Sheryl Friesen

fifth edition

THE VERTEBRATE BODY

SHORTER VERSION

ALFRED SHERWOOD ROMER
Late Alexander Agassiz Professor of Zoology, Harvard University

THOMAS S. PARSONS
Department of Zoology,
University of Toronto

SAUNDERS COLLEGE • Philadelphia

Saunders College Publishing
West Washington Square
Philadelphia, PA 19105

Library of Congress Cataloging in Publication Data

Romer, Alfred Sherwood, 1894–

The vertebrate body.

1. Vertebrates—Anatomy. I. Parsons, Thomas Sturges, 1930– joint author. II. Title. [DNLM: 1. Anatomy, Comparative. 2. Vertebrates—Anatomy and histology. QL805 R763v]

QL805.R65 1978 596′.04 77-11353

ISBN 0-7216-7682-0

Listed here is the latest translated edition of this book together with the language of the translation and the publisher.

Spanish—Third Edition Editorial Interamericana, Mexico

Hindi—Fourth Edition Government of India Ministry of Education, New Delhi, India

The Vertebrate Body—Shorter Version ISBN 0-7216-7682-0

© 1978 by Saunders College Publishing/Holt, Rinehart and Winston. Copyright 1956, 1962 and 1971 by W. B. Saunders Company. Copyright under the International Copyright Union. All rights reserved. This book is protected by copyright. No part of it may be reproduced, stored in a retrieval system, or transmitted in any form or by any means, electronic, mechanical, photocopying, recording, or otherwise, without written permission from the publisher. Made in the United States of America. Press of W. B. Saunders Company. Library of Congress catalog card number 77-11353.

0123 147 9876543

preface

This edition of the "Shorter Version" is based on the fifth edition of the longer work. It contains all of the illustrations and most of the introductory material and appendices of the latter; however, the main part of the text is cut roughly in half, largely by omitting details and comments about exceptions or unusual cases. Such cuts may make this version more suitable for one term courses than the longer one; on the other hand, abridgement makes it less suitable as a reference work.

In the first edition of "The Vertebrate Body," Professor Romer listed in the preface (Apologia) some features he considered desirable in a textbook like this: (1) fairly adequate illustration, (2) a truly comparative treatment, (3) proper paleontologic background, (4) a developmental viewpoint, (5) inclusion of histologic data, and (6) consideration of function. These are equally desirable in this "Shorter Version." However, the shortening inevitably decreases the amount of space that can be devoted to some of these aspects. I suspect that the first three features are retained more successfully than the last three, but others will probably disagree.

This fifth edition differs from the fourth mainly in numerous minor ways. The one major change involves great increase in length of Chapter 3, "Who's Who Among the Vertebrates," and a corresponding decrease in length of Chapter 4, "Cells and Tissues;" the biochemical and physiological information is largely deleted. My rationale is based on the changes in elementary zoology and biology texts. When "The Vertebrate Body" was first written, elementary texts stressed a survey of the animal kingdom; cells, genetics, and physiology were crowded into a chapter or two at one end of the book. Now, most texts are concerned largely with genetic, physiologic, and biochemical principles, with little on the various groups of animals. The change in this book reflects the change in what students entering a course in comparative anatomy may be expected to know.

This edition also introduces a new editor. It does not, however, introduce any change in policy or intent, and I hope that the continuity is more obvious than the break. Professor Romer first asked me to take over this revision some two years before his death. Although all of the actual rewriting was done too late to benefit from his help and advice, we did discuss the major changes in Chapters 3 and 4.

Since this is an abridgement of the longer version, all those people thanked for their help with that also helped, willy-nilly, with the "Shorter Version;" to all of them again my thanks. For assistance, specifically with this version, as well as with the longer, I am indebted to Ludy Djatschenko and Moira N. Loucks in my laboratory in Toronto and to Debbie Patterson at Saunders. Without them and many others both here and at Saunders, revision of this book would have been far more difficult and less pleasant.

Thomas S. Parsons

table of contents

1 Introduction

This work is designed to give, in brief form, a history of the vertebrate body. Basic will be a comparative study of vertebrate structures: the domain of comparative anatomy. This is in itself an interesting and not unprofitable discipline. Of broader import, however, is the fact that the structural modifications witnessed are concerned with functional changes undergone by the vertebrates—changes correlated with the varied environments and modes of life found in the course of their long and eventful history. The evolutionary story of the vertebrates is better known than that of any other animal group, and vertebrate history affords excellent illustrations of many general biologic principles. Knowledge of vertebrate structure is of practical value to workers in many fields of animal biology. To the future medical student such a study gives a broader understanding of the nature of the one specific animal type on which his later studies will be concentrated.

For the most part (Chaps. 6–17) the present volume is devoted to a consideration, *seriatim*, of the various organs and organ systems. In the present chapter is given a "bird's eye" view of vertebrate structure, together with certain introductory matters. Other early chapters discuss general or preliminary topics, including the evolutionary history of the vertebrates and their kin (Chaps. 2 and 3); cells and tissues as the basic structural elements (Chap. 4); and embryonic development (Chap. 5).

THE VERTEBRATE BODY PLAN (Fig. 1)

BILATERAL SYMMETRY. A primary feature of the vertebrate structural pattern is the fact that the members of this group are bilaterally symmetric, with one side of the body essentially a mirror image of the other. Vertebrates share this type of organization with a number of invertebrate groups, notably the annelid worms and the great arthropod phylum, which includes crustaceans, arachnids, insects, and so forth. In strong contrast is the radial symmetry of coelenterates and echinoderms,* in which the body parts radiate out from a central axis like the spokes of a wheel. The degree of activity of animals appears to be correlated with the type of symmetry which is present. The radiate echinoderms and coelenterates are in general sluggish types, slow moving or fixed to the bottom or, if free floating, mainly

*The echinoderms are not truly radial; they start off life as bilateral animals, but later assume a more or less radial form and the habits that go with it (though some may be active predators).

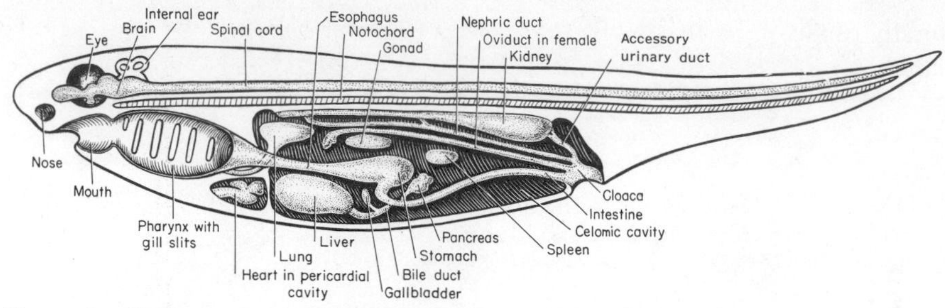

Figure 1. Diagrammatic longitudinal section through an "idealized" vertebrate, to show the relative position of the major organs.

drifters with the current rather than active swimmers. Vertebrates, arthropods, and marine annelids are, on the other hand, generally active animals. Activity would seem to have been one of the keys to the success of the vertebrates and is in a sense as diagnostic as any anatomic feature.

REGIONAL DIFFERENTIATION. In any bilaterally symmetric animal we find some type of longitudinal division into successive body regions—in the annelid worms, for example, a rather monotonous repetition of essentially similar segments, or in insects a pattern in which such segments are consolidated into head, thorax, and abdomen. Vertebrates, too, have well defined body regions, although these regions are not directly comparable to those of invertebrate groups.

There is in vertebrates a highly specialized **head**, or cephalic region; in this region are assembled the principal sense organs, the major nerve centers which form the brain, and the mouth and associated structures. In vertebrates, as in all bilaterally symmetric animals (even a worm), there is a strong tendency toward **cephalization**—a concentration of structures and functions at the anterior end of the body.

In all higher, land-dwelling vertebrate groups a **neck** is present behind the head; this is little more than a connecting piece, allowing movement of the head on the trunk. The presence of a neck region is not, however, a primitive vertebrate feature. In lower, water-breathing vertebrates this section of the body is the stout **branchial region**, containing the breathing apparatus. The appearance of a distinct neck occurs only with the shift to lung breathing and the reduction of the gills.

The main body of the animal, the **trunk**, is the next region; this terminates in the neighborhood of the anus or cloaca. Within the stout trunk are the body cavities containing major body organs, the viscera. In mammals the trunk is divisible into **thorax** and **abdomen**, the former containing the heart and lungs within a rib basket, the latter enclosing most of the digestive tract; there is, however, no clear subdivision here in lower vertebrates.

In most bilateral invertebrates the digestive tube continues almost the entire length of the body. Among the vertebrates, however, we find, in contrast, that the digestive tract and other viscera stop well short of the end of the body; beyond the trunk there typically extends a well developed **tail** or **caudal region**, with skeleton and muscles, but without viscera. The presence of a postanal tail is a basic feature of our group, laid down, it would seem, in an early stage of chordate evolution. The tail is, of course, the main propulsive organ in primitive water-dwelling vertebrates. In land animals it tends to diminish in importance, but is often long, stout at the base, and well developed in many amphibians and reptiles. In mammals it is generally persistent, but is merely a slender appendage. In birds it is shortened and functionally replaced by the tail feathers, arising from its stump; in some forms—frogs, apes, and man—it is, exceptionally, lost completely as an external structure.

GILLS. The presence, in the embryo if not in the adult, of internal gills developed as a paired series of clefts or pouches leading outward from an anterior part of the gut—the pharynx—is one of the most distinctive features (perhaps *the* most distinctive feature) of the vertebrates and their close kin. In higher vertebrates the gills are functionally replaced by lungs, but gill pouches are nevertheless prominent in the embryo. In lower water-dwelling vertebrates gills are the primary breathing organs. Among small invertebrates, many with soft membranous surfaces can get enough oxygen through such membranes to supply their wants. But in forms with a hard or shelly surface, and especially in large forms, in which the surface area is small compared with the bulk of the body, gills of some sort are a necessity. Typical invertebrate gills, as seen in crustaceans or molluscs, are feathery projections from the body surface. The vertebrate gill, however, is an internal development, connected with the digestive tube. Water enters the "throat," or pharynx (usually through the mouth), and passes outward through slits or pouches; on the surface of these passages are gill membranes, at which an exchange of oxygen from the water for carbon dioxide in the blood takes place. Quite in contrast is the function of the gills in certain lowly relatives of the vertebrates. There, as we shall see, the gills and gill slits are of primary importance in food collection—a fact tending to explain the unusual vertebrate condition of an association of the breathing organs with the digestive tube.

NOTOCHORD. In the embryo of every vertebrate there is found, extending from head to tail along the length of the back, a long, flexible, rodlike structure—the notochord. In most vertebrates the notochord is much reduced or absent in the adult, where it is replaced by the vertebral column or backbone. But it is still prominent in some lower vertebrates, and is the main support of the trunk in certain simply built vertebrate relatives (such as amphioxus) in which no vertebral column ever forms. So significant is this primitive supporting structure that the vertebrates and their kin are termed the phylum Chordata, a name referring to the presence of a notochord.

NERVOUS SYSTEM. Longitudinal nerve cords are developed in various bilaterally symmetric invertebrate groups. These, however, are frequently paired and may be lateral or ventral in position. Only in the chordates do we find developed a single cord, dorsally situated and running along the back above the notochord or the vertebrae. Invertebrate nerve cords are generally solid masses of nerve fibers (and supporting cells) running between equally solid clusters of nerve cells, termed ganglia. The chordate nerve cord is, in contrast, a hollow, nonganglionated structure, with a central, fluid-filled cavity. In various invertebrates the process of cephalization is reflected in a concentration of nerve centers in a brainlike structure. Independently, we believe, the vertebrates have evolved a hollow **brain**, with characteristic subdivisions, at the anterior end of the hollow nerve cord—the **spinal cord**. Not exactly matched in any invertebrate group is a series of characteristic **sense organs** developed in the head of vertebrates—paired lateral eyes and, primitively, one or two dorsal, nearly median eyes; nasal structures, usually paired; and paired ears with equilibrium as their primary function.

DIGESTIVE SYSTEM. All metazoans (with degenerate exceptions) have some sort of digestive cavity with a means of entrance to and exit from it. In many of the more primitive metazoans there is but a single opening, serving as both mouth and anus. In vertebrates, as in other more progressive metazoans, there are separate anterior and posterior openings, serving respectively for the entrance of food materials and the exit of wastes. The mouth is situated near the front end of the body, commonly somewhat to the underside. In arthropods and annelids the digestive tube reaches to the posterior end of the body. In vertebrates, however, this is not the case; the anus lies at the end of the trunk, leaving, as we have mentioned, a caudal region in which the digestive tube is absent.

In most vertebrates the digestive tube is divided into a series of characteristic regions serving varied functions—**mouth**, **pharynx**, **esophagus**, **stomach**, and **intestine** (the last variously subdivided). In lower vertebrates the esophagus may be almost nonexistent, and in some groups even the stomach may be absent. In mammals and certain other vertebrates the digestive tract terminates externally at the **anus**. In most groups, however, there is a terminal segment of the gut, the **cloaca**, into which urinary and genital ducts also lead.

A **liver** which performs to some extent a secretory function, but is in the main a seat of food storage and conversion, is present in vertebrates as a large ventral outgrowth of the digestive tube. Somewhat similar but variable structures are present in many invertebrates. In most vertebrate groups a **pancreas** is present dorsally as, primarily, an enzyme-secreting gland.

KIDNEYS. Among invertebrates some type of kidney-like organs for the disposal of nitrogenous waste and the maintenance of a proper composition of the internal fluids of the body is often present, typically as rows of small tubular structures termed **nephridia**. Of chordates below the vertebrate level, amphioxus has nephridia of a special type. In true vertebrates however, the **kidney tubules** serving such a function are of a markedly different type and are characteristically gathered into compact paired kidneys, dorsal in position. **Kidney ducts**, of variable nature, lead to the cloacal region or to the exterior, and a **urinary bladder** may develop along their course.

REPRODUCTIVE ORGANS. Male and female sexes are almost invariably distinct in the vertebrates, as they are in many invertebrate groups. The tissues producing the germ cells—the **gonads**—develop into either **testis** or **ovary**. In all except the lowest vertebrates a duct system leads the eggs or sperm to or toward the surface (frequently by way of the cloaca); in the female, special regions of the duct may be present for shell deposition or for development of the young.

CIRCULATORY SYSTEM. In vertebrates, as in many invertebrates, there is a well developed system containing a body fluid, the blood, with tubular vessels and a pump, the **heart**, to bring about its circulation. The heart in vertebrates is a unit structure, ventrally and rather anteriorly situated. In certain invertebrates the circulation is of an "open" type: the blood is pumped from the heart to the tissues in closed vessels, but is then released and makes its return to the heart by oozing through the tissues without being enclosed in vessels. In the vertebrates, as in some of the more highly organized invertebrates, the system is closed; not only is the blood carried by the **arteries** to the various organs, but the return to the heart, after passing through the tissues in small tubes, the **capillaries**, is also made in closed vessels, the **veins**. In most vertebrates **lymph vessels** form an additional means of returning fluid from the cells to the heart. Many invertebrates contain in their blood streams pigmented metallic compounds in solution, which aid in the transportation of oxygen. Among vertebrates, almost exclusively, the iron compound **hemoglobin** is the oxygen carrier; furthermore, this chemical is not free in the blood, but is contained in red **blood cells** (white cells, with other functions, are also present).

In annelids the circulation of the blood is in general forward along the dorsal side of the body, and backward ventrally in its return to the tissues. The reverse is true of the vertebrates. The blood from the heart passes forward and upward

(primitively via the gills) and back dorsally to reach the organs of trunk and tail, and a major return forward—from the digestive tract—is ventral to the gut (although dorsal veins are important).

CELOM. In certain invertebrates the internal organs are embedded in the body tissues. In others, however, there develop body cavities—**celomic cavities**—filled with a watery fluid, in which most of the major organs are found. This latter condition is present among vertebrates. A major body cavity—the **peritoneal cavity**—occupies much of the trunk and contains most of the digestive tract; various other organs (reproductive, urinary) project into it. Anteriorly there is a discrete **pericardial cavity** enclosing the heart, and in mammals the lungs are contained in separate **pleural cavities.**

MUSCLES. Musculature in the vertebrates is of two types, **striated** and **smooth** (or nonstriated), the two differing sharply in minute structure and in distribution in the body. The former, roughly, includes all the voluntary musculature of the head, trunk, limbs, and tail, and the muscles of the gill region; the smooth musculature, more diffuse, is mainly found in the lining of the digestive tract and blood vessels. The musculature of the heart is in various respects intermediate in microscopic structure. The striated musculature of the trunk develops, unlike most other organ systems, as a series of segmental units.

SKELETON. Hard skeletal materials are present in all vertebrates, and, in all except certain degenerate or (doubtfully) primitive groups, consist in part, at least, of bone. Superficial skeletal parts, the **dermal skeleton**, correspond functionally to the "armor" of certain invertebrates, and are typically bony; internal skeletal structures, the **endoskeleton**, are formed as **cartilage** in the embryo, but are generally replaced by bone in the adult. Cartilage-like materials are found in some invertebrates. **Bone**, however, is a unique vertebrate tissue. It differs in texture and minute structure from the typical chitinous or calcareous skeletal materials of invertebrates. In the fact that the salts deposited in this tissue are mainly calcium phosphate, it differs from most (but not all) invertebrate skeletal structures—in which carbonate is the common calcium compound.

APPENDAGES. Two pairs of limbs, **pectoral** and **pelvic**, are found in most vertebrates in the form of fins or legs, and become increasingly prominent in higher members of the group. They are, however, little developed or absent in the lowest vertebrates, living and extinct, and hence are not absolutely characteristic. They may also be lost in specialized forms. Their structure (in contrast to arthropod limbs) includes internal skeletal elements, with muscles for their movement arrayed above and below.

SEGMENTATION. The great invertebrate phyla of Arthropoda and Annelida are notable for the presence of **metamerism**: a serial repetition of parts in a long series of body segments. In annelids this segmentation is readily apparent; in arthropods the metameric structure may be more or less obscured in the adult, but is clearly seen in embryos or larvae.

Vertebrates, too, are segmented, but the segmentation is limited and has obviously developed independently from that of invertebrate groups, in which all structures, from skin inward to the gut, exhibit segmentation. Among the vertebrates neither skin nor gut is segmented; the metameric arrangement is primarily that of the trunk muscles. In relation, however, to the attachments of these muscles and their nerve supply, much of the skeleton and nervous system has taken on a segmental character.

THE BODY IN SECTION (Fig. 2). We have noted some of the more important body features, with particular regard in many cases to their anteroposterior posi-

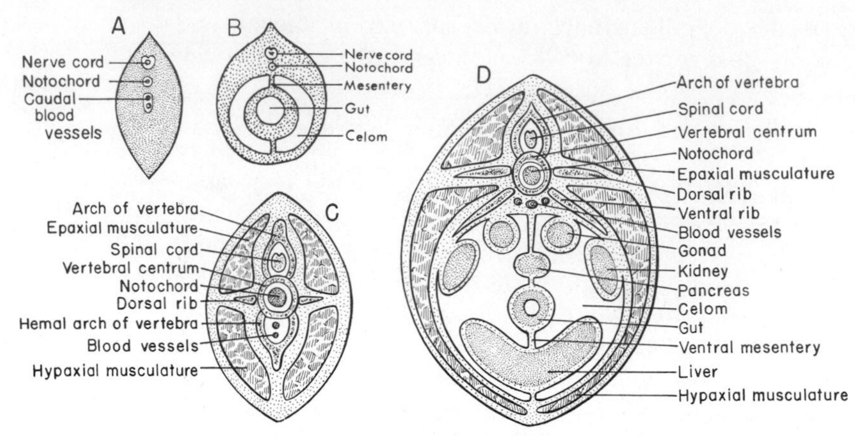

Figure 2. Cross sections through the body of a vertebrate. *A, B,* Much simplified sections through tail and trunk, to show the essential structure of the trunk as a double tube; in the tail the "inner tube" of the gut is absent. *C, D,* More detailed diagrammatic sections of the tail and trunk to show the typical position of main structures.

tions. We may now briefly consider the general organization of the body as seen in cross section.

Structurally, the most simple region of the body is the tail, strongly developed in most vertebrate groups. The section of a tail (Fig. 2 *A, C*) is typically a tall oval, the surface skin-covered. Somewhat above the center is seen the notochord, or the central region of the vertebrae which typically replaces it in the adult, and, above this, a cross section of the nerve tube; the two structures are invariably closely associated topographically. The body cavity and associated viscera are absent in this region; representing them (in a sense) are caudal blood vessels lying below the notochord. Almost all the remainder of the tail is occupied by musculature, usually powerful. This musculature is arrayed in right and left halves, with a median septum dividing them above and below.

A typical section through the trunk is more complicated, even when, as in Figure 2 *B,* this is represented in its most generalized condition. One may consider the trunk as essentially a double tubular system, roughly comparable in structure to the casing and inner tube of an old-fashioned automobile tire. The outer tube in itself contains all the major elements seen in the section of the tail—notochord and nerve cord, and musculature descending on either side beneath an outer covering of skin. Internally, it is as if we had taken the little area below the notochord in the tail, where only the blood vessels were present, and expanded this to enormous proportions as the celomic cavity of the trunk. With the development of this cavity the outer "tube" of the trunk now has an inner as well as an outer surface. The surface lining the body cavity is the **peritoneum**, and that part of this lining which forms the inner surface of the outer tube is the **parietal peritoneum**. The part of the outer tube between celomic cavity and the surface of the body is the body wall.

The "inner tube" is primarily the tube of the digestive tract. The outer lining, facing the celomic cavity, is peritoneum—**visceral peritoneum**. The inner lining is the epithelium lining the digestive tract. Between the two, analogous to the musculature in the body wall, are smooth muscle and connective tissues. In the embryo the gut is connected with the "outer tube" both dorsally and ventrally by **mesenteries**—thin sheets of tissue bounded on either side by peritoneum. The dorsal mesentery—that above the gut—usually persists, but the ventral portion frequently disappears for most of its length.

Although we shall treat the arrangement of the organs in the celom in more detail in a later chapter, we may here go somewhat further in considering the position of the body viscera. In Figure 2 *D*, we have indicated the fact that the digestive tube is not a simple tubular structure, but has various outgrowths—most characteristically the liver ventrally and the pancreas dorsally. These are (in theory and in the embryo, at any rate) median structures, and are developed within the ventral and dorsal mesenteries. Further, we may have other organs projecting into the body cavity, but arising from tissues external to it. The kidneys in many groups project into the abdominal cavity at either upper lateral margin, and the reproductive organs—ovaries or testes—typically project into the cavity more medially along its upper border. It must be noted that the relative size of the body cavity is never so great as represented in this and other diagrams; the viscera actually fill most of the available space.

DIRECTIONS AND PLANES

Although the vertebrate body is essentially a bilaterally symmetric structure, there are many exceptions to this general statement. Organs which primitively lay in the midline may be displaced: the heart may be off center; the abdominal part of the gut—stomach and intestine—is usually twisted and the intestine may be convoluted in a complicated asymmetric fashion. Again, in paired structures those of the two sides may differ markedly; for example, in birds just one of the two ovaries (the left) is functional in the adult. Still greater asymmetry is seen in the flounders, where the whole shape of the body is affected by the substitution of the two sides for the normal top and bottom of the animal.

Either in theory or in practice the body of an animal may be sectioned in various ways at various angles. If the body is considered as sliced crossways, as one would cut a sausage, the plane of section is considered **transverse**. If the line of cleavage is vertical and lengthwise, from snout to tail, the plane is a **sagittal** one. Sometimes this latter term is restricted to a cut actually down the midline—the midsagittal plane—and similar sections to one side or the other are termed parasagittal; but frequently such cuts are considered parts of a series of sagittal sections in a broad sense. The third major plane of cleavage, in the remaining direction, is that of slices cut the length of the body, but horizontally, so as to separate the body into dorsal and ventral parts. Such a plane is termed a **frontal** one—that is, one parallel to the "forehead" of the animal.

Direction within the body is of importance in the description of structural relationships and the naming of the various organs. Terms in this category, fixing a position or pointing out a direction, may be considered.

The head and tail ends of the body are, in most vertebrates, the direction toward which and from which movement of the animal normally takes place. **Anterior** and **posterior** are the common terms of position in this regard; **cranial** and **caudal** are less used, but are essentially synonymous. Upper and lower surfaces—back and belly aspects—are reasonably named **dorsal** and **ventral**. Position in the transverse plane is of course given with reference to the midline; **medial** refers to a position toward the midline; **lateral**, a more removed position.

A fourth pair of terms of less exact meaning but of considerable use are **proximal** and **distal**. The former refers in general to the part of a structure closer to

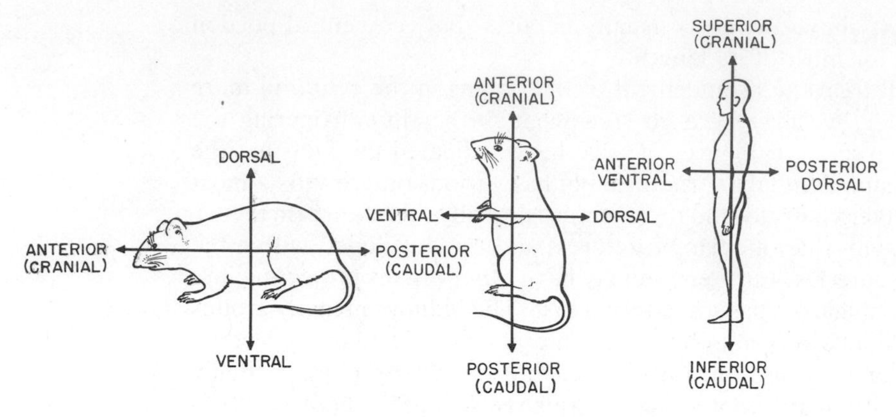

Figure 3. Diagram to show the contrast in positional terms between normal vertebrates and man.

the center of the body or some important point of reference; the latter, to a part farther removed. These terms are clearly available for the limbs and tail. Within the head and trunk their use is less clear, but we may, for example, speak of proximal and distal parts of a nerve with obvious reference to the spinal cord or brain as a center, proximal and distal regions of arteries with reference to the heart as the assumed center, and so on.

For these adjectives of position there are, of course, corresponding adverbs ending in **ly**, and others (rather awkward) to denote motion in a given direction, ending in **ad**, as **posteriorly**, **caudad**.

The major directional terms, anterior, posterior, dorsal, and ventral, apply with perfect clarity to almost all known vertebrates. But in man we have an exception, an aberrant form which stands erect—and hence might have different directional terms applied to him.

It is unfortunate that in the terminology most generally used in medical anatomy this is the case (Fig. 3). The head and "tail" ends of the body are, in the erect human position, above and below rather than fore and aft, and are termed **superior** and **inferior**, rather than anterior and posterior. "Cranial" and "caudal" could, of course, have been used instead, but medical people seem to like these alternatives no better than comparative anatomists. To add to the problem, superior and inferior mean higher and lower in Latin; thus they are occasionally used as synonyms for dorsal and ventral in comparative anatomy. More confusing, however, is the fact that anterior and posterior have generally been used in man—quite needlessly—to replace dorsal and ventral, so that the back side of the human body has been generally termed **posterior** and the belly surface **anterior**. Thus this pair of terms may have contradictory meanings in special human anatomy and in more normal usage, causing needless confusion. For example, each spinal nerve has two roots (cf. Fig. 386). In a dissecting room the two roots in a human cadaver have been generally termed posterior and anterior. But if a neurologist working with (say) rats tries to use the same nomenclature, he is in an obviously absurd position; one root is no more "anterior" or "posterior" than the other. In both rat and man, however, designation of the nerve roots as dorsal and ventral is reasonable and logical. Long established customs, no matter how illogical, are hard to break down; but in a recently adapted revision of human anatomic nomenclature the use of dorsal and ventral has been agreed to in such cases.

THE HOMOLOGY CONCEPT

Even in the early days of zoologic research it was recognized that within each major animal group there was a common basic pattern in the anatomic plan of the body. The same organs could be identified in many or all members of a group, although frequently much modified in size, form, or even function in correlation with changing habits or modes of existence. With the acceptance by biologists of the principle of evolution in the 60's and 70's of the last century, real significance was given to the concept of **homology**: the thesis that specific organs of living members of an animal group have descended, albeit with modification, slight or marked, from basically identical organs present in their common ancestor. For many decades the tracing out of homologies was a leading motif in zoologic research.

Many of the results of such studies were novel and exciting. It was found, for example, that the three little auditory ossicles of our own middle ear (p. 360) were in earlier days part of the jaw apparatus of our piscine ancestors, and appear still earlier to have been part of the supports of the gills of ancestral vertebrates. The muscles with which we smile or frown are derived from those which once helped our fish forebears to pump water through their gills.

Homologous organs are those which are identical—the same—in the series of forms studied. But what do we mean by "the same"? One tends, unthinkingly, to believe that the same actual mass of material, the very same limb or lung or bone, has been handed down, generation by generation, like an heirloom. This is quite absurd, but such a concept has obviously influenced, unconsciously, the minds of many workers. In reality, of course, every organ is re-created anew in every generation, and any identity between homologues is based upon the identity or similarity of the developmental processes which produce them.

The development of the science of genetics has given us a firm base for the interpretation of these processes. They are controlled by hereditary units, the **genes**. These tiny structures are present to the number of thousands, at least, in the chromosomes of every animal cell. The development of the individual is directed by the genes transmitted to the fertilized egg by the parents. Each gene may affect the development of a number of structures or parts of a body; conversely, every organ is influenced in its development by a considerable number of genes. If the genes remain unchanged from generation to generation, the organ produced will remain unchanged (apart from environmental effects upon an individual which may be obvious, but are not inherited by the next or later generations), and the homology is absolute.

Changes, however, do occur in genes, as **mutations**; these mutations produce changes in the structures to which the genes give rise. If the mutations produce effects of small magnitude and occur in only a few of the genes concerned, the organ will be little modified, and its homology with the parent type will remain obvious. If, however, the mutations are numerous and marked in their effects, the organ may be radically modified and its pedigree much less clear. In a sense, a study of organ homology is merely a study of phenomena produced by genes. If the genetic constitution of all animal types were well known, the determination of homology between structures might well rest upon the degree of identity of the genes concerned in their production. But this is not a matter of practical import, for there are few animals whose genetic constitution is at all adequately known, and it is improbable that our range of knowledge will ever be broadened to the necessary degree.

What then are the best criteria for the establishment of homology? Function is no sure guide, for organs which are clearly homologous in two animals may be put to quite different uses. Observation shows that the shape, size, or color of a structure gives little positive evidence of identity. Similarity in general anatomic position and relations to adjacent organs is a more useful clue to identification. Best of all is similarity in developmental history. Embryologic processes in vertebrates tend to be conservative, and organs which are quite different in the adult condition may reveal their homology through similarity in early embryonic stages.

Homology is generally applied to structural identity. Some have proposed that the concept be broadened to include functional identity. This suggestion has not, however, met with general acceptance. The term **analogy** is in some regards a parallel, on the functional side, to homology; analogous organs are those which have similar functions. It is, however, somewhat restricted, for as generally used it implies that the organs concerned are not homologous. A lung and a fish gill, for example, are analogous, for both are used for respiration, but the two are quite different structures.

ADAPTATION AND EVOLUTION

The varied modifications which vertebrate structures have undergone and the varied functions which they have assumed have, of course, come about as the result of evolution. One cannot make a comparative study of the vertebrates without formulating some general concept of the nature of evolutionary processes. Most structural and functional changes in the vertebrate body are quite clearly adaptive modifications to a variety of environments and modes of life. How have these adaptations been brought about? Proper discussion would require a volume in itself; we can here merely indicate the general nature of the problems concerned, and current majority opinion as to their interpretation; the bibliography contains references to more detailed works on this topic.

We sometimes speak, thoughtlessly, of adaptive changes, as if the animal "willed" them or as if its needs or desires in themselves brought new structures or structural changes into being. It would be advantageous, one might say, for a fish to be able to walk on land, and so some fishes made themselves legs; it would be "nice" if the cow's early ancestors developed teeth better able to cope with grain and grass, and so the teeth promptly deepened.

Obviously, such ideas are absurd. They are, however, not far removed from certain theories of evolution which have had, and still have, some vogue. These assume that evolution is an unnatural phenomenon; that changes have been brought about by some "inner urge" within the organism, or are the result of the "design" of some supernatural force. Since such theories are nonscientific, they cannot be scientifically disproved; but we are at liberty to look for more reasonable explanations of evolution, based on known facts. If someone tells us that the operation of an automobile engine is controlled by a little invisible daemon dwelling therein, we cannot prove him wrong. But nothing is gained by adding this hypothetic daemon, and we would prefer an attempt to explain the engine's working in terms of known mechanical principles, the nature of electric currents, and the explosive structure of hydrocarbon molecules.

A more plausible attempt at interpretation of structural evolutionary changes was that first advocated over a century ago by Lamarck—a belief in the inheritance of acquired characters through the effect of use and disuse. If the giraffe's ancestors stretched their necks after foliage on high branches, the effects of stretching, this theory assumes, would be transmitted to their offspring, generation after generation, and an elongate neck gradually developed in the hereditary pattern. If the snake's lizard ancestors ceased to use their legs in locomotion, the cumulative result of disuse would be the eventual loss of the limbs. This attractive theory seems simple, reasonable, and natural. But its present standing is poor indeed. We may summarize by saying that no one, despite repeated efforts, has been able to furnish any valid proof of any instance of the inheritance of an acquired characteristic. Structures useful to an animal may and often do increase in size or complexity in the course of time, and useless or little used structures may diminish. But there is not the slightest evidence that the use or disuse of parts by an individual has any effect whatever upon the build of its offspring.

The science of genetics has in recent decades demonstrated that evolutionary changes are due to mutations. These may produce effects of some magnitude, but most cause only minor modifications: a mutation in a fruit fly may, for example, have no greater visible result than the splitting of a single bristle. We now have considerable knowledge of the chemical structure of the materials involved, and of chemical and physical influences (such as radiation) which play important roles in bringing them about. An understanding of these topics, which are extremely interesting in their own right, is not essential to an understanding of evolution, and discussion of them, therefore, is omitted here. As for evolutionary theories, however, two things stand out clearly: (1) There is no evidence of "design" or "direction" in mutations. They appear to be quite random, rather than tending in any one direction. Some may well be advantageous; most, however, are obviously harmful, and many are lethal. (2) There is no evidence that mutations have any relation whatever to use or disuse of body organs; characters acquired by the individual have no specific influence on the nature of mutations of the genes in its sex cells—mutations whose effect is transmitted to the offspring.

The process of mutation thus seems to be merely one of blind, random change. But vertebrate evolution certainly appears to have resulted in changes both useful and adaptive. How can such results have come about by means of the mutation process?

Our modern ideas on evolution stem from the publication in 1859 of Charles Darwin's On the Origin of Species. Besides giving a mass of evidence that evolution had actually occurred, Darwin presented, for the first time, a convincing theory of *how* it occurred: this is his theory of **natural selection**. He was, of course, quite ignorant of the data now available from genetics, but reasonably assumed that there existed some hereditary mechanisms of the sort with which we are now familiar. Given a supply of random mutations, natural selection will act powerfully to eliminate unfit types and preserve the better-fitted forms in which one or a group of useful mutations have occurred in the germ plasm.

Basically Darwin noted that all species of animals and plants vary and that these variations tend to be inherited. These are obvious and generally accepted ideas: like begets like. Also in nature, species tend to reproduce such that their numbers would increase geometrically if there were no checks (a female cod lays from two to nine million eggs a year; this gives a theoretical potential that boggles the mind). Since numbers of most forms do *not* increase greatly, there must be a very high "infant mortality." Darwin postulated that minor differences would affect an animal's ability to survive. Those differences which are evolutionarily "better," that is, ones that promote survival, will, if inherited, become more common in succeeding generations. Eventually these differences would become the norm, and "new" characters would have evolved.

Although they had, in fact, been discovered by Mendel by 1865, the basic principles of genetics were not generally known in Darwin's day, and he had no real knowledge of the mechanism of inheritance. This is, of course, basic information now, which is included in all elementary textbooks. Major points include the fact that there is no real "blending"; instead, inheritance is by more or less discrete units or genes. For each "character" (in a genetic sense—a character of an animal may depend on many genetic characters), any animal, with various exceptions that need not be treated here, receives one gene from each parent. When that animal produces eggs or sperm, each will contain one of the two original genes. Genes may occasionally change their structure and properties—such a change is a mutation and the mutated form of the gene will be inherited. Mutations of genes or new combinations of existing genes will be responsible for the new characters that may or may not improve the animal's chances of survival.

Obviously things are not simple, and many complications must be considered. First, it is rarely possible to say that a character is simply "good" or "bad." Its value, or lack thereof, to the animal will depend on the animal's environment. A thick fur coat is an obvious advantage to an arctic mammal; in a tropic mammal it could be quite disadvantageous. Some characters are usually beneficial (such as a better means of temperature control) and others not (those that interfere with normal development, for example), but these are probably rare. Thus, as conditions change, characters that were once selected against will become useful and will be selected for.

Second, it is a marked oversimplification to say that a character like blue eyes is inherited if the proper gene is present. The potential to develop blue eyes may be inherited, but whether or not they actually form will depend on many other factors. A shortage of a particular food could reduce the animal's ability to form the proper pigment or, as a more extreme example, if something prevented eyes from forming at all it would obviously not be possible to have blue ones.

Third, what is "selected" is not a specific character. Animals survive or not as a whole, and *all* the characters in an animal must be such that it can survive and reproduce if the "good" genes it possesses are to be passed on. Individual characteristics must be not only beneficial but also able to work together to form a properly integrated whole. Thus, when statements are made to the effect that such a character is advantageous and therefore selected for, remember that several steps are being omitted in the argument. Such short cuts save time and effort and are frequently used in this text and others, but they should be recognized as short cuts.

Fourth, in general genes are carried in duplicate in the cells of every animal (one gene coming from each parent); if the members of a gene pair differ in their potentialities, one tends to dominate over the other in the structures or functions that it controls. It is obvious that selection can have no influence over the "weaker" of such a pair of genes—technically termed recessive—unless by chance both members of the pair of genes concerned are of the same recessive nature. A little consideration makes it clear that as a result of this situation it is practically impossible to eliminate completely a recessive mutant, even if highly deleterious, from an animal stock in which it is once established. It is reasonable to believe that with numerous variables of this sort present in a stock, circumstances might arise (particularly in changed environments) in which certain "suppressed" variants or combinations of them might eventually prove highly advantageous if they should come to light in an individual and result in evolutionary change in the population as a whole. Every species, it would seem, has within its "gene pool" an amount of potential variation which might, to a considerable degree, enable it to adapt to a new situation without the introduction of new mutations.

The mechanism of evolution outlined above can be used to explain many of the observed trends in the phylogeny of vertebrates and other organisms. It is expected that, with time, species will diverge in structures and other characters. Those in different, geographically separate populations will experience different environments so that different qualities will be selected. Moreover, the mutations that occur and are thus available for selection will probably not be the same. After a sufficient (and variable) period of time, the populations will have diverged so that they are different species, families, and so on. **Parallelism** and **convergence**, cases in which two groups of animals change independently but in a similar way or change to become similar from dissimilar ancestors, may reflect selection of similar traits to solve similar problems. Wings are essential for flight in both bats and birds; they show certain similarities but developed from normal front legs quite independently. Such resemblances are, at least in theory, never perfect. Bats remain mammals and birds remain birds; characters not directly involved in the mechanism of the wings are characteristic of their group and quite dissimilar.

This lack of perfect resemblance is also important in the ''law'' that evolution is not reversible. Animals can never return to the ancestral condition because the environment can never be perfectly the same; just the presence of the descendant form makes it different. Also, the chance of the right mutations appearing in the correct order to exactly retrace the previous history in reverse is extremely unlikely. Naturally various structures, having previously evolved, can be lost. Although this is, in a sense, a reversal, it is a very minor one. Some parallelisms can be striking resemblances of reversals; the fusiform shape and the general appearance of sharks, ichthyosaurs (extinct reptiles), and porpoises provide the classic example. However, porpoises have definitely *not* reverted to being fish—they are perfectly good mammals with lungs, not gills, with expanded cerebral hemispheres, not primitive brains, with metanephric, not opisthonephric, kidneys, and so forth. However fishlike a mammal may be or become, it is virtually impossible that it will actually revert to being a true fish.

The differences that are selected may be very minor and the amount of selection that is needed is surprisingly small. Various complex mathematical formulae have been devised to show the changes, given different degrees of selection or starting frequencies, but a few minutes of playing with pencil and paper will show that this is, in fact, true. Evolution does not depend on all-or-none chances of success, so arguments assuming that it does (''this difference is too small to be important'') are simply invalid.

One final note on this theory and its consequences: Selection is nothing more than a measure of differential success in reproduction; it has no teleological or ''purposeful'' component. Therefore selection can only adapt an animal to the existing conditions of the environment. Despite this, the term ''pre-adaptation'' is often used. It describes a situation wherein a character or group of characters was selected because they were of immediate adaptive or selective advantage; later, as conditions changed, these same characters turned out to be even more advantageous and allowed the animal to survive under the new conditions or to adapt to them in a different way. For example, lungs are essential for almost all terrestrial vertebrates and must have appeared before the animals became terrestrial. However, they must have been (and were) adaptive to the original aquatic environment to have arisen in the first place. Their subsequent use by the animal on land was, in a sense, a happy accident—a case of evolutionary serendipity.

SURFACE-VOLUME RELATIONS

It is frequently seen that in any group of animals large and small forms differ notably in the relative size of various organs or parts. The reason for many of these proportionate differences lies in a geometric principle so obvious that it is often overlooked, namely, the fact that *as the size of an animal (or any other object) changes, surfaces increase (or decrease) proportionately to the square of linear dimensions, while volumes change proportionately to the cube of linear dimensions.*

This principle is of wide application, for surface-volume relationships are to be found in a variety of structural and functional features of vertebrates. We cite obvious examples: (1) The strength of a leg (like any supporting column) is proportional to its cross section, which varies as the square of linear dimensions, whereas the weight which it supports is proportionate to the cube of linear dimensions. In consequence an elephant cannot have gazelle-like legs. (2) The amount of food which an active animal needs is roughly proportionate to its volume;* the amount of foodstuffs which its intestine can absorb depends upon the area of the intestinal lining. In consequence, large animals have a disproportionately elongated intestine or one with a complicated structure, resulting in a greater internal surface area for digestion.

ANATOMIC NOMENCLATURE

The student of vertebrate morphology is confronted with a bewildering array of unfamiliar names of anatomic structures. This is unfortunate but inescapable. Vertebrate structures are numerous; for many there are no everyday terms. Even where such names are available, they are often vague and not exactly defined in common usage. Further, it is desirable to have some international system of terms understood in the same sense by scientists of every country.

When anatomy was first studied, all "learned" works were, as a matter of course, written entirely in Latin. In consequence, Latin names where already in existence were applied to anatomic structures, and if no term existed, one was manufactured from Greek or Latin roots and cast in Latin form. Some notes regarding the formation of anatomic terms are given in Appendix 2. Today Latin has ceased to be an international language as far as the general text of scientific books is concerned. Latin anatomic terms, however, are still in vogue. We cannot do without them, although we often use them in a somewhat "anglicized" form—speaking, for example, of the "deltoid muscle" of the shoulder rather than the "musculus deltoideus," or of the "parietal bone" rather than the "os parietalis."

Latin is, of course, an inflected language, and its nouns and adjectives have a variety of endings to express not merely singular and plural numbers, but also a variety of cases and a rather arbitrary system of genders. Until recent decades some knowledge of Latin grammar was part of the equipment of every college student, and the manipulation of Latin terminology presented no difficulty. Today this is not the case, rather unfortunately, for a biologist should at least know enough to avoid such gaucheries as speaking of "humeruses" instead of humeri and "femurs" instead of femora. Fortunately the number of noun and adjectival endings ordinarily used in anatomic terms is limited, and these can be readily learned (cf. Appendix 2).

It is accepted procedure in anatomic nomenclature that where a structure is present in mammals—particularly in man—the name there used be applied to the same structure in other forms.

---

*Emphasis on *active;* basal metabolism in a resting condition is quite another thing.

Thus, for example, man and many mammals have a clavicle, or collar bone, and the equivalent element in the shoulder structure should be called by the same name in reptiles, amphibians, or fishes, even though its appearance is radically different. Sometimes, however, too hasty an identification may be made and a name wrongly applied. Teleost fishes have a bone similar in position to the clavicle, and that name was customarily given to it; we now know, however, that the teleosts lack the true clavicle; the bone present there is a different one (the cleithrum, p. 156). If homologies are in doubt, it is better to use a different name for the structure in question. For example, there is a muscle in the thigh of reptiles which may be homologous with the sartorius muscle of mammals; however, since there is some doubt of the homology, it is customary to give the reptilian muscle a different name—the ambiens muscle (p. 217). If a structure encountered in a lower group has no mammalian equivalent, a new name must, of course, be coined.

Although anatomic terminology has been in general a rather stable and uniform system, there arose, quite naturally, a number of differences in terminology between different schools of work and in different countries. Motivated by the laudable desire to achieve uniformity, the German Anatomical Association, in convention at Basel near the end of the last century, brought forward a comprehensive scheme of terms which members hoped would receive universal adoption in human anatomy. This terminology, usually referred to as the "BNA," was adopted by medical schools and has been widely used in medical work. Quite a number of the terms in this code were, unfortunately, ineptly chosen, and we noted earlier the conflicts in terms of body position. At an International Congress of Anatomists meeting in Paris in 1955, a modified code was drawn up, improving the situation to a considerable degree, and the revisions embodied in the new Nomina Anatomica Parisiensia—"NAP"—are now gradually supplanting the older terms in medical school practice. There is also a "standard" nomenclature in use for veterinary anatomy. All of these systems are based on mammals and become difficult to use for lower vertebrates.

TAXONOMY AND CLASSIFICATION

Obviously, it is essential to have names for different animals if we are to discuss them. It is equally clear that the more information the names give, the more useful they will be, and that there must be general agreement on the names so everybody may understand them. No system is perfect and the one used to name animals is no exception, but it is a workable and generally accepted one dating back over 200 years to the tenth edition of Systema Naturae by the Swedish botanist Linnaeus.

The basic unit is the **species**. The species can be theoretically defined as a group of interbreeding or potentially interbreeding individuals. However, this definition is rarely used in practice. Sometimes it cannot be applied: Many plants and lower animals and even a few vertebrates do not use sexual reproduction, so there is no breeding at all. Fossils too are clearly excluded from any testing by this means. Even with normal, sexually reproducing animals, it is rarely feasible or even possible to test whether animals can interbreed. Finally, the distinction is not clear-cut, and, especially in captivity, forms that everybody agrees are separate species (lions and tigers, for example) may interbreed. A somewhat cynical definition, but one with some degree of truth, is that a species is a group which is considered to be a species by a competent taxonomist. Actually, the problems are not as great as you may think from this—in most cases species are relatively distinct, separate, and recognizable with practice. Lions, tigers, leopards, and jaguars are all big cats, but are obviously different; they are good species.

Species are grouped in various larger or higher units. All species belong to genera (singular, genus), all genera to families, all families to orders, all orders to classes, and all classes to phyla. Thus the domestic dog (*Canis familiaris*) is in the genus *Canis*, the family Canidae (all dogs), the order Carnivora (most carnivorous mammals), the class Mammalia, and the phylum Chordata. Sometimes more subdivisions are needed; in such cases the prefixes super-, sub-, and infra- are used. In a few cases still other ranks are inserted, but these are rare and need not be considered here. There are no nice neat rules for how to distinguish these ranks, or definitions for them. All are subjective, though in many cases they correspond to familiar groupings which have common English names. In the example used above, *Canis* includes what one thinks of as dogs in a loose sense—domestic dogs, wolves, coyotes, and the like. Foxes are dogs in a still broader sense and are in the same family, but not in the genus *Canis*. Similarly, squirrels are a family and rodents are an order.

Since all groups above the specific level are subjective and since new evidence is constantly being discovered, there is no "correct" classification or universal agreement. Ideally, the classification should reflect the phylogeny of the animals. Such a classification is frequently called a "natural" one. In theory, all the members of any taxon (species, genus, or whatever level) share a common ancestor not shared with the members of any other comparable taxon. Again, this is theory; aside from the common ancestor being unknowable without a perfect fossil record which does not exist, there is no agreement about whether the common ancestor is theoretically an individual, a species, or some higher taxon. Nevertheless most people assume that the classification they use is, as far as possible, a "natural" one. Recently other types of classification, based simply on resemblances, have been proposed; most of these resemble the traditional classifications, but the ideas behind them are not always accepted.

Appendix 1 gives a classification of chordates. Like any other, this is arbitrary in many ways and most zoologists would disagree with one or more of its features. However, it does serve as a framework and will be used throughout this book. Unfortunately, confusion can arise easily. Not only are there many synonyms (the order including the turtles can be called Testudines, Testudinata, or Chelonia), but also in many cases the same name is used for different groups (some workers use Crossopterygii for what is here called Sarcopterygii). The classification here is moderately conservative and standard, but in a few places we have used relatively uncommon terms or arrangements that we believe better reflect the relationships of the animals concerned.

Finally, a few practical matters are worth noting. The formal names of taxa are in Latin (or are used as if they were Latin—many are based on Greek roots, personal names, or various unlikely sources). Names of genera and species are controlled by a complex set of rules administered by an International Commission. Basically, these rules concern priority; the first name used, with certain exceptions, must be retained in order to promote stability. The names of higher taxa are not so controlled and may be changed as concepts of the relationships of the animals change.

Generic and specific names are always (or should be) printed in italics (*Canis familiaris* for example); names of higher taxa are not italicized. In writing or typing, italics are indicated by underlining (i.e., Canis familiaris). Generic names are always capitalized; specific ones never, even if they are based on a proper noun (*Sciurus carolinensis,* the gray squirrel of eastern North America). A generic name may be used by itself, but a specific name must always include the generic name as well. The generic name *Sciurus* is used *only* for squirrels, but the specific name *carolinensis* may be used in many different genera (besides *Sciurus,* the mammalian general *Blarina, Castor,* and *Evotomys* include species or subspecies of this name). Names above the generic level are always capitalized and are plural; thus, "the Canidae are . . . ." The names of families always end in -idae; those of subfamilies end in -inae and of superfamilies, in -oidea. There are no standard endings for other ranks, though in some groups many ordinal names end in -iformes. Common names are usually not capitalized, although proper names in them are, and some people, particularly ornithologists, do favor using capitals for common names of species.

2

The Vertebrate Pedigree

Although the present work primarily concerns the vertebrates alone, we must recognize that there exist various animal types lacking a backbone, but closely allied to the vertebrates. Study of these more lowly forms contributes to our understanding of vertebrate structure and history. Further, although (as will be seen) we have little certain knowledge of the early ancestry of the vertebrates, the subject of their pedigree deserves consideration. In this chapter we will first describe, as lower chordates, certain small marine animals definitely allied in some fashion to the vertebrates, then discuss possible relationships of the vertebrates to various invertebrate phyla, and, finally, attempt to plot out a reasonable vertebrate pedigree.

The vertebrates do not in themselves constitute a major division of the animal kingdom. They are considered merely one subdivision—although by far the largest subdivision—of the phylum **Chordata**, the other members of which are to be considered briefly here. The "**lower chordates**" lack the backbone and many other advanced structures of their vertebrate relatives. They do, however, exhibit, to a variable degree, basic features characteristic of the vertebrates and not found elsewhere in the animal kingdom. These features indicate that they are truly related to vertebrates and hence properly included in a common group with them. The term Chordata itself implies that a notochord (or chorda), or some structure thought to be equivalent to a notochord, is generally present. Again, a dorsal, hollow nerve cord is a common feature. Most characteristic of all is the fact that gill slits are almost universally present in chordates.

AMPHIOXUS

Many workers subdivide the phylum Chordata into four subphyla—in roughly ascending order, Hemichordata, Urochordata, Cephalochordata, and Vertebrata. Here we reverse the order and begin our discussion of the lower chordates with the **Cephalochordata**, in which the similarities to the vertebrates are most obvious. The subphylum includes only a few closely related forms, all commonly termed amphioxus, a name replaced in formal taxonomic usage by *Branchiostoma* (Figs. 4, 5). They are translucent animals, fishlike in appearance and proportions, found in shallow marine waters in various regions of the world, and sometimes locally

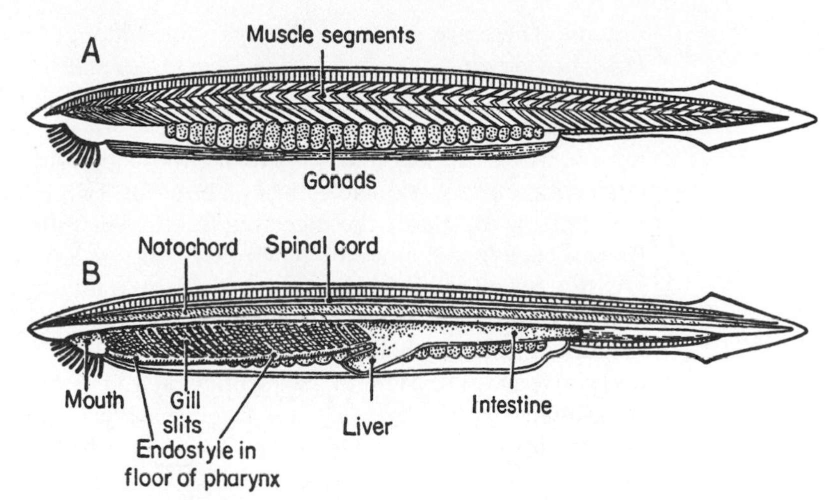

Figure 4. Amphioxus, a primitive chordate. *A,* As seen through the transparent skin; *B,* a sagittal section. (After Gregory.)

abundant. As the shape suggests, they can swim readily, but because of poor development of fins, rather ineffectively; for the most part they spend their time with the body buried in the sands of the bottom, with merely the anterior end projecting.

Despite the piscine appearance, it is obvious that we are dealing with forms far more primitive than any fish. There are no paired fins or limbs of any sort. Cartilage-like materials stiffen the gills, dorsal fin, and mouth parts, but no part of the normal vertebrate skeleton of vertebrae, ribs, or skull is to be found. The main skeletal structure is a highly developed notochord, which persists through life and (in contrast to the vertebrates) extends clear to the tip of the "nose"—a feature to which the group owes its name. The notochord prevents telescoping during vigorous swimming and serves as a convenient central "peg" on which to hang the body

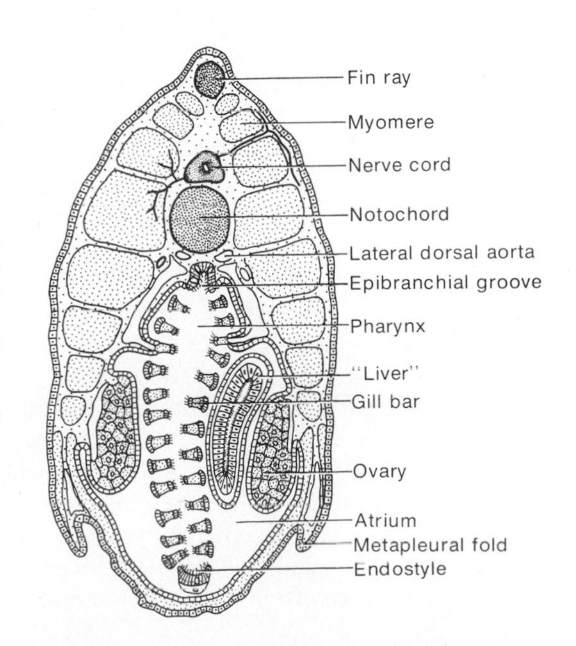

Figure 5. Cross section of amphioxus through the pharynx. The peribranchial space surrounding the pharynx, liver, and so on, is, despite its seeming internal position, actually external to the body and is somewhat analogous to the gill chamber of bony fishes. It is formed by the downgrowth around the pharynx of great metapleural folds meeting one another ventrally, and it connects with the exterior through a posteriorly placed opening. Since the gills lie in a diagonal position, such a vertical section as that shown cuts through a number of successive bars. (After Al-Hussaini and Demian.)

organs. There are nerves serially arranged along a typical, dorsal, hollow nerve cord; but although the cord is somewhat larger anteriorly, there is no true brain, and there are only dubious traces of sense organs which might correspond to nose or eye. Much as in fishes, the major musculature consists of a segmental series of muscle blocks arranged in V's down either side of the body; alternating waves of contraction of these muscles bring about the swimming movements.

For the most part the digestive tract is very simple. There is a mouth cavity (**buccal cavity**) surrounded by a circle of stiffened projecting cirri. The pharynx is greatly elongated, extending about half the total length of the body. Back of the pharynx, the gut is a straightforward tube with little sign of division into successive chambers, although chemical treatment of food appears to predominate in the anterior part of its length, absorption at the back. There is a large pouchlike outgrowth which is generally compared to a liver although the homology is dubious. Within this outgrowth, also called a cecum, there may occur phagocytosis and intracellular digestion, features found in many invertebrates but not in vertebrates. As in vertebrates, the tube ends at the anus, far short of the end of the body, which thus terminates in a true tail.

The **pharynx** is highly specialized for food collecting. Amphioxus lives on particles gathered from the sea water; these are taken in through the mouth by ciliary action and strained out from the water as it passes out of the body through the gill slits. These slits are very numerous, far more so than in any vertebrate, for there may be as many as fifty or more pairs of them, and each gill is essentially a double structure, developing in much the horseshoe-shaped fashion seen in acorn worms (Fig. 9). A pair of great folds grows downward to meet ventrally and enclose the whole gill system, serving to protect these delicate organs when amphioxus is buried in the sand. This forms a pocket, the **atrium**, or peribranchial space, opening to the surface only by a pore at the back of the pharynx. But while the gill slits, by their major development, emphasize the animal's relationships to the vertebrates, they differ from those of typical vertebrates in both purpose and mode of operation. It appears that much of the ''breathing'' of amphioxus is done through the skin—which, in contrast to that of vertebrates, is quite thin—and that here, as in other lower chordates, the gills are primarily feeding devices. Further, water currents through the gills in vertebrates are effected by muscular pumping; in amphioxus ciliary action alone is responsible, and cilia are highly developed in the pharynx. A prominent feature is the development of a longitudinal midventral (hypobranchial) groove termed the **endostyle**, running the length of the pharynx; in this a sticky mucus is abundantly secreted. Ciliary currents carry streamers of this material up the pharyngeal walls, past the gill slits; catching up trapped food particles on the way, the mucus collects in a dorsal (epibranchial) groove. From this point cilia carry back the mucus and the enclosed food particles in a continuous slimy band to the intestine; the animal feeds itself by a conveyor belt system.

The major blood vessels of amphioxus are laid out clearly on the vertebrate pattern (Fig. 326), with the blood coursing forward ventrally and back dorsally after passing upward through the gills. There are, however, no blood cells, red or white, or blood pigments and, further, there is no single heart; movement of the blood is accomplished by wave contraction of some of the principal vessels, together with the contraction of numerous tiny heartlike bulbs situated along the course of the arteries below the gills.

The gonads differ from those of vertebrates—indeed, from those of all other chordates—in being numerous and segmentally arranged. Still more divergent from the vertebrate plan is the nature of the excretory organs. The vertebrate kidney is of a unique type, composed of distinctive water-filtering units which will be described in detail later. In amphioxus, on the other hand, the structures are of a very different sort; they are, as in many invertebrates, segmentally arranged protonephridia, which in amphioxus resemble to a degree those characteristic of many flat worms.

Where does amphioxus stand in relation to the vertebrates? A few theorists, for whose ideas as to vertebrate evolution the existence of amphioxus is inconvenient, deny that it is at all closely related. But the features in which amphioxus resembles the backboned animal are so numerous and so basic that this position is untenable. Strongly in contrast is the suggestion that amphioxus is a degenerate vertebrate. As we shall see, the young of lampreys, quite unlike the adults, live a life as sedentary filter-feeders and are comparable to amphioxus in many ways. Is amphioxus a lamprey which has, so to speak, never grown up and hence retained, as an adult, the simplicity of structure of the larva? It is probable, as we shall see presently, that a factor to be kept in mind in evolutionary studies is the phenomenon known as **neoteny**, in which larval characteristics are retained to a greater or lesser degree while the reproductive organs become functional (**paedogenesis** is sometimes used as a synonym of neoteny for certain cases, or even for some types of parthenogenesis—dictionaries vary and we will use neoteny here in a general sense). But amphioxus shows many features that are different from those expected in the young of a lamprey ancestor—too many to make this suggestion of relationship by degeneration plausible. As a working hypothesis we shall here, like most students of the subject, interpret amphioxus as a specialized and modified survivor of a type of animal ancestral to the vertebrates. Amphioxus is not only more primitive structurally than any vertebrate, but has a mode of life essentially different from and more primitive than that of vertebrates. In general vertebrates actively and aggressively seek large food objects, eat by muscular movements of jaws or analogous structures, and typically use their gills—operated by means of well developed muscles—exclusively for breathing purposes. Amphioxus, although able to swim, is essentially sedentary and, as we have noted, is in contrast a filter-feeder, using cilia rather than muscles in food gathering and utilizing the gills for feeding rather than breathing.

TUNICATES (Figs. 6–8)

In seeking further lowly relatives of the vertebrates we may be well advised to look for other filter-feeders which, even if simpler in structure than amphioxus, show at least some of the basic chordate characters, such as gill slits, notochord, or dorsal, hollow nerve cord. Such a group is that of the **Urochordata**, the **tunicates** or **sea squirts** and their relatives. These are rather common small marine organisms. They are essentially inactive; the adult does not seek its food, but is a highly developed filter-feeder, accepting such particles as it can attract by ciliary action. Many tunicates are found floating freely in the water, singly or in groups, often as tiny barrel-shaped structures; others are attached to the bottom, either as branching colonies or as individuals (Fig. 8 *C–E*). Simplest are the solitary tunicates (Fig. 6 *B*). As an adult, such an animal is an almost formless lump attached to a rock or other underwater object and covered with a leathery-looking "tunic." The only structural features seen externally are an opening at the top, into which water passes, and a lateral opening, through which the water current flows outward. The

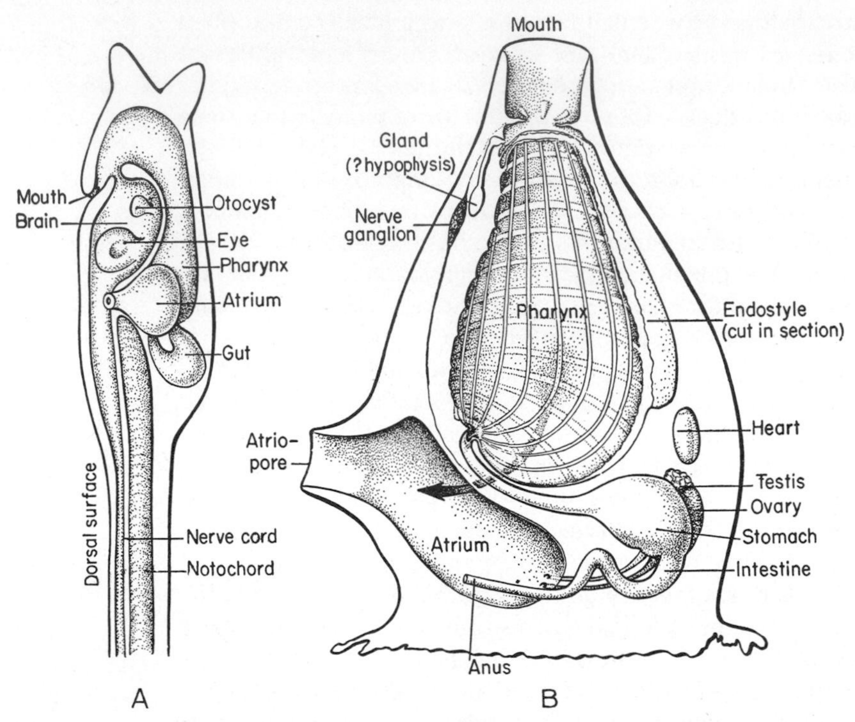

Figure 6. A solitary tunicate. *A,* Diagram of the structures seen in the free-swimming larva (head end above, and only a short section of tail figured). The otocyst is a simple ear structure. *B,* The sessile adult, formed by elaboration of the structures at the anterior end of the larval body. The original dorsal side lies at the left. The large pharynx is attached to the body wall above and below (left and right in the figure); the atrium (corresponding to the peribranchial space of amphioxus) bounds it on either side. Water passing through the latticework gills of the pharynx enters the atrium and, as indicated by the arrow, streams out through the atriopore. (After Delage and Hérouard.)

creature shows no external resemblances to the vertebrates, and internally much of the structure is equally unfamiliar to a student of that group. There is no notochord. Nor is there a nerve cord; instead, there is a simple nerve ganglion with a few nerves splaying out from it.

Much of the interior of the animal is occupied by a barrel-shaped structure which serves as the food-gathering device. The water current, created by ciliary action, is strained through slits in the sides of the barrel into a surrounding chamber, the **atrium**, which leads to the lateral excurrent opening (or atriopore). On closer examination it becomes obvious that the barrel is an exaggerated set of internal gills, constituting the pharyngeal region of the animal; there is even an endostyle comparable to that noted in amphioxus. Below the enormous pharynx, the digestive tube narrows to form esophagus, stomach, and intestine—all of modest size.

We have here, in the pharyngeal gill apparatus, a high degree of development of one of the primary characters to be sought in a relative of the vertebrates. For other chordate characters, however, we must turn to the developmental history. In many tunicates propagation takes place in the main by a process of budding. But in some there is a distinct larval form (Figs. 6 *A*, 7), which appears like an amphibian tadpole. The "head" of the larva corresponds to the entire body of the adult. The tail is a swimming organ, useful in transporting the young tunicate about in its search for a home. Once the animal attaches and "settles down" to its sedentary adult existence, the tail dwindles and is absorbed into the body. In this tail, however, are to be found major proofs of vertebrate relationship of the tunicates. There is in the larval tail (as the group name, Urochordata, implies) a well developed notochord and, above this, a typical hollow, dorsal, nerve cord. These structures are, however, less advanced than those of amphioxus, for there is here no segmentation of the swimming muscles or of the nerves supplying them. Anteriorly, there are in the larva a rudimentary brain and sense organs. At metamorphosis, the notochord and the larval nerve cord (unnecessary in the sessile adult) disappear.

The tunicates, thus, are definitely chordates and definitely related to the vertebrates. How do they fit into the evolutionary story? Those who believe that the vertebrate ancestors were from the earliest times actively swimming animals would regard the tunicates as a degenerate side branch of the vertebrate ancestral line and consider that the common ancestor of vertebrates and tunicates was a free-swimming adult, somewhat like the larval tunicate. From this, it is suggested, the vertebrates "ascended" by an improved continuation of an active mode of life, whereas the tunicates tended to become "degenerate" and lost most progressive structural features except for the food-straining gill barrel; became, in fact, fit subjects for evolutionary sermons on the results of slothful living. There is, however, another interpretation which is more probable, namely, that the chordate ancestor of the vertebrates was, rather, a sessile food-strainer somewhat like an adult tunicate; that the tail first appeared as an adaptation in the larva, rendering easier the search for a suitable place in which the animal could "settle down"; and that the development of higher forms came about by the retention of the tail and the free-swimming habit in adult life with the elimination of a sessile adult stage. It is

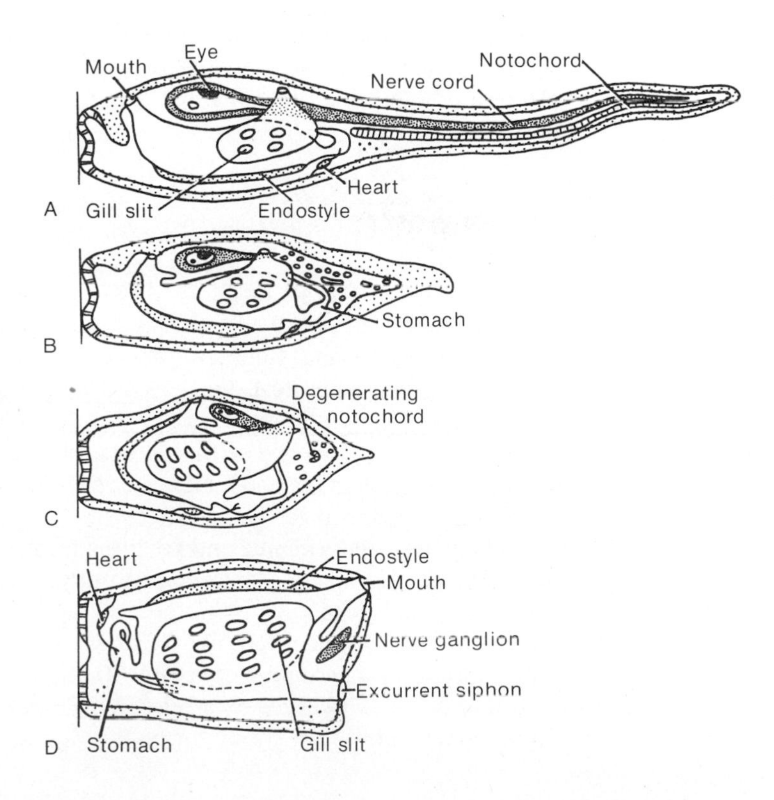

Figure 7. Diagrams of metamorphosis of a solitary tunicate. In *A* the free-swimming larva is attaching to the substrate by anterior suckers. In *B* and *C* the tail is being resorbed with the loss of the notochord and reduction of the nervous system. The internal organs gradually rotate to their adult positions as shown in *D*. (After Storer and Usinger.)

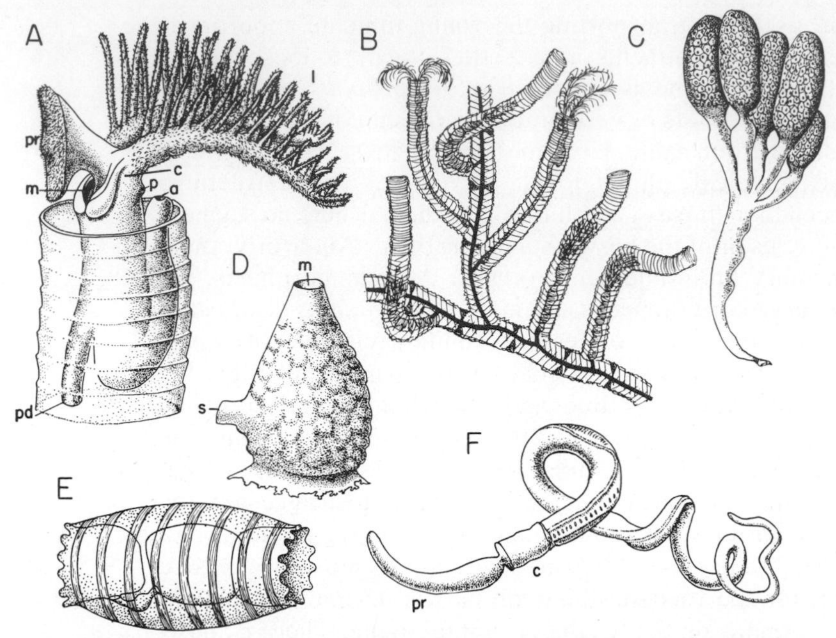

Figure 8. Tunicates and hemichordates. *A,* An individual of the pterobranch genus *Rhabdopleura* projecting from its enclosing tube. *B,* A part of a colony of the same. *C,* A colonial sessile tunicate; each polygonal area is a separate individual of the colony. *D,* External view of a solitary tunicate (cf. Fig. 6 *B*). *E,* A free-floating tunicate, or salp. *F,* An acorn worm (*Balanoglossus*). *a,* Anus; *c,* collar region; *l,* lophophore; *m,* mouth; *p,* pore or opening from celom; *pd,* stalk (peduncle) by which individual is attached to remainder of colony; *pr,* proboscis or anterior projection of body; *s,* siphon which carries off water and body products. (Mainly after Delage and Hérouard.)

reasonable to believe that we have here an example of neoteny—reproduction by larval animals, with the result that in their descendants, the old adult stage was obliterated, and there began a new progressive evolutionary series leading to the vertebrates.

ACORN WORMS

A further series of forms definitely related to the vertebrates is that of the **Hemichordata.** Here the key chordate characters are little developed, and many writers consider the hemichordates (including the pterobranchs, described later) a phylum separate from the Chordata, although related to them.

The best known hemichordates are the **acorn worms**, such as *Balanoglossus* (Figs. 8, 9), termed as a group the **Enteropneusta** and found not uncommonly in tidal flats. The long, slender body suggests that the acorn worms are active animals, as one might hope for in vertebrate ancestors. This is not the case; they are essentially sedentary burrowers in mud and are filter-feeders comparable in their general mode of life to the tunicates. The general body shape is wormlike, but there the resemblance ends, for their structure is not at all comparable to that of ordinary annelid worms. Even externally the acorn worms are distinctive. The body terminates anteriorly in a tough yet flexible and muscular "snout" or proboscis, of variable length, which serves as a burrowing organ. Behind the proboscis a distinct thickened section of the body forms the "collar" region; the name acorn worm is due to the fact that in some forms the proboscis and collar have somewhat the appearance of an acorn in its cup.

In most regards acorn worms show no special resemblance to the vertebrates or other chordates. For a short distance—in the collar region—there is a dorsal nerve cord which is more or less hollow. But over the rest of the body the nerve cells and fibers are rather diffusely distributed in the skin, although there is some development of solid dorsal and ventral strands of nerve tissue. There is no proper notochord, although a stout pouch of tissue at the base of the proboscis has been compared (rather dubiously) to an imperfectly developed structure of that sort.

But—as in tunicates—we find that the vertebrate type of gill system is present in characteristic and highly developed fashion. The gills are not so pronounced as they are in tunicates, but there is, behind the collar, an extended pharynx (partitioned off from the food passage to the stomach), from which on either side open out gill slits quite comparable—even in details of structure and development—to those of amphioxus. Vertebrates are surely not descended from acorn worms as such, but these forms may reasonably be interpreted as a group not distantly removed from our early chordate ancestors—essentially sedentary food-strainers in which, however, some degree of potential motility is present in the adult.

The acorn worms show, in the gill slits at least, definite proof of vertebrate relationships. But in a second type of hemichordates, the **Pterobranchia** (Fig. 8 *A*, *B*), hardly a trace of vertebrate structure is to be found, and were it not clear that they are affiliated with the enteropneusts, one would hardly suspect that they belonged to this general stock. The pterobranchs are tiny, rare marine animals of which only a few genera are known. They form little plantlike colonies, whose individuals project like small flowers at the ends of a branching series of tubes. The short body is doubled back on itself, so that the anus opens anteriorly back of the head. Proofs of relationship to the acorn worms lie in the fact that there is a snoutlike anterior projection beyond the mouth, corresponding to the enteropneust proboscis, and back of this a short collar region. But almost all resemblances to acorn worms—to say nothing of more highly developed chordates—are lacking. There is little development of a nervous system, no trace of a hollow nerve cord, and not the slightest suggestion of a notochord. And the feeding mechanisms are of a very different type. True, these plantlike animals feed, as do more typical lower chordates, on food particles drawn in by ciliary action. But there is almost none of the gill mechanism, which is so important in the filter-feeding of tunicates and amphioxus. One of the two better-known pterobranchs has a single pair of small gill openings, the other none at all. Instead, there project from the collar region large tentacle-like structures, termed **lophophores**; these are supplied with bands of cilia which collect food particles and bring them to the mouth. Even acorn worms, with

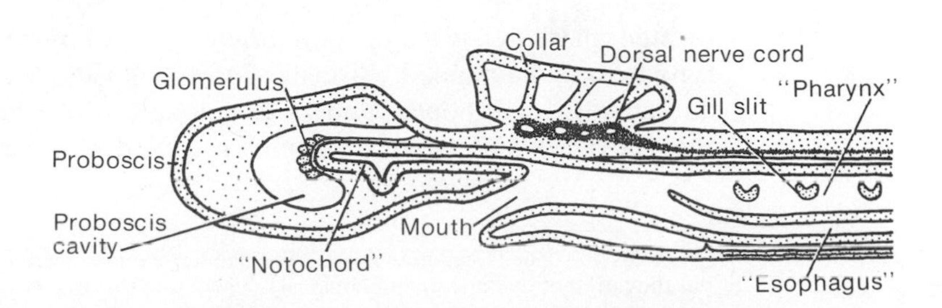

Figure 9. Diagrammatic longitudinal section through the head of *Saccoglossus*, an acorn worm. (After Storer and Usinger.)

good gills and no lophophores, gather almost all their food by cilia and sheets of mucus on the outside of the proboscis and adjacent areas.

So unchordate-like are these small creatures that one is tempted to believe that they are degenerate, perhaps relatively modern in development. But it is believed that they are a very ancient group indeed. Paleontologists have long been familiar with a variety of small tubelike structures termed **graptolites**, which were abundant in the seas long before the appearance of the oldest vertebrates. These tubes are similar to those which shield the modern pterobranchs.

With the pterobranchs we conclude the series of lowly chordates.* What do they teach us about vertebrate origins? As we have said, one (at first thought) might expect the vertebrate ancestral line to lead through active little forms, paralleling, at least, such invertebrate groups as the progressive arthropods like crustaceans and insects. But consideration of the lower chordates tends to dim such expectations. For the most part we see sluggish and sedentary forms and instead of forms actively searching for food, passive filter-feeders. However, before attempting to reach a conclusion let us survey the various invertebrate phyla in search of possible chordate relatives.

INVERTEBRATE PHYLOGENY. In recent decades students of invertebrate zoology have come to agree on many points (but not all!) concerning the phylogeny of animals without backbones (Fig. 10). All forms above the level of protozoans and sponges are termed the Metazoa. The majority opinion is that the ancestral metazoan was a sessile, attached form, probably not unlike in appearance some of the living coelenterates (Cnidaria) such as the sea anemones or little hydras. The living coelenterates, however, are specialized in the possession of stinging cells, which enable them to capture prey of some size. A coelenterate ancestor, lacking such weapons and depending for a living on food particles floating within reach in water currents, seems satisfactory as a truly primitive metazoan.

Despite their specialization in developing stinging cells, the coelenterates appear clearly to represent the ancestral condition in one very important point—the absence of the middle body layer. As presumably was the case in a truly primitive metazoan, the coelenterate body has a simple, two-layered structure, with little between the "skin"—the ectoderm—and the lining, termed the endoderm, of the inner gut cavity. Above the coelenterate level most animals have a third intermediate body layer, the mesoderm, from which muscular, circulatory, and other systems are formed; the mesoderm becomes—in bulk, at any rate—the most important of the three tissues. Further, many invertebrates develop a celom within the mesodermal tissues.

Two contrasting methods of embryonic formation of this third body layer are seen. In one type especially characteristic of echinoderms—starfishes, sea urchins, sea lilies, and the like—the mesoderm arises in the form of pouches growing outward from the gut walls; these pouches remain in the adult as closed body cavities. In a second type the mesoderm arises as solid masses of cells budded off from an area near the posterior end of the body, and the body cavities, when formed, arise by cleavage within the masses of mesodermal cells. To this second type belong the annelid worms and the molluscs. The great group of joint-legged animals, the arthropods, appear to belong to this second group as well, although their developmental pattern is much modified; and certain other forms, such as the

*Possibly allied to chordates are some slender, elongate, deep sea forms termed the Pogonophora, but they are poorly known, and obviously degenerate (the digestive tube is absent), and we may leave them, without loss, in oblivion.

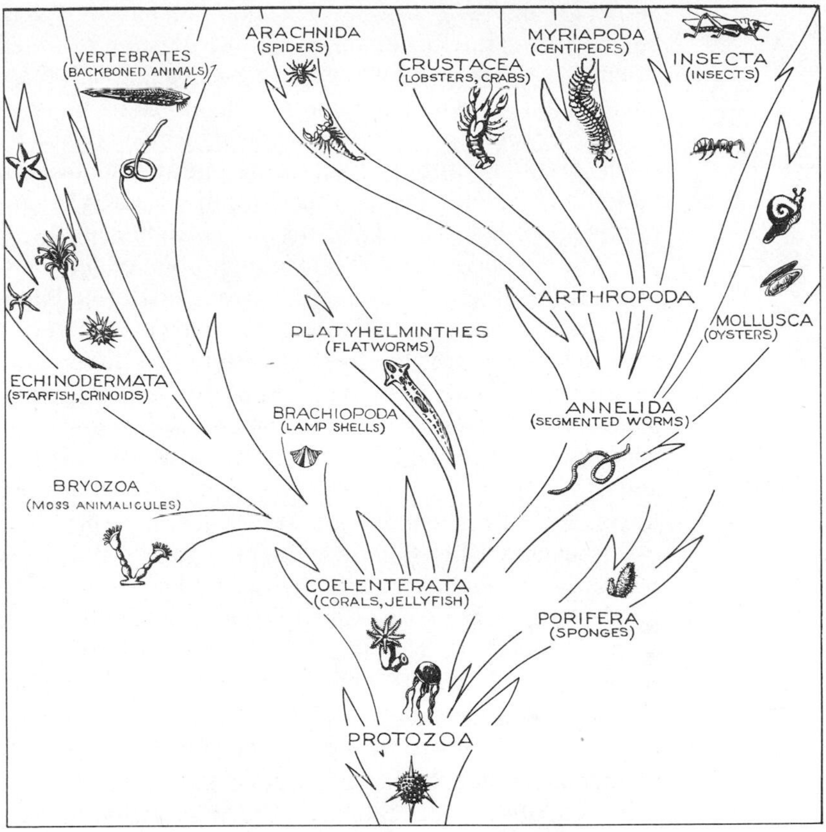

Figure 10. A simplified family tree of the animal kingdom, to show the probable relationships of the vertebrates. (After Romer, Man and the Vertebrates, University of Chicago Press.)

flat worms, appear to be offshoots from the base of this major stock. We thus have the concept that, above the coelenterate level, the invertebrates form two great branches, in Y-fashion, with the echinoderms at the end of one branch and the great host of familiar advanced invertebrates clustered on the other. A few of the less familiar, mainly sessile marine forms, such as the lamp shells (brachiopods) and moss animalcules (bryozoans), do not fit well on either main branch but are perhaps somewhat closer to the echinoderms.

The two stocks contrast not only in the type of formation of the middle body layer, but also in the patterns of cleavage of the fertilized egg, the method of gastrulation, and the larval development. Both echinoderms on the one hand and aquatic annelids and molluscs on the other grow from the egg into tiny larvae of simple structure, with bands and tufts of cilia arranged in characteristic patterns on the surface of the body. The echinoderm larva has an arrangement of these and other features which differ markedly from those of the larvae of annelid worms and molluscs.

From what point on this family tree of the invertebrates does the chordate (and vertebrate) branch arise? Theories on this subject have been numerous, but have given few positive results.

One solution to the problem might be to suggest a direct origin of chordates from the most primitive metazoans. Here there are no great difficulties to overcome, for animals on this level of evolution have few specialized features that must be lost before starting on the path toward the vertebrates. But in reality, advocacy of such a descent would seem to be begging the question. A number of basic advances are common to almost all invertebrate phyla—development of a middle body layer, presence of both mouth and anus in the digestive tract, and so forth. It seems highly improbable that the vertebrates acquired these progressive features entirely independently of other groups. Search seems warranted for possible relatives—if not direct ancestors—at a higher level.

The annelid worms offer a possible point of departure; the theory of vertebrate origin from annelids was warmly advocated during later decades of the nineteenth century. Annelids have bilateral symmetry, as do vertebrates, and, in correlation with this, some are, like typical vertebrates, active animals in contrast with the sessile types common in many invertebrate phyla. Then too, they are segmented forms, as the vertebrates are to at least some extent. As in vertebrates, the central nervous system is composed of a brainlike mass at the anterior end of the body and a longitudinal nerve cord. However, beyond this point the comparison breaks down. Even the segmentation is a weak argument; for the annelid is segmented in every respect, from skin to gut lining, whereas the segmentation of a vertebrate is primarily confined to part of the middle body layer. The annelid has, it is true, a longitudinal nerve cord. But it is solid, not hollow, and it is ventral rather than dorsal in position. This last point is especially troublesome to advocates of this theory. They have "resolved" the difficulties by assuming that a vertebrate is a worm upside down (Fig. 11). This involves further perplexities. The worm's mouth is on the under side of the head, and so is that of a vertebrate. A reversal of surfaces implies that the old mouth of the worm has closed and been replaced, historically, by a new one. Attempts have been made to find traces of the theoretic old mouth opening in vertebrate embryos—it should pass upward and forward through the brain to the top of the head!—but without convincing results.

ECHINODERM AFFINITIES. Unlikely as it seems at first sight, the best clues to chordate relationships are to be found in a study of the echinoderms—the starfish, sea urchins, and the like. Several lines of work suggest that, despite the obvious and strong contrasts, the two phyla are nevertheless related. In most vertebrates mesoderm formation is a complex process, but in amphioxus we find mesoderm

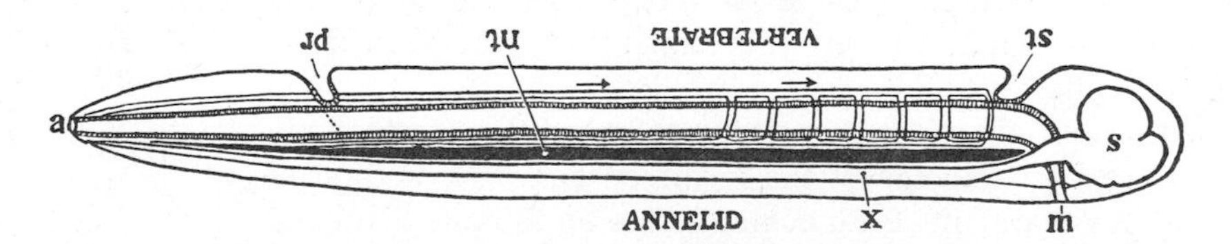

Figure 11. Diagram to illustrate the supposed transformation of an annelid worm into a vertebrate. In normal position this represents the annelid with a "brain" (*s*) at the front end and a nerve cord (*x*) running along the underside of the body. The mouth (*m*) is on the underside of the animal, the anus (*a*) at the end of the tail; the blood stream (indicated by arrows) flows forward on the upper side of the body, back on the underside. Turn the book upside down and now we have the vertebrate, with nerve cord and blood streams reversed. But it is necessary to build a new mouth (*st*) and anus (*pr*) and close the old ones; the worm really had no notochord (*nt*); and the supposed change is not as simple as it seems. (From Wilder, History of the Human Body, by permission of Henry Holt & Co., publishers.)

forming from gut pouches just as it does in echinoderms. Further, whereas amphioxus has a long series of body segments, lower chordate types have only three—and this is also the case with echinoderms. Also, in certain of the hemichordates there is a ciliated larva (Fig. 12) of the same type as that of echinoderms—so similar, in fact, that until the life history was known, the hemichordate larvae were thought to be those of starfish!

Even biochemistry helps to establish the case. The proteins of blood serum vary greatly from form to form, and it appears in general that the more closely related two animals are, the more similar are these proteins. Tests of the sera of acorn worms and other lower chordates show their definite relationships to those of echinoderms, but not to other invertebrates. Again, muscle chemistry tends to link the two groups. The muscles of animals contain phosphorus compounds which speed up the energy release for muscle activity. In vertebrates a material combined with the phosphate is creatine; in most nonvertebrate groups another compound, arginine, is present instead. But some echinoderms have creatine as well as arginine, and arginine is present in some tunicates and present (as well as creatine) in some hemichordates—facts which tend further to link the two groups.

These arguments have been thought convincing by most workers and represent very much the majority view. However, there is not universal agreement, and some workers have recently presented cogent arguments against echinoderm affinities. Possibly in the future we will have to change our minds, but we still prefer the theory just outlined here.

CHORDATE PHYLOGENY. What do these resemblances mean? Surely the vertebrates and their chordate relatives are not derived from echinoderms, with their specialized organs and skeletal plates, and their pronounced (though secondary) radial symmetry. Such a form as a starfish or a sea urchin is obviously far off any line leading to a vertebrate. But one important point may be kept in mind. Most echinoderms are free-living and capable of locomotion to some degree, but the fossil record indicates that the ancestral echinoderms were sessile forms. One group of living echinoderms, the crinoids or sea lilies, is for the most part fixed in habitus. Attached by a stalk to the sea bottom, they spread out above their compact bodies a series of feathery arms; along these arms are bands of cilia, which filter out from the water the food particles upon which the sea lily subsists and carry them down to the mouth.

Here lies, one may believe, the clue to the whole story. This mode of life is precisely that of the little pterobranchs which, we have seen, are unquestionably related to vertebrates despite their simple structure and the absence in them of almost every diagnostic character of vertebrates and even of chordates in general.

Despite the contrasts between primitive echinoderms and pterobranchs, the two can be derived readily from an ancient common ancestor and, except for the little proboscis which tends to tie them to the acorn worms, the pterobranchs are certainly close to the pattern expected in this ancestor.

Clearly the evidence indicates that this ancestor was a sessile bottom dweller,

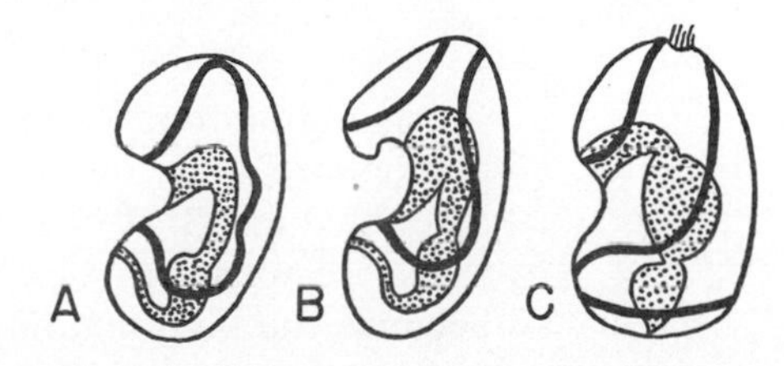

Figure 12. Diagrammatic side views of the larvae of (*A*) a sea cucumber, (*B*) a starfish, and (*C*) an acorn worm, all much enlarged. The black lines represent ciliated bands. The digestive tract (stippled) appears through the translucent body. Views are from the left side; the larvae are bilaterally symmetric. (After Delage and Hérouard.)

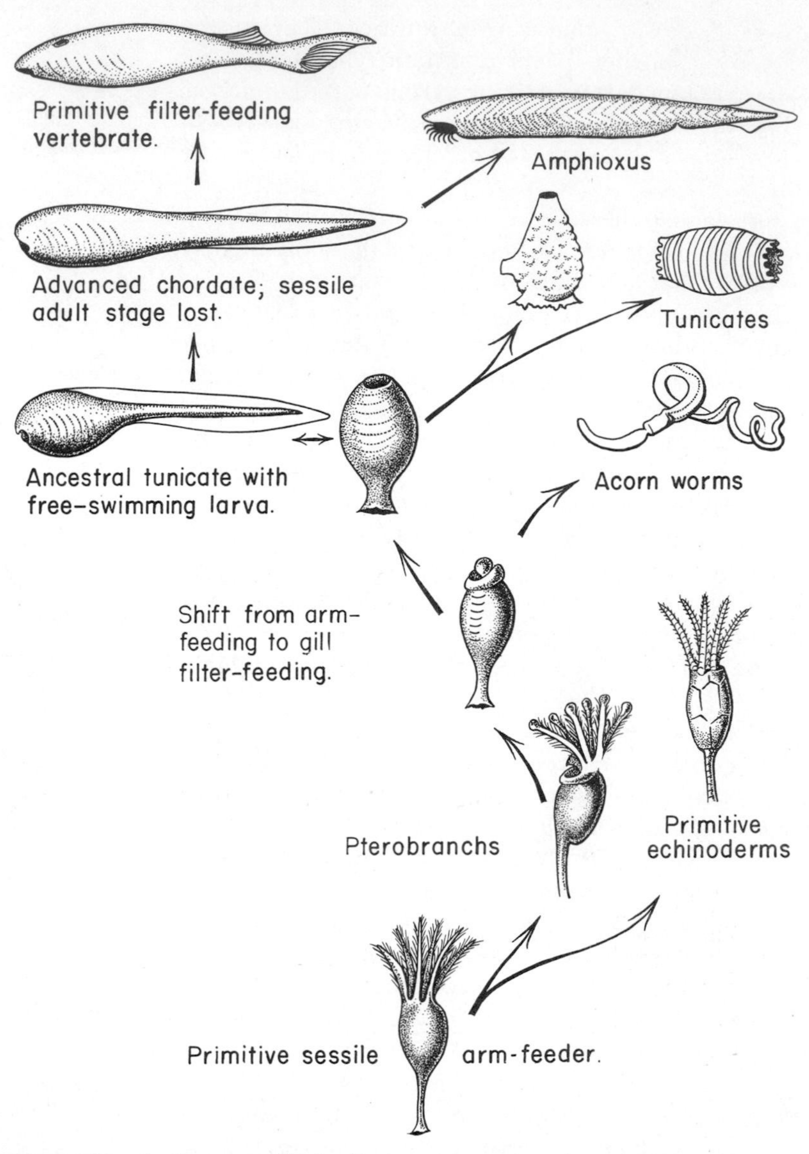

Figure 13. A diagrammatic family tree suggesting the possible mode of evolution of vertebrates. The echinoderms may have arisen from forms not too dissimilar to the little pterobranchs; the acorn worms, from pterobranch descendants which had evolved a gill system but were little more advanced in other regards. Tunicates represent a stage in which, in the adult, the gill apparatus has become highly evolved for feeding, but the important point is the development in some tunicates of a free-swimming larva with advanced features of notochord and nerve cord and free-swimming habits. In further progress to amphioxus and the vertebrates the old sessile adult stage has been abandoned, and it is the larval type that has initiated the advance. (From Romer, The Vertebrate Story, University of Chicago Press.)

subsisting on food particles gathered and brought to the mouth by outstretched lophophore arms. On this basis a reasonable theory of chordate evolution can be erected (Fig. 13). Little animals of this character, collecting food by lophophore arms, are not uncommon today, notable being brachiopods or lamp shells (so called because the body and lophophore are protected by a paired shell) and the tiny bryozoans—the "moss animalcules." These forms are suspected of at least distant relationships to echinoderms and chordates, and the chordate ancestor was rather surely one of a series of such lophophore-bearers. From it, with elaboration of varied specialized organs, may have arisen the echinoderms, and from it, with little change except a thickening of a collar region and a small proboscis, may have come the pterobranchs.

An early development of true chordate characters was a shift in the method of obtaining food particles—the substitution of gill-filtering for lophophores. Even in one of the pterobranchs a single pair of gill slits has developed, apparently aiding in the flow of food materials into the digestive tract. With the increase and elaboration of the gill-filtering system, lophophores were abandoned. The acorn worms, still essentially sedentary, appear to represent a side branch at this stage of evolution.

Further elaboration of the gill-straining device led eventually, in one higher branch, to the typical tunicates, in which the whole animal seems to be little more than an elaborate food-filter. But as this stage was approached, there appeared, it would seem, a new adaptation which was to radically alter the whole picture of higher chordate evolution.

The embryo or larva of a sessile organism must find a proper place on the sea bottom on which to settle down for adult life. How to reach and select it? Some acorn worms, we have noted, have a ciliated larva, but such a larva's powers of locomotion are limited. Much better is the tadpole-like larva which may have developed before the tunicate level was reached. Here a muscular tail is present; a notochord stiffens the structure; a nerve cord and nerves supervise locomotion; sense organs guide the movements of this new swimming structure toward a proper place for fixation and adult existence.

Once this new larval structure was evolved, a radical change of direction in chordate evolution opened out. A new active type of life became possible. Conservative forms specialized as tunicates. But for others, neoteny appears to have entered the picture. The animal ceased to "settle down." The larval locomotor structures were retained throughout life, even though filter-feeding long continued to be the means of sustenance; the pharyngeal filtering apparatus could be transported from place to place as favorable opportunities presented themselves. Amphioxus, although a bit off the direct ascending line, represents a more advanced stage, in which filter-feeding persists but the ancestral fixed adult condition has been abandoned. But for the most part, chordates did not stop at this level. A great burst of evolutionary activity resulted from the development of these new locomotor potentialities and led on to the major story of vertebrate evolution.

"VISCERAL" AND "SOMATIC." In later chapters we will frequently encounter the terms "visceral" and "somatic"—visceral and somatic skeletal structures, visceral and somatic muscles, visceral and somatic nerves. The visceral structures have mainly to do with the gut and its appendages (particularly the pharynx); the somatic structures are those of the "outer" tube of the body (Fig. 2 *B*). One might assume that the two terms had a mere topographic meaning and nothing more. But it is highly probable that there is a long phylogenetic history behind visceral-somatic distinctions (Fig. 14).

Such a lower chordate as a solitary tunicate consists of little else than a pharynx and appended gut tube, plus such necessary additions as the gonads and a very simple nervous system. Except that it is, of course, sheathed externally by skin or tunic, the whole animal represents essentially the visceral component of the vertebrate. The somatic component is the new, added series of locomotor

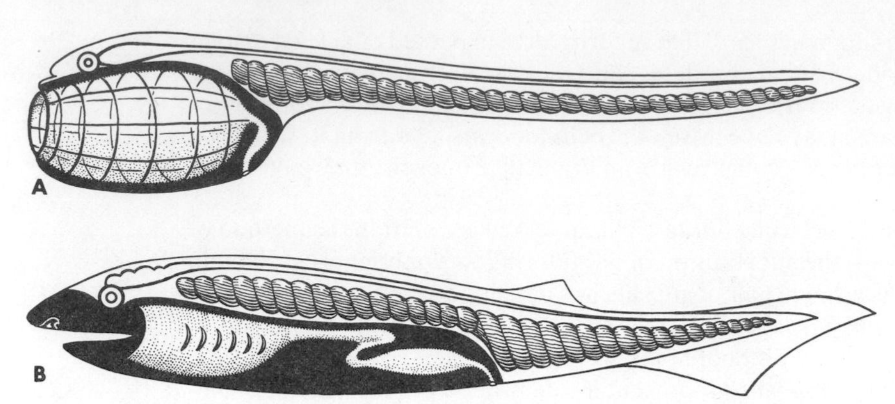

Figure 14. Diagrams to show the contrast between "visceral" and "somatic" components of the chordate body. *A,* A theoretical type of chordate essentially similar to the tunicate larva, but with the somatic component retained in the adult; below, a true vertebrate. The area of the visceral component is outlined in black. In *A* the somatic animal lies posterior to the visceral animal (representing the ancestral chordate), except that sense organs and anterior part of the nerve cord extend forward dorsally. In *B,* the visceral and somatic components overlap to a considerable degree and integration of the two is advancing.

devices—swimming muscles, notochord and, for their direction, a more highly evolved nervous system and sense organs. At first these somatic structures were for the most part appended posterior to the visceral animal and were mainly larval. As vertebrate evolution progressed, the two became more broadly overlapping and coordinated with one another. But even today, as seen in development and in adult structure, the original distinctions tend to persist. In many ways one can regard a vertebrate as two distinct animals, visceral and somatic. The two are welded into a single structure, but some traces of the distinctions between them still persist. The "weld" is an imperfect one.

3 Who's Who Among the Vertebrates

The study of organs and organ systems and their varied forms and functions—the main concern of the present work—gives us but a one-sided account of the vertebrates. What one should know is not merely the discrete parts, but the total animal, its life and its place in nature. Our present study no more gives us a rounded picture of the vertebrates than the dissection of a cadaver and a course in physiology would give us a complete knowledge of mankind. It is to be hoped that the student will read some works on the "natural history" of vertebrates and thus gain an idea of the living animals whose bodies are verbally dissected in this volume. We shall here survey the membership of the vertebrate groups in order to place the forms discussed within a phylogenetic framework.

THE GEOLOGIC RECORD

The fossil record and the extinct animals included in it require attention in this regard. In comparative anatomy one compares often the organs of *existing* members of different groups as if one had descended from the other; as if mammals had descended from the existing reptiles, and these from existing amphibians and fishes. Obviously, however, this is not the case. A turtle is a reptile, but it is not a mammal ancestor; it has had just as much time to diverge from the common primitive reptilian stock as has the mammal. A frog is an amphibian, but it is definitely not the sort of amphibian from which more progressive land vertebrates were derived. Only through paleontology, the study of fossils, can we hope to discover the nature of the actual ancestors from which the varied living vertebrates arose.

In discussing fossils, some notion of the geologic time scale is necessary (cf. Table 1). The earth's history of several billion years is divided by geologists into a few major time units termed **eras**; these are subdivided into a number of **periods**. For the earlier eras there is little adequate knowledge of life of any sort; the fossil record is almost entirely confined to the last three eras, spanning somewhat over half a billion years of earth history.

TABLE 1. **Geologic Periods Subsequent to the Time When Fossils First Became Abundant**

(The Carboniferous is frequently subdivided into two periods, Mississippian [earlier] and Pennsylvanian [later]. The time estimates are based on the rate of disintegration of radioactive materials found in a number of deposits.)

Era (and Duration)	*Period*	*Estimated Time Since Beginning of Each Period (in Millions of Years)*	*Epoch*	*Life*
Cenozoic (age of mammals; about 65 million years)	Quaternary	2+	Holocene (Recent)	Modern species and subspecies; dominance of man.
			Pleistocene	Modern species of mammals or their forerunners; decimation of large mammals; widespread glaciation.
	Tertiary	65	Pliocene	Appearance of many modern genera of mammals.
			Miocene	Rise of modern subfamilies of mammals; spread of grassy plains; evolution of grazing mammals.
			Oligocene	Rise of modern families of mammals.
			Eocene	Rise of modern orders and suborders of mammals.
			Paleocene	Dominance of archaic mammals.
Mesozoic (age of reptiles; about 165 million years)	Cretaceous	130		Dominance of angiosperm plants; extinction of large reptiles and ammonites by end of period.
	Jurassic	180		Reptiles dominant on land, sea, and in air; first birds; archaic mammals.
	Triassic	230		First dinosaurs, turtles, ichthyosaurs, plesiosaurs, mammals; cycads and conifers dominant.
Paleozoic (about 340 million years)	Permian	280		Radiation of reptiles, which displace amphibians as dominant group; widespread glaciation in southern hemisphere.
	Carboniferous	350		Fern and seed fern coal forests; sharks and crinoids abundant; radiation of amphibians; first reptiles.
	Devonian	400		Age of fishes; first trees, forests, and amphibians.
	Silurian	450		Invasion of the land by plants and arthropods; archaic fishes.
	Ordovician	500		Appearance of vertebrates (ostracoderms); brachiopods and cephalopods dominant.
	Cambrian	570		Appearance of all major invertebrate phyla and many classes; dominance of trilobites and brachiopods; diversified algae.

The first of these three, the **Paleozoic Era** or Age of Ancient Life, covered about 340 million years, and is divided into half a dozen periods. The fossil record remaining from the seas of the oldest period (Cambrian) contains abundant representatives of almost every major animal group, except the vertebrates. The first faint traces of backboned animals have been found in the rocks of the following Ordovician period, and a modest number of archaic jawless fishes have been found in the Silurian period that followed. In the Devonian period fishes were abundant and varied in fresh-water and marine deposits. The continental sediments of the Devonian indicate to the geologist that much of the land was subject to marked seasonal droughts, as are certain tropical regions today; times of abundant rainfall alternated with seasons when streams ran dry and pools were stagnant. These conditions appear to have had a major influence on the history of fishes and in the development of terrestrial life.

At the very end of the Devonian appeared the first four-footed vertebrates, the amphibians, and primitive members of that group are common in the swamp deposits that characterize the Carboniferous period, the age during which the major coal seams of Europe and North America were formed. Well before the end of that period the first reptiles had evolved, and early reptile orders were common land animals in the Permian period, with which the Paleozoic Era closed.

The **Mesozoic Era**, the "Middle Age" of the story of life, is frequently termed the Age of Reptiles, for members of that class dominated the land during that era, and many types of reptiles now extinct flourished in the seas and in the air as well. The highest of vertebrate groups, further, had their beginnings in the Mesozoic; the oldest mammals appeared near the end of the Triassic period and the oldest known birds appeared toward the end of the Jurassic, but both groups remained inconspicuous till the end of the era.

The **Cenozoic Era** is the Age of Modern Life or Age of Mammals. At the end of the Mesozoic the reptilian hordes became greatly reduced, leaving that class of vertebrates in its modern impoverished phase. Modern types of birds had appeared by the beginning of the Cenozoic, and, most conspicuously, the mammals rapidly evolved into the varied progressive groups which dominate the land today. Periods may be divided into subdivisions termed epochs; in our geologic chart we have listed the Cenozoic epochs to show, primarily, the stage-by-stage rise of the mammals during Cenozoic times.

VERTEBRATE CLASSIFICATION

The backboned animals constitute the major subphylum Vertebrata of the phylum Chordata (see Appendix 1 for a tabular presentation of vertebrate classification). The next step in classification is a division of the varied vertebrates into a series of classes. The distinguishing features of certain of these classes are obvious to anyone who has the slightest familiarity with nature. The class **Mammalia** includes the mammals, the familiar warm-blooded, hair-clothed animals among which man himself is to be included; the birds, class **Aves**, are readily distinguished by the presence of feathers and wings and by their possession, equally with mammals, of a high, controlled body temperature. The class **Reptilia**, lacking the progressive features of the birds and mammals, represents a lower level, mainly of land dwellers, with lizards, snakes, turtles, and crocodiles as living representatives.

A fourth group is that of the class **Amphibia**, including frogs, toads, and salamanders—four-legged animals, but reminiscent of fishes in many respects.

One commonly lumps the remaining lower vertebrates as "fish," and these forms (or most of them) are sometimes included in a single vertebrate class—the attitude being that, after all, they seem to be built on a common plan, as water dwellers with gills and locomotion performed by fins rather than limbs. But this is a rather personal, human viewpoint. An intellectual and indignant codfish could point out that this is no more sensible than putting all land animals in a single class, since, from his point of view, frogs and men, as four-limbed lung-breathers, are much alike. Actually, a codfish and a lamprey, at two extremes of the fishy world, are as different structurally as amphibian and mammal. The fishes are perhaps best arranged in three classes of lower vertebrates: class **Agnatha** for jawless vertebrates, such as the living lampreys and fossil relatives; class **Elasmobranchiomorphi** for certain extinct armored fishes and the more modern cartilaginous fishes, sharks and their relatives; and class **Osteichthyes**, the higher bony fishes which today constitute most of the piscine world.

If we wish to group these seven classes, we may for convenience consider the four higher land groups as constituting a superclass **Tetrapoda**, or four-footed animals, the fishes as making up a superclass **Pisces**:

Superclass Pisces	Superclass Tetrapoda
Class Agnatha	Class Amphibia
Class Elasmobranchiomorphi	Class Reptilia
Class Osteichthyes	Class Aves
	Class Mammalia

This is but one of several alternative methods of grouping the vertebrate classes. Some, placing emphasis on the development of jaws, would contrast with the Agnatha all the remaining vertebrates as **Gnathostomata**, "jaw-mouthed" forms. Still another grouping is to consider the three highest classes as forming a group termed **Amniota**, the remaining four constituting the **Anamniota**. This is based upon the fact that the lower types generally have a rather simple mode of reproduction, with eggs laid in the water and young developing there, whereas reptiles evolved a shelled egg, laid on land, within which a complex sort of development (described in a later chapter) takes place. Some reptiles and nearly all mammals bear their young alive, but have retained the same general pattern of embryonic development, and the name Amniota is derived from the amnion, one of the membranes surrounding the growing embryo in mammals and birds as well as reptiles.

In its simplest form the phylogeny of the vertebrate classes (Fig. 15) may be diagrammed thus:

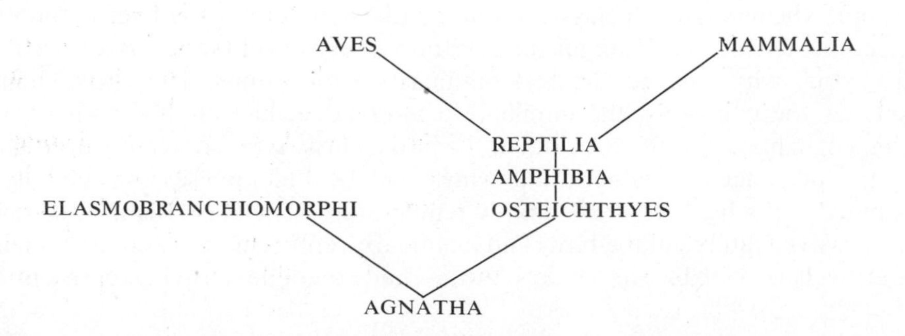

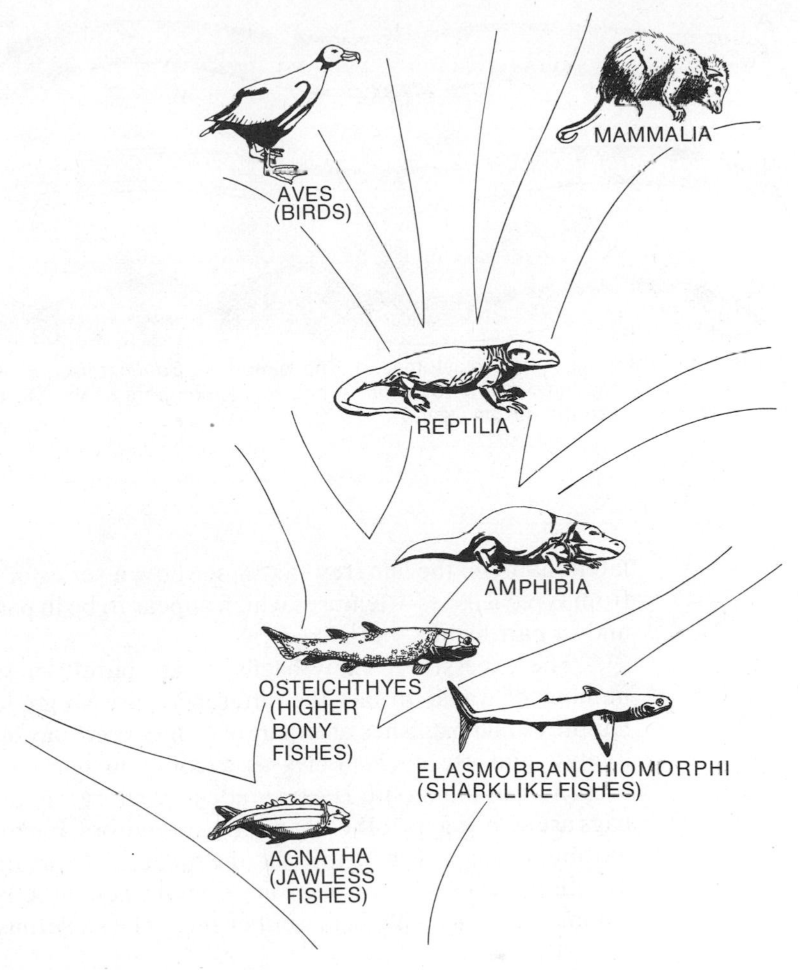

Figure 15. A simplified family tree of the classes of vertebrates. (Mainly after Romer.)

JAWLESS VERTEBRATES

Living **lampreys** and **hagfishes**, together termed **cyclostomes** (Figs. 16, 17), are representatives of a lowly group, the class **Agnatha**—jawless vertebrates. Best known is the marine lamprey (*Petromyzon*). This fish is eel-like in appearance, but much more primitive in its structure than true eels (which are highly developed bony fishes). The lamprey is soft-bodied and scaleless and, though having a feeble skeleton of cartilage, lacks bone entirely. There are no traces of paired fins, and, most especially, jaws are totally lacking. The adult lamprey is predaceous; nevertheless, the rounded mouth cup forms an adhesive disc by which it attaches to the higher types of fishes upon which it preys as a bloodsucker, and a rasping tonguelike structure within the mouth is a fairly effective substitute for the absent jaws. There is but a single nostril, opening high on top of the head, and having a hypophysial pouch (cf. p. 342) combined with it. The gill passages are not slits, as in typical fishes, but spherical pouches, connected by narrower tubes with the pharynx and body surface. In various less obvious structural characters, noted in

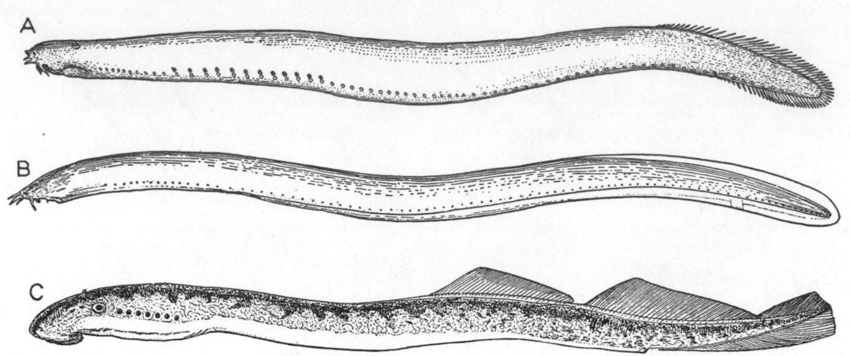

Figure 16. Three types of cyclostomes. *A*, The slime hag, *Bdellostoma; B*, the hagfish, *Myxine; C*, the lamprey, *Petromyzon. A* and *B* are members of the Myxinoidea; *C*, of the Petromyzontia. (From Dean.)

later chapters, the lampreys likewise show a series of features in which they differ from typical fishes—features which appear to be in part primitive, in part aberrant, and in part highly specialized.

The excessively slimy hagfishes are purely marine in habit, and differ in a number of ways. In fact the differences are so great that we prefer to treat the lampreys and hagfishes as separate orders (**Petromyzontia** and **Myxinoidea** respectively), and use cyclostomes as a common name only. The rasping tongue is present, but the mouth is surrounded by short tentacles instead of a sucker. The hags are scavengers rather than active predators, burrowing into the flesh of dead or moribund fishes. The nostril is at the tip of the snout rather than atop the head, and the gill pouches in some hagfishes do not open directly to the surface but join to a common external opening on either side. The skeletons of hagfish and lampreys are also very different.

The hagfish eggs are laid in the sea and the young develop directly there; the marine lamprey, in contrast, has a distinct freshwater larval stage. Every spring, lampreys ascend the streams to spawn, and the developing young spend several years of their lives as little larvae (ammocoetes), which lie nearly buried in the mud of brooks and streams. These larvae are not at all predaceous; there is no rasping tongue or mouth sucker. Instead, they are filter-feeders, which strain food particles

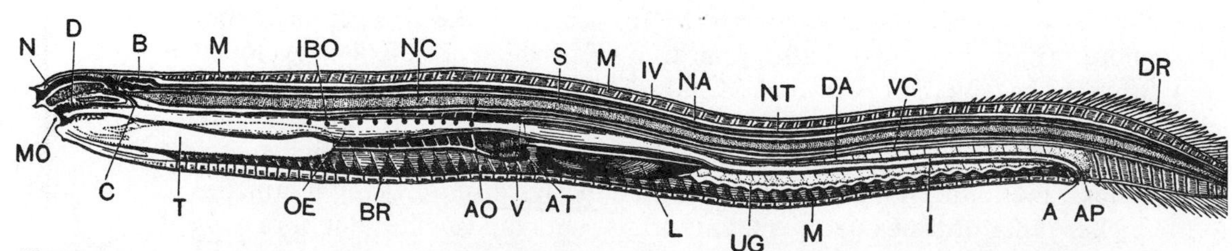

Figure 17. Longitudinal section of a slime hag, *Bdellostoma. A*, Anus; *AO*, ventral aorta; *AP*, abdominal pore; *AT*, atrium of heart; *B*, brain; *BR*, gill pouch; *C*, duct from nasal pit to throat; *D*, horny, toothlike structures; *DA*, dorsal aorta; *DR*, dorsal fin rays; *I*, intestine; *IBO*, internal gill openings; *IV*, septum between muscle plates; *L*, liver; *M*, muscle segments; *MO*, mouth; *N*, nostril sac; *NA*, sheath of spinal cord; *NC*, notochord; *NT*, neural tube (spinal cord); *OE*, pharynx; *S*, sheath of notochord; *T*, extrusible "tongue"; *UG*, urogenital organs; *V*, ventricle of heart; *VC*, posterior cardinal vein. (From Dean.)

much as does amphioxus. A stream of water is brought into the mouth by muscular action (not ciliary, as in amphioxus), passes through a pharynx which even has a structure comparable to the amphioxus endostyle, and thence flows out the gill slits. At the end of the larval period there is a sudden marked change in structure—a **metamorphosis**—and the young lamprey, with adult features fully developed, descends to the sea. It is possible, however, for lampreys to remain in fresh waters for their entire lives—the sea lamprey has successfully invaded the American Great Lakes—and certain small species of lampreys never take up a predaceous life, but reproduce and die in their native streams.

It is generally agreed that the absence of jaws, and, probably, of fins, is a primitive feature of cyclostomes. Other characters, however, are more dubiously primitive. There is considerable reason to regard the absence of a bony skeleton as a degenerate feature; the predaceous or scavenging habits can hardly have been present in ancestral vertebrates (mutual cannibalism is not, to say the least, advantageous), and the rasping tongue is a lamprey specialty. Cyclostomes represent a primitive level of vertebrate development; they are not, however, in themselves ancestral vertebrates.

When we look into the fossil record, we find that the oldest and most primitive of fossil vertebrates, found in Ordovician and Silurian deposits and surviving into the Devonian, were small fishlike creatures known as **ostracoderms** (Fig. 18); these are of several groups (orders) and thus ostracoderm does not appear as a formal term in Appendix 1. Superficially there is little resemblance to the cyclostomes, but study during this century, especially by the Swedish paleontologist E. A. Stensiö, shows that ostracoderms were jawless ancient representatives of the class Agnatha.

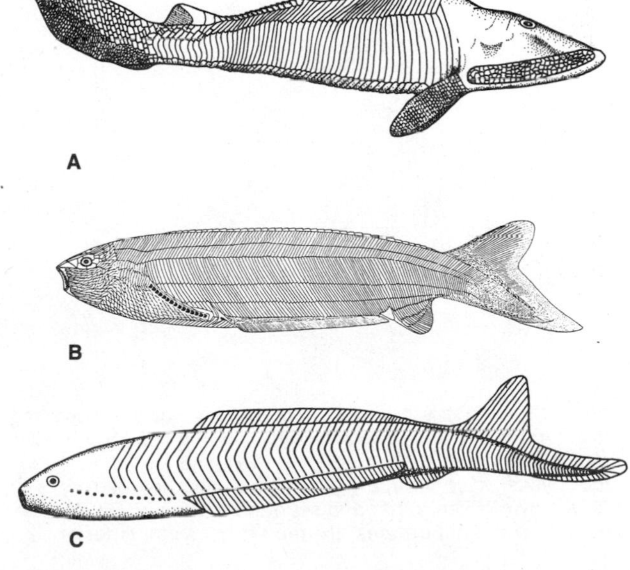

Figure 18. Fossil ostracoderms. *A, Hemicyclaspis,* a member of the Osteostraci; *B, Pharyngolepis; C, Jamoytius,* two members of the Anaspida. In all of these forms the gills had separate openings to the outside and there was a single median nostril on the top of the head. (*A* and *C* after Moy-Thomas and Miles; *B* after Ritchie.)

Many ostracoderms (like cyclostomes) lacked paired appendages, although in some there were aids to navigation in the form of peculiar flaps projecting from the body behind the head rather comparable to pectoral fins, or folds extending out on either side of the trunk as "stabilizers."

A major contrast with the modern cyclostomes lies in the skeleton. All ostracoderms were covered by a good bony armor or at least scales, and in some the head also contained an internal bony skeleton. It was formerly assumed that the primitive vertebrates were (like the living cyclostomes and sharks) boneless forms, with a skeleton of cartilage only. This may have been true of the still older ancestral chordates and immature ostracoderms. But the prevalence of bone in the oldest known fossil vertebrates, and evidence of reduction instead of increase in ossification in the later history of many fish groups, suggest that ancestral vertebrates were armored as adults, and that absence of bone in the lower living vertebrates is a degenerate rather than a primitive characteristic.

As to the reasons for this early development of bone, we are none too certain. One suggestion lies in the fact that we find in association with these ancient vertebrates, in the stream deposits in which many of them are found, remains of eurypterids—ancient water scorpions—and less familiar crustaceans termed ceratiocarids. Both of these arthropod types were voracious and, on the average, considerably larger than the little ostracoderms amidst which they lived. It may be that in their earliest phases the vertebrates were the underdogs; bony armor may have been a defense against these invertebrate enemies. Later, as vertebrates became larger, speedier, and themselves predaceous, the eurypterids vanished from the fossil record and the ceratiocarids shrank to insignificance. Other suggestions are that the armor helped to prevent undue water loss, or represented a store of calcium salts; none of these ideas is really convincing.

Ostracoderms are not one small group of similar animals—they are a very mixed bag of ancient jawless forms, usually (as here) divided into four distinct orders. Best known are the **Osteostraci,** forms such as *Cephalaspis* (Fig. 18 *A*). These had a greatly expanded "head" region (Fig. 19), most of which was occupied

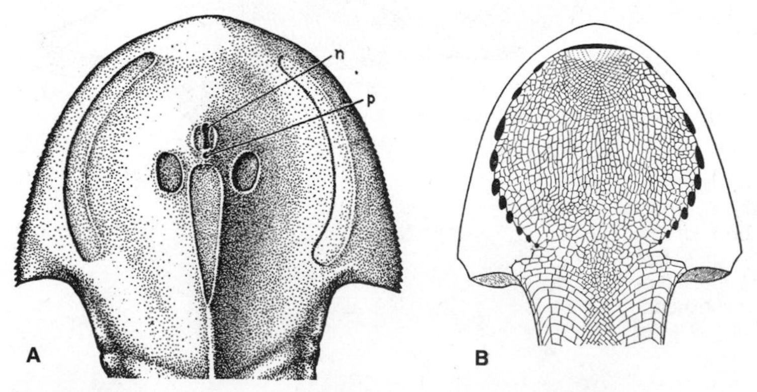

Figure 19. *A,* Dorsal and *B,* ventral views of the head region of a fossil ostracoderm of the *Cephalaspis* type (cf. Fig. 18 *A*). Dorsally are seen openings for the paired eyes, median eye (*p*), and a median slit (*n*) for nostril and hypophyseal sac. Ventrally, the throat was covered by a mosaic of small plates, covering an expanded set of gill pouches. Round openings on either side are for the gill orifices; the mouth is a small anterior slit. (After Stensiö.)

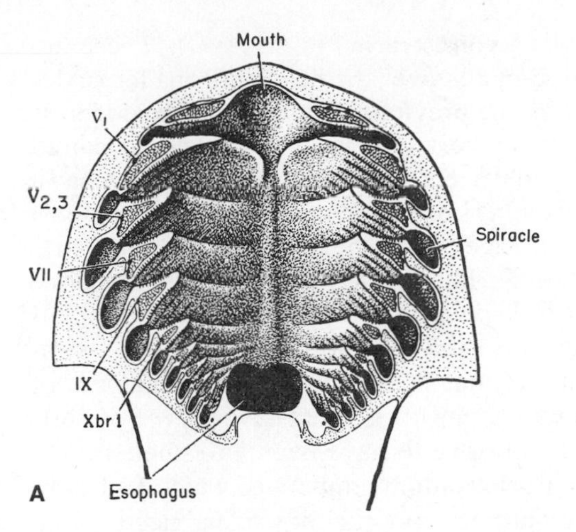

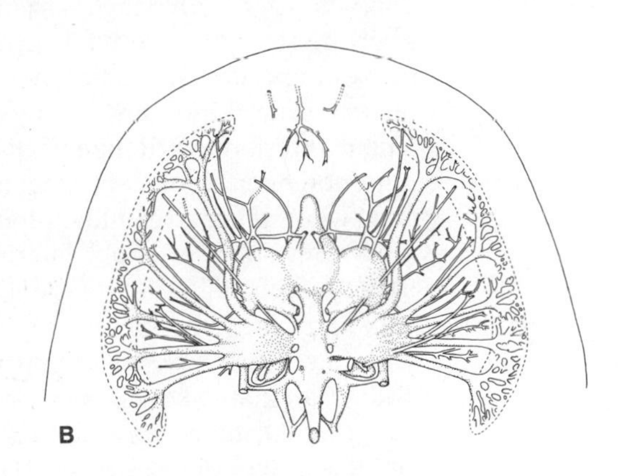

Figure 20. Detailed anatomy of the head of ostracoderms of the order Osteostraci. *A,* A restoration of the pharyngeal region, as seen in ventral view after removal of the bones on the underside of the head. The mouth is very small; the pharynx, an expanded food-straining device. V_1, V_2, V_3, VII, IX, Xbrl, Sections of branchial nerves associated with successive gills (cf. Fig. 396). By this interpretation, two gill pouches appear anterior to that representing the spiracle, but other workers believe the first pouch here to be the spiracle. *B,* Cast of the cranial cavity, the orbits, and canals for the ear, nerves, and blood vessels, as seen in ventral view. The Swedish paleontologist Stensiö has demonstrated the anatomy of these forms in exquisite detail. (After Stensiö.)

by large branchial chambers. Since they lacked jaws or other biting or rasping structures, it seems obvious that these very old vertebrates, like their chordate ancestors and like the larval lampreys of today, made their living by straining food materials through their gill system. Many of them, although capable of locomotion with a fishy tail, were much flattened and must have been relatively sluggish animals. The entire head is enclosed in a solid bony shield; careful dissection of the shields has revealed the detailed structure of the long since rotted gills, brain, cranial nerves, and blood vessels (Fig. 20). In the Osteostraci, as in the modern lampreys, there was a single nostril placed dorsally between the eyes.

A second order, the **Anaspida** (Fig. 18 *B*), shares with the Osteostraci the characteristic of a single median nostril on top of the head. However these were not flattened bottom dwellers, but appear more like active swimmers (although fins were very poorly developed, so we may doubt that they were very agile). The tails tip downwards, the reverse of the shark condition—despite some early pictures that show them restored upside down. Instead of heavy armor, these forms possessed a thinner covering of scalelike plates; internal structures are unknown. The gills opened through a series of small, lateral, circular pores. Although some workers have suggested that the anaspids were free-swimming and lived on plankton near the surface of the water, it seems more probable that they, like the osteostracans, were basically bottom dwellers, filtering out food particles from the mud. Interestingly, at least one anaspid (Fig. 18 *C*) shows a reduced skeleton and appears, for a variety of reasons, to be close to the ancestry of modern lampreys. The relationships of the hagfish are not known; some think them moderately closely related to lampreys and thus, presumably, descended from anaspids, but others claim they are derived from very different forms.

A third and rather large order is the **Heterostraci** (Fig. 21 *A–C*). These are characterized by having an armor of large plates over the head and smaller scales over the rest of the body. They differ from the previously described ostracoderms in lacking a dorsal median nostril; most workers believe they possessed paired nostrils (like us, goldfish, or any other jawed vertebrates), but a few claim they had a single anterior nostril (like hagfish). In members of this group, the gills possessed a common opening on each side, at the back edge of the "head." Typical heterostracans were rather streamlined-looking—almost like cartoon space ships—but paired fins were lacking although lateral and dorsal spines may have provided some stabilizing effects. The eyes were lateral, not dorsal as in bottom-dwelling osteostracans, and the bony plates around the mouth may have allowed a small amount of nibbling. Thus these forms may have been scavengers or even have preyed on soft-bodied animals like worms; they would not have to have been filter-feeders.

Finally there is a small order, the **Coelolepida**, members of which had only small scales in the skin (Fig. 21 *D*). They are very poorly known, and need not be treated here.

Those, then, are the jawless vertebrates—two groups of highly specialized and often degenerate forms still extant and four groups of extinct and thus rather poorly known animals. Unfortunately, their relationships with other groups are a mystery. There are no known intermediates between the lower chordates and any of these forms. The living lampreys appear to be highly modified descendants of the anaspids; the origin of the hagfish is disputed. Presumably the jawed vertebrates arose from some group of ostracoderms. But again, intermediates are completely lacking and we know nothing of the actual stages.

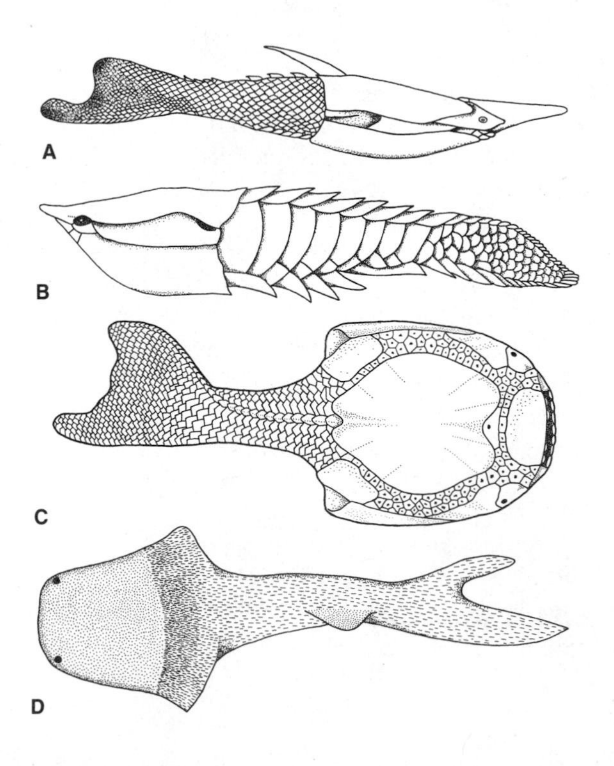

Figure 21. Fossil ostracoderms. *A, B,* and *C* are Heterostraci and *D* is a member of the Coelolepida. The heterostracans were relatively streamlined forms; the true shape of coelolepids is generally not known. *A, Pteraspis; B, Anglaspis; C, Drepanaspis; D, Logania.* (After Moy-Thomas and Miles.)

ELASMOBRANCHIOMORPHS

PLACODERMS. The ostracoderms were at the peak of their development during the Silurian. At the end of that period there appeared somewhat more advanced fish types which were exceedingly prominent in the following Devonian period, but became extinct well before the close of the Paleozoic. These were mostly grotesque forms, quite unlike any fishes living today.

These forms were for a long time considered to form a separate class, the Placodermi, thought to be ancestral to all higher vertebrates. However, more recent work has shown that one group originally placed in this "class" belongs with the bony fish to be discussed later; most of them seem more closely related to sharks and their allies so that the two may be lumped into one class, the **Elasmobranchiomorphi**. The **Placodermi** are, then, reduced to the status of a subclass of the elasmobranchiomorphs.

All had jaws. This represents a major advance over the ostracoderms, one which opened up new avenues of life to fishes and enabled them to become more active and wider-ranging animals. The term **gnathostome**—"jaw-mouthed"—is often applied to placoderms and all higher vertebrates, in contrast with the Agnatha. Originally the jaws of placoderms were thought to be primitive or aberrant in structure—not supported by the next posterior gill arch (the hyoid arch; cf. Fig. 164). This is now doubted by most workers. Unfortunately the hyoid and other gill arches were, presumably, cartilaginous—at least we do not find them in fossils—so we cannot be sure. Paired fins, too, were developing, in connection with the new freedom which fishes were acquiring, but these structures were variable and often oddly designed (from a modern point of view) as if nature were still "experimenting" with them.

Best known of placoderms were the members of the order **Arthrodira**—jointed-necked fishes (Fig. 22 *A*, *B*). In these the head and gill region was covered by a great bony shield, and a ring of armor sheathed much of the body; the two sets of armor were connected by a pair of movable joints (hence the name). Peculiar bony plates served the function of jaws and teeth. The posterior part of the body was quite naked in typical arthrodires but there was a covering of bony scales in the most primitive types. In some forms there have been found true paired fins, but in the most primitive arthrodires we find little but a pair of large, hollow, fixed spines projecting outward from the shoulder region—some sort of holdfast or balancing structures.

Arthrodires were, for the most part, active predators, living on the diverse fish of the Devonian lakes and seas. Indeed, one of them was the largest form of its time, *Dunkleosteus* from the black shales of northern Ohio, a fish similar in shape to the one shown in Figure 22 *B* but reaching a length of nine meters! However, some arthrodires became flattened bottom-dwellers with reduced armor and weak jaws.

The other placoderms are all variants on the arthrodire pattern. The armor that split into separate parts covering the head, lower jaw, and anterior part of the trunk, the presence of some sort of spine in front of the pectoral fins, and tendencies towards bottom dwelling, flattening of the body, and reduction of armor are common, but not universal, characteristics of the different orders. All of the trends we mentioned appear in members of the orders **Phyllolepida** (Fig. 22 *C*), **Petalichthyida** (Fig. 22 *D*), and **Rhenanida** (Fig. 22 *E*); they are a mixed (and poorly

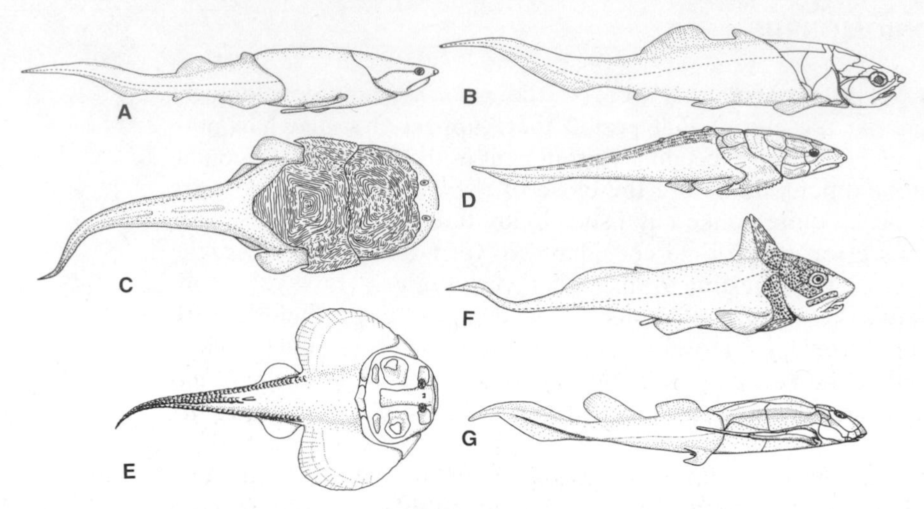

Figure 22. Various placoderms. *A,* The arthrodire *Arctolepis; B,* the arthrodire *Coccosteus; C,* the phyllolepid *Phyllolepis; D,* the petalichthyid *Lunaspis; E,* the rhenanid *Gemuendina; F,* the ptyctodontid *Rhamphodopsis; G,* the antiarch *Bothriolepis. C* and *E* show dorsal views; all others are lateral. (After Stensiö.)

understood) lot, but appear to be early attempts at producing fish comparable to the modern skates and rays. Another order, the **Ptyctodontida**, includes a group that apparently became specialized as crushers of molluscs (Fig. 22 *F*). It is of considerable interest that at least one ptyctodont possessed specialized "claspers" on the pelvic fins of males; these are used in mating and are otherwise known only in the cartilaginous fish or Chondrichthyes. Their presence here is one of the characters linking placoderms and chondrichthyeans.

A last order of placoderms is the **Antiarchi** (Fig. 22 *G*)—grotesque little animals which had two sets of armor like the arthrodires, but small heads, tiny nibbling jaw plates and, for forelimbs, a pair of jointed "flippers" projecting from the body like bony wings. These were pectoral fins, but had dermal armor completely encasing them so that they appeared more like the appendages of a lobster than any sort of fish fins. The heavy armor, flat ventral surface, and dorsally placed eyes indicate that these forms were bottom dwellers and probably relatively inactive.

Most of the placoderms were obviously far off the main lines of vertebrate evolution, and few of the known types can be regarded as actual ancestors of later vertebrates. However, it is highly probable that, with loss of bony armor, certain placoderm types gave rise to the sharks and chimaeras, the second subclass of the Elasmobranchiomorphi.

CHONDRICHTHYEANS. The modern sharks are the typical representatives of a major surviving group of jaw-bearing marine fishes—the subclass **Chondrichthyes**. This name, "cartilaginous fishes," refers to the fact that bone is virtually unknown in any member of the group (there *may* be some bone at the bases of the small scales or denticles). It seems probable that the absence of bone in sharks is due to a process of reduction; the toothlike denticles present in the shark skin and the spines sometimes present on the fins appear to be the last remnants of the armor which once sheathed their placoderm forebears. The main group of cartilaginous fish forms the infraclass **Elasmobranchii**—the sharks, skates, and rays plus some ex-

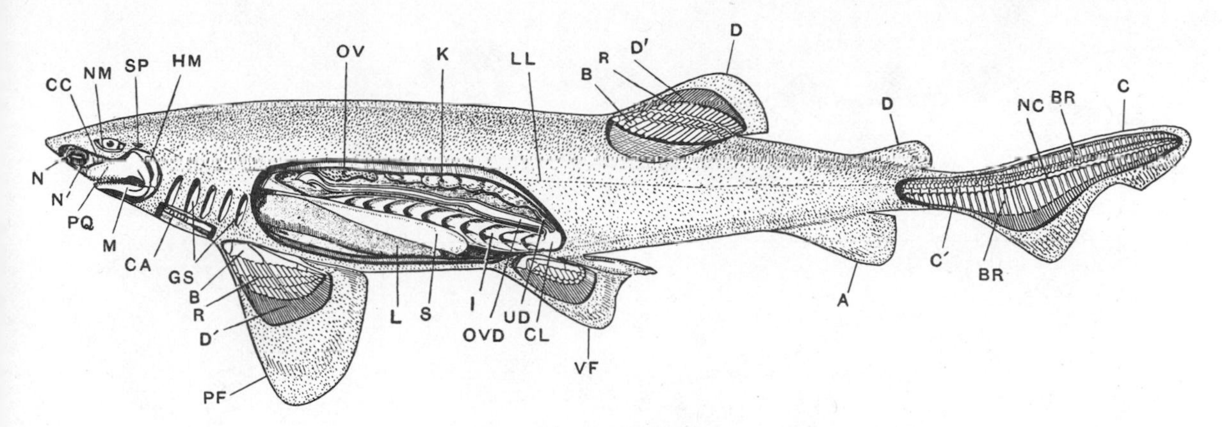

Figure 23. Diagrammatic dissection of a female shark. *A,* Anal fin; *B,* basal elements of fin; *BR,* basal and radial elements, upper and lower lobes of caudal fin; *C,* caudal fin; *C',* centrum; *CA,* conus arteriosus; *CC,* cartilaginous cranium; *CL,* cloaca; *D,* dorsal fins; *D',* dermal rays of fin; *GS,* gill slits; *HM,* hyomandibular; *I,* intestine with spiral valve; *K,* kidney; *L,* liver; *LL,* lateral line; *M,* mandible; *N, N',* anterior and posterior openings to nasal pouch; *NC,* notochord; *NM,* "nictitating membrane" of eye; *OV,* ovary; *OVD,* oviduct; *PF,* pectoral fin; *PQ,* upper jaw cartilage; *R,* radials of fin; *S,* stomach; *SP,* spiracle; *UD,* urinary duct; *VF,* pelvic (ventral) fin. (From Dean.)

tinct relatives (Fig. 23). Present are well formed jaws, although, in the absence of bone, there is no formed skull and the upper jaws are independent of the braincase. The gills border slitlike passages, typically five in number, which open separately to the surface, and there is generally a small accessory anterior opening (the spiracle). Absent are the peculiarities of the cyclostomes and the curious structural "experiments" seen in arthrodires. The nostrils are double, and placed beneath the tip of the snout.

A feature of sharks and their relatives, which is probably not primitive, is the fact that they produce large eggs containing considerable yolk. These eggs are, in many members of the group, encased in a horny shell before they are laid. To effect this, they must be fertilized before leaving the mother's body, and the male sharks have developed "claspers" projecting from the pelvic fins to aid in introduction of the sperm. Internal fertilization allows the possibility of development of the young within the body of the mother. In various sharks and rays the fertilized eggs are retained in the mother's reproductive tract, and develop there, so that the young are born alive (a limited number of reptiles and nearly all mammals have similarly developed this procedure).

Elasmobranchs first appear in the latter part of the Devonian, and *Cladoselache* (Fig. 24 *A*) of that period is a form which may be close to the ancestry of later sharklike fishes. It is representative of the order **Cladoselachii**—a group characterized by broadly attached paired fins and the absence of claspers on the pelvic fins. Both of these traits have been considered primitive, but in both cases this is debatable. An aberrant offshoot of the cladoselachians were the **Pleuracanthodii** (Fig. 24 *B*)—a small group of Paleozoic (and earliest Mesozoic) forms which lived in fresh water.

There was a variety of shark forms in the seas of the late Paleozoic, and toward the end of the Mesozoic we find shark types similar to those in modern oceans (Figs. 23, 24 *C*). All these belong in the large order **Selachii,** the normal sharks.

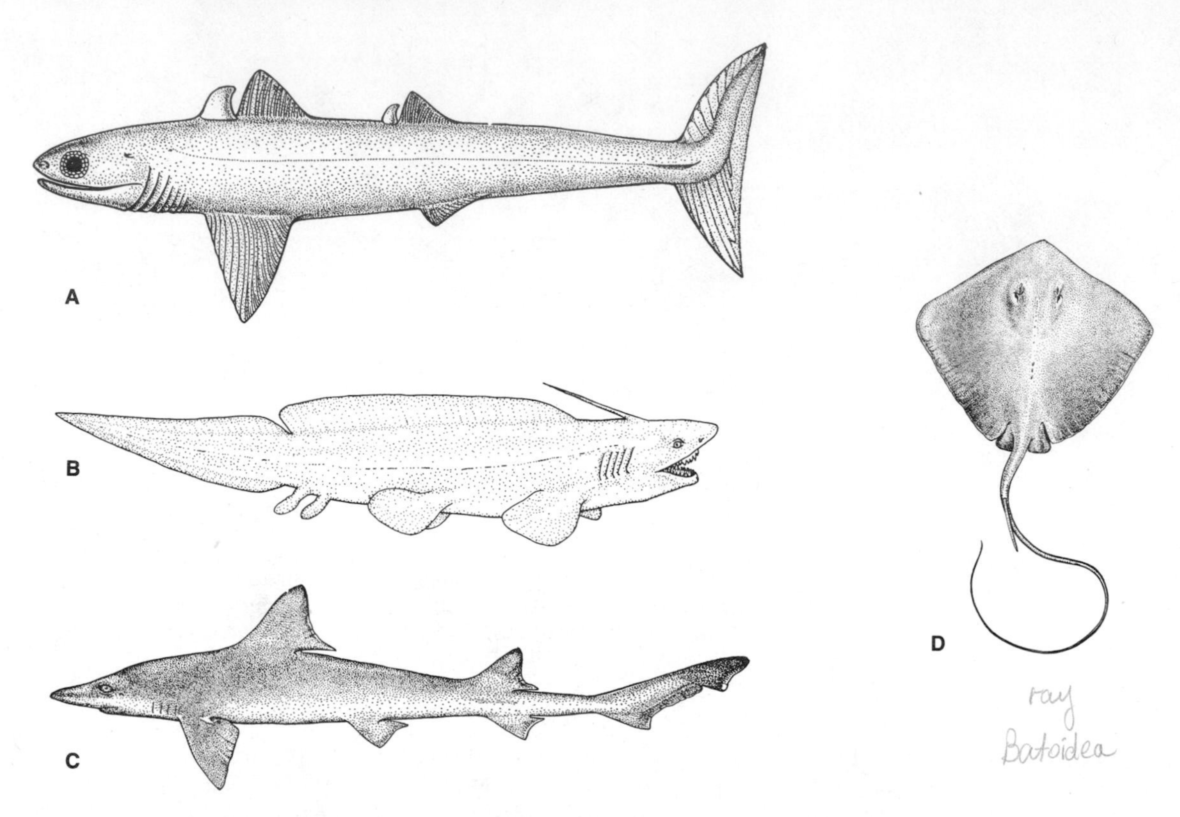

Figure 24. Elasmobranch fishes, members of the Elasmobranchiomorphi which have cartilaginous skeletons in which the upper jaw is not fused to the braincase. *A, Cladoselache,* a primitive member of the Cladoselachii; *B, Xenacanthus,* one of the Pleuracanthodii; *C, Mustelus,* a typical shark of the Selachii; *D, Dasybatis,* one of the greatly flattened forms of the order Batoidea, with greatly expanded pectoral fins and a long whiplash tail. (*A* after Dean and Harris; *B* after Špinar and Burian; *C* and *D* after Garman.)

They are almost purely predaceous in habit, and with few exceptions are purely marine. Except for skeletal degeneration, the sharks appear to be, in general, "proper" fishes of a fairly primitive type. They possess well developed paired fins, and a powerful tail fin with the tip of the body curving into its upper lobe.

In the Mesozoic, too, there appears the first of the skates and rays, members of the order **Batoidea** (Fig. 24 *D*). These are forms derived from sharks; they have taken to a mollusc-eating diet and a bottom-dwelling mode of life with which is correlated their flattened body shape. In typical rays the tail and pelvic fins are much reduced, but the pectoral fins are greatly expanded and, stretching forward above the gill openings, may meet in front of the head; locomotion is accomplished by undulatory movements of these broad appendages. Since in a resting position the mouth may be buried in the mud or sand of the sea bottom, the spiracle (small or even absent in sharks) is here a large opening behind the eyes through which water enters the pharynx.

A distinct group of cartilaginous fishes is that of the chimaeras or ratfish—the **Holocephali** (Figs. 25, 26), which are relatively rare oceanic forms. These are, like the skates, mainly mollusc-eaters; the body is not greatly depressed, and their peculiarities include, among other features, the development of large tooth plates,

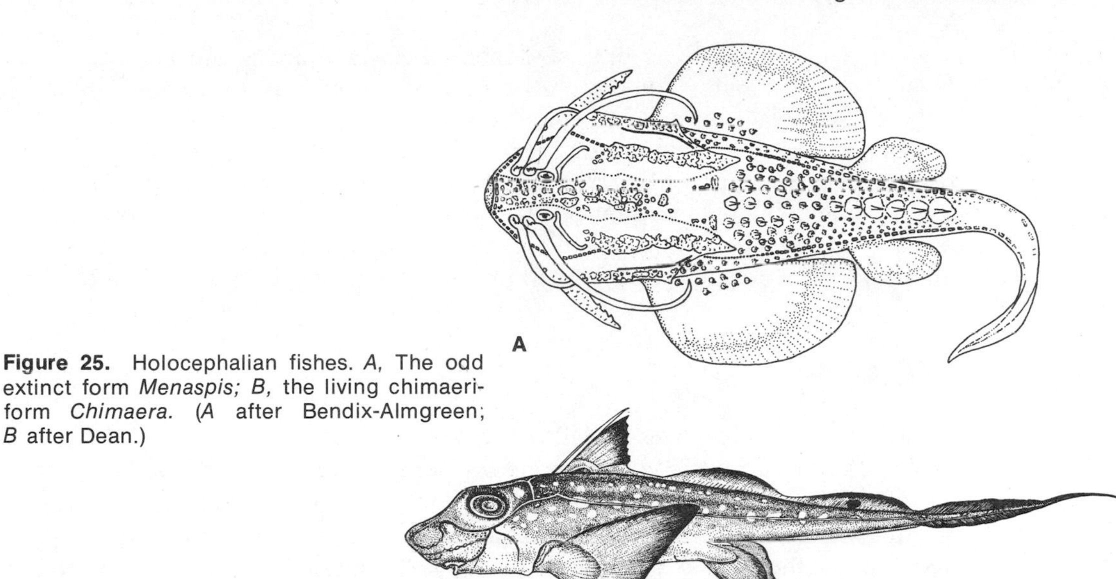

Figure 25. Holocephalian fishes. *A,* The odd extinct form *Menaspis; B,* the living chimaeriform *Chimaera.* (*A* after Bendix-Almgreen; *B* after Dean.)

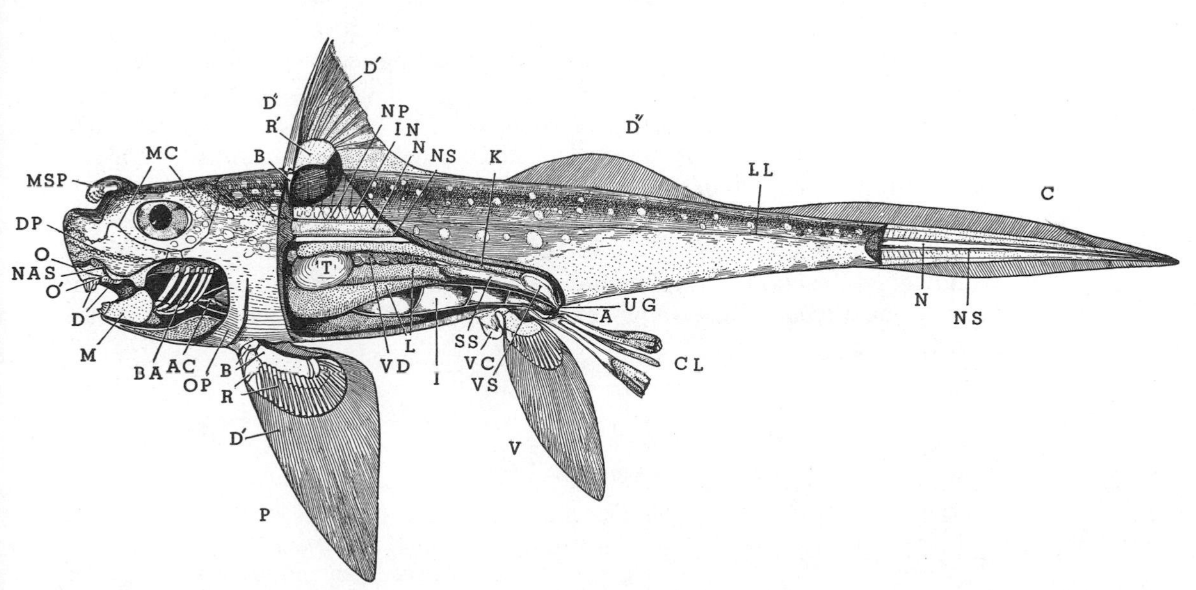

Figure 26. Diagrammatic dissection of a male chimaeroid. *A,* Anus; *AC,* conus arteriosus; *B,* basal element of fin; *BA,* branchial arches; *C,* caudal fin; *CL,* clasper; *D,* dental plates; *D′*, dermal rays of fin; *D″*, dorsal fins; *DP,* small dermal plates of lateral line canal; *I,* intestine with spiral valve; *IN,* interneural cartilages of vertebral column; *K,* kidney; *L,* liver; *LL,* lateral line groove; *M,* lower jaw cartilage; *MC,* lateral line canals (mucous canals) of head; *MSP,* frontal spine peculiar to male; *N,* notochord; *NAS,* nasal pouch; *NP,* neural arch; *NS,* sheath of notochord; *O, O′*, grooves leading to and from nasal cavity, covered by fold of skin; *OP,* operculum; *P,* pectoral fin; *R,* radial cartilages of fin; *R′*, fused radials of anterior dorsal fin; *SS,* sperm sac; *T,* testis; *UG,* urogenital opening; *V,* pelvic (or ventral) fin; *VC,* anterior accessory clasper; *VD,* epididymis; *VS,* sperm vesicle. (From Dean.)

and of upper jaws which (in contrast to those of elasmobranchs) are solidly fused to the braincase. A flap of skin covers the gill region (much as does a series of bony plates in both placoderms and higher bony fishes).

Chimaeras were probably derived from placoderms, as were the sharks; the fossil record, while imperfect, suggests that the chimaeras trace back to the placoderms along a line independent of the sharks. There are quite a few fossil Holocephali—mostly poorly known and bizarre (Fig. 25 *A*); the more normal (or at least familiar) living forms are all included in the order **Chimaeriformes** (Fig. 25 *B*).

BONY FISHES

The class **Osteichthyes** includes the vast majority of fishes. As the name implies, they are forms in which a bony skeleton has been retained, and improved upon. It was once believed that these fishes were descendants of sharklike forms and that bone was in them a new acquisition. It now, however, appears more probable that the bony skeleton here is simply a retention and improvement of that which was present in the ancestral vertebrates.

Below, we shall outline the general evolutionary history of the great groups of fishes which, beginning in the Devonian, are universally agreed to be proper members of the Osteichthyes. But first we must mention the interesting but puzzling little group of very ancient types, the **Acanthodii** (Fig. 27). These were jawed fishes which appeared even earlier than the placoderms, for there are fragmentary remains of acanthodians well back in the Silurian. Most were small, of roughly minnow size. As in ostracoderms and placoderms the head and body were covered with bony plates and scales, and there was a fair amount of internal ossification. The tail was strongly tilted upward in a sharklike pattern. Paired fins were developed, but in most unusual fashion. Each fin was supported by a very stout spine; and in addition to the normal two pairs of paired fins, accessory pairs, up to as many as five, may be present as well. Similar spines lay in front of the dorsal and anal fins. The acanthodians are often called "spiny sharks," but recent detailed studies have revealed features, especially in the skull and branchial skeleton, suggesting their relationships to the Osteichthyes. Because of their peculiar paired fins, however, the known acanthodians do not appear to be directly ancestral to the proper osteichthyans. Thus we place them, with some reservations, as a subclass of the Osteichthyes.

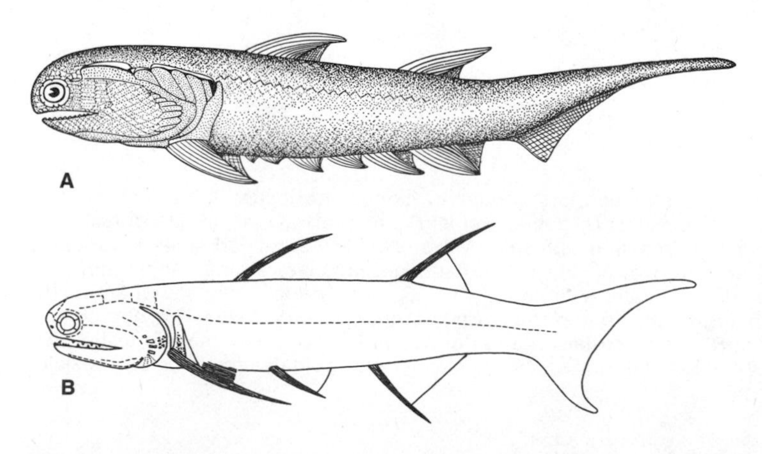

Figure 27. Acanthodians, primitive jawed fishes probably related to the modern Osteichthyes. *A*, *Climatius*, a form with heavy scales and very large fin spines; *B*, *Ischnacanthus*, one with greatly reduced armor and thin spines. (*A*, data from Watson; *B* after Moy-Thomas and Miles.)

As yet the problem of the nature of the truly ancestral jawed vertebrates from which the Placodermi, Chondrichthyes, and Osteichthyes were derived is unsolved. Probably these ancestors arose in the Silurian or even earlier, quite probably in fresh waters; but the preserved record of the rocks of these older times appears to be almost exclusively marine. The biggest gap in the fossil record of vertebrates is our absolute lack of knowledge of the presumed—though quite possibly nonexistent—common ancestor of the various gnathostome groups.

Leaving aside the puzzling "spiny sharks," the first Osteichthyes are found in rock of early Devonian age; the class is, thus, somewhat older than the sharks. By the middle of the Devonian, bony fishes were already the dominant forms in fresh waters, where they remain varied and abundant in later Paleozoic periods. A few bony fishes are present in saltwater deposits in the Paleozoic, and, by the end of the Triassic, marine waters appear to have become the headquarters of the class. Lungs appear to have been present in all primitive bony fishes, although today such structures usually have been lost or converted into a hydrostatic organ, the swim bladder. Lungs were presumably an aid to survival under conditions of seasonal drought; such conditions are believed by many geologists to have been present in the Devonian fresh waters in which the ancestral Osteichthyes lived. Latter, with climatic changes and particularly with the movement of most surviving bony fish types into the sea, the lung lost its importance.

The phylogeny of the bony fishes is complicated, and it is important to keep in mind the position on the family tree of many interesting and anatomically important types (Fig. 28). At the very beginning of their known history the Osteichthyes

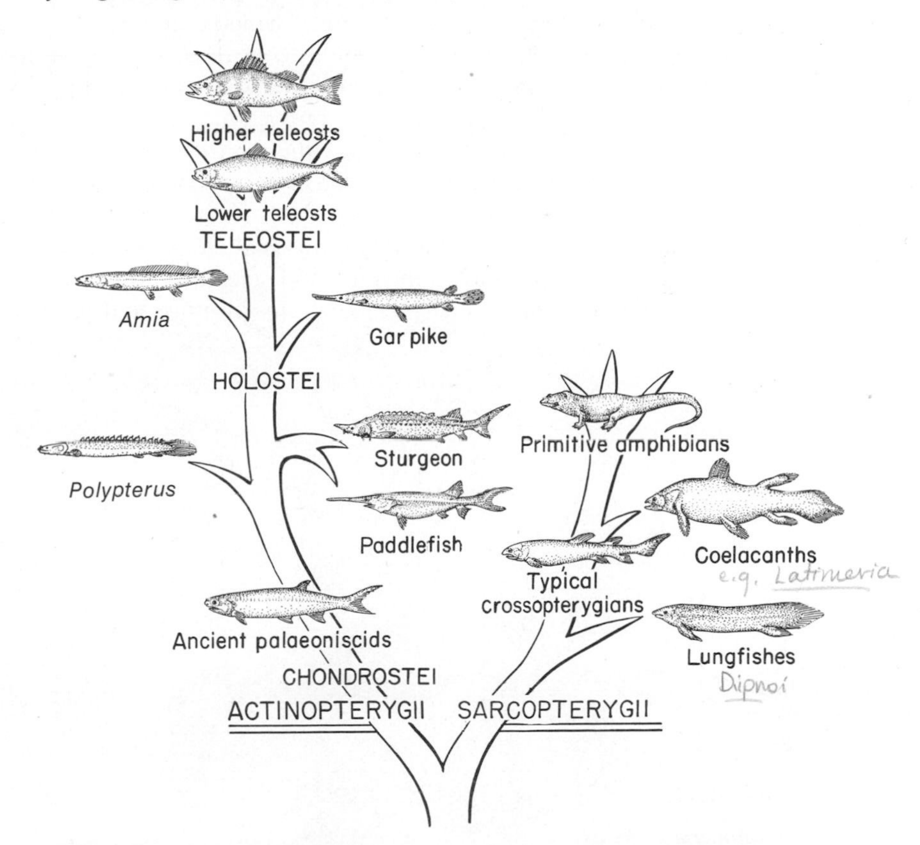

Figure 28. A simplified family tree of the bony fishes (apart from acanthodians), to show their relations to one another and to the amphibians.

(apart from the problematical acanthodians) were already subdivided into two major groups, termed the subclasses Sarcopterygii and Actinopterygii.

SARCOPTERYGII. In considering the descent of land animals, the Sarcopterygii are the most important of the two, for they contain the order **Crossopterygii**, from which land vertebrates appear to have descended, and the order **Dipnoi**, the lungfishes, which are surviving cousins of our piscine ancestors. Some of the sarcopterygians have, in contrast to the other subclass, internal nostrils, as do all land vertebrates. In stronger contrast with the actinopterygians is the fact that there are fleshy-lobed paired fins (a feature to which the group name refers) and, as a technical character, scales which in early forms were of a structure quite distinct from those of the actinopterygians (the cosmoid scale, cf. p. 138).

CROSSOPTERYGIANS. In the Devonian the commonest of bony fishes were crossopterygians (Fig. 29 *A*), aggressive, predaceous fishes which show important structural features of a sort to be expected in the ancestors of the amphibians. In the Carboniferous, however, they became relatively rare, and typical crossopterygians, termed rhipidistians (suborder **Rhipidistia**), were extinct before the close of the Paleozoic.

Meantime, however, a peculiar side branch of the crossopterygians had developed, the suborder **Coelacanthini** (Fig. 29 *B*). These forms, which migrated into the Mesozoic seas, had stub snouts, feeble jaws, and teeth. The last fossil coelacanths are found in Cretaceous rocks, and it was long taught that our crossopterygian relatives had been extinct since the days of the dinosaurs. In 1939, however, to the surprise of science, a strange fish caught off the coast of South Africa proved to be a coelacanth! Since then many further specimens of this form, *Latimeria,* have been obtained from deep waters off the Comoro Islands in the Indian Ocean, and their structure is being studied intensively. Knowledge of the structure of this fish is highly important, since we have here the closest living fish relative of the tetrapods. But although the land vertebrates are derived from crossopterygians, and *Latimeria* is a crossopterygian, one must not expect this survivor to resemble closely the tetrapod ancestors. Even the oldest known fossil coelacanths already differed in various skeletal features from typical members of the order, and the shift from ancestral streams and ponds to deep sea waters has been accompanied by many changes in structure and functions. Coelacanths lack in-

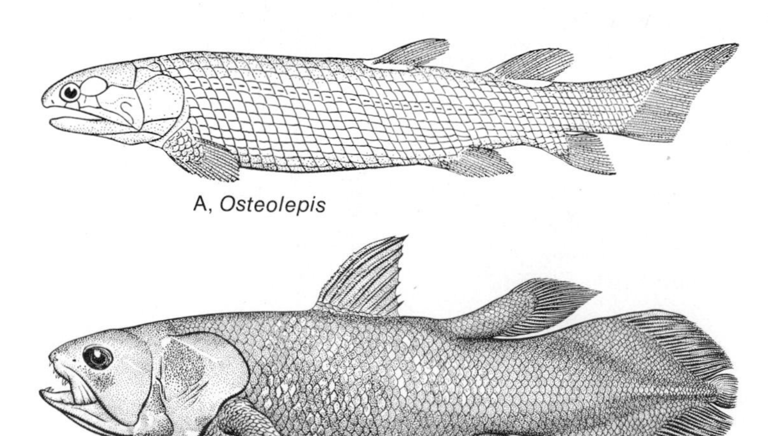

Figure 29. Crossopterygians. *A,* Typical Devonian form; *B,* the only living coelacanth. (*A* after Traquair; *B* after Millot.)

ternal nostrils; there is, instead, a series of canals filled by a jellylike substance in the snout region, which is obviously a sense organ of unknown nature. Lungs, which are of no use in deep marine waters, are represented only by a large sac filled with fat and connective tissue. Bone is much reduced. The heart is constructed in simple fashion; the chambers are nearly linear in arrangement, with little of the folding seen in most vertebrates. The braincase is large, but the brain within is tiny. There is no pineal foramen. The gill septa are more highly developed than in typical bony fish. The intestine retains a spiral valve. Overconcentration of salts internally is prevented as in sharks by retention of urea in the blood (cf. Chap. 13). There is a cloaca in the male. Coelacanths share with the rhipidistians the possession of a transverse joint across the center of the skull, through both the dermal roof and the braincase—a trait known otherwise only in the very earliest of the amphibians.

LUNGFISHES. The Dipnoi, or lungfishes (Figs. 30, 31), are represented today by three genera, living, one each, in tropical regions of Australia, Africa, and South America. In many anatomic features and in their mode of development the lungfishes closely resemble the amphibians, and they were once thought by many to be actual amphibian ancestors. But it is now more reasonable to believe that these features were present as well in their relatives, the ancestral crossopterygians, and that the lungfishes are to be regarded as "uncles" rather than the actual progenitors of land vertebrates. The skull structure of lungfishes, living and fossil, is of a peculiar type obviously unlike that of a proper amphibian ancestor; the transverse joint seen in crossopterygians is never present. During the course of lungfish evolution ossification is much reduced in the skeleton as a whole. In connection with a diet of invertebrates and plant materials, there are present in all lungfishes specialized fan-shaped toothplates. It is of interest that the lungfishes have survived only in regions where we find today conditions of seasonal drought similar to those which we believe to have been present in the Devonian. The Australian form can survive in stagnant water by air breathing; the other two are able to withstand even the complete drying up of the water by digging a burrow in the mud in which they "hole up" until the wet season of the year comes round. So dependent is the African lungfish on air that it will "drown" if kept under water. Both crossopterygians and lungfish are sarcopterygians—they share such features as fleshy fins and cosmoid scales. However, as just noted, they also differ in many

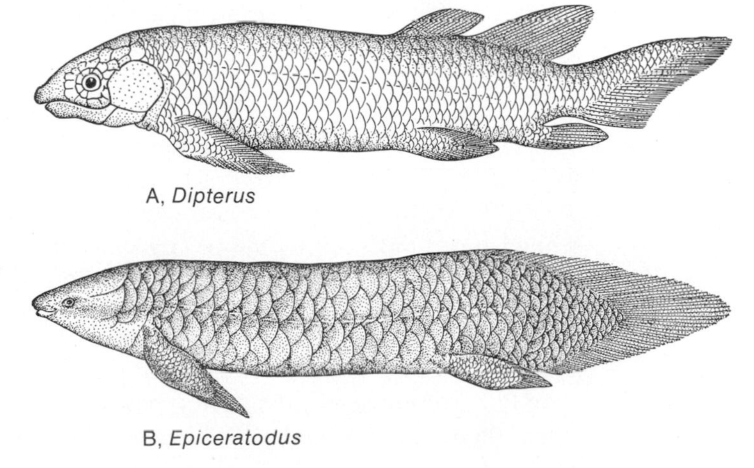

Figure 30. Lungfishes. *A*, An ancient Devonian fossil type; *B*, *Epiceratodus* of Australia. The median fins have changed greatly during the history of the group. (*A* after Traquair; *B* after Dean.)

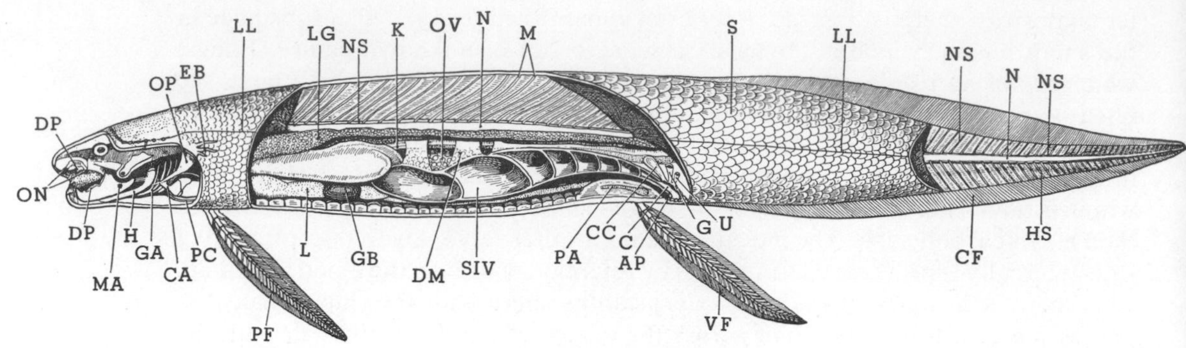

Figure 31. Diagrammatic dissection of a lungfish. *AP,* Abdominal pore; *C,* cloaca; *CA,* conus arteriosus; *CC,* rectal gland; *CF,* caudal fin; *DM,* dorsal mesentery; *DP,* tooth plates; *EB,* external gills; *G,* genital duct; *GA,* gill arches; *GB,* gallbladder; *H,* ceratohyal; *HS,* hemal spine; *K,* kidney; *L,* liver; *LG,* lung; *LL,* lateral line; *M,* muscle segments; *MA,* lower jaw; *N,* notochord; *NS,* neural spine; *ON,* external and internal openings of nostril; *OP,* operculum; *OV,* ovary; *PA,* pelvic girdle; *PC,* pericardium; *PF,* pectoral fin; *S,* scales; *SIV,* spiral valve of intestine; *U,* urinary duct; *VF,* pelvic (ventral) fin. (From Dean.)

ways and no connecting links are known. The differences are great enough that many workers deny a close relationship and consider them to be separate subclasses.

ACTINOPTERYGII. As types ancestral to higher vertebrates, the Sarcopterygii are of major interest; but as successful fishes, the Actinopterygii, or ray-finned fishes, are vastly more important. From Carboniferous times on, these have been the dominant fishes. In contrast with many sarcopterygians, internal nostrils are absent; the scales were primitively of quite another type; and except in a few primitive forms there is never a fleshy lobe to the fins. Instead, as the name implies, the paired fins are webs of skin supported by horny rays. Primitively, but not in most living or extinct forms, the skull was divided by a transverse joint as in crossopterygians, though located more posteriorly.

The actinopterygians have long been divided into three groups—here considered as superorders—which are, in ascending order, the Chondrostei, Holostei, and Teleostei. The names are not particularly appropriate from the point of view of our present knowledge of the evolution of ray-finned fishes, but may be retained for convenience. All of these now appear to be "unnatural" groups—"grades" or "levels of organization" rather than "clades"; in normal words, primitive, intermediate, and advanced forms.

CHONDROSTEI. In the Paleozoic the ray-finned fishes were represented by abundant genera of Chondrostei known as **palaeoniscoids** (Fig. 32 *A*). These were generally fishes of small size, with rather uptilted sharklike tails (the heterocercal type, cf. Fig. 135) and with scales covered by a shiny material known as ganoine (cf. p. 138). In the earliest Devonian days of bony fish history, primitive ray-finned forms were outnumbered by crossopterygians and lungfishes, but in the late Paleozoic they became far more numerous than their early rivals and swarmed in ancient lakes and streams in immense numbers and variety, ranging from long, thin, almost eel-like forms to high, laterally compressed forms shaped like modern angelfish. In the Triassic these ancient actinopterygians were still abundant but mainly represented by advanced types transitional to the Holostei; the palaeoniscoids then rapidly declined and became extinct before the end of the Mesozoic.

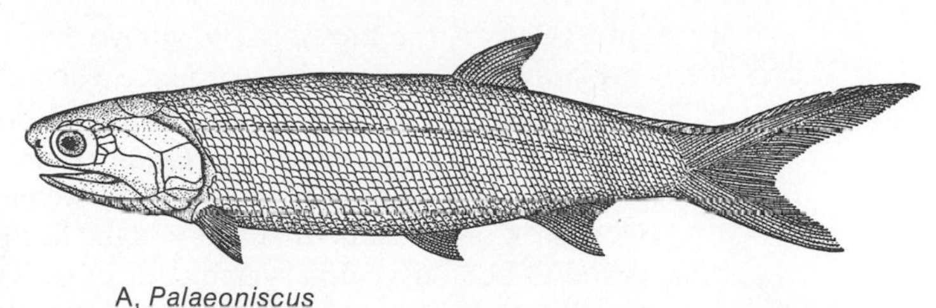

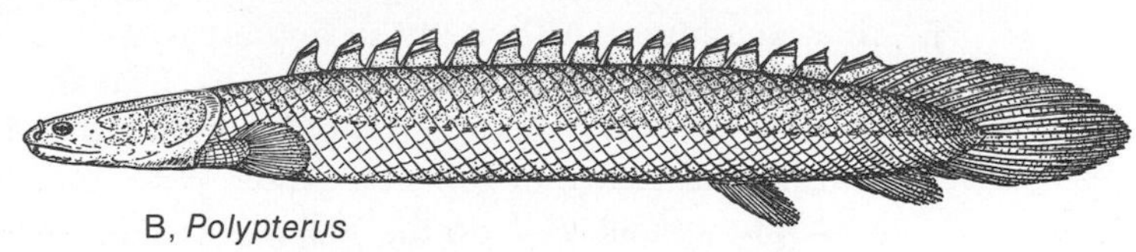

Figure 32. Primitive ray-finned fishes. *A,* An early palaeoniscoid; *B,* a living representative of the ancient palaeoniscoids, with modified fin structure. (*A* after Traquair; *B* after Dean.)

This primitive ray-finned group still survives in the form of three aberrant types. Two, the sturgeons and paddlefishes (both represented in North America), are rather degenerate (Fig. 33). They have lost the ganoid scale covering of their ancestors. Scales may still be present on the tail, but the paddlefish has otherwise only a naked skin and the sturgeon a partial armor of rows of plain bony plates. The internal skeleton, highly ossified in their ancestors, is nearly as degenerate as that of the sharks; little bone remains. Degenerate, too, is their method of feeding. In both sturgeons and paddlefishes the jaws are feeble. In advance of the jaws is a sensitive rostrum which explores for food ahead of them; sturgeons and paddlefishes are bottom-dwelling scavengers or food-strainers. Only their persistently sharklike tail fin recalls the older palaeoniscoids.

The third type of chondrostean survivor is *Polypterus* (Fig. 32 *B*), the bichir of Central Africa, which lives in much the same environment as the lungfish of that continent.* In its fins *Polypterus* is much modified from the ancestral type. Its tail fin has become essentially symmetric; its dorsal fin is split up into a series of small sail-like structures (to which its scientific name refers), and its paired fins, unlike those of any proper actinopterygian, have a short fleshy lobe. Again unique among

***Calamoichthys* is a closely related form, but with a more elongate eel-like shape, from the same region; comments on *Polypterus* usually apply to *Calamoichthys* as well.

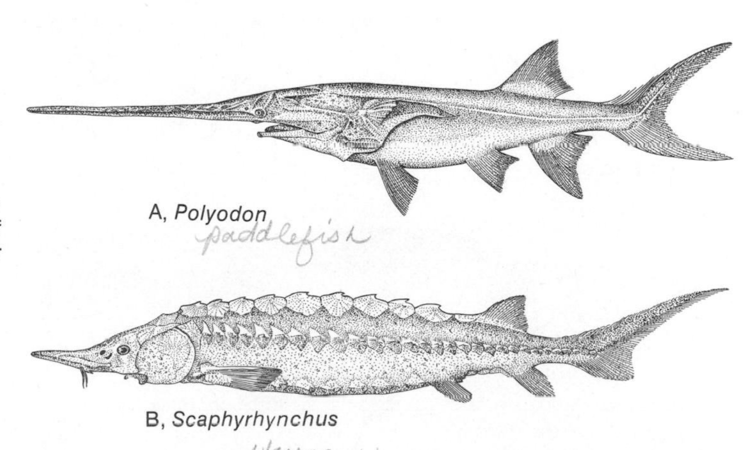

Figure 33. Chondrosteans. *A,* The paddlefish or "spoon-billed cat" of the Mississippi; *B,* a sturgeon. (After Goode.)

actinopterygians is the fact that *Polypterus* has typical (if simple) lungs, whereas other ray-finned fishes have in their place a structure termed the swim bladder (cf. p. 254), which seldom has respiratory functions and is, instead, a hydrostatic organ. The nose has a structure like that of no other actinopterygian, or indeed any other animal except the surviving coelacanth.

Because of the presence of lungs and the fleshy paired fins, *Polypterus* was long considered to be a crossopterygian. But closer study shows this to be incorrect. Lungs were probably present in all primitive bony fishes, and the survival of *Polypterus* in peculiar drought conditions appears to be due (as with the lungfish) to their retention. Although the fins are somewhat fleshy, they differ markedly in pattern from those of crossopterygians. The anatomy of the animal is as a whole in agreement with that of actinopterygians rather than that of the sarcopterygians, and the scales are of the true ganoid type, contrasting strongly with those of the sarcopterygians. *Polypterus* is thus best interpreted as a somewhat modified descendant of the ancient palaeoniscoids, although not all workers agree—some even take it as the type of a separate subclass or class, the Brachyopterygii.

HOLOSTEI. Succeeding the chondrosteans as dominant fishes in the middle Mesozoic were the holosteans. In them the old, long, upturned sharklike tail had become shortened, the jaws tended to have a shorter gape, and the scales, in many cases, tended to lose their shiny ganoid covering. Another trend, too, was apparent at this time: the ray-finned fishes were invading the seas in large numbers. The major center of actinopterygian evolution from the Jurassic period onward appears to have been the ocean. The oceanic holosteans, however, are extinct, and the only two survivors are North American freshwater forms. The gar pikes, *Lepisosteus* (Fig. 34 *A*), are fast-swimming fishes fairly representative of the ancestral holosteans in many ways, but specialized in their elongated jaws associated with predaceous habits. A more advanced type is *Amia* (Fig. 34 *B*), a lake and river fish of the Midwest and South, popularly termed the "dogfish," "mudfish," or bowfin. In these holosteans the internal skeleton is not at all degenerate, but in *Amia* the scales have lost their ganoine covering, and the tail is much like that of the teleosts.

TELEOSTEI. The teleosts, as the name suggests, form the end group of the ray-finned fishes, and are the fishes dominant in the world today. They appear to have originated from the holosteans in the Mesozoic oceans, and before the close of the Cretaceous had replaced the older group as the most flourishing of fish types.

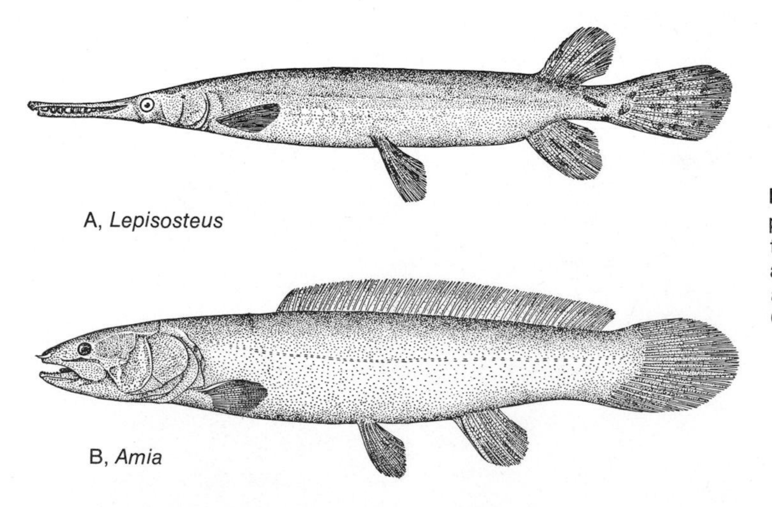

Figure 34. Holosteans. *A*, The gar pike; *B*, the bowfin. Both are inhabitants of North American fresh waters and are the only survivors of a stage antecedent to the teleosts. (After Goode.)

In teleosts the originally sharklike tail has been reduced, and the tail fin has a superficially symmetric appearance although the vertebral axis still turns dorsally (cf. Fig. 122). The paired fins are small; the pectorals are usually well up the sides of the body, and may function as effective brakes; the pelvic fins frequently are found well forward. The scales have lost all trace of the original shiny ganoid covering, and are generally thin, flexible, bony structures. The basic modifications of actinopterygians and the ones that most probably are responsible for their great success are in the locomotor and feeding mechanisms. We have already mentioned the changes in the tail; those in the jaws are more complex and are summarized in Figure 35. In most teleosts the premaxilla becomes independently mobile, so that the upper jaw is protruded as the mouth opens (watch a goldfish—the upper jaw actually slides forward). These bony changes are, of course, correlated with important changes in the muscles, which become larger and very complex. The net result is a system that has become adapted to almost every diet possible (and even some that seem impossible).

In the oceans the teleosts constitute the vast majority of all piscine inhabitants. They have invaded every possible marine habitat from the strand line to the abyssal depths. Further, in fresh waters they constitute almost the entire fish population. Teleosts are unquestionably the most numerous of all vertebrates. It is estimated that there are about 20,000 different species, and the number of individuals of one species alone—the common herring—is probably on the order of a billion billion! The prosperity of the teleosts is certainly due in part to an efficient body organization, but is also at least partially due to extraordinary fecundity. Existing bony fishes of other groups lay but a modest number of eggs; among teleosts the herring, for example, may lay 30,000 in a single season and a female cod is estimated to produce up to 9,000,000 eggs. The individual counts for little; only two eggs need to grow to maturity to keep up the numbers of the race.

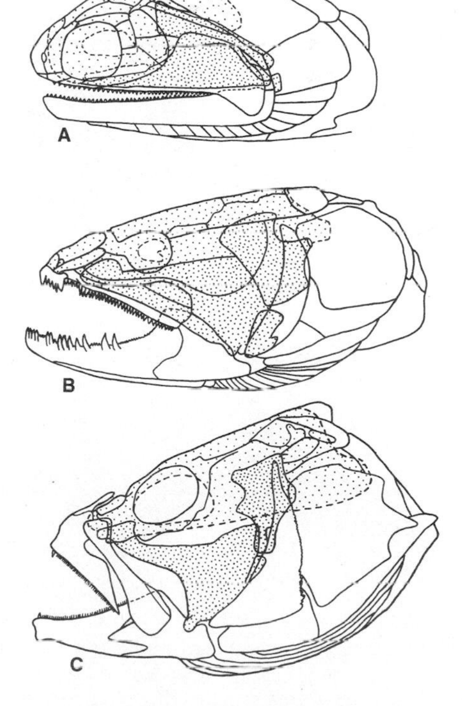

Figure 35. Diagrams of the skulls of actinopterygians in lateral view. *A,* The palaeoniscoid *Pteroniscuius; B,* the holostean *Amia; C,* the advanced teleost *Epinephelus.* The lines show the bones of the skull roof; the braincase is lightly stippled; the jaw suspension and palate are heavily stippled. Note how the jaw suspension keeps its dorsal position near the back of the braincase, but how its ventral end swings anteriorly as the jaw shortens. Note also the freeing of elements at the anterior end of the skull. (After Schaeffer and Rosen.)

The salmon and trout and the herrings and their relatives (Fig. 36 *A*) represent primitive groups of teleosts. The carps and catfishes are characteristic of a major freshwater division of the teleosts. More progressive and numerous, but almost all marine, are the spiny-finned forms—the perch (Figs. 36 *B*, 37) is typical—in which parts, at least, of the fins are supported by stout spines rather than softer rays.

Teleosts are the most versatile of vertebrates. Within both lower and higher divisions of the teleosts there has evolved a great variety of body shapes, a few of which are shown in Figure 38. Equally diverse are their habits. In food they range from eaters of microscopic plant particles to predaceous forms which attack other fishes. And although they have not successfully invaded the land or air, a surprising number of teleosts, such as the "climbing perch," can clamber about on land, and "flying fishes" can glide above the water.

In any comparative study of vertebrate anatomy or physiology in which the teleosts are involved, their ecologic history must be kept in mind. Since land vertebrates come from freshwater fishes, one tends to assume that structures or functions seen in freshwater teleosts may be representative of those once present in the ancestors of the tetrapods. But we must remember that our common fishes belong to a different branch of the fish family tree from that which gave rise to land animals. Further, modern freshwater teleosts have not been, in all probability, continuous residents in that environment since the early days of fish history; between that time and the present there almost certainly intervened a long marine phase.

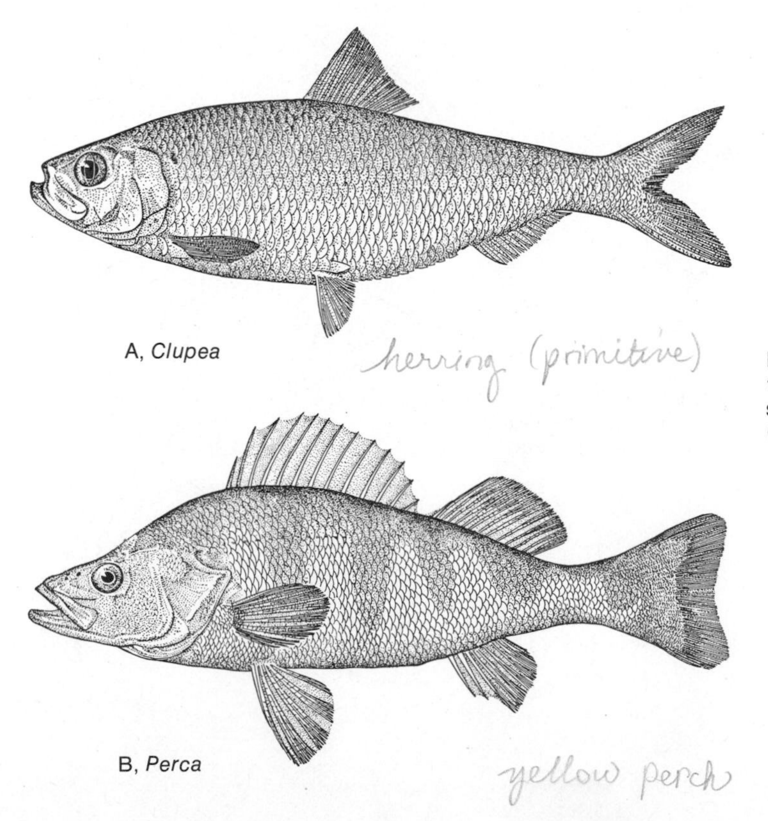

Figure 36. Teleosts. *A,* Primitive type, the herring; *B,* an advanced, spiny teleost, the yellow perch. (After Goode.)

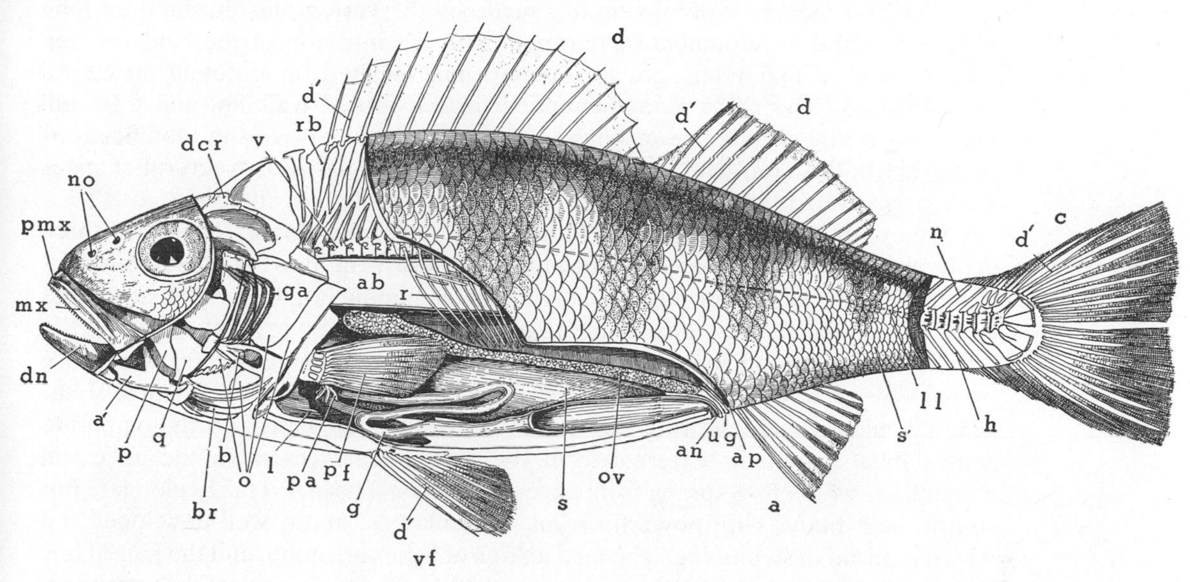

Figure 37. Diagrammatic dissection of a teleost, the perch (*Perca*). *a*, Anal fin; *ab*, air bladder; *an*, anus; *ap*, abdominal pore; *a′*, articular; *b*, bulbus arteriosus (conus arteriosus); *br*, branchiostegal rays; *c*, caudal fin; *d*, dorsal fins; *d′*, dermal rays of fins; *dcr*, dorsal crest of skull; *dn*, dentary; *g*, intestine; *ga*, gill arches; *h*, hemal spines (expanded to hypurals in tail fin); *l*, liver; *ll*, lateral line; *mx*, maxilla; *n*, neural spines; *no*, nasal openings; *o*, opercular bones; *ov*, ovary; *p*, pterygoid; *pa*, pyloric appendices; *pf*, pectoral fin; *pmx*, premaxilla; *q*, quadrate; *r*, ribs; *rb*, basal fin supports; *s*, stomach; *s′*, scales; *ug*, urogenital opening; *v*, vertebral centra; *vf*, pelvic (ventral) fin. (From Dean.)

Figure 38. A variety of teleosts, showing the great diversity of form found in these fishes. For those who may wish to trace these further, the names of the families to which they belong are given. *A*, Pimelodidae; *B*, Syngnathidae; *C*, Fistulariidae; *D*, Scophthalmidae; *E*, Congridae; *F*, Pegasidae; *G*, Linophrynidae; *H*, Diodontidae; *I*, Istiophoridae; *J*, Lophiidae; *K*, Agonidae; *L*, Monodactylidae. (After Greenwood et al.)

AMPHIBIANS

Greatest, perhaps, of all ventures made by the vertebrates during their long history was the development of tetrapods and the invasion of the land—a step which involved major changes in function and resulted in profound structural modifications. The shifts from swimming to four-footed walking, and from gill breathing to the dominance of lungs, are the most obvious of the modifications necessary in this step. But analysis shows that functional and structural changes were necessitated in almost every organ or organ system of the body.

The basic group of land vertebrates is the class **Amphibia**. There are three living orders (Fig. 39): the frogs and toads **(Anura)**; the newts and salamanders **(Urodela)**; and some wormlike burrowers **(Gymnophiona)**. Commonest are the anurans, familiar to us in temperate regions and represented in great variety in the tropics. Their specialized nature is obvious in their jumping habits, which are responsible for many structural modifications, particularly in the skeletal system. The salamanders are retiring but common dwellers in moist north temperate zone habitats. In their external form the salamanders resemble the ancestral amphibians which first sprang from ancestral fishes. There is a fairly elongate but stoutly built body, with powerful trunk musculature, and a well developed tail which is an aid in swimming. The median fins of fishes are gone, and the paired fins have developed into the typical land limbs which are the trademark of the tetrapods.

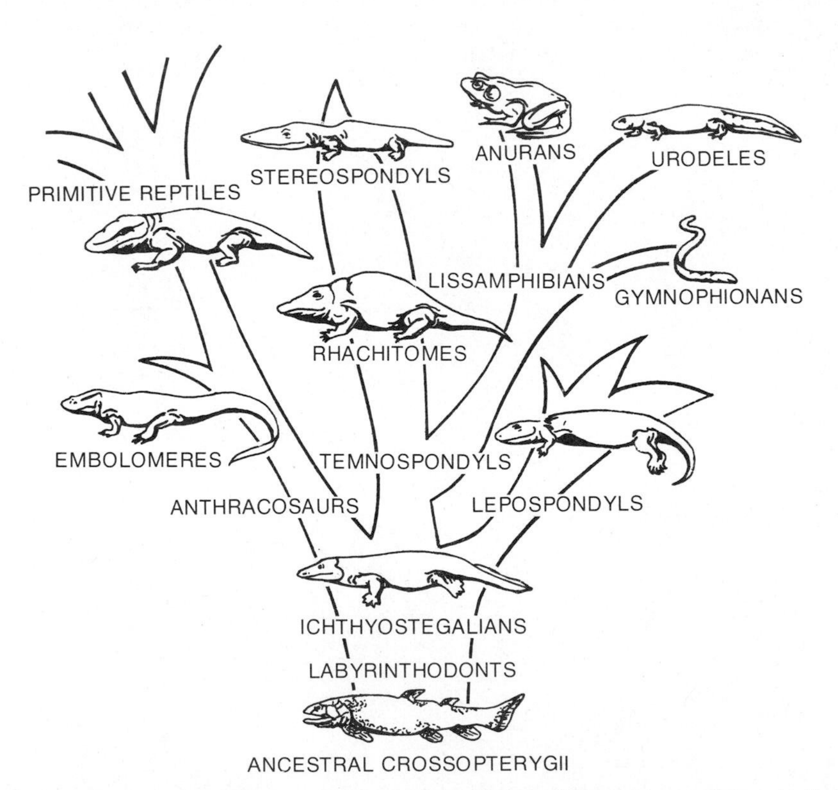

Figure 39. A "family tree" of the Amphibia. The surviving groups, shown in the upper right, probably have a common origin, in the late Paleozoic, from primitive rhachitomes, but this is still open to debate; the arrangement here is merely what we consider most probable.

The Gymnophiona (or Apoda) will not be familiar to many readers of this work, for they include only a few genera of small, blind tropical burrowers which resemble earthworms.

In salamanders, and in the other modern amphibians as well, we find various internal features which bridge structural gaps between the crossopterygian fishes and the higher classes of land vertebrates. Amphibian structure and function are important in comparative studies. But due caution must be exercised. Frogs and salamanders and apodans are amphibians, and the amphibians are the most primitive of tetrapod classes; but we must not complete a false syllogism by concluding that these modern amphibians are really primitive tetrapods. In the modern orders the skeleton is reduced, with considerable loss of bone from the head and, particularly in salamanders, a trend for retention of embryonic cartilages. In all, vertebral structure is apparently quite far removed from that which was present in truly ancestral land forms, and the great shortening of the trunk and reduction of the tail in frogs shows a high degree of specialization. The limbs of newts are seemingly not too aberrant, but those of a frog are certainly highly modified; and the gymnophionans have abolished limbs altogether. Concerning the soft anatomy, we cannot be so sure, but there are indications that here, too, many features are likewise aberrant in these modern orders. A frog is, in many ways, as far removed structurally from the oldest land vertebrates as is a man, and even a salamander must be regarded with suspicion.

For the actual ancestors of the land dwellers as a whole, we must turn to the fossil record of the late Paleozoic when, in the Carboniferous and early Permian, there lived numerous and varied amphibians of a more primitive nature. Two major groups were then distinguishable.

One included a series of small animals, termed **Lepospondyli**, of which a diagnostic feature is the spool shape of the central portions of the backbone segments. Actually the lepospondyls were quite varied (Fig. 40) and may well not

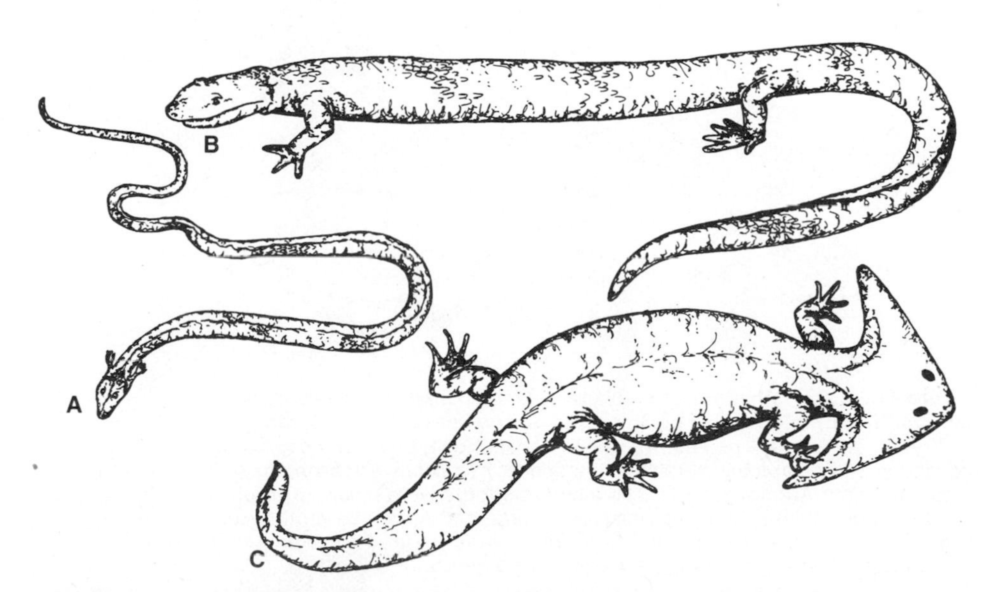

Figure 40. Representations of the extinct lepospondyl amphibians. *A*, *Dolichosoma*, a member of the Aistopoda; *B*, *Cardiocephalus*, one of the Microsauria; *C*, *Diplocaulus*, one of the Nectridea. All were small, more or less salamander-like forms from the swamps of the upper Paleozoic. (*A* after Špinar and Burian; *B* after Gregory et al.; *C* after Colbert.)

have been closely related to one another. Four orders may be recognized (see Appendix 1), but all were more or less salamander-like; some had reduced or even lost the limbs. It seems likely that the older lepospondyls were, despite their antiquity, a side branch, if an early one, of the basal stock of land animals.

The true base is to be sought among a second early group of amphibians, the **Labyrinthodontia** (Fig. 41). These animals of variable size were in general considerably larger than contemporary lepospondyls; some attained the size of crocodiles. Their vertebral construction was a diagnostic feature and was one from which that of reptiles and higher vertebrates could have been readily derived (cf. Figs. 118, 119). Except for the absence of median fins and the presence of short but sturdy legs developed from paired fins, many features of the earliest labyrinthodonts, the **Ichthyostegalia** from the uppermost Devonian, are highly comparable with those of the crossopterygians from which they came. They were the first vertebrates to walk on land.

In the late Paleozoic the labyrinthodonts flourished greatly. One major group, termed the **Temnospondyli**, was especially abundant and varied. Typical ones included semiaquatic forms resembling crocodiles as well as more terrestrial animals like large, clumsy lizards; they survived, in the shape of degenerate, purely water-dwelling forms with flat bodies and small limbs, until the late Triassic. Less abundant and less long-lived, but important in phylogeny, were the **Anthracosauria**, among which were forms closely approaching the reptiles in skeletal structure. It is

Figure 41. Labyrinthodont amphibians. *A, Ichthyostega,* the earliest known tetrapod from the Devonian of Greenland, a member of the Ichthyostegalia; *B, Eryops,* a "typical" member of the Temnospondyli; *C, Metoposaurus,* an advanced and completely aquatic temnospondyl; *D, Diplovertebron,* an early member of the Anthracosauria. The labyrinthodonts were a varied group of primitive amphibians, many of which were large and more like reptiles than the small and "degenerate" modern amphibians. (*A* and *D* after Špinar and Burian; *B* after Romer; *C* after Fenton and Fenton.)

from this group of labyrinthodonts that there arose, in the latter part of the Carboniferous, the first reptiles, and members of it are often taken as "typical primitive tetrapods" (as in Fig. 166 and the accompanying descriptions).

Whence did the three modern orders arise? This is an unsettled problem. The first prefrog is present in the Triassic, and typical, if rather primitive, anurans are present in the Jurassic. In this latter period are found the first urodeles; there is almost no fossil record of the Gymnophiona (Tertiary vertebrae only). There is little in the structure of the modern orders to give a clear clue to their origin from either labyrinthodonts or lepospondyls. The adaptations seen in anurans and urodeles are so divergent that many workers have believed them to have evolved separately from discrete Paleozoic origins. Recently, however, it has been pointed out that there are significant common features in ear apparatus, dentition, and so on. This situation suggests that despite their later divergence the three modern types had a common origin, most probably though very uncertainly from temnospondyl labyrinthodonts, and may be united in a major amphibian subclass, the **Lissamphibia**, equal in rank to the Labyrinthodontia and Lepospondyli.

The development of the early land vertebrates has sometimes been "explained" as the result of some "urge" toward terrestrial life among their fish ancestors. This is, of course, absurd; the evolution of the earliest amphibians capable of walking on land seems to have been essentially a happy accident. The amphibians appear to have evolved from crossopterygian ancestors toward the close of the Devonian, an age during which seasonal droughts were, it seems, common over much of the earth. Lungs, already present in the ancestral bony fishes, are an excellent adaptation for use under stagnant water conditions. But when a stream or pool dries up completely, a typical fish is rendered immobile and dies. Some further development of the fleshy fins already present in crossopterygians would give their fortunate possessor the chance of crawling up or down the stream bed (albeit with considerable pain and effort at first) and enable it to reach some surviving water body where it could resume a normal piscine existence.

Legs, the diagnostic feature of the tetrapod, may thus have been, to begin with, only another adaptation for an aquatic life. The earliest amphibian was little more than a four-legged fish. Life on land would have been the farthest thing from its thoughts (had it had any), for early amphibians were eaters of animal food, and until insects became widespread in the later Carboniferous, there was precious little for an amphibian to eat on land. It was probably only after a long period of time that its descendants began to explore the possibilities of land existence opened out before them through their new locomotor abilities.

The term amphibian implies the double mode of life exhibited by many members of this class. Some toads spend much of their lives on good dry land, but most amphibians do not venture far from the stream banks, and some modern forms are still essentially water dwellers like their ancestors. The typical amphibian mode of development, as exemplified by the familiar frogs and toads of northern temperate regions, is still essentially that of the ancestral fishes. The eggs are laid in the water, and develop there into water-dwelling, gill-breathing tadpoles. Only when adult size is neared do lungs replace gills, do limbs develop and mature, and does terrestrial life become possible. An amphibian is chained to the water by its mode of development and the necessity of returning to that element periodically for reproductive purposes. Although numerous modern amphibians have adapted variously to avoid this complication, none has been a complete success as a fully terrestrial form. Indeed, some salamanders, such as the American mud puppy *(Necturus),* never emerge onto land at any age, retain external gills and water breathing, and reproduce, in neotenic fashion, in essentially a larval condition.

REPTILES

The reptiles are the descendants of the ancient amphibians which, happily, solved this reproductive problem and became the first fully terrestrial vertebrates. The "invention" of the amniote egg (with the associated developmental processes; see Chapter 5) is the major diagnostic feature which distinguishes reptiles from amphibians.

The reptilian egg is laid on land; thus is avoided the necessity of any adaptation for aquatic existence in either young or adult. This egg type is the familiar one preserved in the reptiles' avian descendants. The shell offers protection. A large yolk furnishes an abundant food supply so that the reptilian young (unlike the tadpole) can hatch out, at a fairly good size, as a miniature replica of the adult and thus has no need of prematurely foraging for its food (obviously this apparent advantage can be abandoned—see most birds). Of embryonic membranes developed within the egg shell, one externally sheathes both young and yolk. A second forms a lunglike breathing mechanism for absorption of the oxygen that penetrates the porous shell. A third (the amnion, from which the egg type gets its name) encloses the developing embryo in a liquid-filled space—a miniature replica of the ancestral pond. The development of this new egg type was so important an advance in the evolution of land vertebrates that, as noted earlier, the reptiles, together with the birds and mammals which descended from them, are often styled collectively the amniotes.

Possibly the oldest reptiles were still amphibious in their habits, and the amniote egg was merely an adaptation parallel to, but better than, other adaptations seen in modern amphibians—a device which removed the eggs from the dangers of drought, and of enemies present in the ancestral waters. However, it seems more probable that the first reptiles were small terrestrial forms—things one would consider odd lizards rather than odd salamanders, to use living forms for comparison.

Modern reptiles are moderately abundant in the tropics, but unimportant in temperate zones and absent from cold climates where survival is difficult for these "cold-blooded" creatures. The modern forms, however, are but sparse remnants of the great array of reptiles which, beginning in the late Paleozoic, radiated out into a bewildering variety of forms that long ruled the earth and caused the Mesozoic Era to be popularly known as the Age of Reptiles (Fig. 42). The basic stock from which they sprang, the long-extinct order **Cotylosauria** (Fig. 43 *A*) or "stem reptiles" were, apart from reproductive improvements, still very archaic, with limbs sprawled out sideways from the body, and in most regards little more advanced than their amphibian forebears and cousins. But due, no doubt, to the breaking of the chains that bound them to the water, there presently developed from them the groups which became prominent in the Mesozoic.

A side branch of the stem reptiles was that of the turtles—order **Testudines** (Fig. 43 *B, C*). In their sprawling gait they are reminiscent of their Paleozoic forebears, but turtles have made a conspicuous advance in the development of a protective shell of bone, covered by horn, guarding both back and belly. This shell

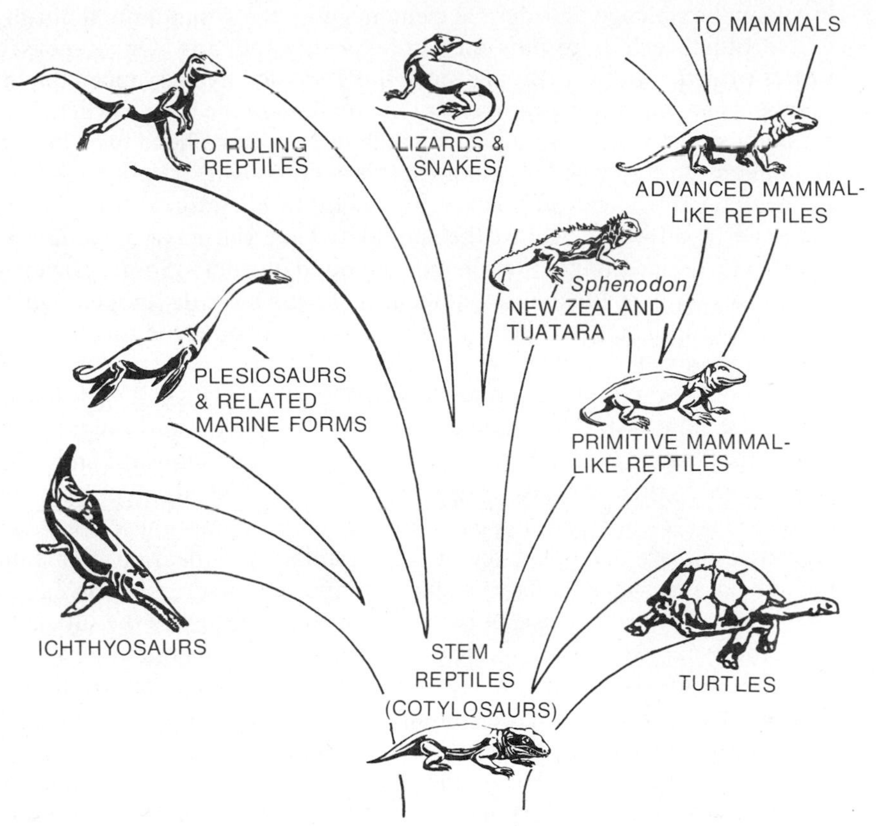

Figure 42. A simple family tree of the reptiles (the archosaurs are shown in more detail in Fig. 47). (After Romer.)

Figure 43. Anapsid reptiles. *A*, The primitive and very lizard-like cotylosaur *Cephalerpeton;* *B* and *C*, two modern turtles. *B*, The snapping turtle *Chelydra*, and *C*, the marine green turtle *Chelonia*. (*A* from Carroll and Baird; *B* and *C* from Young.)

involves the rib cage plus dermal elements with the shoulder and hip girdles inside it. No other vertebrate puts the girdles *inside* the ribs. Once encased in armor turtles turned conservative, and since the Triassic have advanced but little in most aspects. The only later improvement of any note made by the order as a whole was the acquisition of the ability, lacking at first, to pull the head back into the shell. In all familiar types (termed cryptodires) this is done by a straight backward pull, with the neck bent in a vertical S-curve; some odd tropical forms (the pleurodires) tuck the head in sideways against the shoulder. Like the most ancient reptiles, most modern turtles are amphibious marsh and pond dwellers. Some, however, reverted to a purely aquatic life and several marine forms have developed, with paddlelike limbs for propulsion. At the other extreme, one group, the tortoises, has become purely terrestrial.

In both of the reptilian orders mentioned so far, the skull roof, like that of their amphibian ancestors, had openings only for the major sense organs, the nostrils, eyes, and pineal eye. Such a condition is known as "anapsid" and reptiles with it are placed in the subclass **Anapsida** (Fig. 44; see also p. 193). In all other reptiles "extra" openings appear in the temporal region; these are called temporal fenestrae and are diagnostic features in reptilian classification. Some, members of the subclass **Euryapsida**, have a single temporal fenestra high on each side of the cheek—this is the euryapsid or parapsid pattern. Reptiles of the subclass **Synapsida** show the synapsid pattern, a single fenestra on each side but more ventrally placed than in euryapsids. Finally, in the diapsid condition, both pairs of fenestrae occur. Two great subclasses, the **Lepidosauria** and **Archosauria**, have diapsid skulls.

Euryapsid reptiles are all extinct and poorly understood (Fig. 45). Some appear to have been fairly primitive, lizardlike terrestrial forms; these are placed in the order **Araeoscelidia**. Unfortunately, the more these forms are studied, the fewer appear to be euryapsid, so many workers deny the existence of the group completely.

Despite their newly won ability to conquer the land, a number of Mesozoic reptile groups took up (like the sea turtles) a marine existence. Prominent Mesozoic types of this sort, all euryapsids, were the plesiosaurs, the placodonts, and the ichthyosaurs. The plesiosaurs (order **Sauropterygia**) possessed a long neck or long snout or both; the body was short, broad, and relatively flat. Reversion to a truly

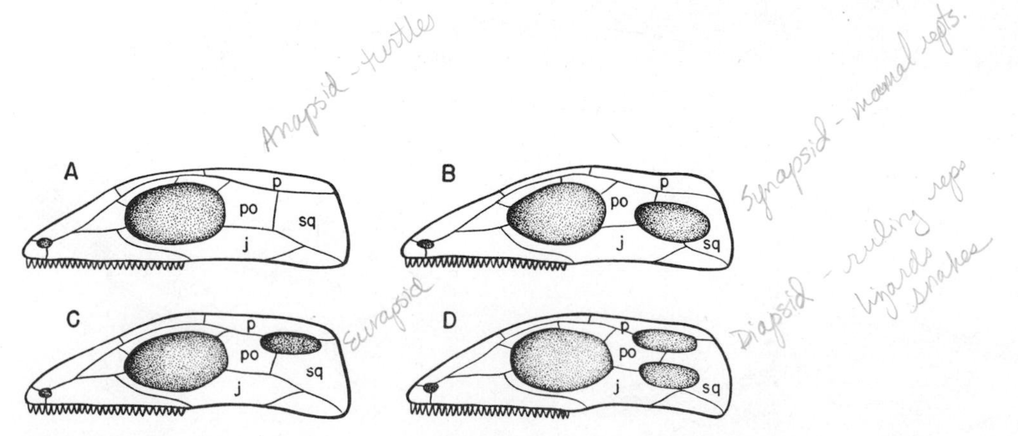

Figure 44. Diagrams to show types of temporal openings in reptiles. *A*, Anapsid type (stem reptiles, turtles); *B*, synapsid type (mammal-like reptiles); *C*, euryapsid type (extinct plesiosaurs, and so forth); *D*, diapsid type (rhynchocephalians, ruling reptiles; lizards and snakes derived by loss of one or both temporal arches). Abbreviations: *j*, jugal; *p*, parietal; *po*, postorbital; *sq*, squamosal.

Figure 45. Euryapsid reptiles, having a single pair of dorsal temporal openings. *A, Araeoscelis,* a small lizard-like member of the Araeoscelidia; *B, Placodus,* one of the mollusc-eating Placodontia; *C, Plesiosaurus,* a member of the Sauropterygia; *D, Stenopterygius,* one of the extremely fishlike Ichthyosauria. (*A, B,* and *C* after Fenton and Fenton; *D* after Špinar and Burian.)

fishlike means of locomotion was impossible, for the trunk was inflexible and the tail short; instead, the limbs were developed into powerful oarlike structures, with which the creature "rowed" its way through the sea.

Placodonts (order **Placodontia**) were Triassic forms which had become eaters of molluscs or other hard-shelled invertebrates. They were heavily built, rather flattened forms, some of which developed rather turtle-like armor. Their relation to the plesiosaurs, like most other things about euryapsids, is currently debated and several workers deny any close affinities between them.

Even more unusual structurally were the **Ichthyosauria**, the "fish-reptiles." Possibly the plesiosaurs were able at least to waddle about on a beach in the fashion of a marine turtle or a seal. The ichthyosaurs, however, had become as completely adapted to a marine life as a porpoise or dolphin; there is fossil evidence indicating that egg-laying on land had been abandoned and that the young were born alive. The body shape was completely reconverted to that of a fish—the neck telescoped to give a fusiform body shape, the limbs shortened into small steering devices. Locomotion was performed, fishlike, by undulations of trunk and tail; a fishlike fin was developed on the back (but, like that of whales, it lacked the skeletal supports found in the dorsal fins of fishes), and the tail became a powerful swimming organ, in appearance like that of a shark. In this last regard, however, there is a notable structural difference; for whereas in a shark the end of the backbone tilts into the upper lobe of the tail fin, that of the ichthyosaur turns sharply down at the back, the fin (as we know from excellent slabs of fossils) expanding above it. Most ichthyosaurs were presumably fish eaters, but some fed on ammonites, large molluscs related to the modern nautilus.

The Lepidosauria (Fig. 46), the first subclass of diapsid reptiles, include a basal ancestral group, the order **Eosuchia**, of extinct, more or less lizard-like (as usual) forms. Another group, which appeared early in the Mesozoic and has persisted to the present day although never prominent, is the order **Rhynchocephalia**, now represented by the tuatara (*Sphenodon*). This creature, lizard-like in appearance, has survived in the relative safety and isolation of New Zealand where, once widespread, it is now preserved on a few small islands.

Descended from ancient forms related to the tuatara is the much more successful order of the **Squamata**, the lizards, amphisbaenians, and snakes. Technically they are readily distinguished by the fact that the cheek and temple region of the skull has been reduced, so as to leave but one temporal arch (most lizards) or none at all (snakes; cf. Fig. 187). The "scaled reptiles" are not only the most flourishing but the most modern of reptile orders, for even lizards amounted to little until late in the Cretaceous and the deployment of most of the various snake types did not take place until Cenozoic times. Lizards (suborder **Lacertilia**) are widespread in the tropics in great variety. Most prominent of American lizards are the iguanas and their relatives, such as the collared lizard or "mountain boomer" of the Southwest and the little "horned toad." In the Old World the largest forms are the monitor lizards (*Varanus*), one of which, in the East Indies, may reach a length of four meters. Relatives of the monitors in the late Cretaceous had a temporary success as a group of giant marine lizards, the mosasaurs. The (true) chameleons of the Old World tropics, with peculiar grasping feet and highly protrusible tongue, are a specialized side branch of the lizard stock. In several lines of lizards there have developed burrowing types, with limbs reduced or absent. Indeed, there are several families entirely composed of such limbless forms.

One such group, often considered a family of lizards, is better taken as a separate suborder, equivalent to the snakes or lizards: the **Amphisbaenia**. They are mainly tropical and limbless (one form has forelimbs), characterized by having the scales arranged in distinct rings on the body—which makes them look even more like earthworms than most other limbless forms do.

Derived from lizards are the snakes (suborder **Serpentes**), which are highly modified in two major regards. As in some lizards and amphisbaenians, the limbs have been reduced and generally lost completely, and locomotion is accomplished

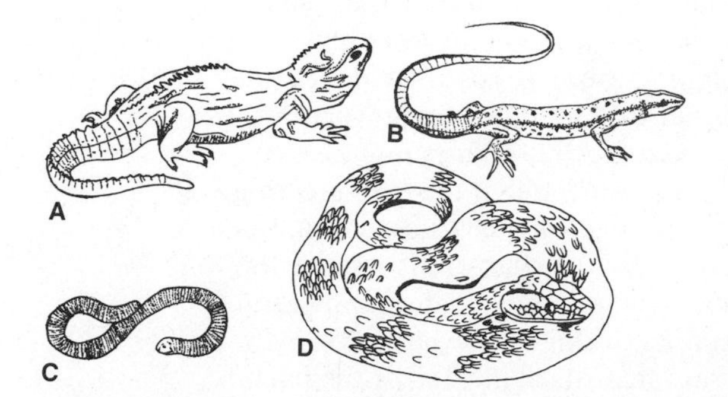

Figure 46. Lepidosaurs, the dominant group of living reptiles. *A, Sphenodon,* the extremely primitive tuatara from New Zealand, the only surviving member of the Rhynchocephalia; *B, Lacerta,* an ordinary lizard or member of the Lacertilia; *C, Amphisbaena,* a limbless member of the Amphisbaenia; *D, Agkistrodon,* a poisonous snake closely allied to the more familiar rattlesnakes, suborder Serpentes. (*A, B,* and *C* after Young; *D* after Gadow.)

by undulations of the body and tail, aided by the projecting horny scales which prevent backslipping; essentially, a snake swims on dry land. More distinctive is the fact that skull parts and jaw apparatus are markedly altered in the direction of flexibility and as a result allow the swallowing of the prey whole. Some snakes of primitive type are burrowers, and it is possible that snake evolution began with such forms, but even among such primitive types as the constrictors—the boas and pythons, some of great size—most now live above the surface. The great majority of snakes belong to an assemblage, often called a family but probably better considered as several (unfortunately not all workers agree on how many or what!), of which the common harmless forms of northern temperate regions are representative. But even within this family many tropical genera have developed poison glands. These, however, are generally small and inoffensive forms and the fangs, situated in the back of the mouth, present little danger to man and other large animals. Two further families include the major poisonous snakes, with highly developed fangs and with powerful and varied venoms which attack the nervous system or cause destruction of tissues: (1) a group, mainly Old World, which consists of the cobras and their kin, including the coral snakes and the sea snakes; (2) the vipers, with erectile fangs, including the adders and other vipers of the Old World, and the pit vipers, mainly American, with such representatives as the rattlesnakes, copperhead, and water moccasins.

An exceedingly important reptile group with the same two-arched type of skull build as *Sphenodon* is the great subclass **Archosauria**, the ruling reptiles (Figs. 47, 48). Today they survive only in the form of the rather aberrant crocodiles and alligators; but most of the dominant land reptiles of the Mesozoic were archosaurs, and the birds are descendants of this group.

The basal stock of the archosaurs is found in the Triassic in the shape of predaceous reptiles forming the order **Thecodontia**. Some early thecodonts were four-footed, but elongate hind legs, a modified hip structure, and other features suggest that others were becoming adapted to a bipedal mode of life. Still other thecodonts became armored or paralleled closely the later crocodilians. From these modest beginnings came the dinosaurs. These are popularly considered as constituting a single group of gigantic reptiles. Actually, although many dinosaurs were large, some were small (one was no bigger than a rooster). There were two major dinosaur stocks, not particularly closely related to one another, although both were descended in common from thecodont ancestors.

In one group, termed the **Saurischia**, or reptile-like dinosaurs, a large proportion of known forms was bipedal carnivores (suborder **Theropoda**). Some of the smaller and more primitive of these bipeds can scarcely be distinguished from their thecodont ancestors. Others grew to immense size; *Tyrannosaurus* was the most ponderous flesh eater the earth has ever seen. Forming a major side branch were the amphibious (sauropod) dinosaurs and their ancestors (suborder **Sauropodomorpha**), which changed to a herbivorous mode of life, and instead of being bipeds were four-footed walkers and grew to such giants as *Apatosaurus* and *Diplodocus* (one is estimated to have reached 50 tons). Both groups had normal reptilian pelvic girdles, the character that gives them their ordinal name.

A second major group was that of the **Ornithischia**, or birdlike dinosaurs, in which the hip girdles (but not other anatomic features) were comparable to those of birds. Like part, at least, of their saurischian cousins, primitive members of this group were bipeds (suborder **Ornithopoda**); but in contrast with that other dinosaur stock, all the birdlike forms were herbivores. Best known of the bipeds of this group are the duckbills (hadrosaurs), which were abundant in the closing days of the Age

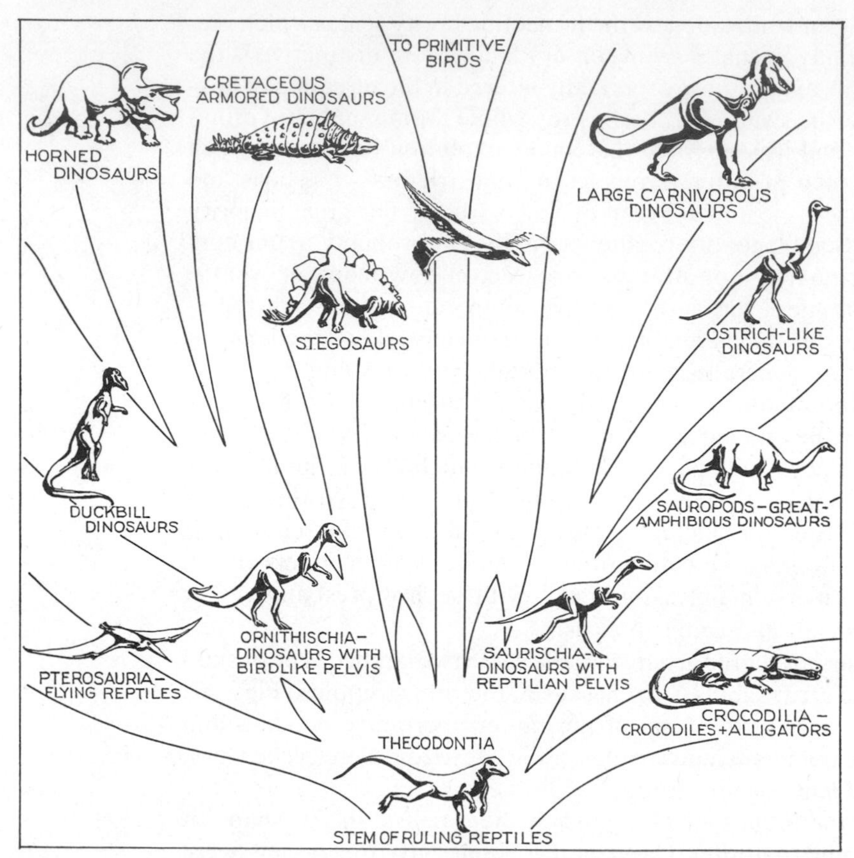

Figure 47. A simple family tree of the ruling reptiles. (From Romer, The Vertebrate Story, University of Chicago Press.)

of Reptiles. In three types of ornithischians there was reversion to a quadrupedal pose, together with some sort of defense against the great carnivores of the day. These types are exemplified by such popular museum exhibits as those of *Stegosaurus* (suborder **Stegosauria**), whose backbone is capped with plates and spines; *Ankylosaurus* (suborder **Ankylosauria**), low and flat and heavily armored on back and tail; and the horned dinosaurs (suborder **Ceratopsia**), such as *Triceratops,* with horns—often a trio of them—and a great frill of bone protecting the neck. The horned dinosaurs, all from the late Cretaceous, were in a sense better "rhinoceroses" than true rhinoceroses.

Dinosaurs flourished during the Jurassic and Cretaceous, and even in the closing phases of the latter period were present in considerable numbers and variety. Then, within a very short space of time (geologically speaking), they disappeared completely. The reason for this abrupt end of the Age of Reptiles is far from fully understood. Geologic events are perhaps basically responsible. The Cretaceous period was one of mountain building during which great ranges, such as the American Rockies, began to emerge from formerly flat country. Many of the low-lying marshy and lagoonal areas where dinosaurs browsed on lush vegetation disappeared. Climatic conditions were radically changed, and new types of plants became dominant with which herbivorous dinosaurs were probably unable to cope.

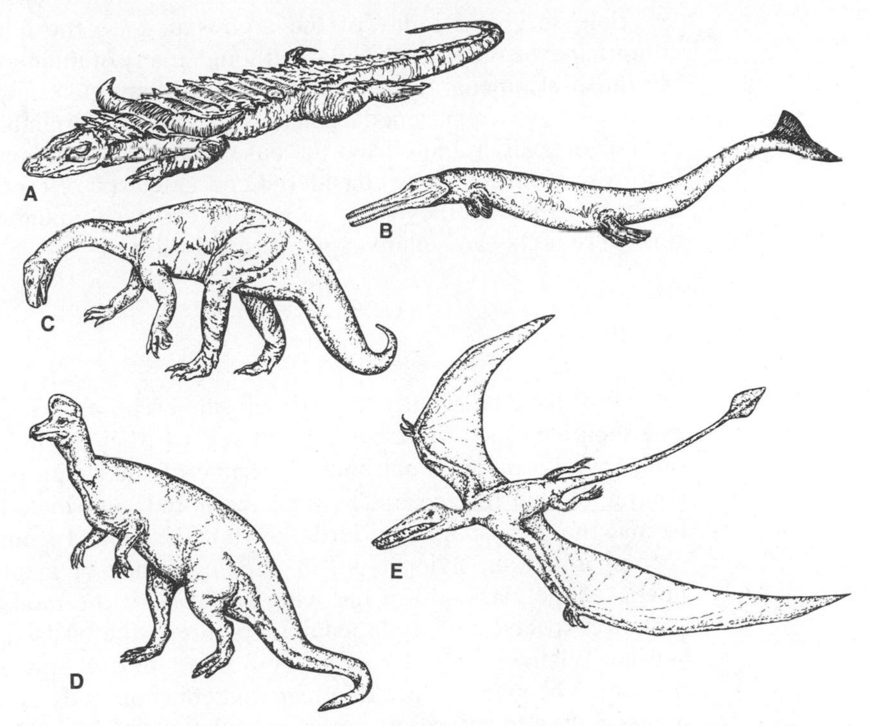

Figure 48. Archosaurs, the most varied and spectacular of the reptiles. *A, Desmatosuchus,* an armored and rather atypical member of the Thecodontia; *B, Geosaurus,* an extinct marine member of the Crocodilia; *C, Plateosaurus,* an early dinosaur of the order Saurischia; *D, Corythosaurus,* a "duck-billed" dinosaur of the order Ornithischia; *E, Rhamphorhynchus,* a flying reptile, a member of the Pterosauria. (*A* after Breed; *B* after Fenton and Fenton; *C* after Colbert; *D* and *E* after Špinar and Burian.)

As the herbivores consequently dwindled and disappeared, their flesh-eating cousins, who preyed upon them, would of necessity follow them to extinction.

Still another group of extinct archosaurs was that of the order **Pterosauria**, the winged reptiles. In them the front limbs had one finger (the fourth) enormously elongated. From it there was extended, in somewhat batlike fashion, a great wing membrane. Manipulation of a wing of this sort would appear to have been an awkward matter, and flight almost certainly consisted mainly of soaring rather than active beating of the wings. The hind legs of pterosaurs, quite in contrast with those of birds, were feeble structures, and it is difficult to see how these creatures could have stood upon them, much less get a running take off, as a bird of any size must do. Probably they perched in batlike fashion. But it is hard to imagine what safe perch could be found for a pterosaur with a wing spread of up to 16 meters (estimated in one Cretaceous form). All in all, it is not difficult to understand why the pterosaurs became extinct after more efficient flying forms—the birds—had evolved. However, this extinction was not rapid—pterosaurs survived and were successful for about a hundred million years!

Sole survivors today of the archosaurs are the alligators and crocodiles, comprising the order **Crocodilia**. Although many of their structural features resemble those of bipedal dinosaurs, the crocodilians, like many of their dinosaurian relatives, have a quadrupedal gait and are, further, amphibious. The crocodiles are phylogenetically remote from the base of the reptilian family tree; their anatomic features are hardly to be considered characteristic of reptiles as a whole, and, as might be expected, they show numerous features found in the birds, whose ancestors were archosaur relatives of the crocodiles.

BIRDS

Birds have been aptly termed "glorified reptiles." We customarily treat them as a separate class, **Aves**, but in many regards they are little farther removed from the general reptilian stock than are some of the ruling reptiles from which they sprang. Within that group, as we have noted, was included one series of flying forms, the pterosaurs; the birds are not descended from pterosaurs, but are a second archosaur flying type, in which, instead of membrane, feathers—diagnostic of the class—form the wing surfaces of the modified pectoral limbs. In certain respects, notably bipedal adaptations, the birds are similar to their dinosaurian relatives, but almost every notable bird characteristic is an adaptation to flight. The maintenance of a high and constant body temperature and improvements in the circulatory system are associated with the need of a high metabolic rate for sustained flight (actually, many workers believe similar adaptations occurred in dinosaurs and pterosaurs). Lightening of the body in various ways (particularly by the development of air sacs and hollow bones) is also associated with flight, as are modifications in the brain and sense organs. The birds are "expert" in wing construction and in the utilization of winds and air currents in gaining distance or elevation. In general birds with an active beating type of flight tend to have relatively short, wide wings; those that rely mainly on gliding and soaring usually have relatively larger wing surfaces with the length greatly increased in most cases. Flying birds (in contrast to ostrich-like ground-dwellers) are seldom of any great size; to maintain flight, wing area must increase, on the whole, proportionately to weight, and too great size of body would require a wingspread so great as to be difficult to manipulate.

Except in primitive Mesozoic birds, teeth were lost and reliance placed on a bill for gathering food. Numerous variations are seen in bill structure, with such extremes as the parrot beak and the effective drilling organ of the woodpeckers. Presumably the food of primitive birds was of some relatively soft nature, insects or the like, and teeth were not necessary. Many modern birds, however, are eaters of grain. For this type of food birds have developed a grinding apparatus in the muscular gizzard containing grit or small pebbles; such a gizzard is also present in crocodilians and may well be a general character of archosaurs.

In the birds we see a class of vertebrates which in many regards is to be considered as on as high a level of organization as the mammals, but (even disregarding differences in aerial vs. terrestrial locomotion) organized in quite another fashion. Birds can, to be sure, be trained, but on the whole seem relatively much less capable of learning by experience than mammals. On the other hand they exhibit innate behavior patterns of a complexity unknown in mammals. Avian "knowledge" of geography is most remarkable. The homing ability of birds is great, and the ability, for example, of young golden plovers to migrate successfully from the Arctic tundras to the Chaco region of South America—unaccompanied by older birds, over a complicated course of many thousands of miles, much of it over open ocean—is an accomplishment apparently verging on the supernatural.

A happy accident of preservation has given us knowledge of five skeletons of an ancestral bird, *Archaeopteryx*, the sole representative of the subclass **Archaeornithes**, from deposits of late Jurassic age (Fig. 49 *A*). In this bird teeth were still present, the wing had clawed fingers, and there persisted a long reptilian type of tail; in fact *Archaeopteryx* might well be called a small dinosaur were not imprints of feathers preserved with the skeletons.

Birds, mainly owing to the delicacy of their skeletons, are relatively rare in the fossil record. However, there is evidence that before the close of the Cretaceous there had evolved forms quite modern in almost every structure. Only one real intermediate form is known—intermediate in that it possesses teeth. *Hesperornis*, the sole representative of the superorder **Odontognathae** (others, sometimes put there, are highly dubious), was an extremely specialized, flightless, fish-eating form (Fig. 49 *B*). Since it is considered closer to living birds than to *Archaeopteryx*, all birds except the latter are placed in one subclass, the **Neornithes**.

Although taxonomists divide the birds into a considerable series of orders, the structural differences between these groups are for the most part small. There is, however, one partial exception to this; for there has been thought to be a distinction between two modern groups, representing primitive and advanced stages in bird evolution. Technically, the two have been defined by details of palatal and jaw structure (with which we need not here concern ourselves) which gave rise to the terms "palaeognathous" and "neognathous" (or superorders **Palaeognathae** and **Neognathae**) to distinguish them. Most birds with which the reader is ordinarily familiar—indeed, the majority of all living birds—are members of the latter group. To the other assemblage (Fig. 50) belong the ostrich and other forms generally termed **ratites** (a term referring to the reduced condition of the breastbone common in flightless birds), such as the cassowary and emu of Australia, the rhea of the

Figure 49. Primitive birds. *A, Archaeopteryx,* the earliest known bird and sole member of the Archaeornithes, had teeth, claws in the "hand," a long bony tail, and other reptilian characters; in fact, were it not for the preservation of feathers on the fossils it would be considered a small saurischian dinosaur. *B, Hesperornis,* a member of the Odontognathae, was primitive in retaining teeth; however, the wings were almost totally lost and it was a highly specialized swimming form, like a flightless loon. (*A* after Heilmann; *B* after Fenton and Fenton.)

Figure 50. Palaeognathous birds, including the ratites. *A,* The kiwi, *Apteryx,* of New Zealand. *B,* The emu of Australia. *C,* The ostrich. *D,* The tinamou of South America. *E,* The cassowary of Australia. *F,* The rhea of South America.

South American pampas, the extinct moas and the little kiwi of New Zealand, and the giant extinct birds of Madagascar.

Most palaeognathous birds have tiny wings and are flightless, a fact that has given rise to the claim that they represent a primitive stage in bird evolution in which flight had not been attained. But anatomic study strongly indicates that the ratites are probably degenerate descendants of once flying types. Most of them are found on islands where there are few terrestrial enemies, or on continents (Australia, South America) where, the fossil record tells us, the same was true at the time that the native ratites evolved. If ground-dwelling enemies are absent, much of the "point" of flying has been lost. The tinamous of South America are birds which have the power of flight (although they are awkward flyers) and yet have the "old-fashioned" type of palate; they may represent an ancestral group from which the ratites have developed. It is quite probable that the various groups of ratites are not closely related—probably the Palaeognathae are not a "natural" group.

Apart from the ratites, most birds (Fig. 51), as has been said before, are rather uniform in basic anatomic features, with differences between orders no greater than those which distinguish the smaller groups, termed families, among mammals. A majority of all birds, including the song birds, are members of the order Passeriformes, or perching birds. These highly evolved bird types are all relatively small; the crows and ravens are giants of the order. Several orders of water birds and oceanic types are customarily considered more primitive, although there is little evidence that they actually are. Of special interest here are the penguins, Southern Hemisphere forms which are flightless but nevertheless have powerful wings which have been transformed into swimming flippers. Here, as in the case of the ratites, it has been argued that absence of flight is primitive. It is, however, more probable that they have descended from flying oceanic birds that also used the wings in swimming.*

*For those of literary tastes, it may be mentioned that the penguins of which Anatole France wrote are the auks of English-speaking peoples, Northern Hemisphere birds which paralleled the penguins in taking up a swimming mode of life.

Figure 51. Thumb-nail sketches of representative members of the major orders of birds above the ratite level. The ordinal name and the popular name of the bird representing it are given in each case.

MAMMALS

MAMMAL-LIKE REPTILES (Fig. 52). The mammals are descended from reptiles; but the fossil record shows that the reptilian line leading to them, the subclass **Synapsida**, diverged almost at the base of the family tree of that class. Their relationship to the existing reptilian orders is thus exceedingly remote.

Oldest of the reptiles antecedent to mammals were members of the order **Pelycosauria** (Fig. 52 *A*), a group that appeared in the Carboniferous and flourished in the early Permian. They were in most regards exceedingly primitive reptiles, but certain characters in skull structure (such as the presence of but a single, rather ventrally placed opening in the temporal region of the skull) indicate that they represent a first stage in the evolution toward mammalian status. Many pelycosaurs, such as *Dimetrodon,* had greatly elongated dorsal spines of the vertebrae to support a large "sail." Succeeding them in late Permian and early Triassic days were the **Therapsida**, progressive mammal-like forms which were the commonest animals of their day (Fig. 52 *B*). The characteristic therapsids were flesh-eaters, active four-footed runners in which, as in their mammalian descendants, elbow and knee had been swung in toward the body, making for better support and greater speed—this in contrast with the sprawled limb pose of primitive land animals. In advanced Triassic members of the group many features of skull, jaw, dentition, and limbs approach closely the mammalian pattern. Other therapsids were, for a time, very successful herbivores.

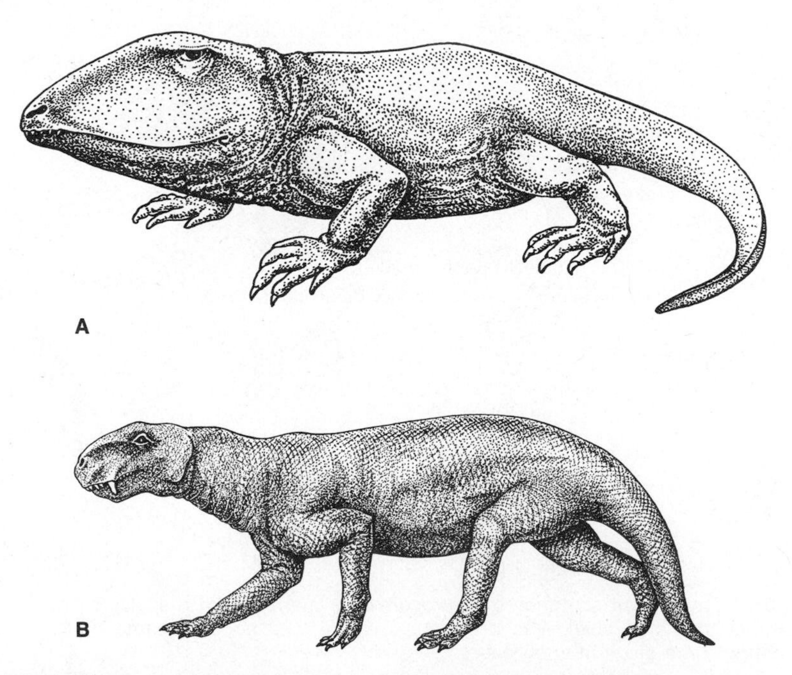

Figure 52. Synapsid reptiles. *A,* The primitive pelycosaur *Ophiacodon* from the early Permian of Texas; *B,* the therapsid *Lycaenops* from the later Permian of South Africa. Note the more mammal-like posture and projecting canine tooth in *Lycaenops.* If hair were drawn on it in place of scales it would appear to be a mammal; this would not be true of *Ophiacodon.* (*B* after Colbert.)

The evolution of mammal-like reptiles was a major feature of early reptilian evolution. But by the end of the Triassic period other reptile groups had become prominent—notably the dinosaurs. It appears that therapsids, for the most part, could not compete successfully with them, and rapidly dwindled and disappeared from the scene. There survived, however, small therapsids from which evolved the oldest mammals, sparse remains of which are found in Mesozoic deposits from the end of the Triassic onwards. Living as they did for tens of millions of years as contemporaries of the dinosaur dynasties, our small Mesozoic mammal ancestors were seemingly insignificant in the life of their times.

Intelligent activity may be reasonably regarded as the keynote of mammalian progress. With activity may be correlated not only the efficient locomotor apparatus characteristic of mammals, but also (as in birds) circulatory improvements and constant body temperature (with which the development of hair is related). In enterprise and ingenuity even the stupidest of mammals is an intellectual giant compared with any reptile. The habit of bearing the young alive—characteristic of all except the most primitive forms—and the development of the nursing habit, with concomitant care and training of the young, are mammalian innovations which have resulted in giving a long period for the development and elaboration of delicate nervous and other mechanisms before the young are sent out into the world. Most of these advanced characters came into existence during the long period of time when mammals lived under dinosaurian dominance. Alertness was necessary for survival, and mammals owe the dinosaurs a vote of thanks for having been unwittingly responsible for the eventual success of higher mammals.

Mammals are properly defined by the presence of mammary glands and, therefore, the habit of suckling the young. Other characters which always apply for living forms are the presence of hair and of a muscular diaphragm. However, all these are "soft" characters—none will be preserved in a fossil. Another character which appears to work and *can* be used is the nature of the jaw articulation; in reptiles (and other submammalian forms) the joint is located between the quadrate and the articular, two parts of the original visceral skeleton, but in mammals these have become ear ossicles and the joint lies between two dermal bones, the squamosal and dentary. Intermediates are known, but this character is still the most widely used criterion for distinguishing mammals from their therapsid ancestors.

The Mesozoic mammals (Fig. 53) are fairly numerous, but are almost all very poorly known, often only from isolated teeth. Their classification has long been debated and there is still no general agreement. We here adopt the idea that all mammals may be divided into two separate lines. One line, the subclass **Prototheria** (a word of warning: this term is frequently used in a much more restricted sense), consists of forms in which much of the lateral wall of the braincase is formed by the periotic rather than the alisphenoid as in therians (see pp. 189–192). Three of the four orders, the **Multituberculata**, **Triconodonta**, and **Docodonta**, are extinct; only one of these, the rather rodent-like **Multituberculata**, survived the Mesozoic and even they disappeared early in the Tertiary. Triconodonts and docodonts are poorly known small animals, comparable in a rough way to some of the eutherian insectivores—small, long-snouted, rather unpleasant carnivores. The prototherians may be divided into two infraclasses, but the evidence for relationships is poor, so these are probably not worth worrying about. Whether prototherians and eutherians form a single "natural" group is debatable. For a long time it was assumed they did, but then opinion gradually shifted and mammals were thought to have arisen independently several times from therapsids; now more workers seem to be returning to the original idea.

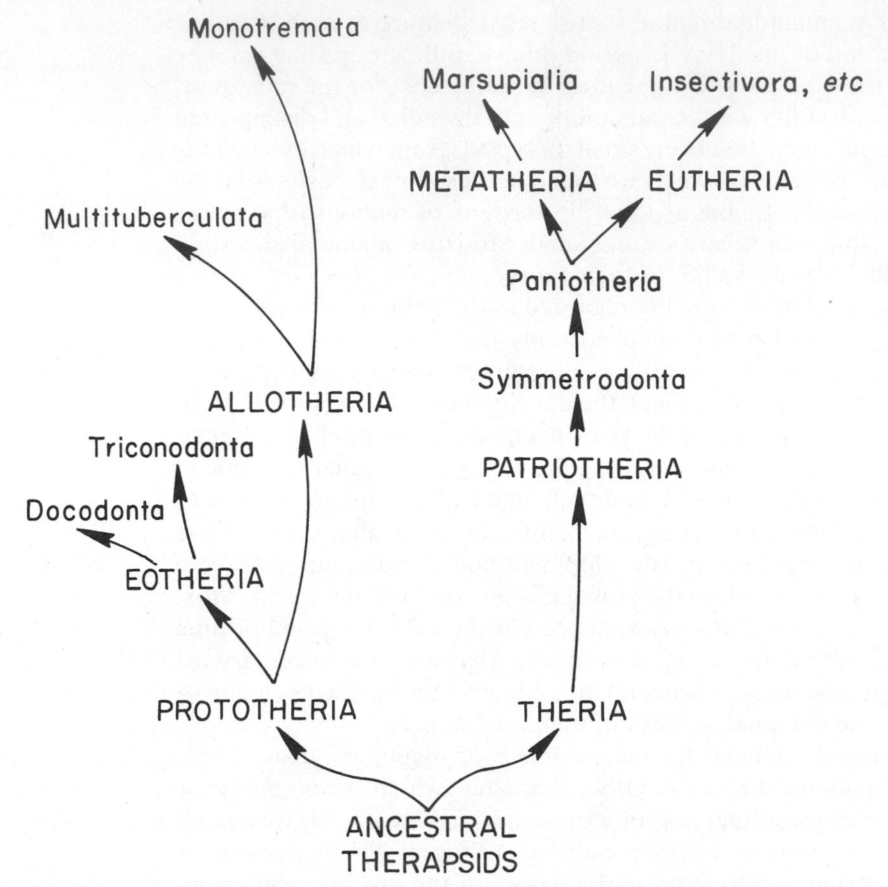

Figure 53. Diagrammatic family tree, to show the evolution and relationships of the primitive groups of mammals. Pictures are omitted because many of the forms are known only from partial specimens and other insufficient evidence.

MONOTREMES. The monotremes, order **Monotremata**, are the only surviving prototherians; as such they are extremely different from any other living mammals. They include only the duckbill and spiny anteaters of the Australian region. These curious animals possess many diagnostic mammalian characters, but retain primitive features in that, alone of mammals, they still lay shelled eggs as did their reptilian ancestors. The duckbill is a semiaquatic, web-footed, fur-covered frequenter of streams in which it finds a food supply of snails and mussels. The spiny anteater, protected from enemies by a coat of spiny hair, subsists on termites; its powerful clawed feet give it phenomenal digging ability. Both types make nests in burrows where the young are nursed after hatching. These two types are so specialized in many ways that they cannot be regarded as in themselves ancestral types. Their survival in Australia may be due to the relative isolation of that area. Unfortunately we know nothing of their history. Much of our knowledge of the relationships of extinct mammals is based on diagnostic characters of molar tooth patterns (discussed in a later chapter); the monotremes are, unfortunately, toothless as adults.

PRIMITIVE THERIANS AND MARSUPIALS. All normal modern mammals belong to the other subclass, the **Theria** (Fig. 53). Aside from a large alisphenoid on the lateral wall of the braincase, they are characterized by molar teeth that start as triangles and become more complex (a story treated in some detail in Chapter 11). Three infraclasses are recognized. The first, **Patriotheria**, consists of two more orders of small, obscure, insectivorous, Mesozoic forms, the **Symmetrodonta** and **Pantotheria.** They differ in dental patterns and the first could be ancestral to the second. From the pantotheres, there appear to have arisen sometime in the Cretaceous the two groups of higher mammals, the infraclasses Metatheria and Eutheria.

The pouched mammals—technically termed the infraclass **Metatheria**, order **Marsupialia** (Fig. 54)—owe their popular name to the fact that although the young are born alive, they are born at a tiny and immature stage; typically, the female marsupial carries on her belly a pouch in which the newborn young are kept and nourished for a further period after birth. The common opossum is characteristic of the group and is a primitive mammal in many regards. In most regions of the world the marsupials have not been able to compete successfully with more progressive mammals, and even the hardy opossum failed to survive except in the Americas. South America proved a haven for many marsupials during Tertiary times, when that continent was long isolated; a variety of marsupials (mainly carnivorous—the suborders **Polyprotodonta** and **Caenolestoidea**) developed there, almost all of which became extinct when the isthmian link was re-established and a host of more advanced mammals invaded that continent. Australia is the one region where marsupials have flourished greatly. The geologic evidence suggests that this latter continent was separated from the rest of the world by Cretaceous times and has since remained isolated. No placental mammals, it appears, had reached Australia at the time of separation, and few have been able to reach it since (bats and whales obviously had no problems, and rats did succeed in getting there). There the marsupials had little opposition, and expanded and diversified to fill almost every type of adaptive niche which placental mammals have occupied in other regions. Fairly directly from the ancient opossum-like, polyprotodont ancestors came such pouched carnivorous forms as the native "cats," the Tasmanian devil, the native "wolf," and pouched parallels to anteaters and even moles of other continents. A second branch of the Australian marsupials (suborder **Diprotodonta**) acquired chisel-like front teeth and developed types comparable to the rodents among placentals. There are varied native squirrel-like types, even a marsupial flying "squirrel." The wombat is analogous to the woodchucks or marmots of other regions; the native "bear," the koala, is a leaf-feeder belonging to this group. The marsupials have failed to parallel the placentals in one respect—there has been no development of hoofed forms comparable in structure to horses, cattle, and antelopes. But the kangaroos—which, like the typical ungulate placentals, are speedy grass-eating plains-dwellers—fill much the same place in nature. Somewhat intermediate between the polyprotodonts and diprotodonts are the Australian bandicoots (suborder **Peramelida**).

PLACENTAL MAMMALS. The major progressive group of mammals includes the host of living forms properly termed the **Eutheria**, the "true mammals," but usually called placentals. The latter name is due to the fact that, in contrast with most marsupials, there is an efficient nutritive connection, the placenta, between mother and embryo; as a result the young can develop to a much more advanced stage before birth. On the extinction of the dinosaurs, these highly developed mammals were already in existence (they are known, albeit rather poorly, from the Upper Cretaceous); they rapidly expanded into a host of types, many of which have continued down to modern times. In some other groups of vertebrates the family "tree" is actually treelike, with a main trunk or at least major branching limbs. That of the placental mammals, however, is comparable to a great bush: the various

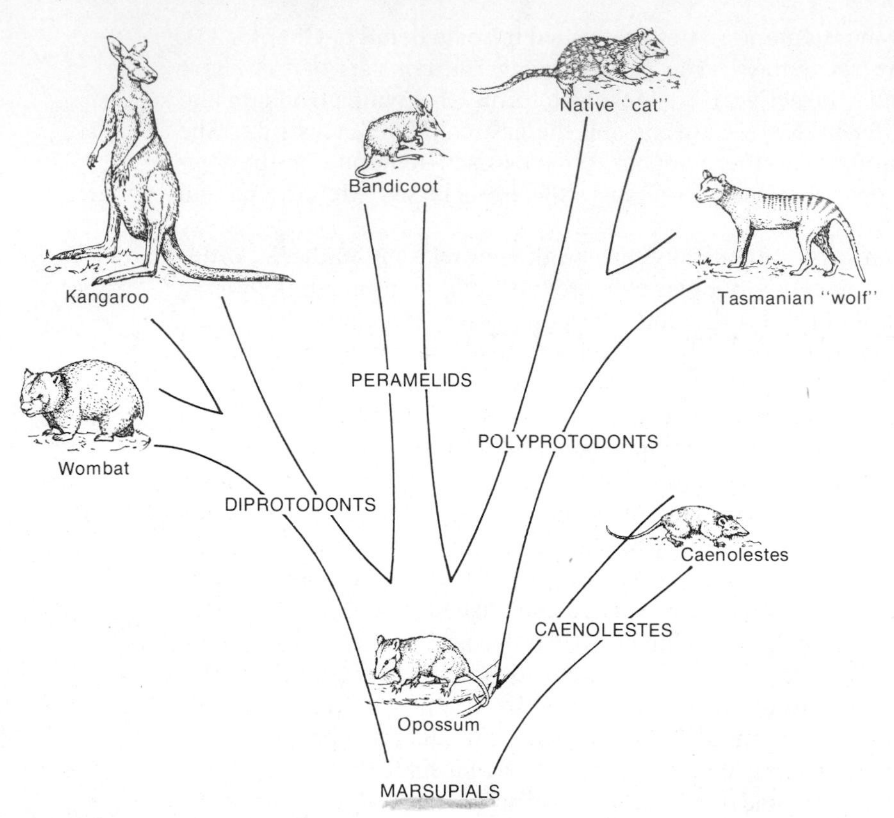

Figure 54. Diagrammatic family tree to show the evolution and relationships of the marsupials. The opossum and *Caenolestes* are basically South American; all the rest are members of the great Australian radiation of this order.

orders are difficult to assemble into groups and appear, for the most part, to have branched out independently of one another in early times. We may briefly note some of the main components of the placental assemblage (Fig. 55). Not all the extinct groups known (and listed in Appendix 1) are discussed here, but those omitted are very obscure and unlikely to be encountered unless one studies paleontology.

The ancestral placentals—indeed, the early mammals as a whole—seem to have been small shy animals which were potential flesh-eaters, but were forced to live, owing to their size, on small prey such as insects, grubs, and worms, and presumably on some of the softer vegetable materials. This phase of mammalian existence lasted for many millions of years before the dinosaurs became extinct.

At the beginning of the Cenozoic there rapidly developed a great radiation of mammals into a variety of orders. Some few forms, however, have remained not too distantly removed in structure and habits from their small insectivorous ancestors; these constitute the order **Insectivora** (Fig. 56). The little shrews closely resemble, in habits at least, their remote ancestors; in many regions shrews are exceedingly abundant in woods and meadows, but so shy are they that they are almost never seen. Other familiar insectivores are the prickly European hedgehog and the moles, which, with powerful digging limbs, have taken up the pursuit of grubs and worms underground. It appears probable that the living mammals least changed from the ancestral placentals are the tree shrews (*Tupaia*) of the Oriental region. It has long been suggested that these attractive little animals are close to the

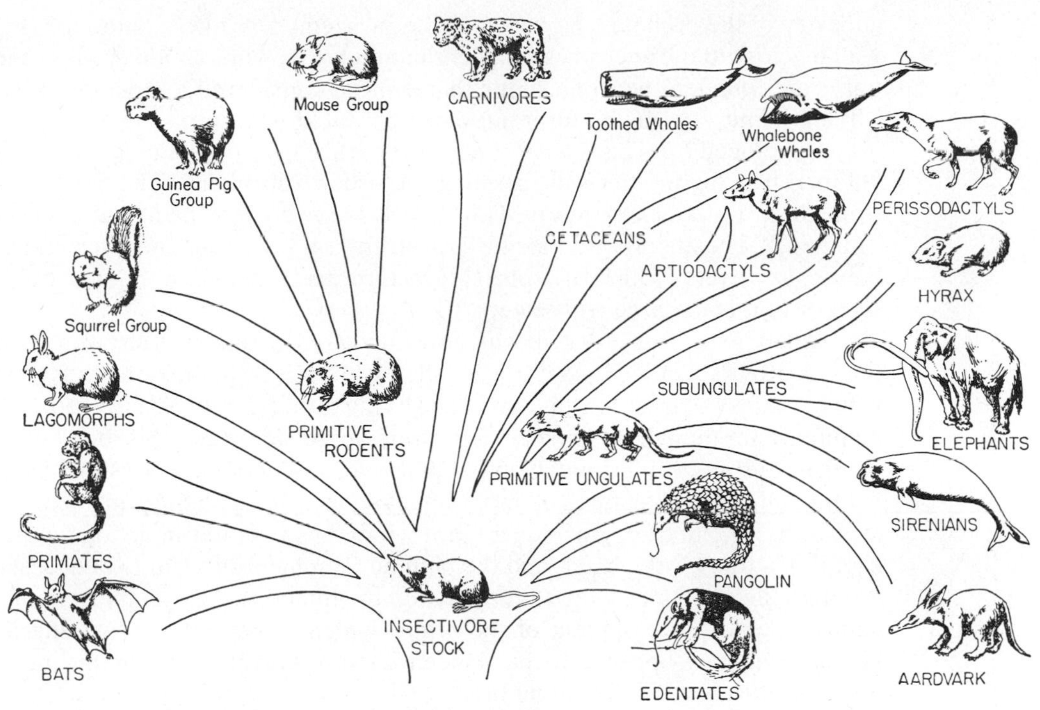

Figure 55. Diagrammatic family tree of the major orders (and some suborders) of eutherian (placental) mammals. Separate diagrams (Figs. 57–61) give in more detail the evolution of primates, carnivores, and odd- and even-toed ungulates.

Figure 56. Insectivores. These rather varied forms are all relatively small and more or less shrew-like. *A, Tupaia,* a tree shrew; *B, Elephantulus,* an elephant shrew; *C, Talpa,* a mole; *D, Erinaceus,* a hedgehog; *E, Potamogale,* an otter shrew; *F, Solenodon,* a large Cuban form. (From Thenius.)

ancestry of the primates; however, this now seems less likely, although they may well be close to the ancestry of higher mammals as a whole. Odd groups of insectivores are found in the tropics; one that is often considered a separate order is the "flying lemur," a gliding form not closely related to lemurs.

Developed from insectivore ancestors is the one group of mammals which has attained true flight—the bats, forming the order **Chiroptera**. The bat wing differs from that of pterosaurs and birds in that it is a web stiffened by four of the five "fingers." The majority of bats (**Microchiroptera**) have remained insectivorous in habits; however, one major group (**Megachiroptera**), abundant in the tropics, consists of relatively large fruit-eaters.

PRIMATES. This order, to which we ourselves belong, was an early offshoot of the insectivores; indeed, so close are the ties that it is an unsettled question as to which of the two orders certain fossil and living forms should be assigned. Primitive mammals are thought to have been to some degree arboreal; this mode of life was emphasized in early primates and appears to have been responsible for the development of many features of importance—general body agility and coordination; the ability to climb by grasping a limb, which has resulted in giving us that most useful of "tools," the hand; and the high development of vision, so necessary for arboreal life (with a concomitant reduction in olfaction). Most important of all, the high degree of development of the brain, which is the outstanding character of higher primates, appears to have been closely correlated with the needs and opportunities found in arboreal life.

Lowest of acknowledged primates alive today are the lemurs (suborder **Lemuroidea**; Fig. 57), still flourishing in the isolation of Madagascar and little changed from early Cenozoic ancestors. These are four-footed arboreal forms, with thick fur, relatively poor eyesight, and a nose forming a typical mammalian muzzle. Certain extinct forms, put in a separate suborder **Plesiadapoidea**, were equally primitive, but had certain rodent-like characters such as gnawing incisors.

A step in advance is represented today by the curious little creature named *Tarsius* (suborder **Tarsioidea**), from the East Indies. The living form is somewhat specialized in such characters as an elongate ankle region—a hopping adaptation to which it owes its name. But while still lemur-like in many ways, *Tarsius* shows advances in such features as excellent eyesight and reduction of the nose to a mere button.

The higher general level of primate evolution is that represented by the monkeys, apes, and man. Although man may pride himself on mental attainments, anatomic differences between the various members of the group are few, and are mostly matters of proportions of various structures or obvious adaptations—such as the distinctive human features of vertebral posture and hind leg structure associated with upright gait. In all these higher primates the cerebral hemispheres are relatively large, eyesight highly developed, the nose reduced, the hands useful grasping organs. Two distinct groups of monkeys appear to have arisen independently from a *Tarsius*-like stock. One, the suborder **Platyrrhini,** is found in South America, where it is represented by a variety of forms such as the common organ-grinder's monkey—the Capuchin—and the little marmosets. A second group of higher primates—the suborder **Catarrhini**—developed in the Old World. Primitive members are the familiar monkeys and baboons of Africa and Asia. More advanced members are the great apes including the gibbon, orang-utan, chimpanzee, and gorilla. All these apes are of relatively large size, in all the tail has been lost, and in the last two forms mentioned the anatomic similarities to man are close indeed, although neither is a human ancestor. The gibbon is an agile arboreal acrobat and

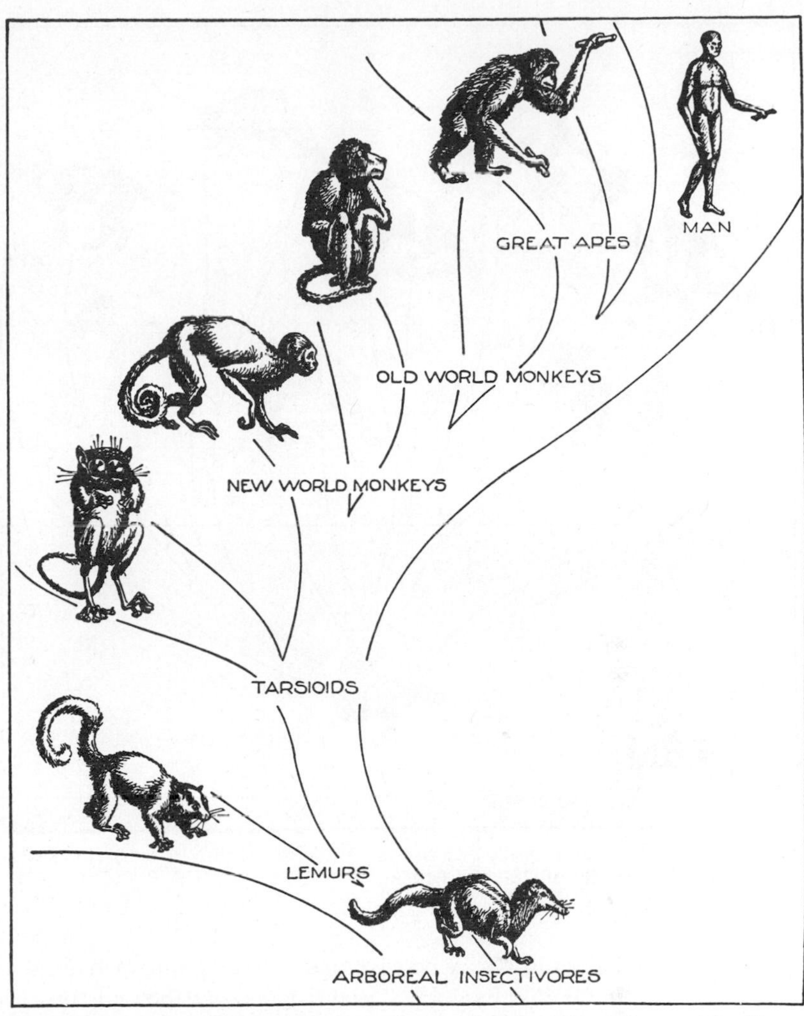

Figure 57. A simplified family tree of the primates. (From Romer, The Vertebrate Story, University of Chicago Press.)

the orang is a good traveller in the trees; but arboreal specialization is less marked in the chimpanzee, and some gorillas have become nearly completely terrestrial, although as quadrupeds. Man is essentially a fifth member of this great ape series; he has become a terrestrial biped, but has, stamped deeply into his structure, features probably acquired during a long sojourn in the trees. We do not fully know our own pedigree, but the fossil "man-apes" of South Africa (the australopithecines) structurally bridge the gap between man and his simian relatives.

CARNIVORES. The insectivores were potential flesh-eaters. With the development of numerous mammals of the more harmless varieties there soon arose from the primitive placental stock varied predaceous types (Fig. 58). All of the living carnivores are included in the order **Carnivora**. However, in the early Tertiary the common flesh-eaters were members of a different stock, which are currently considered as forming a distinct order **Creodonta**. These forms appear, however, to have been relatively slow, clumsy, inefficient and, it would seem, stupid, and soon dwindled and eventually disappeared, giving place to members of the proper stock of Carnivora. The modern terrestrial carnivores—the suborder

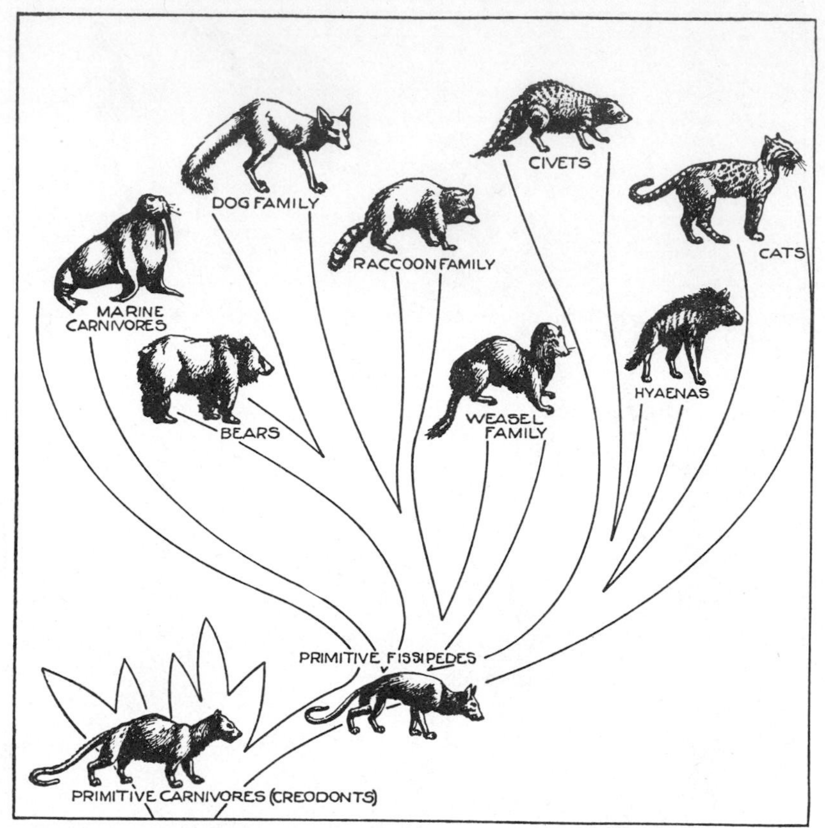

Figure 58. A simplified family tree of the carnivores. (From Romer, The Vertebrate Story, University of Chicago Press.)

Fissipedia—may be essentially divided into two great groups, of which dog and cat are familiar examples, and the weasel tribe and civet group are, respectively, more primitive members. A third group, the infraorder **Miacoidea**, is recognized for various extinct ancestral forms.

The weasels and their close kin, small and short-legged in build and often purely carnivorous in habits, are seemingly primitive members of the general "dog" group (infraorder **Arctoidea**). But within the weasel family there developed a considerable series of forms variant in habits and diet—badgers, skunks, otters, even a marine member, the sea otter of the Pacific. The dog family developed into terrestrial types adapted to running down their prey. The raccoon is related to the dog stock, but is a persistently arboreal animal with an omnivorous taste in food; there are several American relatives, and the lesser panda of Asia is allied. The bears are the members of the general dog group which have departed most widely from ancestral conditions; these clumsy fellows have (except for the polar bear) swung far away from flesh-eating to take up a mixed—but mainly herbivorous—diet. The giant panda, which seems to be an odd bear rather than an odd raccoon, has become completely herbivorous.

Of the "cat" group of carnivores (infraorder **Aeluroidea**), the civets and their relatives appear to occupy a primitive position. These varied forms are mainly Old World tropical forest dwellers, with which the average reader will not be acquainted; the mongoose is the most familiar representative. The unattractive hyenas are an overgrown offshoot of the civets, which have taken up a life of scavenging, although some are highly efficient predators, and one, the aardwolf, has become an eater of termites. The felids form the most highly specialized development in this subgroup of carnivores. The cats are adapted for stalking their prey and making an agile jump onto the victim, rather than running it down as do the dogs; the teeth are highly specialized for stabbing and shearing, and in nature the cats are purely flesh-eating. Most living cat types—such as the lion, tiger, panther, and so forth—are very similar to one another in structure; a variety of extinct sabre-tooths were notable for an exaggerated development of stabbing teeth (cf. Fig. 240 *A*).

Last of the carnivores to be mentioned are the marine **Pinnipedia**—the various seals, which are fish-eaters, and the grotesque walrus, with its digging tusks and blunt molars for crushing the mussels upon which it feeds. These appear to have evolved from land carnivores in mid-Tertiary times; in pinnipeds, among other adaptations, the limbs have been transformed into "flippers"; the hind limbs, turned backward, replace the reduced tail as a swimming organ.

UNGULATES. Notable in Tertiary history has been the development of great series of forms, often of relatively large size, which have assumed an herbivorous mode of existence and developed dentitions with grinding molars for chewing vegetable food materials. The more advanced forms of this sort have tended to become good running types, with their limbs elongated with the addition of an extra joint by lengthening the bones of the "palm" and "sole" region of the feet. They have tended to walk on the tips of the toes, which are generally reduced in number. The claws borne by primitive mammals have generally given place to hooves—hence the term "ungulates" applied to these herbivores.

Although the varied ungulate types have developed some common characters, it is far from certain that all ungulate stocks spring from a common source; there has surely been considerable parallelism in their development. The statement, all too often seen in print, that "the ungulate" has such-and-such physiologic or structural features is valueless unless it is more specifically stated just what type of ungulate is meant; a cow may be as closely related to a lion as to a horse.

Early in the Age of Mammals there rapidly sprang into existence a host of ungulates of varied but archaic types; most passed rapidly out of existence and need not concern us here. One order of these archaic forms, the **Condylarthra**, appears to have been very close to the ancestry of most other ungulates (and even of the carnivores—the two are very hard to tell apart at the start of their history). Some of these extinct orders produced large, clumsy, more or less rhinoceros-like forms. One group of orders developed in South America, which until very recently lacked representatives of the "higher" groups of ungulates. Figure 59 shows a variety of these odd forms.

The dominant ungulates of later Cenozoic and Recent times belong to two very distinct orders, characterized by the horse and cow—the orders Perissodactyla and Artiodactyla, respectively the odd- and even-toed types.

In the **Perissodactyla** (Fig. 60) the key character has been the early reduction of the toes from five to three, and in the case of the later horses further reduction to a single-toed or monodactyl condition (cf. pp. 168–169). Primitive forms such as little *Hyracotherium,* which was not only a "dawn horse" but also close to the ancestry of the entire order, had already reached a three-toed condition in the hind foot but in front had lost merely the thumb. Early perissodactyls were browsers, living on relatively soft food materials in forests and glades. In the progressive series of the horses, mid-Tertiary types occupied the spreading grasslands, developed high-crowned teeth to cope with a diet of grasses and grains, and reduced the toes to

Figure 59. Various archaic ungulates (to very different scales). *A,* The condylarth *Phenacodus; B,* the pantodont *Coryphodon; C,* the dinocerate *Uintatherium; D,* the notoungulate *Toxodon; E,* the pyrothere *Pyrotherium.* (From Kurtén.)

three on each foot. With further approach to the modern genus *Equus* there was attained a single-toed pattern; the middle toe was the sole survivor. The tapirs of Old and New World tropics are persistent browsers which, although larger, have departed little from the mode of life of early perissodactyls. More divergent among extinct odd-toed ungulates were the large and ungainly horned titanotheres and the grotesque chalicotheres, which combined a somewhat horselike body with feet armed with powerful claws (perhaps used for digging tubers). More successful, despite large size, were the rhinoceroses, which during their history developed hornlike defensive weapons; once common and widespread, they are still represented by a few species in the Old World tropics.

The perissodactyls were very successful in the earlier part of the Age of Mammals, but are now reduced to a relatively few species of three types—horses, tapirs, rhinoceroses. Quite in contrast has been the story of the order **Artiodactyla** (Fig. 61). Rare in the early Tertiary, these even-toed ungulates became increasingly abundant in later times and are widespread today. In these forms toe reduction began with the loss of "thumb" and "big toe," giving a four-toed pattern. Among the four, the two side toes tended to reduction or loss, leaving toes three and four to form the so-called "cloven hoof" diagnostic of the group. The pigs of the Old World and the related peccaries of the New are relatively primitive types, omnivorous in habits, as were certain extinct forms that looked like pigs but had skulls up to a meter long. The hippopotamus is a lumbering amphibious cousin of the pigs, a

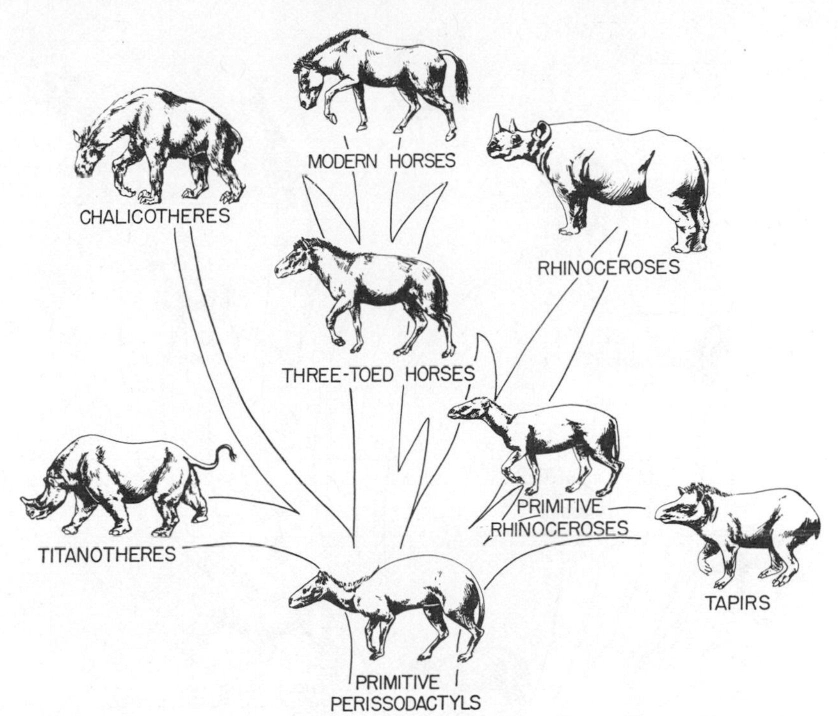

Figure 60. A simplified family tree of the odd-toed ungulates, the perissodactyls.

vegetarian in diet. The more successful artiodactyls became purely herbivorous, developed a series of grinding cheek teeth with characteristic crescentic cusps (cf. p. 246), and perfected a multichambered stomach associated with the cud-chewing habit (rumination, cf. p. 269) for dealing with vegetable foods. The camels (Tylopoda), originating in North America, but now surviving in the Old World and (as the llamas) in South America, are relatively primitive cud-chewers. The most advanced artiodactyls are the Pecora—agile, swift-running ungulates, with highly developed ruminating stomachs and with heads generally armed with some kind of horns or antlers. The deer and giraffe are relatively primitive browsers. Much more abundant are the "cowlike" forms, the bovids, most of which have become (parallel to the horses) grass-eating plains-dwellers. Cattle, sheep, and goats are familiar domesticated bovids; for a host of others, mainly inhabitants of the Old World tropics, we have no familiar specific names and tend to lump them as "antelopes." Finally, in this ruminant assemblage we may mention the prong-buck of the western American plains, survivor of a New World group paralleling the true antelopes.

SUBUNGULATES. Frequently grouped as the subungulates are a series of orders, probably originating in Africa, which are perhaps best regarded as aberrant offshoots of a primitive ungulate stock (Fig. 55). The little hyraxes—order **Hyracoidea**, the "conies" of Scripture—are animals of rather rabbit-like size and habits (though some are larger and others live in trees), but are definitely hoofed ungulates whose pedigree goes back to the fossil records of the early African Tertiary. They have, however, never progressed structurally to any degree and have never ranged beyond Africa and the Mediterranean region. Strange as it appears, these little forms appear to be related to two other quite divergent groups —the elephants and the sea cows.

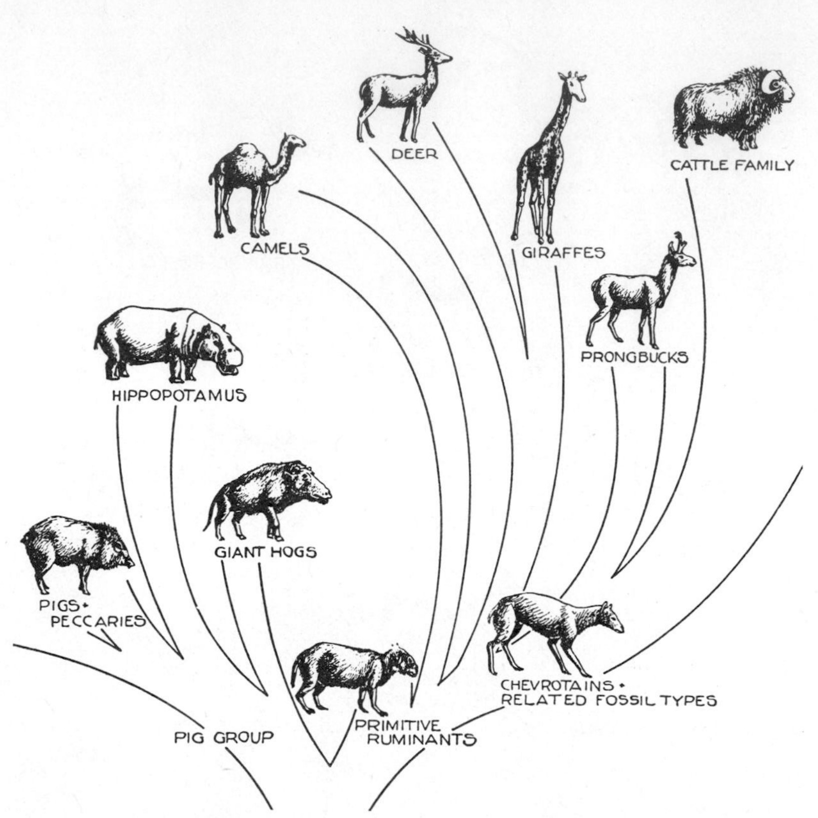

Figure 61. A simplified family tree of the even-toed ungulates, the artiodactyls. A major cleavage is into the pig group, to which the hippopotamus and extinct giant hogs belong, and the cud-chewing ruminants. The little chevrotains of tropical Africa and southern Asia are close to the ancestry of advanced ruminant types. (From Romer, The Vertebrate Story, University of Chicago Press.)

The **Proboscidea** (whose name derives from their trunk) are represented today only by the two elephant types proper to Africa and to southern Asia. Their history, however, has been long and varied. The older types are termed mastodons. The most primitive known form, from the Eocene of Egypt, was already large for its day—much the size of a large pig—with chisel-like front teeth and a good set of grinders in the cheek. Later mastodons increased rapidly in size and developed long jaws, with short tusks both above and below. Later the jaws diminished in length, but the upper tusks, in seeming compensation, tended to elongate and eventually there developed the characteristic head pattern of tusks and trunk seen in the modern elephants. Meanwhile, from mid-Tertiary times on, the mastodons had spread widely over Eurasia and eventually over the Americas; in the Ice Age a variety of true elephants, most of them termed mammoths, roamed every continent except Australia and South America. Toward the end of the Pleistocene nearly all of the proboscideans vanished. The reasons for the sudden reduction of this seemingly flourishing order of mammals are a mystery, but no more of a mystery than the nearly simultaneous extinction of a variety of other large mammals as well. One suggestion is that overhunting by man is responsible.

To class the sea cows—the manatees and dugongs of the order **Sirenia**—with the ungulates would appear superficially to be a gross misuse of terms. These animals, browsers in shallow waters of the tropical Atlantic and Indian Ocean regions, are purely aquatic beasts, with front limbs transformed into paddles, hind legs reduced to concealed vestiges, and the tail redeveloped to form a horizontal swimming fluke. Now relatively rare, they ranged widely over the world for much of the Tertiary. The oldest fossil remains—many from the Eocene of Egypt—show a number of resemblances, especially in the teeth and their replacement, to the most primitive conies and mastodons and add strength to the belief that all three of these curious subungulate groups are varied descendants of some common ancestor in Africa at the dawn of the Age of Mammals.

WHALES. Although both carnivores and subungulates have developed aquatic types, none of these is as highly specialized for marine life as the whales and porpoises which constitute the order **Cetacea**. Much as in the sirenians, the front legs have been transformed into flippers, the hind legs have vanished, and the tail has become a highly developed swimming organ, with horizontal flukes. As in ichthyosaurs, a fishlike dorsal fin may redevelop and the neck is shortened, so that a streamlined fusiform fishlike body is re-attained. The skull is peculiarly altered, for the external nostrils have moved upward to become the "blowhole" atop the head. Most mammals cannot survive long under water, but the physiology of whales has been so modified that some can remain submerged for the better part of an hour. The greater part of the order, including the porpoises and dolphins and a few larger whales (suborder **Odontoceti**), are toothed forms which subsist on animal food—fishes, octopi, and squids. However, the very largest of the "noble cetaceans," the whalebone whales (suborder **Mysticeti**), live on much smaller food materials—the small marine organisms that constitute the ocean's plankton. Teeth are here absent, and instead there hang from the roof of the mouth row after row of thin sheets of cornified skin, the "whalebone." The fringed edges of the whalebone sheets strain the plankton from the water and the resulting "catch" is licked down by the whale's tongue. The oldest known whales, from the Eocene, were already aquatic types, but less specialized in body and skull and suggestive of a descent from some primitive type of land carnivore; however, the relationships of the sorts of modern whales are debated.

EDENTATES. Extremely aberrant in many ways are the members of the order **Edentata**, mainly South American in distribution and history. Characteristics of this group include extra articulations between the vertebrae, a trait responsible for the alternate name of Xenarthra. The living representatives are the tree sloths, dull, sluggish arboreal leaf-eaters; the long-snouted anteaters; and the armadillos, omnivorous feeders that have a well developed back armor. Extinct are two further types which attained large size, the glyptodonts, relatives of the armadillos, with a dome-shaped protective "shell" and an armored tail, and the great ground sloths, elephant-sized and massive herbivores. The anteaters are actually edentulous, but the other forms have at least a good series of molars (although the enamel covering of the teeth is reduced). The group attained its development in South America, but during the Ice Age ground sloths and glyptodonts invaded North America with momentary success. However, only the armadillo has survived in the north; the reason for the extinction of the large edentates is as puzzling as in the case of the mammoths.

We shall mention here two odd types of mammals which in the past were often grouped with the South American edentates, but are now recognized as quite distinct. The aardvark (*Orycteropus*; order **Tubulidentata**) is a grotesque long-snouted African beast; the slender-snouted pangolins (*Manis*; order **Pholidota**) of the Old World tropics are unique among mammals in that they are completely covered by overlapping horny scales which give them somewhat the appearance of animated pine cones. Both make a living by invading termite nests and are armed with powerful claws. The aardvark has retained a few peglike cheek teeth; the pangolin, like its South American anteating analogues, has lost its teeth, useless in this mode of existence. Neither animal has any close relations with any other group.

RODENTS. Most impressive of all mammals from the point of view of both numbers of genera and species and numbers of individuals are the gnawing animals, the order **Rodentia**. The key character of the group lies in the development of an enlarged pair of front teeth in both upper and lower jaws into an effective chisel-like gnawing apparatus. Rodents have never (we may be thankful) developed into flying forms, nor into marine types; but in almost every known terrestrial habitat rodents are the most flourishing group. Rodents have spread out into a great variety of forms, which are difficult to classify into major subgroups—indeed their classification is probably as hotly debated as that of any group of vertebrates. Three "clusters" of living forms, however, stand out rather prominently. (1) A relatively small but familiar group (suborder **Sciuromorpha**) is that of the squirrels, prairie dogs, marmots, and relatives, including, some believe, the beavers. (2) The guinea pig is representative of a great group (suborder **Caviomorpha**) which includes most of the rodents of South America. The North American porcupine appears to be of South American origin; the Eurasian porcupine and a few other Old World rodents are often assigned to this group, but this relationship is now frequently thought to be due to parallelism. (3) Most flourishing of rodents are the rats and mice (suborder **Myomorpha**), ubiquitous in distribution; an indication of their versatility is the fact that they include the only terrestrial placentals that were able to reach Australia before the coming of man.

LAGOMORPHS. The order **Lagomorpha** is a small one, including almost exclusively the hares and rabbits. They were at one time classed with the rodents because, as in that group, there are chisel-like front teeth. But there are few other resemblances; furthermore, the lagomorphs have two pairs of upper chisels rather than one. For decades students of Recent and fossil mammals have recognized the lack of relationship, but despite this there constantly recur instances in which biologists ignorant of the animal world have described structures or functions found in the rabbits or hares as characteristic of "the rodent."

4 Cells and Tissues

Although the anatomic and physiologic studies of the vertebrate body treated in this book generally deal with gross structures—organs and organ systems—it must never be forgotten that these structures are composed of tissues, and these in turn of cells, and that the cells are the basal living units from which and by which the entire complex body is built. A knowledge of cell and tissue structure and function is basic in the study of biology; our purpose here is briefly to review these familiar topics as a background for a better understanding of those structures of greater complexity with which this work is primarily concerned. The functioning of body organs depends basically upon the activities of the cells of which they are composed; conversely, the varied organ systems are nearly all engaged, directly or indirectly, in furnishing the cells with the materials required for their vital processes and in maintaining them in a suitable environment. Each cell lives its own life; however, each is dependent upon other cells and tissues for its continued existence, and each in turn makes its contribution to the welfare of the total organism.

CHEMICAL MATERIALS. Much of vertebrate structure and function is related to the collection, transformation, and transportation to the cells of the basic chemicals needed to form and maintain the protoplasm and enable it to play its proper part in the work of the body and, again, to the disposal of wastes formed by its activity. In consequence, it is important to know the basic chemistry of the various molecules which make up the body. However, this is a broad topic in itself and beyond the scope of this volume.

CELL STRUCTURE AND FUNCTION. The cells of the body are, of course, highly varied in their structures and functions; a liver cell and a nerve cell, for example, appear very different and have very different activities. But despite such differences all cells have in common basic structural features and basic functional activities.

The "gross" features of cell anatomy are readily visible under the ordinary compound microscope, although the electron microscope is needed to see the finer details (Fig. 62). Often centrally situated is the **nucleus**. Within it are amounts of readily stainable—i.e., chromatic—material, to be identified chemically with nucleic acid. Toward a time of cell division the chromatin is compacted into a series of paired rodlike structures, the **chromosomes**; each chromosome splits lengthwise, so that when the cell divides, each daughter cell receives a full complement of chromosomes. Other chromatic material (mainly RNA) may be more or less permanently consolidated into a mass termed a **nucleolus**. Except at times of division, the

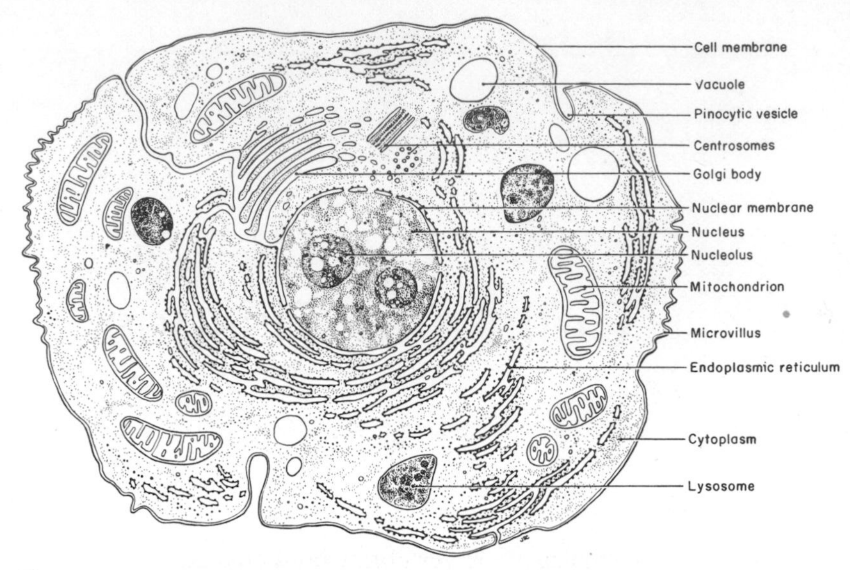

Figure 62. Diagram of a typical animal cell. (From Villee, Biology.)

nucleus is generally seen to be separated by a membrane from the general cell body. The cell in turn is bounded externally by a definite **cell membrane**, including lipids and proteins, and generally of complex structure, which is capable of regulating to a considerable degree the traffic in substances entering or leaving. To the cell materials in general the term **protoplasm** is applied; **cytoplasm** is a more restricted term, designating the contents of the cell body external to the nucleus. With the ordinary microscope certain formed structures can generally be made out in the cytoplasm. Near the nucleus there are often visible small particles, the **centrioles**, which play a part in cell division. Scattered through the cytoplasm are numerous little structures, the **mitochondria**. With the aid of the electron microscope the mitochondria, which appear as mere dots or dashes through the light microscope, are seen to have a definite structure as elongate spheroids with a series of internal cross partitions. The mitochondria are centers of enzyme concentration and activity in which takes place much of the chemical work that supplies energy for the cell. It is in them that sugars and fatty acids are oxidized to yield adenosine triphosphate (ATP), the universal energy-carrier of the cell. Special stains reveal in many cells a **Golgi apparatus** (named for its discoverer), often called a reticular apparatus, a series of tiny vesicles whose function appears to be concentration of proteins which, combined with large carbohydrates manufactured here, are given off from the cell as secretions.

It was long thought that, apart from such structures as those just mentioned, cytoplasm was essentially an amorphous jelly-like colloid, with organic materials in solution in a fluid, watery medium. The electron microscope shows that, on the contrary, the cytoplasm is finely organized, with a network, or **reticulum**, of delicate membranes with vesicles of more liquid material between them. The cytoplasm contains numerous tiny particles termed **ribosomes**, invisible to the ordinary light microscope, embedded in these membranes; the ribosomes are major centers of protein synthesis. When ribosomes are present, the reticulum is a **rough**

endoplasmic reticulum. The endoplasmic reticulum may lack ribosomes and be **smooth**; the functions of this latter sort are not well understood.

Such complexity of cell structure gives the impression of a static condition. This is, of course, the reverse of the actual situation, for the cytoplasm is the seat of constant chemical activities of an exceedingly complex nature.

Every cell must produce energy for its particular function in the body (whether secretion, nervous conduction, muscular movement, or what not), but in addition it constantly undergoes chemical change in the maintenance of its internal economy. A cell, like a body, has, so to speak, its own "basal metabolism," necessary for the support of its vital activities.

CELL ENVIRONMENT; INTERSTITIAL FLUID. An organism must provide the environment, physical and chemical, required by its cellular units. These requirements are rather rigidly fixed for vertebrate cells.

Although the organism as a whole may be able to exist over a fairly wide range of external temperatures, the temperature to which the cells in the interior of the body can be subjected and still survive is limited, although the optimum temperature and permissible departures from it vary considerably from form to form (cf. p. 127). In general, lower, primitively water-dwelling forms have a temperature range pitched lower in the scale than do terrestrial vertebrates. With few exceptions, even fishes are unable to stand internal temperatures much below the freezing point. Thirty degrees Celsius is about the upper limit for fish cells, 45° C or so for the higher, terrestrial vertebrate classes. Beyond such a point "heat death" occurs; some proteins undergo irreparable denaturation.

A cell can live and avoid desiccation only if bathed in a watery liquid medium. Such a material, the **interstitial fluid** (or intercellular fluid), pervades the body. Water itself is not enough: this liquid must contain a considerable amount of material in solution or, through osmotic pressure, swelling and disruption of the cells which it bathes may take place. Further, cells flourish only if the materials in solution, mainly inorganic salts, adhere rather closely to the formula actually present in interstitial fluid normally found in the body of vertebrates. This liquid contains considerable amounts of sodium and chloride ions, lesser quantities of potassium, calcium, and magnesium ions, and small amounts of other elements.

Reasons for the requirement of this salt formula (different from that in the watery constituents of the cells themselves) are poorly understood. It is of interest, however, that with two exceptions (the absence of sulfates and the lesser amounts of magnesium), the "formula" of the interstitial fluid of vertebrates is similar to that of sea water, although it is generally much more dilute. Possibly this is no mere accident; the ancestors of the vertebrates were simple animals bathed in and permeated by the waters of the early Paleozoic ocean, their cellular physiology was evolved with this type of environment as a basic feature, and when, as complex organisms, they developed an independent internal environment, this salty interstitial liquid persisted as, so to speak, a remnant of the ancient seas. The interstitial fluid is in relatively free communication with the similar liquid constituting the bulk of the blood plasma, and it is therefore through the blood circulation that the even distribution through the body of the fluid materials is accomplished.

It is, of course, through the interstitial fluid bathing it that the cell is supplied with the materials that it needs for its existence. Through this same liquid it is relieved of its waste materials. The interstitial fluid must be in intimate contact with the circulatory system so that food materials may be constantly supplied and waste constantly taken away.

The nature of the cell membrane, and the necessity of there being a sufficiency of materials in solution in the interstitial fluid to balance the concentration within the cell, bring to our attention the basic physical-biologic phenomenon of **osmosis**—a phenomenon of importance in a great variety of activities and hence influential in the molding of structure and function in many parts of the body.

If two bodies of liquid—say, of water with materials in solution—are separated by a membrane, man-made or organic, the membrane may be so impermeable that little or no exchange of materials may take place or so tenuous that it may be readily penetrated by all the materials of the liquids present, which may thus exchange freely. Intermediate conditions, however, may exist in which large molecules or even large ions cannot pass through the membrane although small ones may pass readily. Such membranes are termed **semipermeable**.

Cellular membranes of variable degrees of permeability, noted in later chapters, are common in the organs of vertebrates. It is obvious that osmosis is a phenomenon which enters notably into the determination of the architecture of the body. It is to be kept in mind as a process quite distinct from the active transport of specific materials across cell membranes—another phenomenon encountered not infrequently in vertebrate tissues.

EPITHELIA (Fig. 63). The body cells are not isolated entities; they are part of formed **tissues**—organized associations of cells of similar origins and functions. In some cases—notably the connective tissues in a general sense—the aggregation of cells may be relatively diffuse, and sometimes ill defined or amorphous. In the case of cells in the blood stream, the concept that these form a tissue is somewhat of a strain on one's mentality. In many instances, however, the cell association is that termed an **epithelium**—a regular and compact arrangement of cells in a sheet which borders on one of its aspects the surface of the body or one of its cavities.

In the embryo, most epithelia are simple and diagrammatic in appearance. Later, however, by thickening and modification they may lose much of the early epithelial appearance, as in a mass of liver tissue. This differentiation of epithelia during development roughly parallels their probable phyletic differentiation. The earliest and simplest multicellular organisms presumably had but a single epithelium, covering the surface and necessarily serving a variety of functions in protection, sensory reception, and exchange of food and waste materials with the outside world. Later, with the development of internal cavities and internal structures, specialization could and did occur. We may note, however, that in much of

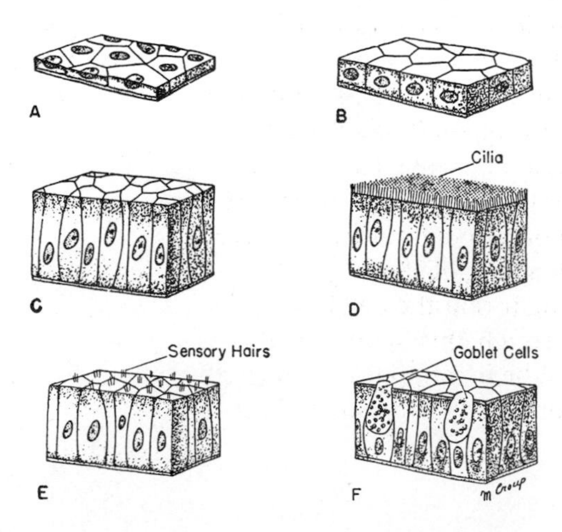

Figure 63. Types of simple epithelial tissue. *A*, Squamous epithelium; *B*, cuboidal epithelium; *C*, columnar epithelium; *D*, ciliated columnar epithelium; *E*, sensory epithelium, showing special sensory processes; *F*, glandular epithelium, including mucus-producing goblet cells. (From Villee, Biology.)

the gut and in other internal structures the epithelium, protected from external harm, may remain persistently thin and delicate in nature, and its cells may have important properties of absorption or secretion. Such epithelia are generally termed **mucous membranes,** since some or all of the cells generally secrete mucus (cf. p. 94); as a result the epithelium is covered and moistened by a thin film of liquid.

For convenience, various names are given to different shapes of epithelial cells. **Squamous cells** are very low relative to their width, appearing in transverse section as thin lines. **Cuboidal cells** are those in which height and width are about equal, so that in section they have a square outline. In **columnar cells** the height is in excess of the width.

Embryologically, epithelia often consist at first of a single layer of cells, and this arrangement is retained in various adult cases as a **simple epithelium**, frequently of the mucous membrane type; such epithelia tend to be present when there is little wear and tear on a surface, or where there is absorption or filtration. In such an epithelium any one of the three cell types noted may be present. Embryonic cells tend to be rounded, but when closely packed they usually assume a cuboidal shape in a simple epithelium; this type is not uncommon in the adult. A thin squamous type is characteristic, for example, of the lining of blood vessels and other areas where osmotic exchange through the membrane is of importance. Superficial cells with secretory or other important functions tend to be of a deeper, columnar type.

When an epithelium shows two or more layers of cells, it is termed a **stratified epithelium**. The stratified types are named from the nature of the cells on their surface; these may be squamous, columnar, or (more rarely) cuboidal. The deeper cells of a stratified squamous epithelium may be cuboidal or columnar, and the underlying cells of a columnar type may be much flattened.

On surface view, or in section parallel to the surface, the cell boundaries of an epithelium may sometimes be irregular, but they are typically polygonal in arrangement, frequently with the hexagonal outline, which gives the best compromise between the ideal circular individual shape and the necessities of a compact arrangement. The adjacent cells in an epithelium may be tightly packed together, but generally are separated by small spaces filled by the interstitial fluid. These spaces may be bridged by numerous tiny bars of protoplasm running from one cell to the surface of its neighbors. As a rule blood vessels are absent in epithelia, and nutrient material reaches the cells by passing upward in the interstitial fluid from vessels beneath the base of the epithelium. In most cases epithelia are bounded below by connective tissue; between the two there is typically a **basement membrane** formed by condensed connective tissue material.

In a stratified epithelium there is frequently a constant loss or destruction of cells at the surface; these are replaced from below, and the lower cells of the epithelium form a **germinative layer** capable of making replacements by cell division (cf. Fig. 89). The major activities of an epithelium are in general associated with its outer, free surface, so it is natural to find that in many stratified epithelia the cells are progressively more specialized as the surface is approached. We further find that in an epithelium of any type, the individual cells are frequently more or less polarized, the proximal (deeper) and distal (superficial) ends being different in nature.

The free surface of an epithelium frequently shows marked specializations. The cells may secrete a **cuticle**, a more or less solid layer of material covering the epithelium. More frequently, however, the surface of the cells themselves may be modified. Among the most specialized of surface structures are **cilia**, found on the surface of a variety of embryonic and adult columnar epithelia of vertebrates as well as invertebrates, and remarkably uniform in structure in all organisms. They are slender, mobile, hairlike "organs," containing 11 slender filaments connected with a complex basal structure. Cilia beat in but one direction; the beat (independent of nervous control) in a ciliated surface usually progresses in wavelike fashion so as to carry mucus or other materials along in a constant unidirectional stream, but in some instances all the cilia may beat in unison.

In certain cases a simple epithelium may present a seemingly stratified appearance—a **pseudostratified epithelium**. If, for example, a simple but tall epithelium contains cells of two sorts, with nuclei at different levels, or with cell bodies expanded at different levels, one gets the impression that two layers are present. The term is better justified where two types of cells are present at the base of a simple epithelium, but only one type reaches the surface. Incidentally, it will be realized that if a simple epithelium is cut at an angle, several cell layers may appear, deceptively, to be present in microscopic section. We may here mention the poorly named **transitional epithelium** found in the amniote bladder or urinary tubes. When the cavity it surrounds is empty or not greatly expanded, this type of epithelium appears to be highly stratified; with distension of the cavity the epithelium expands in area and consequently thins. Other epithelia, such as those lining blood vessels, must also stretch and contract with the surrounding tissues.

In a broad sense the term epithelium may be applied to any tissues of the sort described here, no matter where they are found in the body. By many authors, however, the term is restricted to tissue layers on the outer surface of the body on the one hand, and the gut and its derivatives on the other, or to cavities obviously originating from these outer or inner surfaces. The theoretic consideration behind this is the fact that (as will be seen in the next chapter) such epithelia are derived during individual development from the surfaces of the early embryo. If such a restriction is made, one or more alternative terms are used for tissues lining cavities formed secondarily in the deeper layers of the body. The name **endothelium** is generally used for the epithelial material lining the vessels of the circulatory system. Sometimes this term is applied as well to the lining of the body cavities; **mesothelium** is here a preferable name.

GLANDS (Fig. 64). Secretory activity—the production and discharge of fluid materials—is characteristic of a variety of cells; cells in which this is a dominant activity are termed **gland cells**.

Glands are almost always epithelial in nature, although nerve tissue may also secrete substances. Frequently their nature is obvious in microscopic preparations, owing to the presence of granules of material being secreted or vacuoles filled by it in the cell body. Glands and gland cells may be classified with regard to the effect that the giving off of secreted materials may have upon the cell body. Very commonly a secretory activity may be carried on indefinitely without apparent harm to the cell; this is the condition in the **merocrine** type of gland. But if the secretion is thick or viscous, the cell usually is damaged in discharging it. In the **apocrine** type the superficial part of the cell breaks down as the secretion is released, but no permanent harm is done; this part of the cell is reconstituted and secretion recurs. But in some instances (as in sebaceous glands of mammalian skin) the entire cell is destroyed with the discharge of its contents—the **holocrine** type of secretion.

In many instances gland cells may occur individually or in restricted areas of an epithelium which also serves other functions. Mucus, a thick, slimy lubricating material, moistening and protecting various membranes, may be produced by formed glands, but is frequently produced by individual cells scattered about the

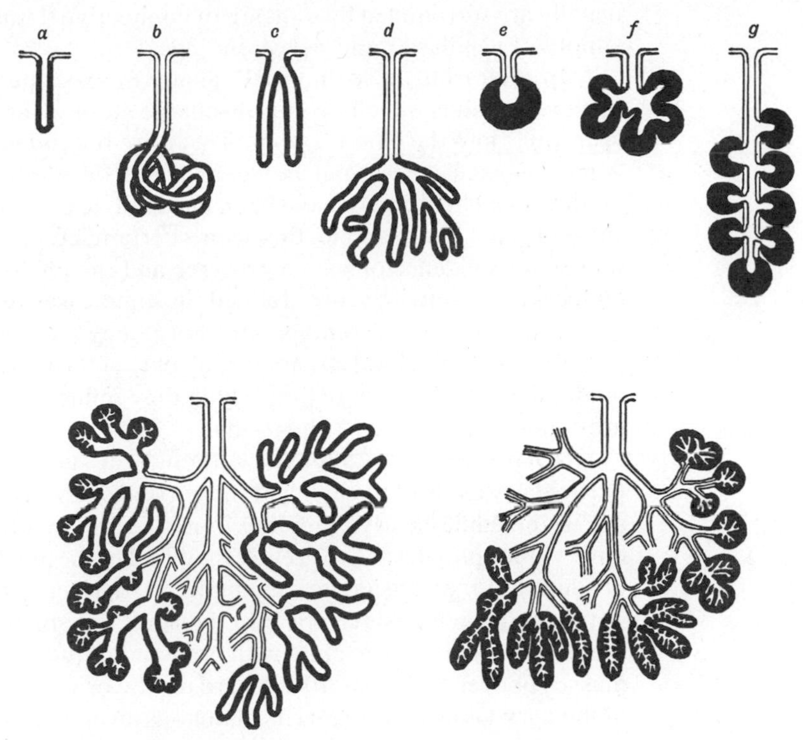

Figure 64. Diagrams of various exocrine gland types. *a–g* (*above*), simple glands; *a,* tubular; *b,* coiled tubular; *c, d,* branched tubular; *e,* alveolar; *f, g,* branched alveolar. *Below,* examples of compound glands. In all diagrams, ducts are in double lines, secretory portions in solid black. (From Maximow and Bloom, Histology.)

surface of an epithelium; this is common, for example, in the skin of various fishes. The secretion frequently accumulates in the cells in the form of large droplets, causing the cells thus distended to be termed **goblet cells.**

Although gland cells or a glandular area may be simply part of the surface of an epithelium, we frequently find that in the course of embryonic development glandular tissues withdraw from the general epithelial surfaces to form discrete glands; the trend for the formation of formed glands is greater among the higher vertebrate groups. In most cases these glands discharge by ducts into some cavity of the body or to the body surface, and are hence termed **exocrine glands**. Withdrawal from the surface serves a variety of purposes. The folding of the glandular epithelium allows a great increase in secretory area; the gland cells are sheltered from the vicissitudes which might be encountered on the general epithelial surface; the presence of a narrow duct (in which the cells generally are nonsecretory) may allow regulation of discharge of the secretion; and provision may be made for storage of considerable quantities of the secretion.

Some of the varied forms that exocrine glands may assume are shown in the accompanying figure. The secretory areas may take the shape of tubules or of rounded pockets (**alveoli** or **acini**), and may be simple or subdivided. Intermediate forms (tubuloacinar, etc.) also occur. Further, in larger glands a compound structure may be brought about by branching of the ducts as well, tending to divide the gland into lobes and lobules. The shape of glands is often quite irregular, for appropriate volume is the desideratum, and form is immaterial. The larger glands usually are surrounded by a sheath of connective tissue; generally there is an ample supply of capillaries and nerve endings.

In contrast to these "normal" glands of exocrine nature, we find in a variety of instances glands which do not discharge to any surface but, instead, pour their secretions into the blood stream. These are the **endocrine glands** (cf. Chap. 17). In some instances these can be seen in the developmental history to arise from epithelia and hence may have been originally exocrine in nature (in the case of the thyroid gland, for example, this seems certain; cf. p. 409). In other cases, however, there is no evidence of such a pedigree and the phylogenetic origin of many of the endocrines is quite obscure. Indeed, in some cases the "gland" is clearly a neural structure. Since the secretions—the hormones—given off by endocrine glands may act, by way of the blood stream, in any part of the body, their position need have no anatomic relation to the organs which they influence, and (as described later) they are found in a variety of situations.

NONEPITHELIAL TISSUES. Although a great series of body tissues—particularly those of the lining of the gut and derivatives, and of the body surface—are in the adult clearly epithelial in nature, many others which are derived from embryonic epithelia are so greatly modified in the adult that their epithelial origin is obscured. Liver tissue, for example, is derived from the epithelium of a gut outpocketing which develops into the adult bile duct; but although the cells of the adult liver retain a connection with branches of the bile duct, the arrangement of the masses of liver tissue has little regard for the original epithelial arrangement. Much of the nervous system begins its history as an epithelium surrounding a neural tube; although cavities representing the tube persist in the adult, there is little trace of typical epithelial arrangement in the nervous tissues. The striated body musculature of amphioxus and the vertebrates can be regarded as of epithelial origin, since in amphioxus it arises as epithelia surrounding a series of paired pouches. But in the adult vertebrate no trace of the ancestral epithelial structure persists, and even in the embryo this musculature shows at the most only fleeting traces of an epithelial origin.

We can still, in a broad sense, regard these tissues as epithelial in nature, since they are ontogenetically or phylogenetically derived from true epithelia. However, it is customary to recognize at least nervous and muscular tissue as separate "basic" sorts, specialized for conduction of impulses and contraction, respectively. Other epithelial derivatives are not as easy to define and so continue to be termed epithelial despite their modifications.

It is quite otherwise with certain other tissues of the vertebrate body—the connective and skeletal tissues and those of the circulatory system. These arise, as will be noted in the next chapter, from a type of embryonic material known as mesenchyme—cells which may be derived from the under surfaces of epithelia, but which are themselves never arranged in a compact epithelial type of tissue. They are, instead, diffusely arranged within a ground substance—a **matrix**—which they have created. In the case of the skeletal elements the matrix becomes the hard material of cartilages or bones; in the case of connective tissue proper, it is gelatinous in nature; in the case of the circulatory system, the matrix is a liquid—the blood plasma. All may be considered connective tissues in a broad sense, this being traditionally the fourth and last of the basic tissues (the others being epithelium, nerve, and muscle).

5 The Early Development of Vertebrates

In later chapters the development of the various organs and tissues will be noted. Here we shall discuss briefly the early developmental history of vertebrates from the egg to the point where the important organ systems have become differentiated and the basic ground plan of the body has been established. In doing so we shall simplify the story in diagrammatic fashion and omit many features of interest to the embryologist.

EGG TYPES

The verbetrate egg varies greatly in size from group to group; this size variation (except in teleosts) is correlated with the quantity of yolk present; the amount and distribution of yolk is, in turn, responsible for major differences in developmental patterns. In consequence it is important for us to distinguish between major types of egg and to follow through, in parallel fashion, the types of development which result from each condition.

In some eggs—those of amphioxus and mammals, for example—very little yolk is present. Such an egg may be termed **microlecithal**.

A second type is that which may be termed **mesolecithal**; the egg is somewhat larger, containing a moderate amount of yolk which tends to settle into the lower hemisphere. Eggs of this sort are found in amphibians such as the familiar frogs, toads, and salamanders, in bony fishes apart from the teleosts, and in lampreys; they are so widespread in lower aquatic forms that it is reasonable to conclude that the mesolecithal egg was characteristic of the ancestral vertebrates.

In the sharklike fishes, on the one hand, and in reptiles and birds on the other, we find eggs of large size—the **macrolecithal** type—with yolk constituting most of the volume of the cell, and with the relatively small amount of cytoplasm concentrated at one pole. In the common modern bony fishes—the teleosts—the egg is sometimes small but is invariably heavily loaded with yolk. It behaves in development somewhat like that of a shark or bird but will not be considered here.

Eggs may also be divided on the basis of the distribution of the yolk within them. It may (as in most microlecithal eggs) be fairly evenly distributed—the oligolecithal (or, linguistically more logically, isolecithal) condition. In most larger eggs the yolk is concentrated in the lower pole—the telolecithal arrangement.

To illustrate the varied patterns seen in the early development of vertebrates we shall select eggs of the three different types discussed above. For a microlecithal egg we shall, in fact, leave the true vertebrates and resort to amphioxus, a lower chordate relative. The frog or urodele egg is a characteristic mesolecithal type. That of the shark or skate is illustrative of an extremely heavy-yolked macrolecithal type, and the bird's egg is similar in nature. We shall, further, note the peculiar early development seen in the mammalian egg. This is tiny and almost devoid of yolk; but mammals have descended from reptiles with large-yolked eggs, and their developmental pattern is one with many "reminiscences" of that of such forms. We shall follow each of these types through three successive major processes: (1) cleavage and the formation of a blastula; (2) gastrulation, with laying down of the main body layers; and (3) formation of neural tube and mesodermal structures.

CLEAVAGE AND BLASTULA FORMATION

AMPHIOXUS. The seemingly inert and relatively featureless egg of a vertebrate or of amphioxus contains within itself all the potentialities required for development of the adult, needing only the proper stimulus—normally the trigger action produced by the entrance of the sperm—to set in motion the developmental story. Yolk distribution brings to light even before fertilization one evidence of organization in the egg. Even in a form with as little yolk as amphioxus, the yolk is slightly more concentrated in a lower "hemisphere," so that we may distinguish an **animal pole** in the relatively clear cytoplasm above, a **vegetal pole** in the yolk region below. In many invertebrates the axis connecting these two poles becomes the anteroposterior axis of the body, the vegetal pole becoming the posterior end. In chordates this is not the case. Related to the greater complexity of development, the adult axis in amphioxus lies about 45 degrees off the egg axis, so that (to put it crudely) the animal pole slants down beneath the prospective chin of the adult, and the vegetal pole slants upward and posteriorly toward the back of the animal.

A first major sequence of events after the entrance of the sperm is the process of **cleavage**, leading to the stage known as the **blastula**. In amphioxus (Fig. 65) the

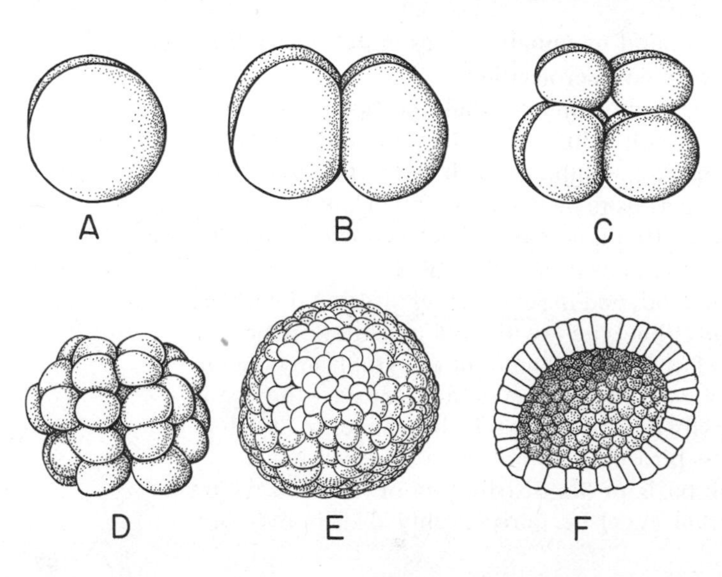

Figure 65. Cleavage and blastula formation in a microlecithal egg—that of amphioxus (cf. Figs. 66 to 68). *A*, First cleavage; animal pole of uncleaved egg is at top of figure. *B*, Second cleavage, to four-celled stage. *C*, Third cleavage; cells of animal hemisphere are somewhat smaller. *D*, After about two further cleavages. *E*, Blastula. *F*, Hemisected blastula, to show segmentation cavity in interior, and single-layered surface. (After Cerfontaine, Conklin.)

first cleavage is longitudinal, extending from pole to pole (much as one cuts an apple into two portions), and results in the formation of two cells destined normally to become the right and left halves of the body—here, and probably in vertebrates generally, the median plane of the future body was already established in the unfertilized egg. A second division is likewise longitudinal, the process being similar to that of cutting an apple into quarters. The third division is at right angles to both those preceding—essentially a cut around the equator of the egg, resulting in an eight-celled stage. Each cell resulting from a division of the egg continues to adhere to its neighbors, but nevertheless tends to assume a spherical shape. In consequence there tends to develop, from this point on, a central cavity inside a sphere of cells—a cavity which becomes increasingly large as division proceeds, and which is known as the **segmentation cavity** or **blastocele**. Another feature also begins to be apparent at this time. A cell tends to divide, not through the center of its gross mass, but through the center of its living protoplasm, without regard for such relatively inert materials as yolk. We have noted that even in amphioxus there is a bit more yolk below the equator than above. In consequence the "equatorial" division just mentioned is not exactly through the equator, but slightly above it, and as a result the four lower cells are slightly larger and more yolky than those above.

Each of the eight cells divides into two, to form a 16-celled stage; each of the 16 again divides, to give 32 cells. Beyond this point, cleavage proceeds in somewhat similar but less regular fashion and, by geometric progression, a few more cleavages result in the formation of a blastula. This product of segmentation is, in amphioxus, a single-layered hollow sphere composed of several hundred cells arranged in a sheet around a central cavity. The cells of the blastula are not too dissimilar from one another, but observation reveals the presence of smaller cells toward the original animal pole and larger and somewhat more yolky cells toward the vegetal pole.

MESOLECITHAL EGGS. The segmentation of the egg of such a mesolecithal vertebrate as a frog or salamander (Fig. 66) follows essentially the same course as that of amphioxus, except for modification caused by the presence of a considerable amount of yolk in the vegetal hemisphere. As in amphioxus the first two divisions are longitudinal, starting at the animal pole, but the presence of masses of inert yolk tends to slow up the cleavage process, so that the second cleavage may be well

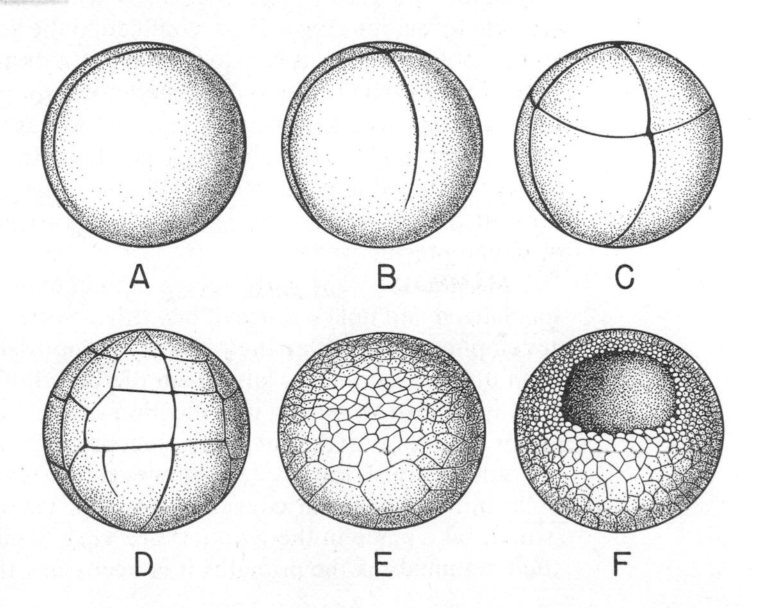

Figure 66. Cleavage and blastula formation in a mesolecithal type of egg found in amphibians. These six figures are comparable to the six in Figure 65 (cf. also Figs. 67 and 68). *A,* First cleavage. *B,* Second cleavage. *C,* Third—meridional—cleavage, with smaller cells in animal hemisphere. *D,* About 36-celled stage; cleavages irregular, but slower and with larger cells in vegetal hemisphere. *E,* Blastula, with strong contrast between cells at two original poles. *F,* Section of blastula, showing segmentation cavity of restricted size; blastula is a number of cell layers in thickness. Yolky mass at vegetal pole has cleaved, but slowly and into a mass of large cells.

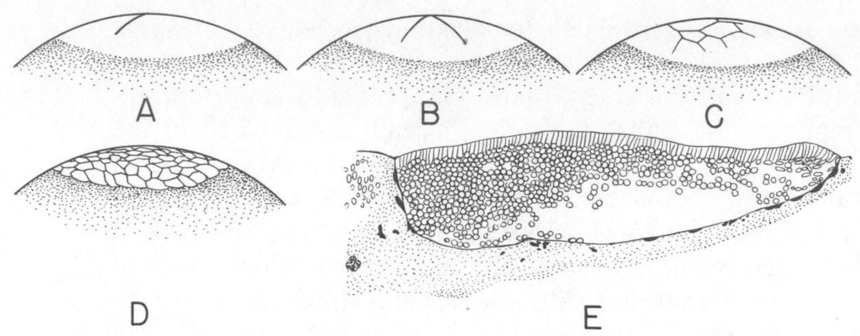

Figure 67. Diagrams to show cleavage and blastula formation in a large-yolked egg, as that of a shark, reptile, or bird (cf. Figs. 65 and 66). In *A* to *D* is figured only the animal pole of the egg, containing an area of clear protoplasm on top of the large, inert yolk mass. *A* to *D* show cleavage stages comparable to those in *A* to *D* of the two preceding figures; the result of cleavage is not a sphere, but a flattened plate of cells. In *E* is shown, at higher magnification, a section through the formed blastula of a shark. The blastula is a flat plate, a number of cells in thickness, with an irregular segmentation cavity lying below it, but above the unsegmented yolk mass. (*E* after von Kupffer.)

under way before the first has been completed to the vegetal pole. This tendency for retardation of division in the lower portion of the egg persists throughout the period of cleavage. Further, the third equatorial division is far from the "equator," because of the disparity in yolk distribution (almost as far north, one might say, as the Tropic of Cancer). In consequence, the cells of the upper ring are far smaller than those below. Because of this unequal division and retardation of cleavage in the vegetal hemisphere, the resulting blastula differs from that of amphioxus. There is a great disparity between the small cells of the animal part of the sphere (here several layers thick) and a mass of large, yolky cells, in part incompletely divided, which make up the vegetal hemisphere and reduce the size of the segmentation cavity.

MACROLECITHAL EGGS. Although the amphibian egg contains a considerable amount of yolk, there is nevertheless a complete cleavage of the egg in blastula formation. In such forms as sharks, reptiles, and birds, however, most of the egg substance is a great, inert yolk mass which hardly cleaves at all; cleavage and blastula formation (Fig. 67) are confined to the small area of clear protoplasm at the animal pole. The result is a disc, which is a number of cells thick, lying above the yolk. The blastula here is a flat sheet, not a sphere. Its margins, bounded all about by yolk, consist of cells which in less yolky eggs would lie in the vegetal pole region, but are here unable to obtain such a position. In cartographic terms, the blastula is a sphere flattened down into a two-dimensional "map" on a north polar projection. Despite their small size, the eggs of teleosts are macrolecithal in structure and mode of development.

MAMMALS. The early stages in mammalian development (Fig. 68) are quite specialized and unlike those of any other vertebrates. Typical mammals carry the developing young within their bodies and nourish them by materials at first derived from uterine secretions, later from the blood of the mother through the placenta. This is formed through a modification of embryonic membranes found in all amniotes, and many of the developmental processes of mammals are similar to those of a reptile or a bird. Early stages, however, are unique. The mammalian egg needs no yolk and has none; in consequence, it is very small, and early cleavage stages (which take place in the oviduct) are very similar to those of amphioxus. But in such mammals as the primates it is necessary that there rapidly develop an outer

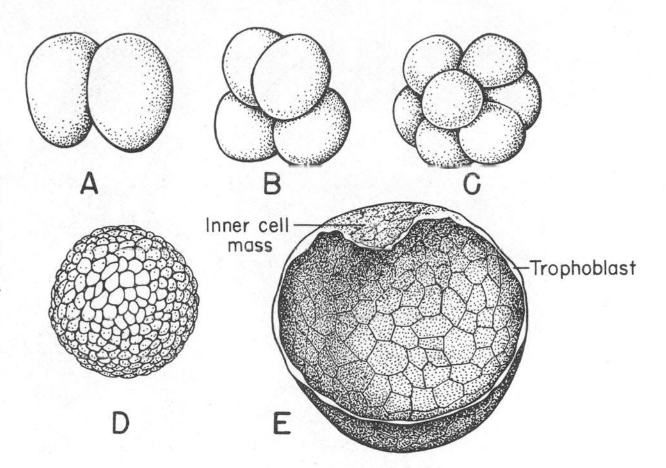

Figure 68. Cleavage and blastula formation in a mammal. The small, almost yolkless egg cleaves (*A* to *D*) in a fashion similar to that of amphioxus. The formed blastula (seen in section at *E*) has a deceptive resemblance to that of amphioxus. Actually, however, the thin external sphere is the trophoblast, which forms a connection with the uterine wall, and the true blastula (or blastoderm) is merely the inner cell mass. This mass is a sheet of cells placed above the internal cavity, much as the blastula of a macrolecithal egg (Fig. 67 *E*) is situated atop the yolk mass. (After Streeter.)

sphere of cells to take part in the formation of a placenta when, very shortly, descent to the uterus has occurred. In consequence, the developing blastula soon comes to be composed of two parts: (1) an **inner cell mass** from which the embryo itself will form and (2) a thin outer sphere of cells, the **trophoblast**—an embryonic membrane whose function it is to make contact with the maternal tissues of the uterus.

GASTRULATION AND GERM LAYER FORMATION

AMPHIOXUS. Cleavage and blastula formation result in an early embryo which consists in most types of a single body layer in the form of a sphere or sheet of cells. In some instances differences in size, pigmentation, or amount of yolk present in different parts of the blastula indicate the differentiation of specific cell areas destined to form one or another major tissue of the later embryo and adult. In other instances a real differentiation is not readily visible, but the future fate of any given area often can be discovered by (for example) applying a stain to cells in the blastula and following the stained area through into later stages of embryonic development. As a result of such observations and experiments it has been found in a considerable number of different chordate types that the future fate of various cell areas is already determined in the blastula stage; "fate maps" of the blastula regions can be drawn, and several of them are given here (Fig. 70 *A–H*).

There now begins a series of movements of specific cell areas toward assumption of the position which they will eventually occupy in the later embryo and adult. A major step is the process of **gastrulation**—the transformation of the single-layered sphere or disc of the blastula stage into a two-layered early embryo, with an outer layer, part of which corresponds to the skin surface of the adult, and an inner layer, part of which will form the adult gut lining; the opening into the interior is the **blastopore**.

In amphioxus gastrulation appears, deceptively, to be a simple process, merely the folding of a sphere into a double-layered hemisphere, with skin outside and gut cavity (the **archenteron**) within (Fig. 69). In such primitive metazoans as the coelenterates this situation actually holds true. All the cells of the original animal hemisphere, which form the outer surface of the coelenterate gastrula, constitute the **ectoderm**, or outer germ layer of the late embryo and adult; the inner cells are

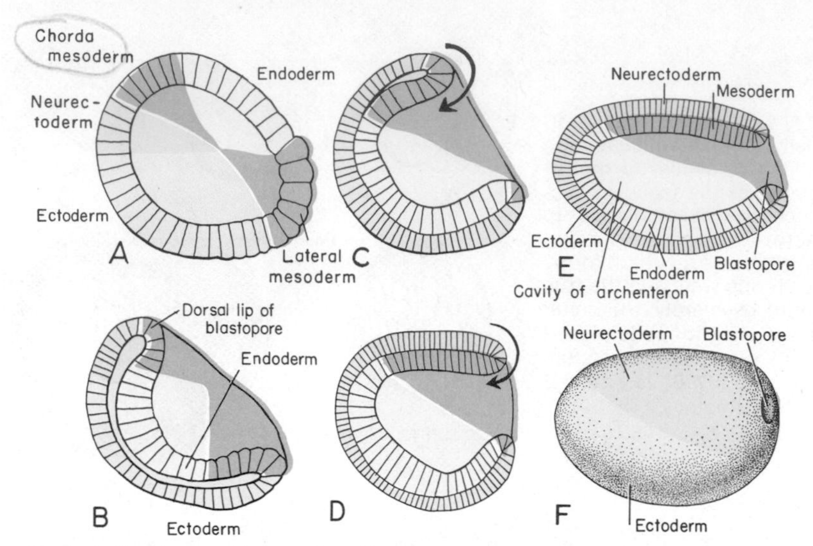

Figure 69. Gastrulation in amphioxus. *A* to *E,* Median sagittal sections showing successive stages. The embryo has been rotated from the original egg position to that of the future adult, with head end at left, the former vegetal pole region posterodorsal. *A,* The endoderm cells are a flattened plate. In *B,* the endoderm has invaginated and the lateral mesoderm cells, originally widely separated from the notochordal materials, are moving upward to join them. In *C* this movement has been mainly accomplished, and the inturning dorsally of chorda materials (arrow) continues. *D* and *E,* Gastrula formation has been completed, and the embryo (particularly notochord and overlying neurectoderm) is elongating. *F,* Surface view of late gastrula, seen from the left. (After Hatschek, Cerfontaine, Conklin.)

In this and subsequent figures in this chapter the following colors are used to distinguish germ layers: skin ectoderm, blue; neurectoderm, green; mesoderm, red; endoderm, yellow.

the **endoderm**, or inner germ layer forming the gut. In the chordates, however, gastrulation is more complex. There is needed a special area of the superficial ectodermal layer for the formation of the complicated nervous system, a **neurectoderm**. There must be formed the materials of the third major germ layer, the **mesoderm**, which constitutes the greater part of the bulk of the chordate body. Still further, there is needed a distinctive mid-dorsal area of mesoderm, the **chordamesoderm**, which forms the notochord and is of especial importance in bringing about by induction the development of the nervous system.

All these areas are already laid out in amphioxus in the blastula stage and are involved in the process of gastrulation (Fig. 70 *A, B*). As gastrulation begins, the large yolky cells at the vegetal pole, which are to become the endoderm, form a flat plate and then bend inward at the ventral rim of the forming blastopore. Above, a sheet of chordamesoderm cells rolls inward over the dorsal lip of the blastopore and pushes forward internally, lengthening in the process and tending to lengthen somewhat the gastrula as a whole. With the infolding of this tissue, the potential neurectoderm comes to occupy a large area on the dorsal surface anterior to the blastopore. On either margin of the blastopore, between chordamesoderm above and endoderm below, cell masses of the mesoderm proper stream in, forward and upward, and align themselves in a sheet on either side of the chordamesoderm.

When gastrulation has been completed, we find the embryo forming a somewhat elongated spheroid, with the only entrance to its interior the posteriorly placed blastopore, now reduced in size. On the outer surface, the future skin ectoderm forms the ventral epithelial covering; the neurectoderm the more posterodorsal area. Internally the endoderm occupies an area essentially comparable to that of the skin ectoderm on the outer surface; farther dorsally and posteriorly is the mesodermal area, with the chordamesoderm occupying a mid-dorsal position along the roof of the archenteron.

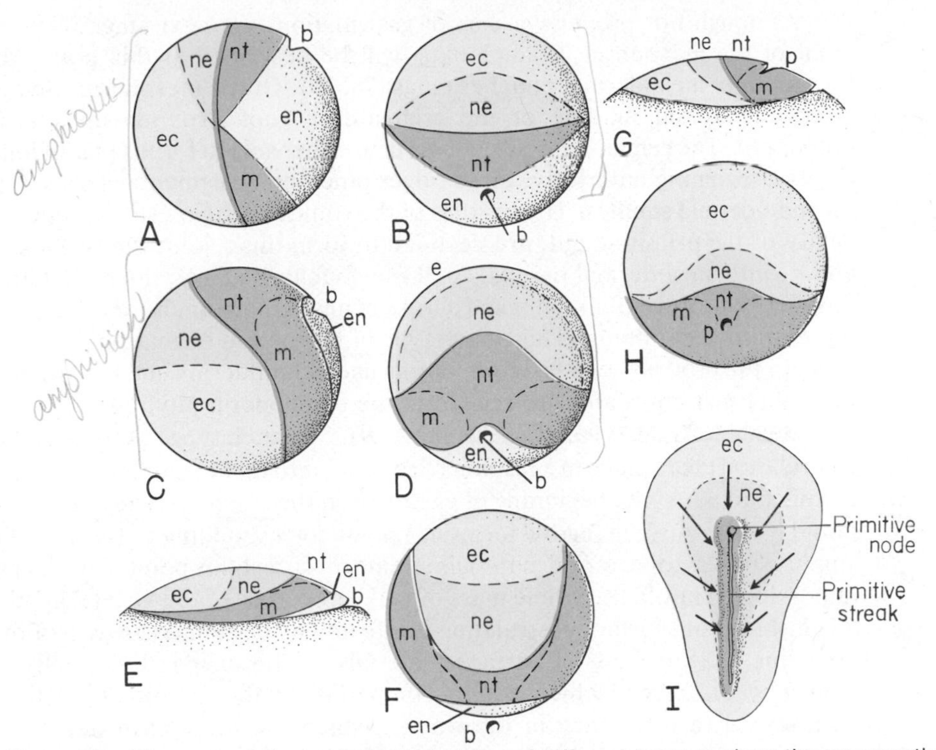

Figure 70. Diagrams of the surface of blastulae of different types to show the presumptive fate of various regions in normal development. *A, B,* Left side and dorsal views of the blastula of amphioxus (cf. Fig. 66 *E*). *C, D,* Similar views of an amphibian egg (cf. Fig. 67 *E*). In these figures the embryo has been rotated from the position in which the egg originally floated to that assumed by the gastrula; the shift is such that the vegetal pole region, originally ventral, has rotated posteriorly and upward to essentially the position from which the endoderm develops. The position at which the blastopore develops is arbitrarily indicated by an indentation. In *E, F, G,* and *H,* similar lateral and dorsal views are shown for the flattened blastula of macrolecithal eggs of shark and bird. Note that in all forms the pattern of potential germ layer areas is similar; however, in the bird (an amniote) the blastopore is replaced, except for endoderm formation, by the primitive streak. *I,* Stage in bird development beyond *H;* the embryo is elongating and mesoderm and neurectoderm are moving inward (as indicated by arrows) to the primitive streak. *b,* Position at which blastopore develops; *ec,* future skin ectoderm; *en,* endoderm; *m,* mesoderm; *ne,* neurectoderm; *nt,* notochordal region of mesoderm; *p,* position in which primitive node and streak make their appearance. (Data from Conklin, Vogt, Vandebroek, Pasteels.)

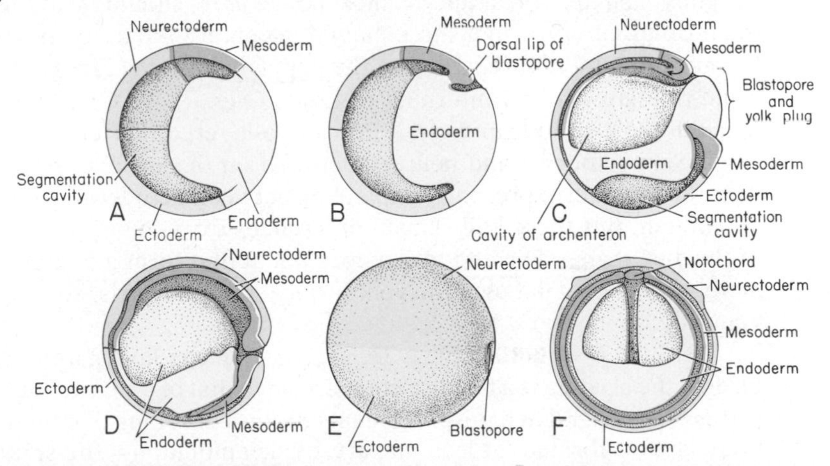

Figure 71. Gastrulation in an amphibian type of egg. *A* to *D* are views comparable to *A* to *E* of Figure 69. The presence, however, of a large mass of yolk restricts invagination to the extent shown at *B;* the remainder of gastrulation is performed by further growth of the blastopore lips, as indicated for dorsal lip by arrow in *C. E,* Gastrula from left side. *F,* Transverse section, looking anteriorly. As can be seen from *C, D* and *F,* the mesoderm folds inward between ectoderm and endoderm (cf. Fig. 80 *A, B*). (After Hamburger.)

Although not properly a part of gastrulation, the next stage in the development of the mesoderm in amphioxus will be described at this point; this is the formation of mesodermal pouches (Figs. 78, 79). Of the mesoderm, that part lying along the dorsal midline of the archenteron roof "rounds up" to form the notochord. The remainder folds outward on either side to form a pair of longitudinal ridges. Beginning anteriorly, these ridges pinch off a segmental series of pouches, the **mesodermal somites**. The cavities of the somites, originally continuous with the cavity of the primitive gut, are destined to form the celom; the walls later differentiate into mesodermal tissues. As these pouches bud off, the endoderm extends upward ventromedial to them and across beneath the notochord to form a continuous definitive gut lining. This pouch type of mesoderm formation is highly comparable to the process seen in acorn worms and echinoderms and is a prime basis for the belief that chordates are related to the echinoderm phylum.

MESOLECITHAL TYPES (Figs. 71; 80*A, B*). In such types as frogs and urodeles, gastrulation takes place in a fashion fairly similar to that of amphioxus, but there are modifications. At the beginning of gastrulation the blastula "tries" (so to speak) to infold itself; a surface furrow forms at a point corresponding to the dorsal lip of the amphioxus blastopore, and materials stream inward at this point. But it is physically impossible to infold the whole mass of yolky materials of the vegetal hemisphere in this fashion, and further gastrulation is effected by continued growth of the blastopore lips and inrolling of surface materials at these lips. Eventually a double hemisphere is formed which is fairly comparable to the amphioxus gastrula. There is, however, a difference in the way in which the mesoderm develops. Pouch formation is suppressed; instead, the infolded mesoderm pushes out between ectoderm and endoderm as a sheet of tissue which eventually reaches the ventral midline; it is only later that segmental divisions and celomic cavities appear.

ELASMOBRANCHS (Fig. 72). Obviously, typical gastrulation cannot occur in the blastula of a macrolecithal type, which is merely a flat sheet of cells, not a sphere; however, the process is basically similar to that of amphioxus or an amphibian.

In normal gastrulation a primary event should be the formation of an endoderm beneath the ectoderm; this should be accomplished, in part at least, by an inrolling of cells at the blastopore lips, particularly the dorsal lip. But where are the blastopore lips in a flat plate? The most reasonable answer is that they lie at the margins of the plate, and the dorsal lip, the most active area, should lie at the posterior end of the forming embryo. In a shark embryo a center of activity at one part of the disc margin can safely be regarded as the dorsal lip region (Fig. 84 *A*). Here there is a rapid overgrowth and inturning of tissue (Fig. 72 *B, C*) which spreads forward beneath the disc to transform it into a two-layered, flattened equivalent of a gastrula. Skin ectoderm and neurectoderm lie on the surface; beneath is endoderm, with the mesoderm presently expanding between ectoderm and endoderm as in an amphibian. But there still remain differences from the mesolecithal type because of the disc shape. The endoderm does not at first form a complete gut tube, but is merely spread out flat over the yolk surface; the embryo is, so to speak, unbuttoned ventrally.

REPTILES AND BIRDS (Figs. 73, 74, 75). In egg-laying amniotes—reptiles and birds—the blastula is a flat disc comparable to that of a shark, but gastrulation is still further specialized in nature. Here there is little inrolling of endoderm; instead, this body layer is formed, at least in part, by delamination—the splitting off of a deep layer of cells from the under surface of the blastoderm (Fig. 73 *A*). There is not yet general agreement on how much of this layer forms in amniotes. Mesoderm, on the other hand, does roll inward through a blastopore; but this blastopore is a highly modified structure, the **primitive streak** (Figs. 70 *I;* 85 *B*). This consists of a pair of longitudinal ridges with a groove between them and a pit at the front end of the groove. On the dorsal surface of the disc there is a steady movement of cells into the margins of the primitive streak. In its walls these cells move downward and then fan outward to interpose themselves between ectoderm and endoderm (Figs. 73 *B*; 74, 75). Anteriorly the central part of the inrolled cells forms the notochord. Laterally the material moves outward to form the somites and other mesodermal structures. As infolding of mesoderm is completed, the primitive streak becomes reduced and vanishes; neurectoderm has moved centrally to occupy its old midline position.

MAMMALS (Fig. 76 *C*). Gastrula formation in mammals is a unique process. In later stages the mammal embryo comes to be identical in major respects with its amniote relatives, but until gastrulation is completed, it is still quite atypical; it has not yet recovered, so to speak, from its early vagaries. The details of gastrulation vary among mammalian groups; described here is that characteristic of primates.

The blastula, we have seen, consisted of an external sheet of cells which made contact with the uterine tissues, and an inner cell mass. In this latter presently appear, above and below, cavities which expand to leave between them a flat, two-layered plate of cells. The upper cavity, lined with ectoderm, is that of the amnion; the lower is a yolk sac with an endodermal lining. The cavities and the materials lining them are parts of the amniote membrane system described later; the two-layered plate between them is a **blastoderm**, in which the embryo is to arise. Since the under surface of the disc is endodermal, the first act of gastrulation is already accomplished. The remainder of the process is the same as that we have seen in a bird or reptile—the development of a primitive streak and the inrolling of mesodermal materials along its margins (Fig. 86 *A*).

NEURAL TUBE AND MESODERM DEVELOPMENT

THE NEURAL TUBE. Following the completion of gastrulation, with the placing in proper relation to one another of the major tissues of the body, early stages in organ formation bring the embryo to a stage termed the **neurula**. Prominent on the outer surface is the formation of a neural tube, its development "induced" by the presence of the notochord beneath it.

In amphioxus the neurectoderm occupies a large oval area on the upper and back surface of the gastrula (Figs. 69 *E, F*). Presently there is an upward folding of the lateral margins of this area. In amphioxus (not in true vertebrates) skin ectoderm and neurectoderm separate as the folds form. The ectodermal margins from the two sides grow medially over the neural region and finally meet to form a complete layer of "skin" over the top of the body. Meanwhile, the lateral margins of the neurectoderm roll upward, meet, and form a neural tube (Fig. 78). For some time the anterior end remains open as a **neuropore**. Posteriorly there is a curious situation in that the neural folds close over the blastopore but leave the hind end of the gut in communication with the cavity of the neural tube by way of a **neurenteric canal** (Fig. 79). In later development, as the tail sprouts out, this canal closes.

Most vertebrates show a type of neural tube development contrasting with that of amphioxus in that the neural folds never completely separate from the ectoderm proper, but a similar end result is obtained (Figs. 75 *C, D;* 77; 83 *B;* 85 *C*). During

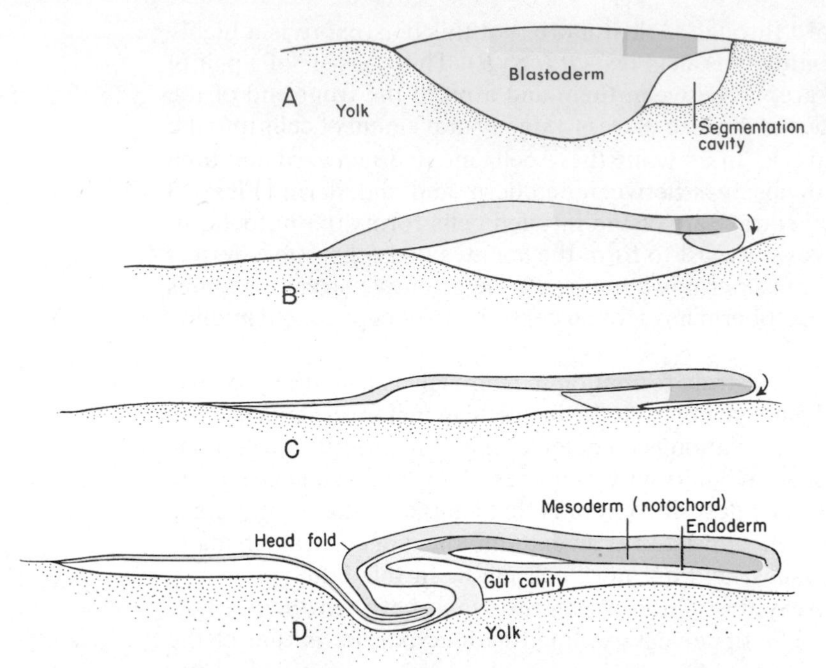

Figure 72. Longitudinal sections of successive stages in gastrulation of an egg of macrolecithal type, as seen in an elasmobranch. Only the disc of the blastula and the neighboring part of the yolk are shown. Anterior end at left. *A*, Blastula (cf. Fig. 68 *E*); *B*, involution of endoderm at posterior end of disc, corresponding to blastopore; *C*, continued process of inturning of mesoderm; *D*, mesoderm separated from endoderm; gut cavity formed, open below, roofed by endoderm. (After Vandebroek.)

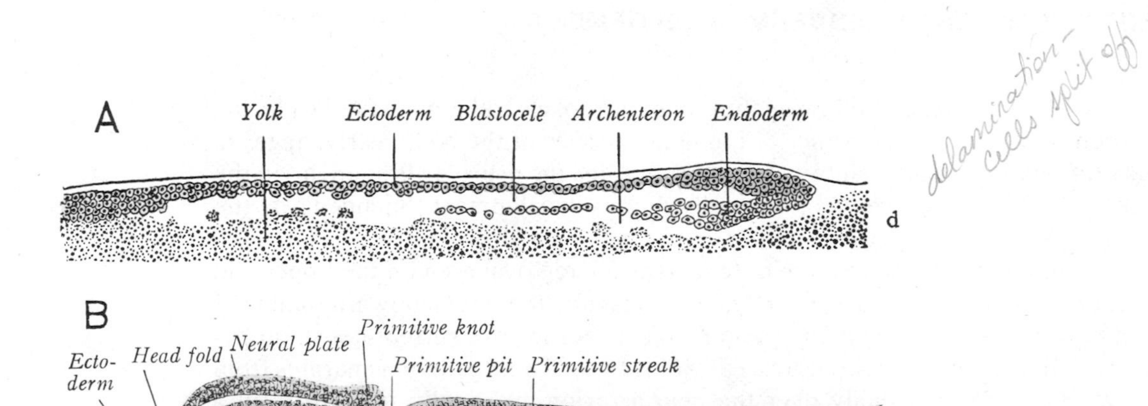

Figure 73. Two successive longitudinal sections of bird embryos at successive stages to show gastrulation. *A*, Figure comparable to the shark of Figure 72 *B*, but the endoderm is delaminating rather than involuting posteriorly to form a roof to the archenteron and, despite the seeming similarity, there is no development of a blastopore posteriorly. *B*, Later stage, comparable to Figure 70 *I*; from the primitive pit backward, cells are turning down inward from the surface (in the plane of the paper) and rolling laterally to form typical mesoderm, and are moving forward from the pit to form the notochord ("head process").

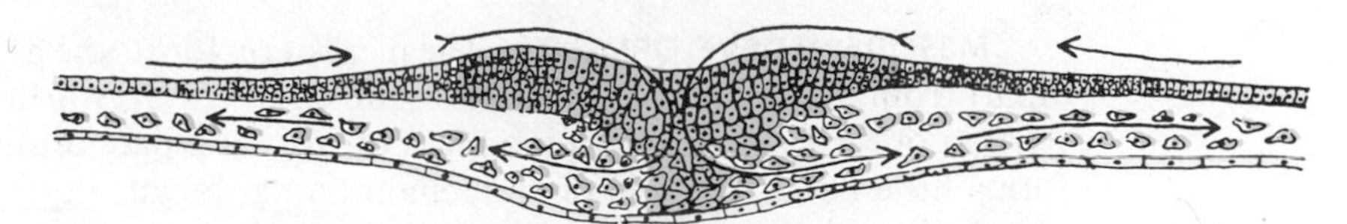

Figure 74. A cross section of the primitive streak at the stage shown in Figures 70 *I* and 73 *B*. Mesoderm, as indicated by arrows, is rolling medially, downward in the primitive streak, and then outward laterally on either side; above, the presumptive neural ectoderm is moving inward toward the midline. (After R. Bellairs, in Marshall, Biology and Comparative Physiology of Birds, Academic Press.)

the folding process, the folds form high **neural crests** on either side, from which masses of cells are pinched off into the interior. Some of these cells form nervous structures; others, as noted in later chapters, have a varied history. In the head region still other future neural elements and sensory structures may arise as **placodes**—thickenings of the embryonic ectoderm lateral to the neural tube region, which detach themselves from the under surface of the future skin.

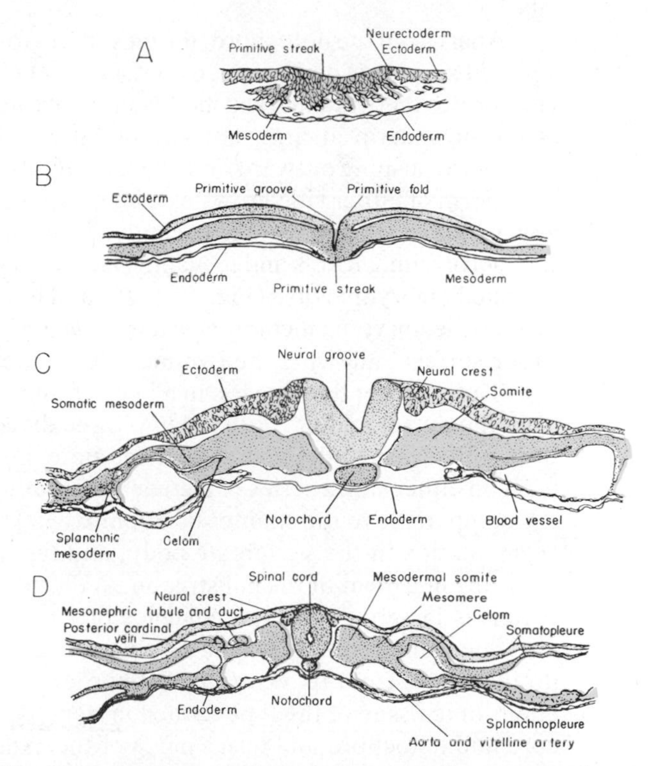

Figure 75. Cross sections of chick embryos to show successive stages in development of mesoderm and neural tube. *A,* Stage in which the inturning of mesoderm in the primitive streak has begun. *B,* The mesoderm has spread widely on either side between ectoderm and endoderm, but has not differentiated further (cf. Fig. 74). *C,* The celom has begun to appear, splitting the lateral part of the mesoderm into outer (somatic) and inner (splanchnic) parts. Centrally, the notochord has separated from the remainder of the mesoderm, and neural folds and crests are appearing on either side of a neural groove. *D,* The neural folds have closed to form a tube, the spinal cord. The mesoderm has divided into somites, mesomeres, and a lateral plate in which a celom separates inner and outer layers of the body–splanchnopleure and somatopleure. (From Arey.)

MESODERM DEVELOPMENT. The mesoderm forms the greater part of the body. Apart from the brain and spinal cord, the ectoderm forms little but the superficial portion of the skin. Except for a mass of liver and pancreas tissue, the endoderm forms little but a thin film of epithelium lining the gut (lungs are endoderm—but mostly air). Practically all the rest of the body is derived from the mesoderm—muscles, connective tissues, skeleton, circulatory, urinary, and genital tissues. If comparison be made with a house, the ectoderm corresponds to the paint on the outside and the wiring system; the endoderm to the floor varnish, wall paper, and perhaps the kitchen stove. All the rest—frame, plumbing, sheathing, even the floor boards, lath, and plaster—is comparable to the mesoderm derivatives.

The **notochord** is sometimes described as a structure discrete from the major germ layers. The chordamesoderm is here considered as a part of the mesoderm, but one which becomes distinct at a very early stage as a longitudinal band of cells lying along the roof of the primitive gut. This rapidly rounds up in section to gain its characteristic as an elongate cylinder (Figs. 75 *C, D;* 78 *C–F;* 79, 82). In higher vertebrates it is reduced or absent in the adults but in every case it is prominent for a long period of embryonic development. It is of primary importance in that its presence is the factor which "induces" the overlying tissues to form the neural tube.

Apart from the notochord, the mesoderm forms in amphioxus, as noted earlier, a paired series of somites, each containing a celomic cavity (Figs. 78, 79). In true vertebrates there is a marked modification of this pattern. There is no segmentation of the mesoderm at first, nor any initial development of a celomic cavity, the mesoderm pushing outward on either side as a solid sheet. In mesolecithal eggs the mesoderm of either side grows as a hemicylinder, following the curve of the body downward and then inward to the midventral line of the belly (Figs. 71; 80 *A, B;* 82). In macrolecithal forms and in mammals, the mesoderm spreads out laterally in the flattened embryonic disc (Fig. 75 *A, B*), and is continued in amniotes outward into the extra-embryonic membranes; it is only at a late stage that the body develops an under surface, allowing the two mesoderm sheets to meet ventrally.

There presently appears, in all vertebrates, a differentiation of the mesoderm from the dorsal midline outward into three divisions, each extending the length of the trunk. Next to the neural tube and notochord thickened masses of mesoderm form on either side a series of **mesodermal somites** (Figs. 75 *C, D;* 80 *C;* 85 *C, D;* 86 *C*), comparable to the somites of amphioxus. These are the first indications of true segmentation in the vertebrate body, and (apart from the independently derived serial arrangement of the gill structures) the segmentation seen in other vertebrate organs is largely due to the influence of the mesodermal somites.

Soon differentiation appears within each somite (Fig. 81). There is a great proliferation, from its ventromedial corner, of cells which form an area of loose embryonic tissue of the type termed mesenchyme. This expands around the nerve cord and notochord and forms much of the axial skeletal structures; in relation to this, the medial part of the somite concerned is termed the **sclerotome**. The external layer of the somite likewise disintegrates; its cells appear to take part in the formation of the connective tissues of the skin, and it is hence termed the **dermatome**. After loss of these two areas, the remaining portion of the somite, the **myotome**, differentiates to form the axial musculature.

Ventral or lateral to the somites, a relatively narrow region of the mesoderm develops in the trunk into the **intermediate mesoderm,** from which form the kidney tubules and their ducts, and the deeper tissues of the gonads as well. This region may develop as an unbroken longitudinal band, but in some cases forms a series of small segmental structures, the **mesomeres** (Figs. 75 *D;* 80 *C;* 288).

Beyond the mesomeric region, extending ventrally or laterally according to the mode of development, is a great sheet of mesoderm, the **lateral plate** (Fig. 80 *C*). This is unsegmented (except in cyclostomes). At first it is a solid sheet of tissue; later, however, it cleaves, and the **celomic cavity**, which in adult life surrounds most of the viscera, develops within it (Fig. 75 *C, D*). The mesoderm external to the celom, plus the adjacent ectoderm, is termed the **somatopleure;** the inner mesodermal layer plus endoderm is the **splanchnopleure.**

During much of embryonic development there exist, between the epithelia and tissue masses of the major organs, relatively empty spaces filled with fluid. Scattered through these spaces is a diffused network of star-shaped cells which compose the **mesenchyme,** the embryonic connective tissue. Much of this is formed by proliferation from the somites, but increments are added from the lateral plate. Both these areas are mesodermal in origin, and mesenchyme is thus a characteristic product of this germ layer. But the ectoderm, as noted elsewhere, and, it seems, the endoderm may also produce tissues of this sort; mesenchyme production is not confined to a single region or a single germ layer.

The mesenchyme is a most versatile tissue. It gives rise in the adult not only to connective tissue, but to the skeleton, the entire circulatory system, and even much of the musculature.

BODY FORM AND EMBRYONIC MEMBRANES

By the attainment of the neurula stage just described, the embryo has laid the foundations for the development of the major organ systems. Their further history will be considered in future chapters. In consequence, we shall merely describe in brief fashion the gradual assumption of definitive body shape and the nature of the embryonic membranes, which are important in the development of large-yolked eggs.

PRIMITIVE TYPES. At the neurula stage the amphioxus embryo had the form of a rather short cylinder. Superficially the remainder of the story is one of body elongation, particularly the budding out posteriorly of the tail, into which neural canal, notochord and somites continue, while anteriorly there develop a mouth and a complex gill structure.

In mesolecithal egg types the neurula is likewise a stubby spheroid, with a rapidly developing nervous system dorsally and ventrally a belly bulging with yolk. With brain growth there is a major development of the head; posteriorly there is a major development of a tail, much as in amphioxus, and it is not long before a body shape recognizable in terms of the adult is attained (Fig. 83).

CARTILAGINOUS FISHES. In sharklike fishes (Fig. 84) the neurula is little more than a pancake-shaped plate atop the massive yolk, with its midline marked by the developing neural tube. With brain growth, the anterior end of the body lifts off the plate, as does the posterior end with tail development. Beneath, the body begins to round off from the yolk, to which it still connects by a stalk containing an extension of the gut cavity. Meanwhile the endoderm (covered externally by thin sheets of ectoderm and mesoderm) has continued, manfully, to grow over the surface of the yolk, which is presently enclosed in a **yolk sac**. The yolk is gradually digested and absorbed, and the sac dwindles and disappears. We may note in passing that the yolk sac of many teleosts appears to be, but is not, similar to that of elasmobranchs.

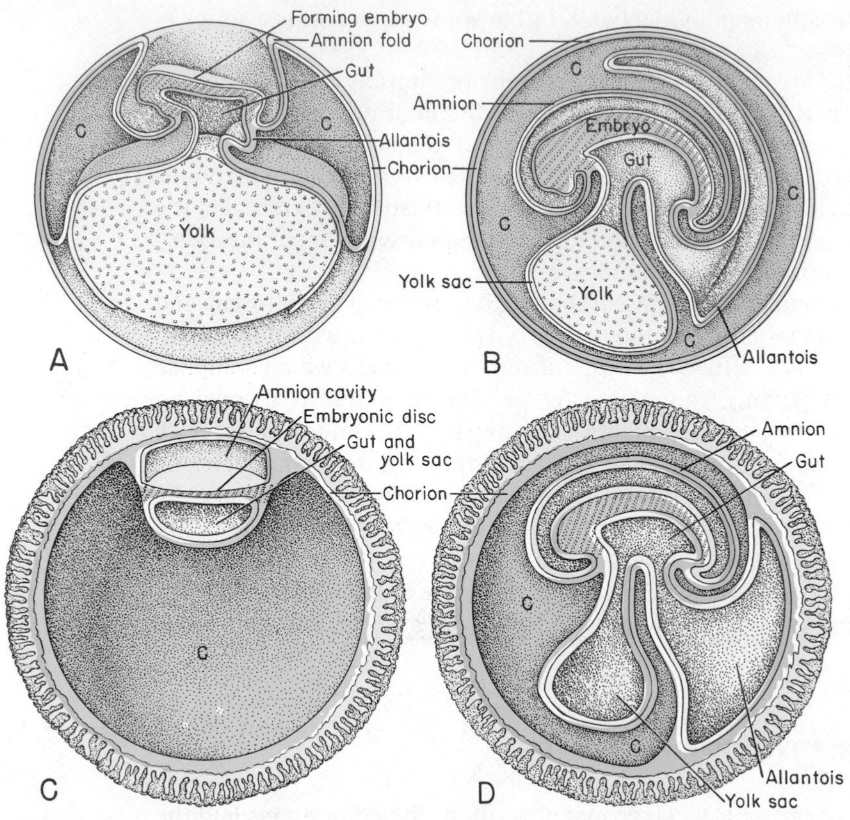

Figure 76. *A, B,* Formation of embryonic membranes in reptile or bird. Anterior end at left. *A,* An early stage. The embryo has been lifted somewhat off the yolk, but gut cavity proper and yolk sac are broadly connected. Yolk sac incompletely formed; amnion folds and chorion incomplete; allantois barely indicated. *B,* Later stage; embryonic membranes formed and yolk already partially reduced. *C, D,* Comparable views of mammalian type of development as seen in primates. *C,* Stage beyond the blastula seen in Figure 68 *E*. The inner cell mass has split ventrally to produce a gut cavity—constituting the major act of gastrulation—and split dorsally to produce an amnion cavity. Between the two cavities the embryo forms a disc in which primitive streak formation occurs much as in a reptile or bird. Mesoderm has already appeared, and chorionic villi are establishing connections with the surrounding uterine wall. *D,* Later stage in mammalian development, corresponding to *B. c,* Celomic cavity.

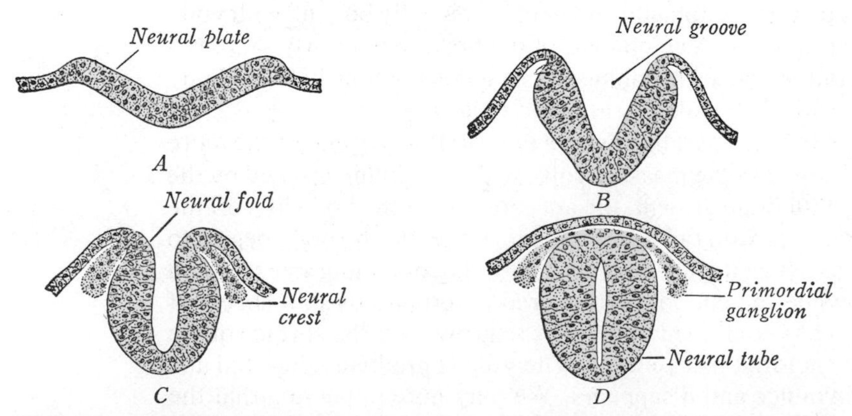

Figure 77. Formation of the neural tube and crest as seen in a typical vertebrate (mammal); a series of transverse sections at successive embryonic stages. (From Arey.)

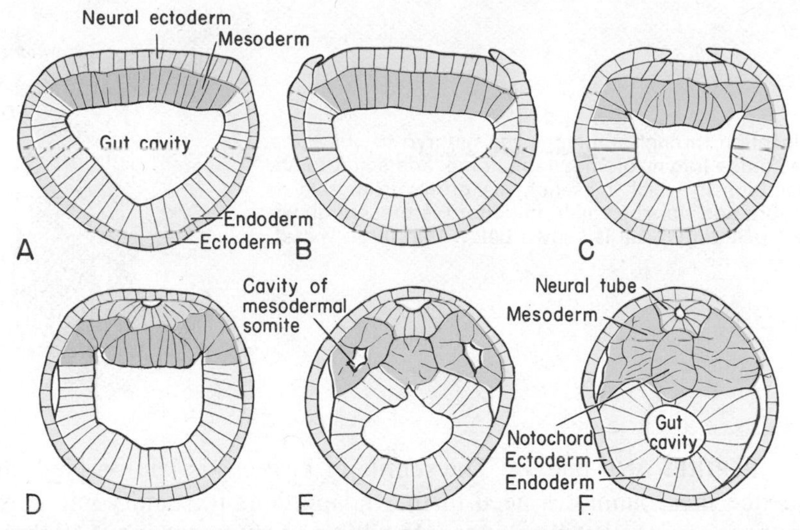

Figure 78. A series of cross sections to show formation of mesodermal pouches and neural tube in amphioxus. (Sections *E, F* somewhat diagrammatic, since the somites of the two sides are alternating in position.) (After Cerfontaine.)

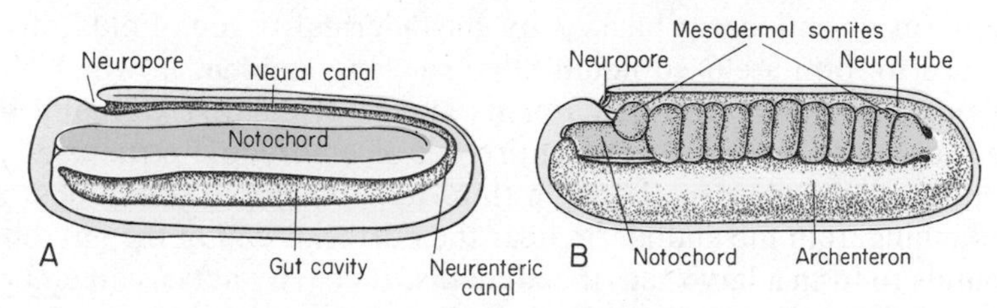

Figure 79. Amphioxus embryos at a stage in which the neural tube has formed and mesoderm is differentiating. *A,* Sagittal section. *B,* Longitudinal view with skin ectoderm sectioned medially, but internal structures preserved intact. (After Cerfontaine and Conklin.)

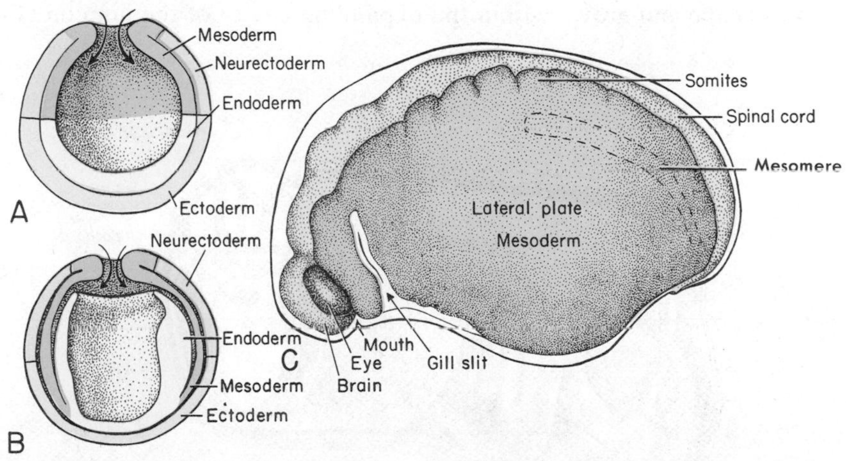

Figure 80. Mesoderm formation in an amphibian. *A,* Section of a urodele gastrula, cut transversely through the blastopore, showing involution of mesoderm into the lateral walls of the archenteron. This is essentially similar to the situation in amphioxus at the stage of Figure 69 *C* or *D. B,* Later stage; the mesoderm, instead of forming hollow pockets, as in amphioxus (Fig. 78), attains its intermediate position by pushing downward and forward between ectoderm and endoderm. *C,* A later embryo of an amphibian, after the neural tube is formed, seen in side view; the skin has been removed. The mesoderm forms a long, continuous sheet on either side of the body. The dorsal part is beginning to subdivide into somites. The part of the mesoderm which will later form kidney tissue is indicated by broken lines. The lateral plate is being broken up anteriorly by the formation of gill clefts. (*A* and *B* after Hamburger; *C* after Adelman.)

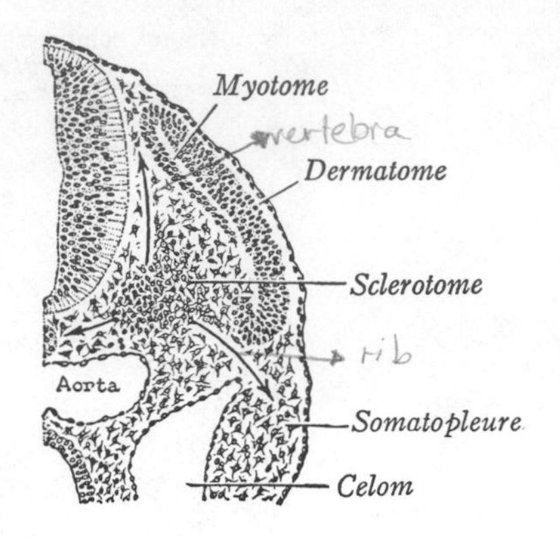

Figure 81. Hemisection through a mammalian embryo to show the subdivision of the somite into myotome, dermatome, and sclerotome. Arrows show directions in which mesenchyme grows from sclerotome to form vertebra and rib. The small notochord is present above the aorta, and part of the gut wall is shown below that large vessel. (From Arey.)

REPTILES AND BIRDS. The reptile or bird egg is laid on land, and in consequence these amniotes need further adaptations for embryonic existence in the medium of air rather than water. In addition to a protective shell there develops a series of membranes which afford the embryo protection and aid its metabolic activities (Fig. 76 *A, B*). A yolk sac is formed much like that of sharks, but before this has gone far in its development further membranes arise, formed of either ectoderm or endoderm backed by mesodermal tissue. Folds of ectoderm grow upward to form a closed liquid-filled sac, the **amnion,** in which the embryo may develop in a miniature replica of its ancestral pond. Externally this ectodermal sheet expands to enclose the entire set of embryonic structures in a protective membrane, the **chorion**. Later, a third new membrane develops, this time as an outpushing from the endoderm near the posterior end of the gut tube. This rapidly expands to form a large sac, the **allantois**. Its cavity acts as an embryonic bladder; much more important, however, is its function as a breathing organ. The combined chorionic and allantoic membranes operate as a lung surface for gas exchange with the air through the porous shell, and the allantoic stalk is richly supplied with blood vessels to aid in this breathing function. With these membranes formed, the embryo takes shape and grows within the expanding cavity of the amnion (Fig. 76 *B*).

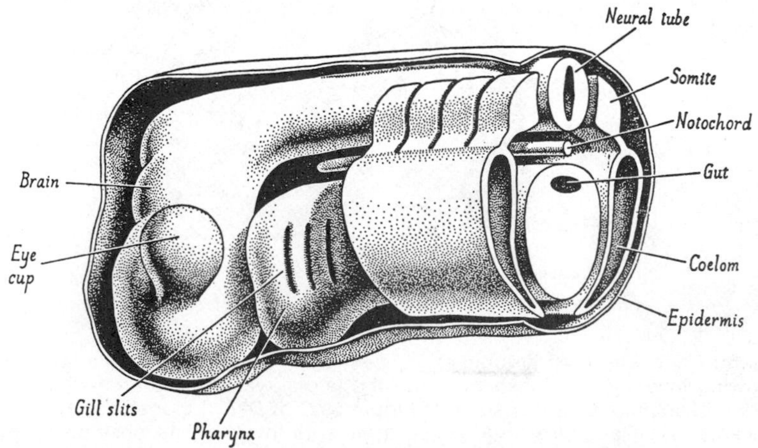

Figure 82. Stereogram of the front part of an embryo (particularly of a mesolecithal type) after partial differentiation of mesodermal components and nervous system. (After Waddington, Principles of Embryology, George Allen and Unwin.)

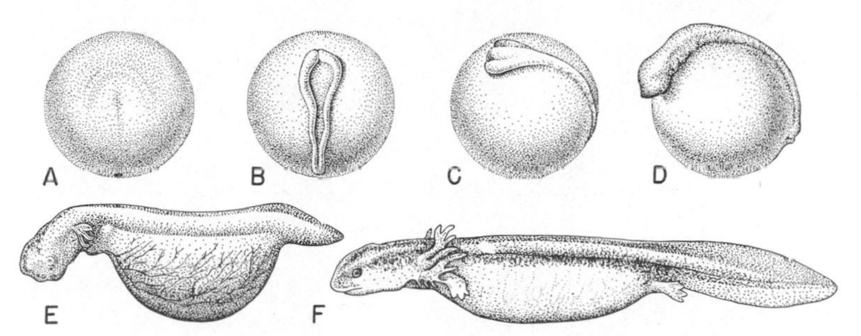

Figure 83. Development of body form in a mesolecithal egg type—the urodele *Necturus* (the mud puppy). *A,* Late gastrula, seen from above, head end at top. *B,* Neural folds forming. *C,* View from left side, neural tube formed, brain bulging upward above sac partly filled with yolk. *D,* Head and trunk taking shape dorsally. *E, F,* Steps in reduction of yolk-filled belly sac and assumption of normal form. External gills and eye appear in *E,* limbs in *F.* (After Keibel.)

MAMMALS (Figs. 76 *C, D;* 86, 87). We have noted the precocious development in such mammals as higher primates of an outer ring of cells, the trophoblast, and, a little later, of epithelia lining cavities both above and below the blastoderm from which the embryo forms. Except that it is not until some time later that these membranes are reinforced by mesodermal tissue, they are exactly equivalent to chorion, amnion, and yolk sac, respectively. Amnion and yolk sac develop much as in reptiles and birds despite the fact that the yolk sac is yolkless. The final member of the amniote membrane series, the allantois, grows out somewhat later from the posterior end of the gut to underlie the chorion. With growth of the embryo and

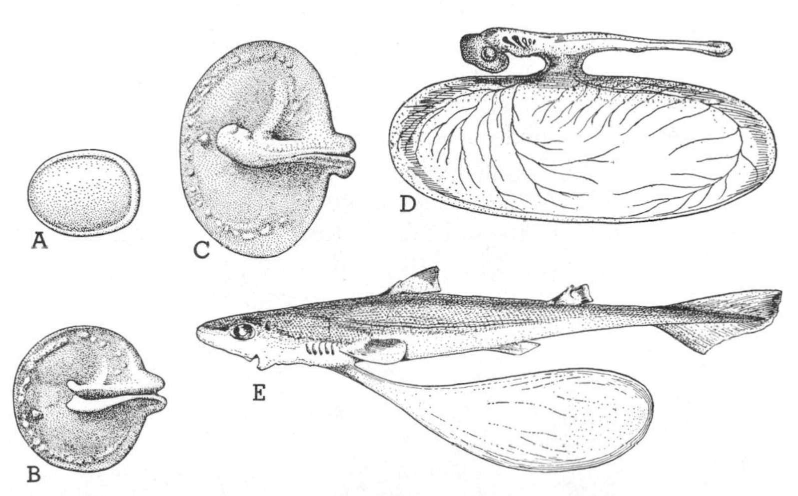

Figure 84. Development of body form in a shark. *A* to *C* are dorsal views of the blastodisc from which the embryo forms; the underlying yolk is omitted in these figures. *A,* The embryonic disc at gastrulation; the endoderm is rolling under at the thickened posterior and lateral margins (cf. Fig. 72 *B*). *B,* The disc is enlarging, and the neural folds are developing on the upper surface. *C,* The neural folds are closed except at the growing posterior end; the body of the embryo is lifting off the yolk, and head region and somites are visible. *D,* The yolk sac is completely formed and the embryo connected with it by a stalk; eyes and gill slits are visible. *E,* Nearly normal shape has developed except for retention of a relatively small yolk sac. (After Ziegler, Dean.)

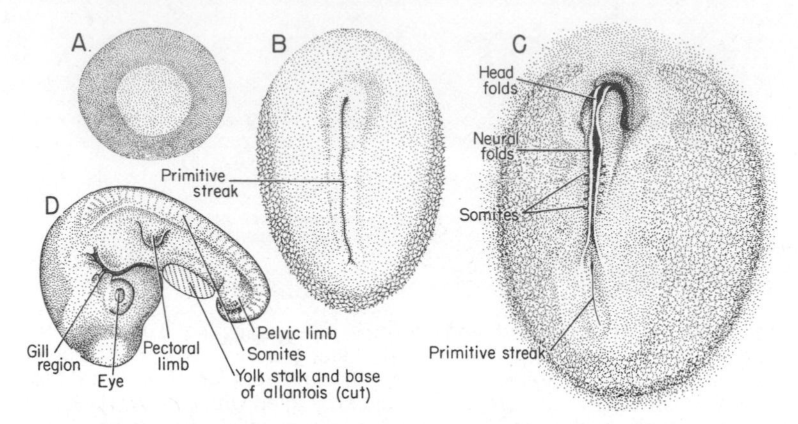

Figure 85. Some stages in amniote development as seen in reptile or bird. *A,* Small germinal disc situated on the upper surface of the yolk. *B,* Formation of primitive streak and elongation of germ disc (cf. Figs. 70 *I,* 73 *B,* 74, and 75). *C,* The embryo is enlarging to cover more of the yolk; the head region is lifting off the yolk surface; neural folds and somites are appearing; the primitive streak, now relatively small, is still active in formation of posterior part of body. *D,* Side view of a considerably later stage, comparable to Figure 76 *B.* The embryo is separated from the yolk except by a stalk (cut). Many head and body structures are formed, and limb buds are appearing. (*B* and *C* after Huettner.)

expansion of the amniotic cavity around it, yolk stalk and allantoic stalk come to be connected with the body only as components of a narrow **umbilical cord**.

The major difference between a placental mammal and its amniote relatives is the development of a **placenta**, which replaces the embryonic lung of reptiles and birds. As in those groups the outer surface of the allantois fuses with the chorion, and the allantoic stalk is richly supplied with blood vessels. The material transported by these vessels, however, is not merely oxygen but the entire food supply for the embryo. The outer surface of the chorion becomes intimately connected with the tissues of the uterine walls, usually by finger-like processes **(villi)**; in these conjoined placental tissues there takes place exchange of materials between the closely apposed (but *not* interconnected) blood vessels of mother and young.

LARVAE. In vertebrates with large yolked eggs, development proceeds rapidly toward adult structure; the young, at birth, is essentially a sturdy little replica of the adult, soon capable of making a living in the fashion of its elders. Not

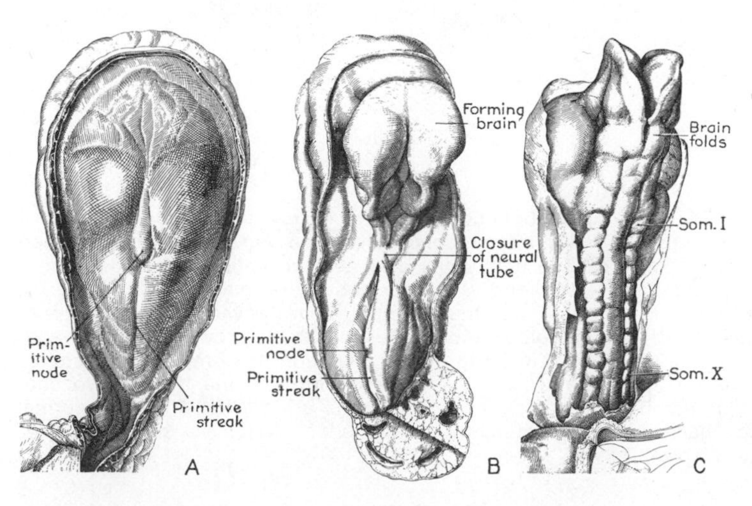

Figure 86. A series of early human embryos, to illustrate stages in mammalian development. All three are dorsal views of the embryo, with the embryonic membranes cut away. *A,* Primitive streak stage, comparable to Figure 85 *B,* for a bird or reptile. *B,* Later stage in which the primitive streak is still active posteriorly, but the neural tube is forming more anteriorly. This stage is comparable to Figure 84 *C,* for the shark, not quite so advanced as Figure 85 *C,* for the bird. *C,* More advanced stage, with neural tube nearly completely closed and somite formation well advanced. (After Heuser, West, Corner.)

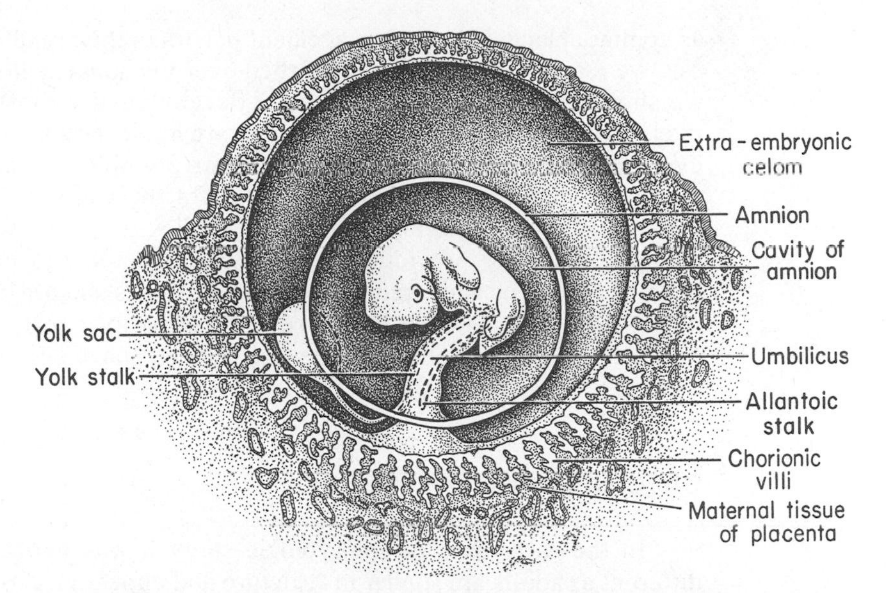

Figure 87. Diagram to show the development of a primate embryo inside its membranes. The stage represented is one at which the embryo, though well formed, is still of small size.

so in many water-dwelling lower vertebrates in which the yolk supply is limited—lampreys, many bony fishes, amphibians. The young, when hatched, is liable to dangers resultant from tiny size and may be incapable of taking up adult feeding habits. In consequence there is often a **larval stage** interjected into the life history, during which the young animal leads a life quite different from that of the adult and may have specialized anatomic structures adapting it to this life; the frog tadpole and the ammocoete larva of a lamprey are familiar examples. With growth, larval features are lost and adult habits and structures assumed—the process of **metamorphosis**.

REGENERATION. In earlier sections we have more or less tacitly assumed that organs and tissues once formed are permanent. But this is far from being universally the case, even under normal conditions—such structures as hair, feathers, epidermal cells of the skin, and blood cells, for example, are normally lost or destroyed and replaced, so that developmental processes of a sort may go on throughout life. Further, when accident or disease causes destruction of tissues there are in every vertebrate certain potentialities of replacement. Even in mammals, which are not notable for regenerative powers, large areas of skin may be renewed, and destroyed liver tissue may be regrown. At the other extreme, urodeles show exceptional regenerative powers, and even a complete limb may be regrown in all essential details from an amputated stump.

DEVELOPMENTAL MECHANICS

In the earlier sections of this chapter we have given a description of the orderly series of events which take place during the development of a vertebrate but we have said little as to the "why?" of these processes. The answer to this question is the major interest of embryologists today. The development of the individual from the seemingly simple egg to complex adult is a miracle so common that we regard it as commonplace. When some accident occurs in this usually well-regulated process, we tend to be puzzled or disturbed over the abnormality that results. Rather, we should marvel that the process of development normally proceeds so effectively. The mechanisms of development are in most regards still a mystery. However, new tools and approaches are making possible the examination of the roles played by the genes in the differentiation of the many and varied types of cells during embryonic development. Also, studies of the movements and other behavior of cells are providing a clearer understanding of how the different cells become organized and arranged to form the tissues and organs of the animal. Work of this sort is considered in any recent embryology text; while this work is interesting and important for anatomical studies, space does not allow us to review it here.

ONTOGENY AND PHYLOGENY

In the early days of embryologic study it was noticed that animals vastly different as adults are similar in structure and appearance as embryos and that the embryos of "higher" vertebrates often exhibit conditions similar to those seen in members of "lower" groups. From such observations came the idea of a **biogenetic "law,"** which proclaimed that individual development—**ontogeny**—repeated the history of the race—**phylogeny**; that an animal in its development climbs its own family tree, successive embryonic stages representing the adult stages of ancestral types.

The main proponent of this extreme version of the biogenetic "law" was the German biologist Haeckel, who believed that it would provide answers to almost all evolutionary problems. His ideas go back to the earlier work of von Baer; the latter, however, made no such exaggerated claims. Von Baer's "laws" noted that more general characters appear before more specific ones during development, that animals progressively diverge more from related forms during development, and that the early stages of young animals resemble those early stages (*not* adult stages) of more primitive forms.

The biogenetic "law" was for decades an important stimulus to embryologic work and in the study of homology. But it is only a half truth. A mammalian embryo at an early stage is fishlike in many regards, as, for example, in the presence of prominent "gills" which are later reduced or lost. But there is actually little resemblance to an adult fish, for the gill pouches do not open to the surface or develop gill membranes. It is the fish embryo, not the adult fish, which the mammal resembles. Development tends to be a conservative process, for departure from the old, tried and true methods will usually result in failure and death. In consequence, the mode of development of an animal may follow well along that which its ancestors pursued, and only toward the end it may diverge to attain an adult condition quite different from the original goal. Ontogeny repeats many important stages in the developmental pattern of ancestral forms. It is especially likely to repeat them if they are structurally or functionally useful in the derived type's own development.

However, embryos and larvae as well as adults must be adapted to the environment in which they live, and in consequence, many structures found in growth stages may never have been present in any adult ancestor. For example, no ancestral shark or amniote ever dragged beneath his body a yolk sac such as is present in the embryo, and it is improbable that the feathery external gills of a larval salamander were ever normally present in an adult fish ancestor. Further, despite the general conservative nature of developmental processes, there may occur striking modifications in the sequence of embryonic events, presumably related to powerful adaptive requirements in embryonic life. A notable example is the method of development of embryonic membranes in mammals. The ancestral method seems surely that present in reptiles and birds. The mammals attain the same end results, but have markedly changed the pattern of their development because of the need for rapid formation of a placenta. However even here we also have retention of odd characters—our yolk sacs do nothing, but still appear and disappear.

THE GERM LAYERS

The theory of the germ layers was an early and most fruitful concept in the study of embryology. In the early embryo, ectoderm and endoderm may be distinguished as inner and outer body layers comparable to those which alone constitute the entire body of a coelenterate; soon there is developed a third, intermediate layer from which, in all animals above the coelenterate level, much of the substance of the body is formed. We have here adhered to this germ layer concept, although emphasizing the early separation between skin and neural portions of the ectoderm and the distinctive nature of the chordamesoderm. In the adult vertebrate, tissue components of organs and organ systems can in general be sorted out readily as regards their derivation from the germ layers. Details and exceptions will be found in later chapters.

From the body ectoderm: the superficial portion (epidermis) of the skin and its extensions into the ends of the digestive tube (mouth, cloacal region); epithelial skin structures, such as hair and feathers.

From the neural ectoderm: the nervous system; the eye retina; certain other derivatives from the neural crest.

From the mesoderm: connective and skeletal tissues; the musculature; the vascular system; most of the urinary and genital systems; the lining of the celomic cavities; the notochord.

From the endoderm: the lining of the gut and the substance of glands derived from it (liver, pancreas); much of the breathing apparatus of gills or lungs.

It was once believed that the origin of various tissue types was absolutely restricted to one or another of the germ layers. In recent years, however, various exceptions have been discovered under both normal and experimental conditions, and there has been a tendency on the part of some to abandon the germ layer concept as meaningless. This, however, is a counsel of despair. In general we find that in normal development the embryonic cells and tissues do follow a consistent pattern of regional movement and arrangement of components. If nothing more, the germ layer terminology is useful as a description of the topography of development. It is, however, more than this. Experimental work has shown that, although in early stages there may be little differentiation between various regions of the embryonic germ layers, there is increasingly, in later stages, a limitation of capacities in different regions. The prospective fate in normal development of the germ layers and subsidiary areas of these layers is in general accord with the experimentally deduced story of their prospective potencies.

The Skin 6

SKIN FUNCTIONS. Forming a covering for the entire body, the skin, with its accessory structures, is an organ system performing varied and important functions. A tough "hide" is a protection against injury and attacks of predaceous enemies. The skin is a continuous line of defense against the invasion of microorganisms and wards off injurious physical and chemical influences. Further, it may play a positive role in many ways—the regulation of body water and of salt content, the intake of oxygen, the elimination of wastes. As the part of the body in immediate contact with the outer world, the skin is the site of important sensory organs, and the nervous system, although withdrawn from the surface in the adult, arises, as we have seen, in continuity with the skin ectoderm.

The skin is not a single structural entity, but consists of two parts, **epidermis** and **dermis**, closely united but differing in nature and origin. The epidermis, ectodermal in origin, is superficial and essentially epithelial in nature; the deep-lying dermis, of mesodermal origin, is primarily a fibrous (connective tissue) structure. The epidermis is thin, the dermis thick; the epidermis gives rise to a host of different structures, such as hair, feathers, various glands; the dermis is relatively simple and uniform in composition.

EPIDERMIS. In amphioxus and hemichordates the epidermis consists of a single layer of columnar cells; in all true vertebrates it is a stratified epithelium. In fishes and water-dwelling amphibians (Fig. 88) it is a persistently simple structure, apart from the presence of glandular elements, and its entire thickness consists of live cells containing a normal protoplasm. There may, however, be present, here and in higher vertebrates, a dark pigment, **melanin** (derived by transfer from dermal pigment cells), and the superficial cells include a certain amount of **keratin** (a waterproof protein abundant in cattle horn sheaths, fingernails, and similar structures). These outer cells tend to be lost by wear or injury; they are constantly replaced from below, and are ultimately derived by the budding off of successive layers of cells from the basal layer of the epithelium. Superficial damage to the epithelium is readily repaired, but if through major injury, such as serious burns, a large area of this basal matrix is destroyed, a recovering of the flesh by skin becomes difficult, if not impossible. The moist epidermis of many lower vertebrates is permeable to some degree and in most modern amphibians is a major respiratory organ, richly supplied by blood vessels deeper in the skin or even, rarely, within the epidermis.

With the assumption of a definitely terrestrial life by certain of the amphibians and by the amniotes, the nature of the epidermis is changed (Fig. 89). Water loss, particularly, is a serious matter, and the surface of the skin becomes dry and

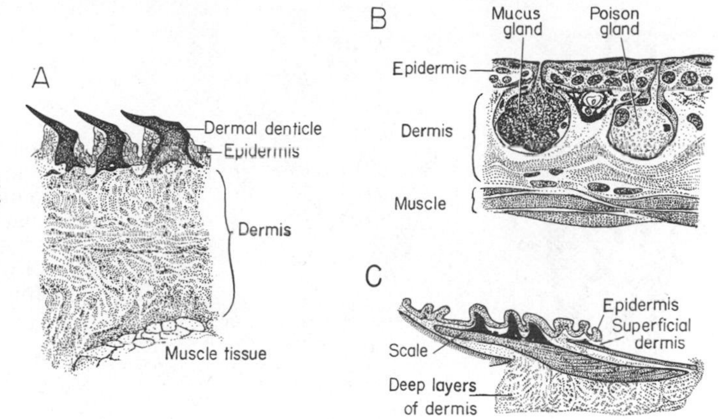

Figure 88. Sections of the skin of *A*, a shark; *B*, a salamander; *C*, a teleost. (After Rabl.)

impervious. The inner cells remain "live" structures, but as the surface is approached, the cells become flatter, lifeless and filled with keratin. The outer cell layers may be rubbed off and lost piecemeal (dandruff is an example) or shed seasonally in reptiles and amphibians. The transition between inner and outer portions may be gradual, as in land-dwelling amphibians, reptiles, and birds. In mammals, however, there may be a sharp contrast (Fig. 89) between a lower zone of live cells, the **stratum germinativum**, and the flattened dead cells of the **stratum corneum** on the surface.

KERATIN SKIN STRUCTURES. Throughout the higher vertebrates the keratin-filled epithelium produces a variety of special structures. Simplest, perhaps, are thickenings or swellings of the stratum corneum, as, for example, in the "warts" of toads, or in the **foot pads** found on the under surfaces of the feet in many land-dwellers. In mammals (Fig. 90 *A*) such pads are characteristically present at or beside the base of each toe, with a pair in addition on the proximal part of palm or sole. In higher primates palm and sole are covered instead by a pattern of **friction ridges** (Fig. 90 *B*) which assist arboreal forms in obtaining a firm grip on the tree limb. In man, the great variation in the arrangement of the loops and whorls on the finger tips affords a ready means of identification.

In reptiles, thickening and hardening of the cornified epithelium results in the formation of **horny scales** or **scutes** (Fig. 91). In lizards and snakes there are generally overlapping scales, highly developed in snakes as an aid to locomotion. In crocodilians and turtles there are, in contrast, flat horny plates. It must be emphasized that these horny epidermal structures are not homologous with the bony dermal scales of fish; in many lizards and crocodilians, however, dermal bony scales underlie the superficial horny structures, and in turtles the horny plates form a superficial sheathing for the body armor.

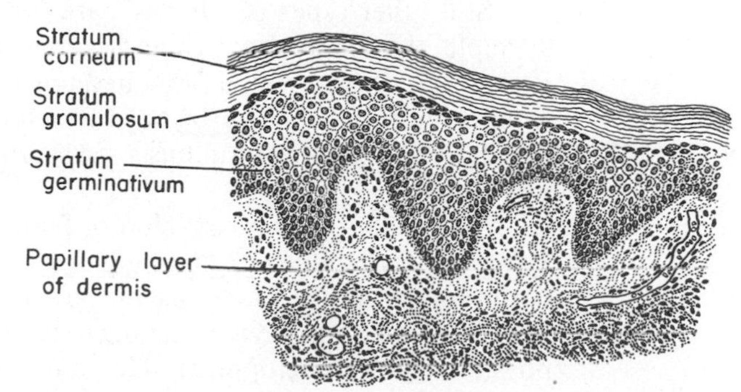

Figure 89. Section of the skin of the human shoulder, × 125. In addition to the germinative layer and stratum corneum, there is (as here) an intermediate granular layer in the epidermis in many areas of mammalian skin; in some situations there is, further, a transparent layer (stratum lucidum) between horny and granular levels. (After Maximow and Bloom.)

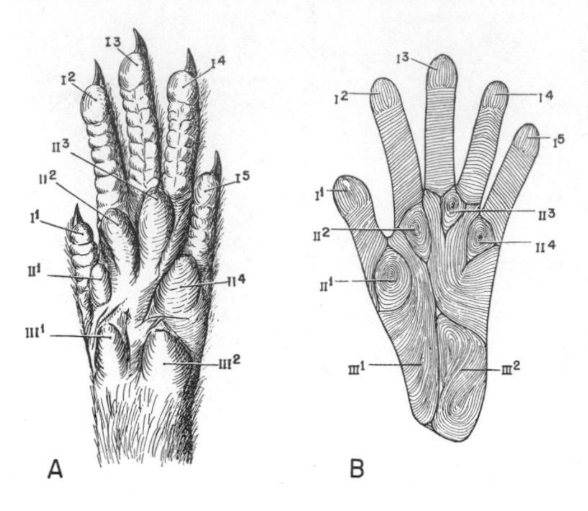

Figure 90. Palmar surface of the manus of *A*, an insectivore; and *B*, a monkey (macaque). The insectivore shows a presumably primitive mammalian structure, with thick pads on either side at the proximal end of the palm (III^1, III^2), pads between the bases of successive digits (II^1 to II^4), and pads at the tip of each toe (I^1 to I^5). In higher primates these pads are replaced by patterns of friction ridges. (After Whipple.)

In mammals and birds horny scales have for the most part disappeared; they persist, however, on the legs of birds, and on the legs and tails of a variety of mammals, notably rodents. The pangolin of the Old World tropics is notable as a mammal which has redeveloped a body covering of large horny scales (Fig. 55).

In birds, in which teeth are reduced or absent, the skin on the jaw margins may cornify as a substitute, producing a **bill** or **beak**. Such structures are also well developed in the turtles, some extinct reptiles, and a few mammals, such as the monotremes.

Claws, nails, and hoofs are keratinized epidermal structures tipping the digits of amniotes (Fig. 92), growing continually outward from a germinative layer beneath or at the base of the structure as the distal end wears off. Beneath the claw or nail tip is a layer of softer, less cornified material, the **subunguis**. The **claw**, V-shaped in section and pointed at the tip, is the basal type; a **nail** is essentially a broadened modification. **Hoofs** are characteristically a development of ungulate mammals which walk on the tips of the digits.

Horns and hornlike structures are widespread in distribution, particularly among ungulate mammals (Fig. 93). A true **horn** is seen in members of the cattle family, including sheep, goats, and antelopes. The core of the horn is a spike of bone arising from a dermal bone of the skull; sheathing this is an epidermal hollow cone of true horn (keratin). Neither core nor sheath is ever shed. Although often called a horn, the **antler** of the deer, almost always confined to males, is quite different. When mature it consists solely of bone; only during growth is it covered by skin in the form of "velvet"; no actual horn substance is present. As further points of contrast we may note that an antler is usually branched—increasingly so in older animals—and is shed annually.

Still other types of "horns" are found among mammals. We may note, for example, the simple, bony, hair-covered unshed horns of the giraffe; the American prongbuck's horn, with a branched horny sheath which is shed and a simple core which is not; the rhinoceros horn which is a fused mass of hairlike horny dermal papillae. Comparable structures are seen, although less commonly, in reptiles and even birds.

FEATHERS. The possession of feathers is the distinguishing mark of a bird. Evolved, it is believed, from reptilian scales, and primarily epidermal in origin, they perform two major functions in avian economy. As a body covering they are effective insulating devices, aiding in temperature regulation; bird flight is rendered possible by the development of large feathers forming the wing surface and the tail "rudder."

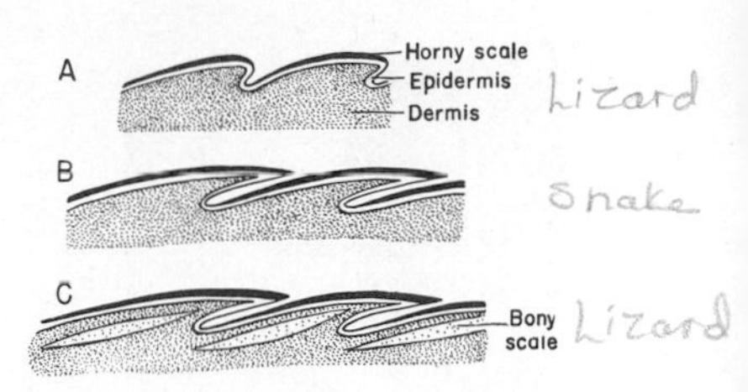

Figure 91. Diagrammatic sections of reptile skin to show scale types. *A*, Lizard skin with simple, horny epidermal scales, gently overlapping; *B*, deeply overlapping horny scales of snake type; *C*, type of scale present in many lizards, with bony scale underlying horny element. (After Boas.)

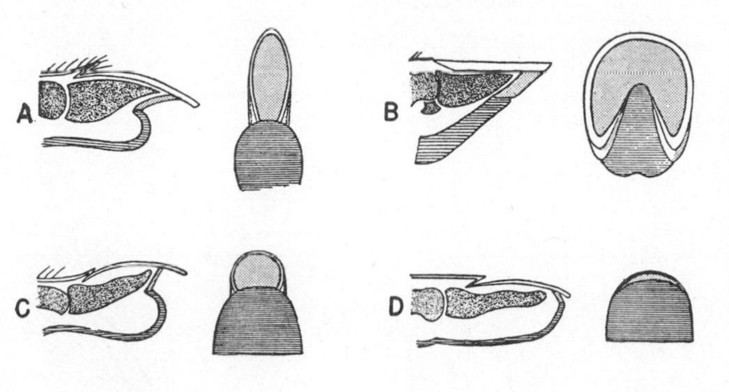

Figure 92. Longitudinal sections and ventral views of terminal phalanges of mammals to show the build of claw, nail, and hoof. Toe phalanges, stippled; subunguis, fine stipple; epidermis of ventral surface of foot, hatched; epidermis of upper surface and horny material of claw, clear. *A*, Claw of carnivore type; *B*, a horse's hoof; *C*, a nail of a typical primate; *D*, a human nail. (After Boas.)

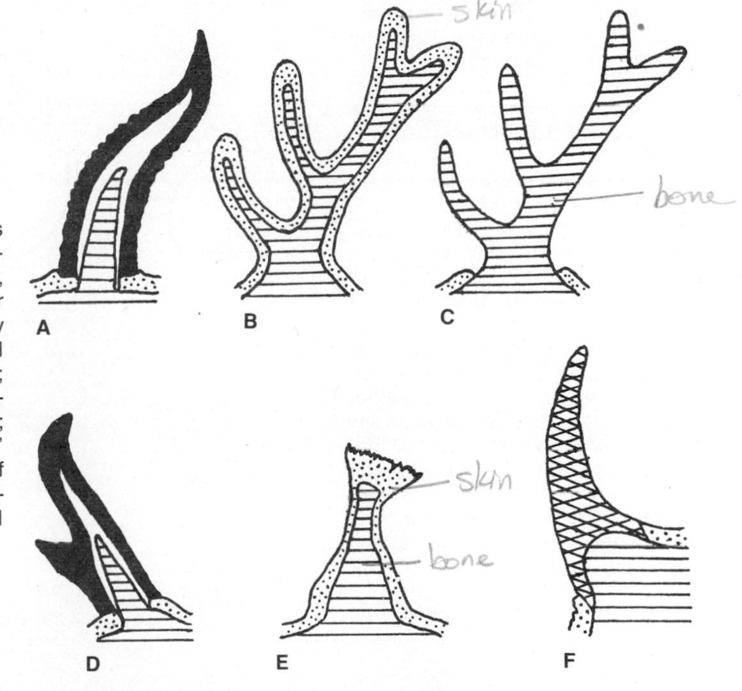

Figure 93. Diagrammatic sections through various horns and related structures. Anterior is to the left. *A*, A true horn, such as that of a cow or sheep; *B*, an antler of a deer in the velvet (the velvet is simply skin over the developing antler—it dies and is rubbed off when the antler is full grown); *C*, the same antler after the velvet is removed; *D*, the "horn" of the prong-buck; *E*, the "horn" of a giraffe; *F*, the "horn" of a rhinoceros. In all cases the bone of the skull is shown by lines, skin by stippling, horn in solid black, and the odd rhino horn by cross-hatching.

Three main feather types may be distinguished (Fig. 94); the down feather, the filoplume, and the contour feather. As the largest and most familar (if most complicated) type, **contour feathers** may be first described. The mature feather is formed entirely from highly cornified epidermal cells. The base is the **quill**, a hollow cylinder with its cavity more or less filled by pith—the remains of mesodermal material present here during the development of the feather. At either end of the quill is an opening—an **umbilicus**. The quill lies in a **follicle**, a cylindric pit extending down into the dermis but surrounded by an epidermal sheath.

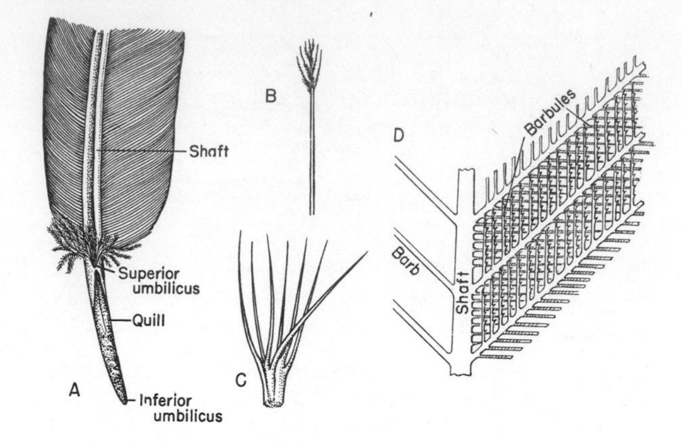

Figure 94. Feathers. *A,* Proximal part of a contour feather; *B,* filoplume; *C,* down feather; *D,* diagram of part of a contour feather, to show interlocking arrangement of barbules. (After Gadow, Bütschli.)

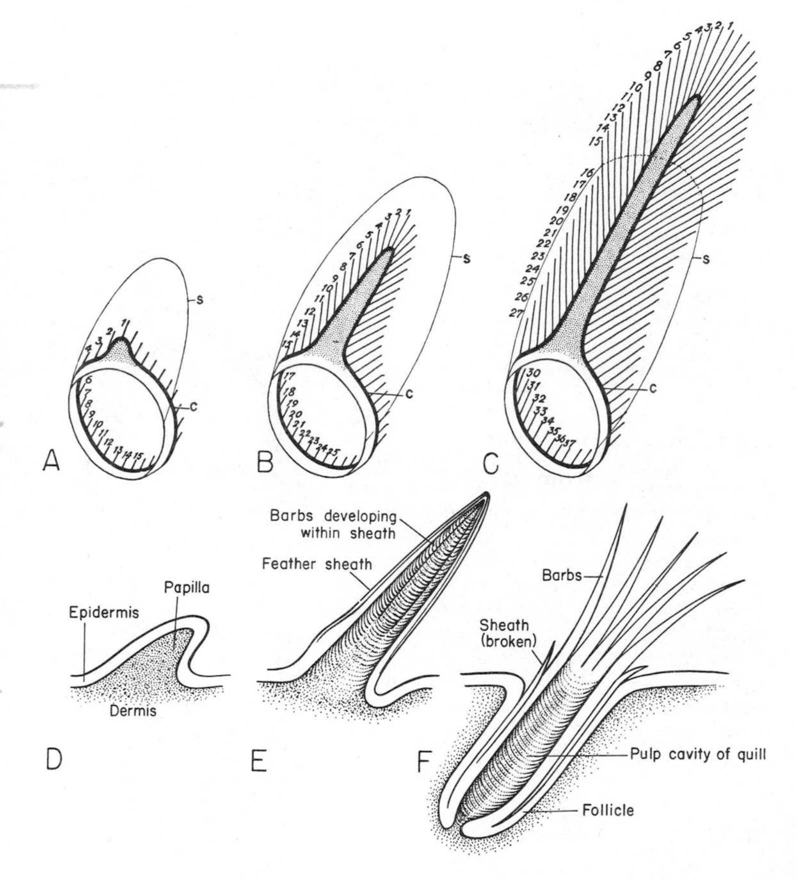

Figure 95. Feather development. *D, E, F,* Diagrammatic sections through successive stages in the development of a down feather. Development begins in the form of a mesodermal papilla; later the structure sinks into a follicle. An outer layer of the ectodermal covering separates as a thin feather sheath. The remainder of the ectoderm forms basally a hollow tube which becomes the quill. More distally it divides into a number of parallel columns. With rupture of the sheath these are freed as the barbs. *A, B, C,* Diagrams to show the development of a replacement contour feather; the basic pattern is comparable to that of a down feather. Feather growth begins at a basal ectodermal collar *(c),* from which develop, as in the down feather, parallel columns of tissue within the feather sheath *(s).* One exceptionally strong upgrowth (stippled) becomes the shaft; the parallel columns of tissue migrate successively (as shown by the numbering) on to this to become the barbs. (*A* to *C* from Lillie and Juhn.)

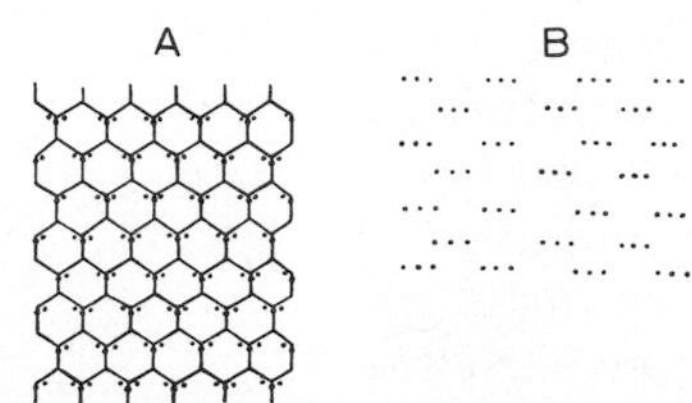

Figure 96. Hair patterns of mammals to show presumed derivation from structures developed in the interstices between scales. *A,* Part of the scaly tail of a tree shrew, with the hair (represented by dots) in this position; *B,* skin of a marmoset, with the hairs arranged in a similar pattern despite the absence of scales. (After De Meijere.)

Beyond the quill is the exposed and expanded portion of the feather, the **vane**. The axis is continued by the **shaft** or **rachis**, which (unlike the quill) is a solid structure. Outward from either side extend the major shaft branches, the **barbs**; each barb is, in most birds, interlocked with its neighbors by tiny hook-bearing branches, the **barbules**. In nonflying ostrich-like birds, where smooth contours are unnecessary for wings or body streamlining, there is little development of barbule hooks, and the contour feathers may be fluffy plumes.

Basically similar but simpler in build are the **down feathers**, or **plumules**, which cover the entire body of the chick and may, as an insulating layer, underlie the contour feathers over much of the adult body. As in contour feathers there is a quill, but distally there is no shaft, simply a splay of slender branches. **Filoplumes** are still simpler, with a single hairlike shaft which, however, may terminate in a tiny tuft of barbs.

In its initial stages embryonic feather development (Fig. 95) is comparable to that of a reptilian scale, for a conical epidermal papilla is formed, within which are mesodermal tissues. Beyond this point, however, feather development follows a course far different from that of a scale, for this papilla sinks inward, with formation of a follicle within which growth of the feather continues. In the formation of a down feather, the portion of the cone enclosed in the follicle becomes the quill, its epidermal covering becoming a cornified cylinder; the contained mesodermal tissues persist as a nutritive pulp until the feather matures. Distally, however, the outer layer of the epithelium separates as a sheath from a deeper layer which divides into a series of thickened longitudinal ridges. When growth is completed the sheath breaks down, and the ridges of epidermis beneath it are freed to become the spreading distal filaments of the down feather.

More complicated, but basically similar, is the development of a contour feather. As in the case of the down feather a cone is formed, with its proximal portion remaining simple in nature as the future quill and with the epidermis of the distal portion separating into a superficial sheath and a deeper series of longitudinal ridges arising from a basal "collar" at the end of the quill region. The entire development of the complex vane takes place within the sheath (Fig. 95 *A–C*). One dominant ridge grows out from the collar as the future shaft; other ridges formed from the collar gradually migrate onto this shaft to form the barbs; the barbules later form by outgrowth from the barbs. When the feather is fully formed within the sheath, this ruptures, and the feather has simply to unroll to attain its mature stage.

Feather replacement continues throughout life, with a basal segment of the papilla persisting at the bottom of the follicle as a feather matrix. Replacement may be a gradual, continuous process, but in many birds, particularly those of temperate and arctic regions, there is a seasonal phenomenon of **moulting**.

HAIR. As an insulating device formed of keratinized epidermis, hair is a mammalian analogue to the avian feather. In other regards, the two structures differ greatly. In contrast to feathers, there is no participation by the mesoderm in the development of hair beyond a basal papilla. Hairs, unlike feathers, are not modifications of horny scales, but are new structural elements of the skin. Hair probably evolved, possibly as special sensory projections, before our reptilian forebears had lost their scaly covering; in such mammals as retain scales, hairs are found growing in definite patterns between the scales; and even when (as usual) scales are absent, the same arrangement of hairs may persist (Fig. 96).

A typical hair includes the projecting **shaft** and the **root** sunk in a pit in the dermis termed the **hair follicle**. Both shaft and root consist (except at the very base) of essentially dead and heavily keratinized epidermal cells; around the root is a

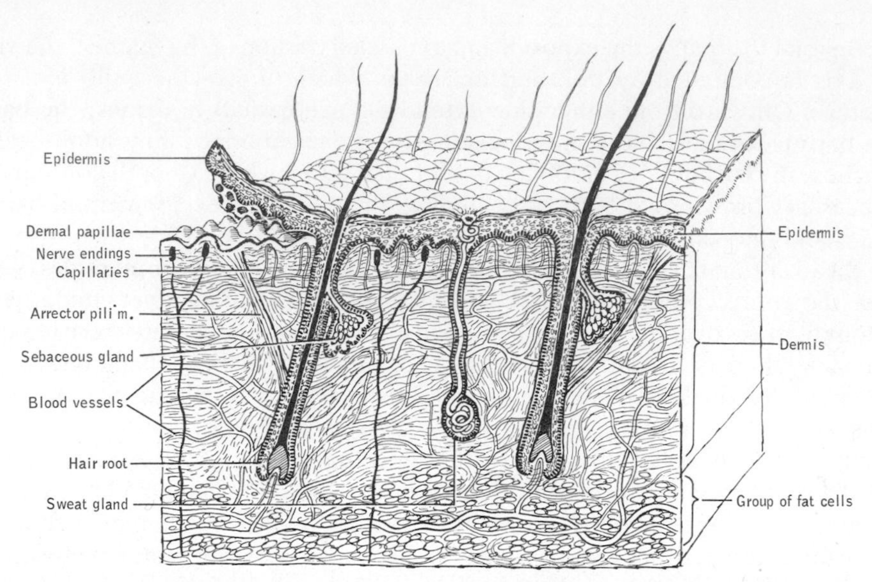

Figure 97. Section of mammalian skin, to show particularly hair, glands, and accessory structures.

sheath which (as shown in Figure 97) may consist of several distinct layers of epidermis and dermis.

At the base, the root expands into a hollow bulb, enclosing a dermal **papilla** containing blood vessels and connective tissue. Surrounding the bulb is a basal layer of "live" cells of epidermal origin, the **hair matrix**, from which are budded off the cells which form the hair root and shaft. Adjacent to the follicle and emptying its oily lubricating material into it may be found a sebaceous gland (cf. p. 125); each hair is further provided with a small muscle which by contraction may erect the hair (and by its pull on the skin cause "gooseflesh").

In the embryonic development of hair (Fig. 98) there is no projecting mesoderm-filled papilla, as there is in a feather; instead a solid column of epidermal cells grows downward. At the base of the column there form a hair matrix and an enclosed mesodermal papilla; the column of epithelial cells above is hollowed out to form the hair shaft. No more than a feather is a hair a permanent structure; most hairs are cast off and replaced throughout life, either as a gradual process or as a seasonal shedding of the coat. Resorption takes place at the bulb; a new hair begins its development from the matrix cells.

Most hairs contain pigment to some degree. Melanin, derived from dermal pigment cells, is the common substance, producing in different concentrations shades of brown and black. A related pigment is responsible for reddish tinges; air bubbles present in the hair may lighten intensity of the color and when abundant may, with pigment reduction, produce gray or white hair.

Mammalian hair is highly variable in numerous regards—in thickness, length, and coarseness, in distribution over body regions, in arrangement of hair tracts, in the slant of hairs in various regions, and so forth. Hairs rounded in section tend to be straight, and if stoutly developed may become sensory **vibrissae** (as in the cat's "whiskers") or protective bristles or spines; hairs oval or flattened in section are more readily bent and may result in a curly or woolly covering.

SKIN GLANDS. Glandular structures develop in the epidermis of every vertebrate class. In fishes and amphibians mucus-producing cells are generally present and widely distributed and in amphibians they take the form of alveolar **mucus glands**. In relatively rare instances **poison glands**, usually associated with spiny

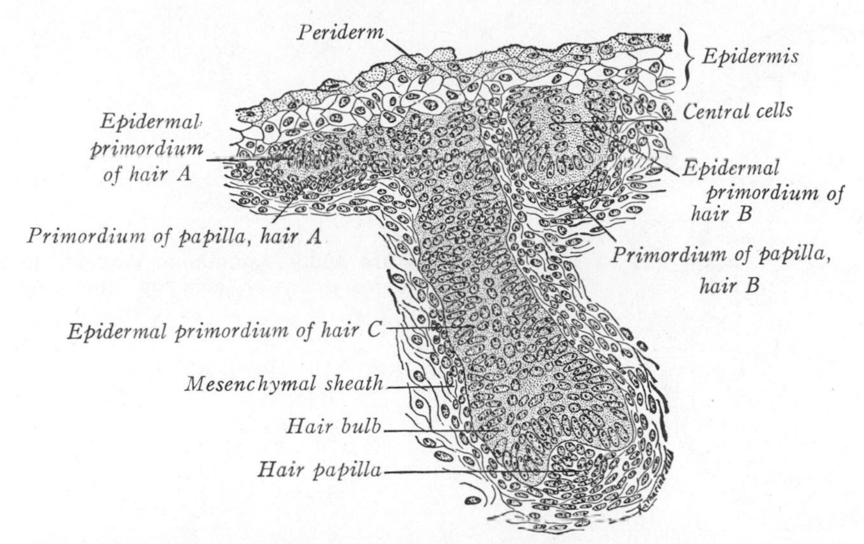

Figure 98. Hair follicles from a 3-month human embryo, showing successive stages of development at *A*, *B*, and *C*. (From Arey.)

structures, are found in fishes. In many amphibians there are glands producing poisons of a variable degree of toxicity; these are termed **granular glands** because of the granulation of the protoplasm of the secreting cells.

An unusual development in some deep sea fishes is that of luminous organs, **photophores** (Fig. 99), which may frighten enemies, lure food, or afford recognition at mating time. These structures appear to be modified mucus glands; accessory organs which may develop include a pigment-backed reflector and a lens, giving much the build of an automobile headlight. Light production may be due either to phosphorescent bacteria or to complicated processes of oxidation in the glandular cells.

In the hard, dry skin of reptiles, glands are little developed, and the same is true of birds, except for the usual presence of an oil-secreting **uropygial** or **preen gland** above the root of the tail. In mammals, however, several new types of glands make their appearance. Oily **sebaceous glands** are associated with hair follicles (Fig. 97) but may persist in regions where hair is absent. **Sweat glands** (Fig. 97) produce a watery secretion containing salts, urea, and other waste products. Through evaporation of their product on the skin they are effective in aiding temperature regulation.

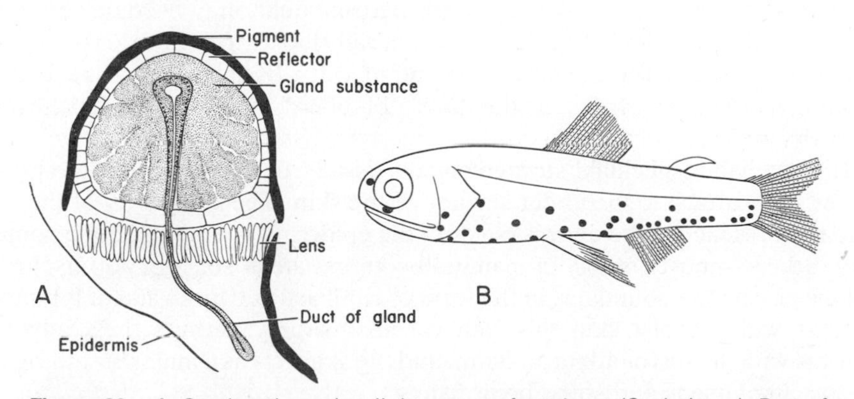

Figure 99. *A*, Section through a light organ of a teleost (*Cyclothone*). Part of the duct leading in from the body surface is seen. *B*, Light organs of a small teleost (*Myctophum*). The organs are in black. (After Brauer.)

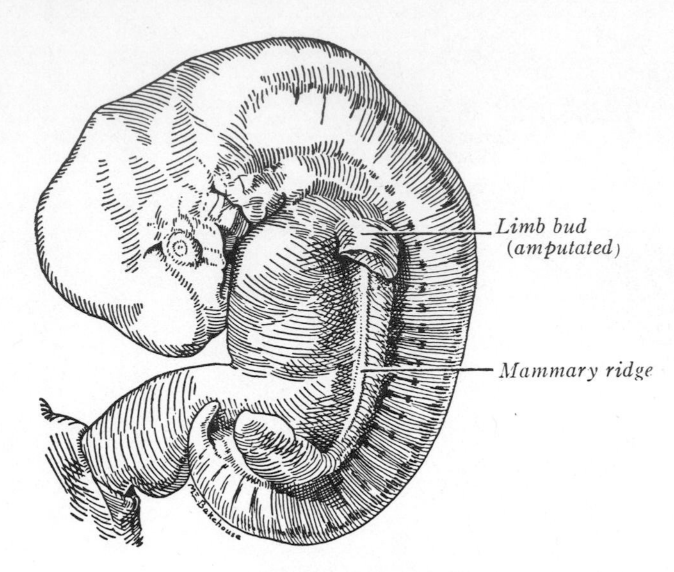

Figure 100. Mammalian embryo, to show mammary ridge or "milk line." (From Arey.)

A further mammalian gland type—one to which, in fact, the class owes its name—is that of the milk-producing **mammary glands**. Thought to be modified sweat glands, they are well developed in the female of every group. In monotremes there are simply two bundles of discrete glands which discharge their secretion into a depression on the belly surface; in other groups the gland openings are concentrated in projecting teats or nipples. The embryonic development of these glands usually begins as a pair of longitudinal swellings, the **mammary ridges** (Fig. 100) running ventrally the length of the trunk; a variable number of concentrations of tissue may occur at a variety of points. The number of teats is generally related to the number of young usually borne. When few in number, they may be either abdominal in position, as in many ungulates, or in the pectoral region, as in higher primates; in such forms as pigs and many carnivores, where litters are large, two long rows of nipples may be present. An active milk gland typically contains clusters of alveoli, from each of which a duct leads toward the surface. Properly, the term **nipple** is applied to types in which the ducts lead directly to the tip; in a **teat** (as in the cow) the ducts empty into a storage reservoir, whence a large duct leads to the surface.

DERMIS. Thicker but less varied in structure than the epidermis is the dermis or **corium** (Fig. 97). This basically consists in most groups of a dense connective tissue—the material which, after appropriate treatment, may become commercial leather. The deep portion of the dermis is generally looser in texture, and, further, is often a major locus for the development of fatty tissues. Fat is an excellent insulating material; in whales, as the thick "blubber," it substitutes for the absent hair in this regard.

In mammals, especially, striated muscle tissues, derived from underlying body muscles, may attach to the under surface of the skin. The sensitivity of the skin is due to the presence in the dermis (seldom the epidermis) of nerve fibers; some end freely, others—most notably in mammals—terminate in sensory corpuscles. Circulatory vessels are abundant, in the form of capillary networks and of lymphatics. In forms with a moist skin this rich vascular supply permits the exchange of materials with the surrounding medium, and the skin acts as a major breathing organ in many amphibians and some bony fishes.

The thick connective tissue of the dermis constitutes a major defense against injury. In most bony fishes, however, this layer is in great measure replaced by stouter defenses in the form of bony scales or plates; these are parts of the dermal skeleton and as such are described in the chapter following. Except for parts of the skull and shoulder girdle this dermal armor is much reduced or absent in most land vertebrates (turtles and armadillos are conspicuous exceptions) and it is likewise absent in cyclostomes and in chondrichthyeans, except for the small denticles embedded in the skin of the latter.

Offhand, one would assume that the more common fibrous condition of the dermis in modern vertebrates is primitive and that the presence of bone in the dermis is secondary. As we have seen, however, the story of vertebrate evolution suggests that the reverse was the case. The oldest known vertebrtates were armored, and it is highly probable that the absence of dermal armor in living jawless forms and sharks is due to bony reduction.

TEMPERATURE REGULATION. Proper functioning of the vertebrate body can take place only at internal temperatures ranging from close to freezing to about 48°C(118°F). In lower vertebrates the internal temperature tends to follow that of the environment, and little (though sometimes important) regulation is possible; but in mammals and birds body temperatures are regulated, under the control of a neural "thermostat" in the hypothalamic region of the brain, so that the internal temperature varies little from a norm which is generally within a few degrees of 38°C. Most heat loss takes place through the skin, which is thus of the greatest importance in regulation. The connective tissue of the dermis and, most especially, its fatty tissues, are insulating in nature, as are hair and feathers. In addition, the skin can act in a positive way in temperature regulation. Hair and feathers are adjustable regulators, and evaporation from sweat glands produces a cooling effect. Highly important is the vascular system of the skin, which is under control of the autonomic nervous system; with distended arterioles and capillaries (and a "flushed" skin), heat is lost rapidly; with constriction of the arterioles (and a blanched skin), heat is conserved.

CHROMATOPHORES. Skin color in vertebrates below the level of mammals or birds is due in great measure to special color-bearing cells, the **chromatophores**, located in the outer part of the dermis (Fig. 101). These are typically stellate in form and contain numerous granules. Common types include (1) **melanophores**, with a dark brownish pigment, (2) **lipophores,** with red or yellow pigments, and (3) **guanophores**, which contain not pigment but tiny crystals which by light reflection may alter the effect of the pigment materials. Nearly all the varied colorations of fishes, amphibians, and reptiles are due to chromatophores of these three types, present in varied numbers and varied arrangements. In many instances striking color changes may occur—the chameleon is proverbial in this regard and flounders are equally remarkable in the variety of color and color patterns which they are

Figure 101. Enlarged surface view of a piece of the skin of a flounder, seen by transmitted light, to show the three types of chromatophores present—pigmented melanophores, lipophores, and crystalline guanophores (or iridocytes). (After Norman.)

able to display. In part these changes are related to shifts in the position of chromatophores of the three types, but in the main they are due to changes in the distribution of the color granules within the individual cell. If the granules are dispersed, the effect is maximum; if concentrated in a tight cluster, little color appears.

Although the chromatophores are situated in the dermis, the embryologic story shows that in many cases—perhaps in all—these cells are not part of the mesodermal tissues. They rise in the embryo from the neural crest and migrate thence to their ultimate peripheral positions.

7 Supporting Tissues—The Skeleton

Most of the functionally "active" tissues of the body are epithelia or tissues derived from epithelia. But were the vertebrate body to be composed solely of such tissues, it would be a flabby and amorphous mass. Materials are needed to back up and reinforce the epithelia and their derivatives, to weld them together into a formed body, to protect this body, and—particularly in nonaquatic forms—to give it strength and support. Such materials are described in the present chapter. The notochord is a distinctive structure of this sort. Connective tissues are widespread and important. Most prominent in vertebrates are the cartilages and bones that compose the skeleton.

NOTOCHORD

The notochord is an ancient structure, present even in such lower chordates as amphioxus and larval tunicates. As has been seen, it arises embryologically from the median portion of the mesodermal tissues, extending in embryonic vertebrates from a point beneath the brain backward along the length of trunk and tail. Its cells are soft and gelatinous; the notochord is, however, surrounded by a sheath and membranes which render it a relatively strong yet flexible structure.

A well-developed notochord persists in the adult in many lower vertebrates—notably cyclostomes, in which the backbone is little developed (Figs. 17, 120). In most fishes and in tetrapods, however, it is progressively replaced by the central elements of the vertebrae, which develop around it and give greater strength if less flexibility. As vertebrae become more highly developed, the notochord is reduced in importance. It is always prominent in the embryo, but in most instances it is soon restricted, during development, by the vertebrae. In many fishes and more primitive tetrapods it may expand between successive vertebral centra but constrict within each segment, so that its contours resemble those of a series of hour glasses set end to end. In most vertebrates it is further reduced, so that in the adult it is represented only by gelatinous materials which may persist between successive centra of the vertebral column.

CONNECTIVE TISSUES

Even in such lowly invertebrates as the coelenterates, which lack a true mesoderm, there is generally interposed between inner and outer layers an intermediate zone of more or less gelatinous material, sometimes fibrous, and containing sparsely distributed cells; such material aids in filling out the form of the body and is comparable to the mesenchyme of the vertebrate embryo's mesoderm. In the adult vertebrate the most direct products of the embryonic mesenchyme are the connective tissues which form the "stuffing" of the body and reinforce the epithelia of many of the body organs. In simple form this tissue (Fig. 102) is of a loose type, with a gelatinous ground substance containing a network of small branching **reticular fibers** and, more prominently, long and slender **collagenous fibers,** flexible but inelastic, which are formed by the spindle-shaped or stellate connective tissue cells **(fibroblasts).** In other cases, such as the dermis of the skin (Fig. 97) the connective tissue is compact, with dense masses of fibers which form a feltlike structure. Most connective tissues contain a small percentage of coarse yellow **elastic fibers,** and in a few instances this fiber type is dominant. **Tendons,** forming the attachment of many muscles, consist of bundles of connective tissue fibers; **ligaments** are comparable structures uniting skeletal elements;* **fasciae** are sheets of connective tissue investing muscles or other objects. Fatty or **adipose tissue** (Fig. 103) is a connective tissue modified for fat storage, commonly developed beneath the skin or in mesenteric folds among the abdominal organs. Present in connective tissues are **mast cells** and various blood cells.

*Certain mesenteries, attached to viscera, are also termed ligaments.

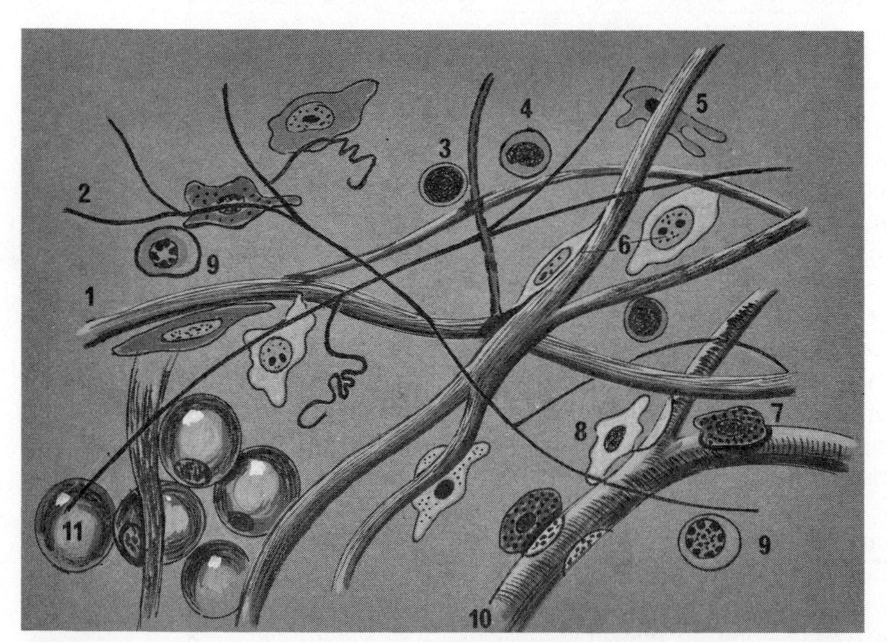

Figure 102. Diagram showing the types of cells present in loose (areolar) connective tissue. (1) Collagenous fiber. (2) Elastic fiber. (3) Lymphocyte. (4) Monocyte. (5) Macrophage. (6) Fibroblasts. (7) Mast cell. (8) Undifferentiated mesenchymal cell. (9) Plasma cell. (10) Capillary. (11) Fat cells. (After Leeson and Leeson, Histology, W. B. Saunders Co., 1976.)

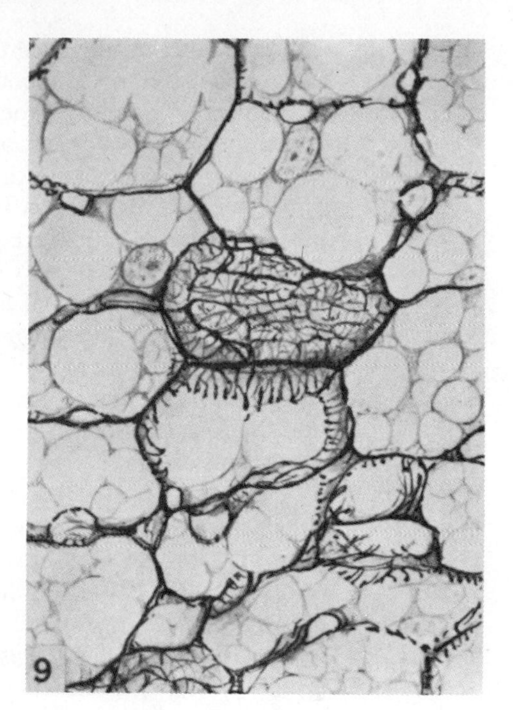

Figure 103. Brown fatty tissue from a rat. Lobules composed of fat-filled cells, with the fat represented by spaces in this preparation, are separated by fibers in the connective tissue. (From Fawcett.)

SKELETAL TISSUES

From a physiologic point of view the skeleton is sometimes (though quite incorrectly) thought of as a relatively inert system. From a broad functional viewpoint, however, it is of the greatest importance. Evolved in phylogeny from the connective tissues and developed in ontogeny from them, or from the mesenchyme which precedes them in the embryo, the hard skeletal structures are vital in welding together and protecting the softer organs and helping in support and in maintenance of body form. Almost all the striated musculature attaches to the skeleton, which is hence the agent through which bodily movement is accomplished.

CARTILAGE. Two skeletal tissues are characteristic of vertebrates—cartilage (''gristle'') and bone. Although both are specialized derivatives of the connective tissues and arise from mesenchyme, they differ markedly in nature and in mode of origin.

Typical **hyaline cartilage** (Fig. 104) is a flexible material with a translucent, glasslike appearance. Its firm ground substance, or **matrix,** is mainly a polysac-

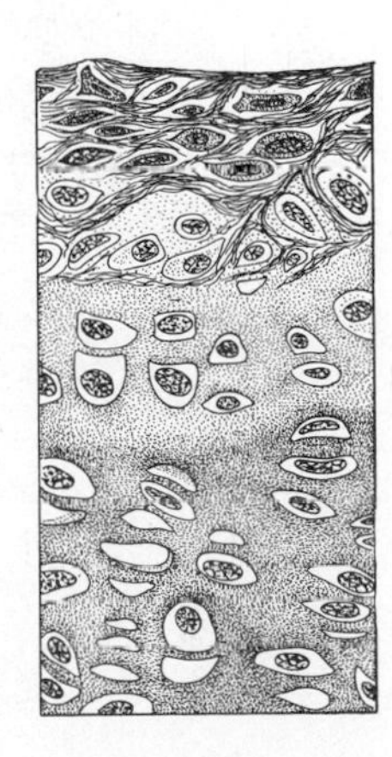

Figure 104. Section through part of a cartilage (from the sternum of a rat). The surface layers (at the top) show a fibrous condition transitional to the perichondrium. (After Maximow.)

charide, forming a firm gel through which is spread a network of connective tissue fibers. Scattered spaces contain cartilage cells, which are generally rounded and without the branching processes characteristic of bone cells. In most cartilages blood vessels are absent, and hence the nutriment received by these cells must reach them by diffusing through the ground substance. The outer surface of a cartilage is covered by a layer of dense, cell-containing connective tissue, the **perichondrium,** the inner cells of which may transform into cartilage cells.

There are numerous variants from this ordinary type of cartilage. Cartilaginous fishes frequently have **calcified cartilage,** which simulates bone in that a deposit of calcium salts is laid down in the matrix. **Elastic cartilage,** seen, for example, in the mammalian ear pinna, gains flexibility through the presence of many elastic fibers in the ground substance. **Fibrocartilage,** frequent in the region of joints and of muscle and tendon attachments, is transitional in composition between dense connective tissue and cartilage.

In cartilage formation, mesenchyme cells round up and develop between themselves the characteristic cartilage ground substance and fibers. Frequent cell divisions are seen in a growing cartilage; we may find as a result cells in pairs or quartets, the members of which gradually separate as more matrix is deposited between them. Cartilage, like bone, may grow by the addition of new cells to its outer surface; but in contrast with bone, it can also grow by internal expansion—a swelling of the ground substance.

Cartilage is essentially an *internal* skeletal material, rarely present on or near the surface of the body. It is always abundant in the embryo and young. In higher living vertebrates and in many of the extinct types of lower vertebrates the adult skeleton is mainly formed of bone, and cartilage is reduced. Only in living lower vertebrates—cyclostomes, Chondrichthyes and a few Osteichthyes—is cartilage a major skeletal material in the adult.

Cartilage is generally a deep tissue, an embryonic tissue, a relatively soft and pliable and readily expandable tissue.

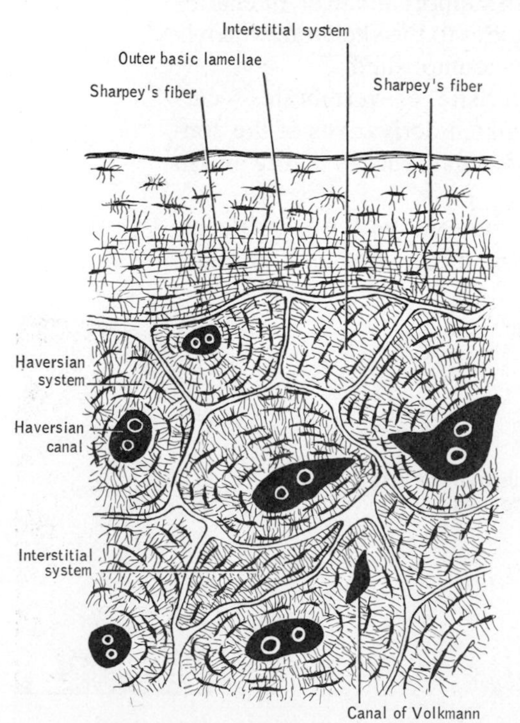

Figure 105. Bone structure. A ground thin section through a mammalian metacarpal. Toward the top (outer) margin are parallel lamellae of bone formed from the periosteum; within are a number of haversian systems cut at various angles. The "interstitial system" includes remains of earlier formed bone layers not destroyed when the present haversian systems were created. A cementing substance binds the different bone areas together. "Volkmann's canals" carry blood vessels from the surface or bone marrow cavity to haversian systems. Sharpey's fibers are connective tissue fibers, which run from the periosteum inward through the bone substance. (After Maximow and Bloom.)

BONE (Fig. 105). Bone is the dominant skeletal material in the adult of most vertebrate groups. Like cartilage it consists of transformed mesenchyme cells enclosed in a ground substance containing connective tissue fibers. Beyond this, however, the two materials differ markedly. The bone matrix rapidly becomes a hard, opaque material containing phosphate and carbonate salts of calcium. The bone cells (and the spaces—lacunae—in which they are enclosed) are irregular, star-shaped; their branching processes continue in tiny canals—canaliculi—to reach neighboring cells.* In contrast with cartilages, bones are penetrated by blood vessels. Since the solid matrix is impervious to nutritive materials, the cells receive sustenance by way of the system of canaliculi. Unlike cartilage, bone cannot expand; it can grow only by the addition of new external layers formed from the dense connective tissue, the **periosteum,** which surrounds it.

Bones have a complex microscopic structure. Many areas—particularly surface regions—consist of compact bone. But in the interior of bones there is often found a spongy type, in which the bone material is reduced to a lattice-like framework, with vascular or fatty tissues forming a **bone marrow** in the interstices. Much of the substance of any bone is laid down embryologically in the form of successive layers, or **lamellae,** but throughout life there continues a process of reworking of bony materials, by absorption of old bone and the redeposition of new. Common is bone destruction by the "eating out" of tubular channels in the bone substance, the destruction perhaps being due to cells termed **osteoclasts.** In these channels bone is redeposited in concentric layers, leaving a small central canal containing blood vessels and nerves; the whole structure formed by this process of redeposition is a **haversian system.** One customarily thinks of bone as a static, inactive material. This is far from the case; not only, as just noted, are bones being constantly reworked, but they are major storehouses of calcium and phosphate.

BONE DEVELOPMENT. Two radically different modes of bone formation—i.e., **ossification**—are to be seen in the embryo. The simpler is the formation of **membrane** (or **dermal**) **bone** (Fig. 106), in which the bone forms directly from mesenchyme. A group of bone-forming cells—**osteoblasts**—lay down between them a thin, irregular plate or membrane of dense matrix in which bone salts are rapidly deposited. This plate gradually expands at its margins and thickens on either surface by the deposition of further layers, the enclosed cells becoming the definitive bone cells, or **osteocytes.** In fishes other than cyclostomes and sharklike forms, **dermal bones** usually form over nearly the entire surface of the body (including the mouth cavity), taking the form of large plates anteriorly and of bony scales over the trunk and tail. In tetrapods, in the usual absence of bony scales or body armor, dermal bone formation tends to be restricted to the head and shoulder region.

Quite different and more complicated is the formation of **endochondral bone** (Figs. 107, 108). This is primarily the replacement of an embryonic cartilage by bone. However, even here much of the bone is laid down directly in membranous fashion external to the cartilage.

In such typical internal structures as long limb bones of tetrapods a cartilage tends to assume the shape of the adult bone at an early stage and tiny size. Presently modification and degeneration of the cartilage begin near the middle of its length. The cartilage cells swell and arrange themselves in columns, and the matrix between them calcifies. Blood vessels break in from the surface; the cartilage of this area is destroyed and bone is laid down in its place. From this central area the process of replacement continues toward each end of the element to form the shaft of the bone—the **diaphysis.**

---

*Cells are lacking in the mature bone in many teleosts and some ancient ostracoderms.

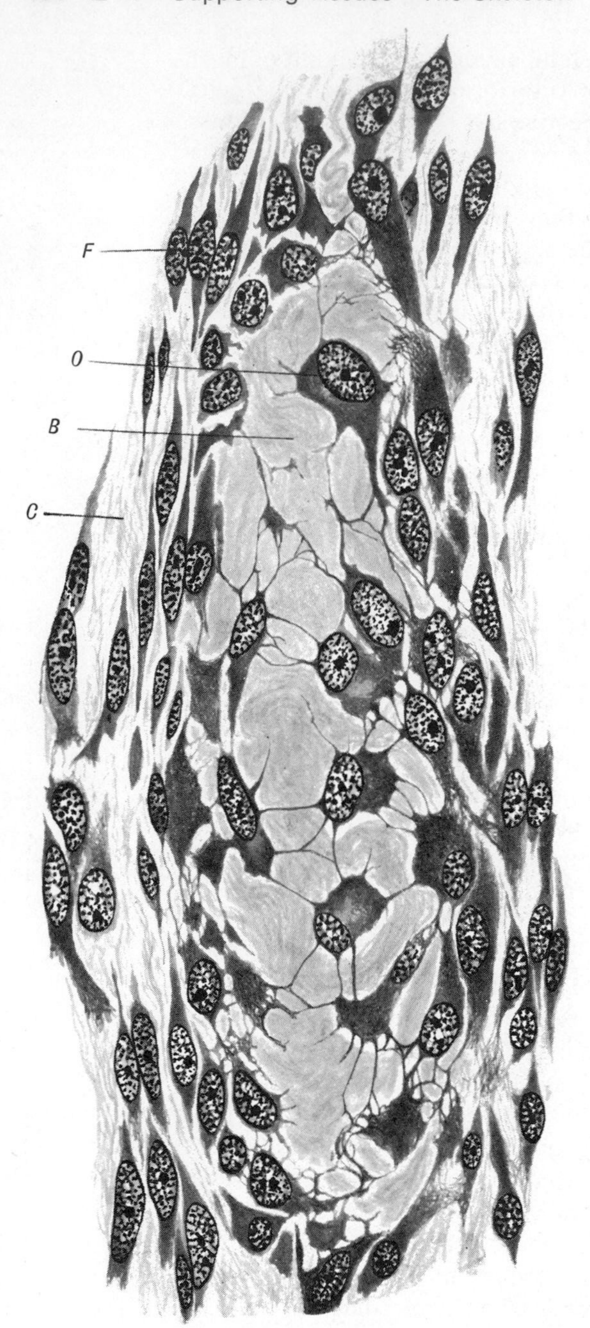

Figure 106. An early stage in the development of a dermal bone in the cat skull. *B,* Homogeneous thickened collagenous fibers which become the interstitial bone substance; *C,* collagenous interstitial substance; *F,* connective tissue cells; *O,* connective tissue cells with processes. These last cells become osteoblasts and later osteocytes. (From Bloom and Fawcett.)

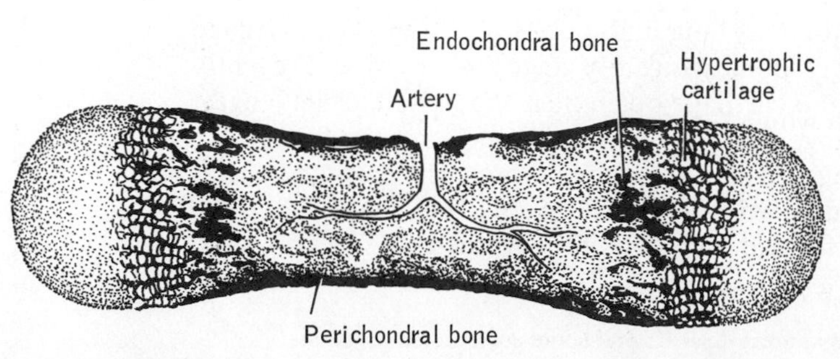

Figure 107. Section of a mammalian embryonic metapodial, in which ossification is taking place in the shaft (bone in black). Perichondral ossification is taking place superficially. At either end is normal cartilage; toward the center of the shaft the cartilage becomes "hypertrophic"; the cells are swollen and arranged in rows; calcification occurs, followed by replacement of the cartilage with dermal bone.

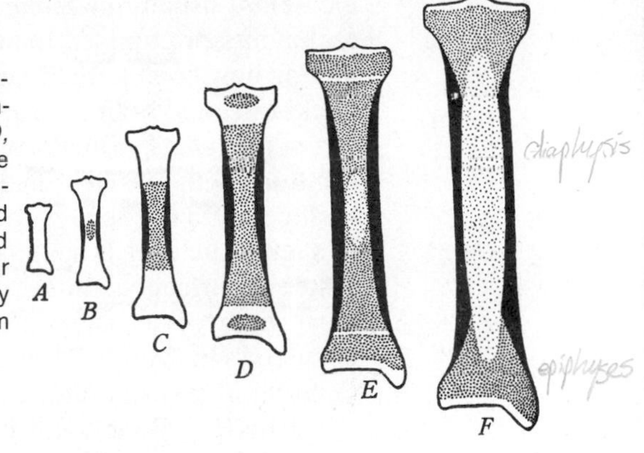

Figure 108. Ossification and growth in a long bone of a mammal. *A*, Cartilaginous stage. *B*, *C*, Deposit of spongy, endochondral bone (stippled) and compact, perichondral bone (black). *D*, Appearance of an epiphysis at either end. *E*, Appearance of the marrow cavity (sparse stipple) owing to resorption of endochondral bone. The lengthwise growth of the bone is confined to the thin strips of actively growing cartilage between shaft and epiphyses. *F*, Union of epiphysis with shaft, leaving articular cartilages on end surfaces; enlargement of marrow cavity by resorption, centrally, as deposition continues peripherally. (From Arey.)

If the embryonic cartilage failed to grow, it would soon be completely replaced by bone. But lengthwise growth of the cartilage, mainly by internal expansion, does continue and proceeds at about the same pace as the ossification within it. The cartilage, so to speak, leads the bone tissue a long "stern chase," which may never be concluded. With complete ossification, growth stops, for internal skeletal elements are usually articulated at their ends with their neighbors and bone cannot be readily added on the articular surfaces.

In lower vertebrates, generally, internal bones ossify from a single center, and their ends are often largely cartilaginous even in the adult. But in mammals (and to a very limited extent in reptiles) accessory ossifications—**epiphyses**—are found (Fig. 108). These are characteristically developed at the ends of long bones or on prominent processes for muscle attachment. These accessory centers may produce ossification (and hence strengthening) of the articular region of the bone long before growth of the shaft is completed. Between epiphysis and shaft there is long a persistent band of cartilage. This band would at first sight seem to be relatively inert and functionless. Actually, of course, it is highly important. This is a zone of growth; the cartilage here is continually growing and is constantly being replaced by bone from shaft and epiphysis. Once this band is eliminated, shaft and epiphysis unite; growth is ended.

Although much of the development of an internal "cartilage bone" takes place by the replacement of cartilage, this is not, as one will readily see upon reflection, the whole story. The original cartilage had but a very small diameter; as growth occurs, the cartilage at the two ends becomes much broader, but if the entire shaft consisted entirely of replaced cartilage, the adult structure would be shaped like an hour glass, with a very thin middle portion. This imperfection is rectified by the direct formation around the shaft of layers of **perichondral** (or, better, **periosteal**) **bone,** formed much in the fashion of membrane bone, thus giving the necessary adult thickness.

The process of endochondral bone formation in higher vertebrates was long thought to be an example of ontogenetic recapitulation of phylogenetic history. It was assumed that the cartilaginous condition of the skeleton seen in cyclostomes and sharks was a primitive one, to which the cartilaginous state of the skeleton in the embryo of higher forms was a true parallel. However, the fossil evidence, reviewed in Chapter 3, suggests that this is not the case, but that the primitive vertebrates had well-ossified skeletons as adults and that no recapitulation is involved.

If so, why this roundabout way of forming bones? The answer may be deduced from the fact that it is only deep-lying bones that are preformed in cartilage. Dermal elements, usually platelike in nature and usually lacking any major muscular attachments or complex joints with other elements, can grow without difficulty by adding new bone to their surfaces and margins. But internal elements of the limbs, backbone, and braincase are usually attached in complex fashion to other skeletal structures and, particularly toward the ends of limb bones, usually have complex relations with muscles. They cannot grow by plastering new layers of bone to their surfaces; and they could not, if formed of bone, expand. What is needed for growth of such structures is some sort of pliable material which can grow without disturbance of surface relations. Cartilage, capable of growth by internal expansion, is an ideal embryonic adaptation for this purpose. Except where reduction has occurred—in various fishes and amphibians—bone is the normal adult skeletal material in a vertebrate, cartilage its indispensable embryonic auxiliary.

JOINTS. Bones and cartilages are joined to one another by structures of varied types. In such cases as bones of the skull where movement is not necessary or desirable, two elements may be firmly connected with one another; lines of separation between two such elements—**sutures**—may remain visible, or the two bones may fuse in the adult. Such an essentially immovable type of union is a **synarthrosis.** A freely movable joint is a **diarthrosis** (Fig. 109); a well-formed, liquid-filled **joint cavity** is frequently developed, lined by compact connective tissue.

CLASSIFICATION OF SKELETAL ELEMENTS. The skeleton includes a wide variety of elements of varied form, structure, function, position, and embryonic origin, associated in variable combinations. Classification of them is difficult; although no method is absolutely satisfactory, we here adopt the following scheme of major subdivision:

- Dermal skeleton
- Endoskeleton
  - Somatic
    - Axial
    - Appendicular
  - Visceral

We have noted above the marked embryologic distinction between membrane bones formed in the dermal layers of the skin and the deeper lying endochondral elements. Among the latter a distinction may be made between two groups of unequal size. The **visceral skeleton** includes the cartilages or bones associated with the gills and skeletal elements (such as the jaw cartilages) derived from them. These, as will be seen, have an embryonic origin quite different from most of the remainder of the internal skeleton, here termed the **somatic skeleton.** This last major division includes in all vertebrates the vertebrae, ribs (when present), and, anteriorly, the braincase. These structures are the axial skeleton. Paired limbs are present in most forms and are prominent in tetrapods; the structures of the limbs and their girdles belong to the somatic group but may be distinguished as the appendicular skeleton.

In certain instances we find that structural units of the adult skeleton contain elements derived from two or more categories. Thus, for example, the shoulder girdle frequently contains both dermal and endoskeletal components; the lower jaw in many forms includes visceral and dermal elements. Most complex of all is the skull, which in bony fishes and tetrapods includes dermal, axial, and visceral structures in its formation.

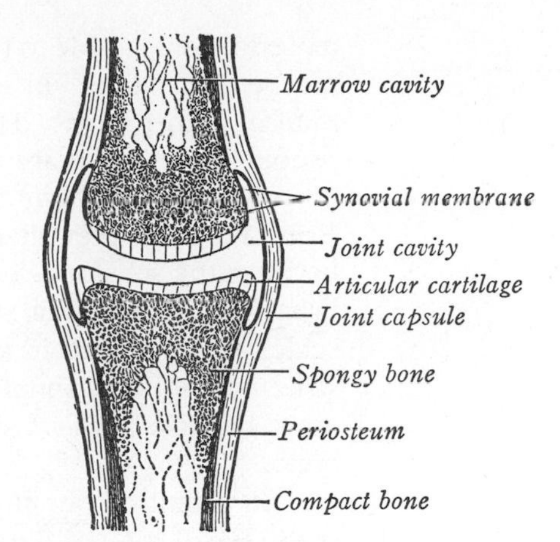

Figure 109. A typical diarthrodial joint. (From Arey.)

DERMAL SKELETON

FISHES. The skin over most of the body of many living vertebrates contains no hard skeletal parts; however dermal bony structures are usually present in the head region, at least, and the fossil evidence leads to the conclusion that ancestral vertebrates were ensheathed in armor, composed in the main of dermal bone. Such armor completely encased the most ancient jawless ostracoderms, covered part or all of the body of the extinct placoderms, and is preserved in most members of the great group of Osteichthyes. Armor is absent in cyclostomes and is found only as denticles in the skin of shark-like fishes; this condition, once thought to be primitive, now seems quite surely to be a degenerate one.

In many ostracoderms and placoderms there was present a pattern of microscopic structure in scales and plates which, with variations, persisted into the primitive bony fish stage (Fig. 110). A middle layer consisted of spongy bone, presumably containing blood vessels; inner and outer layers were compact. The outer surface was frequently ornamented by tubercles or ridges. The substance of these superficial structures was formed of a material much like the dentine of a tooth (cf. p. 238), with a ''pulp'' cavity beneath and with an outer layer of hard, shiny

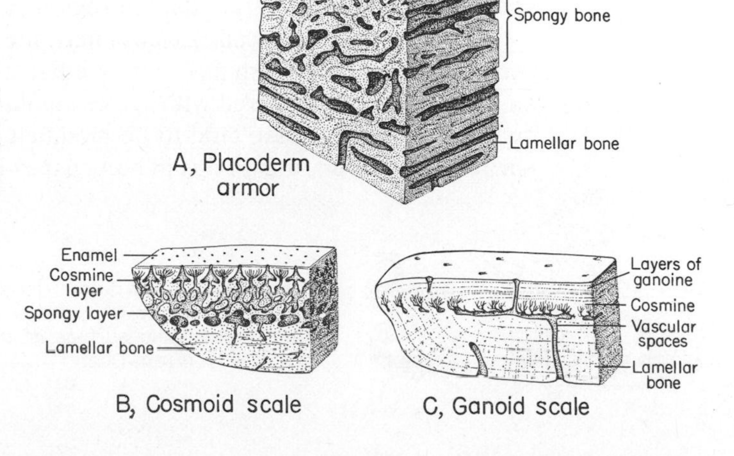

Figure 110. The structure of dermal plates and scales in primitive vertebrates. *A,* Devonian placoderm. *B,* Primitive crossopterygian with a cosmoid scale. *C,* Paleozoic ray-finned fish with a ganoid scale. (After Kiaer, Goodrich.)

material comparable to tooth enamel. However there is much dispute over the true nature of "enamel" in anamniotes, as the tissues are much more variable in them than in higher forms. The whole tubercle, in fact, was toothlike, and it is highly probable that teeth are actually derived from such structures.

Among early bony fishes two types of plate and scale structures were present. Primitive sarcopterygians possessed the **cosmoid scale** (Fig. 110 *B*), so named because the substance of the tubercle consisted of **cosmine,** a form of dentine with branching rather than simple tubules running into it from the pulp cavity. Such scales were present in typical crossopterygians and the earliest lungfish, but the structure became simplified in later members of both groups, and the modern lungfish scale consists simply of a rather fibrous and leathery type of degenerate bone.

In primitive ray-finned fishes there was present, in contrast, the true **ganoid scale** (Fig. 110 *C*), differing from the cosmoid type in that during growth there was laid down on the outer surface layer after layer of shiny enamel-like material termed **ganoine** and, correspondingly, successive layers of compact bone on the inner surface. Today only *Polypterus* and the gar pikes retain the ganoid type, and in modern teleosts the scales (much as in the parallel case of the lungfishes) are reduced to apparently simple thin structures of pliable bone, although it is possible that traces of superficial layers may also be present.

Although we are confident that the cyclostomes are descended from armored ancestors, their skin is absolutely devoid of armor. In the Chondrichthyes there are sometimes fin-spines of a dentine-like nature, but the skin is otherwise bare except for the isolated **dermal denticles** or **placoid** "scales" (Figs. 88 *A*, 111, 236 *A*). These resemble teeth in structure, with a pulp cavity, a tooth substance of dentine, and a shiny enamel-like surface often called vitrodentine. It was once believed that dermal plates and true scales resulted from a fusion of such denticles. It now seems more probable that the reverse is the case: that the dermal denticles are the last superficial remains of ancestral armor, the deeper layers of which have, so to speak, melted away.

In fishes with dermal armor the anterior part of the body is covered not with discrete scales but with large bony plates over the head, gills, and shoulder region. In the ancient ostracoderms and placoderms these plates are arranged in varied patterns. In bony fishes the general pattern is of a more familiar type, although certain of the individual bones are highly variable. There is a well-defined cranial shield which forms part of a typical skull; dermal bony plates are associated with the lower jaw and the inner surface of the mouth; there is, further, a dermal shoulder girdle. These will be discussed later. There may be noted the presence of a series of **opercular elements** covering the gill region (Figs. 122, 169, 171, 172).

We may parenthetically discuss here the nature of the rays which stiffen the peripheral portions of fish fins, both median and paired (Fig. 112). Primitively the fins were probably covered with scales similar to those on the rest of the body. In higher bony fishes these tend to be modified into elongate bony rays, the **lepidotrichia.** In addition, the fin tips in bony fishes may be additionally stiffened by tiny

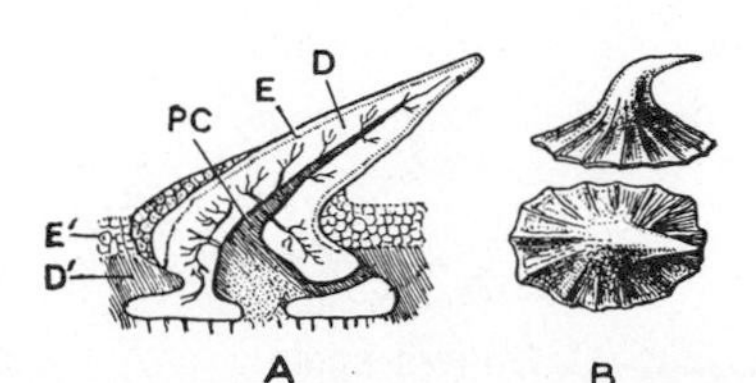

Figure 111. Shark dermal denticles. *A*, Section through a denticle; *B*, Side and surface views of denticle. Abbreviations: *D*, dentine; *D′*, dermis; *E*, hard, enamel-like surface of denticle—vitrodentine; *E′*, epidermis; *PC*, pulp cavity. (From Dean.)

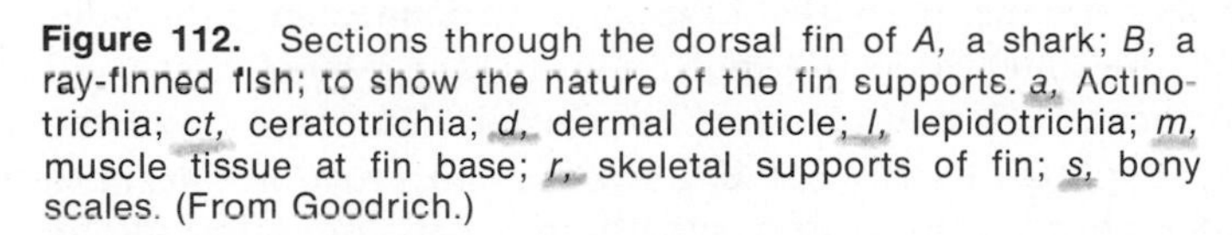
Figure 112. Sections through the dorsal fin of *A*, a shark; *B*, a ray-finned fish; to show the nature of the fin supports. *a*, Actinotrichia; *ct*, ceratotrichia; *d*, dermal denticle; *l*, lepidotrichia; *m*, muscle tissue at fin base; *r*, skeletal supports of fin; *s*, bony scales. (From Goodrich.)

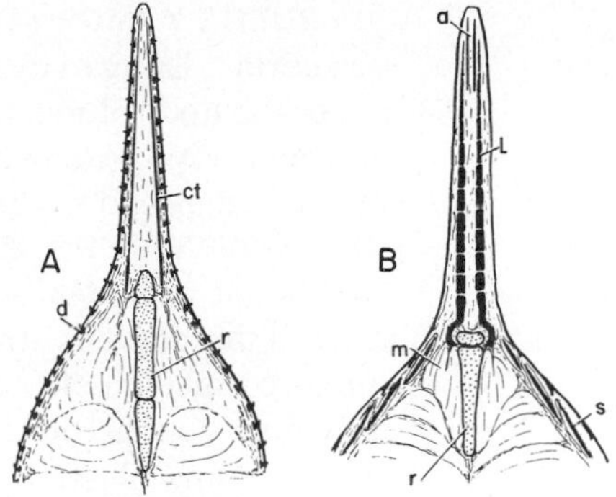

horny rays—**actinotrichia**—and in sharks larger rays of this sort—termed **ceratotrichia**—are the sole supports of the fin web.

TETRAPODS. Of the originally complete dermal covering of their ancestors, tetrapods have retained dermal elements in the skull, jaws, and usually the shoulder girdle. The remainder of the dermal covering tends to be lost. In modern amphibians there is no dermal covering of the trunk and tail save for vestigial scales in the Gymnophiona. In early amphibians and reptiles remains of scales persisted on the belly, and these are retained as V-shaped jointed rods, the **gastralia,** in *Sphenodon*, lizards, and crocodilians. In birds and mammals even these last vestiges of the original armor have vanished.

The skin, however, retains its potentialities of forming dermal bone, and in many reptiles and a few mammals there has been a redevelopment of armor. In lizards bony scales frequently underlie the horny scales of the epidermis; in crocodilians there may be a partial armor of subquadratic bony plates, and certain dinosaurs and other extinct reptiles were armored. Among mammals the armadillos have developed a bony carapace; their extinct cousins, the glyptodonts, had a comparable bony covering developed to a high degree, and ground sloths had small lumps of bone in the dermis.

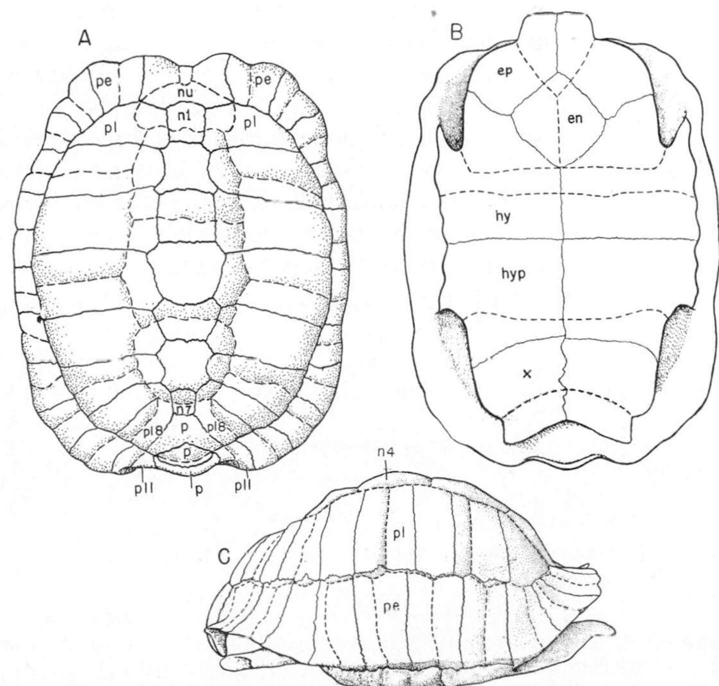

Figure 113. *A*, Dorsal, *B*, ventral, and *C*, lateral views of the shell of a tortoise *(Testudo)*. Sutures between bony plates in solid line, outlines of horny scutes in broken line. Abbreviations: *en*, entoplastron; *ep*, epiplastron; *hy*, hyoplastron; *hyp*, hypoplastron; *nu*, nuchal; *n1–n7*, neurals; *p*, pygal plates (postneurals); *pe*, peripheral plates; *p11*, most posterior (eleventh) peripheral; *pl*, pleural plates; *x*, xiphiplastron.

TURTLE ARMOR (Fig. 113). The most highly developed armor of any land vertebrate, living or extinct, is that of the Testudines. Beneath the horny scutes that cover the body of most turtles there is dorsally a rounded, arched **carapace** of bone and ventrally a flattened **plastron;** the two parts of the shell are connected at the sides by a bridge of bone, while front and back of the shell are open for the head, limbs and tail. Notable in the carapace is a medial series of **neural** elements above the backbone, with **pleurals,** supported by ribs, on either side, and a row of **marginals** around the margins. In the plastron there are typically four pairs of plates and an unpaired median element. The three plates at the anterior end of the plastron are modified dermal bones of the shoulder girdle; all the remaining elements of carapace and plastron are new developments.

AXIAL SKELETON

AMNIOTE VERTEBRAE. The host of skeletal structures remaining for description are endoskeletal, lying (in contrast to the dermal elements) deep within the body and first formed in the embryo in cartilage. Most of these internal structures belong to the system here termed somatic; they are formed (in contrast to those of the visceral system) from mesenchyme of mesodermal origin. Apart from the limb supports, the somatic skeletal elements are classed as axial.

The major axial structure is the **vertebral column,** which usually replaces the notochord in the adult as the main longitudinal girder of the body and extends upward in each segment to enclose and protect the spinal cord. We may first consider the relatively simple and uniform structure of the vertebrae which compose the column in amniotes (Fig. 114) before considering the wild assortment of variants seen in lower vertebrate classes. A major element is the **centrum,** essentially a spool-shaped structure functionally replacing the notochord. Primitively the notochord persisted in the adult amniote in much reduced form, piercing the center of the centrum, but usually this ancient structure has vanished. The centra were hollowed at each end in early reptiles—the **amphicelous** condition (Fig. 115 *A*)—and this condition persists in some reptiles. Usually, however, the centra are more highly developed and their ends are apposed to those of their neighbors (Fig. 115 *B–D*). If the centra are flat-ended, the condition is termed **acelous;** if concave in front, convex behind, **procelous;** if the reverse of this last, **opisthocelous.** In some reptiles there are present in the trunk small **intercentra** wedged ventrally between successive centra. In reptiles and mammals generally these persist in the tail; here they bear **hemal arches,** consisting of a pair of rods extending ventrally to enclose the caudal blood vessels and meeting below in a spine separating the caudal muscles of the two sides.

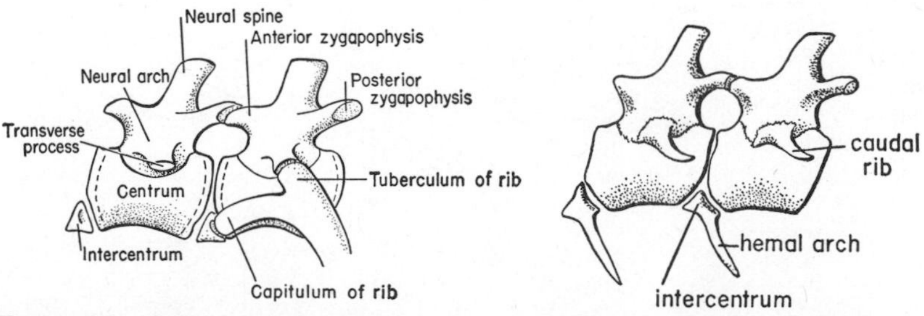

Figure 114. *Left,* two trunk vertebrae of an early generalized reptile (anterior end to the left). *Right,* two caudal vertebrae.

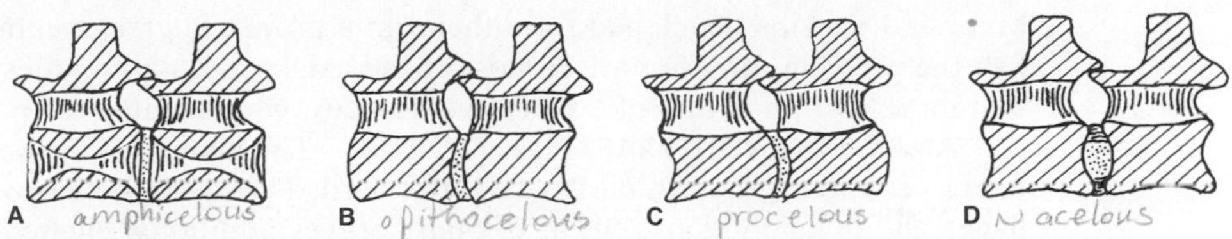

Figure 115. Diagrammatic longitudinal sections through vertebrae to show various types of centra. Anterior is to the left. *A*, Primitive amphicelous type, in this case with the centrum pierced by an opening for a continuous notochord; *B*, the opisthocelous type; *C*, the procelous type; *D*, an essentially acelous type but with the centra slightly biconcave to allow room for an intervertebral disc. (After Romer.)

Above the centrum on either side rises a **neural arch,** sheathing the nerve cord; gaps between successive arches allow space for the exit of the spinal nerves. Above, the arches meet to form a **neural spine.** At either side there is often a prominent **transverse process** for the attachment of a rib. In tetrapods—and in a few fishes—each pair of arches articulates with its neighbors fore and aft by "yoking" processes, the **zygapophyses.** The anterior zygapophyses terminate in surfaces facing inward and upward, which meet corresponding processes facing downward and outward on the opposed posterior zygapophyses of the preceding vertebra (for mnemonic purposes one may recall that in any procession, those at the front of the parade are up and in—those at the rear, down and out).

In the discussion of mesodermal differentiation we noted that there was a proliferation of mesenchyme from the medial side of each somite—the area known as the sclerotome (Fig. 81). It is from this material that the vertebrae are formed. One might assume that each vertebra would form from a single sclerotome, since there is a correspondence in numbers. Actually this is not the case—in amniotes, at any rate. Each sclerotome divides into two parts, and each vertebra is formed from a fusion of adjacent halves of two successive sclerotomes (Fig. 116). On reflection it is seen that this apparently odd development is a logical and functionally necessary one. The muscles of the trunk attach to successive vertebrae and ribs. The trunk muscles are primitively segmental in arrangement, and hence it is necessary that the skeletal elements to which they attach alternate with them. This condition is brought about by the recombination of sclerotome halves, which places the verte-

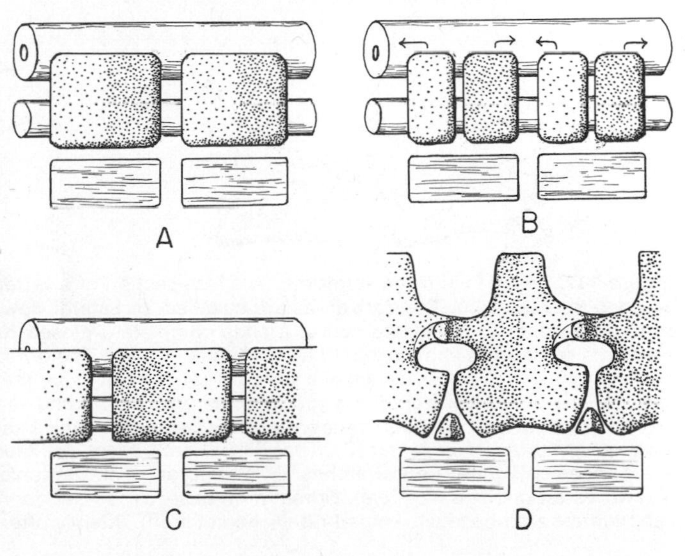

Figure 116. Diagrams to illustrate the manner of utilization of mesenchyme in the formation of amniote vertebrae. Anterior is to the left. *A* is a diagrammatic lateral view of the sclerotomes of two somites; nerve cord and notochord are shown behind them, and below is arbitrarily figured a portion of the musculature belonging to these somites. In *B* the sclerotomes have split into anterior and posterior halves, which move apart and in *C* have fused with the adjacent halves of neighboring sclerotomes. These newly formed sclerotome masses are the materials from which successive vertebrae are derived, and in *D* is diagrammatically represented (by differences in stippling) the dual origin—from parts of two segments—of an adult vertebra. As a result of this seemingly odd developmental process, the vertebrae (as is seen by reference to the blocks of segmental origin) are inter-segmental rather than segmental in position.

brae, and the ribs developed from them, in a proper intersegmental position. In fish the problem does not arise; most of the axial musculature does not insert on vertebrae, but on sheets of connective tissue (myocommata; see page 207).

ANAMNIOTE VERTEBRAE (Figs. 117, 118). The history of the neural arches in lower vertebrate groups is fairly straightforward. Typical arches are present in most fishes and in amphibians, but in Chondrichthyes additional elements are formed between the neural arches and in lampreys one or two pairs of small arches are present in each segment; in hagfishes neural arches are absent altogether except in the tail. Hemal arches comparable to those of amniotes occur in most lower vertebrates.

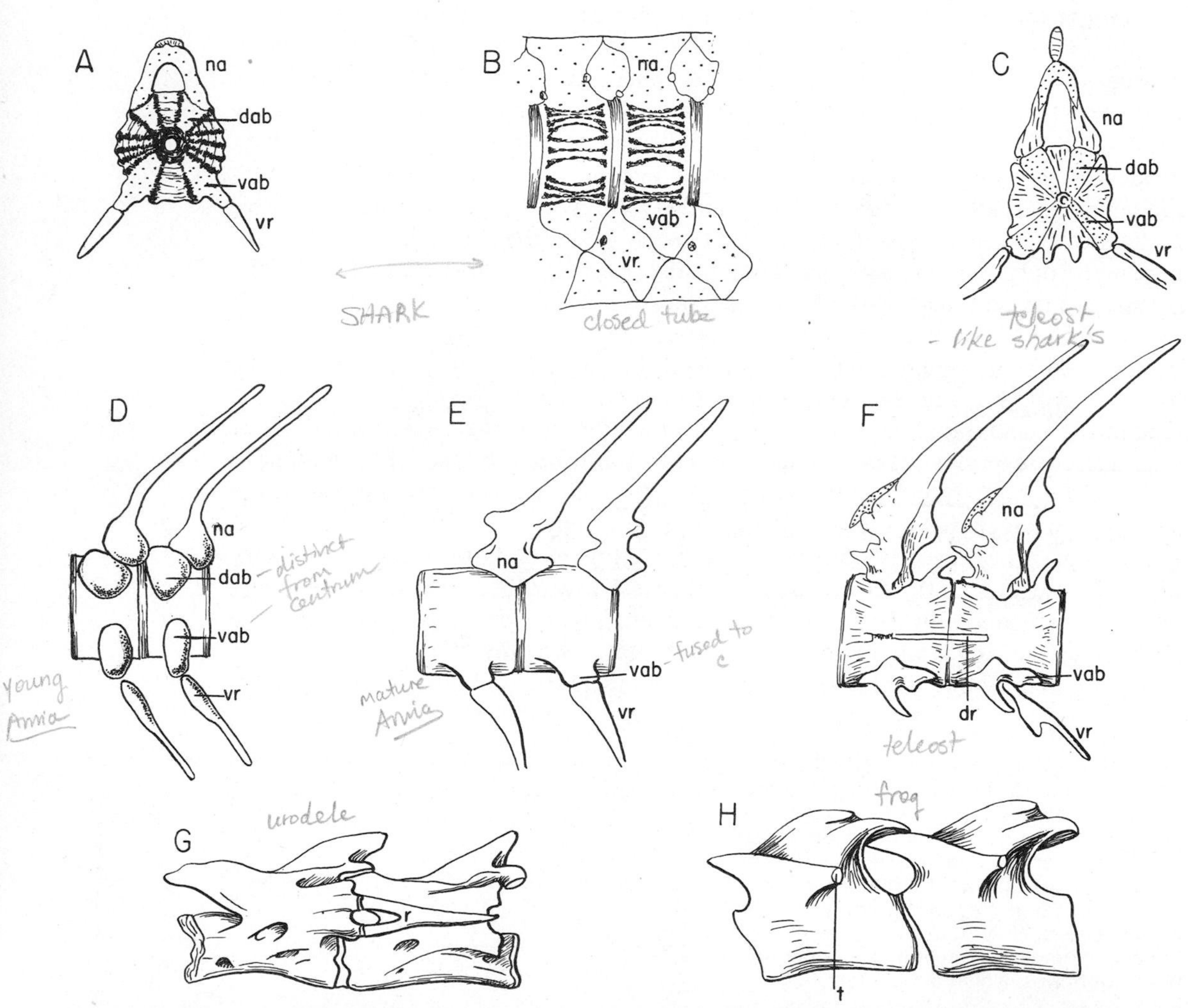

Figure 117. *A–F,* Fish trunk vertebrae. *A,* Cross-section of a vertebra of the shark *Lamna,* showing arch bases wedged into centrum. The dark areas are calcified. *B,* Lateral view of two segments of the column of the same shark. Elements between the neural arches complete a closed tube for the spinal cord; a complex of small elements represents short ventral ribs (= hemal arches). The dorsal arch bases are covered by the base of the neural arches. *C,* A cross-section of a vertebra of a teleost, *Esox;* the structure is comparable to that of a shark. *D,* Lateral view of two vertebrae of a young *Amia;* both dorsal and ventral arch bases are visible and distinct from the centrum proper. *E,* Adult vertebrae of the same; the arch bases are incorporated in the centrum. *F,* Side view of vertebrae of *Esox* (cf. *C*). The ventral arch bases are distinct; as in the shark the dorsal arch bases are concealed by the neural arches. *G,* Vertebrae of the urodele *Necturus,* in which short ribs are present. *H,* Two vertebrae of a frog. *dab,* Dorsal arch base; *dr,* dorsal rib; *na,* neural arch; *r,* rib; *t,* transverse process; *vab,* ventral arch base; *vr,* ventral rib (= hemal arch). (Mainly after Goodrich.)

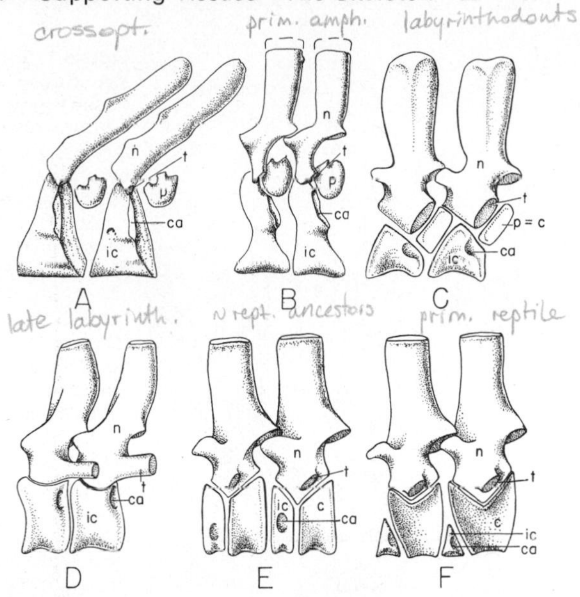

Figure 118. The evolution of vertebral structures from crossopterygians to primitive tetrapods and reptiles. Anterior is to the left. *A,* Two vertebrae of a crossopterygian; the principal central element is a large intercentrum, wedge-shaped in side view, crescentic if seen in end view; there are small paired pleurocentra. *B,* Vertebrae of the most primitive known amphibian type, of similar construction. *C,* The typical rhachitomous type, common in Paleozoic labyrinthodonts. *D,* The stereospondylous type of vertebra, found in late labyrinthodonts; the pleurocentra have disappeared and the intercentrum forms the entire centrum. *E,* The embolomerous type, found in extinct forms related to reptile ancestors; the pleurocentra have expanded and fused to form a complete ring-shaped centrum and the intercentrum forms a similar but thinner structure. *F,* The primitive reptilian type in which (as in *E*) the pleurocentra have expanded to form a large true centrum, but the intercentra are reduced. Abbreviations: *c,* centrum; *ca,* capitular attachment of rib to intercentrum; *ic,* intercentrum; *n,* neural arch; *p,* pleurocentrum (paired); *t,* transverse process on neural arch for attachment of tuberculum of rib. (*A, B,* after Jarvik.)

The history of the centrum is more complex (Fig. 119). Cyclostomes have no centra whatever—merely a large, unconstricted notochord—and hence can be considered vertebrates only by courtesy. In sharks (Fig. 117 *A, B*) the centra are short cylinders, formed in the main of concentric layers of cartilage (frequently calcified). These layers are interrupted, however, dorsally and ventrally on either side, by "plugs" of cartilage which underlie the attachments of neural and hemal arches (the latter represented in the trunk by ventral ribs). In bony fishes these same **arch bases** are present in early stages as distinct structures, but in well-ossified forms such as *Amia* and the teleosts they are fused at maturity with the remainder of the centrum (Fig. 117 *C–F*). In the line of fossil forms leading to land vertebrates ossification in the centrum seems to have been concentrated in these arch bases (Fig. 118). In the ancestral crossopterygians and in many of the ancient labyrinthodont amphibians we find a type of centrum in which bones presumably representing dorsal arch bases are present as a pair of small **pleurocentra** situated high up, close to the base of the neural arch. The fused ventral arch bases of the two sides probably form a large element, which is wedge-shaped when seen laterally, crescentic in front or back view; this structure appears to be identical with the small intercentrum which we have seen in amniotes. As indicated in Figures 118 and 119, most of the ancient fossil labyrinthodonts tended to expand the intercentrum into a still larger structure as the sole central element. In the line leading to the reptiles, on the other hand, the intercentrum underwent reduction toward the amniote condition of small size or complete reduction. In this line toward higher vertebrates, the two pleurocentra apparently become fused and greatly expanded to become the true centrum of amniotes.

Embarrassingly, we cannot readily fit into this otherwise clear story of tetrapod vertebral development the conditions seen in the modern amphibian orders (Fig. 117 *G, H*). The centra are essentially simple cylinders (despite irregularities in outline) in which there is no evidence of separate structures corresponding to pleurocentra and intercentra. In urodeles and gymnophionans the centra ossify directly, with little or no preceding cartilage formation; in most frogs and toads there is formation of a cartilaginous centrum to a variable degree, but the adult centrum ossifies in continuity with the neural arch. Presumably the modern amphibian condition is one of secondary simplification.

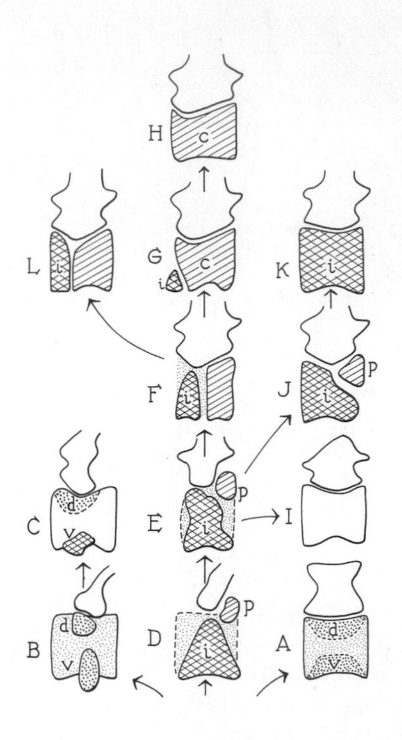

Figure 119. Diagrams to show possible evolutionary sequence in the structure of vertebral centra. *d* and *v* are the dorsal and ventral arch bases found in many fishes, parts of the centrum structure supporting neural arches and hemal arches or ventral ribs, respectively. In crossopterygians and many tetrapods the arch bases appear to be represented by pleurocentra *(p)* and intercentra *(i)*. Cartilage stippled; arch base cartilages, heavy stipple; pleurocentra hatched; intercentra cross-hatched; parts of central ossifications not included in arch bases, plain white. Part or all of the neural arch is shown as well as the central structures. Front end of vertebrae to left.

A, Shark condition; arch bases incorporated in cartilaginous centrum. *B,* Embryonic *Amia* vertebra, showing presumed basic actinopterygian condition, with arch bases forming separately from rest of cartilaginous centrum. *C,* Teleost condition; centrum ossified, and incorporates dorsal arch base and in some cases ventral arch base as well. *D,* Crossopterygian condition, probably basic for tetrapods; much of the centrum was persistently cartilaginous, but small paired dorsal arch bases ossify as pleurocentra, and ventral elements fuse as a wedge-shaped structure. *E,* Oldest known amphibians, similar in structure to crossopterygians. *F,* Type leading toward reptiles among fossil labyrinthodonts; the pleurocentra have grown and fused to form a ring-shaped "true" centrum. *G,* Primitive reptiles; the centrum has grown at the expense of the intercentrum, which is much reduced. *H,* Advanced reptiles, birds, mammals. The intercentrum disappears, and the whole centrum corresponds to the expanded pleurocentra. *I,* Modern amphibians; the centrum ossifies as a simple spool-shaped structure, with no evidence of arch bases. *J,* Structure in rhachitomous labyrinthodonts; the intercentrum is a large wedge-shaped structure, the pleurocentrum persistently small. *K,* Late, stereospondylous labyrinthodonts; the pleurocentra vanish, and the intercentrum constitutes the entire centrum. *L,* Extinct embolomerous labyrinthodonts; both pleurocentra and intercentrum form complete rings.

REGIONAL VARIATIONS IN VERTEBRAE. In higher vertebrates there may be distinguished various regions along the length of the vertebral column, mainly recognizable through the presence and absence of ribs or changes in the nature of the ribs. But in lower vertebrates ribs may be present on every vertebra from neck to tail without break, and we can do little except distinguish posteriorly a **caudal** series where, in contrast to the **trunk vertebrae,** typical hemal arches are present ventrally. In land vertebrates the attachment of the pelvic girdle to the backbone establishes a **sacral** region interposed between **presacral** and caudal segments. In the neck of tetrapods (as noted later) the ribs tend to be short and may be fused or absent, and there is thus established a **cervical** region distinct from the **dorsal** region of the trunk. In the trunk the posterior ribs became shorter, and ribs are absent here in mammals, so that the dorsal vertebrae can be divided in that class into the rib-bearing **thoracic** and the ribless **lumbar** series. Thus there can be progressively established, as we "climb the tree" of the vertebrates, a series of subdivisions as follows:

- presacral
  - cervical
  - dorsal
    - thoracic
    - lumbar
- sacral
- caudal

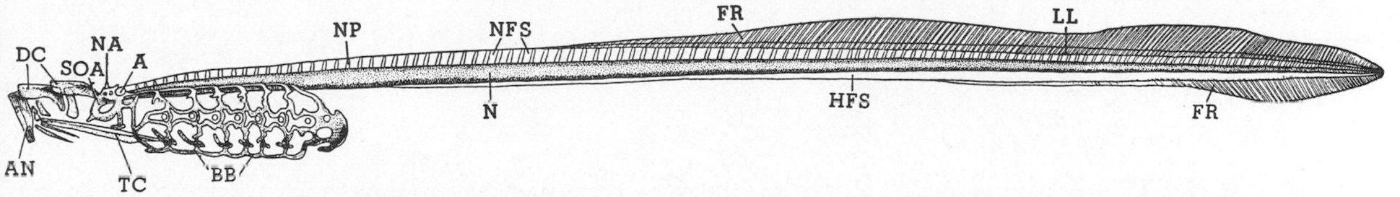

Figure 120. Skeleton of the lamprey, *Petromyzon*. *A*, Otic capsule; *AN*, ring cartilage surrounding mouth; *BB*, cartilages of branchial basket; *DC*, dorsal cartilages of mouth region; *FR*, dermal fin rays; *HFS*, fibrous sheath around dorsal aorta; *LL*, longitudinal ligament connecting tips of neural processes; *N*, notochord; *NA*, openings of nasal capsule; *NFS*, fibrous sheath of spinal cord; *NP*, neural processes; *SOA*, cartilaginous arch around orbit; *TC*, tongue cartilage. The openings in the branchial basket indicate the openings of the gill chambers. (From Dean.)

Numbers of vertebrae are highly variable in fish (Figs. 120–122), but primitive amphibians appear to have had 30 or so presacral vertebrae (of which about seven were cervicals), a single sacral and half a hundred or more caudal elements. Among modern amphibians the worm-like Gymnophiona may run up to 200 or more vertebrae, while on the other hand urodeles (Fig. 123) have a relatively short column, and typical frogs have but nine vertebrae plus a rodlike **urostyle** representing fused caudals. Primitive reptiles (Fig. 124) had about 27 presacrals, two sacrals in most cases, and a long tail; there is great variation within the class, with snakes having a greatly lengthened column. A curious condition is seen in the tail of *Sphenodon* and various lizards, where there is a "breaking point" in the middle of each caudal at which the tail may be shed (it later regenerates). In birds (Fig. 125) the cervical region is distinct and variable in length; the trunk vertebrae tend to be fused, and the sacral region has added to it posterior dorsals and anterior caudals to form an elongate **synsacrum.** The short bony tail of a bird terminates in a **pygostyle,** formed of fused vertebrae, to which the tail feathers are bound.

In mammals (Fig. 126) the cervical region contains almost uniformly seven vertebrae. The number of dorsal vertebrae is usually in the twenties and tends to be stable within many mammalian families or even orders, but the number of ribs—and hence the proportion of thoracics to lumbars—is variable. There are generally three to five sacrals. The mammalian tail is generally a slender and relatively short structure.

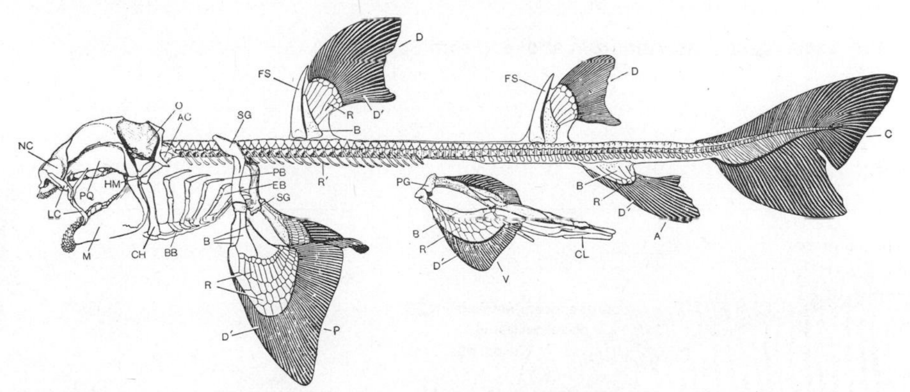

Figure 121. Skeleton of a shark *(Cestracion)*. *A*, Anal fin; *AC*, auditory capsule; *B*, basal elements of fin; *BB*, basibranchial; *C*, caudal fin; *CH*, ceratohyal; *CL*, claspers of male; *D*, dorsal fins; *D'*, dermal rays of fin; *EB*, epibranchial; *FS*, fin spines; *HM*, hyomandibular; *LC*, labial cartilages; *M*, mandible; *NC*, nasal capsule; *O*, orbit; *P*, pectoral fin; *PB*, pharyngobranchial; *PG*, pelvic girdle; *PQ*, palatoquadrate; *R*, radial elements of fin; *R'*, ribs; *SG*, pectoral girdle; *V*, pelvic (ventral) fin. (From Dean.)

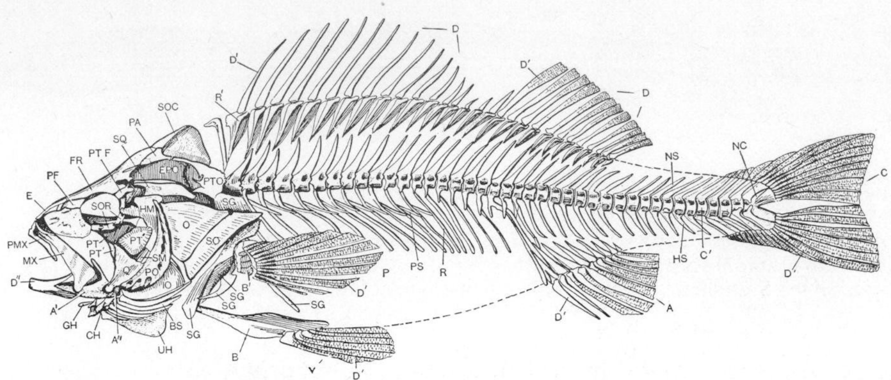

Figure 122. Skeleton of a teleost *(Perca).* The interpretation of certain skull elements differs from that used in Figure 172, but is one commonly used in works on teleosts. *A,* Anal fin; *A',* articular; *A'',* angular; *B,* pelvic girdle; *B',* skeleton of pectoral fin; *BS,* branchiostegal rays; *C,* caudal fin; *C',* centrum of vertebra; *CH,* ceratohyal; *D,* dorsal fins (anterior one supported by stout dermal spines); *D',* dermal rays of fins; *D'',* dentary; *E,* ethmoid; *EPO,* epiotic; *FR,* frontal; *GH,* hypobranchial (glossohyal); *HM,* hyomandibular; *HS,* hemal spine; *IO,* interopercular (part of opercular series); *MX,* maxilla; *NC,* tip of notochord in tail; *NS,* neural spine; *O,* opercular; *P,* pectoral fin; *PA,* parietal; *PF,* prefrontal; *PMX,* premaxilla; *PO,* preopercular (part of opercular series); *PS,* supporting processes of ribs; *PT, PT', PT'',* bones of palatal region; *PTF,* postfrontal; *PTO,* pterotic; *Q,* quadrate; *R,* ribs; *R',* supports of dorsal fins; *SG,* various portions of shoulder girdle; *SM,* symplectic (binding hyomandibular to skull); *SO,* subopercular; *SOC,* supraoccipital; *SOR,* suborbitals; *SQ,* squamosal; *UH,* urohyal (ventral element of branchial skeleton); *V,* pelvic fin. (From Dean.)

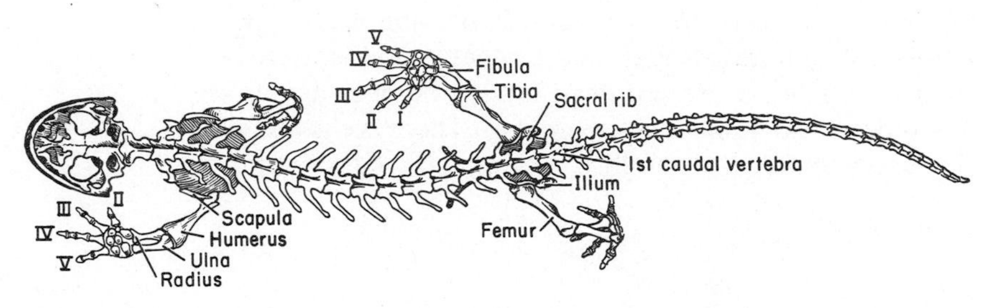

Figure 123. The skeleton of a salamander, as seen from above. (From Schaeffer.)

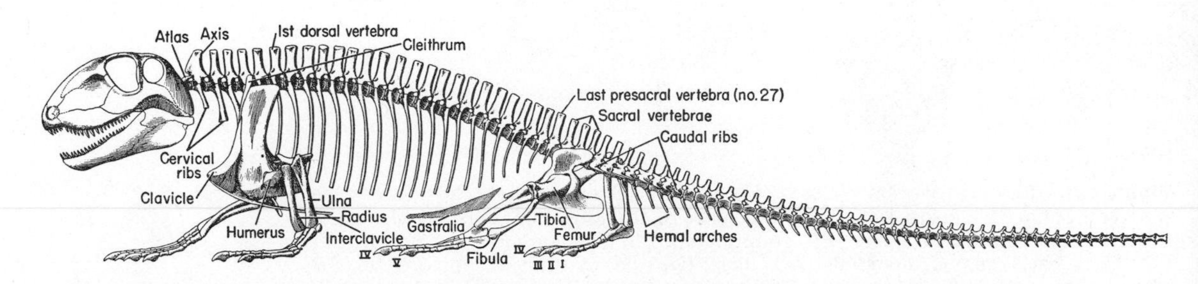

Figure 124. The skeleton of a generalized primitive reptile (the Permian pelycosaur *Haptodus*).

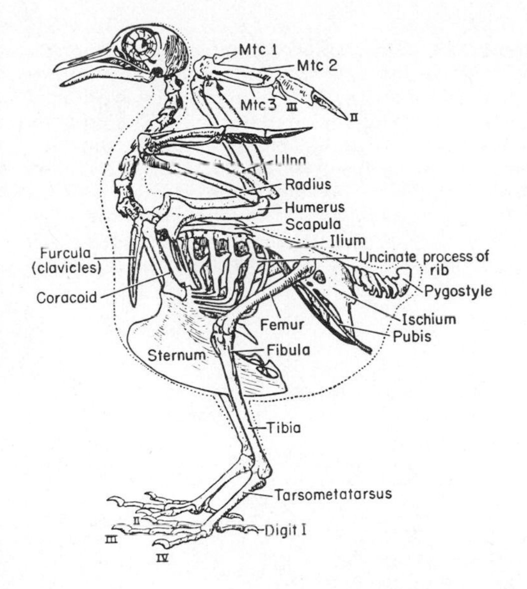

Figure 125. The skeleton of a bird (the pigeon). Mtc, Metacarpal; Roman numerals indicate digits. (After Heilmann.)

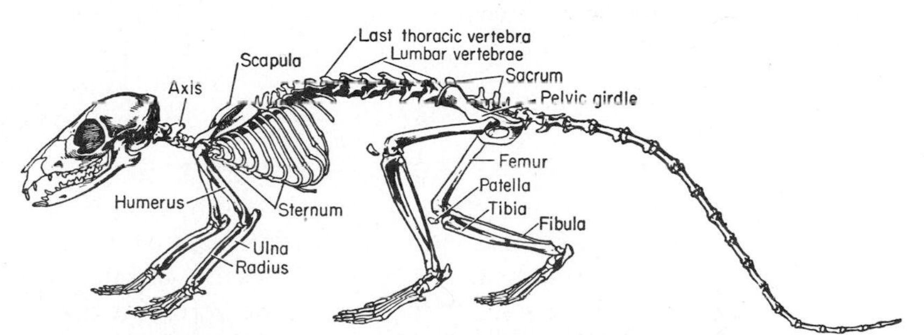

Figure 126. The skeleton of a generalized mammal, the tree-shrew *Tupaia*. (After Gregory.)

ATLAS-AXIS COMPLEX (Fig. 127). In fishes, head and trunk move as a unit, but in land vertebrates independent movement of the head is important, and usually the first two vertebrae—the **atlas** and **axis**—are specialized for this purpose. In typical amphibians the articular surface at the back of the skull—the condyle—is divided into a pair of rounded prominences, one at each side. The atlas correspondingly has a pair of sockets, and the head can swing up and down readily, although there is little facilitation of side-to-side movement. In most reptiles and birds the condyle remains single, but atlas and axis are modified to permit flexibility of movement. The neural arch and intercentrum of the atlas may form a ring on which the head may turn to some degree; and in mammals with two condyles the atlas centrum fuses with the axis as its **odontoid process,** which lies inside the ring of the atlas and aids in rotary head movements. The axis generally has a strong neural spine for attachment of ligaments supporting the head.

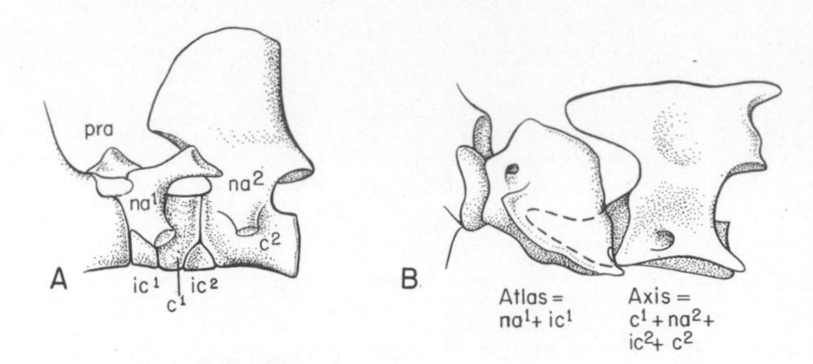

Figure 127. The atlas-axis complex. *A,* The occipital condyle and first two vertebrae in a primitive reptile *(Ophiacodon). B,* The same region in a typical mammal, showing fusion of elements. The proatlas in *A* is the neural arch of a "lost" vertebra of which the centrum has fused with the occiput. The broken line in *B* shows the position of the odontoid process (c^1), which runs forward inside the ring of the atlas. c^1, c^2, Centra of first and second vertebrae; ic^1, ic^2, intercentra; na^1, na^2, neural arches; *pra,* proatlas.

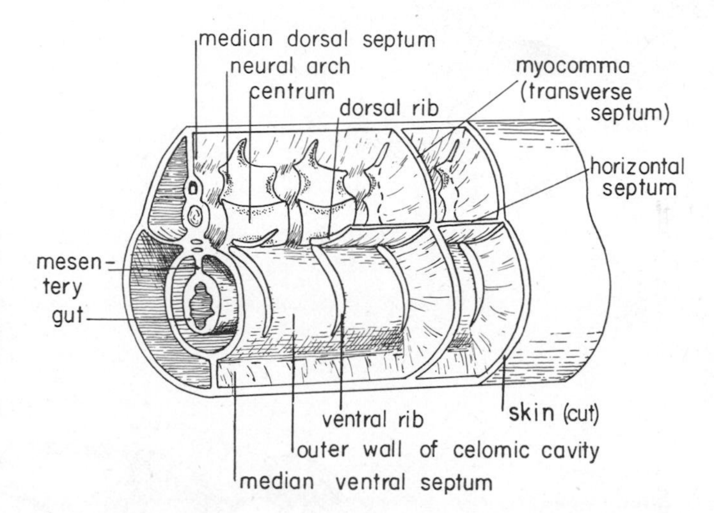

Figure 128. Diagram of a section of the trunk of a vertebrate to show the connective tissue system and the axial skeletal elements. A view from the left side, as if partially dissected out and the muscles removed from between the septa. Vertebral elements develop in the tissue sheath surrounding the spinal cord and notochord; ribs, dorsal or ventral, develop where the transverse septum intersects the horizontal septum or celomic wall. (After Goodrich.)

RIBS. The powerful segmental trunk muscles of fishes are their main locomotor organs, forcing the body ahead by alternate contraction and consequent undulation of trunk and tail. Their force is exerted upon the connective tissue septa—**myocommata**—between successive segments; ribs, formed at strategic points in these septa, connect with the vertebrae and render the muscular effort more effective. In most fishes each muscle segment is divided into dorsal and ventral parts by a longitudinal septum (Figs. 2, 128). A logical place for rib formation is at the intersection of this septum with successive myocommata; in some fishes **dorsal ribs** develop at this point, and the ribs of tetrapods appear to be of this type. A second position in which ribs may form is that where the myocommata reach the walls of the celomic cavity. Such ribs, common in fishes but absent in tetrapods, are termed **ventral ribs.** At the back end of the trunk in fishes (and many tetrapods) the ventral ribs of the two sides approach one another and a bit further back form V-shaped structures, the hemal arches noted above. Rib development is highly variable in fishes. Ribs are absent in cyclostomes and short in sharks; in many bony fishes only ventral ribs are present. The ribs develop as cartilages but, except in the sharklike fishes, are generally ossified, fully or in part.

In tetrapods the ribs were primitively borne, as in fishes, on every vertebra from neck to tail base. The cervical ribs in the more generalized lower tetrapods are relatively short, in relation to the development of a flexible neck; the thoracic ribs, in contrast, are the longest of the body and are generally bound to a median ventral structure, the sternum. In the more primitive tetrapods short lumbar ribs are present; following these are one or two sacral ribs, connecting the backbone with the pelvic girdle. Beyond the pelvis, short ribs, diminishing in size posteriorly, may be present in the base of the tail. In primitive tetrapods the ribs were double-

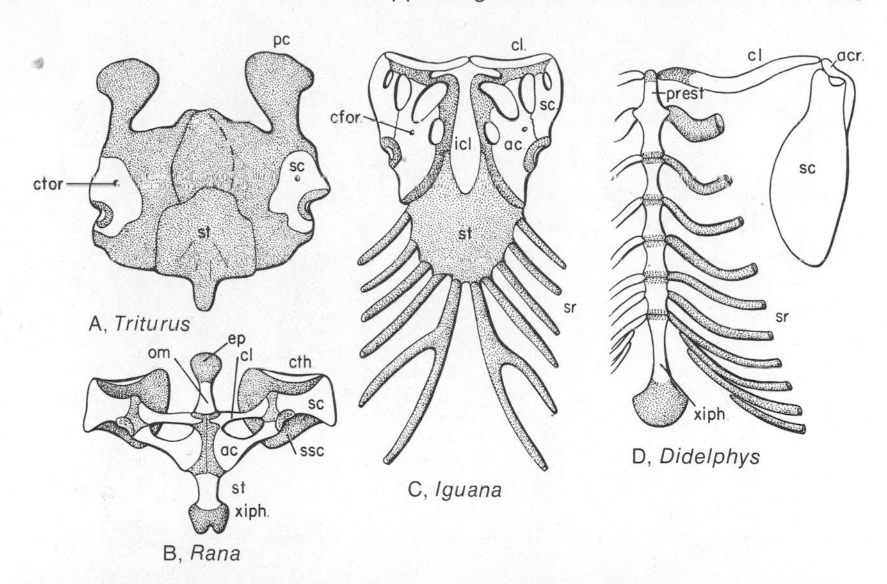

Figure 129. Ventral views of the shoulder girdle and sternal apparatus in various tetrapods. For lateral views, cf. Figures 137 and 138. *A*, Salamander; *B*, a frog; *C*, a lizard; *D*, a mammal (opossum). Anterior end at the top of the figures. In *A* and *C* the dorsally turned scapulae are invisible. In *A* the two coracoid cartilages overlap, as indicated by the broken line. *ac*, Anterior coracoid element; *acr*, acromion; *cfor*, coracoid foramen; *cl*, clavicle; *cth*, cleithrum; *ep*, episternum; *icl*, interclavicle; *om*, omosternum; *pc*, precoracoid region of coracoid plate; *prest*, presternum; *sc*, scapula (in salamanders this ossification extends down into the place of the absent coracoid); *sr*, sternal ribs; *ssc*, suprascapula; *st*, sternum; *xiph*, xiphisternum. Cartilage stippled. (*A*, *C* and *D* after Parker.)

headed, a **capitulum,** or head proper, attaching in early forms to the intercentrum, and the **tuberculum,** an accessory head, to the transverse process of the neural arch (Fig. 114). Posteriorly the two heads became approximated, the capitular attachment shifting upward and backward toward the transverse process. The lumbar ribs, the massive sacrals, and the caudals were generally immovably attached to the vertebra.

Many groups, however, have in many respects departed far from this primitive pattern. In the Recent amphibian orders ribs never reach the sternum, are much reduced in urodeles, and except for a single sacral are generally absent in anurans. In reptiles the rib is frequently single-headed and may attach to either centrum or neural arch. In turtles the ribs are reduced in number; eight of them are firmly fused to the carapace. In snakes the ribs are highly developed and are important in locomotion. In birds the cervical ribs are fused to the vertebrae; free ribs are confined to the short thoracic region, and the pelvic girdle is supported by a long sacral series. In mammals cervical ribs appear in the embryo in the neck region, but in the adult they are fused to the vertebrae; free ribs are absent from both lumbar and caudal regions.

BRAINCASE. The braincase, of cartilage or replacement bone, forms the anterior end of the axial skeleton, here greatly modified in relation to the expanded brain and special sense organs. In most vertebrates the braincase is fused with dermal and visceral skeletal elements to form a definitive skull; in sharklike fishes and cyclostomes, however, it is a separate entity because of the absence of dermal bones—presumably a secondary condition. The cartilaginous braincase of sharks (Fig. 130) shows a structure which, with variations, is repeated in many other lower vertebrates (although in them usually ossified, at least in part); in modified form it is represented in the skull of even the highest of vertebrate groups.

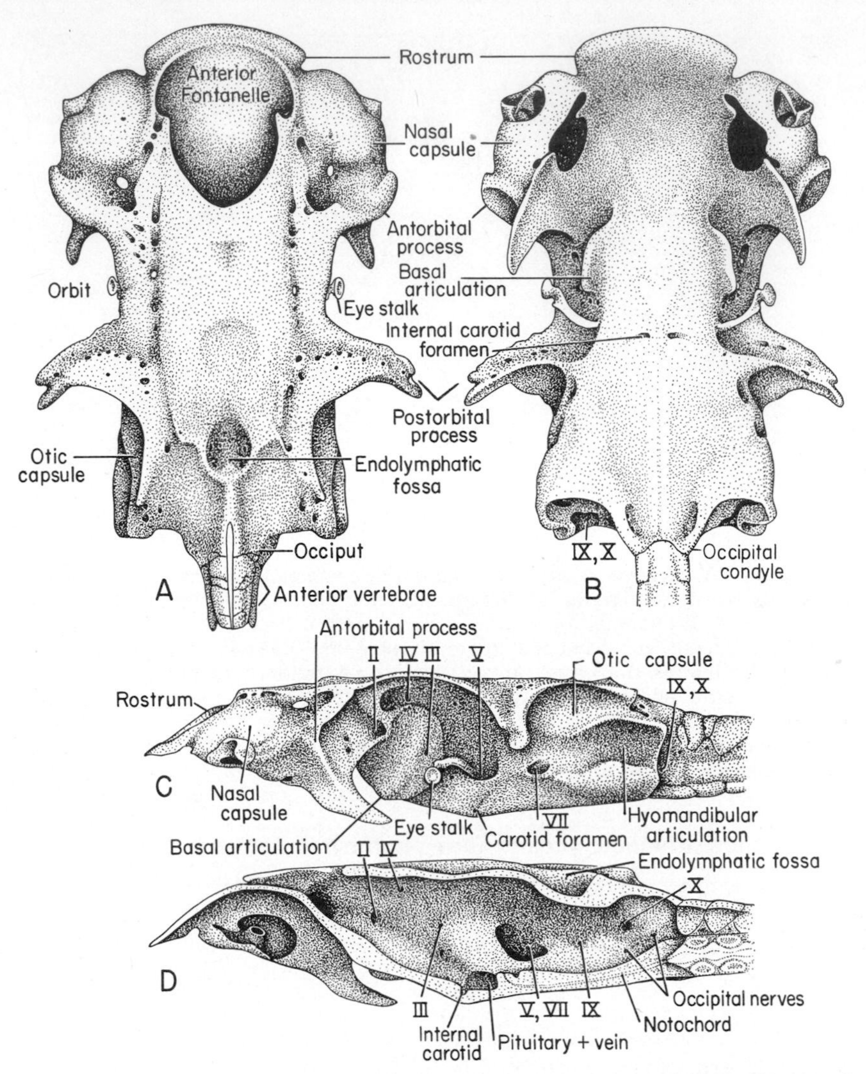

Figure 130. The braincase of the shark *Chlamydoselache; A,* dorsal, *B,* ventral and *C,* lateral views, and *D,* sagittal section. Nerve exits in Roman numerals. (After Allis.)

The most posterior part of the shark braincase is relatively narrow. There is a median opening, the **foramen magnum,** through which the spinal cord passes, and below it a circular **occipital condyle,** which abuts against the centrum of the first vertebra. Anterior to this the braincase expands; there is here incorporated on either side an **otic capsule,** containing the sacs and canals of the internal ear. Farther forward the width decreases to allow the formation of orbital cavities for the eyeballs; here, internally, a median ventral depression lodges the pituitary. Toward the front the braincase again expands to terminate in a rostrum, on either side of which is a **nasal capsule,** containing the olfactory organ. Numerous openings—**foramina**—are present in the braincase for cranial nerves and blood vessels, as may be seen in Figure 130; dorsally the endolymphatic ducts from the internal ear (cf. Chap. 15) open to the surface in elasmobranchs (but not in other groups), and ventrally there enter the carotid arteries which supply blood to the brain. A specialized element of the gill skeleton—the hyomandibular—articulates loosely with the outer side of the otic region and props the jaws on the braincase. The

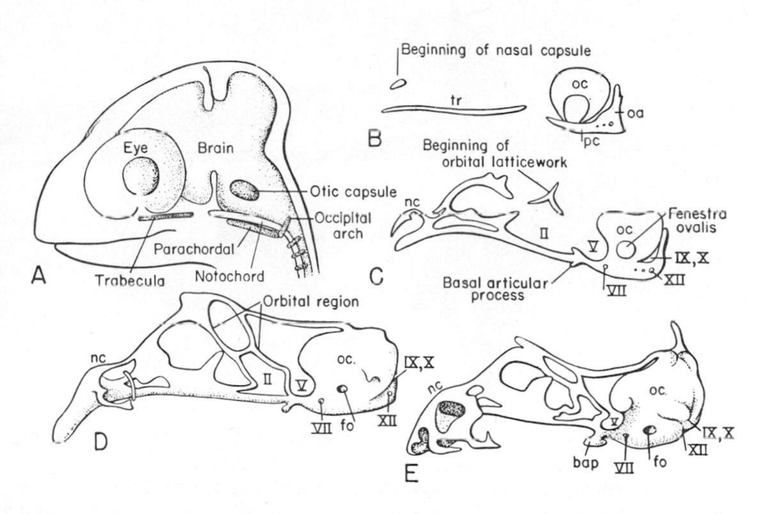

Figure 131. Stages in the embryonic development of the braincase of a lizard. In *A* the outlines of the head, brain, and notochord are given for orientation. The stage shown in *A* is the earliest, that in *E* the latest. The main elements of the braincase structure are appearing—trabeculae, parachordals, otic capsule, and occipital arches (the nasal capsule appears later); further development consists in great measure of the growth and fusion of these elements. In the lizard the orbital region grows as a complicated latticework, rather than a plate. Nerve positions are indicated by Roman numerals. *bap,* Basal articular process with upper jaw cartilage; *fo,* fenestra ovalis (for stapes); *nc,* nasal capsule; *oa,* occipital arch; *oc,* otic capsule; *pc,* parachordal; *tr,* trabecula. (After DeBeer.)

cartilages which form the jaws also articulate with the braincase (Fig. 165). The common jaw connection in modern sharks is one in which the upper jaws are loosely joined anteriorly with the under surface of the braincase; in primitive sharks there is an additional articulation posterior to the orbit. In higher fishes and primitive land vertebrates this latter connection tends to be reduced or absent, and there is, instead, a strong **basal articulation** of the upper jaw and palatal structures with the base of the braincase in the orbital region (Figs. 131 *E*; 166 *B, E*; 167 *A, C, D*; 175*A*).

In the embryo vertebrate (Figs. 131, 132) both the brain and the notochord, which extends forward beneath it to the pituitary region, are already far advanced in development before skeletal structures appear. Basic cartilage elements of the braincase are a pair of **parachordals** which lie on either side of the notochord beneath the brain stem and, farther forward, the **trabeculae**—paired in most vertebrates but a single bar in mammals. Above the parachordals on either side, the otic capsule develops as a shell of cartilage around the internal ear. Behind the otic capsule one or several modified vertebrae form **occipital arches;** in front, between the orbits there may develop a pair of **orbital plates** or a lattice of cartilage bars as a substitute for them. At the front end the nasal capsules presently make their appearance. In later stages all these structures fuse with one another to form the adult braincase, but leaving between the original units spaces for the various nerves and vessels which enter or leave the braincase. In cyclostomes the braincase (Fig. 162 *A*) is highly aberrant. The short, high chimaera braincase (Fig. 133) has as a notable specialization the fusion with it of the upper jaw cartilages; the same peculiarity is repeated in the lungfishes and, in modified fashion, various tetrapods.

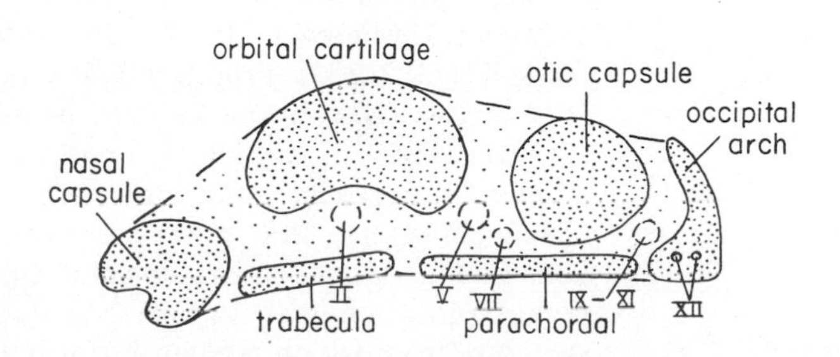

Figure 132. Diagram to illustrate the main embryonic components in the formation of a braincase. Trabeculae and parachordals are the main ventral elements to which an occipital arch (or arches) is added posteriorly. Usually later in development are more dorsal elements: otic capsule, nasal capsule and orbital plate (the last, as in Fig. 131, often develops merely as a latticework). These primary elements are later bound together by a further growth of cartilage, leaving, however, gaps for nerves and blood vessels.

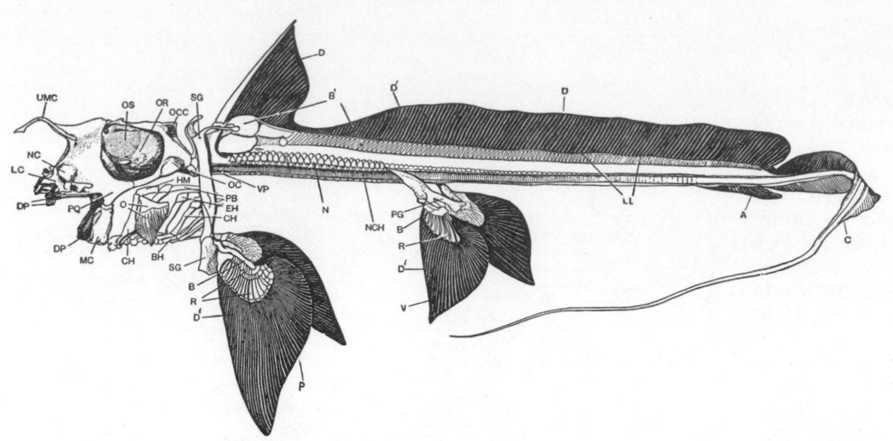

Figure 133. The skeleton of a female chimaera. *A,* Anal fin; *B,* fin basals; *B',* dorsal fin supports; *BH,* basibranchial; *C,* caudal fin; *CH,* ceratohyal, ceratobranchial; *D,* dorsal fin, with spine; *D',* dermal fin rays; *DP,* dental plates; *EH,* epibranchial; *HM,* hyomandibular; *LC,* cartilages in nasal region of "lips"; *LL,* ligament connecting fin supports; *MC,* lower jaw (Meckel's cartilage); *N,* cartilages of neural arch regions; *NC,* nasal capsule; *NCH,* notochord, with ring calcifications of sheath; *O,* cartilages in operculum; *OC,* occipital condyle; *OCC,* crest on occiput; *OR,* orbit; *OS,* interorbital septum; *P,* pectoral fin; *PB,* pharyngobranchial; *PG,* pelvic girdle; *PQ,* palatoquadrate, fused with skull; *R,* radial fin supports; *SG,* shoulder girdle; *UMC,* upper median cartilage of snout (this is not the clasper of the male); *V,* pelvic fin; *VP,* plate formed of fused anterior vertebrae. (From Dean.)

MEDIAN FINS. In primitive aquatic vertebrates the body shape is generally a fusiform one, but with a side-to-side flattening posteriorly. This flattening is associated with the nature of the propulsive force. Forward motion is produced by side-to-side movements of the body effected by the axial muscles (Fig. 134 *A*). Alternating curves, successively produced on opposite sides, travel back along trunk and tail, pushing the body forward as a result of their backward thrust, with accentuation at the expanded **caudal fin** at the posterior end of the body.

Without the aid, in stabilization and steering, of fins other than the caudal, these propulsive movements would tend to be poorly regulated (as are those of a tadpole). Further median fins in addition to the caudal are, above, **dorsal fins**—usually one or two of them—and an **anal fin** below, posterior to the anus. The skeleton of the median fins is formed in the dorsal median septum, in which the neural spines are formed, and in the tail placed in the ventral septum, in which the hemal arches occur. In the caudal fin neural and hemal arches may themselves contribute directly to fin support (Fig. 135), and in teleosts the terminal hemal arches are expanded structures termed the **hypurals.** In the case of dorsals and anals, however, such direct support is not present; the fins are stiffened by **radials** or **radial pterygiophores** (sometimes in two rows), which may articulate at their bases with neural or hemal arches but are often separated by a gap from these structures (Figs. 121, 122, 133). Primitively, it appears, the radials extended well into the fin, as is the case today in sharks; in actinopterygians, however, they extend hardly at all into the free fin, which is supported mainly by lepidotrichia (cf. p. 138). In a number of sharks and in chimaeras the median fins carry, anteriorly, spines which act as cutwaters, and spines were present on dorsal and anal fins in the fossil acanthodians (Fig. 27). In some of the lowly ostracoderms (Fig. 21) we find rows of dorsal spines acting as stabilizers, and it is possible that such spines were the basic structure from which median fins were historically developed.

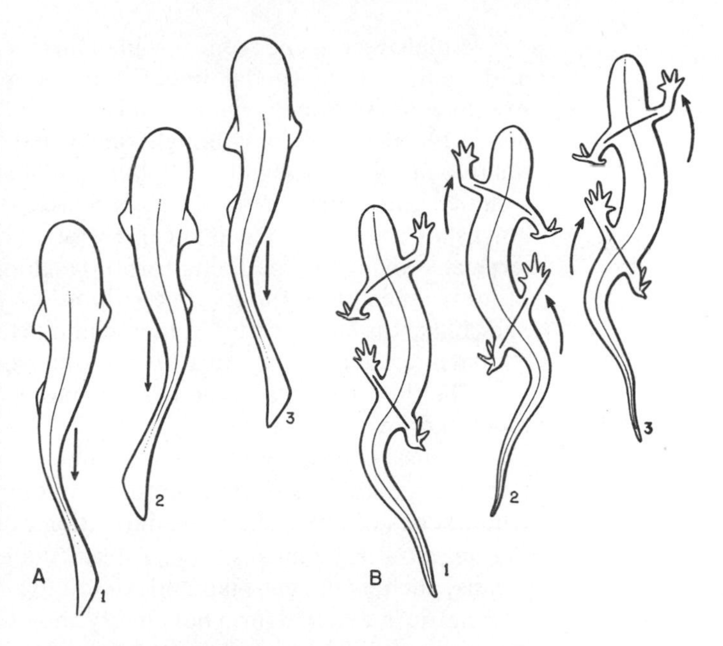

Figure 134. *A,* Dorsal views of a fish swimming, to show the essential method of progression by the backward thrust of the body on the water, resulting from successive waves of curvature traveling backward along trunk and tail. The curve giving the thrust (indicated by arrow) in *1* has passed down the tail in *2* and is replaced by a succeeding wave of curvature; the thrust of this wave carries the fish forward to the positon seen in *3,* and so on. *B,* Dorsal views of locomotion in a salamander. Although limbs are present, much of the forward progress is still accomplished by throwing the body into successive waves of curvature. In position *1,* the right front and left hind feet are kept on the ground, the opposite feet raised; a swing of the body in *2* carries the free feet forward, as indicated by arrows. If these feet are now planted and the other two raised, a following reversed swing of the body will carry them forward another step, as seen in *3.*

Of caudal fins, three major types are to be found in fishes. The **heterocercal** type of tail is that familiar in sharks (Figs. 23; 24 *A, C;* 121; 135 *A*); the tip of the body turns upward distally, and the greater part of the fin is developed below it. This type of tail is also found in many placoderms (Fig. 22), on the one hand, and in all the more ancient bony fishes on the other; it is retained today in Osteichthyes, however, only in sturgeons and paddlefishes (Figs. 29 *A;* 30 *A;* 32 *A;* 33). The dominance of this tail fin structure in so many primitive and ancient fishes strongly suggests that it is the ancestral vertebrate type. Its only rival for antiquity is a **reversed heterocercal** type, seen in some ostracoderms (Figs. 18 *B, C;* 21 *A*) and retained in larval lampreys.

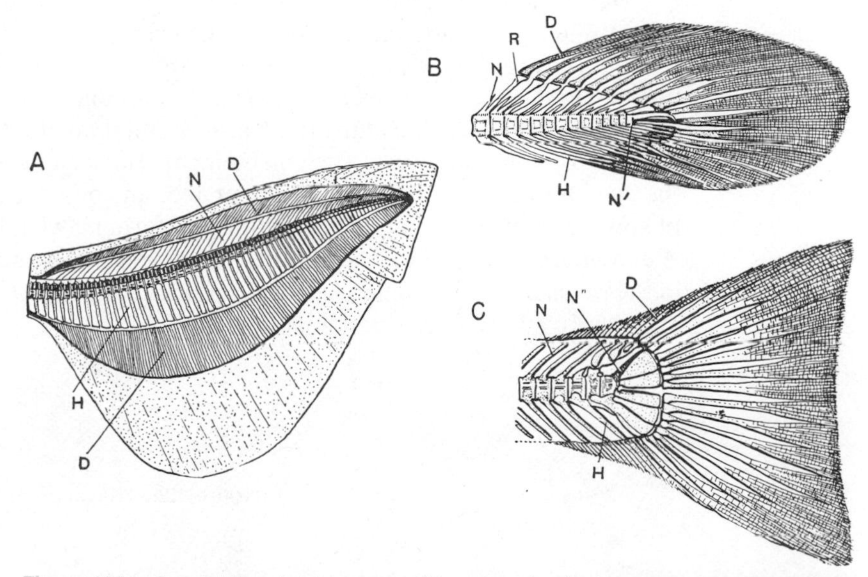

Figure 135. Caudal fins. *A,* Heterocercal type seen in sharks, sturgeons, paddlefish; *B,* diphycercal type, as seen in *Polypterus; C,* homocercal type of teleosts. Abbreviations: *D,* Dermal fin rays; *H,* hemal spines; *N,* neural arches; *N',* tip of notochord; *R,* fin radials. In *C,* enlarged elements beyond *H* are hypural bones. (From Dean.)

A **diphycercal** type is one in which the body axis runs straight outward to the tip of the tail, with the caudal fin developed symmetrically above and below it; good examples are seen in *Polypterus* and the living lungfishes and coelacanth (Figs. 29*B*, 30*B*, 32 *B*, 135 *B*). At first thought one would expect that such a structure would be the true primitive type of tail fin, but in almost every case it can be shown that this symmetric fin is derived from a heterocercal type. Both *Polypterus* and the modern lungfishes come from Paleozoic ancestors with a heterocercal tail; many crossopterygians (including the living form) developed symmetric tail fins, but the most primitive members of that group exhibit the heterocercal structure. Skates, rays, and chimaeras have slender, rather whiplike tails of symmetric build (Figs. 24 *D*, 25 *B*), but intermediate stages are known which connect them with heterocercal types.

The third major tail type is the **homocercal fin** characteristic of the dominant modern fishes, the teleosts (Figs. 36, 37, 122, 135 *C*). This is superficially symmetric, but dissection shows that the backbone tilts strongly upward at the tip; the fin expanse is purely a ventral structure. Among fossil and Recent actinopterygians a whole series of forms show the derivation of this fin from the heterocercal type of the ancestral ray-finned fishes. Intermediate forms are known among the holosteans, such as the gar pike and *Amia* (Fig. 34). In these fishes the tail is nearly symmetric in external form but clearly shows in internal structure that it has arisen by an abbreviation of the uptilted axis of a heterocercal tail.

In tetrapods, except for a few very primitive fossil amphibians, the original median fin structures of fishes have been completely abandoned; even in the tail of a tadpole or salamander, used like that of a fish for swimming, the skeletal supports found in fish ancestors are absent. Many tetrapods have returned to an aquatic existence but often—as in seals, turtles, and the extinct plesiosaurs—limbs rather than a tail act as propulsive organs. Among mammals the cetaceans and sea-cows have a tail "fin," but this consists of horizontally expanded flukes and is not closely comparable to a fish caudal fin. The extinct ichthyosaurs have come the closest to the redevelopment of a fish type of tail, expanded vertically and supported by the axial skeleton; their tail is of the reversed heterocercal type.

HETEROTOPIC ELEMENTS. Embryonic connective tissues are the source of the normal skeletal tissues, and it is hence not surprising that here and there among the various vertebrate types there develop upon occasion cartilages or bones in situations where mere connective tissue is normally present. Small bones—**sesamoids**—may develop along the course of tendons and the mammalian kneecap is an overgrown example of such an element. Bones may develop, for example, in the eyelids of crocodiles, in the hearts of deer and bovids, in the expanded muzzles of some mammals, and in the hand of moles forming what looks like an extra digit. A common example of such a heterotopic bone is the **baculum** (os penis), developed in the penis of many mammals (Fig. 136), including insectivores, bats, carnivores, and nearly all primates (man is an exception).

Figure 136. Baculum of an otter.

APPENDICULAR SKELETON

The skeleton of the girdles and limbs belongs (with the exception of dermal shoulder elements) to the somatic system of internal skeletal structures; their history, however, contrasts with that of the axial elements. Normally two pairs of appendages are present, as paired fins in fishes and as the limbs of tetrapods: the **pectoral appendages,** situated in fishes just behind the gills and in land animals in an equivalent position between neck and trunk, and the **pelvic appendages** placed at the back end of the trunk, just anterior to anus or cloaca.

ORIGIN OF PAIRED FINS. Paired appendages were not ancestral vertebrate structures; they are absent in cyclostomes, are only rarely seen in the early jawless ostracoderms and are poorly developed in many of the ancient placoderms. The mode of origin of paired fins has been much debated. One early theory assumed that they developed from modified gills—that the limb girdles are transformed gill bars and that the limbs themselves developed from gill flaps such as lie outside the gill surfaces in a modern shark. A host of embryologic and morphologic facts show that this idea is purely fanciful; but a reminiscence of this theory remains in the name archipterygium (p. 161) given to a leaf-shaped fin type (Figs. 143 *H*, 144 *D*), which was supposed, under this theory, to be primitive (but apparently is not).

In reaction against this theory arose a rival **finfold theory** of the origin of paired fins. Its advocates pointed out that the paired fins are basically similar to the median fins in structure and hence may have arisen in similar fashion. In either case the fin has a centrally placed set of skeletal structures with a layer of muscle on either side. The median fins appear to have arisen as stabilizing organs in the midline and the paired fins may have been, originally, laterally projecting stabilizing flanges; only later, it would appear, did they become flexible steering organs, and in few forms below the tetrapod level have they any active role in propulsion. In ostracoderms (Fig. 18 *A*) we find some early essays toward paired fin development, such as rows of spines extending out from the base of either flank or as projecting flaps comparable to pectoral fins. In the extinct placoderms and acanthodians paired fins were still "experimental models" in which spines were generally prominent (Figs. 27 *A*, 143). These placoderm structures generally differed as much from the more orthodox fins of later date as did many early (and unsuccessful) flying machines from the modern airplanes. There was even variation in the number of these appendages, with as many as seven pairs in one acanthodian. It is not until we reach the more "modernized" fishes among the sharks and Osteichthyes that there developed the typical two pairs of flexible fish fins.

PECTORAL GIRDLE—DERMAL ELEMENTS (Figs. 137, 138). Each limb has, in addition to the free appendage, an associated girdle lying within the trunk and affording support to the limb skeleton as well as an area of origin for limb muscles. Each girdle consists primarily of endoskeletal cartilage or of bone replacing it. The shoulder, however, lies in the anterior region of the body in which, in early fishes, bony plates were present rather than the small scales covering most of the trunk and tail. Certain of these plates became associated with the limb supports as a **dermal shoulder girdle.** This is present in all jawed fishes, except for the sharklike forms in which all bone has been lost.

Most placoderms, we have noted, had a bony armor surrounding the "chest," and the lateral portions of this armor form essentially a dermal shoulder covering (Figs. 22, 137 *A*), although the elements present are not comparable with those of higher types. In all primitive bony fishes a characteristic pattern of dermal elements is present. Ventrally on either side is a small **clavicle,** homologous with the

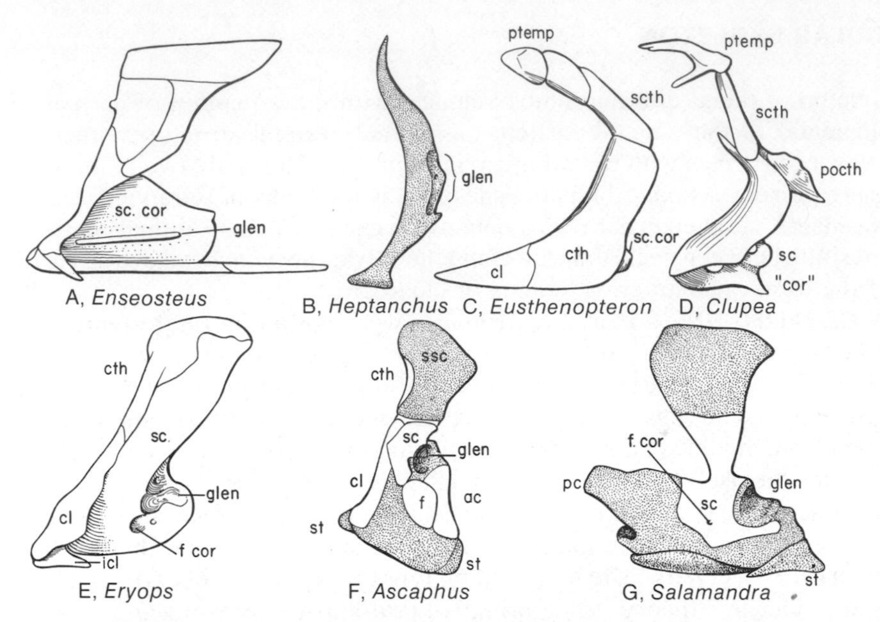

Figure 137. The shoulder girdle in fishes and amphibians. *A,* Devonian placoderm; *B,* a shark; *C,* a Devonian crossopterygian; *D,* a teleost (herring); *E,* a Paleozoic primitive amphibian; *F,* a frog; *G,* a salamander. Cartilage stippled. In all except *B* and *G* a dermal girdle is present; in fishes *A, C, D,* this is the most prominent part of the girdle, including all parts except that labeled scapula and coracoid, and in the placoderm (A) the dermal girdle is the lateral part of an extensive thoracic armor. In amphibians the dermal girdle is reduced or absent. Except in the shark the endoskeletal girdle is relatively small in fishes and partially hidden beneath the dermal elements. In amphibians the endoskeletal girdle is expanded, but generally ossifies from a single center, comparable to the scapula of amniotes; the frog, however, has a coracoid element. Much of the endoskeletal girdle is cartilaginous in living amphibians. *ac,* Anterior coracoid element; *cl,* clavicle; *"cor",* coracoid of teleost (homology with that of land forms doubtful); *cth,* cleithrum; *f,* foramen in coracoid plate of frog; *f. cor,* coracoid foramen for nerve and blood vessels; *glen,* glenoid cavity, point of fin attachment in fishes; *icl,* interclavicle; *pc,* precoracoid process of coracoid plate; *pocth,* postcleithrum; *ptemp,* posttemporal; *sc,* scapula; *sc. cor,* single scapulocoracoid ossification of fishes; *scth,* supracleithrum, *ssc,* suprascapula; *st,* sternum. Cartilage stippled. Anterior is to the left. (*A* after Stensiö, *C* after Jarvik; *D* and *F* after Parker.)

familiar "collar bone"; above it in fishes is a much larger element, the **cleithrum,** and still farther dorsally one or more additional elements extend upward and forward above the gill chamber to attach to the posterior margin of the skull. This type of girdle is characteristic of all crossopterygians and lungfishes. It was also present in the older ray-finned fishes and persists today in the more primitive living members of that group. Advanced actinopterygians, however, lost the clavicle, leaving the cleithrum as the sole external covering for the endochondral girdle beneath.

In primitive tetrapods both clavicle and cleithrum persisted; however the connection with the head was lost (allowing greater freedom of movement), and the bones were no longer broad plates but relatively narrow structures attached to the front margin of the internal girdle for most of its height. There was, however, a new development in the presence of an **interclavicle,** a median ventral plate of bone to which the expanded lower ends of the clavicles attached. In later land vertebrates the dermal elements have had a varied history, as may be seen from Figure 138. On the whole, the story has been one of reduction. The cleithrum is lost in virtually all tetrapods, and today the only relict of this once important bone is a sliver which may be found at the upper front margin of the anuran shoulder girdle. Urodeles, gymnophionans, snakes, and many mammals have lost the entire girdle. The interclavicle persists in many reptiles and in the primitive egg-laying mammals but is otherwise absent. The clavicle has been more persistent; it is found in anurans, lizards, and *Sphenodon* and in a large proportion of the mammals. The fused clavicles form the **furcula,** the "wishbone," of birds. In turtles the clavicles and interclavicle, as we have noted, have been incorporated in the shell.

ENDOSKELETAL SHOULDER GIRDLE (Figs. 137, 138). Functionally the endoskeletal girdle is more important than the dermal component, since it always bears the limb articulation and offers an area of attachment for limb muscles. It is persistently cartilaginous in the Chondrichthyes, but normally is partially or completely ossified in other fishes and tetrapods. The ossifications present in fishes are somewhat variable and need not concern us here, but the general construction of the endochondral girdle seen in bony fishes is comparable to that in tetrapods. Centrally situated on either side is a socket, or series of sockets, for articulation with the skeleton of the appendage; when—as in some fishes and all tetrapods—there is but a single element articulating here, this socket is termed the **glenoid fossa.** Above

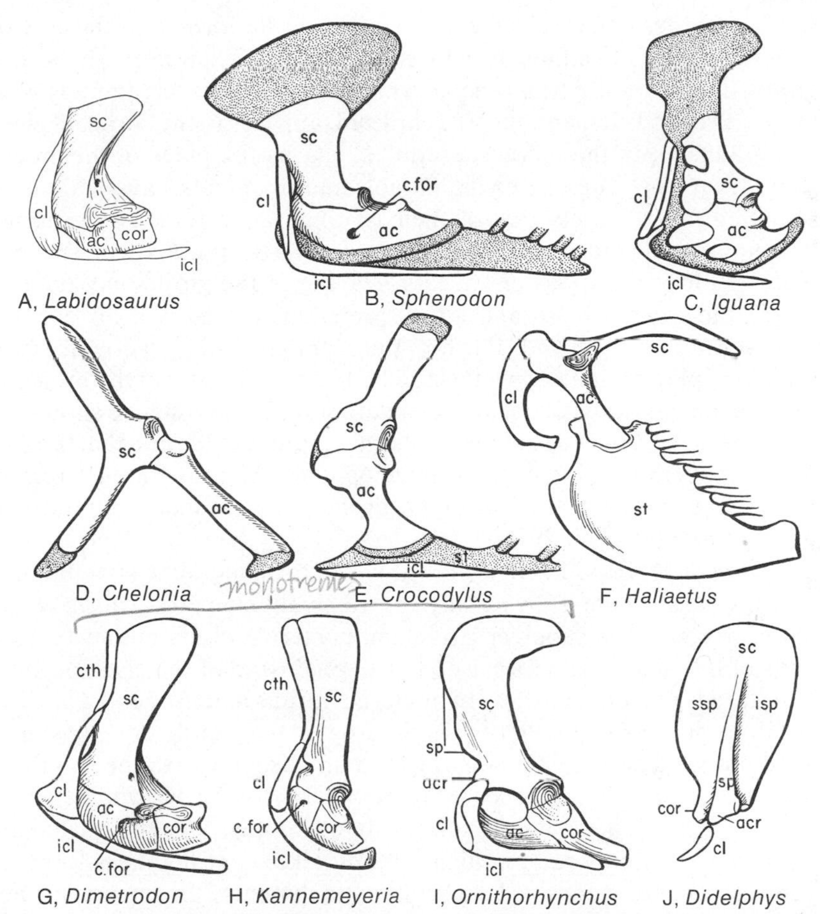

Figure 138. The shoulder girdle in reptiles, birds, and mammals. *A,* "Stem reptile" (cotylosaur); *B, Sphenodon; C,* a lizard; *D,* a turtle; *E,* a crocodile; *F,* a bird; *G,* a pelycosaur (primitive mammal-like reptile); *H,* a therapsid; *I,* a monotreme (duckbill); *J,* opossum. In *B, E,* and *F* the sternum is shown. In most reptiles and birds only one (anterior) coracoid element is present; in mammal-like forms the true coracoid appears and persists, despite the loss of the coracoid plate area. The borders of the scapula and coracoid are often cartilaginous in reptiles. In lizards, scapula and coracoid are often fenestrated at points of muscular origins. The cleithrum has vanished in all living amniotes, but persisted long in mammal-like forms *(G, H);* its position is represented by the scapular spine, which lies at the front edge of the scapula in monotremes, but back of the new supraspinous fossa in higher mammals. *ac,* Anterior coracoid bone of reptiles, birds, monotremes; *acr,* acromion; *c. for,* coracoid foramen; *cl,* clavicle; *cor,* true coracoid; *cth,* cleithrum; *icl,* interclavicle; *isp,* infraspinous fossa; *sc,* scapula; *sp,* spine of scapula; *ssp,* supraspinous fossa; *st,* sternum. (*A* after Romer; *C* and *I* partly after Parker; *H* after Pearson.)

this point is a sheet of bone or cartilage, more or less covered by the dermal girdle. which may be termed the **scapular blade;** below the fin articulation, and deep to the plane of the lower part of the dermal skeleton, is the **coracoid plate.**

In primitive tetrapods this bony fish structure is essentially retained. But whereas, we have noted, the dermal elements are here reduced, the endochondral girdle is greatly expanded—this in relation to the increased size of the tetrapod limb and its need for stronger support and for increased areas of muscle attachment. In primitive amphibians the entire endochondral girdle ossifies as a single element, which comparative studies show is the **scapula** of later types. In anurans and reptiles a second, ventral ossification appears, restricting the scapular ossification to the blade above the glenoid cavity. This lower element in frogs and reptiles is frequently called the coracoid; but, as will be seen shortly, this bone is not the same as the true coracoid of mammals and is better termed **anterior coracoid** or **procoracoid.**

In reptiles and birds an endochondral girdle formed of these two elements is generally present, although with considerable variation in shape from group to group. In the fossil forms leading to mammals, however, there is a new development. A second element, the true **coracoid,** appears in the coracoid plate at the back and gradually, in the therapsid reptiles, usurps the place of the anterior coracoid (Fig. 138 *G–I*). The egg-laying monotreme mammals have a shoulder girdle resembling fairly closely that of their reptilian ancestors. In the transition to the marsupials and placentals, however, there is a striking change. The entire coracoid plate of lower tetrapods disappears, leaving of the girdle only the scapular blade above the glenoid fossa and a tiny projecting "crow's beak" at its lower edge representing the coracoid. Further, the scapular blade is a double structure, with a **scapular spine** running down the middle to end ventrally at the projecting **acromion** for attachment of the clavicle. Consideration of the evidence shows that this spine represents the front edge of the ancestral scapular blade and that the surface in front of (or above) the spine is a new development. The functional "reason" for these major changes in construction of the girdle is to be found in marked changes in limb musculature (cf. Fig. 207).

STERNUM (Figs. 129 *B, E, F;* 138). In the generalized condition seen in many reptiles the sternum is a ventrally placed shield-shaped cartilage which articulates anteriorly with the shoulder girdle and posterolaterally connects with the ventral ends of thoracic ribs to form a complete enclosure of the chest or thorax. No such structure is present in fishes. In modern anurans and urodeles a sternum is present but does not connect with ribs; in the frogs it frequently develops in rodlike form. With limb and girdle reduction, the sternum disappears in the apodous amphibians and snakes; in turtles its absence is associated with the development of the plastron.

In birds the sternum is enormously developed for the attachment of massive chest muscles, important in flight, and in all except flightless forms bears a huge ventral keel for additional muscular attachments. In mammals the sternum is typically an elongate, jointed rod with which the ribs are attached at the "nodes."

Although the sternum has been generally considered part of the axial skeleton because of its close association with the ribs, its embryologic development shows that it arises independently of the ribs. It is, thus, to be regarded as an "appendix" to the shoulder girdle.

PELVIC GIRDLE (Fig. 139–142). In fishes the pelvic girdle is of very modest size; essentially each half of the girdle is a wedge-shaped element, frequently cartilaginous, which lies in the connective tissues of the fish's belly. It is without attachment to the axial skeleton; the girdles of the two sides, however, usually meet one another medially in a **pelvic symphysis.**

In tetrapods, with powerful pelvic limbs, the pelvic girdle increases greatly in size; further, it is imperative that the girdle gain a connection with the axial skeleton to anchor the limbs. The ventral portion of the girdle expands into a large plate, affording a broad area for the attachment of limb muscles; this plate includes two ossifications, the **pubis** anteriorly, the **ischium** posteriorly. A common landmark is the **obturator foramen** piercing the pubis and transmitting a nerve to part of the limb muscles. At the summit of this ventral plate is a socket, the **acetabulum,** into which fits the head of the thigh bone. Above the acetabulum rises a portion of the girdle not known in fishes, which extends upward to gain a firm connection with the backbone through the development of sacral ribs. It usually expands in blade-like fashion to afford an area of dorsal attachment for limb muscles. This dorsal portion of the girdle is formed by a third element, the **ilium.**

Except in cases in which limbs are reduced or lost, the essential structure of the pelvic girdle persists in most tetrapods. In modern amphibians the pubis fails to ossify, and in the anurans the ilium becomes an elongate rod in relation to the shortening of the backbone of these forms. Among reptiles marked modifications are present in most cases in the ventral plate of the girdle. In *Sphenodon,* lizards, and turtles a large opening, frequently termed a **thyroid** (i.e., shield-shaped) **fenestra** develops in this plate for the better functioning of a large limb muscle originating from this area. In the crocodiles and many dinosaurs (Fig. 141 *C, D*) the plate is modified in another fashion. In primitive members of the archosaurs (the major group to which these reptiles belong) there was a strong trend toward a bipedal gait, and it appears to have been advantageous, for better working of the limbs, that muscles attach to the anterior or posterior ends of the plate rather than the middle. In consequence, the ends of pubis and ischium are elongate and turn downward; the middle part of the plate is narrow, giving the girdle a triradiate structure. In one group of dinosaurs (the so-called birdlike forms) and in the bird descendants of the archosaurs there is a further modification (Fig. 141 *E, F*); the pubis has swung back parallel to the ischium. In the dinosaurs with this type of pelvis there has, further, developed a new anterior prong of the pubis which helps support the abdomen; this

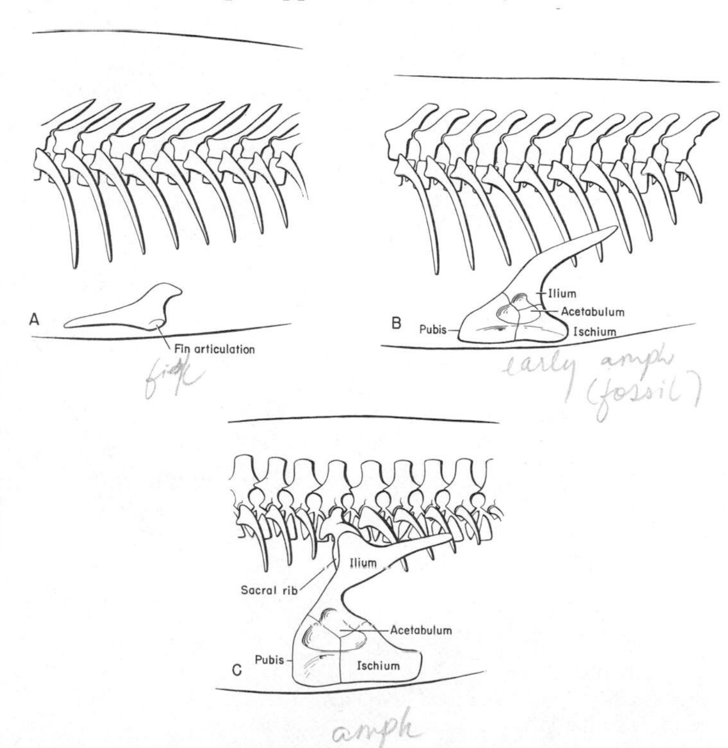

Figure 139. Diagrams to show the development of the pelvic girdle and sacrum in the evolution of amphibians from fishes. *A,* Left lateral view of the pelvic region of a fish, with the vertebral column and ribs above, the small pelvic girdle placed ventrally. *B,* Primitive tetrapod stage, found in some early fossil amphibians. The girdle has expanded, with the three typical bony elements. The ilium extends upward, but was presumably connected with the column only through ligaments binding it to the neighboring ribs. *C,* The girdle has grown further, and the ilium is firmly attached to an enlarged sacral rib. Anterior is to the left.

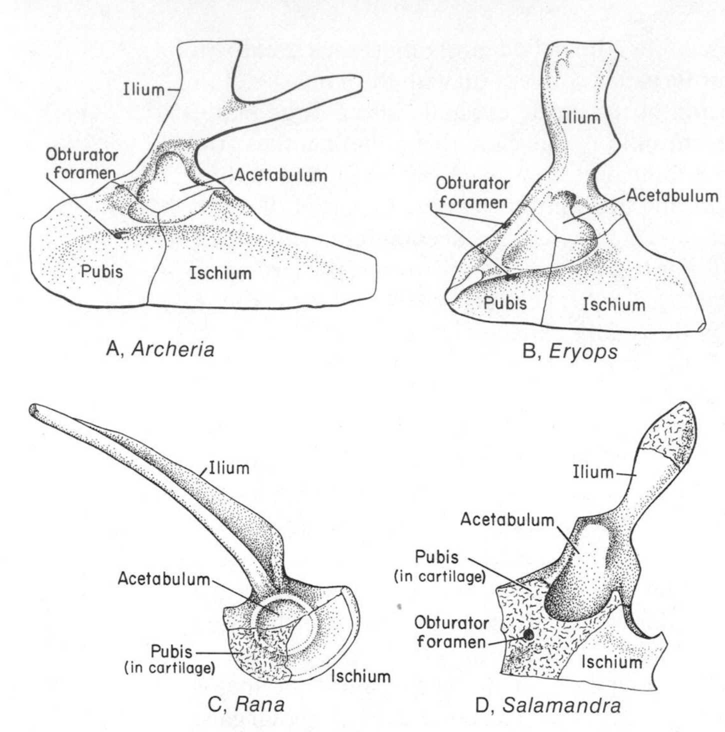

Figure 140. The pelvic girdle in amphibians. *A,* A primitive fossil labyrinthodont; *B,* a typical later labyrinthodont; *C,* a frog; *D,* a urodele. A posterior process of the ilium, present primitively, is retained in many reptiles (cf. Figs. 141 and 142), but is lost in most Amphibia (represented by a prong in *B*). In the anurans the ilium is a specialized and elongate rod. The pubis was primitively well ossified, but remains cartilaginous in many fossil forms and all modern amphibians. Anterior is to the left.

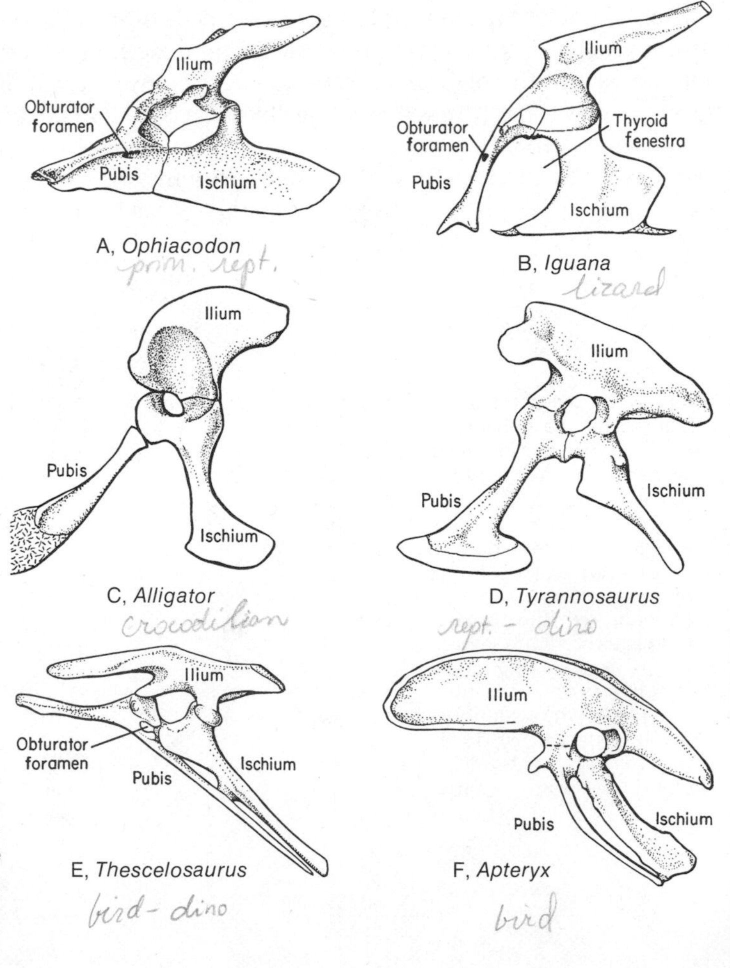

Figure 141. The pelvic girdle in reptiles and birds. *A,* Primitive reptile; *B,* a lizard; *C,* a crocodilian; *D,* a reptile-like dinosaur (Saurischia); *E,* a birdlike dinosaur (Ornithischia); *F,* a bird (the kiwi). In *A* the ilium is a low blade, and the same is true in lizards; in other forms this structure is more expanded. In the dinosaurs and birds, with a bipedal type of locomotion, the ilium has grown forward somewhat as in mammals (cf. Fig. 142). In the archosaurs *(C, D, E)* and the birds descended from them, there is typically an open bottom to the acetabulum for the better reception of the head of the femur. In the primitive reptile the pubo-ischium is a solid plate. In lizards there is a large thyroid fenestra developed between pubis and ischium, from which a large muscle to the femur takes origin (*Sphenodon* and turtles are similar). This fenestra is comparable to one seen in mammals, but there the obturator foramen is concerned in the development of the fenestra. In such archosaurs as the alligators and saurischians there appears at first sight to be a similar structure. Actually, however, this is not the case; the pubis and ischium are twisted downward, and the true ventral margin of the girdle is the curving lower margin of pubis and ischium. In *C* and *D* the pelvis is triradiate, with a simply built pubis; in *E* the pubis is two-pronged, and the pelvis tetraradiate. In the bird the anterior process of the pubis is reduced or absent. In the alligator the pubis is excluded from the acetabulum by the ischium; the pubic region of the girdle extends forward along the belly as a fibrous cartilage. Anterior is to the left.

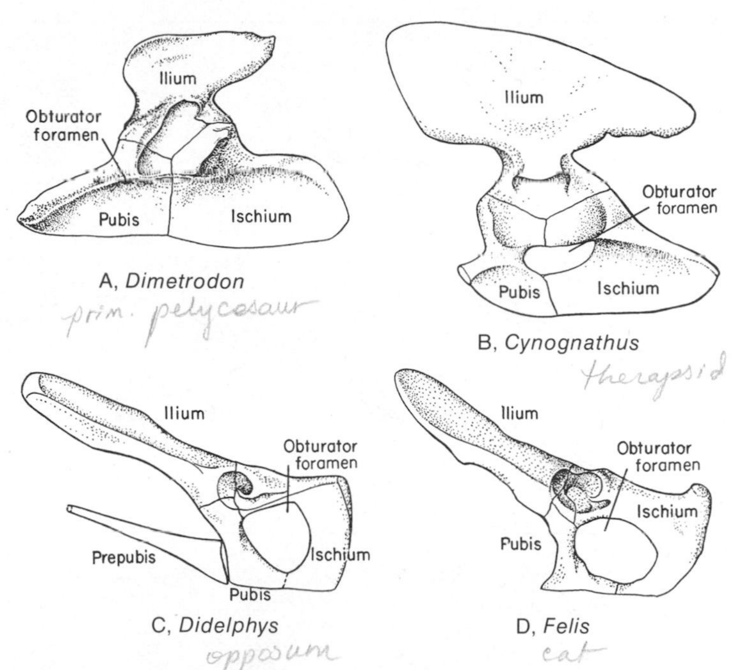

Figure 142. The pelvic girdle in mammal-like reptiles and mammals. *A*, Primitive pelycosaur; *B*, a therapsid; *C*, opossum; *D*, cat. The *Dimetrodon* pelvis is of a primitive reptilian type. In the therapsid the ilium has grown forward dorsally, pubis and ischium have swung back ventrally, and the obturator foramen has expanded (comparable to the situation in many modern reptiles) into an obturator (or thyroid) fenestra. The opossum and cat show a typical mammalian structure, with a large obturator fenestra, a shortened ischium and a slender ilium (secondarily broadened in many heavy mammals, however). The opossum, like other marsupials and the monotremes, has a pair of "marsupial bones" not found in other groups. In the cat, as in certain other mammals, an accessory element is seen in the acetabulum.

is absent in birds, in which support here is rendered unnecessary by the presence of a large sternum (cf. Fig. 125).

In the evolution of mammals (Fig. 142) there has developed a large opening in the ventral plate. This is a parallel to the thyroid fenestra noted in some reptiles; here, however, the opening (in contrast to reptiles) includes the ventral nerve foramen in its area and is termed the **obturator fenestra.** In egg-laying mammals and in marsupials there are present belly supports in the shape of a pair of **prepubic bones;** such structures are virtually unknown in placentals, and they are generally termed "marsupial bones" on the assumption that their function is support of the pouch (marsupium) in which the young are carried.

In the mammalian pelvic limb, as in the pectoral, the musculature has been much modified in connection with a changed posture of the limb; to give better placed areas of muscle attachment, the pubis and ischium have been rotated backward below the acetabulum; the ilium, in contrast, moves forward above it.

PAIRED FINS IN FISHES (Figs. 143, 144). We noted earlier some of the theories connected with the origin of paired fins. Apart from various peculiar early fossil types, we see in the general run of living and fossil fishes two basic types of paired fin skeletons (together with intermediates). One type is the **archipterygium,** well-developed in the lungfish *Epiceratodus* (Figs. 143 *H*, 144 *D*), in which the skeleton of the fin consists of a main axis with side branches. This type of fin is present not only in ancient lungfishes but in some of the most primitive crossopterygians, and hence may be primitive for the sarcopterygians as a whole. In most fossil crossopterygians there was an abbreviated variant of this fin type (Figs. 143 *I*, 144 *E*) which can be interpreted as antecedent to the land limbs of tetrapods. Among fishes, however, the archipterygium is not otherwise known except in some aberrant Chondrichthyes (the pleuracanths).

In strong contrast to the archipterygium is the finfold type of fin seen in ancient Paleozoic sharks such as *Cladoselache* and *Cladodus* (Figs. 143 *B*, 144 *A*). Here the fin has a broad base, and was obviously little more than a horizontal stabilizer with little freedom of movement. In later sharks and in chimaeras (Figs. 143 *C*, *D*; 144 *B*, *D*, *F*) the base has generally become much narrower, thus allowing greater freedom

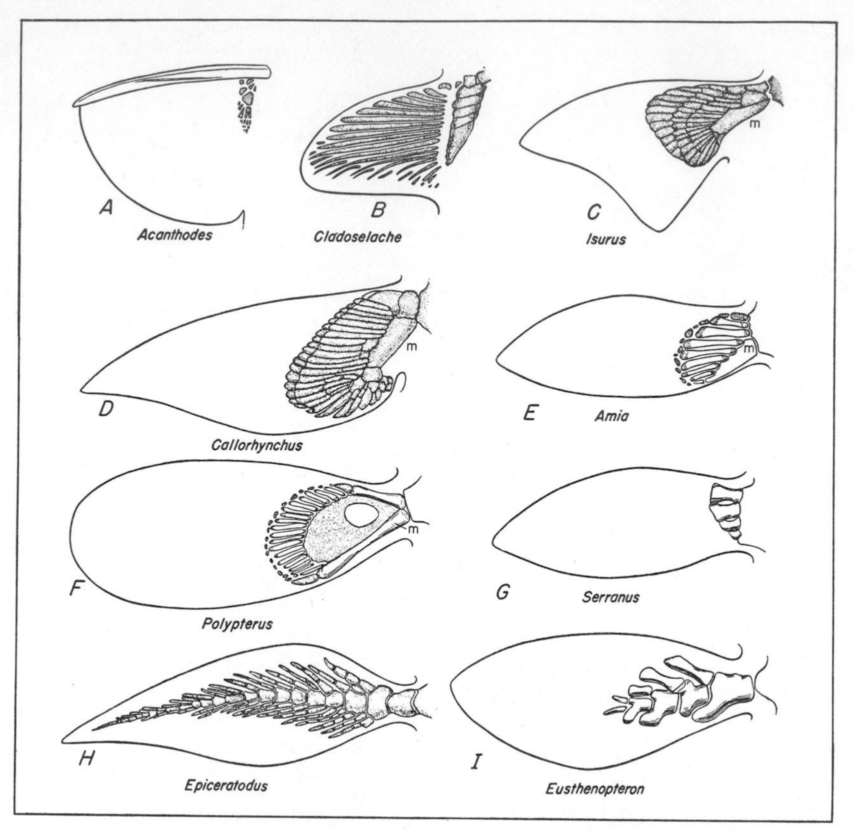

Figure 143. Pectoral fins of fishes. All are fins of the left side, viewed from the upper surface, so oriented that the long axis of the body is vertical on the page, the anterior end above. The outline of the complete fin is indicated, and (except for *A*) the articular region of the girdle is included at the right. *A*, Fossil acanthodian, with the fin skeleton little developed, and a spine forming a cutwater and main fin support. *B*, Primitive fossil shark, with a parallel-bar type of fin. *C*, Modern shark with a narrow-based flexible fin and a basal concentration of bars with formation of a metapterygial axis (the metapterygium is the elongate posterior basal element, *m*). *D*, Comparable type found in chimaeras. *E*, Primitive actinopterygian type, with parallel-bar construction, but metapterygial axis present. *F*, An aberrant modification of the last, found in the archaic actinopterygian *Polypterus*. *G*, Teleost (sea bass) with a much reduced skeleton. *H*, The typical archipterygium of the Australian lungfish. *I*, The abbreviate archipterygium of a fossil crossopterygian. (*A* after Watson; *B* after Dean; *C* and *D* after Mivart.)

of movement; the bases of the supporting bars have been crowded together. There is a tendency for a posterior element, termed the **metapterygium,** to become an axis with which many of the bars articulate; it is probable that the archipterygial fin type has evolved by further development of an axis of this sort.

The actinopterygians are defined as a group in which the fins are mainly supported by horny rays, with the skeleton and flesh restricted to the base of the fin. The fin skeleton in consequence generally consists of short bars of bone or cartilage at its base (Figs. 143 *E–G*, 144 *G–J*). These bars are parallel structures, as in a primitive shark, but (except for such primitive forms as the sturgeons) are few in number, so that the fin usually has a narrow base and flexibility is increased.

In most fishes the pectoral fins are the larger of the two pairs, and in skates and rays they are enormously enlarged. The pelvic fins are reduced or absent in some teleosts, or are moved forward to the shoulder region (Fig. 36 *B*) or even to a position beneath the "chin." Such displacement is usually correlated with a dorsal migration of the pectorals. In the Chondrichthyes the pelvic fins of males bear finger-like **claspers** associated with internal fertilization (Figs. 26, 122, 144 *C*).

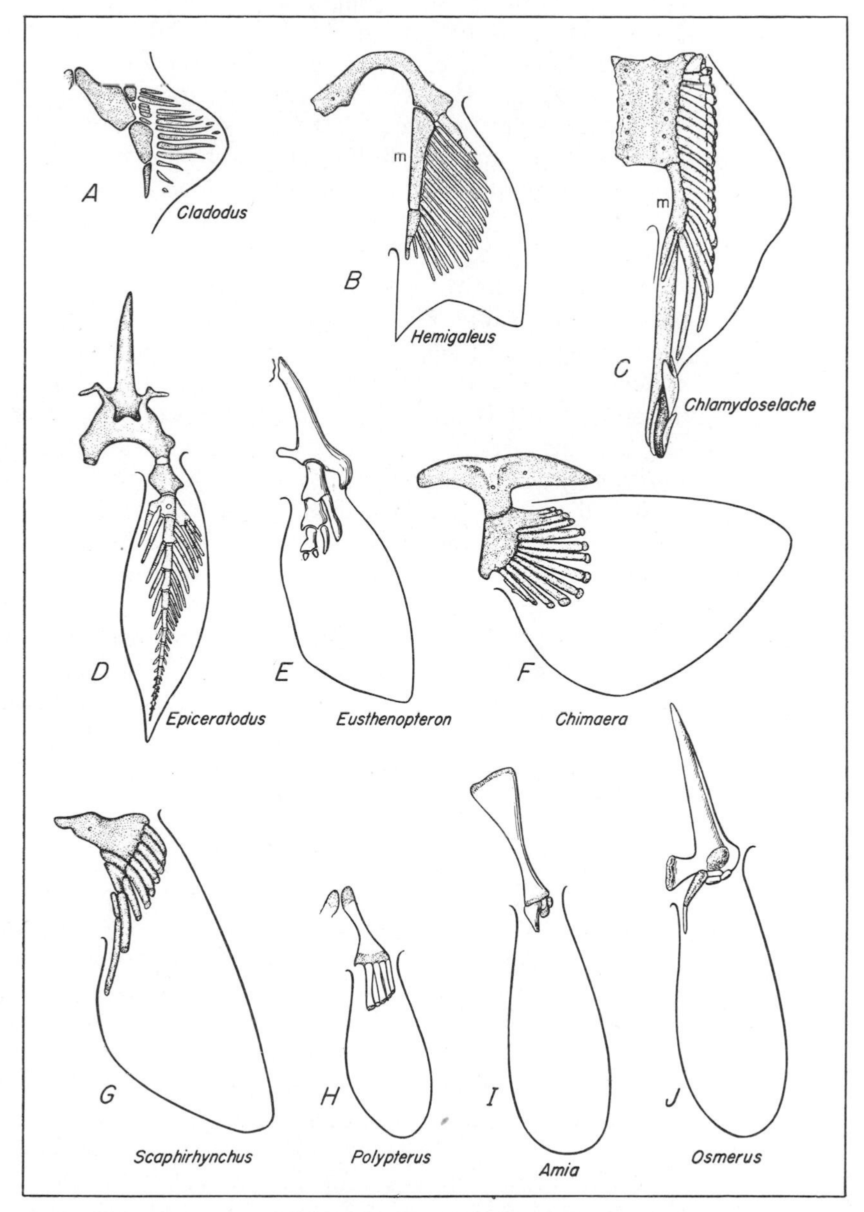

Figure 144. Pelvic fin of fishes. All are of the left side, viewed from below; the midline of the body is at the left, the anterior end above. The fin outline is indicated in each case. The left half of the pelvic girdle is included, or the whole girdle if the halves are fused. *A,* Primitive Carboniferous shark, with a broad-based fin and parallel bars as fin support. *B,* Modern female shark, with similar construction. *C,* Male shark, with additional cartilages supporting the clasper. *D,* The archipterygium of the Australian lungfish. *E,* The abbreviate archipterygium of an ancient crossopterygian. *F,* Female chimaera fin, essentially similar to that of sharks. *G,* A sturgeon, also with a sharklike construction. *H, Polypterus; I,* the bow-fin; *J,* a teleost, showing nearly complete reduction of the bony fin skeleton. *m,* Metapterygium. (*A* after Jaekel; *B* after Garman; *C* after Goode; *D, H, I* after Davidoff; *E* after Gregory.)

THE TETRAPOD LIMB. Although the limbs of terrestrial vertebrates appear, at first sight, strikingly different from fish fins, the two are comparable in basic structure. Tetrapod limb elements are broadly comparable to those of many crossopterygians, and the complex limb muscles of a tetrapod can be analyzed into two opposing muscle groups comparable to those of more simple nature which are present on the upper and lower surfaces of the paired fins of fish.

The limb of a primitive terrestrial vertebrate is composed of three major segments (Fig. 145). In both front and hind limbs—i.e., pectoral and pelvic appendages—the proximal segment includes but a single bone—**humerus** or **femur**—projecting laterally from the body. Beyond the elbow or knee is a second segment which is more or less vertically placed. This contains in either limb a pair of elements—**radius** and **ulna** in the front limb, **tibia** and **fibula** in the hind, the first named in both cases being the more anterior or medial member of the pair. A third segment is that of the foot—**manus** and **pes** respectively. A proximal portion, which more or less flexibly adjusts the foot on the forearm or lower leg is the wrist or ankle region—**carpus** or **tarsus**—while beyond are the toes, or **digits.** The proximal element of a toe (contained in palm or sole) is a **metapodial**—a **metacarpal** or **metatarsal**—while the more distal elements are the **phalanges.**

This arrangement of limb bones is, as mentioned, comparable to that found in the crossopterygian ancestors of the tetrapods (Fig. 146), in which the fin contains, as does the tetrapod limb, a single element in the first joint and two in the second. Beyond this point known crossopterygians show a variable branching arrangement of distal elements, out of which presumably arose the hand or foot of a tetrapod.

From the first there were notable differences between the front and hind limbs of tetrapods, as may be seen, for example, in the contrast between the major joints in the two limbs. The forearm rotates freely on the humerus, but the knee is a simple hinge; on the other hand, the manus cannot be rotated on the forearm, whereas in all tetrapods except mammals the ankle preserves considerable powers of rotation. These differences can, it seems, be traced back to a contrast in the way in which the two land limbs arose from the fish fins (Fig. 147). Although the changes involved are still open to some question, it would appear that the pectoral fin gained contact with the ground by a sharp rotation at the elbow which brought the distal part of the limb forward, so that the toes were immediately placed in a proper forward-pointing position. In the hind limb, apparently, the limb was directed straight outward, reaching the ground by a simple flexure at the knee. As a result, the toes would have pointed laterally if there had not occurred further evolution at the ankle including rotation to enable the toes to be properly directed forward.

LIMB FUNCTION AND POSTURE. In fishes, as we have noted, propulsion is normally accomplished by undulatory movements of the body. In land vertebrates with primitive bodily proportions, such as salamanders (Fig. 134 *B*), body undulations appear still to play a considerable part in locomotion, with the limbs acting in great measure merely as stationary organs through which the push of body undulations may be exerted on the ground. In tetrapods generally, however, the limbs have taken over a positive and dominant role in progression. In primitive tetrapods, as in urodeles, turtles, and lizards today, the limbs were widely sprawled out from the sides of the body, requiring the expenditure of much energy to keep the body off the ground.

Improvements on this situation have been made by numerous tetrapods. In the archosaurs there was a strong trend toward bipedalism, with powerful hind limbs turned forward beneath the body, giving not merely a longer stride but also direct support of the body on the vertically placed limbs. With freeing of the fore limbs from use in terrestrial locomotion there followed their development as wings in

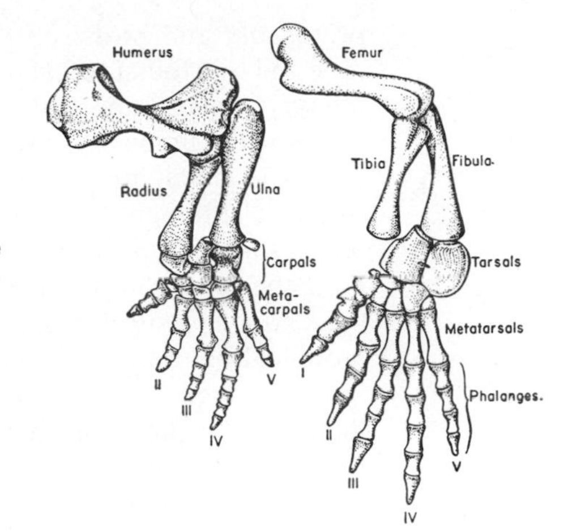

Figure 145. Left front and hind limbs of a primitive reptile (*Ophiacodon*), to show the general pattern of limb construction in early tetrapods. Roman numerals indicate the digits.

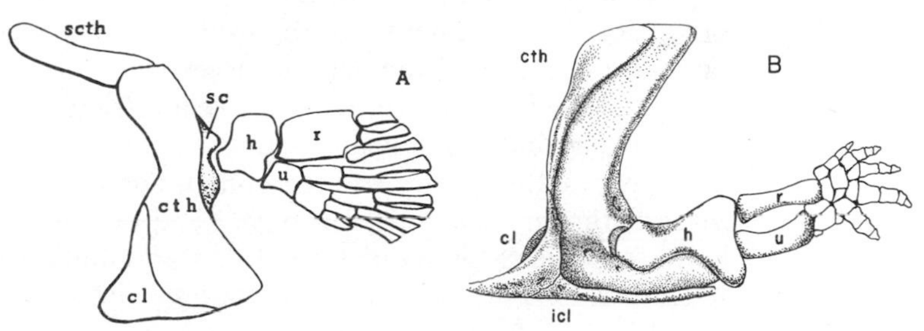

Figure 146. The shoulder girdle and pectoral fin of a crossopterygian, *A*, and the same structures in an ancient fossil amphibian, *B*, placed in a comparable pose to show the basic similarity in limb pattern. *h, r,* and *u,* Humerus, radius, and ulna of the tetrapod and obvious homologues in the fish fin. *cl,* Clavicle; *cth,* cleithrum; *icl*, interclavicle; *sc,* scapula; *scth,* supracleithrum. Anterior is to the left. (*A* after Gregory.)

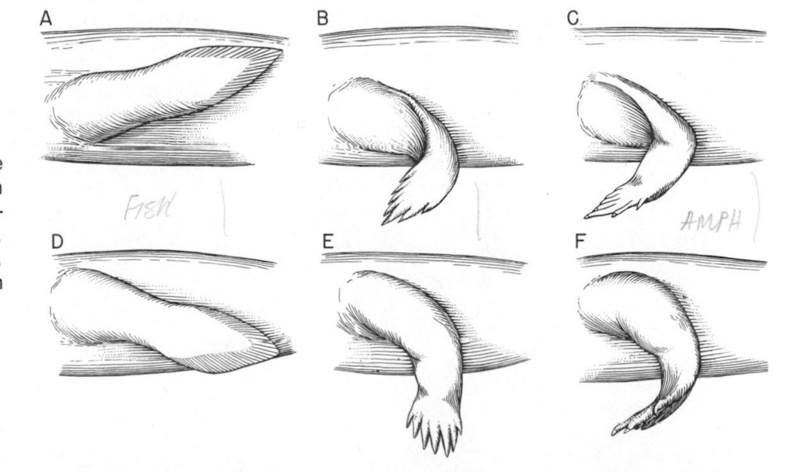

Figure 147. Diagrams to show the postural shift in the paired limbs in the transition from fish to amphibian. *A* to *C*, Pectoral limb; *D* to *F*, pelvic limb; *A, D,* fish position; *B, E,* transitional stage; *C, F,* amphibian position (cf. text).

pterosaurs and birds; on the other hand, many dinosaurs and the crocodilians reverted to a four-footed, quadrupedal, mode of progression. The ancestors of the mammals also improved their gait and gained more effective body support by bringing the limbs into a fore-and-aft pose close to the body. Here, however, both front and hind limbs were concerned; the mammals in general are good four-footed runners, although there have been numerous departures from this locomotor pattern, ranging from flying types to marine forms. Reversion to an aquatic existence has been common among tetrapods, and various reptiles, such as certain turtles and the extinct plesiosaurs and ichthyosaurs, became secondarily marine types with modification of the limbs into flippers (Fig. 148). Still other tetrapods have secondarily reduced or lost their limbs (gymnophionans, various lizards, amphisbaenians, snakes) and have reverted to progression by body undulations, essentially "swimming" on the ground (or beneath it—loss of limbs seems frequently to be correlated with burrowing habits).

MAJOR LIMB BONES (Figs. 149–152). The primitive tetrapod **humerus** was a stout bone, expanded at both ends. In many later forms it assumes a more slender build, but there persist proximal processes for muscular attachments, and there is always some degree of breadth distally, where articular surfaces are present for both radius and ulna and where expansions must be present to furnish areas of origin for the musculature of the forearm. This distally expanded portion is frequently pierced by foramina for nerves to the lower part of the limb, and one of these, on the posterior (or inner) margin is frequently preserved in mammals. The columnar **radius** is the main supporting element of the forearm. The shaft of the **ulna** bears little weight, and the bone tends to be slender and in many mammals fuses with the radius. The proximal end of the bone—the **olecranon** or "funny bone"—projects above the notch for articulation with the humerus and is important as the point of attachment for the main muscle which extends the forearm.

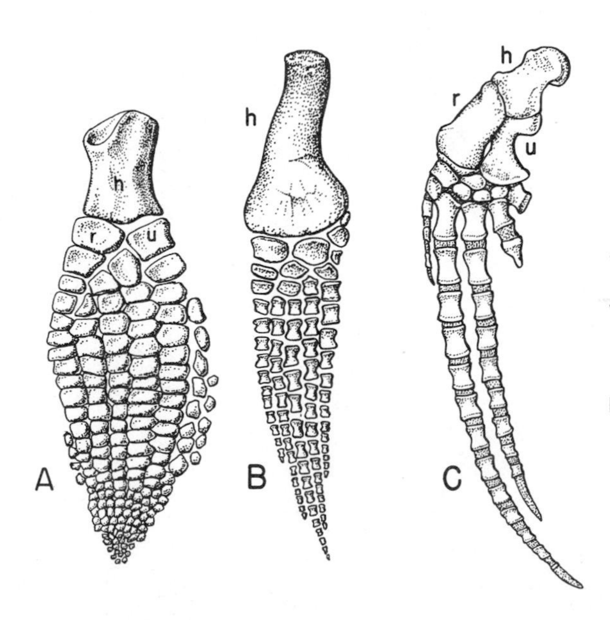

Figure 148. Examples of tetrapod appendages retransformed into fishlike paddles in marine forms. Pectoral limbs of *A*, an ichthyosaur; *B*, a plesiosaur; *C*, a whale. In all there has been a shortening and broadening of the proximal bones and a multiplication of the phalanges. In *A* there has also been an increase in the number of digits. *h*, humerus; *r*, radius; *u*, ulna. (*A* after Huene; *B* after Williston; *C*, after Flower.)

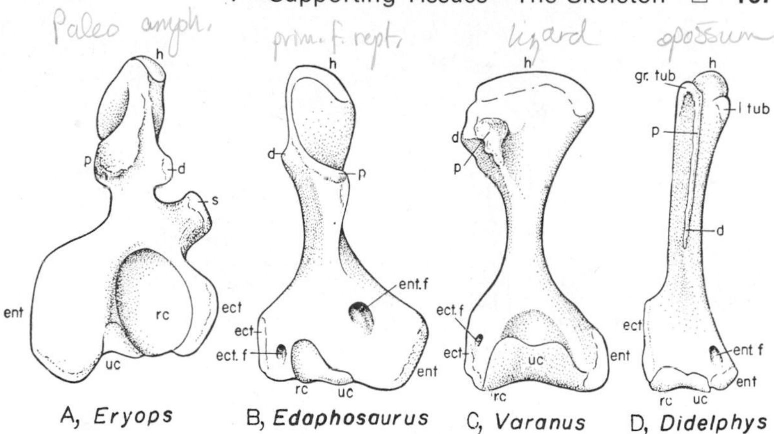

Figure 149. Humeri of *A,* a Paleozoic amphibian; *B,* a primitive fossil reptile; *C,* a lizard; *D,* the opossum, all from the ventral side. *A* shows a left humerus; the others are right humeri. Primitive humeri were short, practically without a shaft, and much expanded at both ends (in *A* and *B* the proximal end is twisted about 90 degrees with the distal and hence appears thin). In all a prominent crest is present to which is attached the pectoralis and deltoid muscles. In later types the bone became relatively long and slender, particularly in small animals. In primitive reptiles, foramina developed distally on the inner or posterior side (entepicondylar foramen, *ent. f*) and on the outer or anterior margin (ectepicondylar foramen, *ect. f*). The former foramen persists in various mammals and in *Sphenodon,* the latter in many reptiles. *d,* Deltoid crest; *ect,* ectepicondyle, for attachment of extensor muscles of forearm; *ent,* entepicondyle, for attachment of flexor muscles of forearm; *gr. tub,* greater tuberosity, for attachment of supraspinatus and infraspinatus muscles; *h,* head; *l tub,* lesser tuberosity, for attachment of subscapular muscle; *p,* pectoral crest; *rc,* radial condyle; *s,* process (supinator) which in reptiles aids in formation of ectepicondylar foramen; *uc,* ulnar condyle, or trochlea.

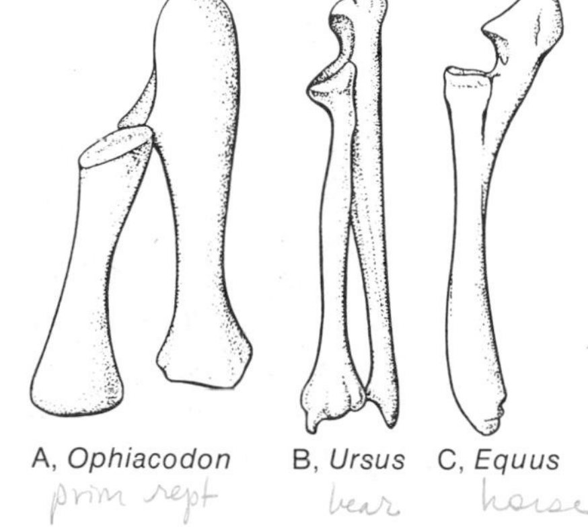

Figure 150. Left radius and ulna, seen from the anterior or extensor surface in *A,* a primitive reptile; *B,* a bear, representing a typical mammalian condition; and *C,* a horse. The humerus articulates with the curved surface of the notch in the ulna and the adjacent head of the radius; above, the projecting olecranon of the ulna serves for the attachment of the powerful triceps muscle, which extends the forearm. In many mammals, as in the horse, the lower part of the ulna is reduced and fused with the radius.

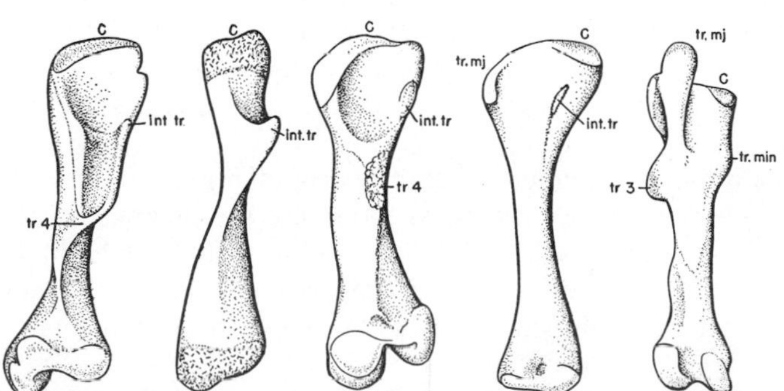

Figure 151. Left femora seen from the ventral surface. *A,* Primitive fossil amphibian; *B,* a urodele; *C,* a primitive reptile; *D,* a mammal-like reptile; *E,* the horse. The proximal end above; distally, the articular surfaces for the tibia. *c,* Head (capitulum); *int. tr.,* internal trochanter of primitive forms, for attachment of obturator externus muscle or equivalent; *tr. 3,* third trochanter of perissodactyls for part of gluteal muscles; *tr. 4,* fourth trochanter, to which are attached tail muscles pulling the femur backward in many amphibians and reptiles; *tr. min.,* lesser (minor) trochanter for iliopsoas muscles of mammal; *tr. mj.,* greater (major) trochanter of mammals, for gluteal muscles.

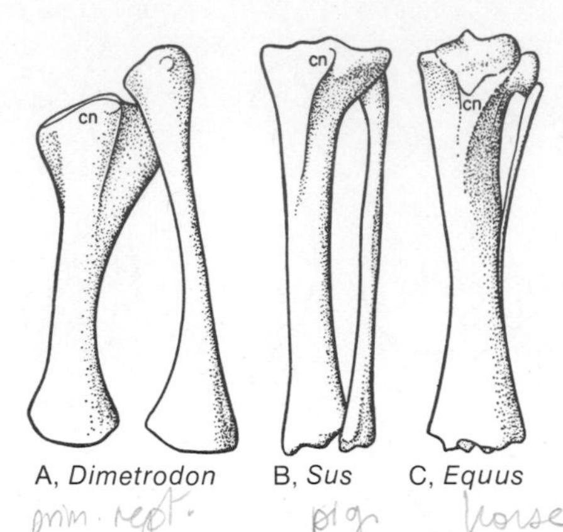

Figure 152. The left tibia and fibula, seen from the extensor (dorsal) surface, of *A*, a primitive reptile (a Permian pelycosaur) with a fibula of good size; *B*, the pig, showing a primitive mammalian condition, with the fibula complete, although slender; *C*, the horse, exemplifying a type with reduced fibula. *cn*, Cnemial crest.

The **femur** in primitive forms was essentially a stout, rodlike structure, with proximal processes—**trochanters**—for muscle attachment, but, in contrast with the humerus, the bone is not greatly expanded either proximally or distally. In primitive forms the femur projected nearly straight laterally from the acetabulum, and the articular surface for the girdle lay, in consequence, directly on the end of the bone. But in some reptiles, and in all mammals, where the limb is rotated to a forward position parallel to the main axis of the body, the head is of necessity turned inward at more or less a right angle to the shaft.

In the lower leg or "shin," the inner bone, the **tibia,** like the radius in the front leg, is the main weight-bearing element, and the **fibula,** on the outer side, is (like the ulna) more slender and mainly serves as an area of origin for musculature. These two bones are, however, quite unlike their counterparts in the forearm in appearance. The head of the tibia is expanded in triangular fashion to underlie most of the distal end of the femur. The fibula, unlike the ulna, has no proximal projection, and the main extensor muscle of the thigh attaches instead to a prominent crest on the tibia; however distally the fibula is important at the ankle, unlike the ulna which takes little part in the wrist in many forms.

FEET (Figs. 153–160). The proximal part of the manus, or hand, consists of a series of small elements, the **carpals,** forming a flexible adjustment between the arm and the digits. In a primitive tetrapod this region (the **carpus**) consists of a dozen bones—three proximal bones termed **radiale, intermedium,** and **ulnare,** four **centralia,** and five **distal carpals,** one for each toe; in reptiles there is added a **pisiform** on the radial side.* The history of these elements in the various classes is a complex and variable one which cannot be followed in detail in any limited space; numerous variants are shown in the figures. In general, there tends to be considerable reduction or fusion of elements—particularly loss of centralia, of which few forms retain more than one or two, and of the fifth distal carpal, an enlarged fourth element generally supporting the two outer digits. The carpal elements of a mammal are readily identifiable in terms of the primitive arrangement, but several sets of names have been applied to the carpal bones. One such mammalian series is shown in Figure 153 *C*.

Beyond the carpus lie the toes, or **digits.** In each a proximal joint lying in the flesh of the palm is termed a **metacarpal;** the distal joints are the **phalanges.** The presence of five digits appears to have been a primitive condition, and except in the "flipper" of some fossil ichthyosaurs is never exceeded. On the other hand, reduction is common. No modern amphibian has more than four digits in the hand;

*A word of warning—the names of the carpals (and tarsals) are not universally agreed upon; there are several different systems in common use.

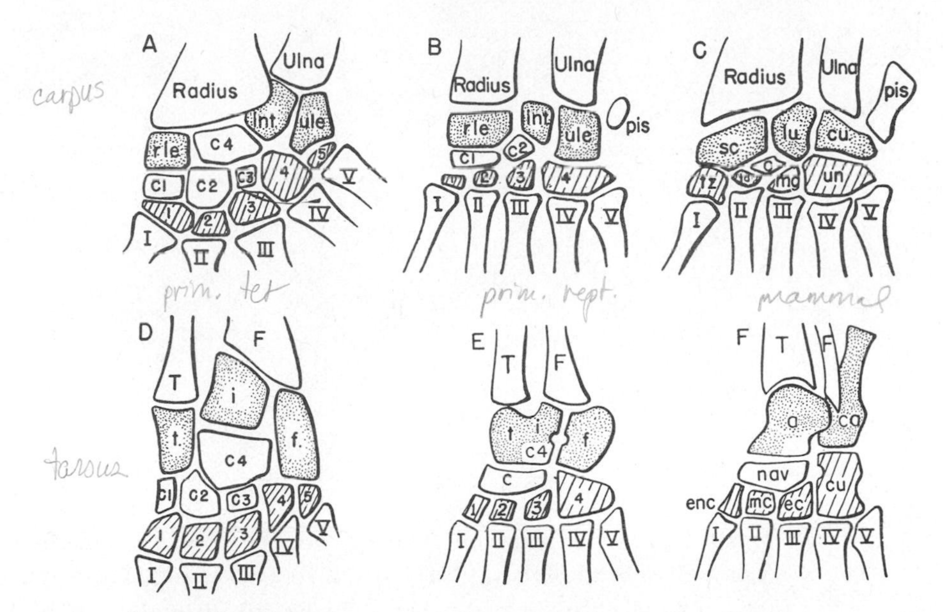

Figure 153. Diagram of carpus *(A* to *C)* and tarsus *(D* to *F)* to show essential homologies between primitive tetrapod *(A, D),* primitive reptile *(B, E),* and mammal *(C, F).* Proximal row of elements stippled; central row (and pisiform) unshaded; distal row hatched. Digits indicated by Roman numerals; distal carpals and tarsals by Arabic numerals. *a,* Astragalus; *c, c1* to *c4,* centralia; *ca,* calcaneum; *cu,* cuneiform in carpus, cuboid in tarsus; *ec,* external cuneiform (ectocuneiform); *enc,* internal cuneiform (entocuneiform); *f,* fibulare; *F,* fibula; *i, int,* intermedium; *lu,* lunar; *mc,* middle cuneiform (mesocuneiform); *mg,* magnum; *nav,* navicular; *pis,* pisiform; *rle,* radiale; *sc,* scaphoid; *t,* tibiale; *T,* tibia; *td,* trapezoid; *tz,* trapezium; *ule,* ulnare; *un,* unciform.

dinosaurs and their kin tended to reduce the number by loss of the outer digits, and birds have remains of only the three inner fingers in the skeleton of the wing. In primitive mammals the first toe tended to be set off from the others as an aid in grasping—the thumb or **pollex;** this is reduced in many running forms. In the hoofed mammals, the ungulates, there is frequently further loss. In the even-toed ungulates—the artiodactyls—digits II and V are reduced and may be lost, leaving

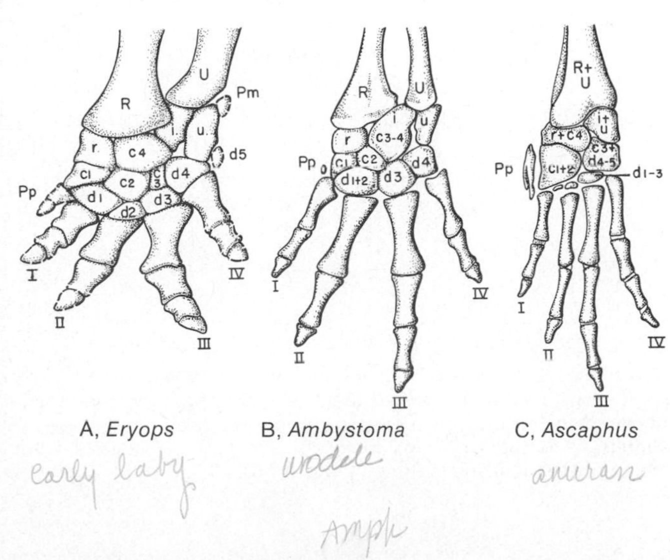

Figure 154. The left manus of amphibians, including an early labyrinthodont, a urodele, and an anuran. Restored elements in *A* in broken line. All twelve elements thought to have been present in the primitive carpus are found in *Eryops;* in modern amphibians fusions of various sorts have occurred. As in most amphibians, four toes are present in the forms figured, although all have some development of a digit medial to the pollex, and an element in *Eryops* may have been the stub of an extra toe beyond the reduced fifth digit. The phalangeal count of two or three is usual in amphibians. *c1* to *c4,* Centralia; *cu.,* cuneiform; *d1–d5,* distal carpals; *i,* intermedium; *l,* lunare; *m,* magnum; *m1, 3, 5,* metacarpals; *p,* pisiform; *Pm,* postminimum digit; *Pp,* prepollex; *R,* radius; *r,* radiale, *s,* scaphoid; *td,* trapezoid; *tm,* trapezium; *U,* ulna; *u,* ulnare; *un,* unciform; *I* to *V,* digits. (After Gregory, Miner, and Noble.)

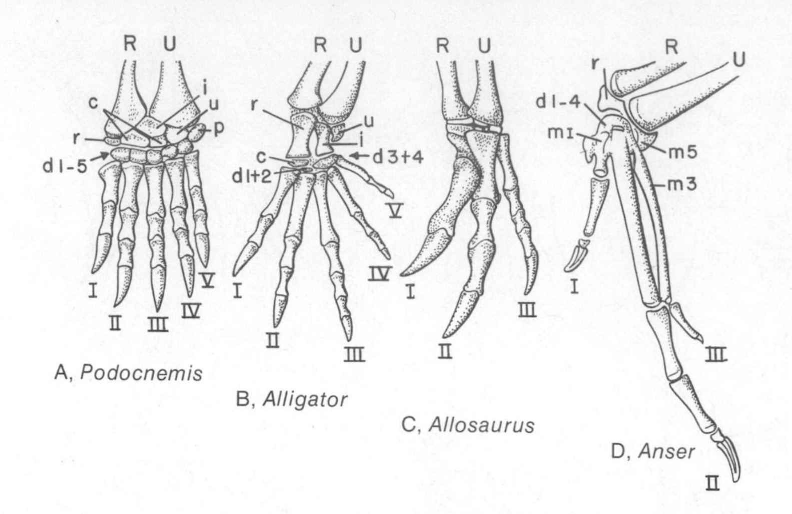

Figure 155. The manus in *A*, a turtle; *B*, the alligator; *C*, a carnivorous dinosaur; *D*, the goose. Abbreviations as in Figure 154. The bird manus includes the fused elements of the first three digits only; a somewhat comparable structure is seen in certain dinosaurs. (*A* after Williston; *C* after Gilmore; *D* after Steiner.)

toes III and IV to form the "cloven" hoof of a cow or deer. In perissodactyls toe V as well as the pollex was early lost, giving a three-toed foot and in modern horses only the central digit—toe III—remains. In ungulates toe reduction is usually paralleled by an elongation of the metacarpals, which has the effect of adding a third major segment to the limb and a correlated increase in speed.

A **phalangeal formula** gives in brief form the count of phalanges in each digit from the first toe outward. The number of joints present in primitive tetrapod toes is uncertain; in amphibians there are seldom more than three. In primitive reptiles, however, the formula is typically 2.3.4.5.3, and this count is maintained in a variety of reptiles, notably the lizards. In the transition from reptiles to mammals there tended to be an "evening up" of the toes and the development of the formula of 2.3.3.3.3. This is still the digital count common in primates (such as man); it also occurs in non-mammalian forms, such as many turtles. In mammals in which the number of toes is reduced, those present generally retain the typical number of joints. Very seldom in amniotes is there any secondary increase in the number of phalanges in a toe, the only exceptions being in the "flippers" of such aquatic forms as the extinct ichthyosaurs and plesiosaurs and the whales.

The structure of the **tarsus,** or ankle region, in primitive tetrapods was very similar to that of the carpus. Twelve elements were present here as in the case of the

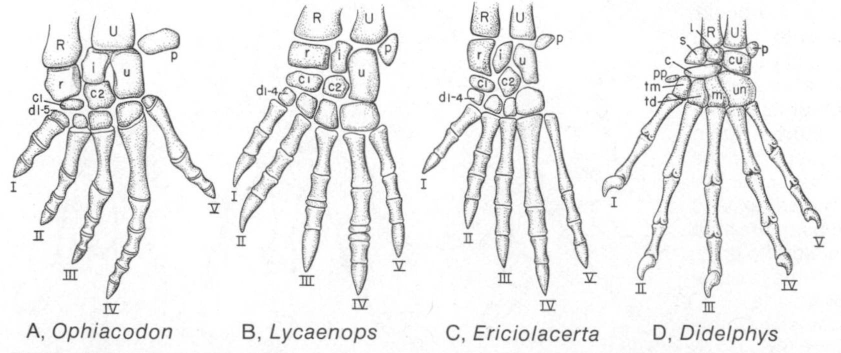

Figure 156. Evolution of the mammalian manus. *A*, Primitive reptile; *B*, a primitive therapsid; *C*, an advanced therapsid; *D*, a primitive mammal, the opossum. Abbreviations as in Figure 154. In the carpus there is a loss of the fifth distal element, a reduction from two centralia to one; distally, "supernumerary" phalanges are lost from digits III and IV. (*B* after Broom; *C* after Watson.)

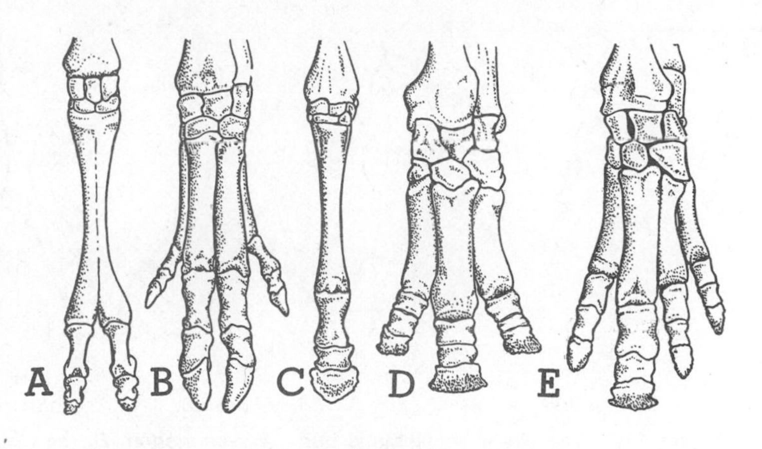

Figure 157. Left front feet of ungulates—in order, camel, pig, horse, rhinoceros, and tapir. The first two are artiodactyls, in which the axis of symmetry lies between the third and fourth toes. In the pig *(B)* lateral toes, 2 and 5, are complete but small. In most artiodactyls, as in the camel *(A)*, the two main metapodials are fused into a cannon bone. The three remaining forms are perissodactyls, in which the axis runs through the third toe. In the tapir *(E)* the pollex is lost, but the other four toes remain; in modern rhinoceroses *(D)* the fifth toe has disappeared; in modern horses *(C)* the second and fourth are reduced to splints. (After Flower.)

wrist. There were three proximal elements—**tibiale, intermedium,** and **fibulare;** four **centralia;** five **distal tarsals.** As in the case of the manus, the centralia tended to be reduced in number and the last distal tarsal lost. More striking, however, is a marked change in the proximal region in all amniotes. The fibulare persists, but on the inner side of the ankle the reduced tibiale fuses with the intermedium and a centrale to form a large bone on which the tibia moves freely. This compound bone is usually called by its mammalian name of **astragalus,** and the fibulare, its lateral companion, is generally termed the **calcaneum,** as in mammals. This proximal modification is apparently associated with the need, noted earlier, for rotating the hind foot forward on the shin to bring it into a proper forward pointing position. A further change in the tarsus is the development in mammals of a projecting heel on the calcaneum. As noted in Chapter 9 this is associated with the development here of an attachment of the major calf muscle through the "tendon of Achilles."

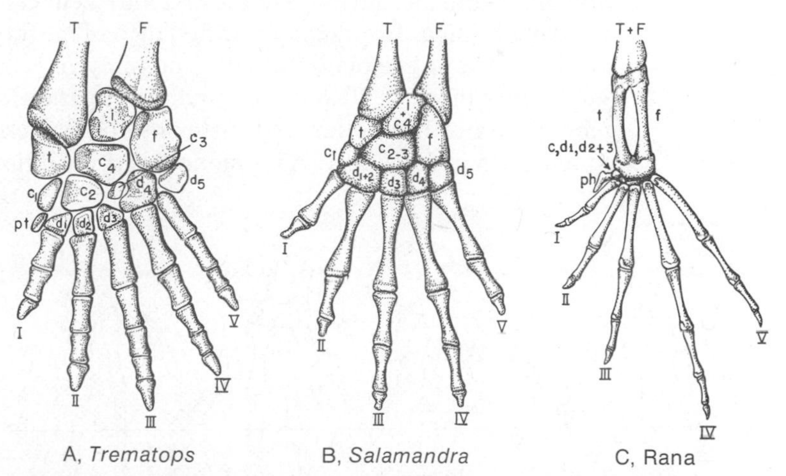

Figure 158. The pes of amphibians, including *A,* an early labyrinthodont; *B,* a salamander; and *C,* a frog. The tarsus of *Trematops* includes all elements presumed to be present in ancestral tetrapods, as well as an additional pretarsal bone. In the urodele some fusion of tarsal elements has occurred. In the frog, tibiale and fibulare are so elongated as to constitute an additional limb segment; except for a few small distal elements, the fate of the other tarsal bones is not clear. Five toes are typically present, as contrasted with four in the manus, and the count of phalanges is generally higher in the pes than in the amphibian manus (cf. Fig. 154). The frog has a developed prehallux. *c1* to *c4,* Centralia; *d1* to *d5,* distal tarsals; *F,* fibula; *f,* fibulare; *i,* intermedium; *ph,* prehallux; *pt,* pretarsal element; *T,* tibia; *t,* tibiale. (*A* after Schaeffer; *B* after Schmalhausen; *C* after Gaupp.)

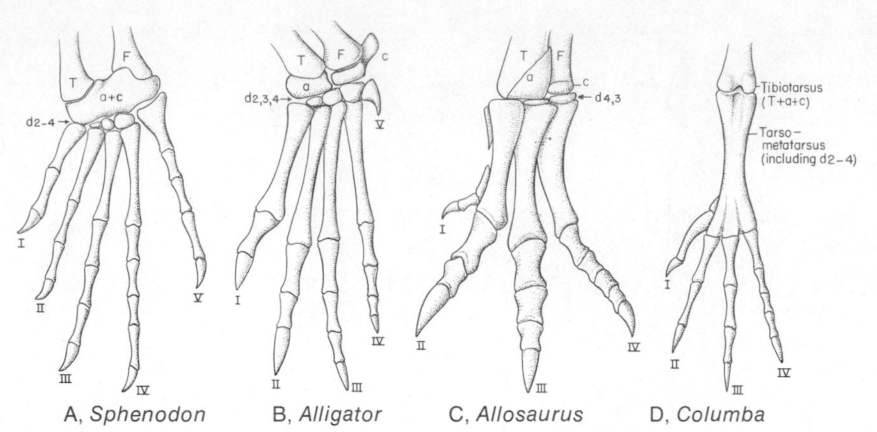

Figure 159. The pes in reptiles and birds. *A, Sphenodon; B,* the alligator; *C,* a carnivorous dinosaur; *D,* the pigeon. In all forms there have been various types of reduction and fusion of the tarsal elements and a trend toward the development of a main joint within the tarsus between proximal and distal elements. In the birds all the tarsals are fused with the tibia proximally or with the fused metatarsals distally. In archosaurs, such as the alligator and dinosaurs, there is a strong trend toward the loss of the fifth toe, and it has disappeared in the dinosaur figured and in the bird. Except for the lack of fusion of the metatarsals, the *Allosaurus* foot is quite similar to that of birds. *a,* Astragalus; *c,* calcaneum; other abbreviations as in Figure 158. (*B* after Williston; *C* after Gilmore).

The build of the toes of the hind foot is very similar to that of the fore foot; the proximal joints, however, are (naturally) termed **metatarsals** rather than metacarpals. Five digits in the pes appears to be the primitive number, and this number is retained in most amphibians and reptiles. In dinosaurs and birds there was a trend toward the development of a foot with three toes, digits II–IV pointing forward, with the central one the longest; the first toe is reduced or turned backward as a prop or aid in perching. In mammals the first digit of the foot—the **hallux**—early became specialized as a grasping structure, possibly in relation to arboreal life. Digital reduction in the pes of fast running mammals paralleled that seen in the manus with general loss of the pollex and, in many forms, further reduction to a two-toed condition in artiodactyls and a monodactyl condition in horses.

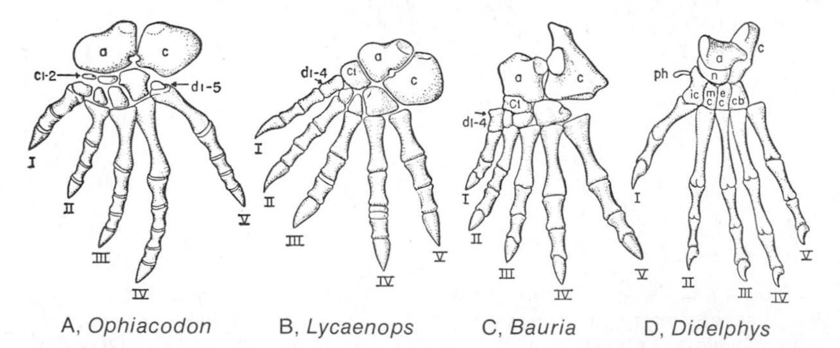

Figure 160. Evolution of the mammalian type of pes. *A,* Primitive reptile (of the early Permian); *B,* a primitive therapsid (of the late Permian); *C,* an advanced Triassic mammal-like reptile; *D,* the opossum. Principal changes include development of a pulley surface on the mammalian astragalus, development of a heel on the calcaneum (in *C* and *D*), loss of two tarsal elements, and reduction in phalangeal count (a transitional stage in *B*). *a,* Astragalus; *c,* calcaneum; *c1* to *c2,* centralia; *cb,* cuboid; *d1* to *d5,* distal tarsals; *ec,* external cuneiform; *ic,* internal cuneiform; *mc,* middle cuneiform; *n,* navicular; *ph,* prehallux (exceptional in mammals). (*B* and *C* after Schaeffer.)

As in the manus, amphibians tend to have but few phalanges per digit in the foot, but in early reptiles there appeared a phalangeal formula of 2.3.4.5.4, almost identical with that of the hand. This count was retained in most reptilian groups (although turtles tend to reduction in the hind as well as fore feet). We noted, above, the symmetric build of the toes in many dinosaurs and in birds. It is of interest that this symmetry has been attained without modification of phalangeal numbers, for the long third toe commonly has four phalanges, and the two shorter ones, inner and outer, have three and five joints respectively. Mammals have reduced the phalangeal count in the pes, as in the manus, to 2.3.3.3.3. Again as in the hand, the phalangeal count tends to be persistently constant under conditions of toe reduction. Only in aquatic reptiles—again as in the manus—is there any increase in the number of phalanges in a given digit of the pes.

VISCERAL SKELETON

Between the gill openings of water-breathing vertebrates lie cartilaginous or bony bars, which are the basic components of a set of structures termed the visceral skeleton. The visceral elements form but a modest part of the skeleton, but they are versatile structures in their potentialities. Primitively this entire series acted as gill supports, but early in vertebrate history anterior visceral arches were transformed into jaws; and although gills have disappeared in amniotes, persistently surviving visceral elements are to be found, even in mammals, in such varied areas as the skull, auditory ossicles, and larynx.

We shall note elsewhere that the gill region is unique in its musculature and nerve supply; its skeletal elements are unique, also, particularly as regards their embryologic origin. Like other cartilages, those of the visceral system are derived from mesenchyme; but this mesenchyme, most exceptionally, is not of mesodermal

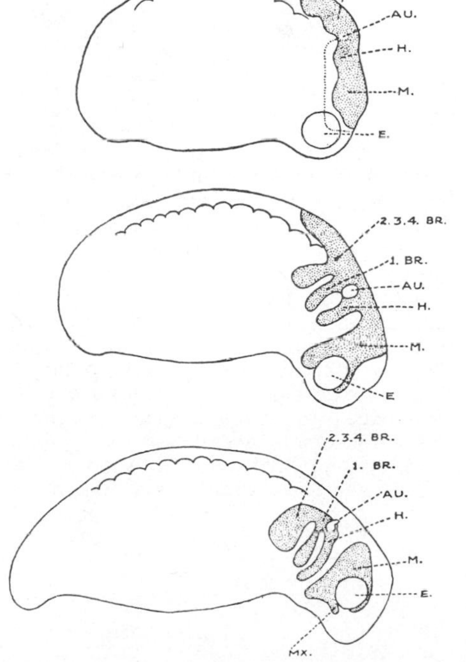

Figure 161. Side views of salamander embryos to show the downward growth of the neural crest (stippled) to form visceral skeletal structures in the head and throat region. *E*, eye; *AU*, auditory capsule; *MX, M, H, BR*, neural crest downgrowths in maxillary, mandibular, hyoid, and branchial arch regions. (From Stone, L. S., Journal of Experimental Zoology, 44:95–131, 1926.)

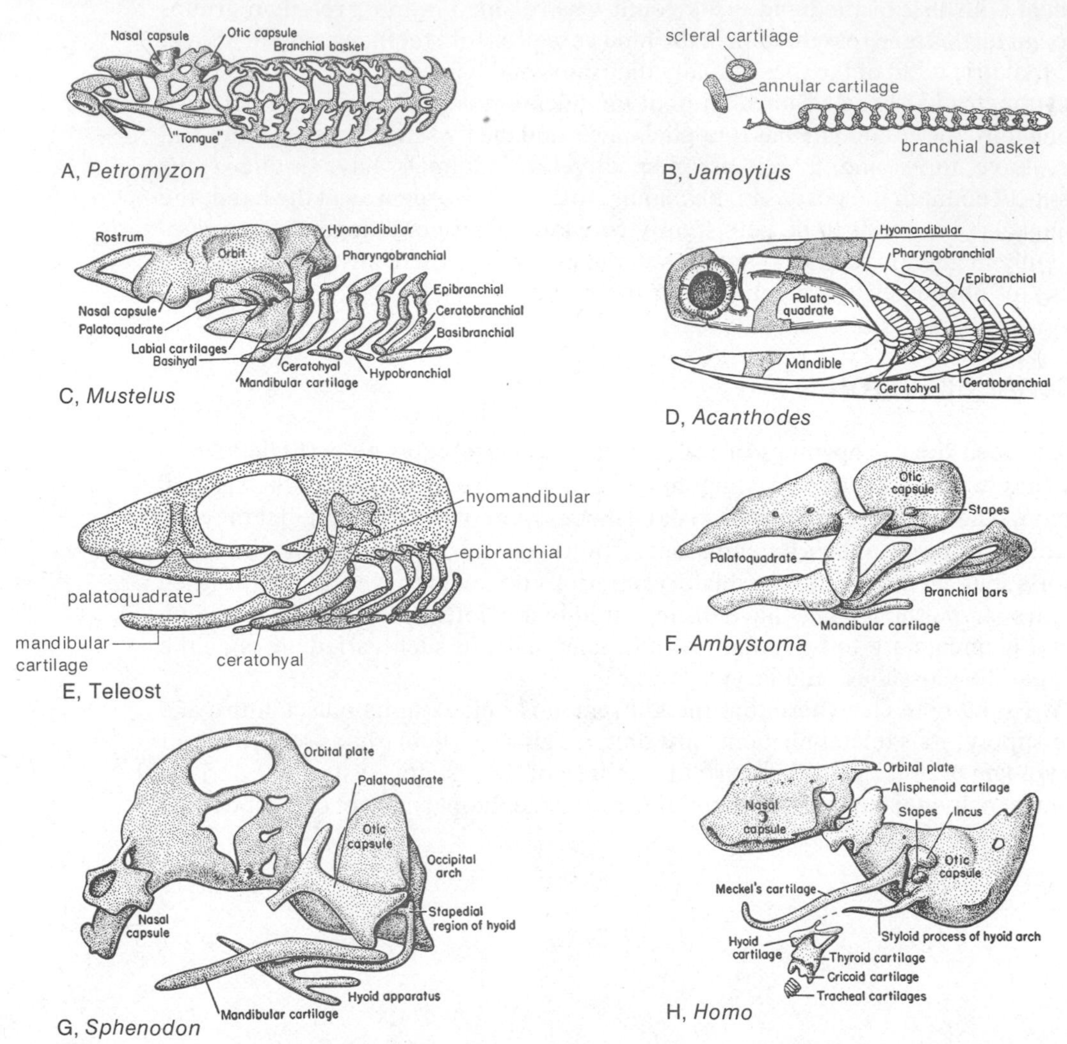

Figure 162. Visceral skeleton and braincase in representative vertebrates. *A* to *D,* Adult forms; *E* to *H,* embryos; cartilage stippled, bone unstippled. *A,* The lamprey, with a peculiar braincase and associated cartilages not readily comparable to those of other groups; the branchial basket is fused anteriorly to the braincase. *B,* An anaspid, one of the extinct ostracoderms, showing a branchial basket somewhat like that of the lamprey; the braincase is not preserved. *C,* A modern shark, with hyostylic jaw suspension. *D,* A Paleozoic acanthodian, a very primitive bony fish; the sclerotic plates of the orbit are included as is a slender dermal bone beneath the mandible. *E,* A rather diagrammatic teleost embryo; in the adult almost the entire visceral skeleton, as well as the braincase, would be encased in dermal bones. *F,* A urodele; the palatoquadrate is reduced, and the branchial arches are reduced even in the embryo or larva. *G,* The reptile *Sphenodon;* the braincase is incompletely formed (cf. lizard in Fig. 131 *D*), and hyoid and branchial bars are reduced to a hyoid apparatus and stapedial cartilage. *H,* Human foetus; the braincase develops only ventrally and anteriorly around the brain; the palatoquadrate is reduced to alisphenoid and incus (= quadrate); the lower jaw (Meckel's cartilage) is reduced, the proximal part becoming the malleus in the adult; other visceral elements include the hyoid and its styloid process, stapes, laryngeal, and tracheal cartilages. (*A* and *C* after Goodrich; *B* after Ritchie; *D* after Watson; *E* after Daget; *F* after Hörstadius; *G* after Howes and Swinnerton; *H* after Gaupp, Macklin).

origin—it is, instead, derived from the ectoderm of the neural crest. The main derivatives of this structure are elements of the nervous system. In the head, however, neural crest cells migrate downward and form the mesenchyme from which the visceral arches (and in addition a fraction of the braincase) are formed (Fig. 161).

THE GILL SKELETON. In cyclostomes (Figs. 120, 162 *A*) part of the gill skeleton is specialized to form the supports of the peculiar rasping "tongue." The remainder forms a lattice-work surrounding the gill pouches. Here and, as far as we can tell, in the ostracoderms, the visceral skeleton lies *lateral* to the actual gills and immediately beneath the skin, in marked contrast to its deeper position in jawed vertebrates. Whether or not this is a primitive condition is uncertain; in all more advanced fishes, at any rate, the gill supports consist (except for cases of secondary fusion) of a number of series of joined bars, following one another in sequence along the walls of the pharynx between successive gill slits. These **visceral arches** are serially arranged, as are the gill openings, but this serial arrangement may not be related to the segmentation seen in other parts of the body, which is founded on segmentation of the mesodermal somites; this topic is further discussed later (see pp. 381–383). Each typical gill arch of a jawed fish (Figs. 162 *C, E;* 163 *A;* 211) includes on either side a major dorsal element, an **epibranchial,** and a major ventral one, a **ceratobranchial.** There is frequently a **pharyngobranchial** at the top of the arch and a short **hypobranchial** below the major ventral element. Ventrally, median structures—**basibranchials** or **copulae**—tie the arches of the two sides to one another and connect successive arches. The gill bars generally bear a row of short **gill rakers** on their inner margins, and **gill rays** which extend outward and stiffen the gills (Fig. 248). In most jawed fishes five typical arches are present. The gill arches of bony fish typically bear dermal tooth-bearing bones on the side toward the pharynx.

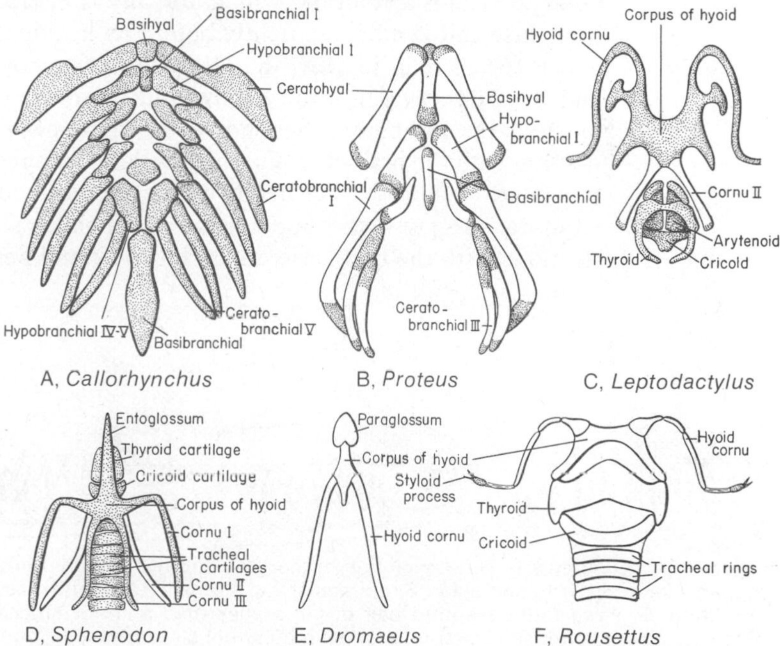

Figure 163. Gill bars and their derivatives in tetrapods, in ventral view, in *A,* a chimaera; *B,* a water-dwelling salamander; *C,* a frog; *D,* a reptile, *Sphenodon; E,* the cassowary; and *F,* a bat. In *A* the dorsal arch elements are not included. In *C, D,* and *F* laryngeal cartilages are included. The entoglossum of *D* and paraglossum of *E* are anterior, tongue-supporting developments from the body of the hyoid.

DEVELOPMENT OF JAWS. The development of jaws was one of the greatest advances in vertebrate history—one which brought about a revolution in feeding habits and mode of life. The visceral arches played a leading role in this process, for transformed gill arches are basic structures in the formation of jaws. Indeed, they form the entire skeleton of the jaws in sharklike fishes. It appears that with the development of an expanding mouth cavity, a pair of gill bars lying adjacent to it enlarged; the epibranchial element of this arch became the **palatoquadrate cartilage** forming the shark upper jaw; the corresponding ceratobranchial became the **mandibular cartilage** of the lower jaw (Figs. 162 *C*, 164). Various lines of evidence suggest that the jaws are not the most anterior of the original gill arch series, but that one or possibly two sets of anterior bars were crowded out of existence as the mouth expanded.

In bony fishes and tetrapods, dermal elements play a major part in jaw formation, and the role of the visceral cartilages or the bones that replace them is much reduced. However, at least that part of the primitive cartilages which forms the articulation between upper and lower jaws persists as high up the scale as reptiles and birds, and relics of these elements persist even in mammals, although, as will be seen, in curiously modified form.

JAW SUSPENSION. It is believed that in the ancestral fishes the jaws articulated with the braincase without additional support—the **autostylic condition.** This articulation was primitively by movable joints, but in some fishes—chimaeras and lungfishes—there is a firm fusion of the upper jaw with the braincase. Forms in which autostyly is present are, however, the exception in fishes, for the next visceral arch behind the jaws, termed the **hyoid arch,** is generally called in to aid in suspension of the jaws on the braincase. The major ventral element of this arch, the **ceratohyal,** is little specialized; however, the main dorsal element, the **hyomandibular,** becomes a stout bar which dorsally is braced against the otic region of the braincase and ventrally is tightly bound by ligaments to the region of the jaw joint (Figs. 164 *C*, 165). In a few sharks the jaw is supported both by the hyomandibular and by a direct connection of jaw and braincase—the **amphistylic condition.** In most sharks and bony fishes, however, the upper jaw loses any major direct connection with the braincase and the jaws are propped solely by the hyomandibular; this mode of support is termed **hyostylic.** In tetrapods the hyomandibular no longer supports the jaws; the upper jaws and palatal structures form their own connections with the braincase and skull roof, and hence are autostylic in construction.

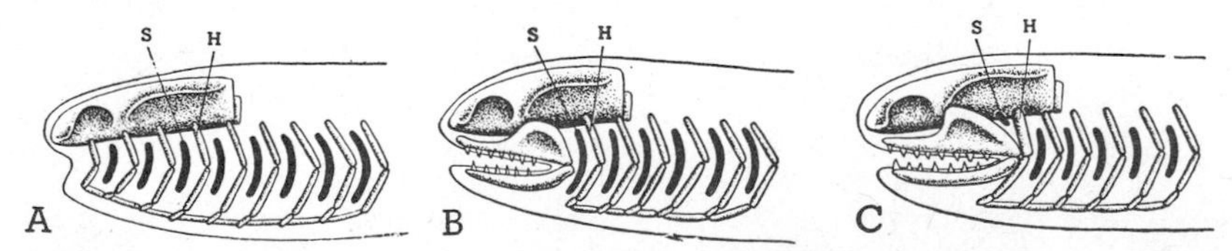

Figure 164. Diagrams to show evolution of the jaws and hyoid arch. Gill openings in black. *H*, Hyomandibular; *S*, spiracular gill slit. In *A*, a primitive jawless condition. *B*, Jaws formed from a pair of gill arches (two anterior arches and slits may have been lost in the process); spiracular gill slit unreduced, and hyomandibular not specialized. *C*, Condition seen in primitive jawed fishes; the hyomandibular has become a jaw support, and the intervening gill slit reduced to a spiracle.

TETRAPOD GILL ARCH DERIVATIVES (Figs. 162 *F–H,* 163 *B–F*). Except in larval amphibians the gills have lost their function in the tetrapod classes and in correlation with this the gill bars are reduced and modified in structure. The history of the elements associated with the jaws and that of the hyomandibular (which becomes an auditory ossicle) is discussed elsewhere. The remainder of the old system of visceral arches becomes associated with two structures prominent in tetrapods but little developed in fish groups—the tongue and the lungs.

The main body of the tongue is a mass of muscle; embedded in its base, and extending from it backward and upward around the sides of the pharynx, is the **hyoid apparatus.** The main body of this structure—the **corpus**—is formed from one or more of the median ventral arch elements present in the fish. Expanding outward and upward from the corpus are slender "horns"—**cornua**—which represent the main ventral elements of the hyoid and succeeding gill bars of the ancestral fishes; sometimes small dorsal elements are appended to the tips of the horns. Three such pairs of "horns" are frequently present in amphibians and reptiles; mammals have but two pairs and birds one.

The windpipe, leading to the lungs, is supported in tetrapods by cartilages which are modified parts of the visceral arches. Just beyond the entrance to the windpipe there typically develops an enlargement of the tube, the larynx (cf. Chap. 11). The basal part of the hyoid apparatus generally lies close to the front end of the floor of this cavity, but in addition a series of special cartilages tend to form a complex **laryngeal skeleton** around it. In most tetrapods the windpipe, the trachea, is strengthened by ring-shaped **tracheal cartilages,** which are not directly comparable to any specific elements seen in fish, but are reasonably to be regarded as new developments of the visceral skeletal system.

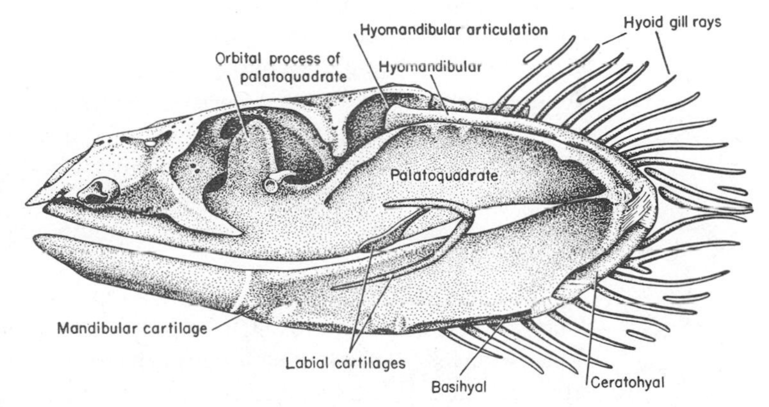

Figure 165. The cranial skeleton of a shark, *Chlamydoselache:* braincase, jaws, hyoid arch. (After Allis.)

The Skull 8

The term "skull" is used in somewhat variable fashion. In a broad way it refers to any type of skeletal structure found in the head. In this sense one may consider the lamprey or shark as having a skull composed of a braincase and other isolated cartilages. But in common parlance the term has a somewhat different meaning. The familiar skull of every form from a bony fish to a mammal is a unit, in which braincase and upper jaws are welded together by a series of dermal bones; the lower jaw is not included.

The old ostracoderms and placoderms appear to have possessed some sort of fused cranial skeleton forming a skull in this sense, but in most of these older fishes the structure is too aberrant and too incompletely known to be considered in detail here. In higher bony fishes and all tetrapods there is a well-constructed skull, with many common features throughout. Even so, it is difficult to give a generalized description based on any one living type, for most have become so specialized or degenerate that the true phylogenetic story is obscured.

But in the case of the skeleton, in contrast with other organ systems, we have the advantage that in many instances there are preserved as fossils the bones of actual ancestors. We have a fairly complete knowledge of the skulls of the ancient labyrinthodont amphibians of the late Paleozoic, the very types from which those of all later tetrapods have been derived. Further, their skulls are not very different from those of the crossopterygian fishes from which tetrapods are descended, and through them this ancestral tetrapod skull pattern can be "tied in" with that of other bony fishes. We shall, therefore, give at this point a fairly full account of this central skull pattern and then discuss the major modifications seen in the various groups of fish and tetrapods.

SKULL COMPONENTS. To avoid (as far as possible) mental indigestion in the description of this complex structure, it is best that it be resolved into components (Figs. 166, 167). The skull includes both dermal bones and cartilages, or their bony replacements, of both somatic and visceral nature (cf. p. 136). One possible approach would be to consider as units the elements derived from each of these three embryologic sources. It is, however, preferable to recognize that a primitive skull seems to be composed of three major functional units, one purely dermal, the other two composites, as follows:

A. **Dermal skull roof.** This shield of membrane bone covers thoroughly the top and sides of the head and extends down to the jaw rims, where elements of the shield bear the marginal teeth. The roof is unbroken save for openings for the nostrils **(external nares)**, eyes **(orbits)**, and a small **parietal foramen** for a third, median eye. On either side the shield is notched behind the orbit in early tetrapods for the eardrum opening, which replaces the spiracle present here in bony fishes.

B. **Palatal complex.** This includes ossifications in a palatoquadrate cartilage of visceral origin (which in sharks forms the entire upper jaw), and in addition a series of membrane bones formed in the roof of thc mouth below this cartilage and in great measure replacing it. The anterior portion of this complex forms a broad palatal plate with, antcriorly, latcral gaps for the **internal nares,** or **choanae.** Posteriorly the palatal complex on either side is separated from the edge of the shield by the **subtemporal fossae,** through which descend the adductor muscles for closing the jaws.

C. **Braincase.** This is formed in cartilage (mainly of somatic endochondral origin) but usually ossified to a considerable degree; to its lower surface, in early

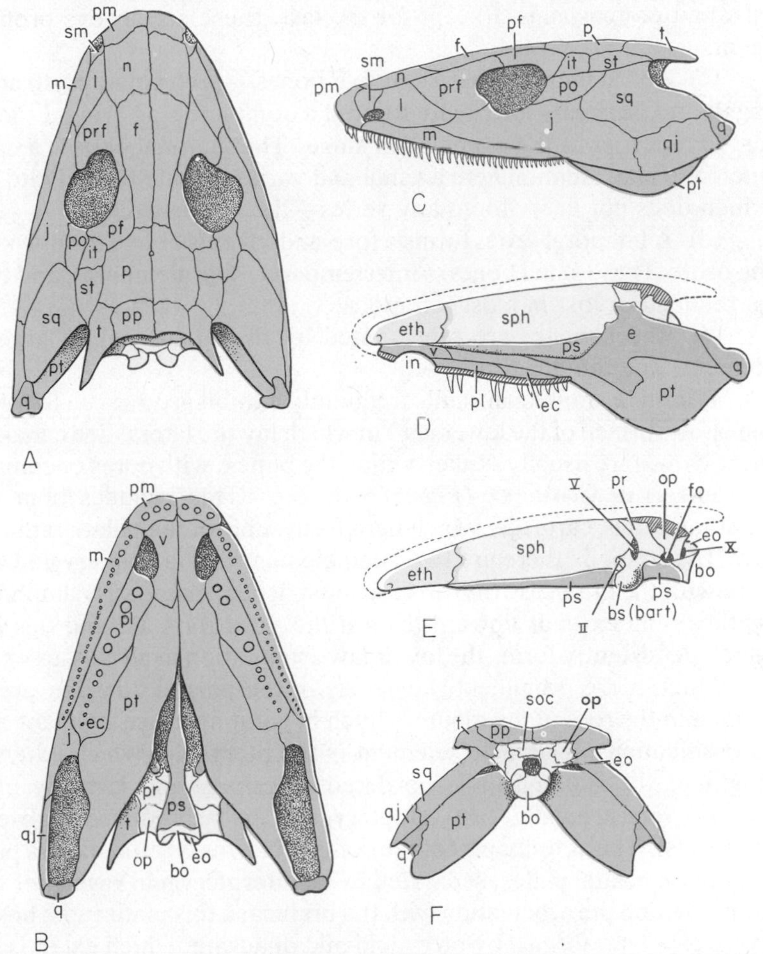

Figure 166. The skull of an ancestral land vertebrate, based primarily on the Carboniferous labyrinthodont *Palaeogyrinus. A,* Dorsal view of skull; *B,* palate; *C,* lateral view; *D,* lateral view with dermal skull roof removed (outline in broken line); palatal bones—dermal and endoskeletal—of left side are shown; deep to them, the braincase. The ventral hatched area is the sutural surface of palatal bones and maxilla. *E,* Lateral view of braincase; *F,* posterior view. Dermal bones, red; endochondral bones of palate, orange; endochondral bones of braincase, yellow; cartilage, blue. (After Watson and Panchen.)

Abbreviations: *bart,* basal articulation of braincase and palate; *bo,* basioccipital; *bs,* basisphenoid; *con,* condyle; *e,* epipterygoid; *ec,* ectopterygoid; *en,* external naris; *eo,* exoccipital; *eth,* ethmoid region; *f,* frontal; *fo,* fenestra ovalis; *ic,* foramen for internal carotid; *in,* internal naris; *iptv,* interpterygoid vacuity; *it,* intertemporal; *j,* jugal; *l,* lacrimal; *m,* maxilla; *n,* nasal; *nc,* nasal capsule; *on,* otic notch; *op,* opisthotic; *or,* orbit; *p,* parietal; *paf,* parietal foramen; *pf,* postfrontal; *pl,* palatine; *pm,* premaxilla; *po,* postorbital; *pp,* postparietal; *pr,* prootic; *prf,* prefrontal; *ps,* parasphenoid; *pt,* pterygoid; *ptf,* posttemporal fenestra; *pv,* pituitary vein; *q,* quadrate; *qj,* quadratojugal; *sm,* septomaxilla; *soc,* supraoccipital; *sph,* sphenethmoid; *sq,* squamosal; *st,* supratemporal; *stf,* subtemporal fossa; *t,* tabular; *v,* vomer. Roman numerals, foramina for cranial nerves.

tetrapods and bony fishes, is applied a sheet of dermal bone formed in the central area of the mouth vault.

THE PRIMITIVE AMPHIBIAN SKULL. DERMAL ROOF. In a primitive tetrapod the dermal skull roof includes a considerable number of elements, mainly paired, which are suturally united to form a practically solid shield. Some of these elements are lost in variable fashion in later tetrapods; many, however, are important in every group of bony vertebrates. It is a strain to commit to memory this long series of names. As an aid we may arbitrarily group them in several series (Fig. 168):

(a) Tooth-bearing marginal bones are the small, anteriorly placed **premaxilla** and the large **maxilla.**

(b) Paired elements along the dorsal midline include **nasals, frontals, parietals,** and **postparietals.** Except for the last, these are always prominent skull elements.

(c) A circumorbital series of five bones—**prefrontal, postfrontal, postorbital, jugal,** and **lacrimal**—originally formed a complete ring around the orbit. Only the last two persist to the mammalian stage. The lacrimal carries a canal for the tear duct. We may mention here a small and variable bone tucked into the nasal cavity which does not fit well into any series—the **septomaxilla.**

(d) A temporal series forms a fore-and-aft row above the otic notch and behind the orbit. These small bones—**intertemporal, supratemporal,** and **tabular**—tend to be reduced or lost in most tetrapods.

(e) Cheek bones are represented by the **squamosal,** a large and persistent element, and the **quadratojugal.**

Primitive amphibian skulls frequently exhibit grooves on the skull roof (and on the outer surface of the lower jaw) in which lay the lateral line canals. In bony fishes these canals are usually sunken within the bones, with pores opening to the surface.

PALATAL COMPLEX (Fig. 166 *B, D*). This includes bones formed in the palatoquadrate cartilage, which here forms part of the palate rather than the upper jaw. In tetrapods there are two such elements. The **epipterygoid** articulates at its base with the braincase (movably in most fishes, many early amphibians, and some reptiles) and extends upward toward the skull roof. Behind this is the **quadrate,** which persistently forms the lower jaw articulation in all classes except mammals.

Much more prominent, however, in the palatal complex are dermal bones, formed in the roof of the mouth, which in great measure supplant the cartilage and its ossifications. The major element is the **pterygoid,** which extends much of the length of the skull. This is bordered anteriorly and laterally by three smaller bones—**vomer, palatine,** and **ectopterygoid** (the last reduced or absent in many later tetrapods). The anterior part of the complex forms on either side a pair of essentially horizontal palatal plates, separated by an **interpterygoid vacuity** of variable proportions. Behind the articulation with the braincase this plate ends; posteriorly there is a vertical plate, formed by pterygoid and quadrate, which extends back and out to the jaw articulation, medial to the subtemporal fossa.

THE BRAINCASE (Fig. 166 *B, E, F*). In most groups of bony vertebrates a median dermal element, the **parasphenoid,** is closely applied to the under surface of the braincase; it is often difficult to distinguish it from the rest of the braincase (and in mammals and birds it has ceased to exist as a separate element). The braincase itself is generally well ossified, but the region of the nasal capsule never ossifies in typical land forms, and in modern amphibians ossification of the entire braincase is much reduced. In well ossified forms the bony elements of the braincase frequently fuse in the adult, making interpretation of the individual bones difficult.

The occiput includes a ring of four bones—the paired **exoccipitals** on either side of the foramen magnum, and the median **supraoccipital** and **basioccipital** above and below; the condyle (primitively single) is carried by exoccipitals as well as the basioccipital. The hypoglossal nerve (XII), except where posterior to the skull in the specialized modern amphibians, pierces the exoccipital; the vagus complex (X and XI) and a vein usually emerge through a foramen (jugular) just in front of the exoccipital, and nerve IX has its exit here or by a separate opening just ahead of this point. On either side, in front of the occipital bones, is the region of the otic capsule containing the internal ear. In primitive tetrapods this ossifies as two elements, **prootic** and **opisthotic,** and in tetrapods there is present an external opening, the **fenestra ovalis,** into which fits the base of the sound-transmitting stapes (evolved from the fish hyomandibular). Nerve VIII, of course, enters the inner surface of the capsule; nerve VII penetrates to the outer surface of the braincase by a canal in the prootic, and nerve V emerges through one or more openings at or near the anterior margin of that bone.

Forward from the otic region the braincase rapidly narrows to the sphenoid region lying between the orbits. A major element here is the **basisphenoid,** a median ventral ossification, sheathed below by the parasphenoid. It contains a pocket (or fossa) for the pituitary (not seen in the figures) and laterally sends out a basal process for articulation with the palatal complex. Ventrally the internal carotid arteries pierce the bone to enter the cranial cavity; the side walls are variably developed but contain at least a major opening for the optic nerve (II) and allow emergence of the small nerves to the eye muscles (III, IV, VI). The anterior end of the primitive tetrapod braincase, as far as ossified, is formed by a single large median element, the **sphenethmoid,** which is still present in rather primitive fashion in the living anurans. This contains the olfactory nerves, running back from the cartilaginous nasal capsules.

THE SKULL IN BONY FISHES. Having seen the general plan of a vertebrate skull in a representative and rather primitive type, we may now proceed to study variations on the "theme" which it presents, beginning with a comparison of its structure with that of bony fishes. Logically we should begin with the crossopterygians (now extinct except for one specialized form), since they include the ancestors of tetrapods. In these forms (Fig. 169) most major elements of the skull and palate can be readily identified in terms of tetrapod bones. They differ, however, in certain regards; in front, the nasal and "rostral" regions include a rather variable mosaic of small bones (probably a primitive condition), and posteriorly there is an extra row of "extrascapular" bones which are not part of the skull proper. Notable, however, are differences in proportions; the crossopterygian has a very short facial region in front of the orbits and a long postorbital region, so that the posterior bones of the roof are much elongated. The braincase of the ancient crossopterygians was highly ossified—so completely so that sutures are obscured and individual elements cannot be made out. In many features the braincase (Fig. 169 *D*), too, is like that of a typical early amphibian. There is, however, one seemingly major stumbling block in the comparison of braincases in the types—a stumbling block which at one time led to the belief that crossopterygians were too specialized to be ancestors of the tetrapods. The braincase is formed in two discrete pieces, front and back, which can move slightly on each other; the hind half is perforated by a great ventral tunnel for an enormous notochord. These unusual structural features are not found in ordinary tetrapods, even in amphibians as primitive as that here used as a type, and it was long felt that the crossopterygians possessed very specialized braincases. Very recently, however, there have been found braincases of the very oldest amphibians from the late Devonian. In them (to our astonishment and delight) there is seen a large notochordal canal, and, while there does not appear to have been motility between halves of the braincase, there is distinct evidence of its ossifi-

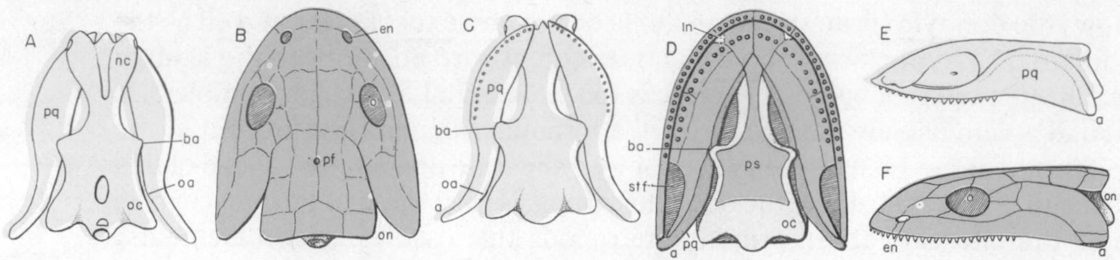

Figure 167. Diagrams to show the components of the skull. *A, C, E,* Dorsal, ventral, and lateral views of endoskeletal structures of braincase and palate (or upper jaw) as present in a shark or an embryo of higher jawed vertebrates; *B, D, F,* similar views with dermal elements added. In *B* and *F* a dermal roof shield binds together and covers braincase and endoskeletal jaw structures; in *D,* dermal palatal structures are seen reinforcing the endoskeletal palate, and a further dermal element underlies the braincase. Dermal bones, red; endochondral bones of palate, orange; endochondral bones of braincase, yellow; cartilage, blue. *a,* Articulation with lower jaw; *ba,* basal articulation of braincase with palatoquadrate; *en,* external naris; *in,* internal naris; *nc,* nasal capsule; *o,* orbit; *oa,* otic articulation of palatoquadrate; *oc,* otic capsule; *on,* otic notch; *pf,* parietal foramen; *pq,* palatoquadrate; *ps,* parasphenoid; *stf,* subtemporal fossa for jaw muscle.

cation in two units. Thus, here (as in many other instances) structures once thought to be "aberrant" or "specialized" apparently represent ancestral conditions.

The dipnoans are related to crossopterygians (and hence to tetrapod ancestors); but in their skull structures (Fig. 170) they have branched off in an entirely independent direction. The roof has lost much of its original extent (particularly in the living genera) and is represented by variable plates, which cannot be readily identified. Most of the bones of the palate (except for large pterygoids) have been lost or remain cartilaginous; the braincase has many features comparable to those of a tetrapod, but remains purely cartilaginous in the modern lungfishes and even in most fossil genera. In the skull as in the rest of the body the lungfish skeleton is thus highly reduced.

The ray-finned fishes (except for the sturgeons and paddlefishes) have a well-ossified skull (Figs. 122, 171, 172), which is comparable as to main components and their arrangement with that of early tetrapods and crossopterygians (although the braincase is more rarely—and differently—subdivided than in the latter group).

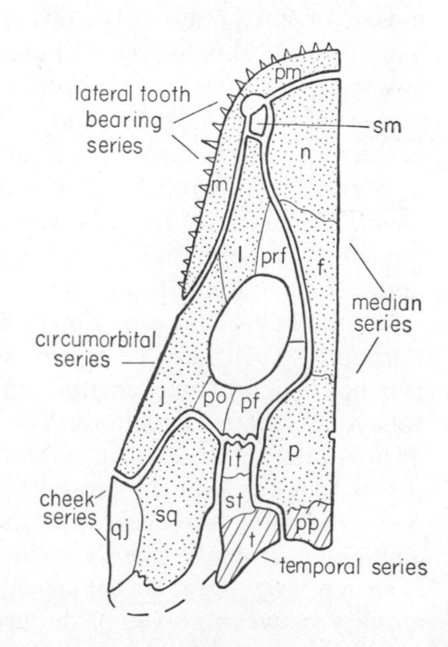

Figure 168. A diagram of the skull roof of a primitive tetrapod with the elements grouped (rather arbitrarily) into regional series. The little septomaxilla does not readily fit into any series. The stippled elements are retained as such in the skull of typical mammals; hatched elements at back appear to be fused into the mammalian occipital bone; elements unshaded are lost in mammals. *f,* Frontal; *it,* intertemporal; *j,* jugal; *l,* lacrimal; *m,* maxilla; *n,* nasal; *p,* parietal; *pf,* postfrontal; *pm,* premaxilla; *po,* postorbital; *pp,* postparietal; *prf,* prefrontal; *qj,* quadratojugal; *sm,* septomaxilla; *sq,* squamosal; *st,* supratemporal; *t,* tabular.

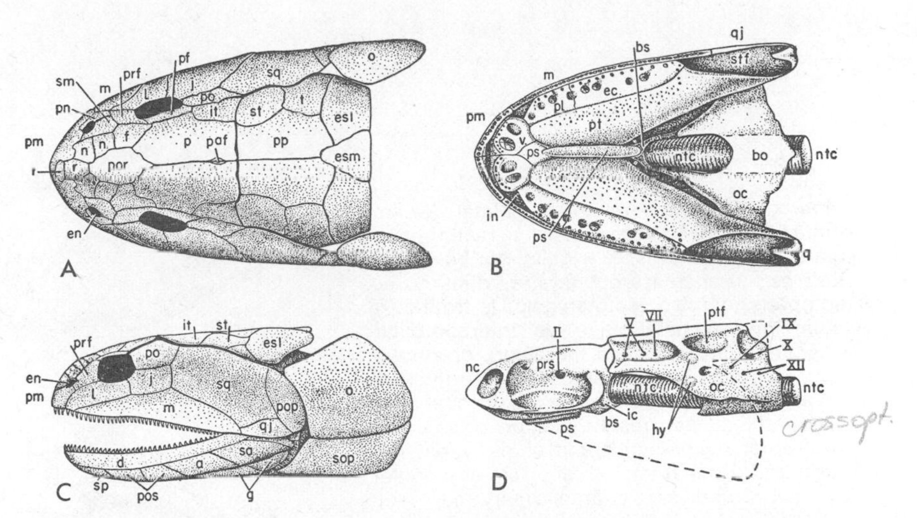

Figure 169. The skull of a Paleozoic crossopterygian (composite), for comparison with that of a primitive land vertebrate (cf. Fig. 166). *A*, Dorsal view; *B*, palatal view; *C*, lateral view; *D*, lateral view of braincase. From the crossopterygian dermal covering that of a labyrinthodont differs primarily in loss of opercular elements, relative reduction of length of posterior part of skull and elongation of "face" region, and reduction of small elements in rostral and nasal region; the "extrascapulars" at the back of the skull are not true skull bones but enlarged neck scales. The palate is similar in the two. The labyrinthodont braincase is less completely ossified, formed in one piece instead of the two present in crossopterygians, and lacks the greatly expanded notochord of the latter.

a, Angular; *bo*, basioccipital; *bs*, basisphenoid; *d*, dentary; *ec*, ectopterygoid; *en*, external naris; *esl*, *esm*, lateral and medial extrascapulars; *f*, frontal; *g*, gulars; *hy*, hyomandibular articulation; *ic*, foramen for internal carotid; *in*, internal naris; *it*, intertemporal; *j*, jugal; *l*, lacrimal; *m*, maxilla; *n*, nasal; *nc*, nasal capsule; *ntc*, notochord (restored); *o*, opercular; *oc*, otic capsule; *p*, parietal; *paf*, parietal foramen; *pf*, postfrontal; *pl*, palatine; *pm*, premaxilla; *pn*, postnasal; *po*, postorbital; *pop*, preopercular; *por*, postrostral; *pos*, postsplenial; *pp*, postparietal; *prf*, prefrontal; *prs*, presphenoid (sphenethmoid); *ps*, parasphenoid; *pt*, pterygoid; *ptf*, posttemporal fenestra; *q*, quadrate; *qj*, quadratojugal; *r*, rostrals; *sa*, surangular; *sm*, septomaxilla; *sop*, subopercular; *sp*, splenial; *sq*, squamosal; *st*, supratemporal; *stf*, subtemporal fossa; *t*, tabular; *v*, vomer. Roman numerals, foramina for cranial nerves. (*A* and *B* based on *Eusthenopteron; C* based on *Osteolepis; D* based on *Ectosteorhachis;* data from Jarvik, Romer, Säve-Söderbergh, Stensiö.)

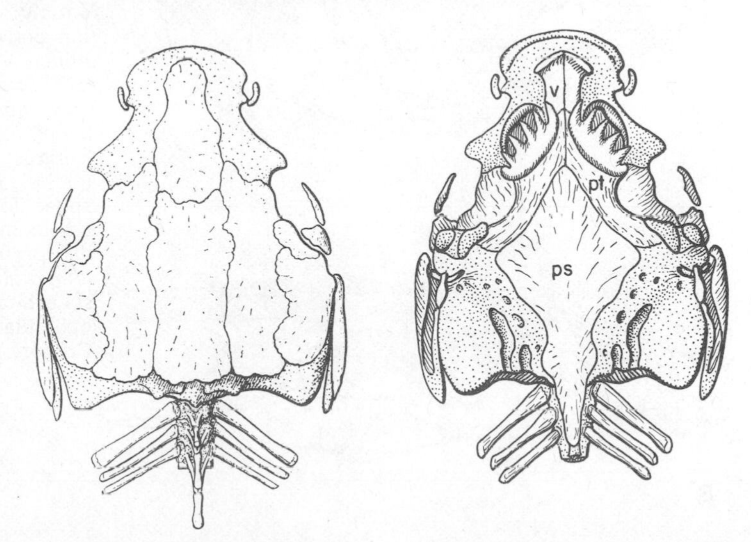

Figure 170. Skull roof and palate of the Australian lungfish, *Epiceratodus*. The braincase (stippled) is cartilaginous, but is underlain by a large parasphenoid *(ps);* the dermal roof includes only a small series of large plates which cannot be readily homologized with those of other forms. The upper jaws are fused to the braincase, and the only ossifications in the upper jaw or palate are large pterygoids *(pt)* and vomers *(v)*. Large fan-shaped toothplates are borne by the pterygoids, and a small cutting tooth is present on each vomer. (After Goodrich.)

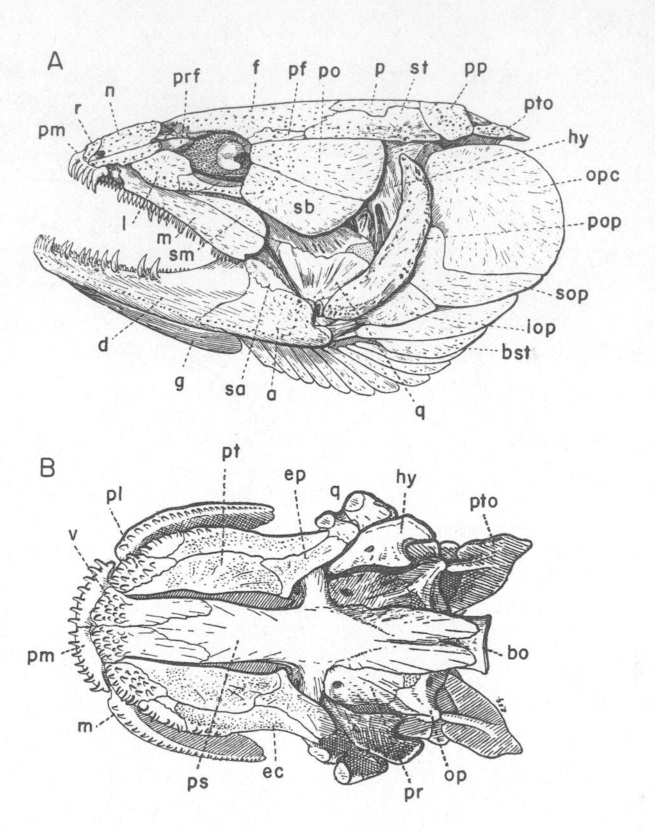

Figure 171. *A*, Lateral and *B*, palatal views of *Amia*, least specialized of living actinopterygians. Abbreviations: *a*, angular; *bo*, basioccipital; *bst*, branchiostegal rays; *d*, dentary; *ec*, ectopterygoid; *ep*, epipterygoid; *f*, frontal; *g*, gular; *hy*, hyomandibular; *iop*, interopercular; *l*, lacrimal; *m*, maxilla; *n*, nasal; *op*, opisthotic; *opc*, opercular; *p*, parietal; *pf*, postfrontal; *pl*, palatine; *pm*, premaxilla; *po*, postorbital; *pop*, preopercular; *pp*, postparietal; *pr*, prootic; *prf*, prefrontal; *ps*, parasphenoid; *pt*, pterygoid; *pto*, pterotic; *q*, quadrate; *r*, rostral; *sa*, surangular; *sb*, suborbital; *sm*, supramaxillary; *sop*, subopercular; *st*, supratemporal; *v*, vomer. The identity of certain of these elements with similarly named bones in crossopterygians and tetrapods is doubtful. (After Goodrich.)

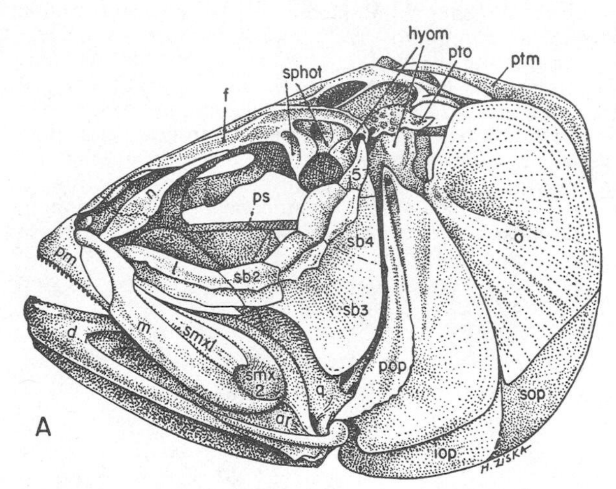

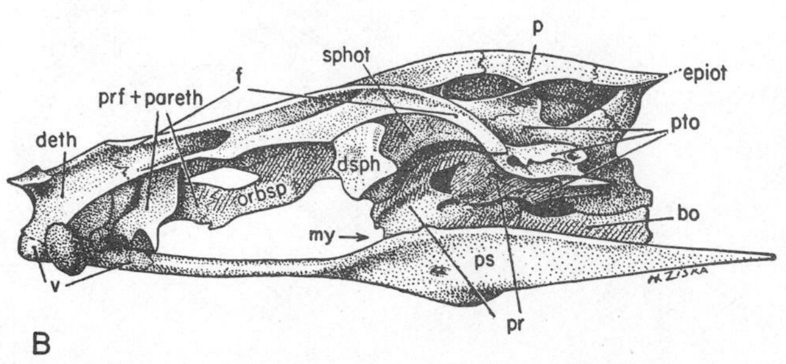

Figure 172. The teleost *Clupea* (herring). *A*, Lateral view of skeleton of head including the operculum; *B*, lateral view of braincase. The structure of teleost skulls is highly specialized, but derived from the general pattern seen in *Amia*. Abbreviations: *ar*, articular; *bo*, basioccipital; *d*, dentary; *deth*, dermethmoid; *dsph*, dermosphenotic; *epiot*, epiotic; *f*, frontal; *hyom*, hyomandibular; *iop*, interopercular; *l*, lacrimal; *m*, maxilla; *my*, opening of myodome (arrow); *n*, nasal; *o*, opercular; *orbsp*, orbitosphenoid; *p*, parietal; *pareth*, parethmoid; *pm*, premaxilla; *pop*, preopercular; *pr*, prootic; *prf*, prefrontal; *ps*, parasphenoid; *ptm*, posttemporal; *pto*, pterotic; *q*, quadrate; *sb 1–5*, suborbitals; *smx 1, 2*, supramaxilla; *sop*, subopercular; *sphot*, sphenotic; *v*, vomer. (After Gregory.)

When, however, we try to compare and name individual bones, we are in grave trouble. It is probable that the actinopterygians diverged at an exceedingly early date from the line leading to the crossopterygians and their tetrapod descendants, and that in consequence their bone patterns have little relation to those seen in the other group. Most living actinopterygians are teleosts, in which, among other specializations, there is great reduction of the cheek region, a shortening of the jaw gape, and a loosening and reduction of the upper jaw elements (Fig. 35).

HISTORY OF THE MAMMALIAN SKULL. In *Alice in Wonderland* the King of Hearts advises the White Rabbit that the proper way to tell a story is to "begin at the beginning and go on till you come to the end; then stop." We have already violated that advice by describing a primitive amphibian skull before discussing that of the fishes, and here we shall violate it again. Orthodox procedure would be to discuss, seriatim, the cranial changes seen in the various amphibians, reptiles, and birds, and only at the end come to mammals. However, we suspect that the major interest of most readers of this book lies in the evolution of the interesting but complex skull of mammals from primitive beginnings. Consequently, we shall now deal directly with the evolution of the mammalian skull, and only after that return to describe the variations seen in other tetrapod groups.

THE SKULL ROOF (Figs. 173–177). In the development of early reptiles from ancestral amphibians, the skull tends to become higher and narrower and, more importantly, the primitive otic notch is eliminated; the ear drum presumably lay behind the cheek bones and rather lower toward the jaw articulation. Further, there begins a reduction of the row of three small bones in the temple region—intertemporal, supratemporal, and tabular. The first is lost at the beginning of reptilian history, the supratemporal is much reduced at an early stage, and the tabulars (together with the postparietal, lying between the tabulars) are (so to speak) pushed off the skull roof to survive for the time as elements along the upper margin of the occiput (Fig. 182).

The line leading toward mammals split off early from other reptiles as the Pelycosauria of the late Carboniferous and early Permian, the lower of the two orders constituting the subclass Synapsida. The pelycosaurs were still very primitive reptiles but showed strong indications of a drift in a mammalian direction. Most notable is the development of an opening, a **temporal fenestra,** in the dermal bones of the cheek. In primitive tetrapods the strong mandibular adductor muscles, which close the jaw, are shut inside the solid dermal covering of the cheek (or temporal) region of the skull. When the jaws close, these muscles shorten but do not, of course, decrease in bulk, and hence must expand in breadth *within* the skull. A "solution" to this difficulty is brought about in most reptiles by the development of one (or two) openings in the primitively solid cheek covering (Figs. 44, 175). As noted later, such openings—fenestrae—are variable in reptiles; in pelycosaurs and higher synapsids the characteristic type is a **lateral temporal fenestra,** primitively well down on the side of the cheek.

The advanced synapsids are the Therapsida of the later Permian and Triassic, in which there is a much closer approach to mammalian conditions in the skull. The two temporal openings may enlarge and reach toward or almost to the mid-dorsal line, leaving below only a narrow band of bone, a **zygomatic arch** (Fig. 175). The bar of bone between orbit and temporal fenestra disappears in some advanced types; the maxilla (bearing the large canines) expands, reducing the size of the lacrimal, and the two external nares crowd toward the midline to form a single common bony opening. Various primitive elements tend toward reduction and to loss in advanced members of the group; these include septomaxilla, pre- and postfrontals, postorbi-

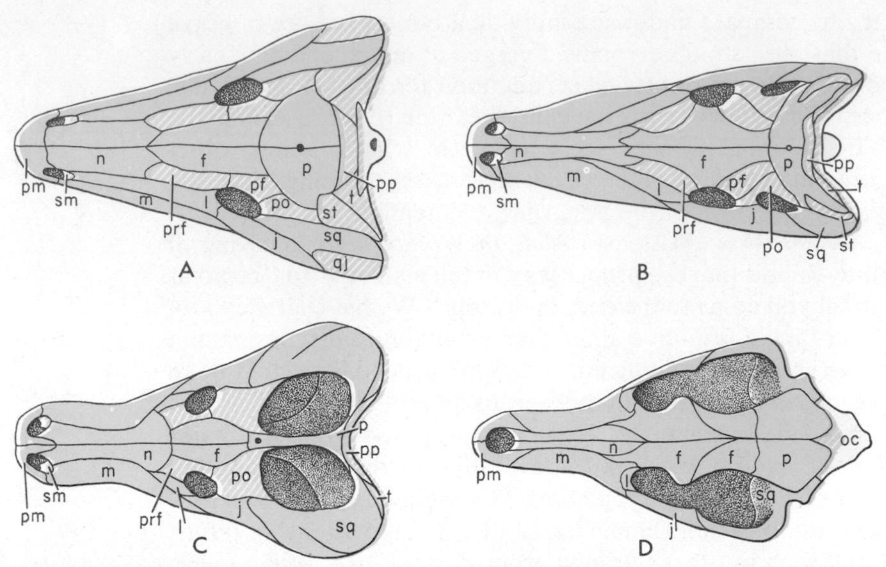

Figure 173. Diagrammatic dorsal views of skull roof to show the evolution of the mammalian condition. The principal changes are: (1), loss of many primitive elements; (2), development of a temporal opening, and reduction of the area between the openings to a sagittal crest, followed by an expansion of the braincase beneath. Colors and abbreviations as in Figure 166; however, dermal skull roof elements lost or fused with others in mammals are shown as hatched here and in Figures 174 and 176. *A,* a stem reptile (cotylosaur); *B,* a primitive mammal-like reptile (pelycosaur); *C,* an advanced mammal-like reptile (therapsid); *D,* a placental mammal. The forms shown in Figure 176 are representative of stages *B–D.*

tal, supratemporal, and quadratojugal. The originally paired postparietals fuse to a single median element, which in most mammals fuses with the occipital complex (as apparently do the tabulars; Fig. 182).

The primitive pattern of the mammalian skull roof is preserved, in the main, in such a form as the dog (Figs. 176 *D,* 177 *D*). There are, however, numerous variations in proportions and disposition of parts in mammals. For example, the bar behind the orbit, lost in ancestral mammals, may be rebuilt; the zygomatic arch may

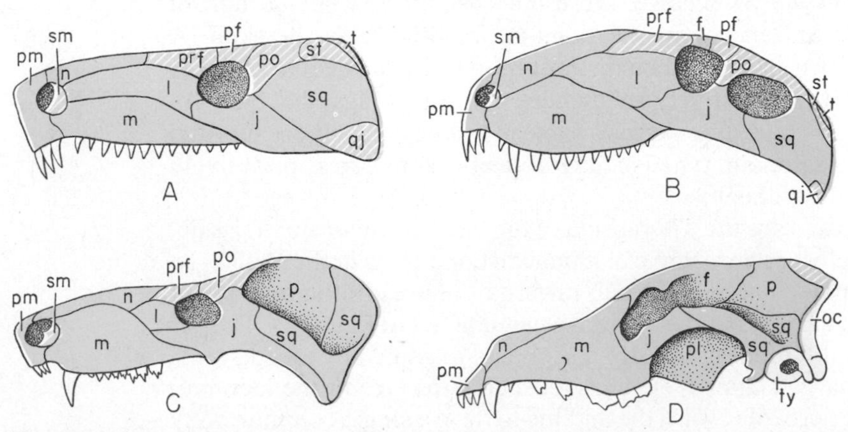

Figure 174. Diagrammatic views of the stages of the mammalian skull roof shown in Figure 173, seen in lateral view. Colors as in Figure 173; abbreviations as in Figure 177. The forms shown in Figure 177 are representative of stages *B–D.*

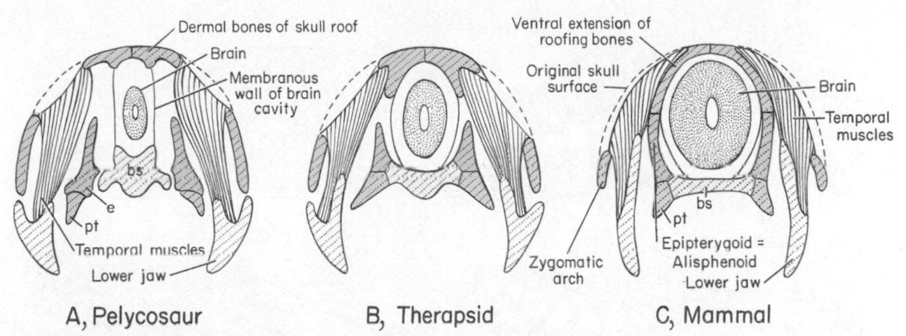

Figure 175. Diagrammatic cross sections of the skull and jaws of *A*, a primitive mammal-like reptile; *B*, an advanced mammal-like form; and *C*, a mammal; to show features in the development of the skull roof and braincase. (1) Part of the lateral walls enclosing the expanding brain were originally membranous. This lateral area comes to be ensheathed by ventral extensions of the roofing bones and by the incorporation of the epipterygoid of the palatal structure as the alisphenoid. (2) This extension of the roofing bones downward around the brain gives the appearance in mammals of being the original skull surface; the original surface, however, lay external to the temporal muscles, as indicated by the broken lines. (3) The upper jaw elements (*e*, epipterygoid; *pt*, pterygoid) originally articulated movably with the braincase (*bs*, basisphenoid); the two structures are fused in *B* and *C*. Dermal bones of skull roof and palate, red; endochondral bone of braincase, yellow; endochondral bone of upper jaw (epipterygoid = alisphenoid), orange.

be lost, notably in many insectivores and edentates; the face is greatly elongated in many insectivores and anteaters of several other orders, but, on the other hand, is much shortened in higher primates. The external nares may move upward and backward in forms with a flexible snout or trunk (tapirs, elephants) and, most

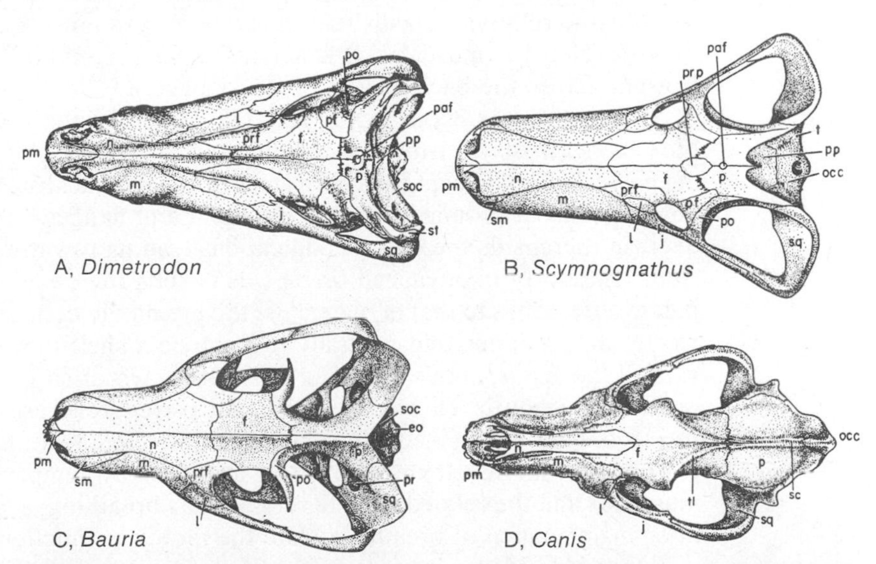

Figure 176. Dorsal views of skulls to show the evolution of the mammalian skull roof. *A*, Primitive early Permian mammal ancestor (pelycosaur); *B*, a later Permian therapsid; *C*, a progressive Triassic therapsid; *D*, the dog. Abbreviations: *eo*, exoccipital; *f*, frontal; *j*, jugal; *l*, lacrimal; *m*, maxilla; *n*, nasal; *occ*, occipital bone; *p*, parietal; *paf*, parietal foramen; *pf*, postfrontal; *pm*, premaxilla; *po*, postorbital; *pp*, postparietal; *pr*, prootic; *prf*, prefrontal; *prp*, preparietal; *q*, quadrate; *qj*, quadratojugal; *sc*, sagittal crest; *sm*, septomaxilla; *soc*, supraoccipital; *sq*, squamosal; *st*, supratemporal; *t*, tabular; *tl*, temporal lines. (*B* after Watson; *C* after Boonstra.)

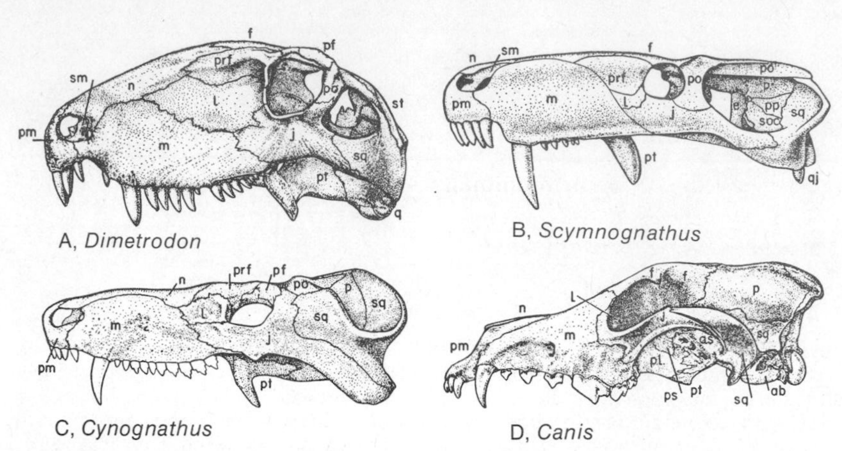

Figure 177. Lateral views to show the evolution of the mammalian skull. *A*, Primitive early Permian mammal ancestor (pelycosaur); *B*, a late Permian therapsid; *C*, a progressive Triassic therapsid; *D*, the dog. Abbreviations: *ab*, auditory bulla; *as*, alisphenoid; *e*, epipterygoid; *f*, frontal; *j*, jugal; *l*, lacrimal; *m*, maxilla; *n*, nasal; *p*, parietal; *pf*, postfrontal; *pl*, palatine; *pm*, premaxilla; *po*, postorbital; *pp*, postparietal; *prf*, prefrontal; *ps*, presphenoid; *pt*, pterygoid; *q*, quadrate; *qj*, quadratojugal; *sm*, septomaxilla; *soc*, supraoccipital; *sq*, squamosal; *st*, supratemporal. (*B* after Watson; *C* after Broili and Schroeder.)

notably, in the aquatic sirenians and whales. In rodents (cf. p. 227) the masseteric muscles of the jaw are highly developed, with correlated excavation of the side of the face for their accommodation. In small mammals with a relatively large brain, the dorsal surface of the skull appears swollen and smooth-surfaced; in contrast, in forms with a relatively small brain and powerful jaw muscles, a median **sagittal crest** may develop for muscular attachments, and a transverse **nuchal crest** may be present across the back margin of the skull roof.

The Palatal Complex (Figs. 178–180). In the palate there is relatively little change in the transition from primitive amphibians to primitive reptiles (including pelycosaurs), except for a narrowing of the palatal structure as a whole and the development of prominent tooth-bearing lateral flanges on the pterygoids. However, in therapsids there are modifications leading toward the mammalian condition. Anteriorly in advanced therapsids is seen the development of a secondary palate antecedent to that of mammals; the premaxillae, maxillae, and, more posteriorly, the palatines fold ventrally to produce a shelf that lies below the primary roof of the snout, above which air from the external nares passes far back before entering the mouth. The vomers, originally paired, form a single bone that supports the secondary palate medially. The palatal development may be reasonably associated with the development of the constant body temperature characteristic of mammals and the related need for continuous breathing; a secondary palate facilitates maintenance of breathing while the mouth is functioning in eating.

In therapsids (as in many lower tetrapods) the originally movable articulation of palate with braincase has been abandoned, and the pterygoids become immovably joined to the base of the skull (Fig. 175). These bones are already somewhat reduced in extent in therapsids, and in mammals are little more than small wings of bone projecting ventrally below on either side of the sphenoid region. Above the pterygoids, the epipterygoid bones persist as vertical plates on either side of the cranial cavity and in mammals fuse into the braincase walls as alisphenoids (p. 189).

In mammals a new jaw joint, between squamosal and dentary, replaces the old quadrate-articular joint. Transitional stages are seen in therapsids, for the quadrate becomes much reduced in size (see p. 200).

THE BRAINCASE (Figs. 181, 182). In early reptiles the braincase retains much of the structure already described for primitive amphibians, but tends to become relatively higher and narrower, and the walls of the interorbital region are rarely ossified. There tends, on the whole, to be little change in the mammal-like reptiles (in which there was little advance in brain size or structure). By the time true mammalian conditions are reached, however, there are notable changes. Posteriorly, the series of occipital elements have fused into a single **occipital bone** with which, we have noted, the postparietal and possibly the tabulars are also incorporated. In lower tetrapods the otic capsule was formed of two elements (pro- and opisthotic); in mammals it is formed from a variable number of centers of ossification which in the adult form a single **periotic** element. Posteriorly this is often exposed as a projecting **mastoid process** (medial to which there is frequently a **paroccipital process** of the occipital bone).

A new addition to the skull in almost all mammals is an **auditory bulla** (Figs. 180 *D;* 184), which surrounds the cavity of the middle ear and the delicate auditory ossicles that it contains; at its outer margin lies the ear drum (cf. p. 360). Apparently no bulla was present in the ancestral mammals, and there is much variation in the mode of formation of the bulla. Frequently present is a **tympanic** bone (which is thought to be a modification of the angular bone of the lower jaw). In its simplest form this is merely a bony ring surrounding and supporting the ear drum; in other cases it extends inward to form part or all of the bulla. In many groups of mammals there is a second, **entotympanic,** bone. This is formed in cartilage and is (most unusually) a new addition to the skull. It may form the entire bulla or fuse with the tympanic when the latter is present. Bulla and periotic may remain separate in the adult; frequently, however, they fuse with one another and with the adjacent squamosal to form a compound **temporal** bone.

Forward from the ear region, the basisphenoid continues to form the floor of the braincase in the region of the pituitary. Above this region, as we have noted, the side walls of the braincase were open in early reptiles; in mammals, however, the gap here is filled by the **alisphenoid** bone; this is the old epipterygoid bone of the palatal complex, modified to play a new role. Forward of the basisphenoid, in the region of the primitive sphenethmoid, we find its mammalian homologue, the **presphenoid;** its ascending lateral processes are sometimes distinct and are termed the **orbitosphenoids.** All the "sphenoid" bones may be discrete (particularly in embryos and young) but are frequently fused to form a single adult **sphenoid** bone—small but highly complex in structure (it may also include the pterygoid and even the parasphenoid). As noted earlier, there is seldom any ossification at the front end of the braincase in lower tetrapods. In some groups of mammals, however, an extra anterior element, a **mesethmoid,** is present in the nasal region. In mammals the highly developed nasal cavities lie close to the front end of the expanded cranial cavity, so that instead of forming an olfactory nerve stem, the nerve branches perforate the bar of bone between brain and nose, the **cribriform plate,** in a series of small foramina. In the nasal cavity there may develop variable scrolls of cartilage or bone—**turbinals**—which fuse with adjacent elements (maxilla, nasal, mesethmoid). Covered with mucosa, they increase the surface area of the nasal passages (see p. 343).

The bones named thus far compose the entire roster of elements of the original braincase. But if we examine the actual "braincase" of a typical mammal (Fig.

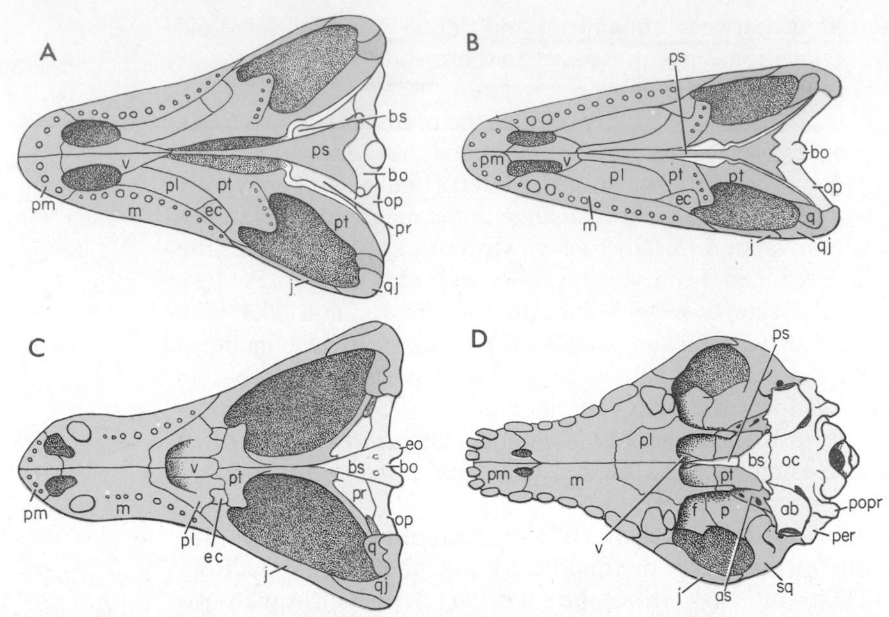

Figure 178. Diagrams to show evolution of the mammalian palate as seen in ventral view. *A,* a stem reptile (cotylosaur); *B,* a primitive mammal-like reptile (pelycosaur); *C,* an advanced mammal-like reptile (therapsid); *D,* a placental mammal. The principal changes are: (1) development of a secondary palate; (2) fusion of palatal structures with braincase at mid-length of skull; (3) reduction of pterygoid and quadrate, and loss of latter from skull structure. Colors as in Figure 166; abbreviations as in Figure 180. The forms shown in Figure 180 are representative of stages B–D.

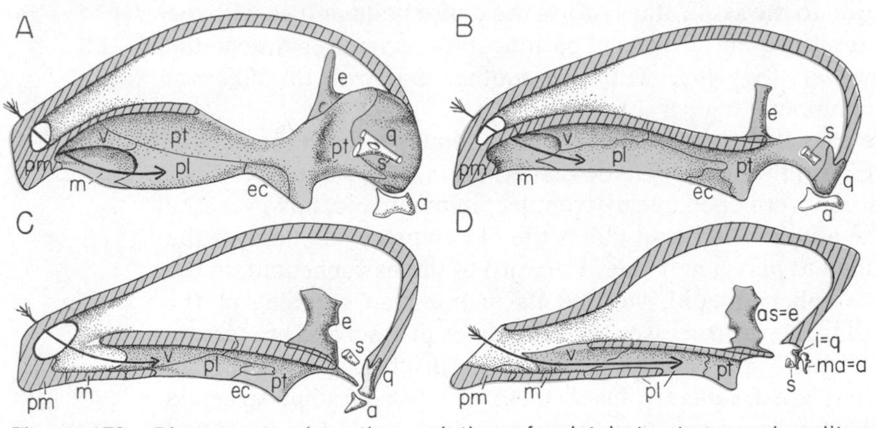

Figure 179. Diagrams to show the evolution of palatal structures and auditory ossicles from primitive reptiles to mammals. Longitudinal sections of skulls, with bones of skull roof and palate represented as if cut vertically just to the right of the midline (hatched surfaces). The left half of the skull and the entire braincase removed, so that the palatal structures of the right side are seen in medial view. In addition, the stapes and the articular element of the lower jaw are figured to show the evolution of the auditory ossicles. *A,* A primitive pelycosaur *(Dimetrodon); B,* a primitive therapsid; *C,* an advanced mammal-like reptile; *D,* a mammal. In *A* and *B* incoming air *(arrow)* enters the mouth cavity directly through anteriorly-placed internal nostrils; in *C* and *D* a secondary palate is developed and there is some degeneration of the primary palate. In *A* the palate articulates movably with the braincase at a socket on the epipterygoid; in *B,* the palate becomes fixed to the braincase; the epipterygoid loses its original function but survives as the alisphenoid bone of mammals. In *B–D* the pterygoid is reduced; the quadrate and articular, forming the articulation between upper and lower jaws, are reduced in size and lose their original function but survive as auditory ossicles. Abbreviations: *a,* articular; *as,* alisphenoid; *e,* epipterygoid; *ec,* ectopterygoid; *i,* incus; *m,* maxilla; *ma,* malleus; *pl,* palatine; *pm,* premaxilla; *pt,* pterygoid; *q,* quadrate; *s,* stapes; *v,* vomer. Colors as in Figure 166.

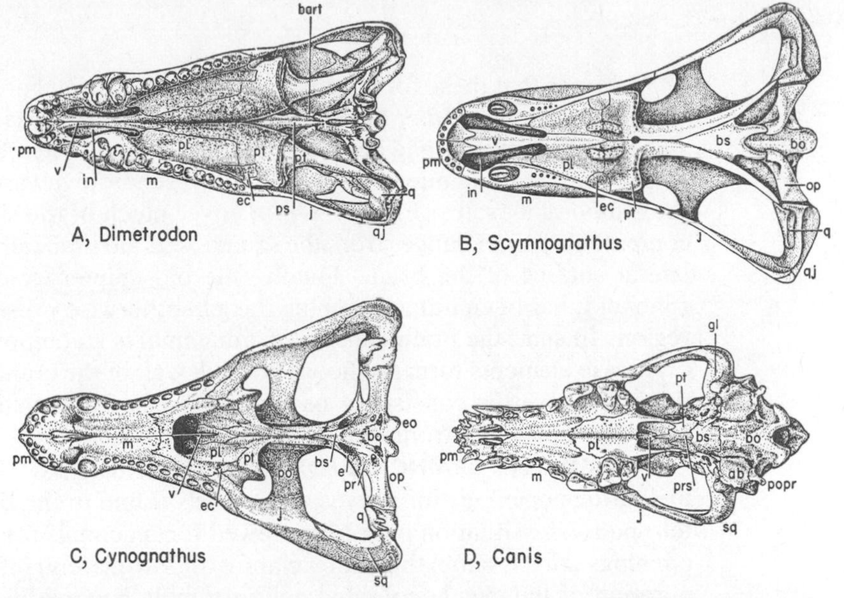

Figure 180. A series of skulls in ventral view to show the evolution of the mammalian palate. *A*, Primitive early Permian pelycosaur; *B*, a late Permian therapsid; *C*, an advanced Triassic therapsid; *D*, the dog. Principal changes include development of secondary palate in *C* and *D;* loss of movable basal articulation and fusion of braincase and palate in *B–D;* reduction of pterygoid; loss of quadrate from skull structure and development of new jaw joint in *D;* addition of auditory bulla in *D.* Abbreviations: *ab,* auditory bulla; *bart,* basal articulation of palate and braincase; *bo,* basioccipital; *bs,* basisphenoid; *e,* epipterygoid; *ec,* ectopterygoid; *eo,* exoccipital; *gl,* glenoid cavity; *in,* internal nares; *j,* jugal; *m,* maxilla; *op,* opisthotic; *pl,* palatine; *pm,* premaxilla; *po,* postorbital; *popr,* paroccipital process; *pr,* prootic; *prs,* presphenoid; *ps,* parasphenoid; *pt,* pterygoid; *q,* quadrate; *qj,* quadratojugal; *sq,* squamosal; *v,* vomer.

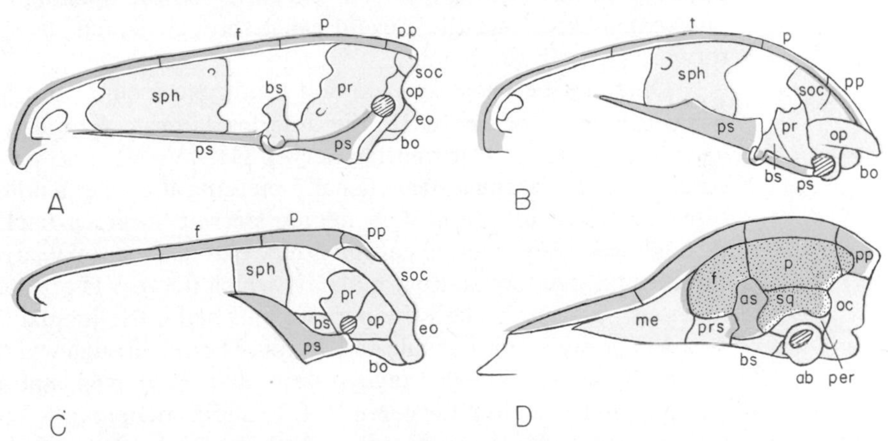

Figure 181. Diagrams to show evolution of mammalian braincase. *A,* primitive reptile (cotylosaur); *B,* primitive mammal-like reptile (pelycosaur); *C,* advanced mammal-like reptile (therapsid); *D,* placental mammal. Colors as in Figure 166. The principal changes are: (1) development anteriorly of presphenoid from sphenethmoid and in some mammals an added mesethmoid; (2) disappearance of parasphenoid; (3) addition of alisphenoid to braincase by transformation of epipterygoid of upper jaw; (4) fusion of otic elements into a periotic; (5) addition of tympanic element(s); (6) fusion of occipital elements plus postparietals into a single occipital bone. Except in mammals, the brain is confined to the posterior part of the cranial cavity (essentially between the otic and occipital elements) and the sphenethmoid serves merely for passage of the most anterior nerves. In the expanded mammalian brain, not only are sphenethmoid and epipterygoid (as presphenoid and alisphenoid) pressed into service to aid in brain coverage, but also flanges (stippled) from frontal, parietal, postparietal, and squamosal form parts of the lateral wall. Abbreviations: *ab,* auditory bulla; *as,* alisphenoid; *bo,* basioccipital; *bs,* basisphenoid; *eo,* exoccipital; *f,* frontal; *me,* mesethmoid; *occ,* occipital bone; *op,* opisthotic; *per,* periotic; *pr,* prootic; *prs,* presphenoid; *ps,* parasphenoid; *soc,* supraoccipital; *sph,* sphenethmoid. Other abbreviations as in Figure 182.

183), we see that these form little more than a floor and back wall for the expanded mammalian brain. Other elements help to completely sheathe the delicate brain. Parietals and frontals had been much reduced in surface area with the enlargement of the temporal openings. However, they send downward new flanges, deep to the temporal muscles (Fig. 175 *C*), to cover much of the side walls of the brain; a comparable pair of flanges from the squamosals aid in sheathing the back part of the lateral surface of the brain. Finally, the old epipterygoid, now termed the alisphenoid, has been utilized to plug the gap otherwise present above the sphenoid region. In sum, the brain capsule of a mammal is a composite affair. The original braincase elements form the floor and back wall of the cranial cavity; all the rest of the brain covering consists of new growths of dermal elements from the original skull roof, plus one from the erstwhile epipterygoid.

MAMMALIAN BRAINCASE FORAMINA (Fig. 184). We have mentioned earlier the various openings for nerves and vessels found in the braincase of a primitive tetrapod. The situation may be reviewed for mammals for, although many of the openings are the same, the nomenclature (unfortunately) differs; further, the incorporation of the alisphenoid and auditory bulla has modified the structure of the braincase.

Certain of the foramina and canals seen on the surface of the skull do not enter the braincase. In this category we may note: the **incisive foramina** in the front of the palate in some mammals, which connect the mouth with the vomeronasal organ (cf. p. 344); the **infraorbital foramen** (sometimes enlarged to a canal) which carries nerves and vessels forward from orbit to snout; the **nasolacrimal canal,** containing the tear duct; the **alisphenoid canal** piercing the bone of that name in some forms and carrying a branch of the internal carotid artery forward onto the palate; the **external auditory meatus** leading out from the eardrum; the opening into the bulla for the eustachian tube; and the **carotid canal** through which the internal carotid runs forward beneath the bulla.

Openings between surface and braincase include the **optic foramen** in the orbitosphenoid for nerve II; the **anterior lacerate foramen,** in front of the alisphenoid, typically transmitting nerves III, IV, VI, and part of V; the **foramen rotundum** and **foramen ovale** usually piercing the alisphenoid bone and carrying other branches of nerve V; a **middle lacerate foramen** back of the alisphenoid through which the internal carotid artery enters the cranial cavity; the **stylomastoid foramen** back of the auditory bulla, by which nerve VII reaches the surface after a tortuous passage through the periotic and bulla; the **jugular** or **posterior lacerate foramen** between otic capsule and occipital bone, through which emerge nerves IX, X, and XI and the internal jugular vein; and the **hypoglossal foramen** (sometimes multiple) in the occiput for nerve XII. In addition, openings from the cranial cavity which do not reach the surface include the multiple openings piercing the cribriform plate for nerve I and the **internal auditory meatus,** through which nerve VIII leaves the braincase and enters the otic capsule and through which nerve VII begins its outward journey.

THE SKULL ROOF IN LOWER TETRAPODS (Figs. 185–187). Having followed through the story of mammalian skull evolution, we will now present an outline of the evolution of the skull in other tetrapod types. Rather than discuss the changes rung on the primitive pattern in group after group of nonmammalian tetrapods, we shall, as in the case of mammal evolution, follow through separately the history of the major components.

The later history of the elements of the dermal skull roof is almost exclusively one of loss and degeneration. Almost never is there any development of a new element; always there is, in time, a greater or lesser degree of reduction. No living tetrapod has retained in full the pattern of its early ancestors, and few have preserved a solid roof covering.

Although a great variety of Paleozoic and early Mesozoic amphibians retained the primitive pattern with little change, the modern amphibians show a greater reduction of the skull roof than do many of the amniotes. In the broad, flat skull of a modern frog or salamander (Fig. 185 *B, C*) only a small proportion of the ancient dermal roof elements is preserved. Premaxilla, maxilla, nasal, frontal, parietal, and squamosal form the usual complement of bones. In urodeles a prefrontal is also present, but in anurans the parietal (as seen from embryos) appears to be fused with the frontals, and in some urodeles maxilla and nasal are absent. All other elements have vanished, and much of the upper surface of the head is bare of dermal bone. In the Gymnophiona (Fig. 185 *D*) the skull roof is a solid structure, but there is a similar reduction in the number of bones present.

In stem reptiles (Figs. 186 *A*, 187 *A*) there was little loss of the original elements, except that reduction of the temporal series had begun (these little elements and the postparietals disappear completely in most reptile groups). In later reptiles (Figs. 186 *B, D;* 187 *B–F*) there are numerous and varied modifications in the pattern of the skull roof. To some extent losses of elements occur, but major modifications are mainly associated with the development of openings in the cheek region of the skull—the **temporal fenestrae,** which offer valuable clues to reptile classification and relationship (Fig. 44). These openings, which afford release to the mandibular adductor muscles during their contraction, may develop high on the side of the cheek, in a low position, or in both—the last being termed the **diapsid** condition, because of the two bars of bone left on the cheek after fenestration. Stem reptiles have no temporal openings, and the same is true of turtles, although the roof may be "eaten away"—i.e., emarginated—to give a result comparable to fenestration. Such extinct reptiles as the plesiosaurs and ichthyosaurs have an upper opening. A great majority of reptiles, alive and extinct, are diapsid types. This is true of the tuatara—*Sphenodon*—and of the whole host of archosaurian reptiles with the crocodilians as living representatives; the birds (Fig. 188 *A*) are diapsid in pedigree, but the cheek bars, external to the swollen braincase, have largely disappeared. The Squamata are descended, too, from diapsid ancestors, but in lizards at least the lower of the two bars has gone, and in snakes both have vanished.

THE PALATAL COMPLEX IN LOWER TETRAPODS. In modern amphibians (Fig. 189 *B–D*) the original dermal elements of the palatal complex are retained except for the posterior lateral element—the ectopterygoid—but the quadrate may be incompletely ossified and the epipterygoid is absent; further, the originally movable joint between braincase and palate is lost and the two are immovably fused. In frogs the anterior end of the braincase is narrow, and there are very large interpterygoid vacuities. In the other two orders the floor of the braincase, covered by the parasphenoid, is very broad and flat and the vacuities on either side less developed in consequence.

In reptiles (Fig. 190), early types held fairly close to the primitive conditions. The Squamata and *Sphenodon* have modestly developed interpterygoid vacuities and retain the original motility of palate on braincase, but in other groups the two structures fuse together in the sphenoid region, and the palate tends to become a solid plate for its entire width. The reduction of temporal arches in the Squamata has enabled the quadrate to move freely on the remainder of the skull, and in snakes this, combined with flexibility of other jaw structures, gives them an enormous gape for swallowing large prey whole (Fig. 187 *D, E*). In some turtles the internal nostril

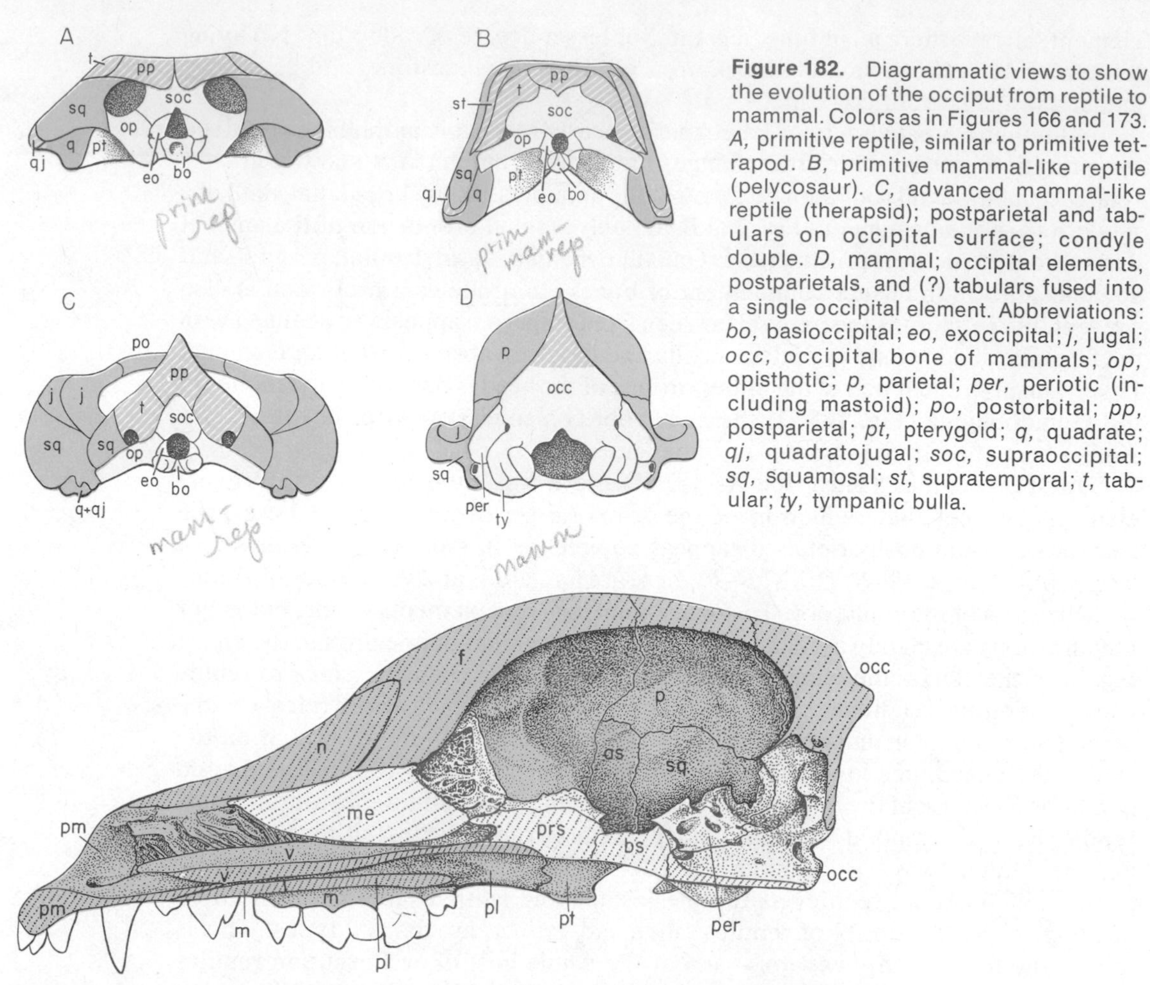

Figure 182. Diagrammatic views to show the evolution of the occiput from reptile to mammal. Colors as in Figures 166 and 173. *A*, primitive reptile, similar to primitive tetrapod. *B*, primitive mammal-like reptile (pelycosaur). *C*, advanced mammal-like reptile (therapsid); postparietal and tabulars on occipital surface; condyle double. *D*, mammal; occipital elements, postparietals, and (?) tabulars fused into a single occipital element. Abbreviations: *bo*, basioccipital; *eo*, exoccipital; *j*, jugal; *occ*, occipital bone of mammals; *op*, opisthotic; *p*, parietal; *per*, periotic (including mastoid); *po*, postorbital; *pp*, postparietal; *pt*, pterygoid; *q*, quadrate; *qj*, quadratojugal; *soc*, supraoccipital; *sq*, squamosal; *st*, supratemporal; *t*, tabular; *ty*, tympanic bulla.

Figure 183. Median sagittal section of the dog skull. Colors as in Figure 166. Diagonal lines indicate sectioned bones. It will be seen that most of the bone enclosing the brain cavity is derived from dermal elements *(f, p, sq)*, and even part of the occipital is embryologically of dermal origin. Abbreviations: *as*, alisphenoid; *bs*, basisphenoid; *f*, frontal; *m*, maxilla; *me*, mesethmoid; *n*, nasal; *occ*, occipital; *p*, parietal, *per*, periotic; *pl*, palatine; *pm*, premaxilla; *prs*, presphenoid; *pt*, pterygoid; *sq*, squamosal; *v*, vomer.

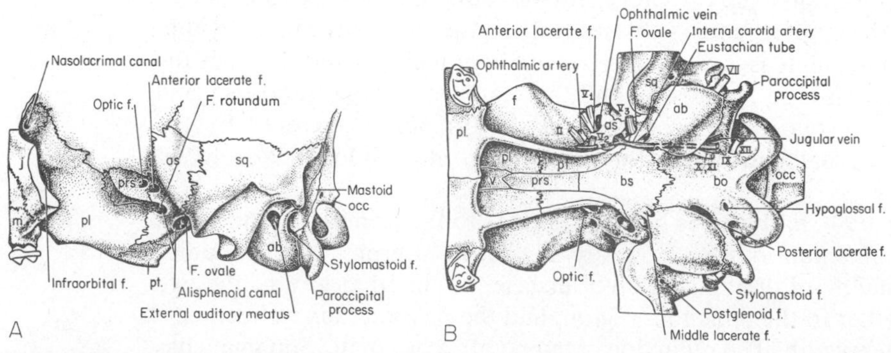

Figure 184. Braincase region of a dog in *A*, lateral view, *B*, ventral view, to show foramina. In *B* are indicated the main nerves, the course of the internal carotid artery and its palatine branch, and the jugular vein. Abbreviations: *ab*, auditory bulla; *as*, alisphenoid; *bo*, basioccipital; *bs*, basisphenoid; *f*, frontal; *j*, jugal; *m*, maxilla; *occ*, occipital; *pl*, palatine; *prs*, presphenoid; *pt*, pterygoid; *sq*, squamosal; *v*, vomer.

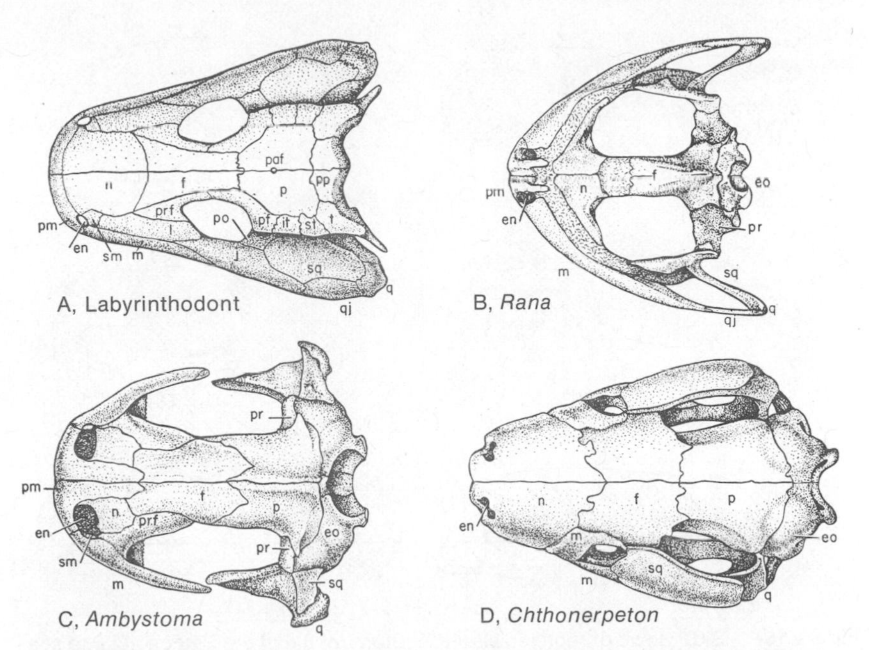

Figure 185. Dorsal views of amphibian skulls. *A*, A primitive labyrinthodont; *B*, a frog; *C*, a salamander; *D*, a gymnophionan. Abbreviations: *en*, external naris; *eo*, exoccipital; *f*, frontal (fused with parietal in frogs); *it*, intertemporal; *j*, jugal; *l*, lacrimal; *m*, maxilla; *n*, nasal; *p*, parietal; *paf*, parietal foramen; *pf*, postfrontal; *pm*, premaxilla; *po*, postorbital; *pp*, postparietal; *pr*, prootic; *prf*, prefrontal; *q*, quadrate; *qj*, quadratojugal; *sm*, septomaxilla; *sq*, squamosal; *st*, supratemporal; *t*, tabular. (*A* after Watson; *D* after Marcus.)

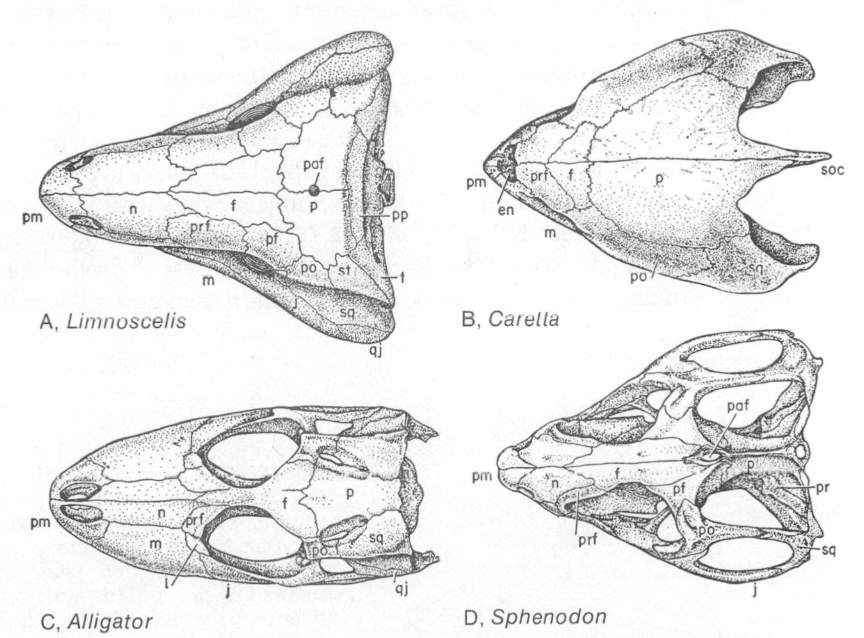

Figure 186. Dorsal views of reptilian skulls. *A*, Stem reptile of the Paleozoic; *B*, a sea turtle; *C*, a young alligator; *D*, *Sphenodon*. Abbreviations: *en*, external naris; *f*, frontal; *j*, jugal; *l*, lacrimal; *m*, maxilla; *n*, nasal; *p*, parietal; *paf*, parietal foramen; *pf*, postfrontal; *pm*, premaxilla; *po*, postorbital; *pp*, postparietal; *pr*, prootic; *prf*, prefrontal; *qj*, quadratojugal; *soc*, supraoccipital; *sq*, squamosal; *st*, supratemporal; *t*, tabular.

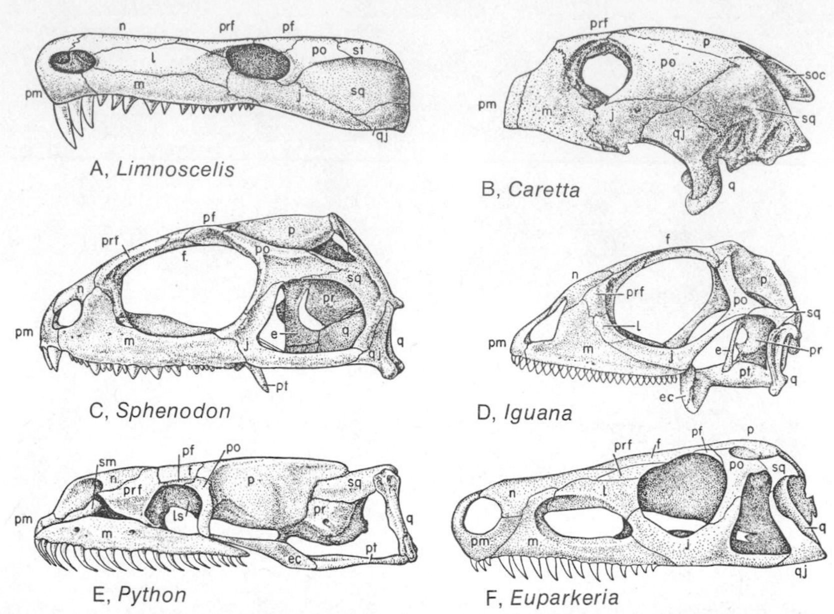

Figure 187. Side views of reptilian skulls. *A,* Stem reptile of the Paleozoic; *B,* a sea turtle; *C, Sphenodon; D,* a lizard; *E,* a python; *F,* a primitive ruling reptile of a type from which birds, dinosaurs, and crocodilians have descended. Abbreviations: *e,* epipterygoid; *ec,* ectopterygoid; *f,* frontal; *j,* jugal; *l,* lacrimal; *ls,* laterosphenoid; *m,* maxilla; *n,* nasal; *p,* parietal; *pf,* postfrontal; *pm,* premaxilla; *po,* postorbital; *pr,* prootic; *prf,* prefrontal; *pt,* pterygoid; *q,* quadrate; *qj,* quadratojugal; *sm,* septomaxilla; *soc,* supraoccipital; *sq,* squamosal; *st,* supratemporal. (*F* after Broom.)

openings, placed far forward in primitive forms, lie in a pocket in the roof of the mouth with some development of a secondary shelf of bone below them. We see here the beginning of a **secondary palate.** This is developed to a much higher degree in crocodiles and alligators, in which the secondary shelf of bone is so extensive that the air channels extend far toward the back of the skull before opening into the mouth—a feature useful to these aquatic animals when dealing with prey under water. In birds (Fig. 188 *B*) the palatal structures are lightly built and flexible, with a movable articulation with the braincase and with freely movable quadrates.

THE BRAINCASE IN LOWER TETRAPODS. In modern amphibians the braincase has tended to become broad and flat—more so in urodeles and gymnophionans than in the frogs. With the skeletal reduction seen generally in the three living orders, it

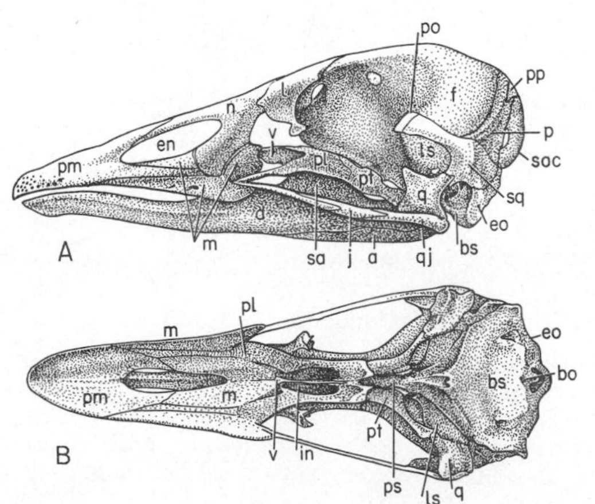

Figure 188. *A,* Lateral, and *B,* ventral views of the skull of a duck *(Anas).* Abbreviations: *a,* angular; *bo,* basioccipital; *bs,* basisphenoid; *d,* dentary; *en,* external naris; *eo,* exoccipital; *f,* frontal; *in,* internal naris; *j,* jugal; *l,* lacrimal; *ls,* laterosphenoid; *m,* maxilla; *n,* nasal; *p,* parietal; *pl,* palatine; *pm,* premaxilla; *po,* postorbital; *pp,* postparietal; *ps,* parasphenoid; *pt,* pterygoid; *q,* quadrate; *qj,* quadratojugal; *sa,* surangular; *soc,* supraoccipital; *sq,* squamosal; *v,* vomer. (After Heilmann.)

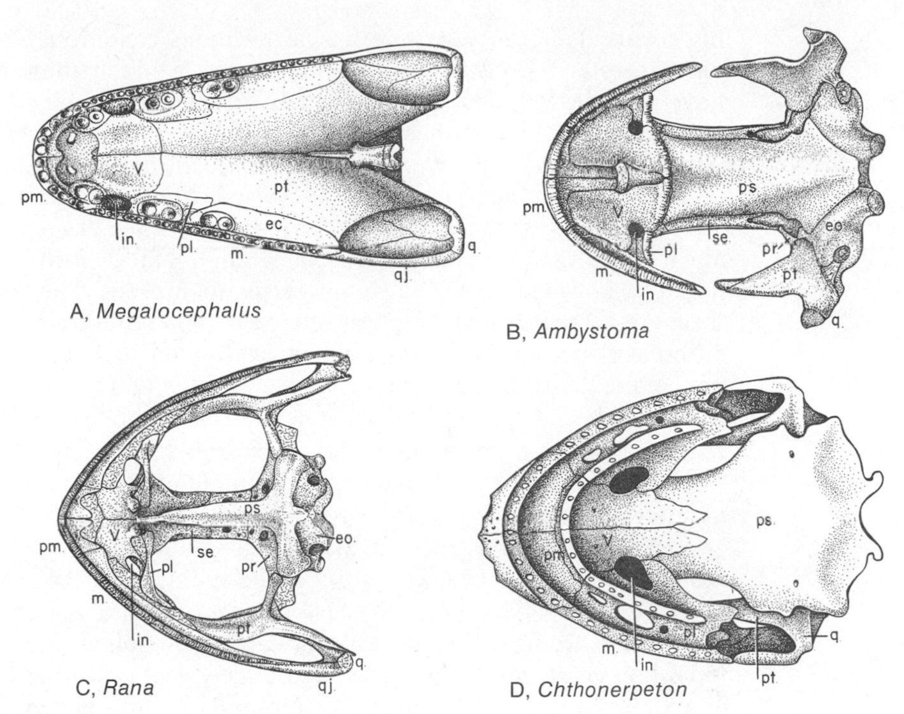

Figure 189. The palate of amphibians. *A,* Paleozoic labyrinthodont; *B,* a salamander; *C,* a frog; *D,* a gymnophione amphibian. Abbreviations: *ec,* ectopterygoid; *eo,* exoccipital; *in,* internal naris; *m,* maxilla; *pl,* palatine; *pm,* premaxilla; *pr,* prootic; *ps,* parasphenoid; *pt,* pterygoid; *q,* quadrate; *qj,* quadratojugal; *se,* sphenethmoid; *sq,* squamosal; *v,* vomer. (*A* after Watson; *D* after Marcus.)

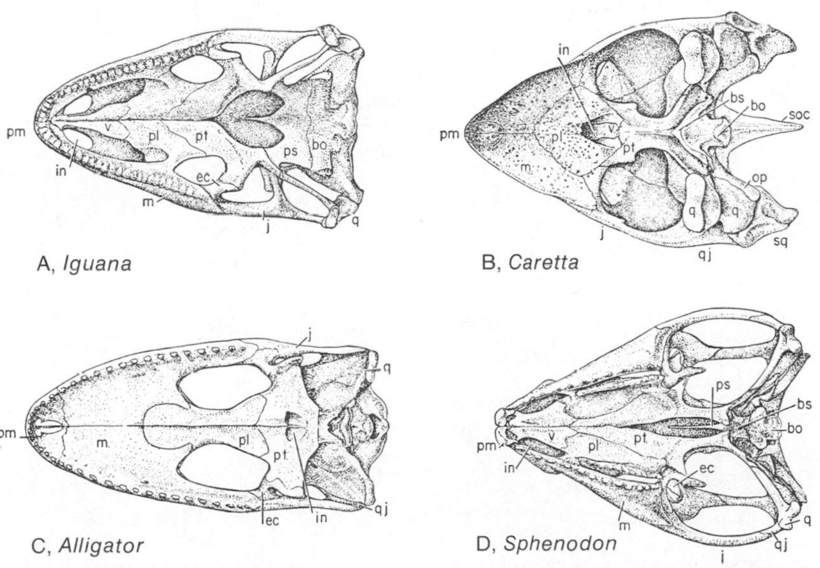

Figure 190. The palate of reptiles. *A,* Lizard; *B,* a sea turtle; *C,* a young alligator; *D, Sphenodon.* Abbreviations: *bo,* basioccipital; *bs,* basisphenoid; *ec,* ectopterygoid; *in,* internal naris; *j,* jugal; *m,* maxilla; *op,* opisthotic; *pl,* palatine; *pm,* premaxilla; *ps,* parasphenoid; *pt,* pterygoid; *q,* quadrate; *qj,* quadratojugal; *soc,* supraoccipital; *sq,* squamosal; *v,* vomer.

has reverted in great measure to a cartilaginous condition. There are paired exoccipitals, often a prootic bone and, anteriorly, a sphenethmoid; other ossifications have disappeared. The occipital condyle, originally single, is paired in living amphibians, and the skull has apparently shortened posteriorly, for in the occipital region a foramen for the twelfth cranial nerve (the hypoglossal) is no longer present.

Reptiles, on the other hand, show in many instances a much better retention of primitive features: there is usually no flattening of the braincase, and ossification is much more extensive than in modern amphibians; with one exception, every primitive bony element of the braincase is still present. The condyle remains single in all typical reptiles, and reptiles retain a hypoglossal foramen. There is one major difference from a primitive amphibian: presumably because of the relative narrowness of the skull, the two orbits are close together and the region between them fails to ossify; the old sphenethmoid ossification is reduced or absent (although partially relief) to the relatively simple story of the evolution of the lower jaw. As we have (Fig. 188) the braincase is swollen and completely ossified; it is partly formed by a new development of bony plates comparable to that described earlier for mammals.

LOWER JAW (Figs. 191, 192). To conclude this chapter we may turn (with relief) to the relatively simple story of the evolution of the lower jaw. As we have noted, this may have begun its history as a part of a visceral arch, which in the sharklike fishes forms the **mandibular cartilage.** In all vertebrates with bony skeletons, however, this is reinforced and largely replaced functionally by a series of dermal elements. The cartilage develops fully in the embryo, but usually produces

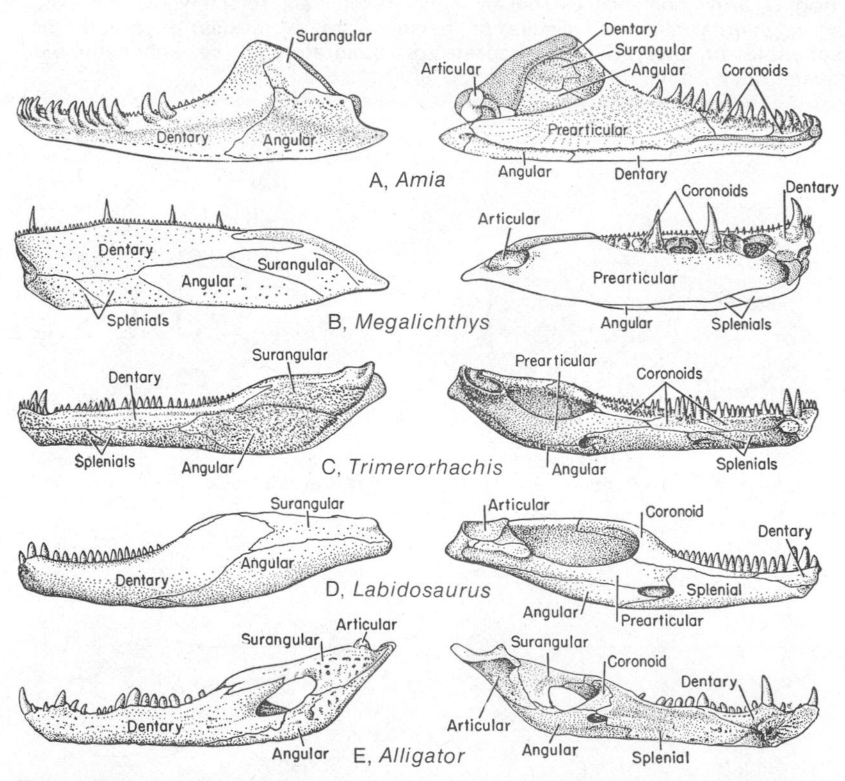

Figure 191. Left lower jaws, outer views at left, inner views at right, of *A*, a ray-finned fish, the bow fin; *B*, a primitive crossopterygian; *C*, a primitive labyrinthodont; *D*, a primitive reptile; *E*, an alligator. The jaws of modern teleosts, amphibians, and reptiles, in which the number of elements is reduced, have been derived from the types shown in *A*, *C*, and *D*, respectively.

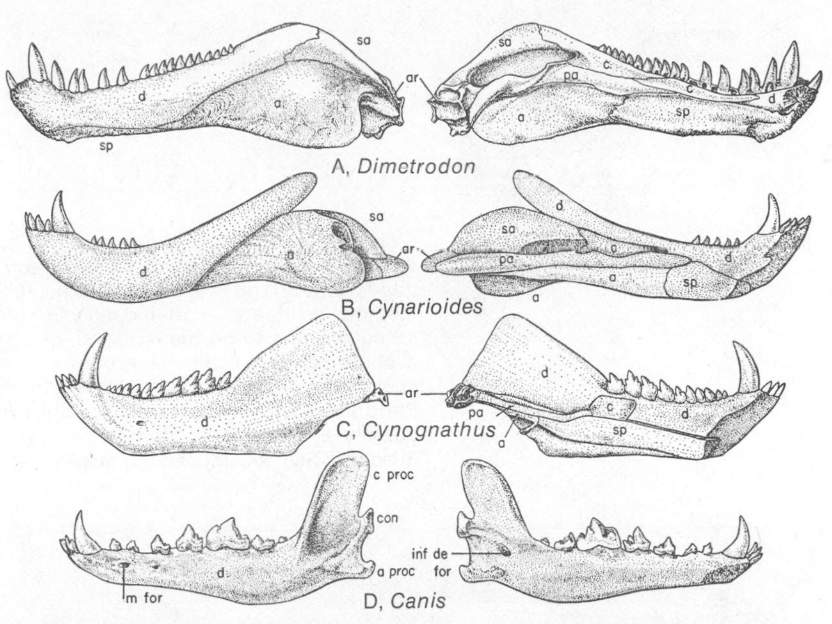

Figure 192. Left lower jaws of mammal-like reptiles and mammals, illustrating the reduction of jaw elements. Outer views *(left)* and inner views *(right)*. *A,* Primitive mammal-like reptile (pelycosaur); *B,* a primitive therapsid; *C,* an advanced therapsid; *D,* a typical mammal (dog). Abbreviations: *a,* angular; *a proc,* angular process; *ar,* articular; *c,* coronoid; *con,* condyle; *c proc,* coronoid process; *d,* dentary; *inf de for,* inferior dental foramen; *m for,* mental foramen; *pa,* prearticular; *sa,* surangular; *sp,* splenial.

in the adult only a single bony element, the **articular,** situated at the back of the jaw and bearing, as the name implies, an articular surface for the quadrate bone of the skull.

Of sheathing bones on the outer surface, the most important and largest is the **dentary** bone, which bears the marginal tooth row and forms part or all of the **symphysis** uniting the two sides of the jaw. Below and behind the dentary in many primitive fishes and early tetrapods is a whole series of dermal bones on the outer jaw surface: two **splenial** elements, an **angular** and a **surangular.** On the upper surface of the jaw, in front of the short articular bone, there is a fossa into which the major muscles closing the jaw insert and into which enter blood vessels and nerves serving the jaw. Below this fossa, on the inner side, there primitively runs forward a long **prearticular** bone; below the dentary there is primitively, on the inside of the jaw, a series of three slender **coronoid** bones, often bearing teeth. The lower part of the inner surface of the jaw is often sheathed by extensions of the bones of the outer surface.

From this primitive structure there are numerous variations; almost all involve reduction in the number of elements present. The dentary is almost invariably retained, and the articular, because of its function, is present in all groups except mammals (although it fails to ossify in some skeletally degenerate fishes and amphibians). On the other hand, there is seldom more than one splenial; the coronoids are usually reduced in number and may be lost; angular, surangular, and prearticular are more constant, but even these may be absent or fused with neighboring elements. A few variants in jaw structure among fishes and lower tetrapods are shown in Figure 191.

In the evolution of mammals (Fig. 192), we find in the fossil therapsids a steady

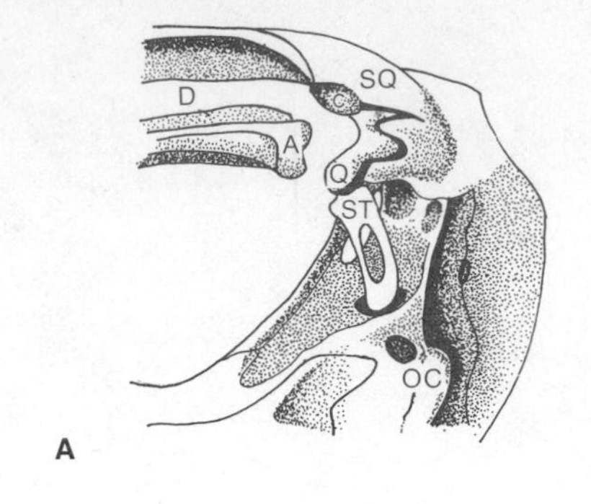

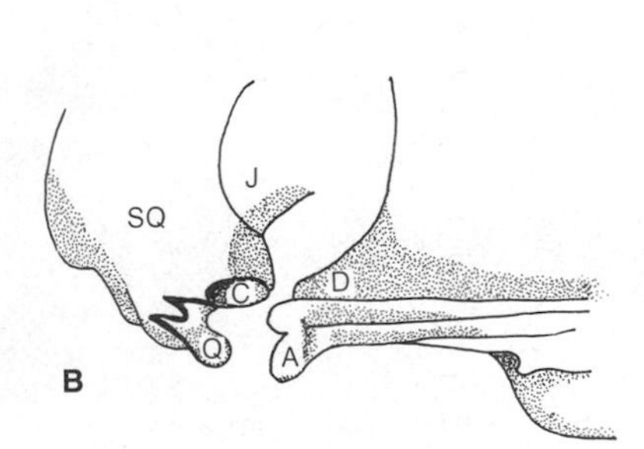

Figure 193. The jaw articulation of an advanced therapsid, the cynodont *Probainognathus,* in *(A)* ventral and *(B)* medial views. In this form both the old, reptilian articular-quadrate joint and the new, mammalian dentary-squamosal joint are functional. Note how the articular (soon to be the malleus), quadrate (incus), and stapes form a chain leading from the region of the tympanic membrane (not shown; posteroventral to the jaw articulation) to the otic capsule of the braincase. *A,* Articular; *C,* condyle in the squamosal with which dentary articulates; *D,* dentary; *J,* jugal; *OC,* occipital condyle; *Q,* quadrate; *SQ,* squamosal; *ST,* stapes. (After Romer.)

increase in the size of the dentary and a corresponding decrease in the size and strength of other elements. In advanced therapsids the enlarged dentary has developed an ascending **coronoid process** to which much of the jaw musculature is attached, and posteriorly reaches back to a point close to that at which the articular element gains contact with the skull. The remaining jaw elements are small and feeble structures plastered on to the inner surface of the dentary. With the transition to the mammal condition, these elements disappear from the jaw, which now consists of the dentary bone alone. The old elements are not, however, entirely abandoned. As will be seen later the articular (and quadrate) takes up a new if modest career as a tiny ear ossicle (cf. p. 360). The transitional stages in advanced therapsids are surprisingly well known; indeed, we now have numerous forms in which both the old reptilian joint (articular-quadrate) and the new mammalian one (dentary-squamosal) are simultaneously developed and functional (Fig. 193). The system will work as long as all the joints are on, or very close to, one transverse line—a door can have as many hinges as desired along its side as long as they are all in a straight line. Finally the angular bone has been, it seems, incorporated into the skull as the tympanic bone of the auditory bulla.

9 Muscular System

The muscular system, judged quantitatively, at least, should loom large in any study of the present sort, for muscle tissue constitutes from a third to half of the bulk of the average vertebrate. Functionally, too, the musculature is of the highest importance. From locomotion to the circulation of the blood, the major functions of the body are caused by or associated with muscular activity. Movement—of trunk, limbs, jaws, or any organ—is the major product of this activity, but muscular work may be expended without obvious movement in maintaining body stability and the production of body heat. Indeed, the activity of the nervous system—even the highest functioning of a human brain—has little mode of expression other than the contraction of muscle fibers.

MUSCLE FIBER TYPES (Fig. 194). Histologically, two major categories of muscular tissue may be distinguished—smooth and striated fibers. **Smooth muscle fibers** are the simpler and smaller of the two types. They are normally derived from embryonic mesenchyme, in association with the connective tissues. The main site of smooth muscle fibers is in the lining of the digestive tract, or of its embryonic outgrowths such as the trachea and bronchi of the lungs, and the bladder. Still other loci, however, are independent of the gut—notably the walls of circulatory vessels. A typical smooth muscle fiber is a slender, spindle-shaped body, averaging a few tenths of a millimeter in length. There is a single centrally situated nucleus; with special stains the seemingly homogeneous protoplasm exhibits tiny fibrils running the length of the cell. As the name implies, the fibers in this type of muscle lack the cross-banding seen in striated muscle cells. Smooth muscle fibers may be scattered, but more generally are arranged in bands or bundles, with interspersed connectve tissue fibers uniting them into an effective common mass (cf., for example, Figs. 270, 278).

In the heart is found a special type of **cardiac muscle** not present elsewhere. Embryologically heart muscle is of common origin with smooth muscle, but in correlation with its important functions and constant activity, it has developed a cross-banding similar to that seen in striated fibers. In contrast with that tissue, however, heart muscle does not consist of individual fibers, but is a continuous network of dividing and recombining strands; at intervals along them prominent cross bands, termed **intercalated discs,** separate successive cellular units.

Striated muscle fibers form the "flesh" of the body, the voluntary muscles, derived in great measure from the myotomes of the embryo; they generally attach to and move skeletal structures. These fibers are large multinucleate cells, with lengths which vary from about a millimeter to a number of centimeters. As in

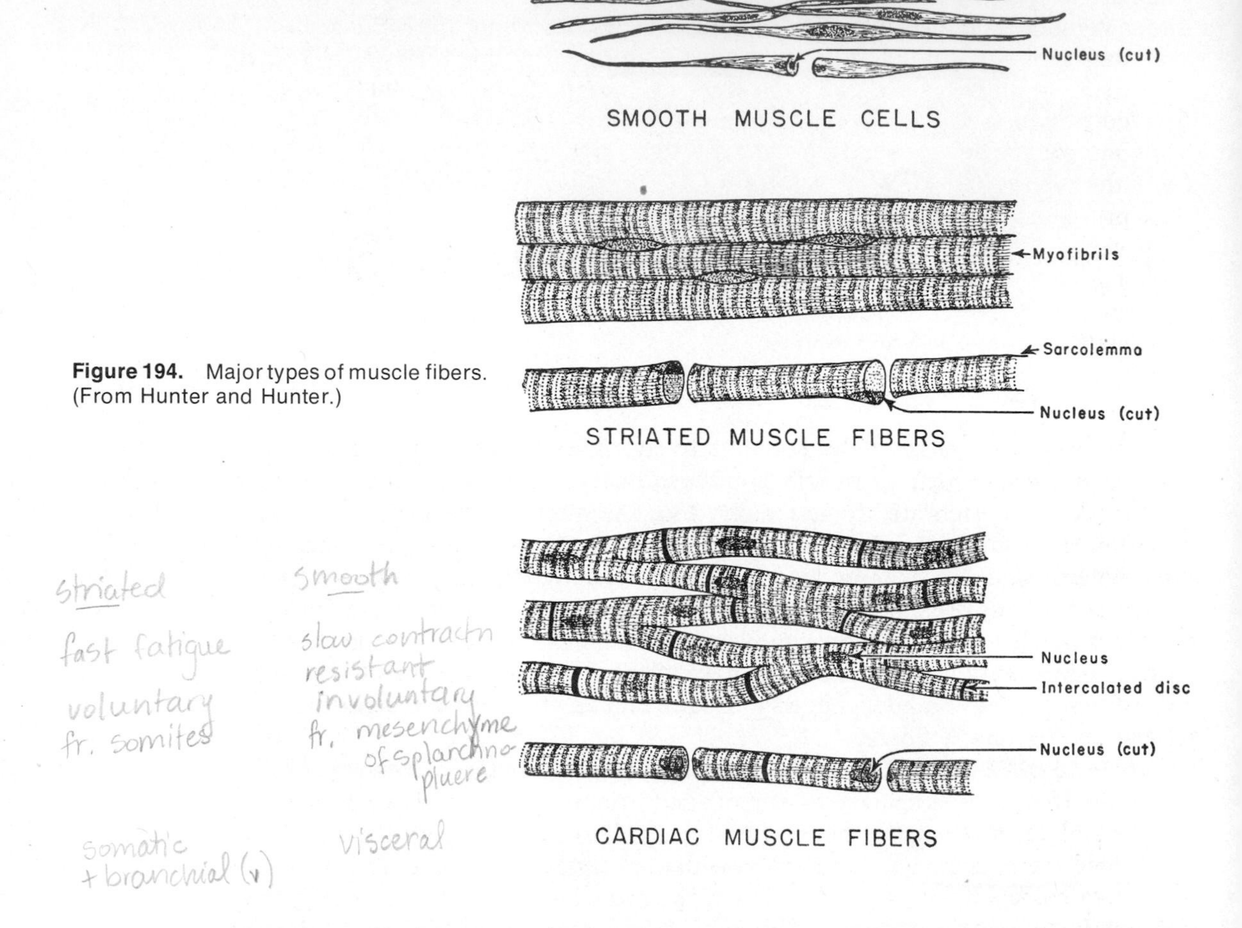

Figure 194. Major types of muscle fibers. (From Hunter and Hunter.)

smooth muscle, the fiber contains a large number of closely packed longitudinal fibrils. The striated appearance is due to the fact that the fibers consist of alternating light and dark portions, which occur at the same point on each fibril; changes in the banding occur between relaxed and contracted phases of a fiber. Striated fibers are arranged in parallel fashion to form muscles. Connective tissues run between the fibers and bind them together, form sheaths for fiber bundles and, further, form an external sheath for the muscle as a whole.

The force of a muscle is exerted through a contraction of its fibers. In smooth muscle the contraction is relatively slight and slow, but may be long sustained; striated muscle may be stimulated rapidly and contract vigorously, but is more readily fatigued. A muscle as a whole may contract slightly or strongly, briefly or for a considerable period of time; the result varies according to the number of fibers stimulated by the nerves and the rapidity of the stimuli. Individual fibers, however, work on an "all or none" basis; each fiber either contracts as fully as possible or fails to contract at all. The sharp contraction of a striated fiber and the more gradual relaxation which follows take altogether but a tenth of a second or so. Normally, however, muscle contraction is due not to a single stimulation but to a continuous tattoo of nerve impulses.

CLASSIFICATION OF MUSCLE TISSUES. How may the varied muscle tissues of the body be classified? One obvious suggestion is to do this on the basis of histologic structure, with striated and smooth (including cardiac) muscles forming the two main divisions. This at first sight seems reasonable. Striated muscles are generally under voluntary control, whereas smooth muscles are under the influence of involuntary nerves; striated muscles are mostly formed in the "outer tube" of the body, smooth muscles are associated with the gut; most striated musculature is derived from somites, while smooth muscles come from mesenchyme.

One prominent group of muscles, however, is in many ways anomalous and ruins the seeming simplicity of such a classification: the branchial system of muscles, primitively associated with the gill bars and prominent in the head and "neck" of all vertebrates. These muscles are striated and under voluntary control. But they do not come from myotomes; they arise, rather, from mesenchyme of the splanchnopleure, like smooth muscles (Fig. 199); they are associated with the gut, as are typical smooth muscles; and their innervation (as will be seen in Chapter 16) is by nerves more closely comparable to those supplying the smooth muscles than to those running to typical striated musculature.

All this suggests that the primitive vertebrate had two discrete sets of musculature. The first, which we may term the **somatic musculature,** forms the muscles of the "outer tube" of the body, is universally striated, is typically developed from myotomes, is innervated by somatic motor neurons (p. 371), and is associated functionally with the adjustment of the organism to its external environment. To this group belong the typical muscles of trunk, tail, and limbs (and, incidentally, those of the eyeball). The second group is the **visceral musculature,** connected mainly with the gut tube, derived not from myotomes but from mesenchyme, innervated by visceral motor nerves, and mainly associated with digestion and other functions of the animal's internal economy. In this second group the musculature in the posterior part of the gut has retained a simple structure of smooth muscle fibers; in the head and pharyngeal region, however, functions in eating and breathing have necessitated the development of the more vigorously acting striated type of muscle.

It appears, then, that despite the seeming impropriety of associating in one major group all smooth and some striated elements, the most natural classification of muscles is as follows:

- Somatic ....
  - Axial
    - Trunk and tail
    - Eyeball
  - Appendicular
- Visceral ....
  - Branchial (striated)
  - Smooth (gut, and the like)

The smooth muscles are, in general, component parts of various organs and need no separate consideration here. We shall in this chapter discuss only the formed muscles of striated type—the various somatic muscle groups and the branchial muscles of the visceral system.

MUSCLE TERMINOLOGY. Several terms are frequently used to describe muscles, particularly limb muscles, according to the type of action they perform. An **extensor** muscle is one which acts to open out a joint; a **flexor** closes it. An **adductor** draws a segment toward the midline of the body; an **abductor** does the reverse. A

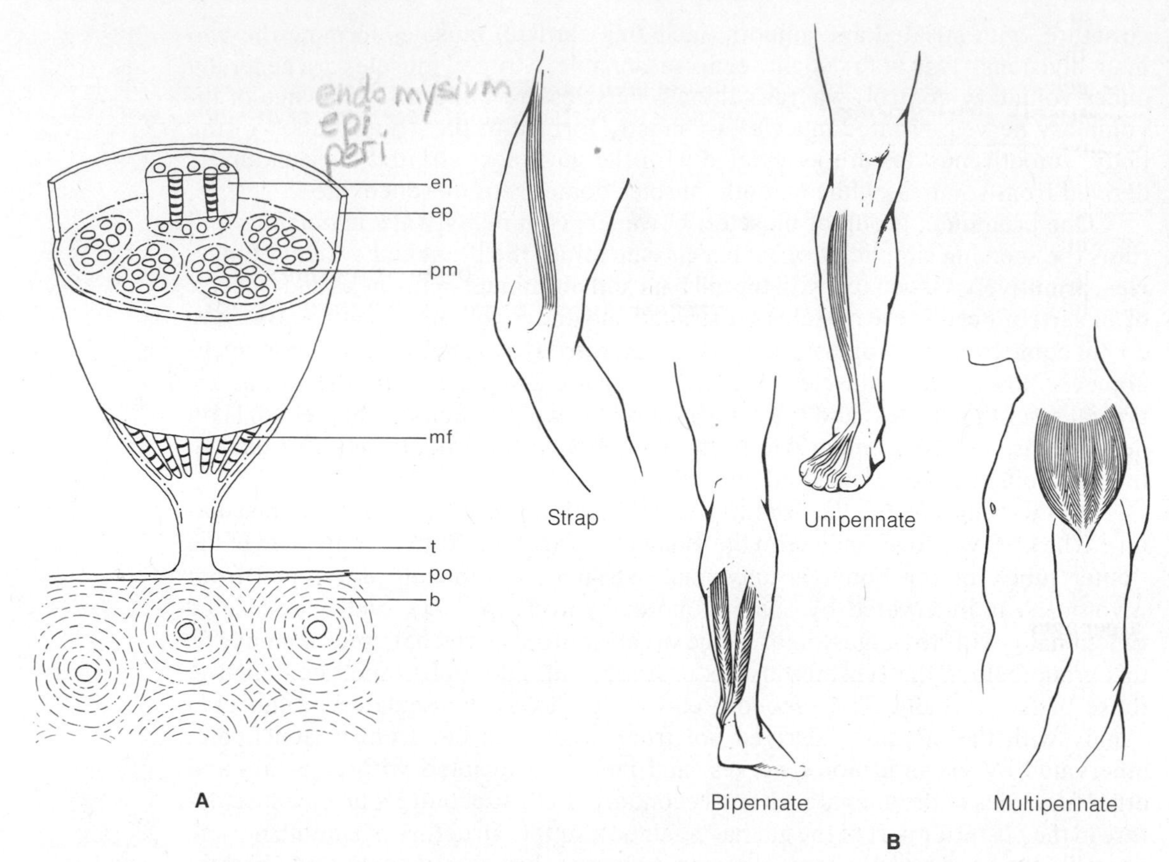

Figure 195. *A,* Diagram of the end of a muscle, showing the various layers of connective tissue. Although they are shown as partially separate, they are in fact continuous and indistinguishable except by position. The connective tissue would be fibrous and the muscle fibers would occupy more volume than they appear to in the drawing. *en,* Endomysium; *ep,* epimysium; *pm,* perimysium; *mf,* muscle fiber; *t,* tendon; *po,* periosteum; *b,* bone.

B, Diagrams showing various arrangements of muscle fibers, based on human examples. The strap or parallel-fibered muscle is the sartorius; the unipennate muscle is one of the extensor digitorum longus muscles; the bipennate muscles are the flexor digitorum longus and flexor hallucis longus; and the multipennate muscle is the deltoid. (*B* after Young.)

levator raises a structure, in contrast to a **depressor.** A **rotator** twists a limb segment; a **pronator** or **supinator** rotates the distal part of a limb toward a prone or supine position of the foot (i.e., with palm or sole down, or vice versa). **Protractors** draw parts forward or out; **retractors,** naturally, retract them. **Constrictor** or **sphincter** muscles are those which surround orifices (as gills, anus) and tend to close them when contracted; they may be opposed by **dilators.**

Muscles usually attach to skeletal elements at either end. The more stable attachment is considered the **origin,** the other the **insertion.** By convention, in limb muscles the proximal end is always considered to be the point of origin; this does *not* mean it moves less (in locomotion it frequently does not). A muscle with multiple heads may be termed bicipital, tricipital and so forth.

Muscle fibers never attach directly to a bone or cartilage; the attachment is always mediated by connective tissue fibers. The muscle as a whole and bundles within it are surrounded by thin sheets of connective tissue (Fig. 195 *A*). Fibers from these sheets are continuous with those around skeletal elements and form the actual attachment. In many cases the muscle fibers approach the bone closely, in a "fleshy" attachment; in others, the muscle terminates in a **tendon,** a ropelike structure of connective tissue fibers. Some tendons form thin, flat sheets; these are termed **aponeuroses.** Tendons and aponeuroses may lie along the surface or penetrate the substance of a muscle; when this occurs, there may develop complex arrangements of muscle fibers, producing **pinnate muscles** (Fig. 195 *B*).

MUSCLE HOMOLOGIES. The comparative study of musculature is difficult because of the variability of muscles and the apparent ease with which their function may alter. A single muscle in one animal may be split into two or more distinct muscles in another, and there appear to be other cases in which originally separate muscles have fused. The general pattern of their arrangement may aid in sorting out individual muscles, but the pattern may be obscured by shifts in the muscle's attachments. Muscles are given, as far as possible, names used in human anatomy. Unfortunately, we are in doubt in many cases as to the homologues of human muscles in lower vertebrates. Students in elementary zoology, for example, frequently dissect the thigh of a frog and find applied to its muscles the names of those found in the human thigh; but it is doubtful that many of the muscles given like names are really homologous. A better procedure is to give muscles of lower vertebrates of which the homologies are in doubt names which are simply descriptive of their general position or attachments.

Embryologic origin is here, as ever, an important criterion for identification of homologies. Often groups of individual muscles in the adult can be traced back to larger aggregations of muscular or premuscular tissue in the embryo (cf. Fig. 104 *B*); the mode of breaking up these masses gives valuable evidence of homologies. Unfortunately, little work has been done on muscle embryology and the information usually indicates homologies of groups of muscles more than of individual ones.

The motor innervation of muscles provides valuable clues. Many workers believed that there is an unalterable phylogenetic relation between a given nerve and muscle—a proposition known as the Fürbringer hypothesis, after the anatomist who first explicitly stated it. Embryology, however, gives no indication that there is any mysterious affinity between specific nerve fibers and the specific muscle fibers which form a given muscle, and in a few cases it seems quite certain that the innervation of a muscle *is* different in different animals. Nevertheless, actual practice indicates that the nerve supply to a particular mass of muscle does tend to remain constant, and that innervation affords an important clue to the identity of muscles.

MUSCLE FUNCTION. The action of muscles has, of course, always been of great interest to anatomists. Often workers have tried to deduce the actions of

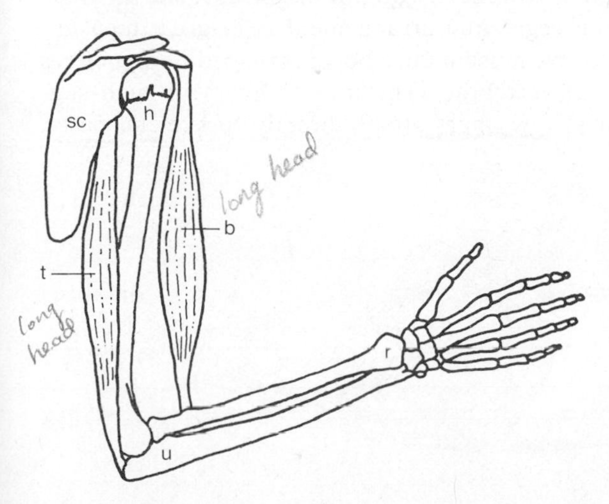

Figure 196. Diagram showing muscles acting on the human elbow. The long head of the biceps *(b)* flexes the elbow while the long head of the triceps *(t)* extends it. *h,* Humerus; *r,* radius; *sc,* scapula; *u,* ulna.

muscles simply from dissections of preserved animals—a probably impossible task. Sometimes this has required simplifying assumptions, the absence of other muscles or the absolute immobility of one end of the system. Such assumptions are reflected in the familiar diagrams illustrating actions, like Figure 196. The biceps is shown as the flexor of the elbow, the triceps as the extensor. Even in this simplified system, flexion does not depend only on the contraction of the biceps; the triceps must relax at the same time. And why should contraction of one or the other of these muscles bend the elbow at all? Both cross the shoulder joint as well as the elbow; why should they not rotate the humerus or the scapula instead? And what are all the other muscles, conveniently omitted here, doing?

This problem of oversimplification also arises when we try to show muscle action as a system of simple levers. In Figure 196, contraction of the biceps always works as a third class lever system because the force is between the load and the fulcrum; however, when a book is being lifted, the hand is the load and the elbow joint is the fulcrum; the reverse is true if the subject is chinning himself on a bar. Use of the triceps will involve either a first or a second order lever system, depending on which end moves. The basic physical principles apply, of course, but not necessarily in diagrammatically simple or unchanging ways.

Thus to discover and describe the actions or functions of muscles more than casual inspection is needed. First, behavioral studies are important to see what the animal actually does. Then, dissection may show what muscles might be involved in the activity being studied. Finally, more physiological techniques, such as electromyography, may indicate which muscles actually are contracting when a given action is being performed. Naturally, all these stages may overlap and there will almost certainly be complexities not even hinted at here. However, many techniques are now being developed and used, and functional analysis of muscular systems has become an important aspect of anatomy.

AXIAL MUSCLES

TRUNK MUSCULATURE IN FISHES. The major part of the somatic muscle division, in fishes, is the axial musculature, arranged for the most part in segmental masses along the flanks (Figs. 197; 198*A, B*). It forms the major locomotor organ of a fish; by rhythmic, alternate contractions of the muscles of the two sides the fish's body is thrown into propulsive waves (cf. Fig. 134).

The axial musculature is, in fishes, of direct myotomic origin, as is shown in the embryo of a shark (Fig. 199). The segmental arrangement is in great measure retained in the adult, most of the trunk musculature being arranged in **myomeres** which correspond in number to the vertebrae. The muscle fibers in each segment are oriented anteroposteriorly; few fibers attach directly to skeletal parts,

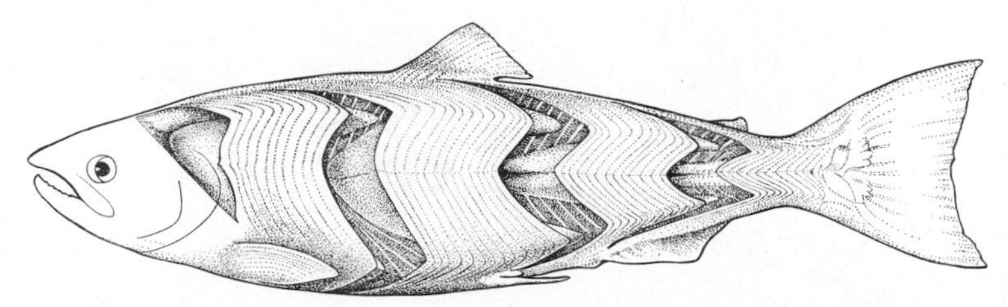

Figure 197. Dissection of a salmon to show axial musculature. In four places a series of myomeres has been removed to show the complicated internal folding of these segmental structures. Within the body each V projects farther anteriorly or posteriorly than it does at the surface. The horizontal septum is visible, cutting the main, anterior-pointing V. (After Greene.)

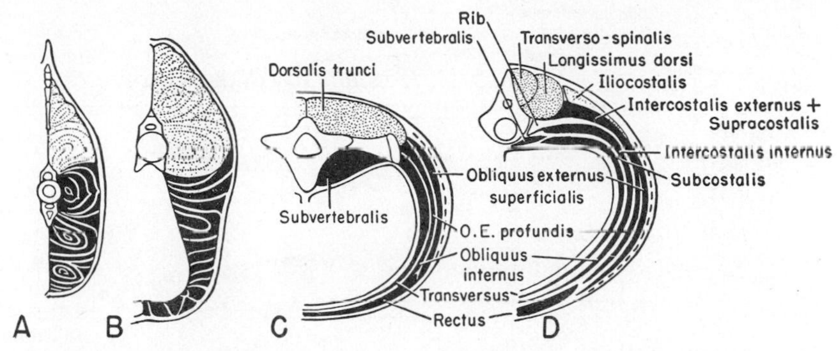

Figure 198. Diagrammatic sections to show the divisions of the trunk musculature in *A*, a shark tail; *B*, the shark trunk; *C*, a urodele; *D*, a lizard. The epaxial muscles are stippled, hypaxial muscles in black. In *D* a rib is assumed to be present dorsally, and the adjacent parts of the hypaxial muscles are labeled as in the rib-bearing region; more ventrally the names are those of the corresponding abdominal muscles. (Mainly after Nishi.)

but insert into stout sheets of connective tissue, the **myocommata** (Fig. 128), which lie between successive myomeres and reach inward to tie into the vertebral column; it is in the myocommata that the ribs (and the extra intermuscular bones of teleosts) are developed. In the embryo the myomeres begin as simple vertical bands, but in the adult (above the cyclostome level) they are folded in a zigzag fashion which appears to promote muscular efficiency. In amphioxus each fold is a **V,** with the point turned forward along the flank. In most jawed fishes there is greater complexity, to give the shape of a **W** with its upper edge turned forward; deep to the surface each myomere may run some distance fore and aft, overlapping and underlapping its neighbors (Fig. 197).

Jawed fishes develop a **horizontal septum** of connective tissue running fore and aft just below the tip of the anterior-pointing V's; it is at the points of intersection of this septum with successive myocommata that dorsal ribs develop (Figs. 2 *D,* 128).

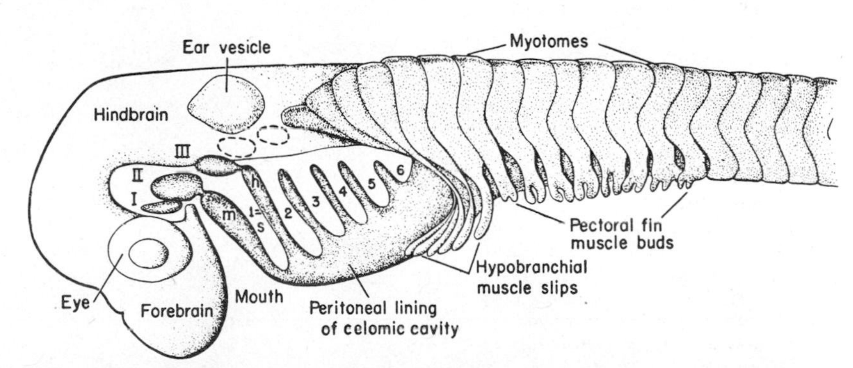

Figure 199. A diagrammatic view of a shark embryo to show the development of the muscles. Skin and gut tube removed; the brain and eye and ear vesicles are included as landmarks. Posteriorly, the myotomes have extended downward to form myomeres; in the region of the pectoral fin, paired buds are forming from neighboring myotomes as potential fin muscles. Anterior to this, buds from anterior myotomes extend ventrally to form hypobranchial muscles. In the ear region, myotomes (broken lines) are rudimentary or absent, but farther forward three myotomes (I to III) persist to form eye muscles. The position of the spiracular slit *(s)* and normal gill slits (2 to 6) is indicated. These interrupt the continuity of the celom and its peritoneal epithelium. Buds of this project upward between the gill slits; from the splanchnopleure there arise the visceral muscles of the mandibular arch *(m)*, the hyoid arch *(h)*, and the more posterior gill arches. (In part after Braus.)

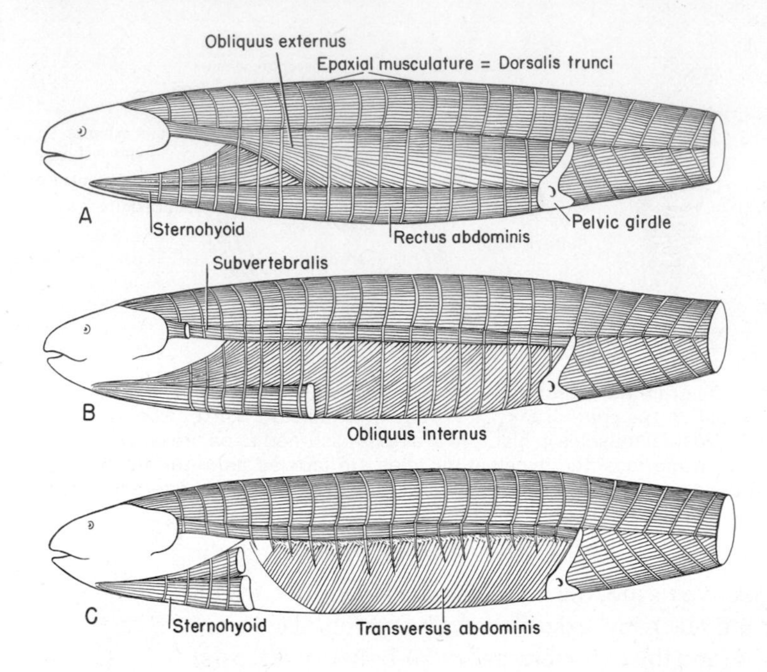

Figure 200. Lateral views of the axial musculature of a urodele. *A,* Surface view (a thin superficial sheet of the external oblique, however, has been removed). *B,* The external oblique and rectus have been cut to show the internal oblique and subvertebral muscles. *C,* The internal oblique has been removed to show the transversus. (Modified after Maurer.)

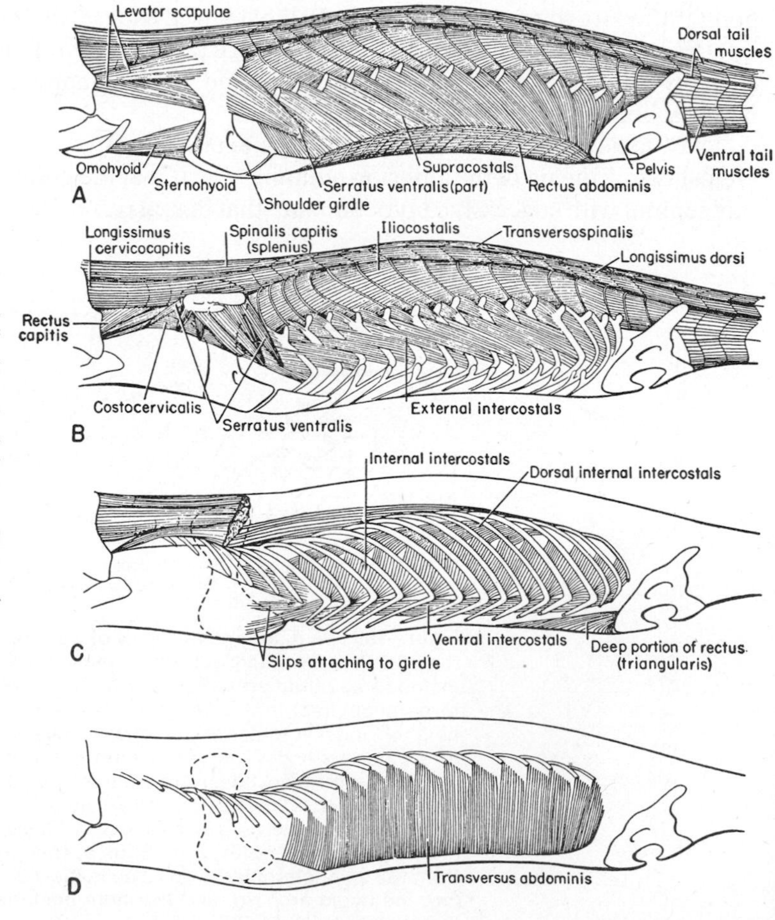

Figure 201. A series of diagrammatic dissections of *Sphenodon* to show the anatomy of the axial muscles. In *A* a thin superficial sheet of the external oblique has been removed. In *B* the supracostals, rectus, throat muscles, and more superficial muscles to the scapula have been removed. In *C* the epaxial muscles are cut posteriorly, and the internal intercostals and triangularis (not shown in the last figure) are indicated. In *D* the ribs are cut, and all other muscles removed to show the transversus. (After Maurer and Fürbringer.)

The axial muscles of gnathostomes may be divided into two major groups: the **epaxial musculature** lying above the septum and above (or external to) the dorsal ribs, and, below the septum, the **hypaxial musculature** (Fig. 198). Apart from the main mass of fish axial musculature, lesser muscles may be developed in connection with the median fins. Further, regional specializations, most highly developed in land vertebrates, may be found in the tail and, particularly, in the region from the shoulder girdle forward. Also in tetrapods, as noted on page 141, the myomeres come to *alternate* with the vertebrae. Below, we shall first follow the history of the main trunk muscle groups—epaxial and hypaxial—from fishes upward through the tetrapods, and then return to pick up the story of more specialized anterior and posterior regions.

EPAXIAL TRUNK MUSCLES. The dorsal musculature has led a relatively uneventful phylogenetic career. In fishes (Figs. 197, 198 *A, B*), it is generally a massive column of segmented musculature which shows little signs of subdivision and may be termed as a whole the **dorsalis trunci.** In tetrapods, where (except in urodeles) body undulation is of little importance in locomotion, it is generally reduced in extent and restricted to a dorsal channel lying above the transverse processes and the vertebrae. Simple in urodeles (Fig. 198 *C*), it tends in amniotes to be divided transversely into several longitudinal subdivisions such as those seen in Figure 198 *D*. In turtles, with development of the shell, these dorsal trunk muscles (and the ventral ones also) are much reduced, and they are reduced in birds as well; in snakes, on the other hand, resumption of major locomotor functions by the axial muscles has resulted in a high development of these dorsal muscles.

HYPAXIAL TRUNK MUSCLES. In fishes, the hypaxial musculature of the trunk is essentially a unit, extending downward from the horizontal septum on either flank around the body wall (Fig. 198 *B*). In terrestrial vertebrates the thickness of this wall—and consequently of the hypaxial musculature—is much reduced; these muscles are, however, complex in structure (Figs. 198 *C, D;* 200; 201). We may distinguish three major subdivisions:

1. Subvertebral muscles, dorsally and medially;
2. A lateral series of muscular sheets along the flanks;
3. A rectus group ventrally.

The **subvertebral musculature** is generally small in volume; it functions in opposition to the dorsal musculature in dorsoventral movements of the spinal column.

The **flank muscles,** extending over the region from the transverse processes down to the ventral territory held by the rectus system, are complicated and varied. There are typically three superimposed major sheets of muscle (each of which may be subdivided, however, in various regions and areas). They are segmentally arranged in fish, but usually continuous sheets in tetrapods. In urodeles, which lack ribs, the three layers are an **external oblique** muscle, whose fibers run essentially anteroposteriorly, but slant somewhat upward anteriorly; an **internal oblique** whose fibers, on the contrary, slant upward posteriorly; and, deepest of the three, the **transverse** muscle, whose fibers, in contrast to those of the obliques, run in a dorsoventral direction. In amniotes a comparable series of simple muscle sheets may be present in the lumbar region, where ribs are short or absent. More anteriorly the transverse muscle usually persists, but the presence of the ribs breaks the two outer layers into a bewildering series of small muscles—intercostals, supracostals, subcostals, and so forth (Fig. 201). To attempt to describe them in detail would be as tiring to the author as to the student.

The **rectus abdominis** primitively ran along the belly from shoulder region to pelvis as it does today in urodeles. At its lateral margins the rectus may be more or less continuous with the oblique muscles, particularly the internal oblique. In tetrapods with a highly developed sternum the rectus is shortened and in mammals may be restricted to the abdomen.

Mammals are notable for the development of the diaphragm, a partition separating thoracic and abdominal cavities and important in breathing (cf. p. 233). The diaphragm is moved by a series of thin muscle sheets which converge from its boundaries toward its center; these appear to be specialized derivatives of the anterior end of the rectus musculature.

TRUNK MUSCLES OF THE SHOULDER AND HEAD REGION. As might be expected, the forward extension of the trunk musculature is largely interrupted in the region of the shoulder girdle, and the specialized structures in the branchial region or neck make for exceptional conditions in the anterior portions of the axial musculature. The dorsal epaxial muscles run forward with relatively little interruption past the shoulder to terminate at the occipital region of the skull; special slips may develop to support the head. In addition slender elements of the subvertebral system run forward just beneath the vertebrae.

The flank muscles—obliques and transverse—or their intercostal equivalents terminate anteriorly at the shoulder region. From the oblique series there develops, however, a very special group of muscles supporting the shoulder girdle in tetrapods. Quite in contrast to the condition at the tetrapod pelvis, where the girdle is firmly fused to the backbone, the shoulder girdle has no direct connection with the column. The body is, instead, suspended between the two scapular blades in elastic slings formed by special flank muscle elements, the **serratus** and **levator scapulae** muscles. These run on either side from the top of the scapular blade downward, in fan-shaped fashion, to attach to the anterior ribs or transverse processes (Figs. 201 *A, B;* 202); through this pliable type of connection (analogous to the use of springs in automobile construction) the body is eased of much of the jolts and jars of locomo-

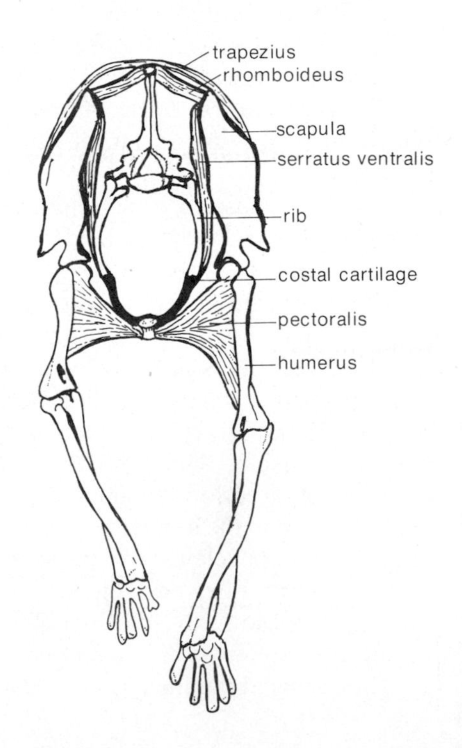

Figure 202. Diagram showing how the body is supported on the forelimbs of a mammal by muscles. The main support is by the serratus ventralis, the two of which form a sling suspending the trunk between the scapulae. The trapezius and rhomboideus dorsally and the pectoralis ventrally hold the scapula and leg in the proper position and prevent the system from collapsing. (After Walker.)

tion. In mammals there is, in addition, a **rhomboideus** muscle, placed more dorsally, which tends to keep the upper end of the scapular blade in position by a pull toward the midline. Ventrally an appendicular muscle, the **pectoralis,** holds the base of the forelimb in place.

Although the flank muscles cease at the shoulder region, the ventral musculature, forming part of the rectus system, continues forward beneath the throat from the shoulder girdle. Such muscles in fish are known collectively as the **hypobranchial musculature** or **coracoarcuales,** for elements of this series typically originate from the coracoid region of the girdle and attach to the ventral end of the branchial arches. In tetrapods various slips of this musculature persist as the **sternohyoid, omohyoid,** and so forth (Figs. 200, 201 *A*). The tongue of tetrapods develops in the floor of the mouth from the region of the branchial arch bases. As it expands, it carries with it a mass of the hypobranchial muscle fibers present in this region; these constitute the flesh of the tongue (Fig. 229).

The embryology and nerve supply of the hypobranchial muscles are not without interest. As axial muscles, they are of myotomic derivation; however, the development of the gills separates the throat region from direct connection dorsally with the myotomes of the occipital and cervical regions from which we would expect them to come. In some embryos, slips from these myotomes are seen migrating circuitously backward above the gills, down behind the gill chamber, and then forward in the throat to form the hypobranchial (and tongue) muscles (Figs. 199, 398). As noted earlier there tends to be a constant nerve supply to a given mass of muscle, even if it has migrated far from its original position. In correlation with this fact we find that the hypobranchial muscles in fishes are innervated by nerves from the occipital region of the skull and the anterior part of the cervical region, which follow the same path of migration as did the muscle tissue, around the back of the branchial chamber and forward along the throat. In amniotes comparable nerves form the hypoglossal nerve and cervical plexus; in the embryo these nerves follow the ancestral route back and down behind the embryonic gill pouches, and even in the adult they pursue a roundabout course to the throat and tongue (cf. p. 379).

CAUDAL MUSCLES. In fishes, the axial musculature continues with little interruption past the cloacal or anal region into the tail. The epaxial musculature is simply a continuation of that of the trunk. Ventrally, however, in the absence here of the body cavity and its contained viscera, the hypaxial muscles change from a series of sheet-like structures to a compact pair of ventral bundles similar to the epaxial muscles above (Fig. 198).

In tetrapods the great development of the pelvic girdles and the limb muscles arising from them has tended to break the continuity of the axial muscles between trunk and tail. The epaxial muscles may be little disturbed, but the interruption of the hypaxial elements at the girdle is complete, or nearly so. The tail, while not as important as in fish, is persistently stout and muscular in urodeles and many reptiles; ventrally, however, part of its volume is made up by muscles (the caudifemorales, described later) which run anteriorly to the femur and hence are limb muscles rather than true caudal muscles. Obviously the caudal musculature is reduced in forms—such as anurans, birds, mammals, and turtles—in which the tail as a whole is reduced in importance. From the ventral tail musculature just behind the girdle there usually develops a sphincter muscle closing the cloaca or anal opening.

EYE MUSCLES. The muscles which move the eyeball form a far-flung anterior outpost of the axial musculature. Except in cyclostomes the series of mesodermal somites found the length of the trunk is interrupted anteriorly by the expanded

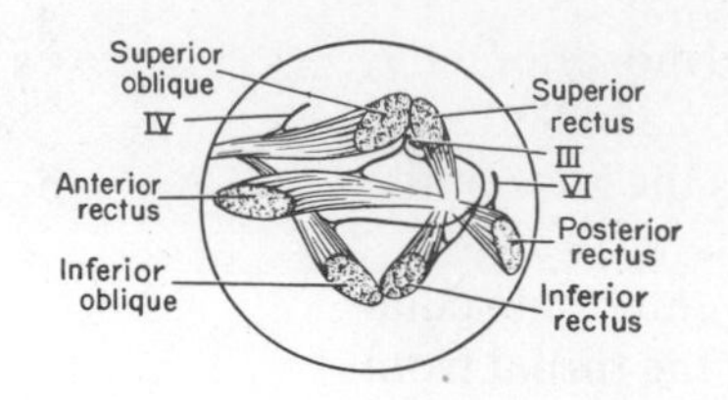

Figure 203. Eye muscles. A lateral view, with the eyeball (in outline) removed; the ovals are the muscle attachments. The three eye muscle nerves are shown (III, IV, VI). Anterior is to the left. (After Goodrich.)

braincase in the ear region; more anteriorly, small somites, usually three in number, persist in members of every vertebrate class in the region of the eye (Fig. 199). These play little part in the development of skeletal or connective tissues, but from them develop the muscles of the eyeball. Associated with these three somites, and innervating the muscles which they form, are three small cranial nerves—III, IV, and VI of the numbered series (pp. 375, 379).

In most vertebrates six typical straplike muscles develop from these somites (Fig. 203). They take origin from the surface of the braincase and fan outward to attach to the eyeball; in varied combinations their pull will rotate the eye in any desired direction. Four of them, the **rectus muscles,** arise posteriorly close to the eye stalk or optic nerve; the other two, the **oblique muscles,** generally spring from the anterior part of the orbit. Four of the six muscles are innervated by nerve III, the superior oblique by nerve IV, the posterior rectus by nerve VI. As this would lead us to suspect, we find that in the embryo four of these muscles usually arise from the first of the three eye somites, one each from the other two.

In addition to the six normal muscles, accessory ones may be present. In most tetrapods (birds and primates are exceptions) there is a **retractor bulbi** muscle, which tends to pull the eyeball deeper into its socket; in most amniotes there is a **levator palpebrae superioris** raising the upper lid, and rather variable slips which move the nictitating membrane of the eye.

LIMB MUSCLES

The musculature of the paired appendages is derived, historically, from the general myotomic musculature of the trunk and hence is part of the somatic system. The limb muscles, however, are so distinct in position and nature and so important in higher vertebrates that they deserve special treatment. In tetrapods, axial musculature declines in volume; the limbs and their musculature grow in most cases to relatively enormous bulk. To use a homely example, fish as food is axial musculature; in a steer, lamb, or hog, the meat is almost entirely limb muscle, with little of axial origin remaining.

As derivatives of the somatic system, limb muscles should, in theory at least, originate in the embryo from myotomes. In some lower vertebrates—specifically sharks—this origin appears to be demonstrable (Fig. 199), the paired fin muscles being derived from buds extending from the tips of a series of myotomes. In tetrapods, however, such an origin has not been demonstrated. The limb muscles arise from masses of condensed mesenchyme; possibly this mesenchyme is of ultimate myotomic derivation.

In the fins of fishes the musculature is simple in construction (Fig. 204 *B*). Two opposed little masses of muscle are generally discernible, a dorsal mass serving to elevate or extend the fin, a ventral one to depress or adduct it. In addition small slips may develop from either group which give rotary or other special fin movements.

TETRAPOD LIMBS. In terrestrial vertebrates we meet with a different situation.

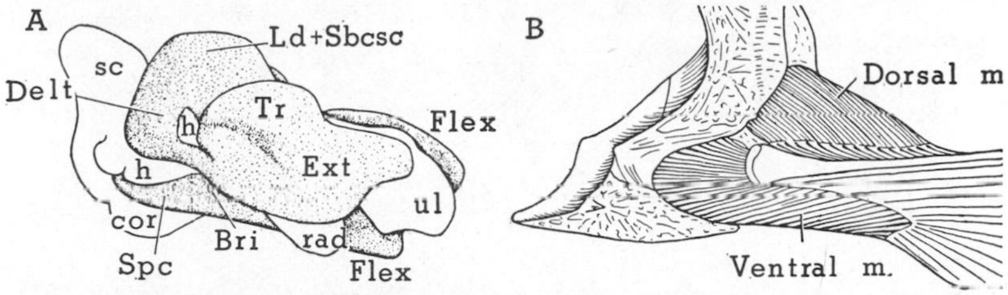

Figure 204. *A,* External view of the left pectoral girdle and limb and its musculature in a lizard embryo (the skeleton unshaded, the muscle tissue stippled); *B,* the pectoral girdle and fin in a fish (sturgeon). In the fish fin the musculature consists simply of opposed dorsal and ventral muscle masses. In the adult land vertebrate the limb has a large number of discrete muscles, but in the embryo these are arranged in two opposed masses comparable to those of the fish fin. At the stage figured the two masses are barely beginning their differentiation into the muscles of the adult (cf. Fig. 205, *A*). The dorsal mass is well shown; the ventral mass is mostly concealed beneath the limb (in which the foot is not yet developed). *Bri,* brachialis inferior; *cor,* coracoid region; *Delt,* deltoid; *Ext,* extensor muscles; *Flex,* flexor muscles; *h,* humerus; *Ld,* latissimus dorsi; *rad,* radius; *Sbcsc,* subcoracoscapularis; *sc,* scapula; *Spc,* supracoracoideus; *Tr,* triceps; *ul,* ulna.

The limb musculature is not only more bulky but more complex. The mode of development, however, affords a clue to a natural classification of the muscles present. Early in ontogeny, while the tetrapod limb is still a short bud from the body, a mass of premuscular tissue (Fig. 204 *A*) is formed on both the upper and lower surfaces of the developing skeleton; it is clear that these two opposed masses are comparable to the dorsal and ventral muscle masses of the fish fin. From these masses arise all the complicated muscles of the mature limb; these muscles can, in consequence, be sorted out into two main series—dorsal and ventral or (roughly) extensor and flexor series. In the distal part of the limb the distinctions between members of the two series generally remain clear; in the proximal regions of shoulder and hip, however, various modifications would make a sorting out of these groups difficult were their embryonic origins not known.

To describe and compare in detail the highly varied musculature of all the tetrapod groups would require a volume in itself, one as exhausting as exhaustive. We will here confine ourselves to picturing and describing, very briefly, the major features of the musculature of a lizard, as representative of a fairly generalized primitive tetrapod condition, and of an opossum as representing a basic mammalian type.

PECTORAL LIMB. *Dorsal Muscles* (Figs. 205, 206 *A, D*). In all tetrapods a number of dorsal muscles attach to the humerus near its head and are responsible for much of the movement of that bone on the shoulder girdle. Two prominent superficial fan-shaped muscles of this sort seen in both reptile and mammal are the **latissimus dorsi** arising from the fascia of the flank and back, and the **deltoideus,** taking origin (often in two parts) from scapula and clavicle. In mammals a slip of the former muscle has gained contact with the scapula as the **teres major.** In reptiles a small external dorsal muscle, the **scapulohumeralis anterior,** is present deep to the deltoid; in mammals this has been pushed to the back margin of the scapula as the **teres minor.** In both reptiles and mammals a broad muscle (only partly showing in the figures) runs from the inner side of the shoulder girdle to insert on the humerus near the latissimus; this is the **subcoracoscapularis** of reptiles, the **subscapularis** of mammals.

More distally the dorsal surface of the humerus is covered by the **triceps,** which arises from the humerus and by one or more heads from the adjacent parts of the

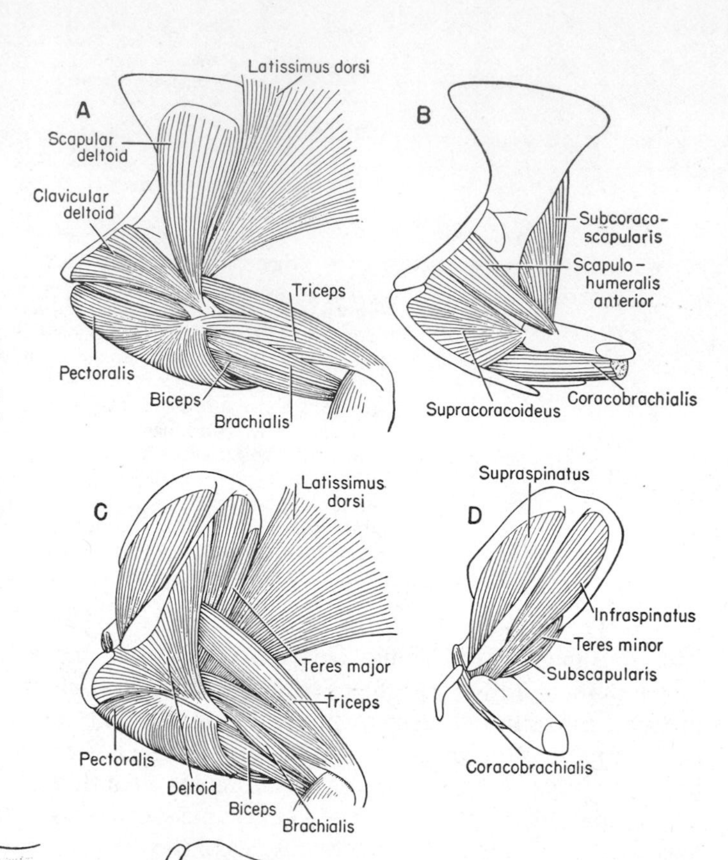

Figure 205. Shoulder and upper arm muscles in the lizard *(A, B)* and opossum *(C, D),* lateral views; the right hand figures in each case are comparable deep dissections with latissimus, deltoid, pectoralis, and long muscles (triceps, biceps, brachialis) removed. Notable is the upward migration of the supracoracoideus to become the two "spinatus" muscles.

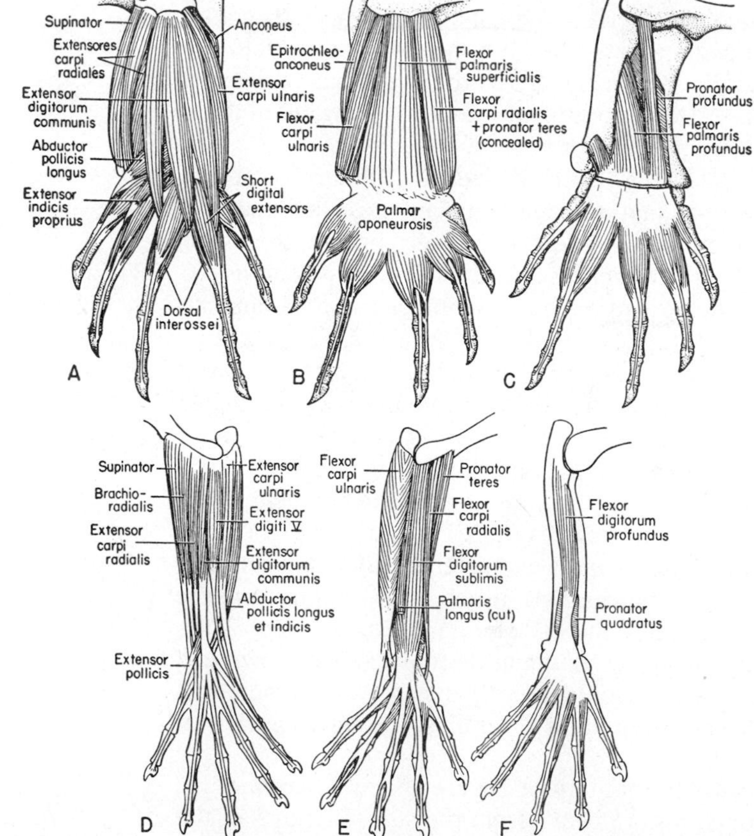

Figure 206. Muscles of the forearm and hand in the lizard *(A* to *C)* and opossum *(D* to *F),* somewhat diagrammatic and simplified. *A* and *D* are views of the extensor surface; *B* and *E* superficial, *C* and *F* deeper dissections of the flexor aspect. On the extensor surface the most prominent change from reptile to mammal is the reduction of short muscles on the manus and the development from the common extensor of tendons to the toes. Long special muscles have developed for movement of the "thumb" and fifth digit. On the flexor aspect a prominent feature in reptiles is the presence of a stout and complex aponeurosis on the "palm" with which connect the long flexors proximally and a variety of tendons and short muscles to the toes. In mammals this is broken up; the palmaris longus inserts into a superficial aponeurosis over the wrist, which is cut away in the figure, and the two deeper flexors present here have each developed a broad palmar tendon. Deeply placed in the hand are various short muscles of the digits, which are not shown.

girdle; the muscle attaches distally to the olecranon of the ulna—this attachment is in fact the reason for the existence of the process—and serves to extend the forearm. Below the elbow the dorsal series is continued by the extensor series of the forearm. Most prominent is a complex sheet of muscles running downward from the elbow and fanning out to the bones of the forearm and hand; a series of short extensors is present in the region of the "wrist" and digits. Lizard and mammal show much the same arrangement, except that in the latter a long extensor from the elbow has tendons, lacking in reptiles, running out directly into the digits.

Ventral Muscles (Figs. 205; 206 *B, C, E, F*). On the under side of shoulder an important superficial muscle, the chest muscle or **pectoralis,** gives a strong pull backward and downward on the humerus; this spreads fanwise far back over the sternum and ribs and inserts on a powerful process beneath the proximal end of the humerus. As noted later (p. 228), it is also important in the formation of mammalian dermal muscles. A deeper, smaller ventral muscle is the **coracobrachialis,** running from the coracoid bone to the lower side of the humerus. Ventral muscles with a flexor action opposing the triceps are the **biceps** and **brachialis,** extending along the humerus to insert on the forearm bones near their heads.

These four proximal muscles are present in fairly comparable form in both reptiles and mammals. But a fifth reptile muscle in this region appears, at first sight, to have no homologue in a mammal. This is the **supracoracoideus,** a large fleshy muscle which runs from the coracoid plate to the under side of the humerus. In the primitive sprawled tetrapod posture this muscle is important in keeping the body from sagging downward between the limbs. In mammals there is no muscle present in this position; indeed there is no coracoid plate from which it could arise.

The muscle, however, is actually present and prominent in the form of the **supraspinatus** and **infraspinatus** muscles on the scapula (Fig. 207); this major muscular migration is presumably responsible for the reduction of the coracoid region of the girdle, which afforded the muscle origin in reptiles, and for the development of the spine and supraspinous fossa of the mammalian scapula. The reptilian muscle has pushed its way upward beneath the deltoid (as can be seen repeated in the mammalian embryo) and has (a) taken over the old scapular blade as

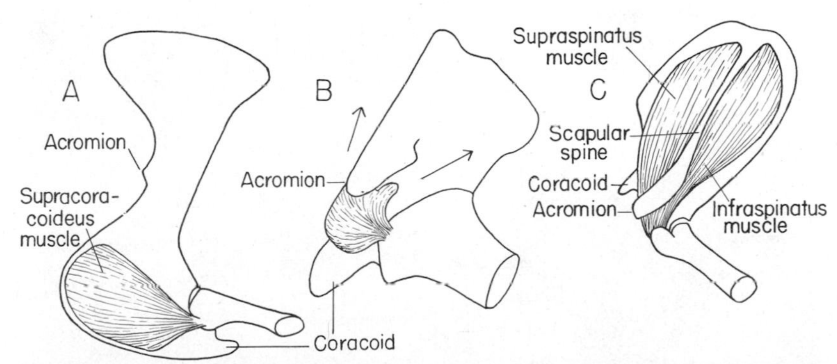

Figure 207. Diagrams of the shoulder region of *A*, a lizard, *B*, an embryonic opossum and *C*, an adult opossum, to show a major shift in shoulder musculature between reptiles and mammals and the consequent modification of shoulder girdle structure. In the lizard the supracoracoideus is a large ventral muscle running from coracoid plate to humerus. In the embryo opossum a comparable muscle is found, but this is, at the stage figured, tending to split and grow upward *(arrows)* on either side of the acromion. In the adult mammal this muscle mass has become the dorsally situated supraspinatus and infraspinatus muscles; a new portion of the scapula has developed for the reception of the supraspinatus muscle, while the coracoid has been reduced to a nubbin. (*B* after Cheng.)

the infraspinatus (and restricted the deltoid origin to the spine), and (b) occupied a new shelf (supraspinous fossa) built for reception of the derived supraspinatus muscle in front of the old anterior margin of the scapula. With the changed limb posture of mammals the supracoracoid ceased to function in its old position. It has, however, retained its supporting function. The two homologous mammalian muscles insert at the very tip of the humerus in front of the glenoid; the resulting lever action tends to swing the limb downward and forward or, conversely, to pull the body up and back on the arm.

In the distal part of the limb, the main propulsive effort is a backward push of the forearm and digits accomplished by the muscles of the ventral, flexor surface, which are hence powerful. A series of long flexors fans out to the forearm and wrist somewhat in the fashion of the opposed extensors. But flexion of the digits is rendered difficult by the fact that muscle serving this function would have to pass round the curve on the underside of the wrist if they were to extend directly to the digits. This is avoided by the development of an aponeurosis, a pad of connective tissue beneath the wrist; to this attach the long flexors proximally and certain of the short toe muscles and tendons distally (a similar structure in the hind leg is shown in Fig. 210 *A*). In mammals this pad of tissue is subdivided into several superimposed tendinous sheets.

PELVIC LIMB. *Dorsal Muscles* (Figs. 208, 209 *A, D*). Certain of the dorsal muscles or muscle groups in the hip and thigh region are readily comparable

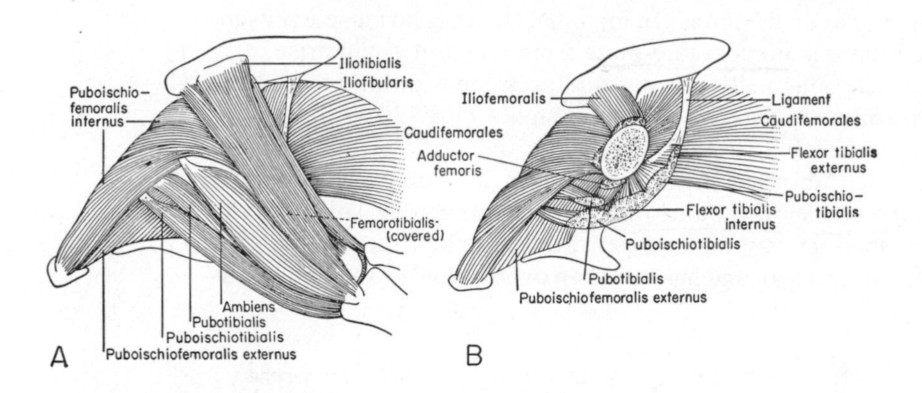

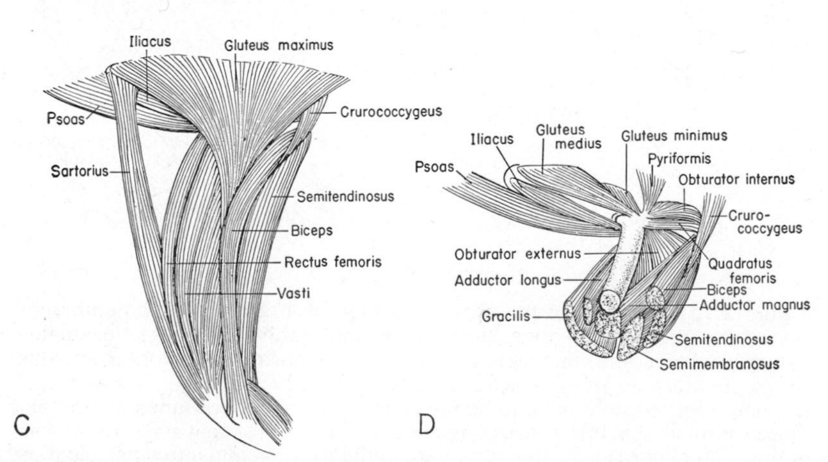

Figure 208. Limb muscles of the pelvis and thigh in a lizard *(A, B)*, and opossum *(C, D)*, lateral views. *A, C,* Superficial views; *B, D,* dissections to show deeper layers of musculature. Anterior is to the left.

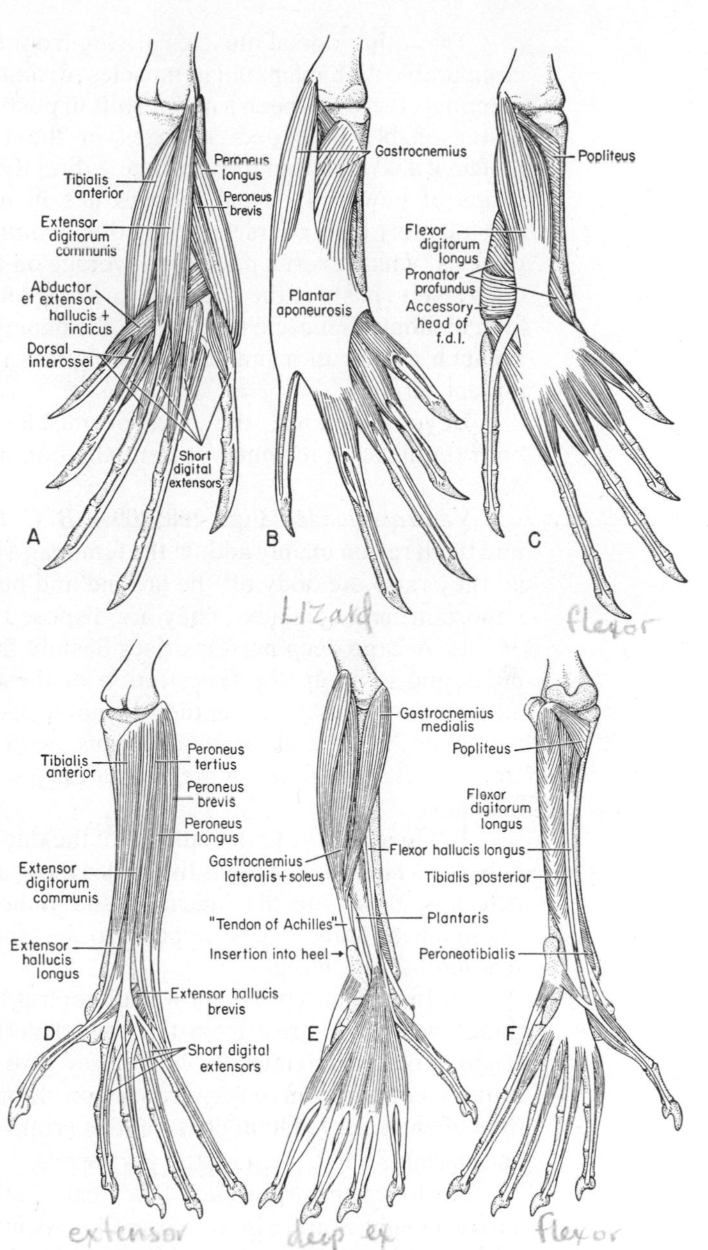

Figure 209. Muscles of the lower leg and foot in a lizard *(A* to *C)* and opossum *(D* to *F),* somewhat diagrammatic and simplified. *A* and *D* are views of the extensor surface; *B* and *E* superficial, *C* and *F* deeper dissections of the flexor aspect. The extensor surface of the lizard hind leg is comparable to that of the forearm and manus except for a lesser development of individual muscles on the inner (tibial = radial) side. In the change to mammals the modifications are similar to those seen in the front leg, including the development of separate tendons from the common extensor and the development of long muscles working on the first and fifth toes. The flexor aspect of the lizard hind leg resembles that of the front in many regards, including the development in reptiles of a stout plantar (sole) aponeurosis, into which much of the musculature attaches both proximally and distally. It differs, however, in the absence of the flexors running down either side and the development of a powerful two-headed "calf" muscle, the gastrocnemius. In mammals the calf musculature has (except for the plantaris) changed to a new attachment on the "heel bone." The long flexor of the digits, however, runs on to the reduced plantar aponeurosis. From this extend, as in reptiles, distal tendons and toe muscles (in *F* the superficial muscles of this set have been removed to show the deeper tendons and muscles). In both reptile and mammal there are, on the flexor surface, deep, short toe muscles not shown in the figures.

in lower tetrapods and mammals. Reptiles have a powerful fleshy **puboischiofemoralis internus** (what names these muscles have!) which arises from the lumbar region and the inner surface of the girdle and inserts on the femur near its head; in mammals this develops as the **iliacus** and **psoas** muscles. Frequently grouped as the **quadriceps femoris** in both reptiles and mammals is a group of muscles which runs down the femur to insert by a common stout tendon on the head of the tibia; this tendon contains the patella of mammals. The quadriceps extends the leg in much the fashion of the triceps in the "arm." The **vasti** muscles are heads of this complex arising from the femur; heads arising from the girdle, however, are differently named in the two cases, for while it is probable that the iliac head of reptiles, termed the **iliotibialis,** is the same as the **rectus femoris** of mammals, it is far from sure that the **ambiens** of reptiles is the same as the **sartorius,** the "tailor's muscle" of mammals.

Two other dorsal muscles arising from the reptile ilium are none too readily comparable with mammalian muscles arising in this region. We have noted that in mammals there has been a major shift in posture of the thigh, and in consequence it is reasonable to expect changes in the related musculature. In reptiles an **iliofemoralis** runs from the iliac blade directly outward to the femur; in mammals a series of powerful **gluteal muscles** lies in much the same position, but differs considerably in its course and function in limb movement—particularly the deeper gluteals, which exert a powerful leverage on the femur in pulling the knee back or, conversely, pushing the body upward and forward on the leg. In reptiles there is a long **iliofibularis** muscle running (as the name implies) from ilium to fibula. There is no such muscle in mammals; possibly it is represented by a long member of the gluteal series.

Beyond the knee, the extensor muscles of the pelvic limb show a pattern in both reptiles and mammals essentially comparable to that of the pectoral appendage.

Ventral Muscles (Figs. 208; 209*A, B, C, E, F*). The ventral muscles of the hip and thigh region mainly adduct the femur and flex the knee joint; in locomotion, that is, they raise the body off the ground and push it forward. They are hence large, important, and complex. They are disposed in three main groups:

1. A large deep muscle arises fleshily from much of the outer surface of the pubis and ischium (the fenestration of these elements is related to this muscle attachment). This has in reptiles the noble name of **puboischiofemoralis externus;** in mammals it is the **obturator externus,** with the **quadratus femoris** as a split-off fraction. Related to these are deeper muscles including the mammalian **obturator internus.**

2. Covering the under surface of the thigh is a large and complex group of long muscles which flex the tibia. In reptiles these are the **puboischiotibialis, flexor tibilias externus, flexor tibialis internus,** and **pubotibialis** (more lovely names!); their mammalian homologues appear to be the **gracilis, semimembranosus, semitendinosus,** and **biceps.**

3. In typical reptiles, powerful ventral limb muscles, the two **caudifemorales** (long and short), arise from the caudal vertebrae and run forward to insert by tendons onto the femur, to which they give a powerful backward pull with consequent contribution to forward motion. In mammals, however, with reduction of the tail and changed limb posture, this group of muscles has been reduced to small and variable slips, such as the **pyriformis.**

The long ventral muscles of the "calf" are mostly concentrated into a powerful **gastrocnemius** muscle in all tetrapods. As in the case of the front leg flexors, the problem of "rounding the turn" at the ankle is a major structural problem. Reptiles generally have solved this, as in the fore foot, by the development of an aponeurosis. In mammals, however, a new type of foot-raising device has been evolved by having the calf muscles insert on a heel tuber of the calcaneum. The major gastrocnemius heads no longer extend to the underside of the foot, but insert by the "tendon of Achilles" onto this tubercle (Fig. 210).

BRANCHIAL MUSCULATURE

Markedly different from the striated musculature so far considered is the branchial or branchiomeric musculature, highly developed around the gills of the ancestral vertebrates, and persistently prominent in much modified form in even the

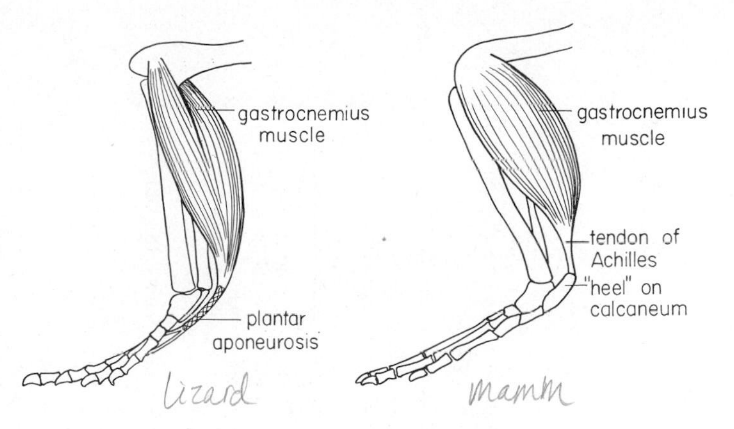

Figure 210. Side views of the hind leg of a lizard *(left)* and a typical mammal *(right)* to show the contrast in action of the main calf muscle, the gastrocnemius, in extending the foot. In lower tetrapods it rounds the ankle region to act on the under surface of the foot by attaching to a sheet of connective tissue, the plantar aponeurosis, which in turn makes connections with the toes (cf. Fig. 209 *B*). In mammals its action is simplified by the development of a heel on the calcaneum (cf. Fig. 209 *E*); attachment of the muscle tendon here raises the foot off the ground.

highest groups. As noted elsewhere, the skeleton and nerves of the pharynx are highly distinctive in nature. The pharyngeal musculature is equally noteworthy, for, in contrast to all other striated musculature, it arises, not from the myotomes, but from mesenchyme derived from the peritoneum of the lateral plate (Figs. 80 *C*, 199). The smooth musculature of the gut proper, posterior to the pharynx, arises in similar fashion.

Gut muscles, striated or smooth, are but anterior and posterior parts of a single great visceral system of muscles whose primary locus is the walls of the digestive tract. In the stomach and intestine, the slow movement of smooth musculature suffices; for the vigorous movements needed in the pharynx of primitive vertebrates—for breathing and, still more primitively, food straining—striated muscle is needed. Generally the boundary, along the gut tube, between striated and smooth muscle lies at the posterior end of the pharynx. But this is not a fixed point, and both in fishes, on the one hand, and in mammals on the other, striated visceral musculature may extend back into the esophagus. In higher vertebrates the pharynx is reduced in size and importance, but the striated pharyngeal musculature is persistently prominent, for portions of it have assumed a variety of natures as facial and jaw muscles and even a fraction of the shoulder musculature.

Branchial musculature is well developed in cyclostomes, as sheets of muscle constricting the gill pouches and as specialized muscles operating the peculiar "tongue." The construction of the lamprey musculature is, however, quite unlike that of other vertebrate groups and will not be considered further here.

In the sharks (Fig. 212 *A*) the branchial muscles show a pattern which may be considered basal to that of other gnathostomes. Anterior elements are, however, already specialized for operation of the jaws. We shall, hence, follow the history of the muscles connected with the typical branchial arches from the sharks upward through the higher vertebrates before returning to consider the muscles farther forward in the region of the hyoid and mandibular arches.

MUSCLES OF THE TYPICAL GILL BARS AND THEIR DERIVATIVES (Figs. 211, 212 *A*). Behind the hyoid arch there are typically, in fishes, five gill slits with four intervening arches, each with its own proper musculature as well as its own skeletal bars. Even when, in tetrapods, the gills themselves have disappeared as landmarks, muscles derived from various parts of the branchial system can be readily traced because of their innervation. These muscles are supplied by a special series of cranial nerves, numbers V, VII, IX, and X (cf. pp. 378–379, and Fig. 396). The mandibular arch is supplied by nerve V, the hyoid by nerve VII; the first typical branchial arch is the territory of nerve IX; and the further arches are innervated by branches of nerve X (which continues onward far down the gut).

TABLE 2. **Muscle Homologies in Various Tetrapod Groups**

No attempt has been made to include all the variants present in different members of the groups tabulated, and many details are omitted. For each group only one common name is given for each muscle; often there are synonyms. In many instances homologies are doubtful.

Pectoral Limb, Dorsal Musculature

Mammal	*Reptile*	*Urodele*	*Frog*	*Bird*
Latissimus dorsi, Teres major	Latissimus dorsi	Latissimus dorsi	Latissimus dorsi	Latissimus dorsi
Subscapularis	Subcoracoscapularis, Scapulohumeralis posterior	Subcoracoscapularis	———	Coracobrachialis posterior, Subcoracoscapularis, Scapulohumeralis anterior
Deltoideus	Dorsalis scapulae, Deltoideus clavicularis	Deltoideus scapularis, Procoracohumeralis longus	Dorsalis scapulae, Deltoideus	Deltoideus + Propatagialis
Teres minor	Scapulohumeralis anterior	Procoracohumeralis brevis	Scapulohumeralis brevis	Scapulohumeralis anterior
Triceps	Triceps	Triceps	Anconeus	Triceps
Supinator	Supinator	Supinator longus	Extensor antibrachii radialis	Extensor antibrachii radialis
Brachioradialis, Extensores carpi radiales	Extensores carpi radiales	Extensores carpi radiales	Extensor carpi radialis	Extensor carpi radialis
Extensor digitorum communis, Extensor digiti quinti	Extensor digitorum communis	Extensor digitorum communis	Extensor digitorum communis	Extensor digitorum communis
Extensor carpi ulnaris	Extensor carpi ulnaris, Anconeus	Extensor carpi ulnaris	Extensor carpi ulnaris, Epicondylocubitalis	Extensor carpi ulnaris
Anconeus				Extensor antibrachii ulnaris
Abductores pollicis	Abductor pollicis longus	Supinator manus	Abductor indicis longus	Abductor pollicis
Extensores digitorum 1-3	Extensores digitorum breves	Extensor digitorum brevis	Extensores digitorum breves	Extensores digitorum breves
	Dorsal interossei	Dorsal interossei	Dorsal interossei	

Pectoral Limb, Ventral Musculature

Mammal	*Reptile*	*Urodele*	*Frog*	*Bird*
Pectoralis	Pectoralis	Pectoralis	Pectoralis	Pectoralis
Supraspinatus Infraspinatus	Supracoracoideus	Supracoracoideus Coracoradialis	Coracohumeralis Coracoradialis	Coracobrachialis anterior Supracoracoideus
Biceps brachii	Biceps brachii	(Not formed)	(Not formed)	Biceps brachii
Coracobrachiales	Coracobrachiales	Coracobrachiales	Coracobrachiales	Coracobrachialis
Brachialis	Brachialis inferior	Brachialis	———	Brachialis
Pronator teres Flexor carpi radialis	Pronator teres Flexor carpi radialis	Flexor carpi radialis	Flexor antibrachii medialis Flexores carpi radiales	Epicondyloradialis
Flexor digitorum sublimis Palmaris longus	Flexor palmaris superficialis	Flexor palmaris superficialis	Palmaris longus	Flexor digitorum sublimis
Epitrochleoanconeus	Epitrochleoanconeus	Flexor antibrachii ulnaris	Flexor antibrachii laterales	
Flexor carpi ulnaris	Flexor carpi ulnaris	Flexor carpi ulnaris	Flexor carpi ulnaris Epitrochleocubitalis	Flexor carpi ulnaris
———	———	Ulnocarpalis	Ulnocarpalis	———
Flexor digitorum profundus	Flexor digitorum profundus	Flexor palmaris profundus	Flexor palmaris profundus	Flexor accessorius
Pronator quadratus	Pronator profundus	Pronator profundus	Pronator profundus	Pronator profundus
Superficial short digital flexors: palmaris brevis, contrahentes, lumbricales, etc.				
Deep short digital flexors, digital interossei, etc.				

Table continued on the following page

TABLE 2. **Muscle Homologies in Various Tetrapod Groups** (*Continued*)

Hind Leg, Dorsal Musculature

Mammal	*Reptile*	*Urodele*	*Frog*	*Bird*
Sartorius	?Ambiens	?Iliotibialis	?Tensor fascia latae	?Ambiens
Rectus femoris	Iliotibialis	Ilioextensorius	Crureus glutaeus	Iliotibialis Sartorius
Vasti	Femorotibialis	———	———	Femorotibialis
Gluteus maximus	?Iliofibularis	?Iliofibularis	?Iliofibularis	?Iliofibularis
Iliacus Psoas Pectineus	Puboischiofemoralis internus	Puboischiofemoralis internus	Iliacus Pectineus Adductor longus	Iliofemoralis internus
Gluteus medius, minimus	Iliofemoralis	Iliofemoralis	Iliofemoralis	Iliofemoralis externus Iliotrochantericus
Tibialis anterior	Tibialis anterior	Tibialis anterior	Extensor cruris brevis	Tibialis anterior
Extensor digitorum longus Extensor hallucis longus Peroneus tertius	Extensor digitorum communis	Extensor digitorum communis	Tibialis anticus longus	Extensor digitorum communis
Peroneus longus Peroneus brevis	Peroneus longus Peroneus brevis	Peroneus longus Peroneus brevis	Peroneus	Peroneus longus Peroneus brevis
Extensores digitorum breves	Extensores digitorum breves	Extensores digitorum breves	Tibialis anticus brevis	Extensores digitorum breves
———	Dorsal interossei	Dorsal interossei	———	———

Hind Leg, Ventral Musculature

Mammal	*Reptile*	*Urodele*	*Frog*	*Bird*
Obturator externus, Quadratus femoris	Puboischiofemoralis externus	Puboischiofemoralis externus	Adductor magnus (pt.)	Obturator externus
Obturator internus, Gemelli	Ischiotrochantericus	Ischiofemoralis	Gemelli, Obturator internus	Ischiofemoralis
Adductores femoris brevis et longus	Adductor femoris	Adductor femoris	Obturator externus, Quadratus femoris	Puboischiofemoralis
Adductor magnus	?Pubotibialis			———
		?Pubotibialis	?Adductor magnus (pt.)	
Crurococcygeus	Caudifemoralis longus	Caudifemoralis longus	Caudalipuboischiotibialis	Coccygeofemorales
Pyriformis	Caudifemoralis brevis	Caudifemoralis brevis	Pyriformis	———
		Puboischiotibialis		
Gracilis	Puboischiotibialis		Semitendinosus, Sartorius	———
Semimembranosus, Semitendinosus	Flexor tibialis internus	Ischioflexorius	Semimembranosus, Gracilis	Ischioflexorius
Biceps	?Flexor tibialis externus	———	———	Caudilioflexorius
Gastrocnemius medialis, Flexor hallucis longus	Gastrocnemius internus	Flexor digitorum sublimis, Flexor digitorum longus	Plantaris longus	Gastrocnemius internus, Flexor hallucis longus, Flexor profundus
Flexor digitorum longus	Flexor digitorum longus			
Tibialis posterior	Pronator profundus	Pronator profundus		
Popliteus	Popliteus	Popliteus, Fibulotarsalis (?)	Tibialis posticus	Tibialis posticus, Popliteus, Gastrocnemius externus
Gastrocnemius lateralis, Soleus, Plantaris	Gastrocnemius externus			
Interosseus	Interosseus	Interosseus	———	———
Superficial short digital flexors: flexor digitorum brevis, contrahentes, lumbricales, etc.				
Deep short digital flexors, digital interossei, etc.				

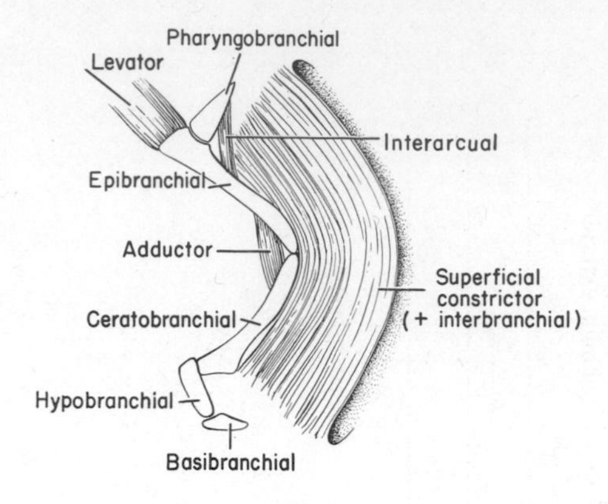

Figure 211. A single gill arch of a shark and its musculature. In many sharks, including the familiar *Squalus* or spiny dogfish, the levators are fused to form a cucullaris associated with the scapula rather than the epibranchials, and the fibers of the superficial constrictors are not parallel to those of the interbranchial, but form a shallow V with the apex anterior (to the left here).

Although there is frequently a fusion of muscle tissues above and below the gill openings, each typical shark gill has a characteristic series of muscle slips proper to it. The most prominent element is the **superficial constrictor,** a broad, thin sheet whose fibers generally run vertically in the flap of skin extending outward in the gill septum. Above and below, most of the constrictor fibers terminate in sheets of fascia on the back and throat, but deeper slips may attach to the gill bars and may form separate **interbranchial** muscles.

In addition there are deeper muscles. The **adductors** of the arches run from epibranchials to ceratobranchials and tend to bend the two together; dorsal **interarcual** muscles function similarly in connection with pharyngobranchials and epi-

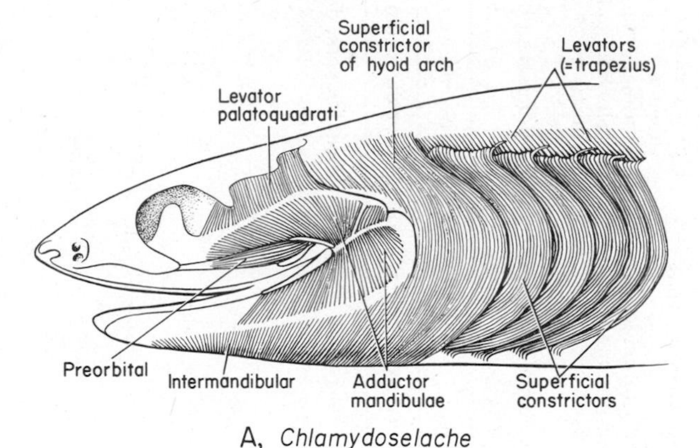

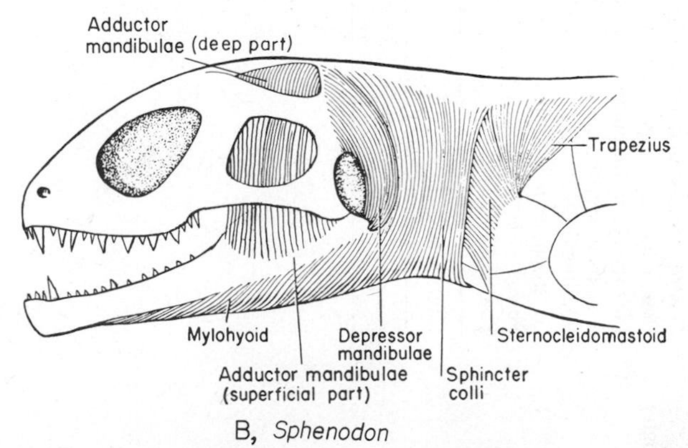

Figure 212. Lateral views of the branchial arch musculature and its derivatives in a shark and the reptile *Sphenodon.* (*A* after Allis; *B* after Adams and Fürbringer.)

branchials of the same or neighboring arches. Dorsally, fibers from the dorsal fascia run backward and downward to insert on successive branchial arches as **levators,** but in many sharks most or all of these fibers run farther back to insert on the shoulder girdle.

In bony fishes the development of the gill musculature is more restricted. Since branchial septa are lost, the superficial constrictors are absent, although ventral slips may persist as **subarcual muscles;** in teleosts the levators are lost and the remaining small muscles are reduced or absent.

Among tetrapods, gill-breathing larval amphibians retain a series of branchial muscles resembling those of bony fishes, but in true terrestrial vertebrates the muscles moving the typical gill bars have disappeared except for small slips associated with the hyoid apparatus and larynx. There is, however, one conspicuous if aberrant relic of the gill series in tetrapods—the **trapezius** musculature (Fig. 212 *B*). This is derived from slips of the levator musculature which, we noted, tend in sharks to run back above the gills to attach to the shoulder girdle. In tetrapods these form a thin sheet of muscle which arises from the occiput and dorsal fascia and inserts along the anterior margin of the shoulder girdle. Primitively this trapezius sheet attached to the clavicle and cleithrum, but with reduction or loss of these elements the attachment may be on the front margin of the scapula (or the corresponding spine in mammals) and may reach the sternum ventrally. Anterior and ventral slips may become separate muscles, such as the **sternomastoid** and **cleidomastoid,** and with reduction of the clavicle in many mammals may fuse with slips from the deltoid to form long, slender compound muscles extending directly from head to front limb.

MUSCLES OF THE HYOID ARCH. Presumably in the ancestral jawless fishes muscles of the hyoid arch were comparable to those of typical branchial arches. But in all living jawed vertebrates this arch, as we have seen, has become highly modified, and its muscles—which may be identified through their innervation by nerve VII—are modified as well. Even in sharks the only surviving muscle of the arch is the superficial constrictor. This muscle may be variously subdivided in fishes, however, with deep slips connecting hyoid arch elements with one another and with the jaw joint. Certain of these slips may persist as tiny elements in the hyoid and ear region of tetrapods.

Facial nerve

In contrast with the trend to reduction of much of the hyoid musculature, the dorsal part of the hyoid constrictor sheet is persistently prominent. In bony fishes this constrictor is highly developed to control movements of the bony operculum covering the gills. In tetrapods, with loss of the operculum, this muscle spreads around the neck in a thin sheet, generally adherent to the skin, as the **sphincter colli** (Fig. 212 *B*). In mammals it expands in spectacular fashion to form the **facial muscles** or muscles of expression. These are especially concentrated about the orbits, outer ear, and lips (Fig. 213).

Mechanisms for opening the mouth have apparently not been taken seriously (so to speak) by vertebrates (mouths tend, rather, to open by themselves), and various makeshift devices are seen in different groups. As noted earlier, the ventral axial muscles run forward along the throat and may attach to the ventral portions of branchial arches and jaws; a backward pull on these muscles can open the jaws of some fishes. In most tetrapods except mammals there is substituted a **depressor mandibulae** (Fig. 212 *B*), an anterior slip of the hyoid constrictor which runs downward from the back of the skull behind the eardrum to attach to the back end of the lower jaw.

In mammals, as we have seen, the lower jaw has been refashioned; the

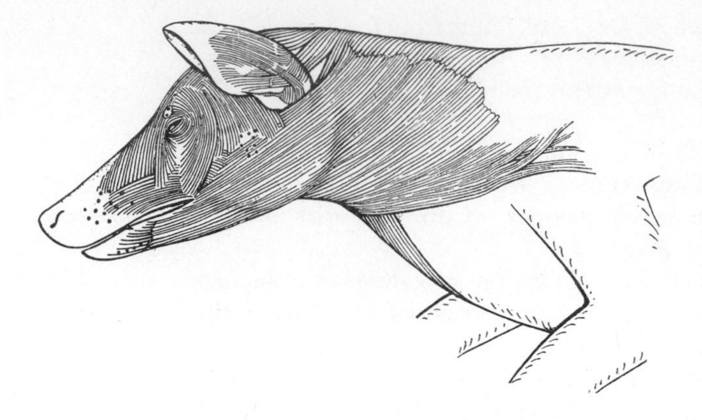

Figure 213. Facial musculature; the head and neck of an opossum. (After Huber.)

elements about the region of attachment of the depressor muscle are lost, and the muscle with them. As a substitute, another slip of the hyoid muscle emerges to take part in the formation of a new jaw-opener, the **digastric** (Fig. 214 *B*). As its name implies, this has two bellies. The posterior one is a hyoid slip which runs downward from the ear region of the skull; the anterior belly is formed by fibers derived from the proper musculature of the jaws. The two may be at a sharp angle to one another, but between them successfully perform the none too arduous task of depressing the jaw.

JAW MUSCLES (Figs. 212, 214). With the modification in gnathostomes of the elements of an anterior gill arch to form basic jaw structures, the muscles of this arch (innervated by nerve V) have become highly modified to serve special functions. In sharks, the jaw musculature consists of three parts: (1) The upper jaw in sharks is but loosely attached to the braincase; connecting these two skeletal elements, between eye and spiracle, is a **levator palatoquadrati,** rather comparable to the levator muscles of ordinary branchial arches. (2) The major muscle mass of the mandibular segment is the **adductor mandibulae,** roughly comparable to the adductor of an ordinary gill, but of vastly greater size, since it performs the important function of pressing the jaws together in the biting or grinding motions essential to feeding. The main mass of the adductor is arranged in simple fashion, running between palatoquadrate and mandibular cartilages. A specialized **preorbital** muscle slip in sharks runs forward to aid in anchoring the jaws to the braincase. (3) Less important is a ventral **intermandibular** muscle, a thin sheet of fibers connecting the two mandibular rami and having fibers from the hyoid as well as the mandibular arch.

In higher vertebrates the first and third of these components amount to little. The levator persists as one or more slips in forms in which palate and braincase

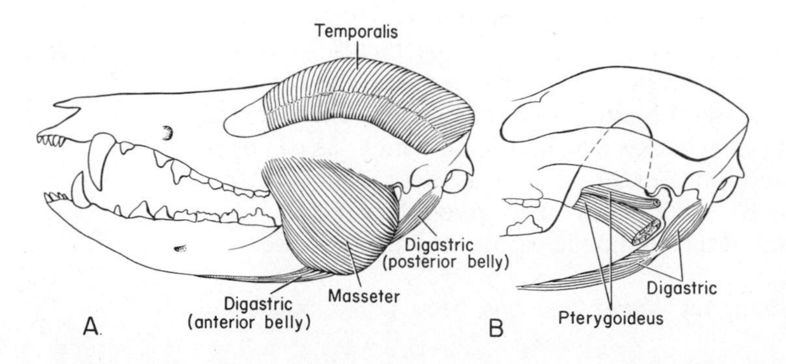

Figure 214. Jaw musculature of the opossum. *A,* Superficial view; *B,* deeper dissection. The jaw is represented as transparent in *B* to show the pterygoid muscles, which attach to its inner surface.

retain some degree of independent movement, but is vestigial or lost in groups in which the skull is a solidly fused structure—chimaeras, lungfishes, modern amphibians, turtles, crocodiles, mammals. The ventral sheet between the jaws persists as a **mylohyoid** muscle and, as noted above, may contribute to the formation of the mammalian digastric.

The adductor mandibulae and its derivatives remain prominent throughout the gnathostomes. In bony fishes and tetrapods the ''cheek'' region in which it lies is covered by dermal bones of the skull; the origin of the adductor is not confined, as in sharks, to the palatoquadrate or the bones which replace it, but spreads onto these dermal bones and may extend upward and inward to attach to the braincase as well. In lower tetrapods the adductor musculature is divided into three main parts separated by the major branches of the trigeminal nerve. In many forms the muscle is extremely complex and heavily subdivided—one recent work recognizes up to a dozen parts in some snakes and lizards. In amniotes, as discussed in connection with the reptilian skull, fenestration of the skull roof allows greater freedom of action for the adductor musculature. In mammals the original adductor is divided into three parts. The **temporalis** has much the original position of the muscle and inserts into the coronoid process of the mandible. A second muscle, the **masseter,** is more superficial in position. With its fibers running at a considerable angle to those of the temporalis, it pulls the jaw forward as well as upward; it is particularly developed in rodents. Finally the little **pterygoideus** muscles form a deep division of the adductor mass; they typically originate from the pterygoid region of the palate and insert on the inner or back surface of the jaw.

DERMAL MUSCULATURE

Although in tetrapods (in contrast to fishes) the skin usually lies relatively loosely over the surface of the trunk muscles, there are often applied to its undersurface thin sheets or ribbons of muscle derived from the underlying layers and functioning in movements of the skin. Such dermal musculature is little developed in amphibians and reptiles, except for slips derived from the pectoral muscles. In snakes, however, a dermal muscle is attached to each of the large scales, aiding them in serving as holdfasts, which prevent a backward slip in the undulatory motion. In birds dermal muscles are prominent in the skin of the wing.

The greatest development of dermal musculature is in mammals. In many forms almost the entire trunk and neck are enveloped in a continuous sheath of dermal muscle, the **panniculus carnosus** (Fig. 215); the twitch of a horse's skin where a fly has settled is evidence of the presence and functioning of this muscle

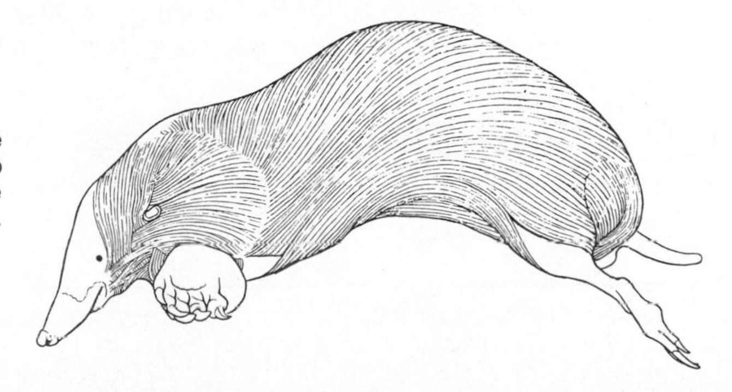

Figure 215. Dermal muscles sheathing the body of a mole. Those anterior to the forelimb are parts of the facial musculature; the large one posterior to it is the panniculus carnosus. (After Nishi.)

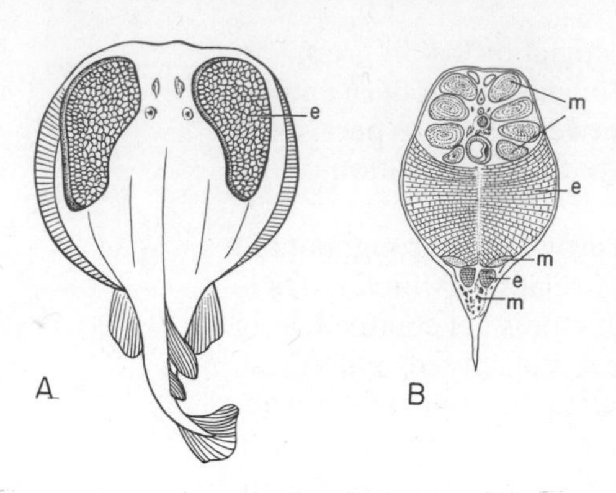

Figure 216. Electric organs. *A, Torpedo,* a ray in which musculature of the expanded pectoral fins has been transformed into electric cells *(e).* The skin is dissected away to show the electric organs. *B,* Section of the tail of the electric "eel" of South America *(Gymnotus).* Typical axial musculature *(m)* is present above and below, but much of the tail muscle has been transformed into electricity-producing tissue *(e).* (*A* after Garten; *B* after duBois-Reymond.)

sheet. Most of the panniculus is derived from the pectoralis, an appendicular muscle, but in the neck the sphincter colli is, as we have noted, a part of the visceral musculature, innervated by the facial nerve (VII). In mammals this anterior part of the dermal musculature undergoes a striking development.

ELECTRIC ORGANS

Several types of fishes, such as rays of the genus *Torpedo,* the electric "eel" *Gymnotus,* and the electric catfish *Malapterurus,* have developed special organs capable of producing a heavy electric shock. Weaker electric organs are present in a number of other fishes, and in some cases it seems clear that they are used as aids in navigation in much the fashion of radar. The bulk, at least, of these electric organs (Fig. 216) are modified muscular tissues. Muscle fibers are structures chemically adapted, as we have seen, for the rapid release of energy; in the present instance the energy is utilized for the production of electricity rather than for muscular contraction.

The modified muscle fibers which appear to form the electric organs in most of these fishes develop as flattened plates of multinucleated protoplasm, each innervated by a nerve fiber, and arranged in series of piles, comparable to the old fashioned voltaic pile, famous in the history of electrical discovery. They form essentially an organic battery, made effective by plus and minus differences between the two surfaces of each plate. In *Torpedo,* strengths of more than 200 volts and 2000 watts have been recorded.

Despite the usual basic similarity of their construction, the electric organs vary greatly in position and appearance in different fish. In *Torpedo* they are present as two large groups located on either side of the head in the expanded pectoral fins. The organ of the electric "eel" is formed from much of the caudal musculature. In the electric catfish of the Nile the electric tissue encircles the whole body just beneath the skin; in this instance its origin from muscular tissue is not certain.

10 Body Cavities

In the vertebrates, as in all the more highly organized invertebrate types, most of the body organs are not bedded in solid tissues or mesenchyme, but are situated within the bounds of fluid-filled body cavities, more properly **celomic cavities.** The viscera are thus placed in a situation where they are more or less at liberty to move freely during their functional activity and to change the more readily in size or shape during growth. Before proceeding to the description of the organs which are enclosed in these cavities or border them, we may here give a brief résumé of the arrangement.

DEVELOPMENT OF THE CELOM. The celomic cavities are formed in the mesodermal tissues and their linings are a mesodermal epithelium, the **peritoneum.** Something of their early embryologic development was noted in Chapter 5. In amphioxus they are formed in the mesoderm in segmental fashion and are at first continuous with the gut cavity (Figs. 78, 79). This condition (probably primitive) does not hold in true vertebrates; ephemeral cavities may appear in the somites or kidney-forming tissue, but permanent celomic cavities develop only in the lateral plate of the mesoderm and generally show no indication of a segmental arrangement. In forms developing from a mesolecithal type of egg the lateral plate at an early stage extends down the flanks of the body and the two sheets presently meet, or come close to meeting, in the ventral midline (Fig. 80); in large-yolked types the lateral plates at first spread out widely to the sides (Fig. 75), and only at a relatively late stage do the two sheets meet ventrally. At first each lateral plate is a solid sheet of tissue. Presently, however, it splits into inner and outer layers with an intervening liquid-filled cavity, the embryonic celom, continuous on either side for the length of the trunk; its outer and inner walls (apart from giving rise to connective tissues, muscles, and other materials) are destined to form the parietal and splanchnic layers of the peritoneum (Fig. 217). The **parietal** (somatic) **peritoneum** forms the inner surface of the great "external tube" of the body (the somatopleure of the embryo); the **splanchnic** (or visceral) **peritoneum** forms the outer wall of the gut and its outgrowths (the splanchnopleure).

As development proceeds the cavities of the two sides come to lie close to one another and below the digestive tract (Figs. 2 *B, D;* 220 *B*). They are separated only by thin sheets of tissue, suspending the abdominal organs—the **dorsal mesentery** above, a **ventral mesentery** below. The former is usually a persistent structure; the latter generally disappears for most of its length by the time the adult stage is reached. On either side, the celomic cavity, in the embryo at least, extends far forward ventrally (Fig. 218 *A*) to reach the floor of the pharynx (the development of

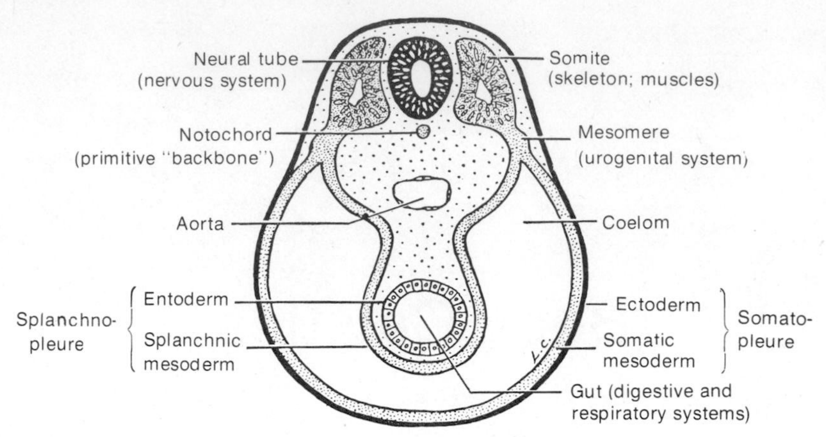

Figure 217. Diagrammatic transverse section of a mammalian embryo to show the relations of the mesoderm. (From Arey.)

gill pouches prevents its dorsal development in this region; Fig. 199). The celomic cavities are almost completely separated from other cavities or the exterior. However, the funnels of the oviducts open from them; in some lower vertebrates some anterior kidney tubules may tap the celom; a pair of small pores open posteriorly to the surface from the celom in most fishes except teleosts.

The celomic structure of an early embryo is simple and readily understood; that of the adult is not. It is complicated, owing to three factors: (1) a pushing into the celomic spaces of other organs than those of the digestive tract—heart, gonads, kidneys, lungs; (2) longitudinal subdivision into compartments (Fig. 203)—in most cases into a cavity for the heart and general body cavity, but with further subdivision in mammals and birds; (3) elaboration and twisting of the gut and outgrowth of its appendages—liver, pancreas—with a consequent complicated folding of mesenteries.

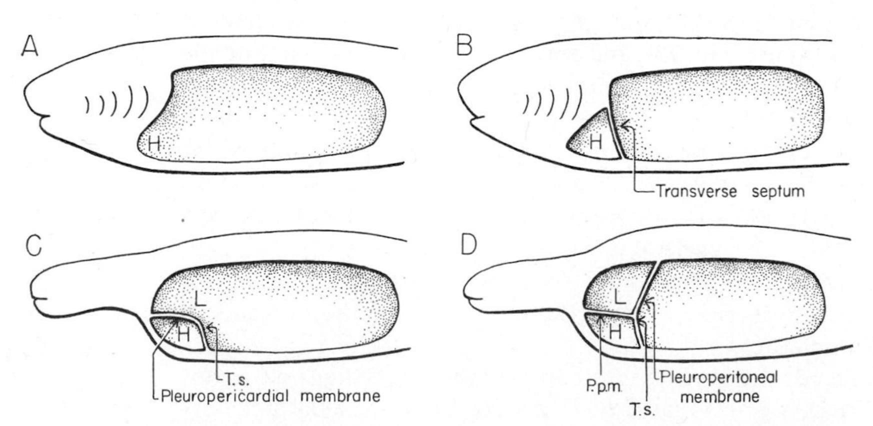

Figure 218. Diagrammatic longitudinal sections of the body to show the evolution of the celomic cavity. *A,* A primitive condition, with the entire celom forming a single cavity. *B,* Typical fish condition; the pericardial chamber is separated from the main cavity. *C,* Typical amphibian and reptilian condition; the lungs are developed but the spaces in which they lie are in continuity with the main celom. *D,* Mammalian condition, with a formed diaphragm (see text). *H,* Heart; *L,* lungs; *P.p.m.,* pleuropericardial membrane; *T.s.,* transverse septum.

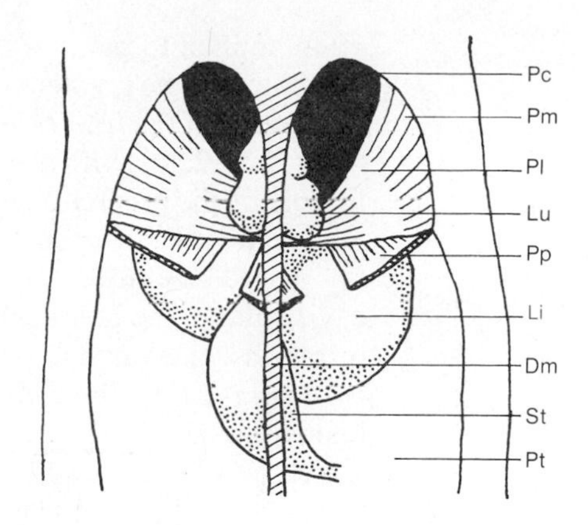

Figure 219. Diagram of the developing celom of a mammal showing its division, here incomplete, into various parts; dorsal view. *Dm,* Dorsal mesentery; *Li,* liver; *Lu,* lung; *Pc,* pericardial cavity; *Pl,* pleural cavity; *Pm,* pleuropericardial membrane; *Pp,* pleuroperitoneal membrane; *Pt,* peritoneal cavity; *St,* stomach. (After Nelson.)

PERICARDIAL CAVITY. The heart in all vertebrates lies in the most anterior and ventral region of the embryonic celom, primitively in the floor of the "throat" below the gills. At an early stage in development there forms behind it a vertical **transverse septum,** separating this cavity from the celom of the trunk (Figs. 218 *A;* 219). This septum becomes a complete partition in most vertebrates, but a communicating opening persists in some fishes, notably sharks and hagfishes. The liver, as it grows, attaches to the posterior surface of the septum; this attachment persists in the adult, although in many cases it is restricted to a relatively narrow ligament. The upper wall of the pericardial cavity is in fishes the floor of the pharynx. When, in tetrapods, the pharynx is reduced in size, the lungs occupy the area dorsal to the heart, and a **pleuropericardial membrane** develops to separate the two organs (Fig. 218 *C*).

GENERAL BODY CAVITY. With the separation of the pericardial cavity, there remains in most vertebrates a single great celomic cavity occupying, with its enclosed organs, most of the trunk. The ventral mesentery, we have noted, tends to disappear for the most part in most groups; anteriorly, however, a portion of it, the **lesser omentum,** connects the stomach with the liver (morphologically ventral to the gut), and a **falciform ligament** may persist below the liver (Fig. 220 *B*). The dorsal mesentery remains a continuous structure in mammals and reptiles, but tends to be broken into segments with intervening gaps in other groups. Distinct names are

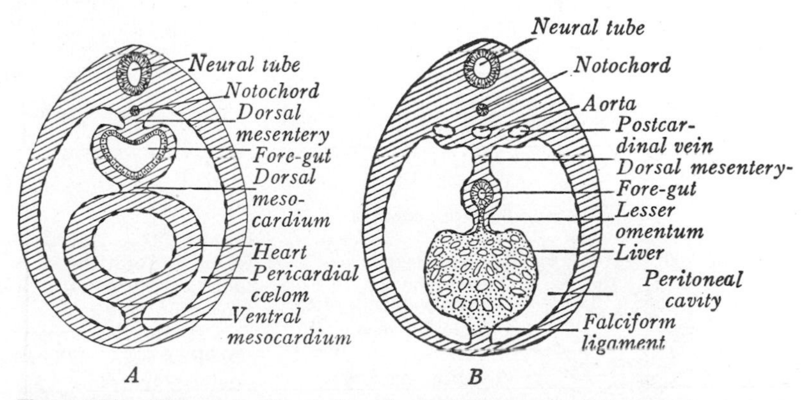

Figure 220. Diagrammatic sections through the heart and liver regions of an amniote embryo to show the relations of the mesenteries. (From Arey, after Prentiss.)

often applied to parts of the mesentery which connect with one organ or another; prominent is the **greater omentum** supporting the stomach. In teleosts and tetrapods the coiling of the intestine results in a confusing folding of the attached mesenteries (Figs. 221, 224). With the generally sigmoid curvature of the stomach is associated a folding of the greater omentum in such fashion that part of the right celomic cavity may come to lie in a pouch above and to the left of the stomach—the **omental bursa.**

LUNG POCKETS. When lungs are developed they push backward into the body cavity above the heart and on either side of the esophagus; in lungfishes and lower tetrapods, they are supported by folds of tissue forming little mesenteries of their own (Fig. 222). The right pulmonary fold, which ventrally touches the liver, has been utilized in lungfishes and tetrapods in the formation of the posterior vena cava, which plunges down through this fold in its course toward the heart (cf. p. 326). In fishes, amphibians, and many reptiles the lungs merely lie within the anterior part of the general celom and in some reptiles the lungs may be more or less buried in the body wall; in other reptiles, however, the lung pockets may be closed off as separate **pleural cavities,** much as in birds and mammals.

THE CELOM IN BIRDS AND MAMMALS. The great development of air sacs in connection with the lungs in birds is associated with a complicated subdivision of the body cavity (Fig. 223). The original abdominal cavity is divided into a pair of pleural cavities and two pairs of cavities surrounding the abdominal viscera; air sacs develop between pleural cavities above and the sets of abdominal cavities below.

In mammals no air sacs are present, but the arrangement of celomic cavities is

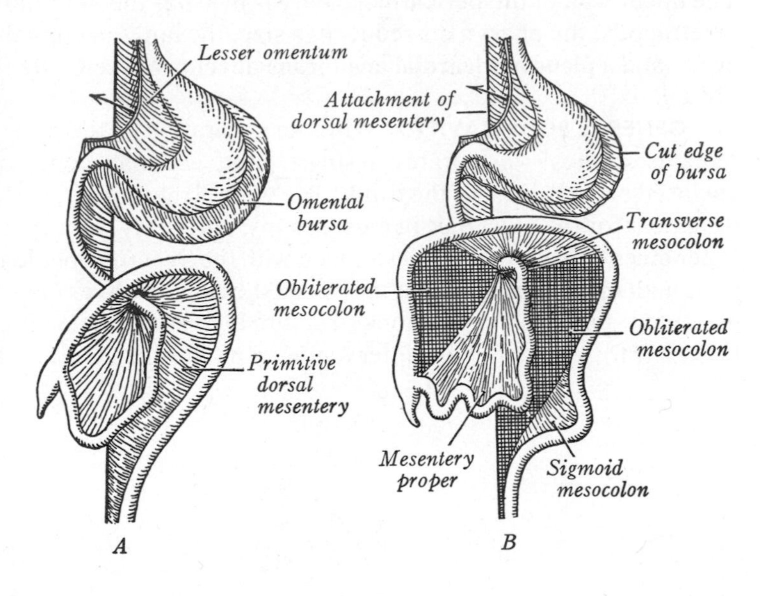

Figure 221. Diagrammatic ventral views of the gut and mesenteries of a mammal in *A*, an embryo, and *B*, essentially adult conditions. The small intestine is greatly simplified in *B*. *A* shows the general type of folding which the mesenteries must undergo because of the asymmetrical position of the stomach and twisting of the intestine. As shown in *B*, this folding may result in obliteration or fusion of parts of the mesentery. In *A* the bursa is a structure of minor dimensions; in many mammals the enlarged bursa would extend down, covering over much of the intestine, but has been cut off short in *B*. The opening from the bursa to the celom of the right side (epiploic foramen) is shown by an arrow; the double line "above" the arrow (i.e., ventral to it) is the cut attachment of the lesser omentum to the liver. For lateral views of the same structures, see Figure 224. (From Arey.)

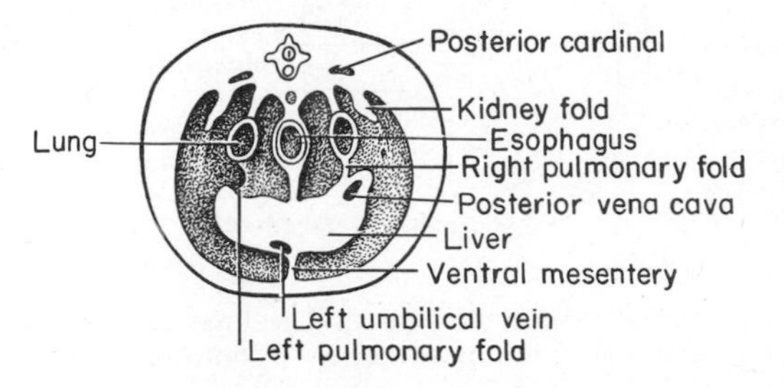

Figure 222. Diagrammatic cross section, looking forward, through the body of a lizard in the region of the lungs, to show the pulmonary folds, between which and the esophagus lie pulmonary recesses. The left pulmonary fold is little developed; it is down the right fold that the posterior vena cava makes its way from the region of the posterior cardinal vein to the liver (cf. p. 326). (After Goodrich.)

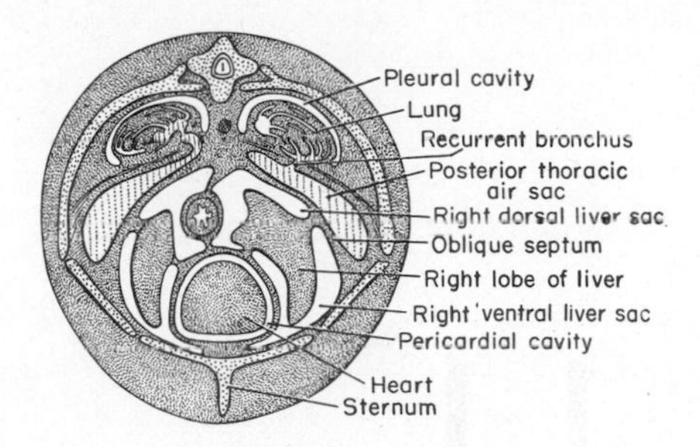

Figure 223. Diagram of transverse section through the thorax of a bird to show especially the subdivision of the celomic cavity. In addition to the pericardial cavity, the pleural cavities and the sacs dorsal and ventral to the liver are shown; the main intestinal celom is too far posterior to be included in this section. The thoracic air sacs (hatched) are also present in this section. (After Goodrich.)

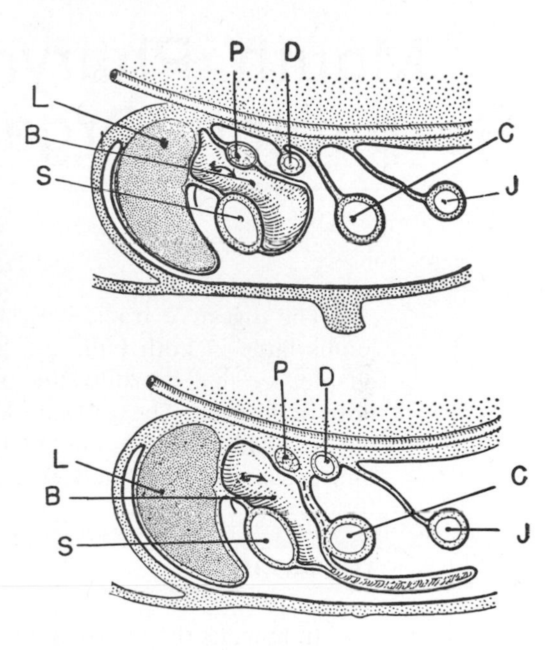

Figure 224. Diagrams of embryonic and adult conditions of the omentum and bursa in a mammal; longitudinal sections, seen from the left side, with head end at left. In the embryo (upper figure) the bursa is small; the entrance (epiploic foramen) is indicated by an arrow. In the adult of many mammals (lower figure) the dorsal mesentery, as the greater omentum, has become a long fold. The diagrams further show how the mesenteries of the gut may be fused together (as is here that of the transverse colon to the greater omentum), or obliterated (as is that of the duodenum, in the diagram): *B,* omental bursa; *C,* transverse colon; *D,* duodenum; *J,* jejunum; *L,* liver; *P,* pancreas; *S,* stomach. For comparable ventral views, see Figure 221. (After Arey.)

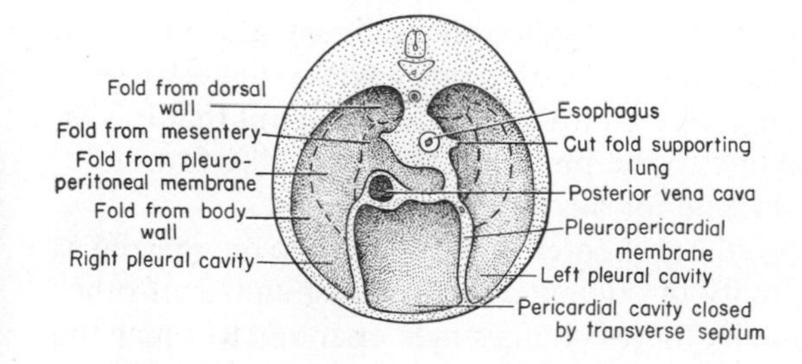

Figure 225. Anterior view of the diaphragm region of an embryonic mammal, to show the various elements which make up the diaphragm. Heart and lungs have been removed to show the posterior walls of the pleural and pericardial chambers. At this stage, the pericardial cavity still extends to the ventral wall of the chest, whereas in the later stage the pleural cavities extend below it. (After Broman, Goodrich.)

nevertheless complicated. The omental bursa is generally well developed as a large and nearly closed pouch (Figs. 221, 224), and the greater omentum may extend down from it ventrally as a great apron over the abdominal viscera. Further, as in birds and some reptiles, the lungs are enclosed in separate pleural cavities which are closed off from the rest of the celom by the development of the **diaphragm** (Figs. 218 *D,* 219, 225). The transverse septum, separating heart from abdominal cavity, is the principal ventral component of this complex structure. Above this, folds from the body wall extend inward on either side of the body, and are continued medially by the **pleuroperitoneal membranes.** These lateral growths meet folds growing out from either side of the mesentery which complete the diaphragm as a solid wall of tissue separating the abdomen from the chest. Into the diaphragm grow muscles from the axial system, so that it becomes a positive element in expansion of the pleural cavities and is important in mammalian breathing.

11 Mouth, Pharynx, Respiratory Organs

The digestive tract, with its various outgrowths and accessory structures, looms large in both bulk and importance in body organization. In the present chapter we shall consider the mouth and pharyngeal region, "introductory" sections of the tract. These play little role in alimentation beyond the reception of food, but are highly important in other regards—notably as the place of origin of respiratory organs and of important glandular structures.

THE MOUTH

In humans the mouth appears to be a well-defined structural unit, with fixed, uniform features such as the lips, tooth row, tongue, and salivary glands. But a broad survey of vertebrates shows that these structures vary widely; every one of the familiar landmarks may be absent in one group or another. Except that it is an inturned area—a **buccal cavity**—leading to the pharynx, we can make few statements about the mouth that will hold true for all vertebrates.

The embryonic gut of vertebrates, the archenteron, was long ago recognized as comparable to the adult digestive cavity of coelenterates and a number of other simple invertebrate types; into this cavity there is but a single opening, to which the vertebrate embryonic blastopore and adult anus are broadly comparable. In the more highly organized invertebrate phyla, however, a second opening develops at the opposite end of the gut as a progressive feature. That vertebrates followed a similar evolutionary course is suggested by the ontogenetic history. In the embryo the archenteron ends blindly at its anterior end, the region of the future pharynx (Figs. 226*A*, 268). In front of this area the head turns downward over the surface of the yolk-swollen body or yolk sac, producing beneath it an inturned fold or pocket of ectoderm, the **stomodeum.** This is the primitive mouth cavity, at first separated by a membrane from the adjacent pharynx. Later the membrane breaks down; mouth and pharynx are placed in continuity and the gut has thus acquired an anterior opening. The epithelia of the two regions concerned blend with one another and in later stages are difficult or impossible to distinguish; broadly, however, the pharynx continues to be lined, mainly if not entirely, with endoderm, but the oral epithelium, in contrast, is ectodermal, essentially a continuation of the epidermis.

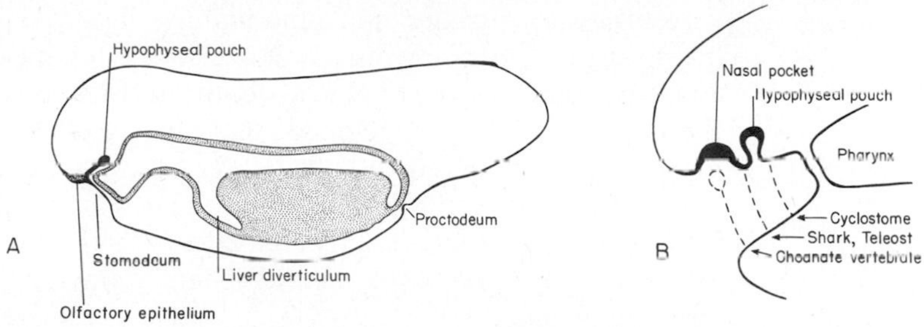

Figure 226. *A,* Diagram of a larval amphibian (at about the stage of Fig. 83 *E*), in longitudinal section, to show the extent of the endoderm (stippled) and its relation to structures of the mouth region. *B,* Diagram to show the comparative position of the oral margins in various types of vertebrates (cf. text).

The extent of the mouth in the vertebrate groups is highly variable (Fig. 226 *B*). One would, *a priori,* expect that the mouth of the adult would include in general the same stomodeal area in all groups. But this is far from the case.

Two good landmarks are always present in the roof of the embryonic stomodeal region. Near the outer end of this funnel, beneath the swelling forebrain, is the embryonic nasal region—a pair of ectodermal pits or thickenings in embryonic vertebrates. Farther back in the roof is a median pit—the **hypophyseal pouch** (Rathke's pouch) —the epithelium of which is to form much of the adult pituitary body. In most sarcopterygians and all tetrapods the jaw margins form a transverse bridge beneath the nasal pockets, so that internal as well as external openings are present, and the site of the hypophyseal pouch lies far back of this point; the mouth cavity is extensive. In ray-finned fishes and sharklike forms, however, the nasal sacs are external to the margins of the jaws; the mouth is, hence, less developed in these groups. Still less inclusive is the peculiar cyclostome mouth; embryonic development shows that it corresponds only with the inner recesses of the oral cavity of the higher vertebrates. In the larval cyclostome (Fig. 227) a nasal pit (single in all but earliest embryonic stages) and a hypophyseal pouch are found in the stomodeal depression, somewhat as in other vertebrates. But as the embryo develops, forces of differential growth cause a rotation of both nostril and pit (the two are closely connected) forward and upward on to the outer surface of the head, as in the hagfish (Fig. 17) and, in the lamprey, to a position high on the dorsal surface, far removed from the adult mouth (Fig. 253). Most of the outer surface of the lamprey head is thus covered by ectoderm, which in gnathostomes lies within the mouth.

In most vertebrates the oral margins are formed by **lips,** soft pliable skin structures. In cyclostomes the mouth is rounded (as the group name implies); its margins bear sensory tentacles in hagfishes, and in lampreys form an effective sucker by which the animal attaches itself to its prey (Fig. 253). In many other

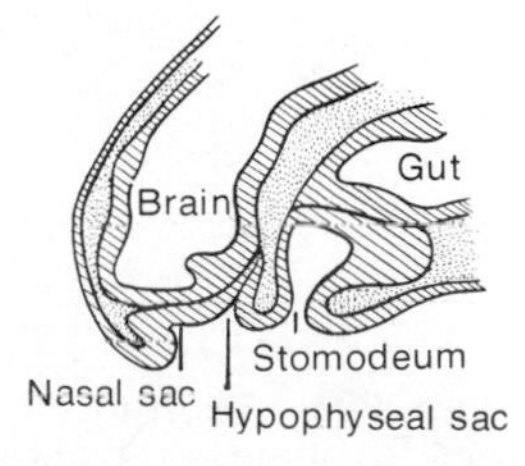

Figure 227. Section of the head of a larval lamprey. At this stage nasal and hypophyseal sacs are still ventral in position (cf. Fig. 253).

vertebrates the lips are small and unimportant folds of skin, and in such forms as birds, turtles, and a few mammals are converted into a bill or beak which functionally replaces a reduced or absent dentition. In mammals, in general, the lips are, on the contrary, highly developed; they are separated by deep clefts from the jaw margins and rendered mobile by the presence of the facial musculature. The opening of the mouth generally terminates in mammals well forward of the jaw articulation, and there is thus created a skin-covered **cheek** region, which, in such forms as Old World monkeys and rodents, may expand into pouches useful in the carriage of food.

In the oral roof, the **palate,** of the typical crossopterygians and tetrapods there develop paired **choanae,** the internal openings of the nasal passages. In amphibians the palate is nearly flat; in reptiles, however, it is vaulted, giving an improved passage for air back from the internal nares to the pharynx and lungs. In crocodilians and mammals (Figs. 180, 190 *C*, 228 *C*, 229) this air channel has been shut off from the mouth by the secondary palate noted in our discussion of the skull; this "hard palate" is extended backward in mammals by a thick membrane, the soft palate (Figs. 228 *C*, 229). The vomeronasal organs (p. 344) may gain openings to the mouth independently of the choanae in some reptiles and mammals. In modern amphibians the tissues of the palate (and of the oral cavity in general) are richly vascularized and the mouth functions as an important respiratory organ. In mammals there are often cornified transverse ridges on the palate which aid in the manipulation of food. In the toothless whalebone whales (Mysticeti) these have developed into long parallel plates of "whalebone" hanging down into the mouth; fringes on their margins strain small marine organisms which are licked off by the tongue and form the food supply of these giant mammals.

In fishes the lower ends of the gill bars, with accompanying musculature, slant forward below and between the jaws into the floor of the mouth. In lampreys, derivatives of these structures form an extrusible structure armed with horny "teeth" wherewith to rasp the flesh of the animal's prey. This is called a tongue, but it is obviously not homologous with one, and in fishes generally there is little in the way of a true tongue. This organ is essentially a development in tetrapods for the better manipulation of food in the absence of water; reduction of the gills enables terrestrial animals to put to this new use the gill bars and their musculature. The musculature of the tongue, we have noted, is derived from the hypobranchial system, and the tongue is anchored at its base by modified visceral arches—the hyoid apparatus.

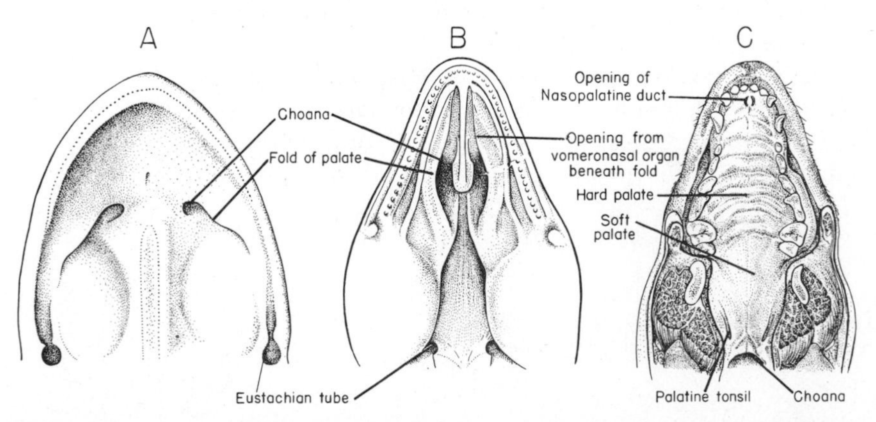

Figure 228. The roof of the mouth in *A*, a urodele; *B*, a lizard; and *C*, a mammal (dog); to show particularly the position of the choanae.

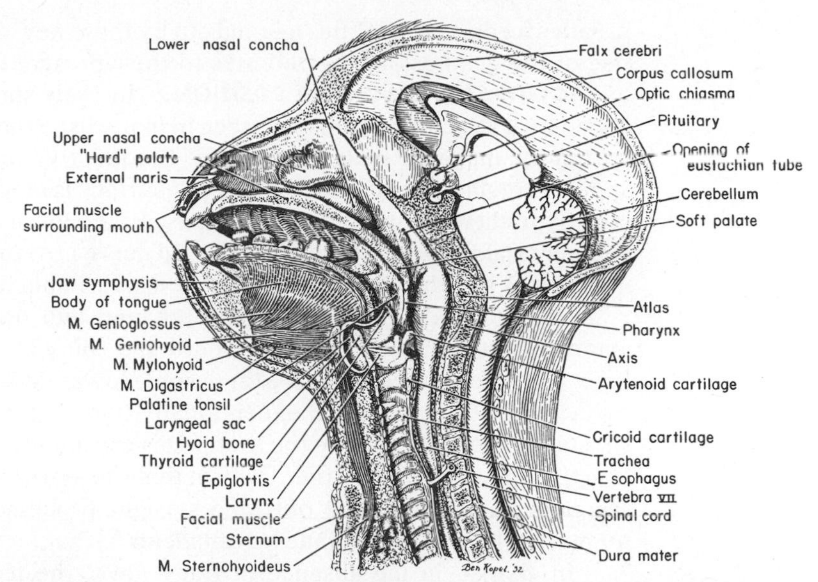

Figure 229. Median section of the head and neck of a rhesus monkey. (After Geist.)

There are numerous specializations in lingual structure among tetrapods. Thus, while some anurans are tongueless, common frog and toad types have a tongue readily protrusible to pick up an insect with its sticky tip; chameleons have a comparable structure; woodpeckers and several mammal termite-eaters have very elongate tongues. Mechanisms for tongue extrusion are varied; in cases of rapid extrusion, muscular action is always involved, but modest expansion or extrusion may be due to filling blood sinuses or lymph sacs.

Apart from a specialized gland in the lamprey, which prevents coagulation of the blood of its prey, fishes generally have little in the way of oral glands except scattered mucous cells. In land vertebrates, in the absence of a water medium, **salivary glands** make their appearance as aids in moistening and swallowing food, secreting mucin and a more watery material to produce the saliva. In amphibians, reptiles, and birds such glands are usually small, if sometimes numerous. In amphibians, however, there is generally a large median **intermaxillary gland** in the anterior part of the palate, and in many snakes and in the Gila monster (the one poisonous lizard) there develop special oral glands which produce venoms which pass by way of a groove or duct in adjacent fangs into the flesh of the prey (their use in defense is secondary). In mammals salivary glands are highly developed, notably as prominent **parotid, submaxillary,** and **sublingual glands.** For the most part salivary glands lack chemically active materials; but in many mammals (including man), many birds, and even a few anurans, enzymes may be present.

DENTITION

The **teeth,** although modified parts of the dermal skeletal materials, may be appropriately discussed here as "inhabitants" of the oral cavity. They are unknown in lower chordates and in jawless vertebrates, living or fossil, although cyclostomes (and larval anurans) have functionally comparable horny structures. With the advent of jaws, it seems, teeth simultaneously developed as biting structures; the broader feeding possibilities opened out by these new devices paved the way for the rise of the gnathostome vertebrates to their present estate.

TOOTH STRUCTURE AND POSITION. In their simplest form (Fig. 230) teeth are conical structures of the type seen in many fishes and reptiles and in the anterior part of the mammalian dental battery. Frequently, however, more complex types appear, notably teeth in which the upper surface is more or less broadened to form a **crown** for chewing or crushing purposes. In the tooth interior is a **pulp cavity,** with soft materials including blood vessels and nerves. At the base may be present one or more **roots** by which the tooth may be firmly implanted in the jaw.

The major dental materials are enamel and dentine. **Enamel,** as found in mammalian teeth, is exceptionally hard and shiny in appearance and forms a thin layer over the surface of the tooth; in some lower vertebrates enamel appears to be replaced functionally by an exceptionally hard outer layer of **dentine.** This latter substance forms the bulk of the tooth. It is essentially similar in chemical composition to bone, but differs structurally in that the associated cells have their bodies in the pulp cavity and send out long straight processes into the dentine through numerous fine parallel tubules—**canaliculi.***

In sharks, in the absence of bony jaws, the teeth are attached by fibrous connective tissue; fibrous attachments are also characteristic of snakes and some lizards, and ligamentous attachments may be found in teleosts. In general, however, teeth are firmly attached to the underlying bony elements, often by a spongy, bonelike material, **cement.** Teeth are often inserted in sockets—the **thecodont** condition (Fig. 231); in other cases (as in *Sphenodon* and most teleosts) they may be fused to the bone surface—the **acrodont** type; in many lizards a variant is the **pleurodont** condition, in which the tooth is attached by one side to the inner surface of the jaw bones.

The most important element of the vertebrate dental equipment is usually a marginal row of teeth along each upper and lower jaw. Such teeth, although sometimes reduced or absent, are to be found in representatives of every class of jawed vertebrates. Teeth are not, however, confined to the edges of the jaws; teeth or denticles may develop from the ectoderm at any point, and since the mouth is lined with ectoderm, it is not astonishing that in many groups of vertebrates (barring birds and mammals) teeth are present on the dermal bones of the palate (cf. Figs. 166 *B,* 169 *B,* 170 *B*), and are to be found on the inner surfaces of the lower jaws in bony fishes and extinct amphibians; in actinopterygians teeth may even develop in the pharynx.

*These definitions really apply only to mammals; in lower vertebrates there is considerable histological variation in dental tissues.

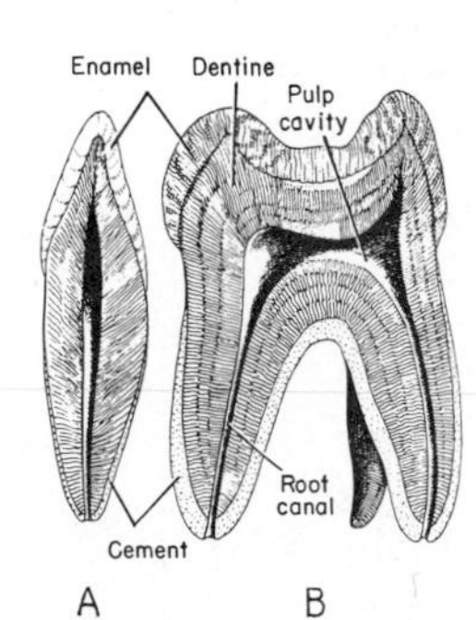

Figure 230. Sections through a mammalian incisor, *A,* and molar, *B.* (After Weber.)

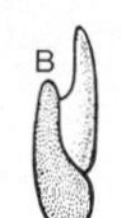

Figure 231. Diagrammatic sections through reptilian lower jaws to show the distinction between thecodont, *A*, pleurodont, *B*, and acrodont, *C*, tooth attachments.

It has long been recognized that teeth and the dermal denticles of the shark skin (cf. Figs. 111, 236 *A*) are essentially similar structures, derived from ectoderm and underlying dermis, and hence are presumably homologous. It was once believed that teeth originated directly from discrete dermal denticles which lay along the oral margins. Our current conceptions of the nature of the primitive fish as having a bony skeleton demands a modification of this idea. Shark denticles, as we have seen, appear to represent superficial tubercles on the bony plates of ancestral fishes, the remaining layers having, so to speak, melted away during the course of evolutionary history. Teeth, we believe, similarly represent such dermal tubercles which lay on the surface of plates situated at the jaw margins. We may thus continue to regard teeth and dermal denticles as homologous, for although they have not descended one from the other, both have a common ancestry as superficial "ornaments" on the dermal bony plates of ancestral vertebrates.

TOOTH DEVELOPMENT AND REPLACEMENT. Embryologically, only the enamel is an ectodermal product, the remainder of the tooth being mesodermal in origin. The first embryologic indication of teeth is an infolding of the epidermis. In marginal tooth rows, there develops as a continuous furrow the length of the jaw, a **dentinal lamina,** from which individual **tooth germs** develop (Fig. 232). If, as is usually the case, a succession of teeth is to develop, the germinal material is not used directly, but, while remaining deep within the jaw, buds off a bit of tissue which gradually moves toward the surface. These ectodermal tooth buds usually form as hollow cones which outline the future surface of the tooth. The bud typically secretes the enamel of the tooth on its inner surface, hence the name **enamel organ** frequently applied to it. Meanwhile cells of mesenchyme origin gather within the cavity of the enamel organ; they deposit the dentine of the tooth, but their cell bodies remain in a persistent pulp cavity and become associated with small blood vessels and nerves. As growth takes place, the tooth works toward the surface and finally erupts; meanwhile one or more successional buds may have formed from the germinal material and pushed outward for replacement.

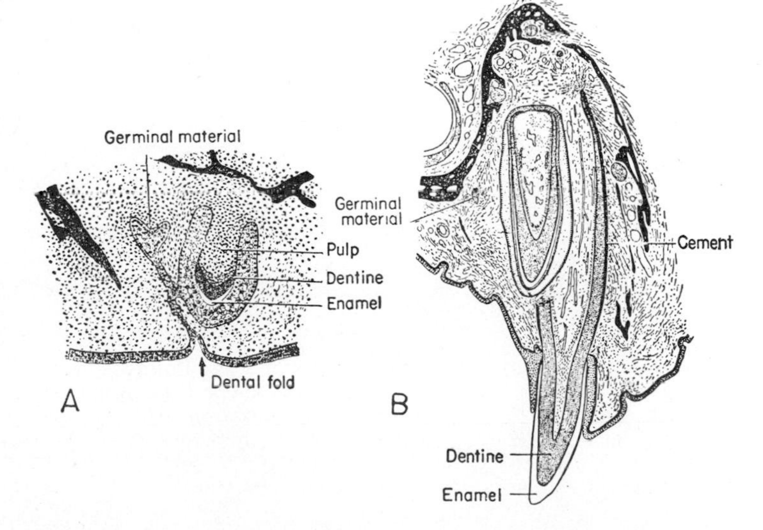

Figure 232. Two sections of a crocodilian upper jaw to show tooth development in lower tetrapods. *A*, Embryo with a first tooth developing and germinal material in reserve at the base of the dental fold. *B*, Mature animal; a tooth is functioning, but is being resorbed at the inner side of the root, where the next successional tooth is already in process of formation. Germinal material for successional teeth persists. (After Röse.)

In mammals (as we are well aware) there is no tooth replacement, apart from the succession of "permanent" teeth for a "milk dentition" in the front of the mouth. Far different is the situation in most other vertebrates. Little is known of the replacement of teeth found on the palate and inside of the jaw in many bony fishes, amphibians, and reptiles; and certain of the large tooth plates present in various fishes and reptiles appear to undergo little if any replacement. However, in the major, marginal dentition of fishes, amphibians, and reptiles, tooth replacement continues through life. Teeth are constantly being formed deep within the tissues of the jaws; they grow in size, erupt, and function and presently, through resorption at their bases or loosening of their connections with the jaw elements, are shed and replaced by a new generation of teeth.

In many fishes, amphibians, and reptiles the tooth row has a seemingly irregular appearance with old teeth, mature teeth, and newly erupted teeth scattered along the jaw in apparently random fashion (Fig. 233). Actually there is method in this seeming madness. Replacement is taking place in an interesting way, which guarantees a continuous function of the dentition despite frequent renewal of individual teeth. The teeth and tooth germs appear to be arranged in two series, the "odds" and "evens" in each tooth row. One may find a condition in which, for example, in a given area of the jaw, the odd-numbered teeth are functional; between them, in place of the even-numbered ones, are sockets within which new teeth are forming. Later this region may show a condition in which both sets of teeth are in place at the same time, but the odd set shows evidence of age and wear. Still later, the odd-numbered teeth drop out, leaving the even series as the functional set, and so on. This neat device guarantees that at least half the teeth in any region will be functioning at any time. This replacement appears to take place in waves that travel

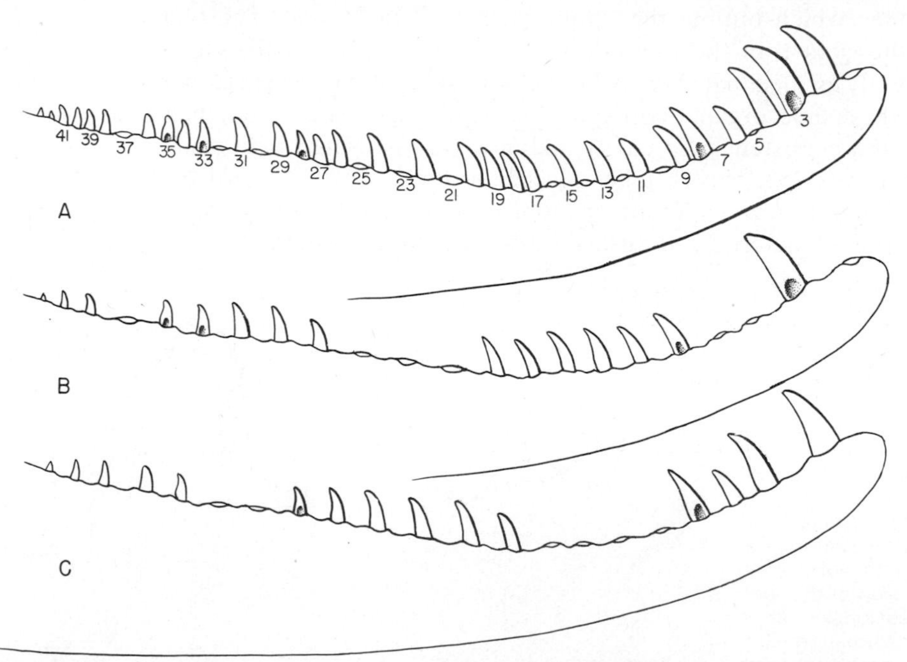

Figure 233. Inner surface of the lower jaw of a primitive fossil reptile *(Ophiacodon)* to show the alternation in tooth replacement between odd and even tooth series. The gaps in the tooth row seem haphazard at first sight; however, if the two series are considered separately, as in *B* and *C,* it will be seen that each includes several waves of replacement arranged in a regular alternating manner. Certain teeth–3, 8, 28, 33, 35–show resorption at the root and are at the point of being lost.

along the jaw rami; at a given time the "even" set may be functioning in some areas of the jaw ramus, the "odd" set in others.

This type of tooth replacement appears to be due to waves of stimulation that proceed, one after another, from the front of the tooth row along strands of tissue connecting the series of tooth germs; the alternating appearance of "odds" and "evens" is due to the spacing between successive waves (Fig. 234). In *a* and *b*, the horizontal lines represent the persistent strand of tissue connecting successive tooth germs in many vertebrates, the vertical lines the distance from the level of the tooth germs up to the surface of the jaw. In *a1* it is assumed that an impulse for tooth formation has begun at the most anterior position, and a tooth rudiment (or Anlage, represented diagrammatically by a black dot) has formed. In *a2* the impulse has travelled to the next position and started a second rudiment; meanwhile the first embryonic tooth in position one (represented by a small circle) has grown somewhat and moved toward the oral surface. The stimulus continues to travel backward, initiating tooth development at successive positions until, in the diagram at *a8*, this first stimulus has produced a row of teeth of which the first (represented by a large circle) has reached maturity. But meanwhile a second stimulus, spaced somewhat more than two tooth germs behind the first, has followed, and a third has just made its appearance.

If this sequence be continued, the mature jaw will obviously come to have the structure diagrammatically shown in *b*, and, as seen in side view, in *c*. Here there is, at first sight, a seeming jumble of old, mature, and young teeth and tooth germs of various sizes; but, as we have just seen, the underlying principle is a simple and orderly one. In *c* one series of alternate teeth is shown in black; we see here an alternation between black and white teeth—between "evens" and "odds"—quite similar to the alternation found in the actual jaw of Figure 233. If the spacing between impulses had been exactly two tooth germ intervals, there would be an exact alternation of odd and even series the length of the jaw. But if the interval is somewhat greater than two tooth spaces (as in the example shown) or somewhat less, the appearance will be seemingly irregular but basically orderly. This relatively simple system may reflect the true situation; however, different mathematical models could account for the observed patterns in other ways.

TEETH IN LOWER VERTEBRATES. No teeth are present in lower chordates or fossil ostracoderms, most of which are—or were—food strainers, nor in cyclostomes. In all jawed vertebrates, with the possible exception of some placoderms,

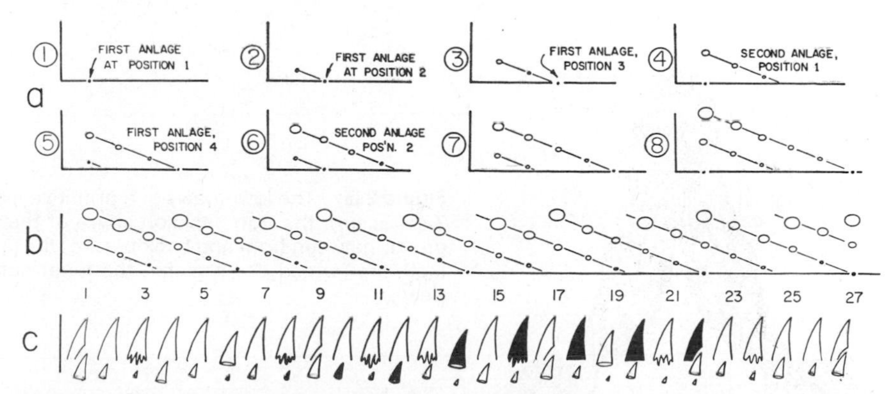

Figure 234. Diagrams (discussed in the text) to show the mode of tooth replacement in typical lower vertebrates. (After Edmund.)

teeth are universally present except where secondarily lost, as in turtles and modern birds. The most generalized type of of tooth is that of a simple cone, present in many elasmobranchs although often modified by accessory cusps or by flattening to give a triangular shape. In some sharks and most skates and rays the teeth may be flattened to form a crushing battery for feeding on molluscs (Fig. 235). In chimaeras, likewise primarily shellfish eaters, the dentition is reduced to a pair of crushing plates in both upper and lower jaws and an accessory pair of small upper plates (Fig. 237 *A*). An analogous development is present in lungfishes (Fig. 237 *B*); the marginal teeth are lost, and the dentition typically consists of four fan-shaped compound dental plates (plus an accessory upper pair). Among actinopterygians the conical tooth shape is primary. In teleosts the cement-like material joining tooth and jaw bone may form a massive base for the tooth as essentially a separate "bone of attachment" (Fig. 236 *B*). In many instances large "fangs" are attached to the jaw by an elastic hinge, bending inward to allow the prey to enter the mouth but opposing its escape. In higher actinopterygians the maxilla becomes toothless, and in some teleosts the marginal dentition disappears entirely; reliance is placed on palatal and pharyngeal teeth, often numerous. Many ancient crossopterygians were notable for the presence of longitudinal grooves on the teeth; these represent infoldings of the enamel, often of complicated pattern, giving the tooth in cross section a labyrinthine appearance (Fig. 237 *C, D*). This crossopterygian tooth type is repeated in many of the oldest amphibians; it is so prominent a feature that it gives the name Labyrinthodontia to the ancestral amphibian group. In modern amphibians, living on small and soft materials, the teeth are small and simple in structure, and are entirely absent in some toads. A curious feature is the usual presence of a zone of weak calcification near the base of each tooth.

In most reptiles the teeth are essentially of the simple conical type. Palatal teeth were present in ancestral reptiles (Fig. 238 *B*) and are preserved in *Sphenodon,* snakes, and lizards; such teeth are absent in crocodilians. Turtles have lost their teeth, relying instead on a horny bill. Ancestral reptiles generally had teeth inserted in shallow sockets, and typical thecodont teeth are seen in crocodiles. Most lizards, however, have a pleurodont type of tooth attachment, and some lizards (including chameleons) are acrodont, as is *Sphenodon;* in snakes the teeth are attached only by fibrous tissue and may form hypodermic-like fangs (Fig. 238 *C*). The primitive bird *Archaeopteryx* was toothed, but the bill (plus gizzard stones) has functionally replaced the dentition in later birds.

In the fossil mammal-like reptiles are seen stages in the initiation of the mammalian type of dentition. The most ancient members of this series, the

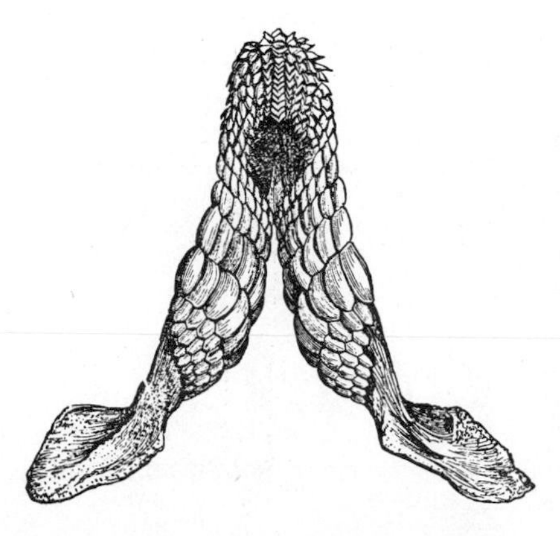

Figure 235. The lower jaws of a primitive living shark, *Heterodontus (Cestracion),* the Port Jackson shark of the Pacific. The teeth differ greatly between front and back parts of the jaw. Rows of successional teeth are formed down within the inner surface of the jaws. (From Dean.)

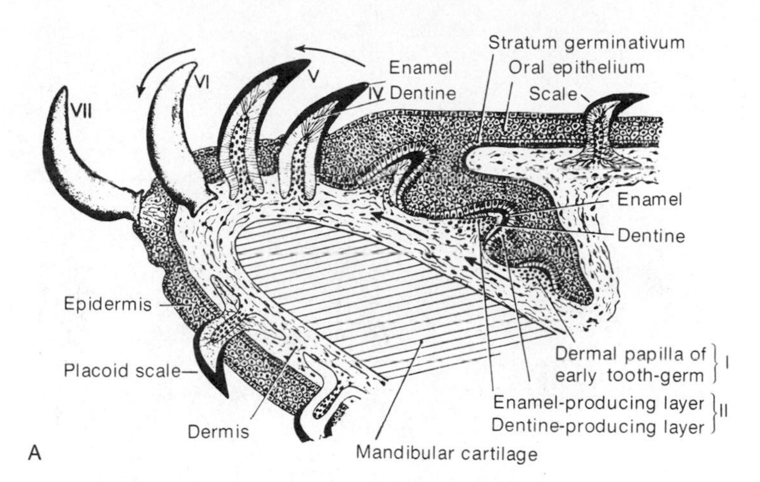

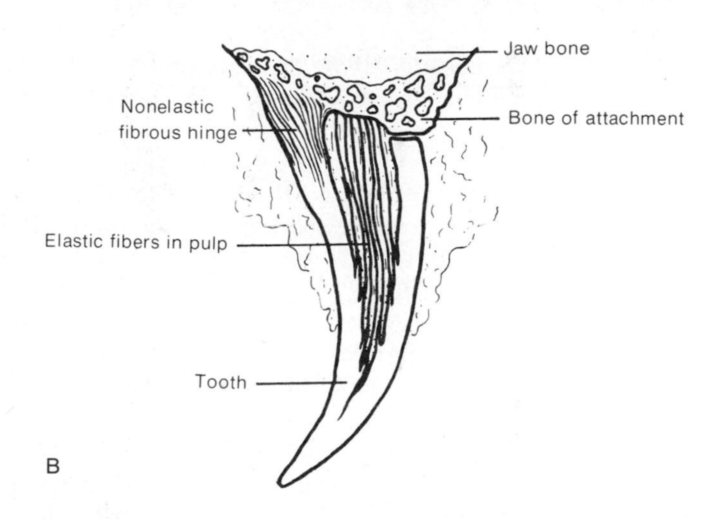

Figure 236. Attachment of teeth in fish. *A,* Section through the jaw of a shark, to show stages (I–VII) in the development of a tooth to its functional stage and to the point where it is about to be shed (VII). For comparison, placoid scales (denticles) in the adjacent epithelium are shown as well. (From Rand, The Chordates, The Blakiston Company.) *B,* Section through a tooth of a type found in many teleosts, in which the tooth may be bent back into the mouth on a fibrous hinge. (From Scott and Symons, Introduction to Dental Anatomy, The Williams & Wilkins Company.)

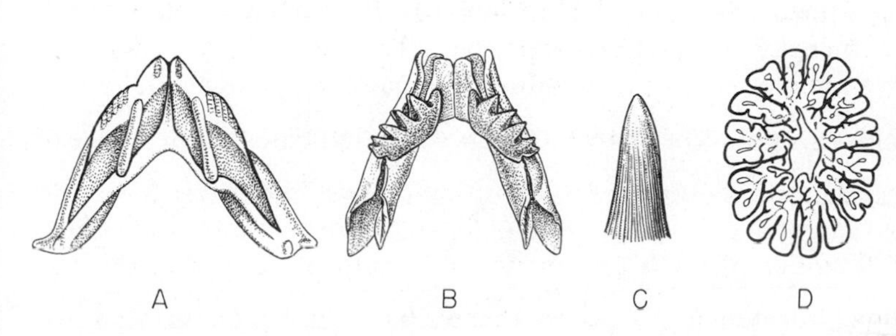

Figure 237. *A,* Dorsal view of the lower jaws of a chimaera, to show the pair of large tooth plates, of complex shape, which cover most of their surface. *B,* Similar view of the bony elements of the jaws of the lungfish *Epiceratodus* to show the pair of fan-shaped toothplates. *C,* External view of a grooved labyrinthine tooth characteristic of crossopterygians and ancestral tetrapods. *D,* Section through such a tooth to show the complicated folding of the enamel layer (heavy black line). (*A* after Dean; *B* after Watson; *C* and *D* after Bystrow.)

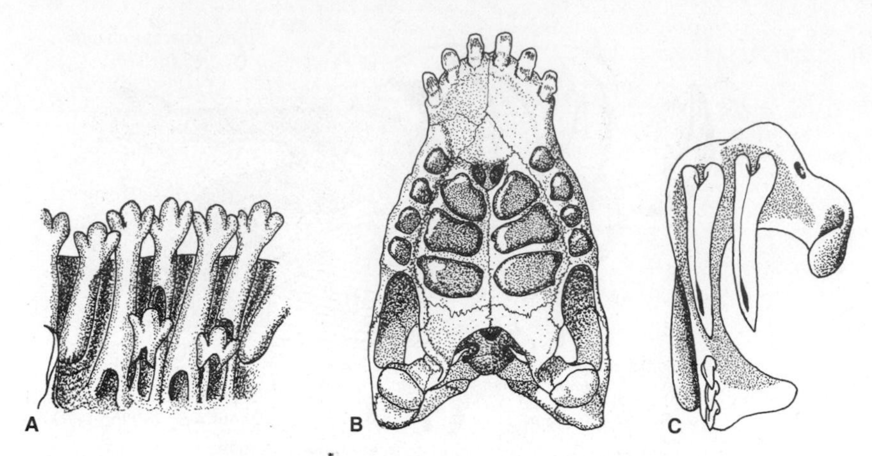

Figure 238. Teeth of various reptiles. *A*, The lizard *Amblyrhynchus;* in many herbivorous lizards the teeth are not simple pegs, but have multiple cusps. *B*, The extinct placodont *Placodus;* the anterior teeth formed a clam rake and the more posterior ones, including palatal teeth, were flat plates for crushing hard molluscs. *C*, The maxillary fangs of a king cobra, *Ophiophagus;* the teeth are hollow hypodermic needles serving to inject the venom into the prey. (After Edmund.)

pelycosaurs, show the development of large upper stabbing teeth comparable to mammalian canines, separating incisors anteriorly from a series of cheek teeth (Fig. 177 *A*); various therapsids show further stages in the evolution of a mammal-like dentition (Fig. 177 *B, C*).

THE MAMMALIAN DENTITION. The most primitive of living mammals, the monotremes, are aberrant in having entirely lost their teeth as adults, and reduction or loss of teeth is found in various other forms, notably anteaters and whalebone whales. Typically, however, mammals have a short marginal series of teeth in each of the quadrants of the jaws, in which four tooth types can be distinguished in anteroposterior order (Fig. 239 *A*). Most anterior are **incisors,** nipping teeth with a simple conical or chisel-like build; next a single **canine,** primitively a long, sharp stabbing structure. Following the canine is a series of **premolars** (the "bicuspids" of the dentist), which frequently show a degree of grinding surface on the crown, and finally, a **molar** series, the members of which generally assume a chewing function and develop a complex crown pattern. In early fossil mammals the number of teeth in the various categories appears to have been variable and rather high; primitive placentals, however, settled down to a count in each half of each jaw of three incisors, a canine, four premolars, and three molars.

There has been devised a simple "shorthand" for the nomenclature of placental molar teeth and for the number of teeth of each type present in a given mammal. The letters I, C, P and M followed by a number in upper or lower position will define any tooth in terms of the original placental formula: I^1, for example, refers to the most anterior upper incisor, M_3, the last lower molar. The number of teeth of each type present, above and below, in the dental equipment of any mammal, can be formulated in succinct fashion. The **dental formula** $\frac{3.1.4.3}{3.1.4.3}$ indicates the presence of the primitive placental formula on either side of both upper and lower jaws (the number of incisors, canines, premolars and molars are represented by successive figures). The total number of teeth present in the mouth of such an animal is forty-four. The human formula is $\frac{2.1.2.3}{2.1.2.3}$; that is, we have lost an incisor and two premolars from each quadrant of the jaw and reduced our teeth to a count of 32.

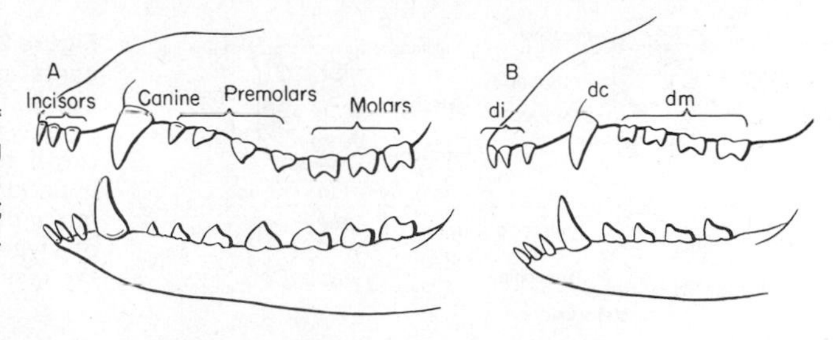

Figure 239. Left side view of the dentition of a generalized placental mammal, showing permanent dentition in *A; B,* deciduous teeth. *dc,* Deciduous canine; *di,* deciduous incisors; *dm,* "milk molars" (= deciduous premolars).

Incisors are retained in most mammals, although the ruminants, for example, have lost the upper incisors (Fig. 240 *B*) and must crop grass by a bite between lower incisors and gums. The elephant's tusks are greatly enlarged upper incisors. In rodents a pair of upper and of lower incisors are developed as gnawing chisels which grow persistently at their roots as their tips are worn away; in this development they are paralleled by a variety of other mammals. Canines are prominent in carnivores, and reach their maximum in the extinct sabertooths (Fig. 240 *A*); in nonpredaceous mammals they may be retained as defensive weapons or, in males, for use at mating time, but are generally reduced and often lost.

The cheek teeth—premolars and molars—have a varied history. In carnivores, in which there is little chewing of the food, they tend to be reduced in number and often in size, except that in most terrestrial carnivores a single pair on either side are specialized as shearing teeth—the **carnassials** (Fig. 240 *A*). In herbivores, on the other hand, the cheek teeth are usually retained (except for the frequent loss of the first premolar) and develop as an efficient grinding battery which is usually separated from the cropping teeth by a gap in the tooth row—a **diastema.**

The crowns of the mammalian cheek teeth—particularly the molars—develop a varied and complicated pattern of cusps which are valuable in diagnosis of relationships and consequently have been studied in detail by systematists, paleontologists, and anthropologists. Only the basic features of these molar patterns, as found in placental mammals, will be noted here (Figs. 241–243). The nomenclature of the cusps is simple. Each is termed a **cone;** a small cusp is a **conule;** a lower jaw element has the suffix **-id.** Typical cusp patterns and their nomenclature are illustrated in Figure 241.

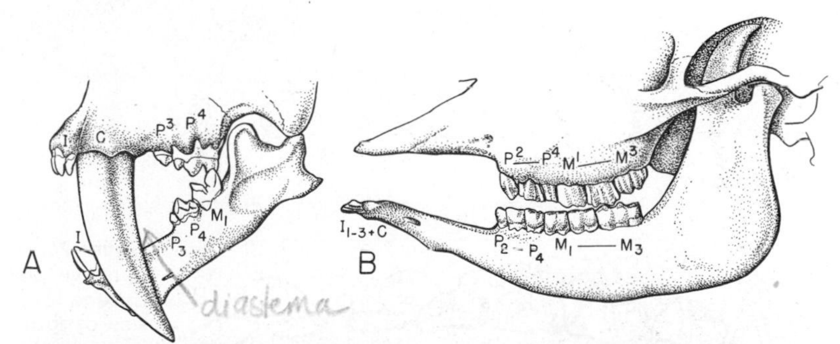

Figure 240. Two specialized mammalian dental types. *A,* Felid; *B,* the cow. For the felid, the extinct sabertooth, with extreme specialization, is shown. In the felids the dentition is much reduced; there are modest incisors and lower canine (concealed in the figure) and stabbing upper canine. In the cheek little remains except the carnassials $\frac{(P^4)}{(M_1)}$. In the ruminants the cheek teeth are expanded into a grinding battery; they are separated by a diastema from the cropping teeth–here consisting only of lower incisors and canine, working against a horny pad on the upper jaw.

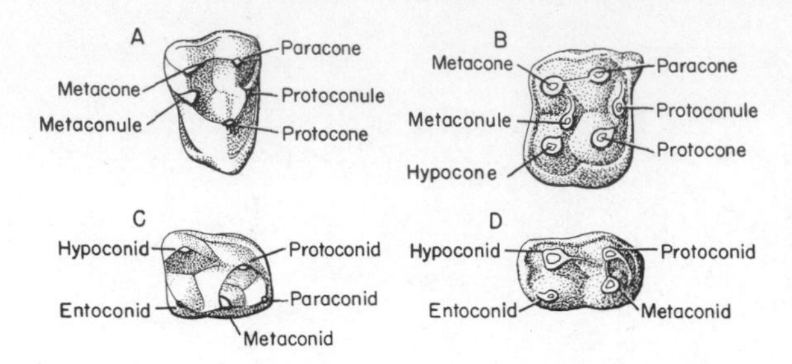

Figure 241. Diagrams of molar tooth patterns of placental mammals. In all cases, outer edge of tooth above, front edge to right. *A,* Right upper molar of a primitive form. *B,* The same of a type in which the tooth has been "squared up" by addition of a hypocone at the back inner corner. *C,* Left lower molar of a primitive form with five cusps. *D,* The same of a type in which the tooth has been "squared up" by the loss of the paraconid.

In generalized primitive placentals the upper molars were triangular, with three major cusps, one at an internal apex, two at the external base. It is believed that the lower teeth were originally similar triangles (but with internal bases and external apices) with a similar trio of cusps. However, such teeth, if meeting one another on closing of the jaws—**occlusion**—would tend to pass rather than meet one another. With resulting greater efficiency, each lower molar developed a "heel" with two additional cusps, into which fitted the apex of the triangular upper molar (as shown in Figure 242 *B*). This type of lower molar, already present in early placentals, is shown in Figure 241 *C*. Exactly how this pattern arose from the simple conical teeth of reptiles is still a subject of much controversy. Two of the most widely accepted theories are diagrammed in Figure 243; both may be correct—one for the uppers and the other for the lowers! A further step in molar development taken by many herbivores or mixed feeders is to "square up" both upper and lower grinders to oblongs, as shown in Figure 241 *B, D*. As can be seen, this is accomplished in the upper teeth by adding a fourth major cusp at the back inner corner and in the lower by losing one of the original three; in either case there results a tooth with four major cusps.

The cusps were primitively sharp-pointed; this, however, is ineffective for chewing. In mixed feeders (such as men and swine) they tend to be rounded "hillocks"—the **bunodont** type. In many ungulates the cusps may connect to form ridges, the **lophodont** condition, or the individual cusps may assume a crescentic outline in a **selenodont** pattern (Fig. 244).

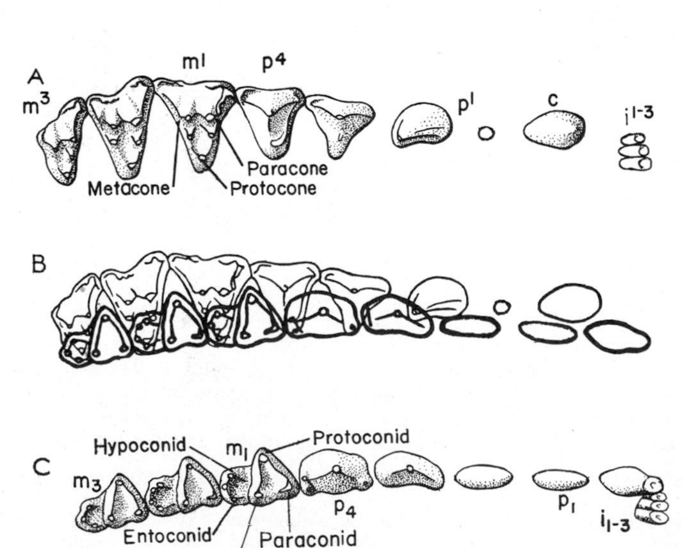

Figure 242. Diagrams of the dentition of a primitive placental mammal (based on the Eocene insectivore *Didelphodus*). *A,* Crown view of right upper teeth; *C,* crown view of left lower teeth; between, in *B,* the teeth are placed in occlusion, the outlines of the lower teeth (heavy lines) superposed on those of the uppers. (After Gregory.)

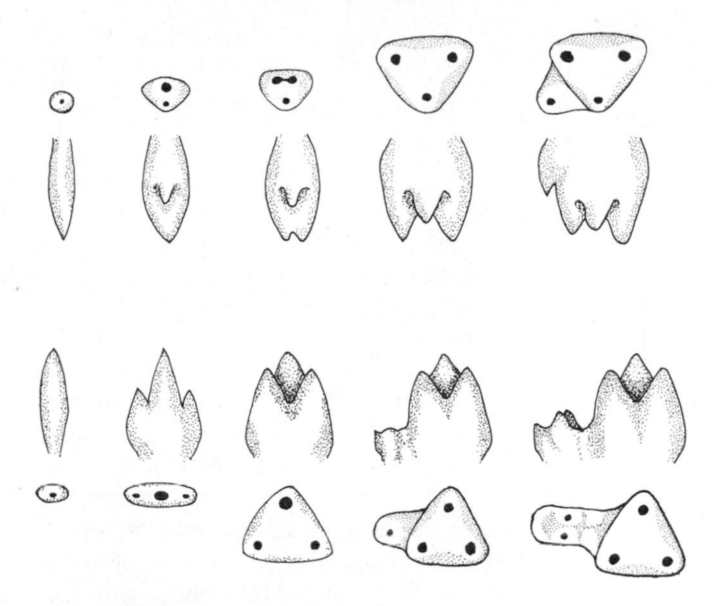

Figure 243. Diagrams to show the origin of mammalian molar structure. The top and bottom rows show occlusal views with the major cusps as black dots; in all, lateral (buccal) is to the top and anterior (mesial) is to the right. The two middle rows show medial views of the teeth, again with anterior to the right. The evolutionary sequences move from left to right. In the lower teeth, starting with the single reptilian cusp (protoconid), anterior and posterior cusps (paraconid and metaconid) are added, the tooth becomes triangular, and finally a heel (talonid) with two additional cusps (hypoconid and entoconid) is added posteriorly; this sequence is that suggested by Cope and Osborn. The upper teeth here show another theory, the amphicone theory. Here the original cusp (the eocone) divides to form the paracone and metacone; the protocone is a new development and the heel is small or absent.

Grazing presents a serious "problem" to an ungulate, for grass is a hard, gritty material which would wear a low-crowned tooth to the roots in short order. In relation to this fact we find that such forms as horses and cattle have developed a high-crowned—**hypsodont**—tooth type (Fig. 245). One could, in imagination, develop a high-crowned tooth by elongation of the dentine-filled bulk of the tooth body, leaving the cusps in their original shape on the grinding surface. This actually occurred in some early fossil mammals, but proved unsuccessful for, once the hard enamel surface of such a tooth is worn away, the bulk of the wear falls on the relatively soft dentine. Successful forms have hypsodont teeth built on quite another plan. The height is attained by a skyscraper-like growth of each cusp or ridge on the tooth; these slender peaks are fused together by a growth of cement over the entire surface of the tooth while it is yet beneath the gums. As wear takes

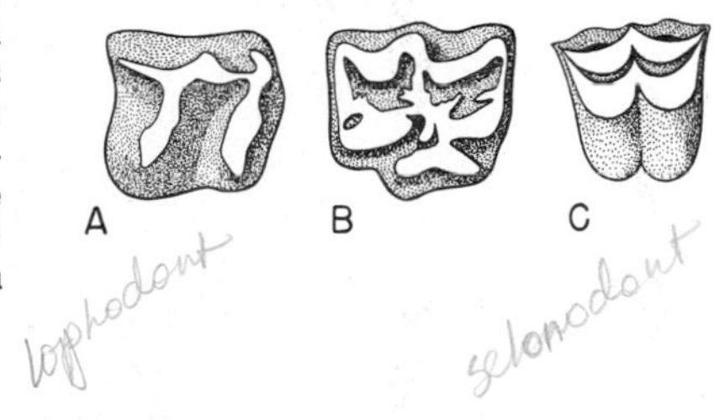

Figure 244. Crown views of molar teeth of *A*, rhinoceros; *B*, horse; *C*, ox; to show types of molar patterns. The white areas are worn surfaces showing dentine; the black line surrounding the white is the worn edge of the enamel; stippled areas, cement covering or unworn surface. *A* is a simple lophodont type, with an external ridge (above) and two cross ridges; the development of this pattern by connecting up the cusps seen in Figure 241 *B* can be readily followed. The horse *(B)* shows a development of the same pattern into a more complex form in which the primitive lophodont pattern is obscured. *C* shows a selenodont pattern characteristic of the ruminants; each of four main cusps takes on a crescentic shape.

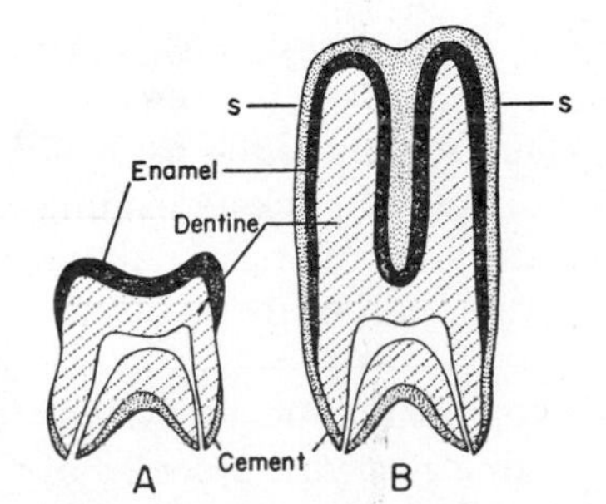

Figure 245. Diagrams to show the development of a hypsodont tooth. *A*, Normal low-crowned tooth (cf. Fig. 230). *B*, Hypsodont tooth, with cusps elevated and the whole covered by cement. As a tooth of this sort wears down to any level, such as that indicated at *s-s*, it will be seen that no less than nine successive layers of contrasting materials are present across the crown surface.

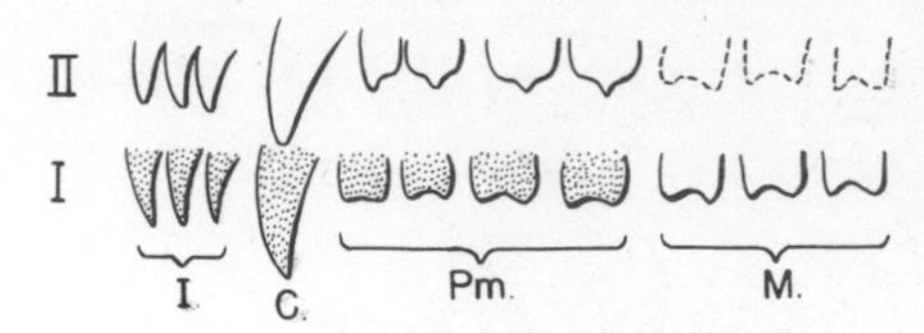

Figure 246. Diagram to show the composition of the "permanent" dentition of a mammal; left upper teeth of a generalized placental mammal. One complete set of teeth (I) develops from incisors back to molars. All of these except the molars, however, are shed (as indicated by stipple). A second set of teeth develops (II), but never produces molars. Hence the "permanent" dentition includes portions of two tooth series (cf. Fig. 239).

place, it grinds down through a resistant complex of layers of all the tooth materials—enamel, dentine, and cement. Teeth may also remain open-rooted and continue to grow after eruption.

In mammals, tooth renewal is a much reduced process; there is only one replacing set, and that is incomplete, for the molars have no successors. We usually think of the mammalian dentition as composed of a "milk" or **deciduous** dentition—which lacks molars, and a complete permanent set (Fig. 239). If, however, we consider the development of the teeth we gain a different concept (Fig. 246). The "milk" teeth usually develop successively in anteroposterior order from incisors to deciduous premolars (the canines, with long roots, may lag in eruption). After the appearance of the last premolar, the successive eruption of the three molar teeth begins. Meanwhile, without waiting for the completion of this first wave, the anterior teeth are replaced, but in the succession the molars are not replaced. Obviously, we have one complete set of teeth and a partial second set; the molars, although permanent, belong in series to the set which, farther forward, is deciduous in nature. That is why your own first molars were in place long before you replaced either premolar—indeed, in man the first molar is the first permanent tooth to appear. In various mammals, replacement may be modified (elephants, for example, are odd) or may be greatly reduced as in marsupials.

GILLS

The **pharynx** is a short and unimportant segment of the digestive tract in the higher vertebrates—a minor connecting piece between mouth and esophagus where ventrally the **glottis** opens to the lung apparatus and dorsally the paired **eustachian tubes** lead to the middle ear cavities (Fig. 229). In mammals it is merely the place where air and food channels cross one another in awkward fashion, and where there accumulate masses of lymphoid tissues, the **tonsils.** But from both phylogenetic and ontogenetic viewpoints the pharynx is an area of the utmost importance; it is the region in which are developed the gill pouches, basic in the construction of respiratory devices in lower vertebrate classes, and persistently important in the developmental story in higher groups.

In small animals with a permeable skin, sufficient oxygen can be readily obtained by exchange of gases through the skin. But in forms of greater size, in which the area of the skin may become insufficient to supply the need, and in forms with relatively impermeable skins, special respiratory structures—**gills**—are needed. These are formed in a variety of ways among invertebrates; in the vertebrates and their close chordate relatives, as we have seen, they take the form of **internal gills**—respiratory structures located in a series of slits or pockets leading from the pharyngeal region to the surface of the body. As the locus of origin of these slits, the pharynx is a highly developed and highly important part of the digestive tube in the lower vertebrate classes.

THE GILL SYSTEM IN SHARKS. A characteristic development of the gill system is seen in sharks (Fig. 247 *C*). On either side of the long pharynx a series of

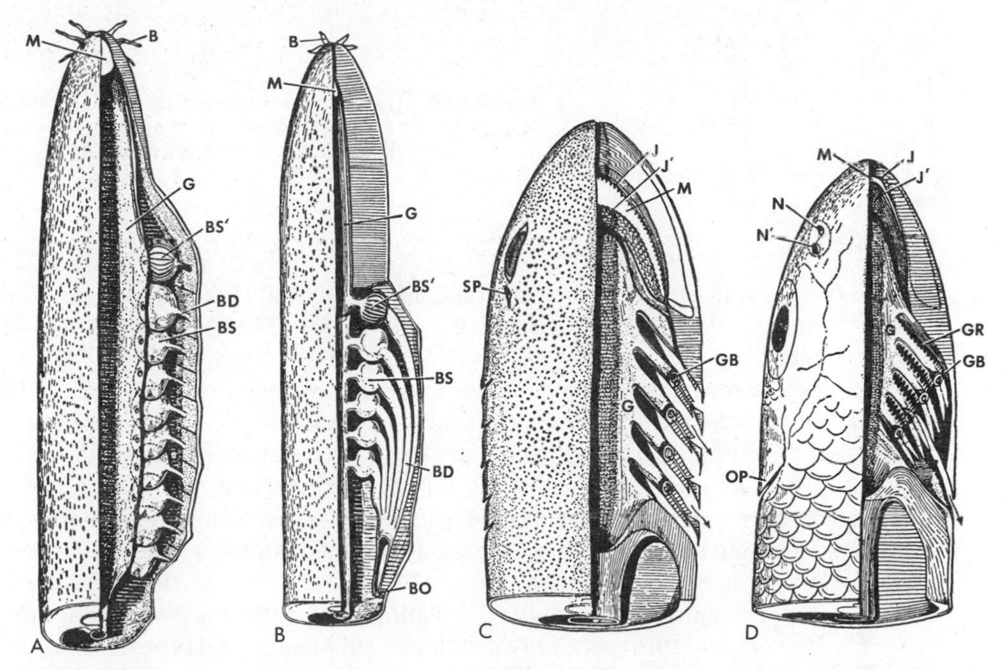

Figure 247. Heads of various fishes to show gill arrangement. *A*, The slime-hag *Bdellostoma; B*, the hagfish *Myxine; C*, a shark; *D*, a teleost. The right half of each is sectioned horizontally through the pharynx. Abbreviations: *B*, barbels around the mouth; *BD*, ducts from gill pouches; *BO*, common outer openings of gill pouches; *BS*, gill sacs; *BS'*, sacs sectioned to show internal folds of gill; *G*, gut (pharynx); *GB*, cut gill arch; *GR*, gill rakers; *J,J'*, upper and lower jaws; *M*, mouth; *N,N'*, anterior and posterior openings of nasal chamber; *OP*, operculum; *SP*, spiracle. (From Dean.)

openings runs outward to the surface of the body. Anteriorly, there is generally present a small and specialized paired opening, the spiracle (described later); back of it, on either side, are found the typical **gill slits,** five in number in most sharks (six or seven in a few special cases). In these slits are located the respiratory organs, the gills themselves. Water, brought in through the mouth to the pharynx, flows outward through the gill slits; in its passage past the gill surfaces, the respiratory exchange takes place. The term **gill bar*** is applied to the tissues lying between successive openings. The region between mouth and spiracle is the **mandibular bar;** that between the spiracle and the first normal gill slit, the **hyoid bar;** more posterior gill arches are generally referred to by number.

Each gill bar includes a series of characteristic structures (Fig. 248*A*). We have earlier described the skeletal elements, including elements of the gill arches, gill rakers, and gill rays. In addition there are branchial muscles, blood vessels (pp. 315–316); and for each gill a special cranial nerve or nerve branch (cf. Figs. 396, 397). Stiffening the gill and extending outward in sharks is a **gill septum** of connective tissue, which externally becomes a fold of skin overlapping and protecting the gill next posterior to it. The gill itself is a richly vascularized structure covered by a thin epithelium, folded into numerous parallel **gill lamellae.**

---

*There is a source of confusion here, for the term "arch" is often used in three different senses in connection with the gill region. As noted in Chapter 7, it may refer to the series of bars which form the skeleton of each gill segment. The word is also used to describe the arterial blood vessels—aortic arches—which traverse each gill (cf. Chapter 14). Finally it is used in a broader, inclusive sense, to describe the total structure lying between two successive gill slits. To reduce this problem, we will use bar rather than arch for the last of these three structures.

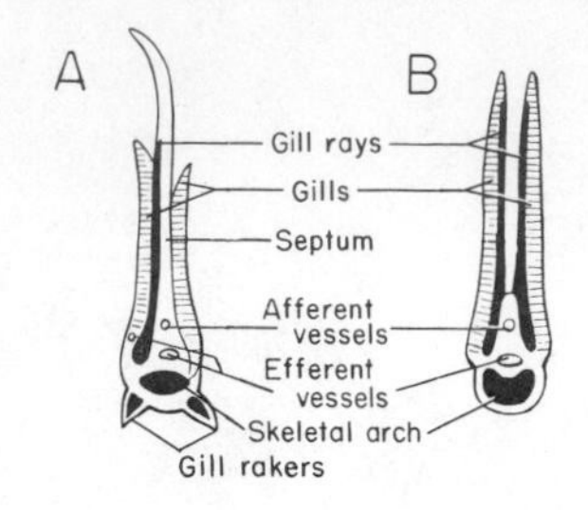

Figure 248. Horizontal sections of a gill bar of *A,* a shark; *B,* a teleost. External surface above, pharyngeal margin below; cartilage or bone in black. In the shark there is a projecting septum, lost in the teleost.

A gill may develop on either side of any gill slit or, stated in another fashion, may develop on either surface of a gill bar (Fig. 249 *A*). In most cases a bar carries a gill on both surfaces, and is hence considered to be a "complete" gill, a **holobranch.** Less commonly in fishes, a gill may be developed on only one surface of a bar, which is hence termed a **hemibranch.** No vascular arch is commonly present behind the last gill slit and in consequence there is almost never any gill development on the posterior surface of the last slit. All other typical slits in sharks, however, bear gills on both sides; in terms of gills (rather than slits) there are thus four holobranchs. There are no gill lamellae on the posterior side of the spiracle; hence the hyoid bar behind it is a hemibranch.

The **spiracle** is a small gill opening lying between mandibular and hyoid bars. Presumably this was a fully developed slit in ancestral types, but in all living jawed vertebrates the hyomandibular, as we have seen, comes to be connected with the jaw joint, and any opening between mandibular and hyoid arches is in consequence restricted in development. Even in its reduced state the shark spiracle bears a small gill on its anterior margin. However, the blood which reaches it comes from the next gill behind it, where it has already been aerated, and the spiracular gill is thus thought of as a "false" gill, a **pseudobranch.** In the skates and rays the spiracle is enlarged and serves a useful purpose. These forms are bottom dwellers, in which the mouth is for much of the time buried in mud or sand; the spiracle is put to use as a substitute place of entrance for the water current.

The gill slits of elasmobranchs arise embryologically in a fashion characteristic of vertebrates generally (Fig. 250 *B*). Early in development paired pouches push out from the endodermal lining of the pharyngeal region and come in contact with infoldings of the superficial ectoderm. Presently the intervening membranes break down, and the two epithelial layers join to form a continuous lining for the gill slits

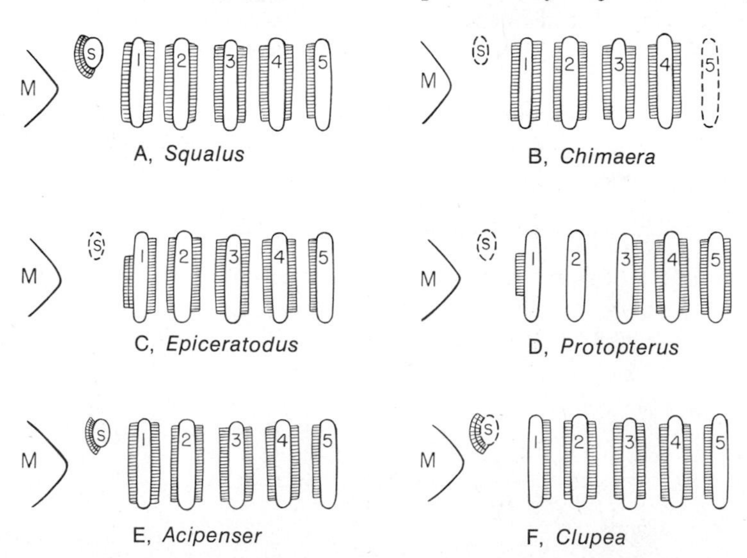

Figure 249. Diagrams to show the gill arrangement. *A,* Shark; *B,* a chimaera; *C,* the Australian lungfish; *D,* the African lungfish; *E,* sturgeon; *F,* a teleost (herring). Broken line indicates a closed slit. Hatched area adjoining slit indicates gill surface; vertical line through hatching indicates pseudobranch. The presence in *Protopterus* of a gill surface on the posterior side of the last slit is unique. *M,* mouth; *s,* spiracle; postspiracular slits numbered.

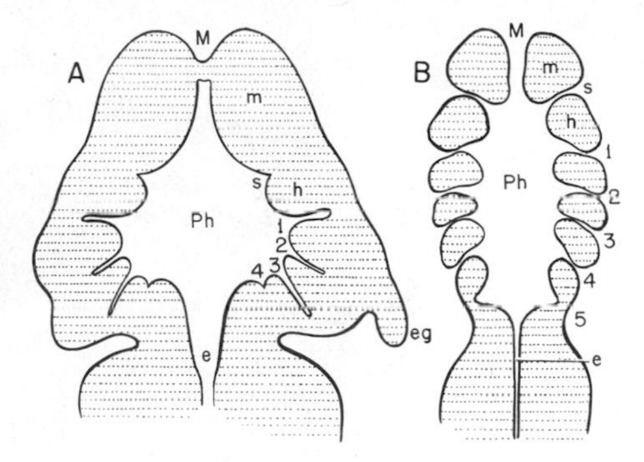

Figure 250. Horizontal sections through the head and pharynx in embryos of *A*, a frog (*Rana*); and *B*, an elasmobranch; to show development of gill pouches. In *A* the pouches are developing as narrow slits, which have not as yet opened to the surface. The mouth is also closed. In *B* all the slits except the last one are open. *e*, Esophagus; *eg*, external gill developing; *h*, hyoid bar; *M*, mouth; *m*, mandibular bar; *Ph*, pharynx; *s*, spiracular gill slit; 1 to 5, post-spiracular slits.

and, later, the folded gill lamellae. In most cases the ectodermal part of the lining is responsible for the formation of the lamellae. The gill pouches form in paired longitudinal series and are basically responsible for a segmental arrangement of the associated nerves and skeletal structures. Many workers have attempted to correlate the gill segmentation with that seen in the myotomes and the skeletal and nervous elements associated with them. This is discussed later (pp. 381–383).

The gill slits of elasmobranchs open directly and independently to the surface, but in the related chimaeras there develops a fold of skin, an **operculum,** which, extending backward from a point behind the jaws, covers and protects the gill series. We may note that chimaeras (Figs. 25 *B*, 26, 249 *B*) differ from their shark cousins in two further respects—the spiracle has disappeared (as it has in a few sharks as well), and the last gill slit is closed.

GILLS IN JAWLESS FISHES. Descending the evolutionary scale from the sharks, we find that in lampreys and hagfishes (Fig. 247 *A, B*) gills are equally well developed, but in a rather different fashion. The gill passages are not slits but spherical pouches, connected by narrow openings with the pharynx, on the one hand, and the exterior on the other; and the gill lamellae, in the adult, form a continuous ring around each pouch. Lampreys have well-developed muscles which effect a pumping movement on the sacs, so that water can be drawn in through the external tubes and forced out again through them, allowing the gills to function even

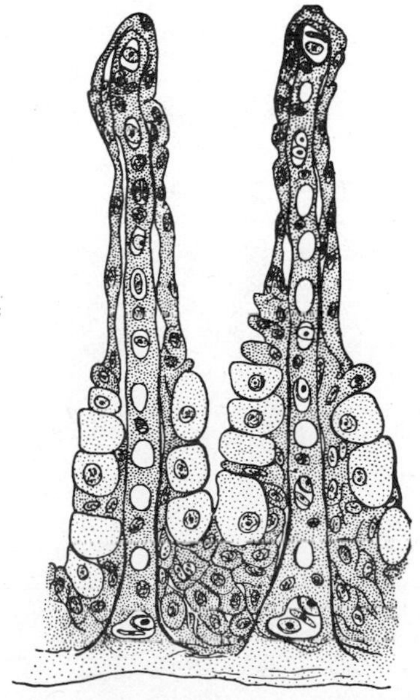

Figure 251. Two lamellae from the gill of an eel; the large cells at the base of the lamella are salt-secreting. The open (or largely open) circles in the center of the lamellae are the capillaries of the gill. (After Keys and Willmer.)

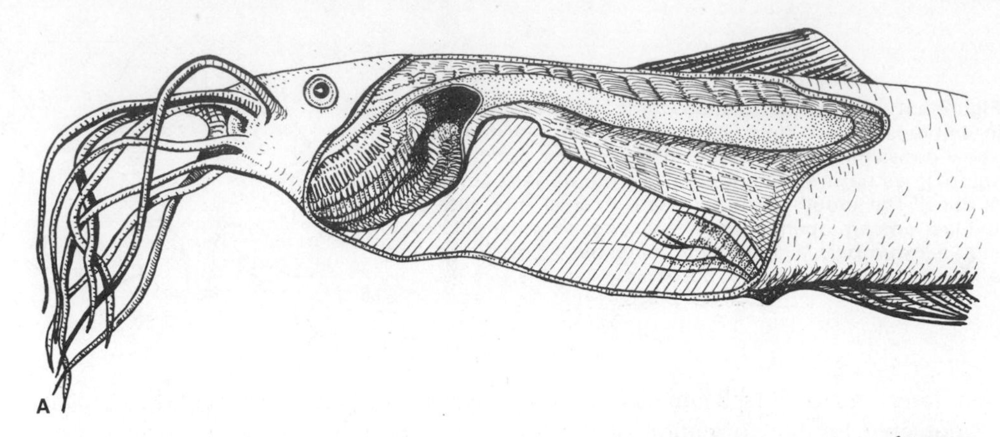

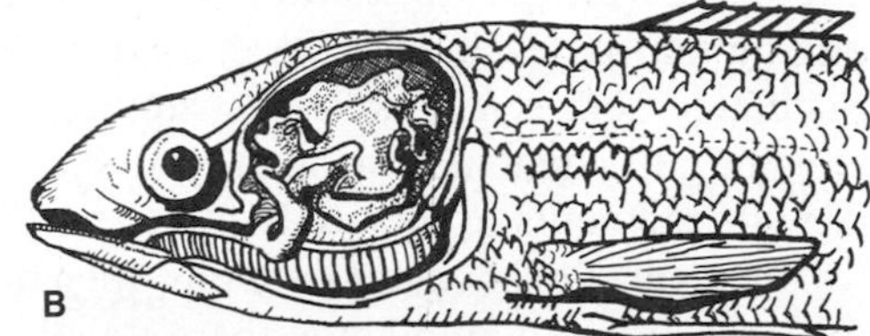

Figure 252. Accessory respiratory organs in teleosts. *A, Saccobranchus,* a tropical catfish; a rather lunglike sac extends posteriorly from the branchial chamber. *B, Anabas,* the "climbing perch"; here a complex folded structure, the labyrinth organ, projects into the branchial chamber. (After Giersberg and Rietschel.)

when the animal's mouth is occupied with its prey. Further devices to aid simultaneous feeding and breathing are present in both lampreys and hagfishes. In the adult lamprey the pharynx splits into two tubes; a small dorsal duct ("esophagus") leads directly back from mouth to esophagus, by-passing the gills; these lie in a larger, ventral pharyngeal sac ("respiratory tube") which ends blindly at its posterior end (Fig. 253). In hagfishes an equally peculiar device is present. The combined nostril and hypophyseal pouch breaks through posteriorly into the roof of the pharynx.

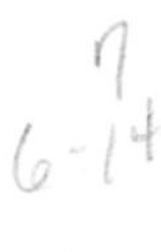

In most cyclostomes we find a higher number of gill pouches than the normal gnathostome count of five slits plus a spiracle. *Petromyzon* has seven pairs of pouches; hagfishes from six to fourteen. There is variation in external openings as well. In lampreys and the slime hag *Bdellostoma* each gill pouch opens by a separate orifice; in the common hag, *Myxine,* in contrast, the whole series of tubes of either side fuse externally to form a single outer opening.

The ammocoete larva of the marine lamprey, we have noted, spends its existence half buried in the mud of streams and ponds; it feeds on minute food

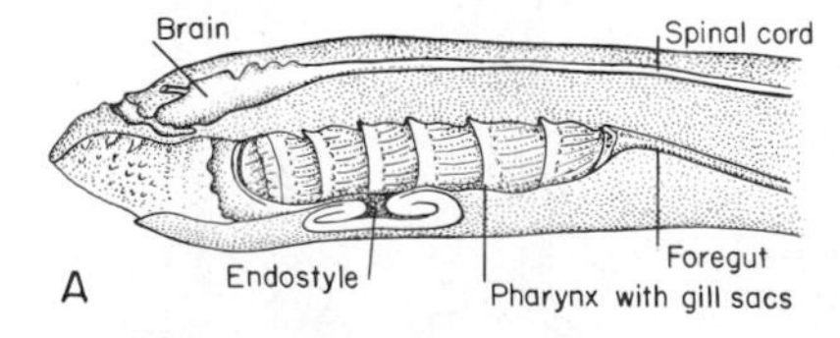

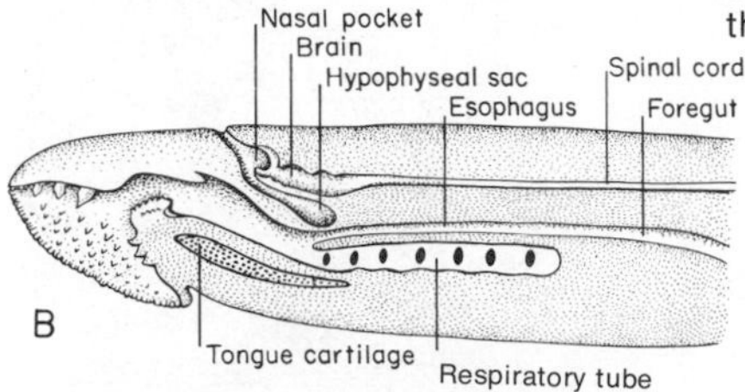

Figure 253. Sagittal sections of *A,* the ammocoete larva of a lamprey; *B,* adult lamprey; to show especially the division of the adult pharynx into two parts. The endostyle of the larva becomes the thyroid gland of the adult (not shown). (After Goodrich.)

particles which are collected from the water passing through the pharynx and gills in a fashion similar to that seen in amphioxus and the tunicates. This food-collecting function of the gill system is here a more important one than that of respiration, which could be readily carried on in the skin of these small animals. This situation, reinforced by the facts discussed in Chapter 2, leads to the concept that food collecting was the primary raison d'etre of the gills and respiration was at first only an accessory function.

Further evidence that food gathering was a major function of the gill system not only in their lower chordate ancestors but in ancestral true vertebrates as well can be adduced from a study of the most ancient of vertebrates, the fossil ostracoderms (Figs. 18, 19, 21). In such forms as the cephalaspids (Fig. 20 *A*) the mouth was small and apparently without means of aggressively gathering food, but the gill pouches (here about ten in number) occupied an enormous chamber on the under surface of the head and were obviously far larger than need have been the case were they concerned with respiration alone. It is obvious that these early vertebrates were still filter feeders; not until jaws (or the cyclostome substitute for them) evolved was this mode of alimentation abandoned and the gills became purely respiratory organs.

PHARYNX AND GILLS IN BONY FISHES AND TETRAPODS. In the bony fishes the gills (Fig. 247 *D*) are basically similar to those of sharks; however, there are various differences, due in part to the fact that in all members of this class of fishes there is a highly developed operculum, here (in contrast to chimaeras) reinforced by bony plates (Figs. 122, 169 *C*, 171 *A*, 172 *A*). Beneath the operculum there is present a considerable **branchial chamber.** In sharks, breathing is accomplished by (1) expansion of the pharynx with gill slits closed and the mouth open, with the effect of drawing in a volume of water, and (2) constriction of the pharynx with the mouth closed, thus forcing the water out the open slits, past the gills. In most bony fishes the action is similar, except that it is the opercular opening that is closed and opened rather than individual slits. Since the gills are protected by the operculum, the flaplike development of the gill septa present in sharks is reduced (Fig. 248 *B*).

Normally teleosts have five pairs of typical slits as do sharks, but aberrant conditions are occasionally found. There tends to be some reduction in the number of gills present (Fig. 249 *C–F*). In teleosts, for example, no gill is developed on the anterior margin of the first slit, and gills are much reduced in African and South American lungfishes. That the spiracle was present in primitive bony fishes is attested by its presence in the most primitive ray-finned forms—*Polypterus*, sturgeons, and paddlefishes. The spiracular opening is lost, however, in all other living Osteichthyes, although (curiously) a small gill may persist in the teleost pharynx near the site of this lost opening.

The gills of teleosts have excretory as well as respiratory functions. Glandular cells of the gill membranes (Fig. 251) appear to excrete nitrogenous wastes, supplementing the work of the kidneys. An important function of such cells in marine teleosts is their ability to excrete salt, further aiding the kidneys in maintaining a proper "internal environment" in which too high salt concentrations are harmful (cf. pp. 278–280). Although gills are the primary respiratory organs of almost all fishes, various teleosts have evolved different ways of surviving in poorly aerated water or even in air. Some respire through their skins, oral epithelium, or lining of the gut. Others have more elaborate organs, usually associated with the branchial chamber (Fig. 252).

In the larvae of a few fishes—*Polypterus* and the African and South American lungfishes—and in most amphibians as well, there are present accessory respira-

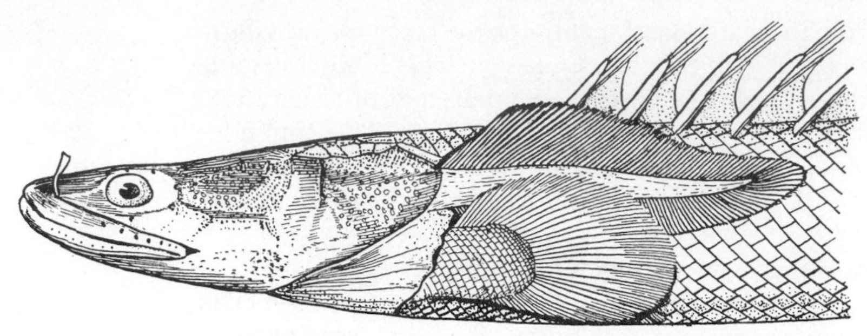

Figure 254. Larval form of the primitive African ray-finned fish, *Polypterus;* the large external gill arises from the top of the hyoid arch and extends back above the pectoral fin. (From Dean.)

tory organs in the form of **external gills.** As seen in the fishes mentioned (Fig. 254) and in urodele amphibians, they are feathery processes, from one to four in number, which grow out of the side of the neck above the gills; in life they are reddish owing to the presence of an abundant blood supply. The citation of the specific fishes which have these organs is in itself sufficient to bring to mind the most probable reason for their existence—that they aid in obtaining the utmost in oxygen from waters stagnant because of drought. It is, of course, during the rapid growth characteristic of embryonic and larval periods that oxygen demands are at their height. Probably these larval structures were characteristic, as were lungs, of the ancestral bony fishes, only to be abandoned later by most members of the group. In the Amphibia they are present in typical fashion in urodeles and the Gymnophiona. In anurans external gills begin to develop at an early stage, but later the gills of both sides are enveloped by a great fold of tissue, termed an operculum but not homologous with that structure in a fish. Beneath it there is a mass of filamentous gill tissue which appears to be derived in part from external gills, in part from outgrowths of the internal gill structures.

With the appearance of lungs, followed by the evolution of life on land, we find in adult amphibians and in all amniotes a great reduction in size and importance of the pharynx. In amphibians gill slits break through to a variable degree in the larvae, but internal gills never function (except in the aberrant condition we have noted in frogs), and apart from persistently larval forms the slits disappear in the adult. In amniotes, gill pouches are always formed in the embryo and push outward, as in fishes, to meet superficial furrows of ectoderm; but gill slits open only in a transitory manner if they open at all. The amniote embryo, in conservative fashion, repeats the embryologic processes of countless ancestral generations; never, however, is there true ontogenetic repetition of the development of the full-blown adult gills seen in fish.

In the adult amniote, as was said in the introduction to this section, the pharynx retains hardly a vestige of its former importance. The lungs developed from the pharynx in fishes, and the entrance to them is still to be found in the pharyngeal floor. Of the gill pouches, the spiracle persists in modified form in connection with the ear; in the adult amniote the other pouches have disappeared completely.

THE SWIM BLADDER

Characteristic of most ray-finned fishes is the presence of a swim bladder, an elongate sac arising as a dorsal outgrowth from the anterior part of the digestive tube; this is usually distensible and is filled with air or other gases (Figs. 255, 256

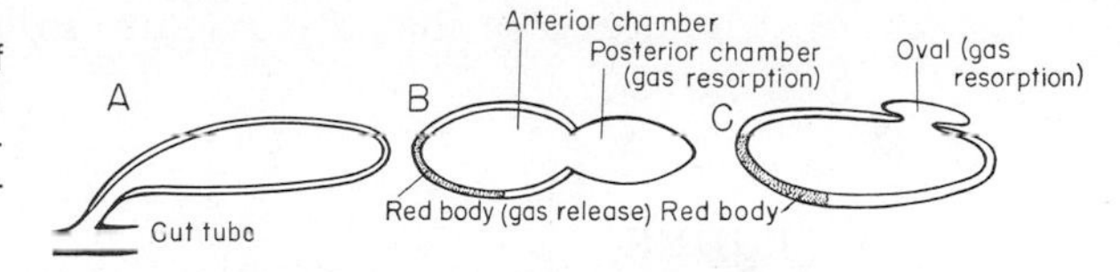

Figure 255. Diagrammatic longitudinal sections of the air bladder in various teleosts. *A*, Primitive type, open to the gut; *B*, *C*, closed types, with gas-producing red body and other areas for gas resorption.

A–C, 258*C*). Its major function is that of a hydrostatic organ; filling or emptying this sac alters the specific gravity of a fish and aids it in keeping at a depth in the water proper to its habits. It is thus obvious that this structure is most useful to fishes living in deep-water bodies—most especially the ocean. *Polypterus*, most primitive of actinopterygians, has no swim bladder as such but has, instead, ventral paired lungs; all other actinopterygians, however, have this air sac or appear to have secondarily lost it. It seems certain that the swim bladder is a specialization of the ray-finned fishes, developed at an early stage in their history; it is found in no other fish group whatsoever.

In its most primitive condition the air bladder connects with the pharynx by a **pneumatic duct** through which, when the fish rises to the surface, air can be taken into the sac, and through which the air can, as well, be discharged. In more specialized teleosts, however, this tube may shift to a more posterior connection with the digestive tract. In many advanced teleosts such connection has entirely disappeared; in such forms the walls of the swim bladder become specialized as a retort which can fill the bladder with gas, and a specific area is also present in which the gas so formed can be resorbed (Fig. 255 *B–C*).

In some cases—notably the relatively primitive actinopterygians *Amia* and the gar pike—the air bladder is rather lunglike in texture (Fig. 256 *B*) and is an auxiliary respiratory mechanism. This suggests the possibility that air-breathing may have been the original function of the bladder. In its dorsal position and single rather than paired nature the swim bladder contrasts strongly with the lungs. However, most anatomists believe that the two are at least partially homologous

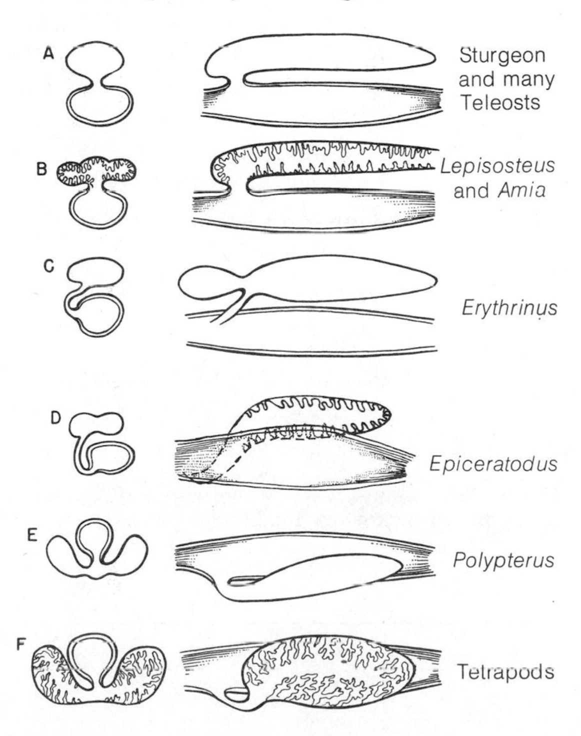

Figure 256. Diagrammatic cross sections and longitudinal sections of the air bladder or lung in various fishes and in tetrapods. *A*, Typical dorsal air bladder of actinopterygians; *B*, a bladder type found in holosteans with a folded inner surface capable of some breathing function; *C*, an unusual teleost type with a lateral opening suggestive of a transition from lung to air bladder; *D*, the Australian lungfish, with the bilobed lung rotated dorsally, although the opening remains ventral in position; *E*, the archaic actinopterygian, *Polypterus*, with a ventral lung, probably primitive and antecedent to all other lung or air bladder types; *F*, the type of lung developed in land vertebrates, with complex internal structure. Transverse sections viewed from the front and longitudinal ones from the left. (After Dean.)

structures. We shall discuss this further after a consideration of the history of the lungs.

LUNGS

LUNG STRUCTURE. Structurally distinct from the gills, although similar in basic function, and, like them, pharyngeal derivatives, are the lungs, which in typical air-breathing vertebrates replace the gills as the structure through which oxygen is brought to the blood and tissues. In most tetrapods the air-breathing apparatus has a characteristic pattern which may be sketched at this point (Fig. 257). Entrance to the air duct is gained by a median ventral opening in the pharynx, the **glottis.** Immediately beyond the glottis the duct enlarges to a chamber termed the **larynx;** beyond this a ventral median tube, the **trachea,** extends backward to divide into primary **bronchi,** one leading to each lung. The paired **lungs,** primitively ventral, may expand to occupy a lateral or even dorsal position in the anterior part of the celomic cavities. Embryologically the lungs appear as a median ventral outpocketing in the floor of the posterior end of the pharynx. This is in many cases distinctly bilobed at an early stage; the lung buds grow backward, accompanied by connective tissues which strengthen their walls and may produce skeletal elements related to the larynx, trachea, and bronchi.

In addition to an internal epithelium of endodermal origin, the lungs include in their substance variable amounts of connective tissue and smooth muscle fibers, and, of course, an abundance of blood vessels of the special pulmonary system; externally they are covered by epithelium of the body cavities within which they lie. The efficiency of a lung depends in the main upon the area of the internal surface present for gas exchange. Birds and mammals, with greater activity and greater needs for oxygen, increase the respiratory area not so much by increasing lung size as by increasing the complexity of internal subdivision. Here, as usual, we must keep in mind the problem of surface-volume relationships; in large animals the lungs must be disproportionately increased in size or attain greater complexity of subdivision so that their surface area may keep pace with the volumetric growth of tissue demanding oxygen.

Lungs are not the only possible organs for obtaining atmospheric oxygen; any thin moist membrane may effect gas exchange. In the modern amphibian orders the skin, as has been mentioned, is an important breathing structure and is richly

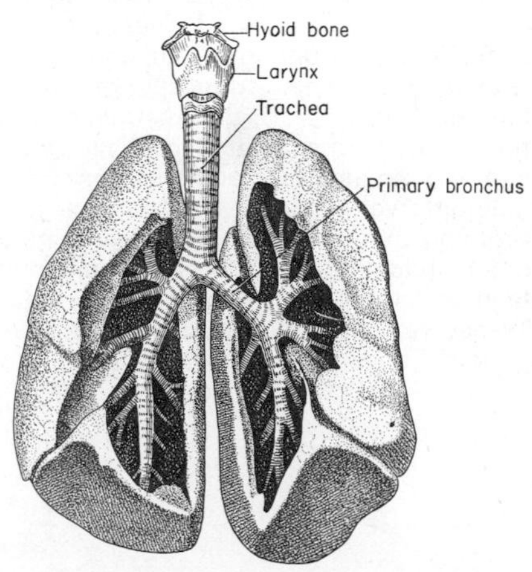

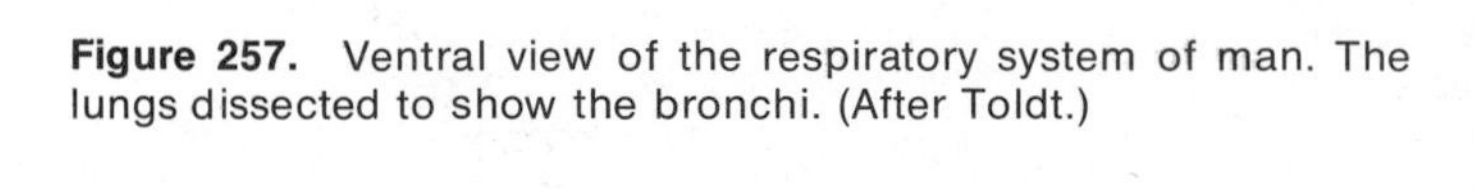
Figure 257. Ventral view of the respiratory system of man. The lungs dissected to show the bronchi. (After Toldt.)

supplied with blood vessels; in frogs and toads much of the breathing takes place through the moist membranes of the mouth. As already noted, many teleost fishes, such as the "tree climbing perch" *(Anabas)* of the East Indies, can breathe air by structures developed in or from gill chambers, kept moist beneath the operculum.

LUNGS IN FISHES. Although lungs are most characteristically developed in tetrapods, they are phylogenetically older structures. Lungs are present in the dipnoans, who gain their popular name of lungfishes from this fact, and surely were present in their close relatives, the ancient crossopterygians, from which land animals are descended. Still further, lungs are present in *Polypterus,* the most primitive living member of the ray-finned group of bony fishes.

Fish lungs (Figs. 256 *D, E*; 258 *A, B*) are simple in construction. In *Polypterus* an opening in the pharynx floor leads to a bilobed sac; the lobes pass back on either side of the esophagus. Except for the elongation of one lobe they resemble the lungs of amphibians and appear to be quite primitive. In dipnoans there is some modification, for although the opening, as always, lies in the floor of the pharynx, the duct curves upward around the right side of the pharynx, so that the lungs (which may be partially or entirely fused to form a single unit) are dorsal in position. The course of the duct suggests that this dorsal position is a secondary one. This supposition is reinforced by consideration of the blood supply; part of the lung is supplied by an artery (Fig. 258 *B, la*) which follows the presumed phylogenetic migratory path in looping ventrally around the esophagus from the left side of the body to reach the lung.

The embryonic development of lungs suggests that they arose in phylogeny as pockets of the moist pharyngeal epithelium which became specialized for absorption of atmospheric oxygen. But when and why did they arise? Since they are found in both the Sarcopterygii and the most primitive of actinopterygians, they obviously date in time from a very early stage in fish history; there is even evidence suggesting their presence in the very ancient placoderms. If they did have lungs, then they may well have been common to all primitive jawed fishes. Presumably the conditions of

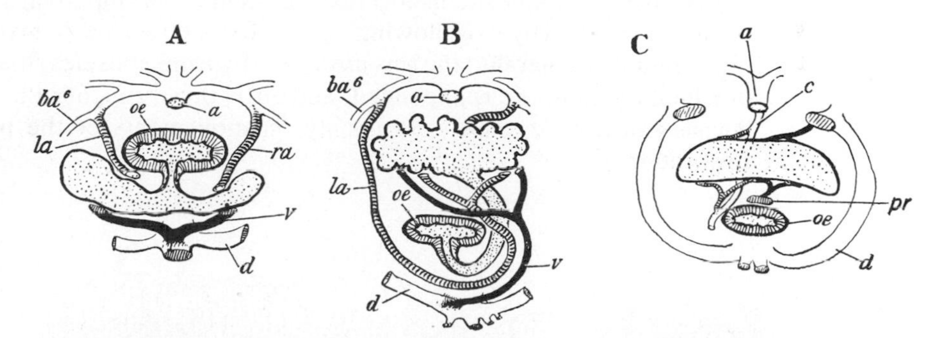

Figure 258. Diagrammatic cross sections of bony fishes (seen from behind) to show the position of the lung or swim bladder and the blood vessels connected with it. *A, Polypterus,* a primitive actinopterygian, with a ventral paired lung; *B, Epiceratodus,* a lungfish, with a single lung dorsal in position, but with a ventral opening from the gut; *C,* a teleost with air bladder (duct lost). In the lung, the arterial supply is from the last aortic arch; in *B* the curve of the vessel from the left arch below the gut indicates the path of dorsal migration of the lung. In the teleost the arterial supply to the air bladder is from the dorsal aorta (by way of the celiac artery). In the lung the venous return is directly to the heart region. The *Epiceratodus* veins show an asymmetrical condition comparable to that of the arteries. From the air bladder the blood returns to the heart via the normal venous system. *a,* Dorsal aorta; ba^6, sixth aortic arch; *c,* celiac artery, whence branches run to teleost air bladder; *d,* common cardinal vein (duct of Cuvier); *la,* left pulmonary artery; *oe,* esophagus; *pr,* hepatic portal vein, draining part of teleost air bladder; *ra,* right pulmonary artery; *v,* pulmonary veins. (After Goodrich).

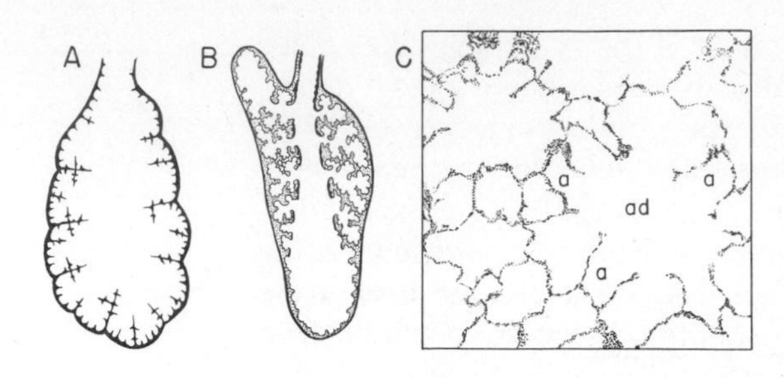

Figure 259. Diagrammatic sections through the lung of *A*, a frog, and *B*, a lizard. *C*, Section of a small area of a human lung, × about 50, showing its complex construction. *ad*, An alveolar duct, smallest component of the duct system; *a*, one of the alveoli to which it leads. (*A* after Vialleton; *B* after Goodrich.)

seasonal drought which we have previously noted were probably widespread in the early days of fish evolution may have been responsible for their evolution as accessory breathing organs when waters became stagnant or streams dried.

If, as we assume, lungs and swim bladder are homologous, which is the ancestral type? Early writers took it for granted that, since the swim bladder is a fish structure and lungs primarily characteristic of tetrapods, the swim bladder was the progenitor of the lungs. But consideration of the evidence now available suggests that the reverse is the case: the lung is primitive, the swim bladder a derivative evolved during the evolution of the ray-finned division of the bony fishes, the only group in which it is found. This is readily correlated with known facts in their history. The lung has been retained as such in modern fishes only in a few forms still living in tropical regions of seasonal drought. Most actinopterygians do not live in regions where such conditions now occur; most, indeed, are marine. A lung is of little or no use to such fishes; but if converted to a dorsal bladder, it could—and has—become a hydrostatic organ, especially useful in the marine environment in which most later actinopterygian evolution occurred.

THE LUNGS IN TETRAPODS. In amphibians the lungs remain relatively simple structures, with little internal subdivision (Fig. 259 *A*). Many reptiles show little further advance (Fig. 259 *B*), but in some lizards and in turtles and crocodilians there is multiplication of septa and internal partitions to give a more complex structure and a rather spongelike texture. In modern amphibians, in default of ribs, the lungs are filled by "swallowing" air—a force pump mechanism. In most reptiles (and amniotes generally) the ribs are raised by trunk muscles, the abdominal cavity in which the lungs lie is expanded, and air is thus brought in by suction. Turtles, encased in a solid shell, rely mainly on movements of the pectoral girdle for respiration.

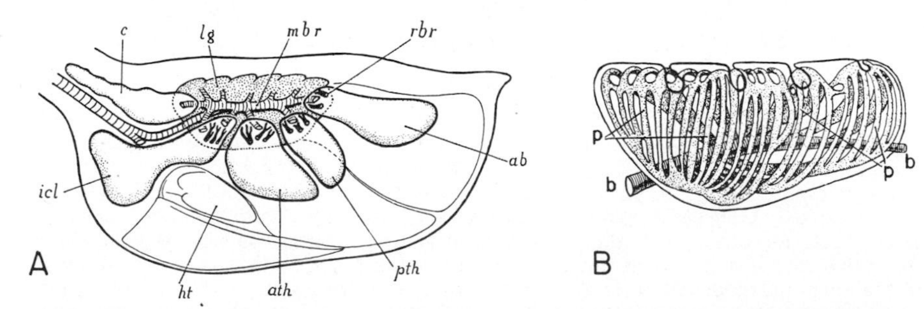

Figure 260. *A*, Diagram of the trunk of a bird, left side view, to show the position of the respiratory sacs. *ab*, Abdominal air sac; *ath*, anterior thoracic sac; *c*, cervical air sac; *ht*, heart; *icl*, interclavicular air sac; *lg*, left lung; *mbr*, bronchus running through the lung and leading from trachea to various air sacs; *pth*, posterior thoracic air sac; *rbr*, recurrent bronchi which return air from sacs to respiratory areas of lung. *B*, Lateral view of the left lung of a bird. The primary bronchus, *b*, traverses the length of the lung to the abdominal sac. Branches are given off, leading to other air sacs and to parabronchi *(p)*. These connect at both ends with the bronchi. Notches in the dorsal outline of the lung are rib impressions. (*A* after Goodrich; *B* after Locy and Larsell.)

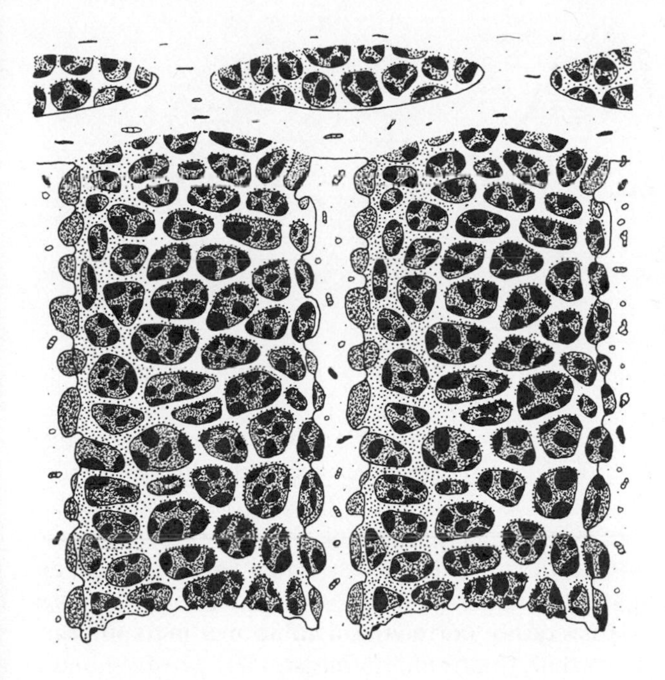

Figure 261. Diagram of the fine structure of a bird's lung, based on scanning electron micrographs. Two parabronchi are cut longitudinally and others, at the top, are shown in cross section. The respiratory exchange takes place in the many small pockets and spongelike areas along the wall of the parabronchi; airflow through the parabronchi occurs in one direction only. (After Schmidt-Nielsen.)

In birds the lungs themselves are small and compact, but the respiratory apparatus is complex in structure and function. Out beyond the lungs there develop some four pairs of **air sacs** which invade every major part of the body (Fig. 260), and from these sacs there may be air passages invading elements of the skeleton. The sacs themselves absorb little oxygen but nevertheless play a major role in respiration. Air is drawn in by raising the ribs and consequently expanding the volume of the trunk; it passes inward through the lungs, by channels continuous with the bronchi, to the air sacs. Respiration mainly takes place on the "return trip" from the sacs; small channels lead from the air sacs to the respiratory passages and sur-

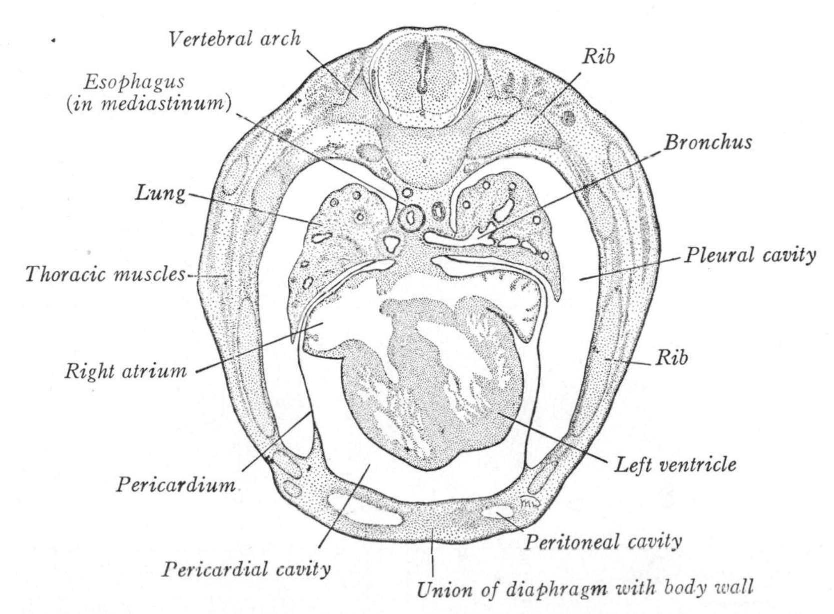

Figure 262. Cross section of a human embryo (at eight weeks) to show the development of the lungs and of pleural and pericardial cavities and heart. (From Arey.)

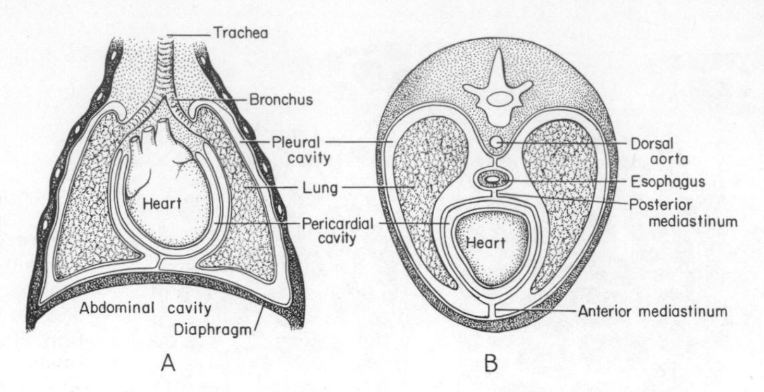

Figure 263. Diagrammatic longitudinal, *A*, and cross section, *B*, of the thorax in mammals to show the position of heart and lungs and of pleural and pericardial cavities.

faces within the lungs and thence outward to the bronchi. The internal structure of the bird lung is likewise unique (Fig. 260 *B*). In other amniotes the respiratory membranes are in "dead-end" alveoli. In birds, every passage, large or small, is open at both ends, so that there is a true circulation of air. The respiratory exchange occurs in spongy tissue surrounding the tubes, called **parabronchi,** passing through the lungs (Fig. 261).

In mammals the lungs are large, occupying a considerable area in the thorax, but show a rather simpler pattern than that found in birds. They are finely subdivided into tiny and exceedingly numerous **alveoli** (Fig. 259 *C*) to which the air

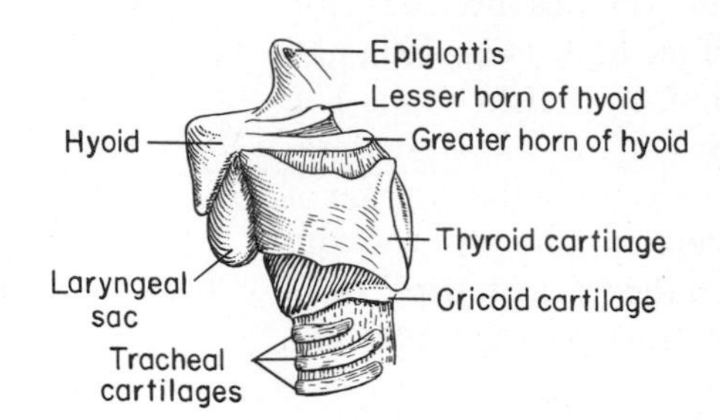

Figure 264. The larynx of the rhesus monkey, in side view. The hyoid apparatus, with its greater and lesser horns, is seated above and anterior to the larynx. The rhesus monkey has a resonating chamber in the laryngeal sac. Shown are the thyroid and cricoid cartilages, tracheal cartilages, and a membrane connecting the hyoid and thyroid cartilages. A muscle runs from thyroid to cricoid and covers most of the latter; there are several other small muscles proper to the larynx more deeply situated and not shown. (After Hartman and Straus: Anatomy of the Rhesus Monkey. Williams & Wilkins Company.)

passes by means of a branching series of larger and smaller bronchi and, finally, **bronchioles;** the alveoli cluster about the terminal ducts like grapes on the stem. In the embryo mammal (Fig. 262) the primary lung bud invades a compact mass of mesenchyme, within which, by repeated budding and subdivision of cavities, the adult structure is eventually attained. Breathing is accomplished by a highly efficient suction mechanism. The lungs no longer lie in the general abdominal cavity as they do in primitive tetrapods; instead each lies within an individual **pleural**

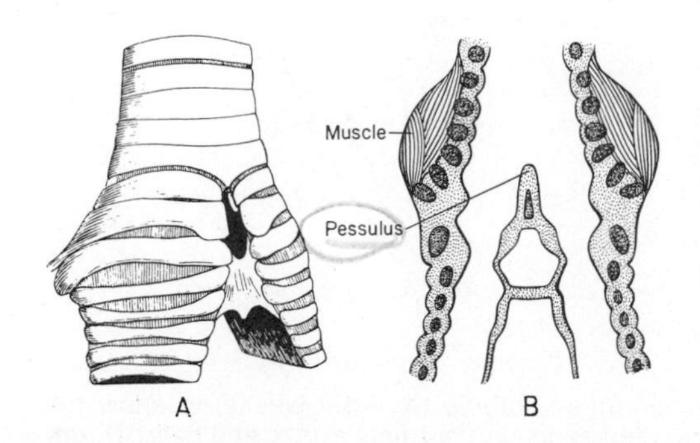

Figure 265. The syrinx of a songbird (magpie). *A,* External view; *B,* in section. Vibratory membranes on the inner aspect of the two bronchi meet at the base of the trachea to form the median pessulus; further membranes may develop between the expanded rings at the tracheal forks. The syrinx is controlled by musculature of the hypobranchial group, which can be very complex and varies greatly among different groups of birds. (After Haecker.)

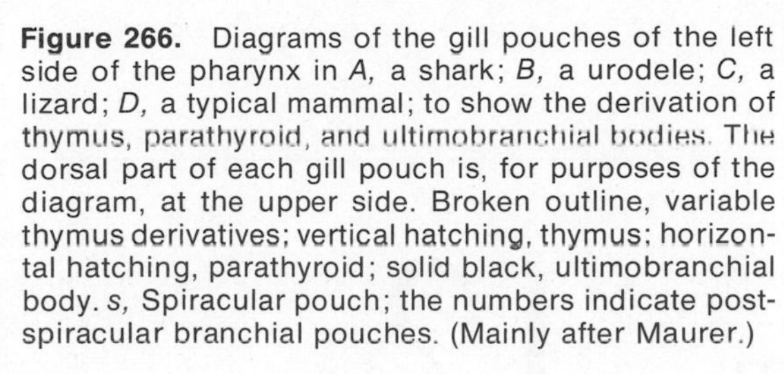

Figure 266. Diagrams of the gill pouches of the left side of the pharynx in *A*, a shark; *B*, a urodele; *C*, a lizard; *D*, a typical mammal; to show the derivation of thymus, parathyroid, and ultimobranchial bodies. The dorsal part of each gill pouch is, for purposes of the diagram, at the upper side. Broken outline, variable thymus derivatives; vertical hatching, thymus; horizontal hatching, parathyroid; solid black, ultimobranchial body. *s*, Spiracular pouch; the numbers indicate postspiracular branchial pouches. (Mainly after Maurer.)

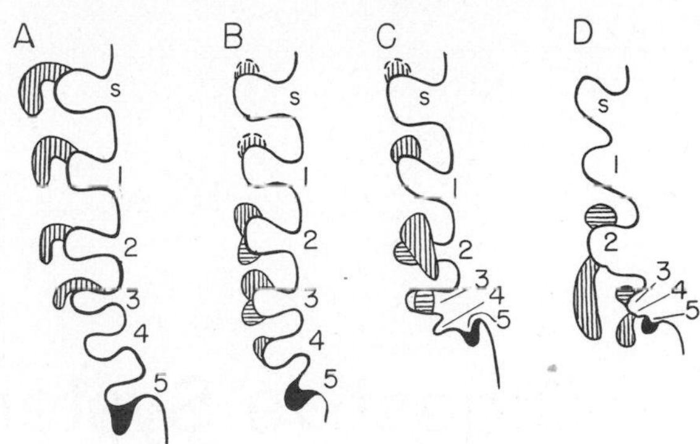

cavity, closed off posteriorly by the development of the diaphragm (Fig. 263; cf. p. 233). Expansion of the cavities, and consequently of the lungs, is attained as much or more by downward movement of the diaphragm as by movement of the rib basket.

With the rise in importance of the lungs in land vertebrates, the **larynx** (Figs. 229, 264), surrounded by a complex of cartilages or bone, arose at the entrance to the system into the pharynx; and a flap of skin, the **epiglottis,** developed to cover its entrance in mammals. Frogs and toads, a few lizards, and, notably, most mammals have acquired a voice through the development of elastic ridges, the **vocal cords,** stretched across the larynx. In birds no such cords develop, but voice production occurs in an organ, the **syrinx** (Fig. 265), typically developed at the point of subdivision of the trachea. With the development of the neck in tetrapods the originally short duct from throat to lungs becomes the elongate **trachea,** its walls generally stiffened by cartilages which in amniotes are ringlike structures. In amniotes the trachea divides into two primary bronchi before the lungs are reached.

PHARYNGEAL DERIVATIVES The pharynx is important in every group of vertebrates as the embryonic source of glandular structures. For the most part they are derived from the epithelial walls of the gill pouches (Figs. 266, 267), and form varied masses of tissues present in the neck of the adult.

Certain of these glands, notably the parathyroids and thyroid, are glands of internal secretion and are described in Chapter 17; the thymus, concerned with lymphocyte formation and the immune reaction, is discussed in connection with the circulatory system.

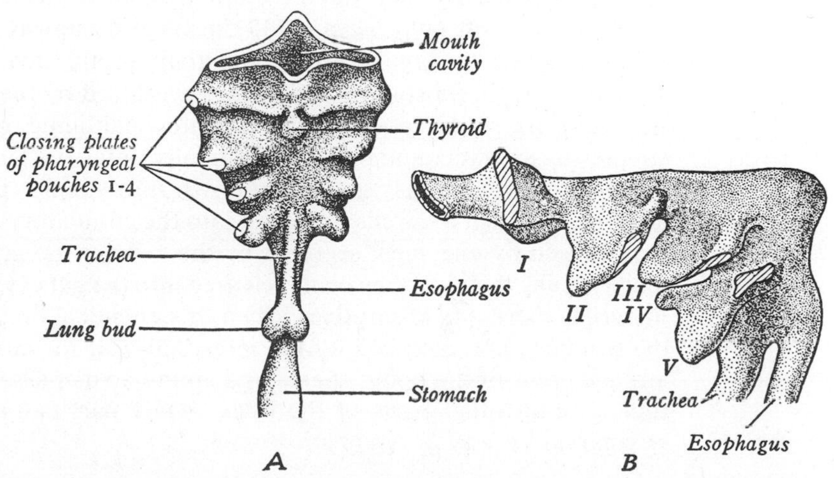

Figure 267. *A*, Ventral view of a model of the anterior part of the gut tube and its outgrowths in an embryo mammal (*Homo*). *B*, Lateral view of the pharynx in a slightly later embryo. Closing plates separating tips of gill pouches from the surface are hatched; pouches, including the spiracular pouch (middle ear), indicated by Roman numerals. (From Arey.)

Digestive System 12

The mouth and pharynx, described in the last chapter, are the forward outposts of the digestive tube, with gathering of food materials as their original major duty. The business of digestion is the function of the remainder of the digestive tract, to which the ancient and simple Anglo-Saxon term ''gut'' may properly be applied. The gut, so limited, together with its outgrowths—liver and pancreas—will be considered in the present chapter.

GUT FUNCTIONS. The functions of the gut may be considered under four heads: (1) **Transportation.** Once food materials are gathered, they must be carried along the ''dis-assembly line'' of the successive sectors of the gut, and wastes must be ultimately disposed of as feces. Although cilia are present in the epithelium of the gut in some instances, transportation is mainly the function of the visceral musculature which surrounds the digestive tube for its whole length—mostly smooth muscle, but often some striated muscle at the anterior end of the gut. The smooth musculature, usually arranged in sheets of both longitudinally and circularly arranged fibers (Fig. 270), is influenced by nerves of the autonomic (involuntary) system but in great measure operates independently of central nervous control. The major muscular activity causing movement of food materials is **peristalsis**—successive waves of muscular contraction causing constrictions of the gut which travel backward and push the food before them. (2) **Physical treatment.** Food may enter the digestive tract in large masses which must be reduced in size before an efficient chemical attack on them can be made; effective here (in contrast to peristalsis) are rhythmic contractions of the gut musculature. This ''squeezing'' of the food masses, together with the addition of mucus from glands in the gut, reduces the food to a soft pulp, **chyme.** (3) **Chemical treatment.** This is digestion in the technical sense, the breakdown of potentially useful ''raw materials'' in the food to relatively simple substances which can be utilized by the body. Water and necessary salts are readily absorbed by the intestinal lining; other needed materials—simple sugars, fats, amino acids—generally enter the gut in the form of complex compounds which must be broken down into simpler units before they can be absorbed through the walls of the gut into the circulatory vessels. This breakdown is produced by enzymes, secreted by the cells of the gut and its outgrowths but mainly doing their work by being released into the gut cavity and attacking the food materials there. (4) **Absorption.** When this chemical breakdown is accomplished, the products are absorbed by the intestinal wall for circulation to the cells and storage areas of the body; sugars and amino acids pass via veins to the liver and thence to all other parts of the body, while part of the fats reach the general circulation by way of lymphatic vessels.

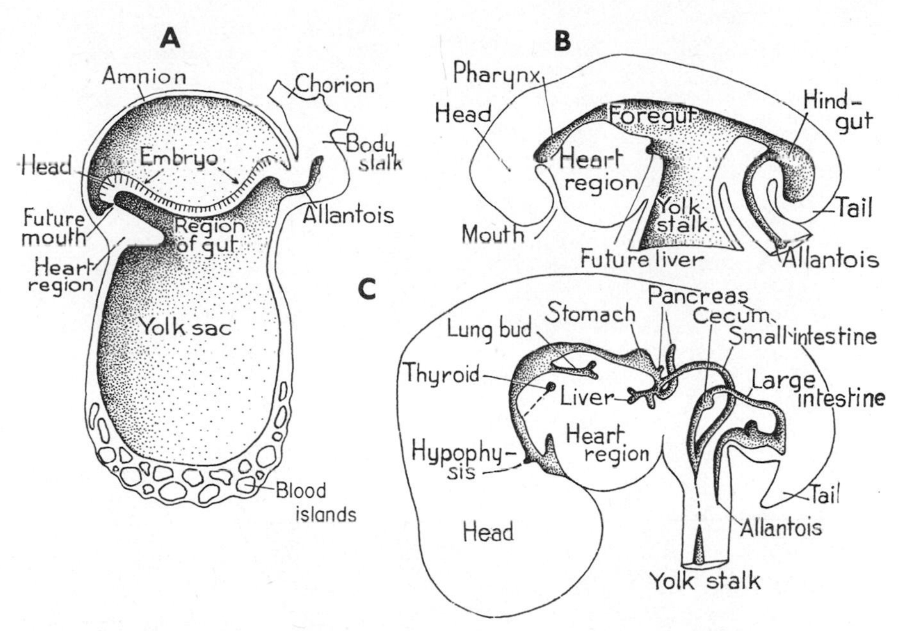

Figure 268. Diagrams to show the development of the digestive tract in a mammal *(Homo). A,* Stage somewhat later than that of Figure 76 *C*; *B,* stage slightly later than that of Figure 76 *D; C,* an embryo of about the age of that in Figure 87. (After Arey.)

DEVELOPMENT. In Chapter 5 was described the formation, in one fashion or another, of an archenteron, or primitive gut. In amphioxus (Figs. 69, 79) this is at first a simple cylindric pouch; in mesolecithal eggs, such as those of amphibians, the early gut is similarly constructed except that the gut floor is distended in paunchy fashion by a mass of endodermal cells rich in yolk (Fig. 71). In macrolecithal eggs, however, the presence of a great amount of yolk causes a radical change in early development of the gut. In early stages the gut lining does not form a closed tube but is spread out over the yolk, which it eventually encloses in a saclike extension of the gut; only later does the entrance to the sac become constricted and the gut assume a tubular shape (Figs. 76 *A, B;* 84 *D, E*). Mammals, despite the absence of yolk, follow this same pattern of development and produce a yolk sac, although an empty

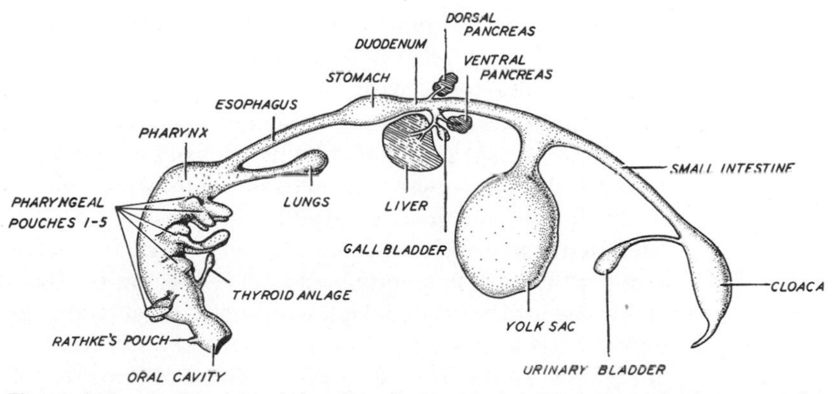

Figure 269. A diagram of the digestive tract and its outgrowths in an amniote embryo similar to that of Figure 268 *C*, but with the structures concerned shown as solid objects rather than in section. The first pharyngeal pouch here is the spiracle; the successive pouches are those numbered 1 through 4 elsewhere. (After Turner.)

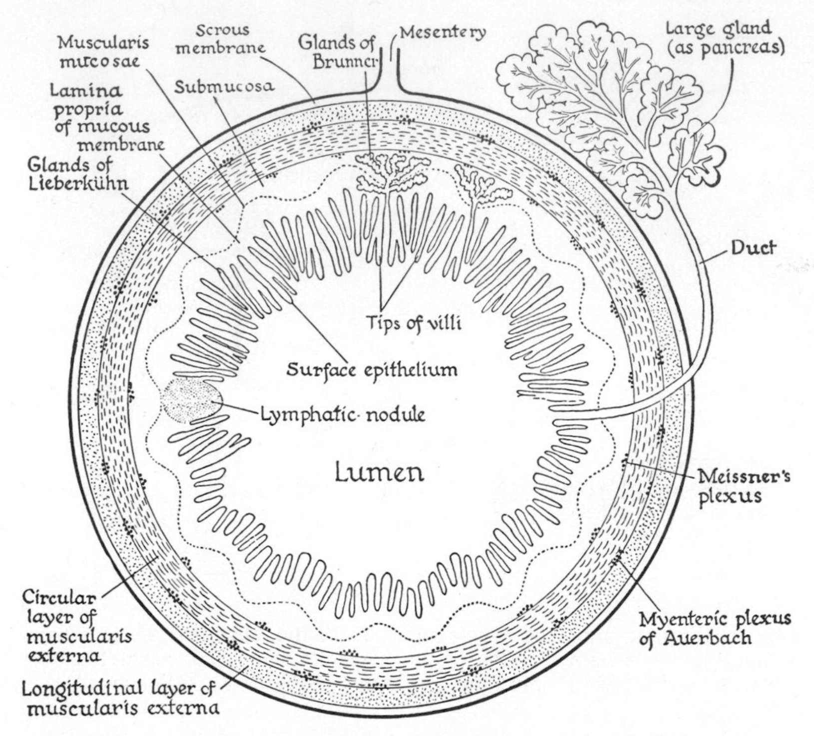

Figure 270. A diagram of a generalized cross section of the mammalian gut. In the upper half of the drawing the mucous membrane is provided with glands and villi; in the lower half it contains only villi. Meissner's plexus (submucous plexus) and the myenteric plexus are part of the autonomic nervous system; the former consists of sympathetic fibers, the latter, cells and fibers of the parasympathetic system. The tunica mucosa extends outward as far as the muscularis mucosae; it is followed outward in the section by the submucous tissue, the muscular tunic, in two layers, and finally, the serous tunic. The glands of Brunner and of Lieberkühn are characteristic of the mammalian small intestine. (From Maximow and Bloom.)

one; in all amniotes there is a second, more posterior diverticulum from the gut, the allantois, from which in the adult arises the urinary bladder (Figs. 76 *B, D*; 87; 268; 269).

In all cases, however, the gut eventually becomes a tubular structure. In most forms it remains for some time closed at both ends. As noted in the last chapter the anterior end eventually breaks through to connect with the oral cavity. Posteriorly, many vertebrate types, we have seen, have in the gastrula stage a posterior opening, the blastopore, which is in approximately the region of the anus or cloacal opening of the adult, but in large-yolked types this opening is at best transitory in nature. In any case the blastopore soon closes, so that the gut ends blindly posteriorly, as anteriorly, for much of embryonic life. Analogous to the development of the stomodeum anteriorly is the appearance posteriorly of the **proctodeum** (Fig. 226 *A*), an indipping pit of ectoderm beneath the tail; this is separated from the distal end of the gut by a membrane which eventually disappears. Meanwhile, within the gut there occurs a lengthening of the tube and a differentiation of successive regions into adult organs (Figs. 268, 269). Of these regions the most anterior, the pharynx, has been described previously and the most posterior, the cloaca, is so intimately associated with the urinary and genital as well as digestive systems that its consideration is best postponed. Subtracting these regions, there remain for consideration

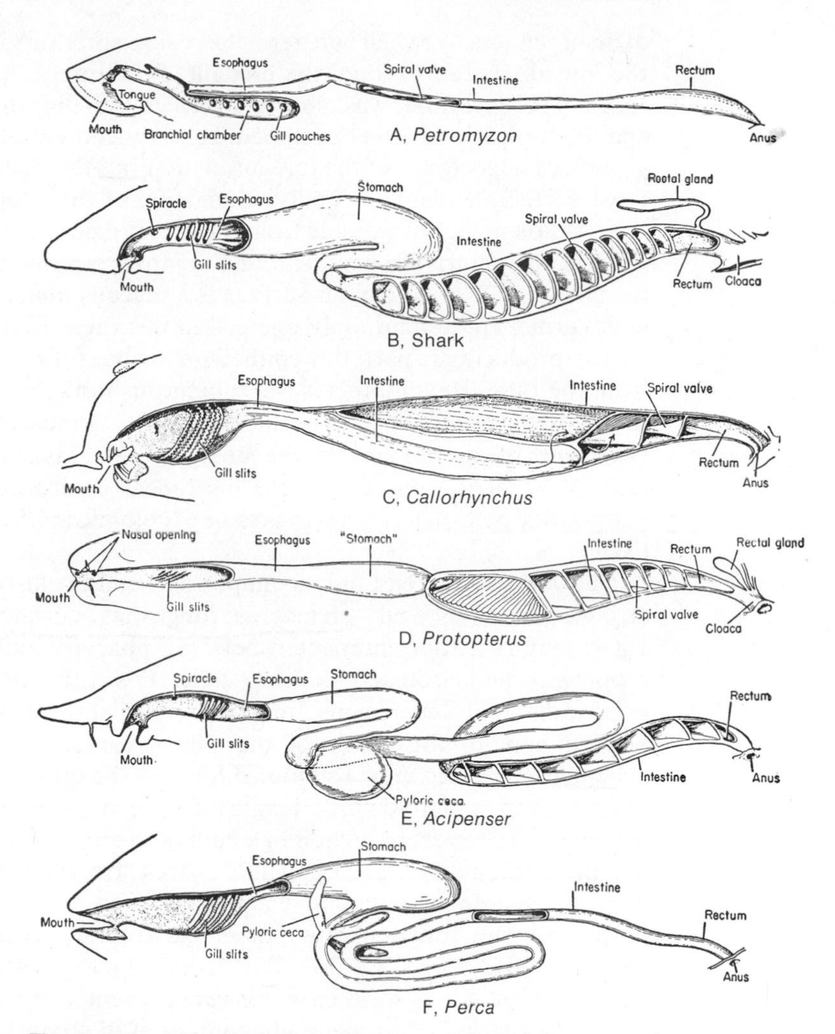

Figure 271. Digestive tracts of *A,* a lamprey; *B,* a shark; *C,* a chimaera; *D,* a lungfish; *E,* a sturgeon; *F,* a teleost (perch). The "stomach" of the lungfish is nonglandular and is simply a somewhat enlarged section of the esophagus. (From Dean.)

the major segments of the endodermal tube to which the term gut is for present purposes restricted.

GUT REGIONS AND STRUCTURE. A study of higher vertebrates gives one the impression that the succession of structures along the gut is consistent and uniform: esophagus, stomach, small intestine, large intestine, rectum, and anus. But when we extend our view to lower groups the picture becomes confused (Fig. 271). The distinction between large and small intestine breaks down in primitive vertebrates; the esophagus may be vestigial or absent; in some fishes there is no stomach. In amphioxus and cyclostomes the postpharyngeal gut is essentially a single tubular unit; most familiar landmarks are absent.

It is, however, possible at least to distinguish two successive regions of this tube in almost all vertebrates. Between stomach and intestine there is generally found a distinct **pylorus,** a constriction, usually muscular, which guards the gateway into the intestine. When this is poorly developed, the fact that the bile duct from the liver enters just back of the pylorus enables us, even so, to establish a line of demarcation. We shall here term the region anterior to the pylorus the **foregut** and the intestinal area beyond, the **hindgut.** Primitively, it would seem, the hindgut, forming the intestine, was alone concerned with digestion and absorption of food, and the foregut was little developed. In advanced vertebrates we see an increasing growth of importance of the foregut, with, first, the development of the stomach in most vertebrates and, secondly, elongation of the esophagus in tetrapods.

Although highly variable from region to region and from form to form, certain general characteristics may be found in most regions in the histologic structure of the gut (Fig. 270). The inner surface is a **mucous tunic;** this consists mainly of the endodermal epithelium, only one cell in thickness in many regions and generally mucus-producing in part; this epithelium further forms varied glands extending out from the tube. Beyond this is the **submucous tunic,** mainly formed by connective tissue, with numerous included blood vessels. A **muscular tunic** generally includes two layers of smooth muscle, the inner a circular layer and the outer a longitudinal one. Most of the gut lies in the peritoneal cavity of the celom, and hence is surrounded externally by a **serous tunic** of celomic epithelium backed by connective tissue.

ESOPHAGUS. Not only in amphioxus and cyclostomes, but in a fair number of jawed fishes as well—chimaeras, lungfishes, certain teleosts—the entire foregut is simply a tube interjected between pharynx and intestine. Although both esophagus and stomach develop from this tube in more progressive vertebrates, the term "esophagus" may, in default of the development of a stomach, be applied here to the whole of the simple foregut. In all other living fishes—elasmobranchs and most ray-finned forms—the development of a stomach leaves only a short and ill-defined length of tube anterior to it as an esophagus. In tetrapods, however, reduction in length of the pharynx (with loss of gill breathing) and the concomitant development of a neck, frequently of some length, bring into being a well-defined tubular esophageal segment. Only exceptionally has the esophagus any function other than that of transportation of food. Except for mucous cells, glands are little developed; the epithelium is generally of a tough stratified type, but in some cases cilia are present, even in the adult. There is a stout muscular coat which is generally composed of smooth musculature, but in many fishes and again in mammals (particularly ruminants) may be striated musculature, although this is of visceral origin (cf. p. 219). Two unusual developments may be noted. In lampreys the pharynx, we have seen, splits into two parts, the lower becoming a blind pocket for the gills, the upper giving a long anterior extension to the esophageal tube (Figs. 253 *B*, 271 *A*). In birds a distensible sac, the **crop** (Fig. 272 *C*), develops part way down the esophagus. It serves as a place for temporary storage of grain and other food and in doves exudes a milky material with which, regurgitated, the young are fed.

STOMACH. Although we are accustomed to think of the stomach as a normal part of the gut, it is absent in various lower forms, not only amphioxus but a number of different fish—cyclostomes, chimaeras, lungfishes, and some teleosts. When present, as in all other vertebrates, it serves three functions: storage of food awaiting entrance into the intestine, physical treatment of this food, and initial chemical treatment of proteins. In ancestral vertebrates, as today in fishes lacking a stomach, the last two of these functions were presumably served by the intestine alone, and the primary function of a stomach, when first developed in phylogeny, was for the storage of food masses which could not all be accommodated at once by the intestine. Such a situation would have first arisen when there developed, from early filter-feeders, predaceous jawed forms, such as the sharks, which would bolt

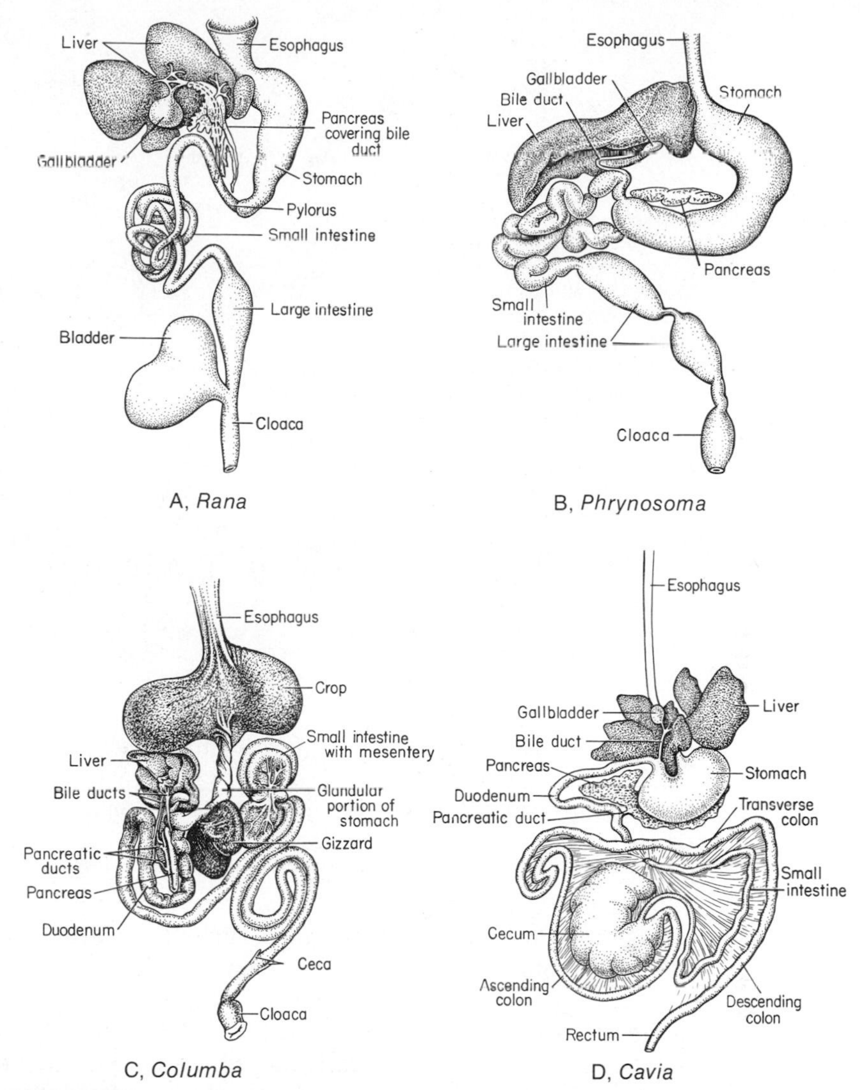

Figure 272. Diagrams of digestive tract and appendages, seen in ventral view, in *A*, a frog; *B*, a lizard; *C*, a bird (pigeon); *D*, a mammal (guinea pig). (*A* after Gaupp; *B* after Potter; *C* after Schimkewitsch.)

chunks of food in large quantities at irregular intervals. A pouch for food storage and physical treatment would thus become necessary; the introduction of digestive enzymes was presumably a phylogenetic "afterthought."

As regards external shape (Figs. 271–273), the central type of stomach structure, seen in various groups from shark to man, is one in which at the end of the esophagus the gut curves to the left, "descends," as a **cardiac region** of the stomach, to a major sac, the **fundus,** and then "ascends" to the right as the **pyloric region;** it thus forms a J. It should be noted that the shape of a stomach in a dead, dissected animal may be quite different from that in life, and that marked changes may take place from moment to moment in an active stomach (Fig. 274).

Highly important is the nature of the epithelial lining of various regions of the stomach. An **esophageal** type of epithelium may be present in a proximal segment of the stomach which is presumably an area "borrowed" from the esophagus. In

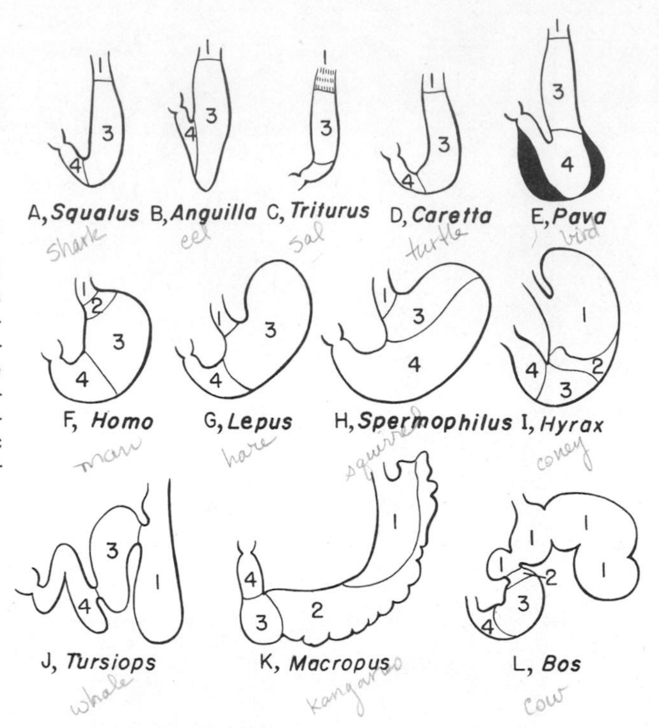

Figure 273. Diagrams to show stomach form and nature of internal lining in *A,* a shark; *B,* a teleost (eel); *C,* a salamander; *D,* a turtle; *E,* a bird (peacock—the thickened wall of the gizzard is indicated); *F,* man; *G,* a hare; *H,* a ground squirrel; *I,* the coney of Africa; *J,* a whale; *K,* a kangaroo; *L,* a cow. *1,* Epithelium of esophageal type (ciliated in *C*) which may penetrate into stomach, particularly in mammals; *2,* cardiac epithelium (found only in some mammals); *3,* fundic epithelium; *4,* pyloric epithelium. (After Pernkopf.)

mammals alone there may be a transitional **cardiac** type of epithelium proximally which contains glands, but glands of little chemical importance. The **fundic** type of epithelium is characterized (Fig. 275) by the presence of numerous tubular glands which usually contain two types of cells—the so-called **chief cells,** producing **pepsin,** an enzyme aiding in protein breakdown, and the **parietal cells,** which furnish hydrochloric acid and give the pepsin a more favorably acid medium in which to work. The **pyloric** epithelium resembles that of the cardiac region. Mucous cells are present throughout and the walls are almost always very muscular.

There are numerous variations among vertebrates in gastric form and in the distribution of epithelial linings (Fig. 273). The types of epithelial linings have no necessary relation to the similarly named morphologic regions of the stomach; for example, the stomachs of man and ground squirrel are much the same in shape but differ radically in the distribution of epithelia within them (Fig. 273 *F, H*). In most vertebrates the stomach is a relatively simple structure which may vary in shape from a straight, cigar-shaped tube, as in a variety of fishes and amphibians and the snakes, to a form, as in most teleosts, in which the fundus is a sharp V. Subdivisions of the stomach are found in some groups. In birds and crocodilians there develops a distinct rough-walled muscular compartment, the **gizzard** (Figs. 272 *C,* 273 *E*), in

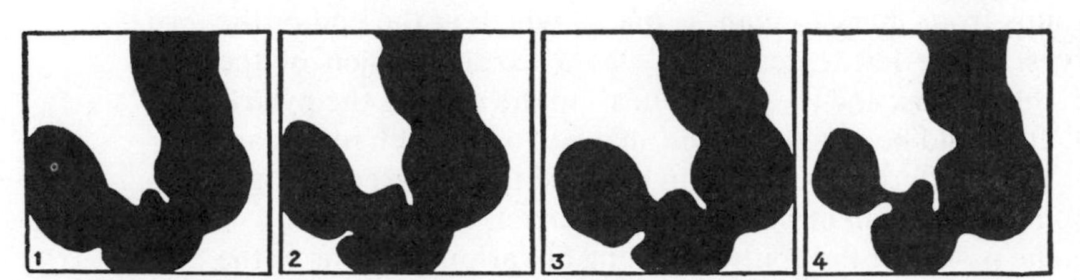

Figure 274. Skiagrams of a human stomach taken at intervals after food, showing its variable shape when active. (After Cole, from Fulton.)

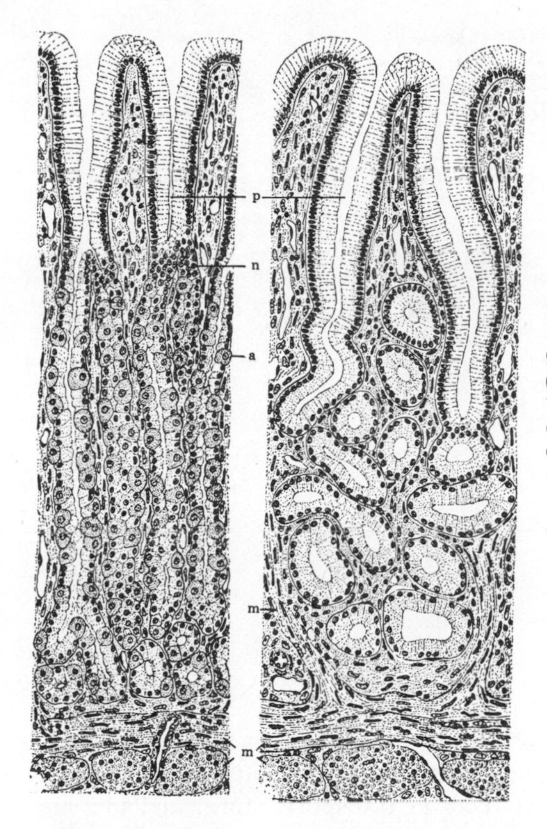

Figure 275. *Left,* a section of a portion of the fundic region of a mammalian stomach, showing the gastric pits *(p)* and the glands deep to the pits and opening into them. Mucous cells *(n)* are present in the neck region of the glands; below, two cell types can be distinguished: parietal cells *(a),* secreting hydrochloric acid, and smaller and more lightly staining chief cells, producing pepsin; *m,* smooth muscle cells. *Right,* a section through the pyloric epithelium, showing pits and glands of an essentially mucous type. (From Windle, Textbook of Histology, The McGraw-Hill Company.)

which are contained small stones; this makes a grinding mill which functionally replaces the teeth which are lost in birds. In the ruminating artiodactyls among mammals—such as cow, sheep, goat, deer, and so forth—four distinct compartments are present in the stomach (Fig. 276). The first two, **rumen** and **reticulum,** are storage pouches where vegetable food is kneaded to a more workable pulp by action of muscular walls and subjected to the action of microorganisms, which break down complex materials (especially cellulose) and manufacture useful organic substances (including some vitamins), some of which are absorbed in the rumen. At leisure the animal regurgitates the "cud" for chewing—or rumination, if you will—and then sends it, by a bypass, to the **omasum,** where there is further physical reworking and, at long last, to a final compartment, the **abomasum.** Here alone (Fig. 273 *L*) are found the three types of epithelium distinctive of the mammalian stomach; obviously the three preceding compartments are not part of the original stomach, but essentially elaborations of its proximal part in which, as we have noted, an esophagus-like epithelium is present in many mammals.

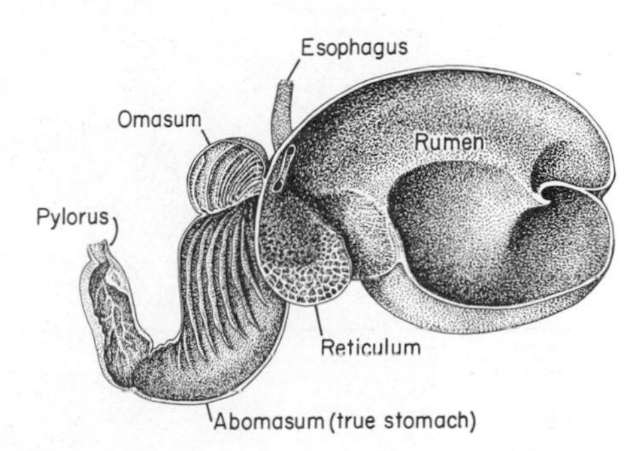

Figure 276. The stomach of a sheep, sectioned to show the four compartments characteristic of higher ruminants. (After Pernkopf.)

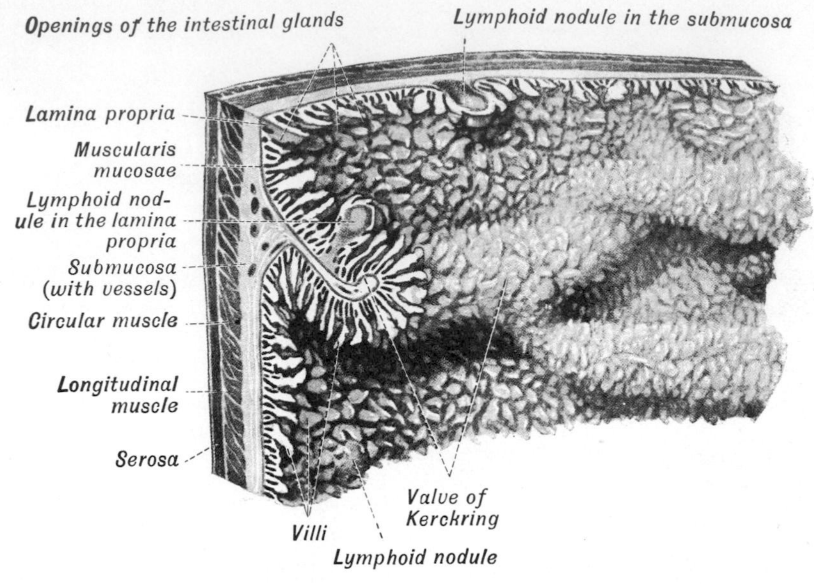

Figure 277. Part of the wall of the mammalian small intestine. (After Maximow and Bloom.)

THE INTESTINE (Figs. 271, 272, 277, 278). The major stages in the true digestive process normally occur in the hindgut—the variably built intestine. More anterior segments of the digestive tract receive, transport, store, and prepare food materials. In the intestine, however, occur most—and primitively, all—of the chemical processes of digestion. And here alone occurs, in most cases, the final crucial step, the absorption of food materials for body use.

In living vertebrates the intestine has come to share the production of digestive enzymes with other organs of the digestive tract. Many are formed in a special glandular outgrowth of the intestine, the pancreas; in most vertebrates the stomach

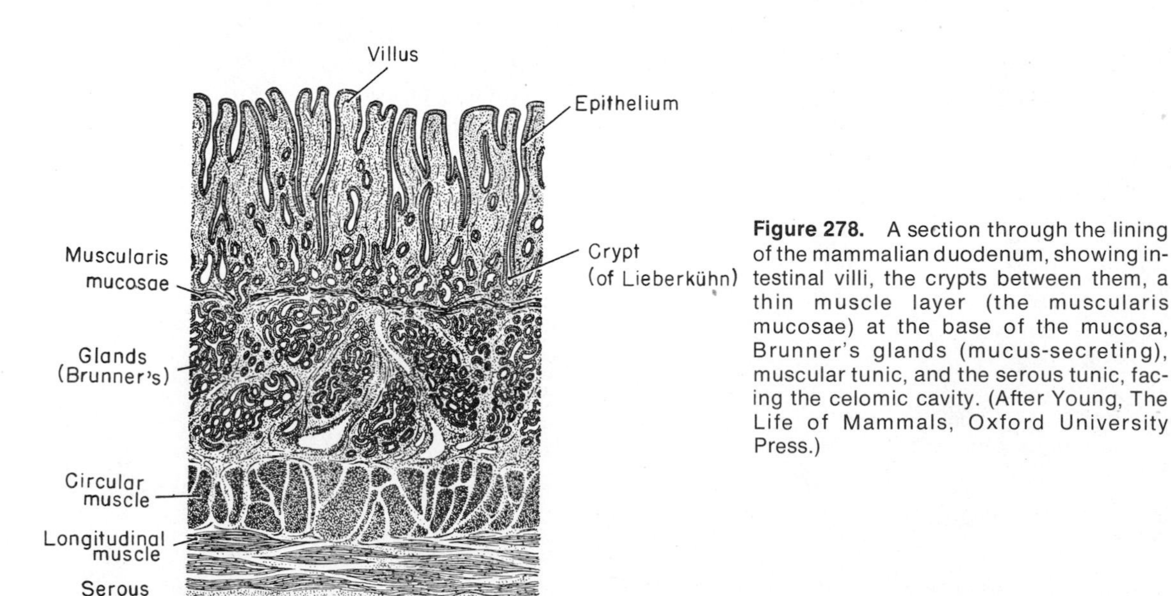

Figure 278. A section through the lining of the mammalian duodenum, showing intestinal villi, the crypts between them, a thin muscle layer (the muscularis mucosae) at the base of the mucosa, Brunner's glands (mucus-secreting), muscular tunic, and the serous tunic, facing the celomic cavity. (After Young, The Life of Mammals, Oxford University Press.)

has become a minor center of enzyme production, and even the salivary glands may take on such a function. Despite this, the intestine itself is still of importance, secreting a series of enzymes in numerous small glands embedded in its walls and to some extent even in the superficial epithelium itself. This epithelium, of simple columnar type, has, however, as its dominant function the absorption of the materials prepared from the food by the work of the digestive enzymes; once this epithelial wall is passed, they are discharged into the capillaries and lymphatics with which the intestine is richly supplied.

To allow sufficient absorption, a larger area of intestinal epithelium is required than could be provided if the intestine were a straight, smooth-walled tube. Methods of increasing this surface are effected by all vertebrates—particularly large forms, because of surface-volume relations between the absorptive surface and the bulk of tissues for which food must be absorbed. Such increases of surface have been brought about by structural features on three levels of magnitude. (1) Of microscopic size are countless small folds of the mucous tunic, which appear to have been primitively a network of tiny ridges but in mammals are generally in the form of countless small finger-like **villi** (Figs. 277, 278). Ridges or villi occur in all vertebrates. (2) In numerous instances there may be somewhat larger folds in which the submucous layer is involved as well as the epithelium (as in the mammalian folds of Kerkring, Fig. 277). (3) Major structural developments occur which greatly increase the area of the intestinal surface. Two are most notable: the spiral valve characteristic of primitive vertebrates and the slender but long and highly coiled tubular intestine developed by teleosts and tetrapods.

THE SPIRAL INTESTINE (Figs. 23, 26, 31, 256 *A–E*, 271). In members of every major group of fishes we meet with a type of intestine seemingly primitive for vertebrates—the spiral intestine. This is present in all bony fish except the teleosts; it is, however, most characteristically developed in the sharklike fishes. There is in such forms no division of the intestine into "small" and "large" regions. Except for a short proximal tube connecting with the stomach and a short distal rectal region (where there may be an accessory gland, which in sharks secretes sodium chloride), the entire length of the hindgut is occupied by the spiral intestine, a large, cigar-shaped body running fore-and-aft for most of the length of the peritoneal cavity. Its internal structure is complex; in addition to minor folds, its surface area is greatly increased by the presence of the **spiral valve**—a fold of epithelium and connective tissue extending in spiral fashion from one end of the intestine to the other, somewhat as if a carpenter's augur, or bit, were enclosed in a tube. In a few sharks the spiral valve has a different form which is, however, equally effective: the base of the valve twists but little and hence is relatively short; the valvular fold, however, is highly developed and rolled up into a great scroll running the length of the intestine. In either case, the surface area inside the gut is greatly increased. Cyclostomes are more problematic; the lamprey has a slightly spiralled ridge along its intestinal wall which may represent the spiral valve. Hagfish lack even this ridge.

THE INTESTINE IN HIGHER VERTEBRATES. Although among bony fishes the dipnoans and lower actinopterygians persistently keep the spiral valve structure, the teleosts, on the one hand, and tetrapods as well, have abandoned it for a new type of hindgut structure in which the intestine is a slender tube without major internal folds (Figs. 271 *F*, 272), but which, in compensation, may be greatly elongated. In teleosts most of the length of the tube is a highly active digestive region, beyond which a short segment leads to the anus. A special teleost development is that of **pyloric ceca**—pouches, sometimes very numerous (up to 900!), in the proximal end of the intestine into which food materials can enter and be absorbed.

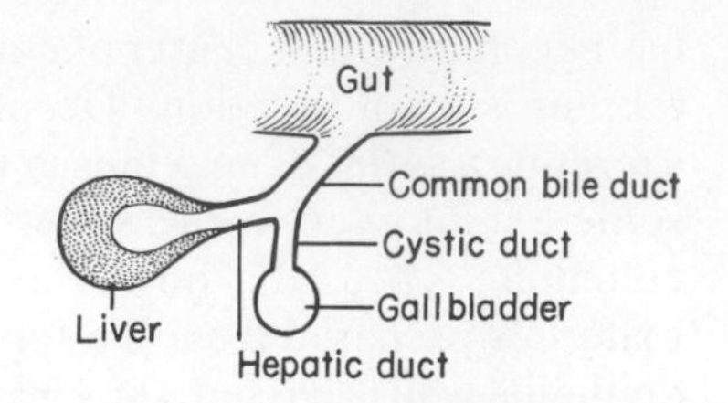

Figure 279. Diagram of an early stage of liver development to show the relationship of the ducts.

In both teleosts and tetrapods there are enormous differences from form to form in the length of the intestinal tube and consequently in the area of absorptive surface. Two factors influence intestinal length: food habits and absolute size. Plant foods often include masses of complex carbohydrates which are difficult to digest and absorb, and hence the intestine is in general longer in herbivores than in carnivores. As regards the differences due to size of the animal, we encounter here again the question of volume-surface relationships. With increasing size, the volume of the body requiring nutriment increases faster than does the intestinal surface, and disproportionate elongation of the intestine is necessary to keep the absorptive surface in line with the demands made on it. From teleosts to mammals, small flesh eaters tend to have the shortest intestines, large herbivores the longest.

No trace of the spiral valve persists in any tetrapod, and the greater part of the hindgut universally consists of a slender, **small intestine,** coiled to a variable degree—in general more complexly so in birds and mammals than in the lower classes. This is the major seat of digestion and absorption. In some cases (as in man) subdivisions may be named, but the differences between them are very slight. Beyond the small intestine there is usually, in lower tetrapods, a short but broader segment which appears to be homologous, at least in a broad sense, with the **colon,** or **large intestine,*** which is a highly developed terminal area of the gut in mammals, generally separated by a valve from the small intestine. There is frequently a small outpocketing at the proximal end of the colon in lower tetrapods. In amniotes this becomes the variably developed **cecum,** which in man and various other forms terminates in the **vermiform appendix.** In most mammals the cecum is a midventral structure embryonically, that of reptiles is usually dorsal, and that of birds paired and lateral. The appendix is sometimes claimed to have evolutionary significance; but this does not seem to be the case. The colon, when well developed, is not merely a place for the collection of feces pending evacuation, but a terminal digestive area where bacteria, plentifully present in colon and cecum, may make a final attack on cellulose or other difficult carbohydrates and a certain amount of absorption of water and other materials takes place. In lower tetrapods and birds the intestine opens distally into the cloaca; in typical mammals the short **rectum,** leading to the anus, is derived from the embryonic cloaca.

THE LIVER. In concluding this chapter we treat of two organs, liver and pancreas, derived embryologically from the endoderm, which are important both for secretions which they furnish to the intestine and for their functions in the metabolism of food already digested.

Amphioxus (Fig. 4) has a saclike outgrowth of the gut which is similar in position to the liver but of dubious homology; a large liver is, however, universally

*The "large" and "small" refer only to the diameter of those parts of the intestine: for other measures they are reversed.

present in all true vertebrates. Embryonically, it develops as an outpocketing of the gut below the stomach in the ventral mesentery, extends forward to attach to the transverse septum behind the heart, and then expands backward into the abdominal cavity. It has no constant form, and needs none, for its proper functioning rests on an appropriate internal arrangement and sufficient bulk. In general, we have noted, the endoderm forms only the thin lining of the gut and the glands developing from it; only in the liver (and to a lesser extent in the pancreas) does endodermal material bulk large in body composition.

Primitively, it would seem, the liver evolved as a gland, and in its early embryonic development a glandular structure is prominent (Figs. 226 *A*, 268 *C*, 269). As its main mass of tissue pushes forward, it remains connected by a duct system with the gut. As one of its functions the adult liver secretes a liquid, the **bile,** which consists partly of wastes but includes certain salts useful as an aid in fat digestion. This material is collected by tiny tubules from the liver cells and leaves via the **hepatic duct** (Fig. 279); the bile may be stored in a **gall bladder,** which usually (but not always) develops along a side branch, the **cystic duct;** eventually the bile reaches the gut through a **common bile duct.**

The major duties of the multitudinous liver cells, however, are not those directly connected with digestion but with the treatment of food materials after their digestion and absorption into the body. In part its activity is that of a storage depot, particularly for carbohydrate storage as glycogen and for fat storage (as in the cod). Further, the liver cells act as complex chemical factories; proteins may be synthesized here, fats altered in composition, proteins and fats transformed into carbohydrates, nitrogenous wastes such as ammonia transformed into less harmful substances such as urea or uric acid. For its major functions as storage depot and manufacturing plant, the liver is strategically situated along a "main line" of the circulatory system. As will be described later (cf. pp. 323–324), all the food-carrying veins from the intestine collect as a hepatic portal system which filters through the liver tissues in a series of sinusoids before reaching the general body circulation (Fig. 280). From these small vessels the liver cells thus have the first opportunity to select for storage or modification food materials newly arrived in the body.

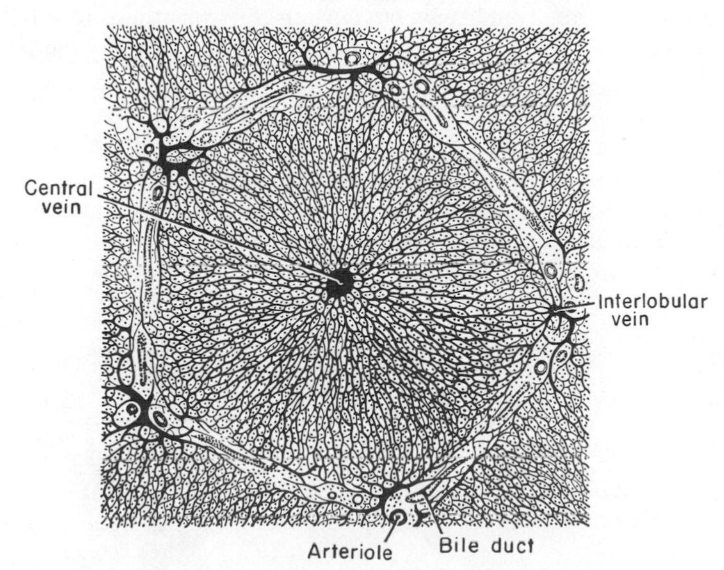

Figure 280. Section of a mammalian liver, showing a lobule and portions of others. The portal system of veins has been injected (black), showing the course of the blood from interlobular veins *via* a multitude of sinusoids through the sheets of liver cells to central veins of the lobules; branches of the bile duct and of the hepatic artery are seen in the interlobular septa. (After Young, The Life of Mammals, Oxford University Press.)

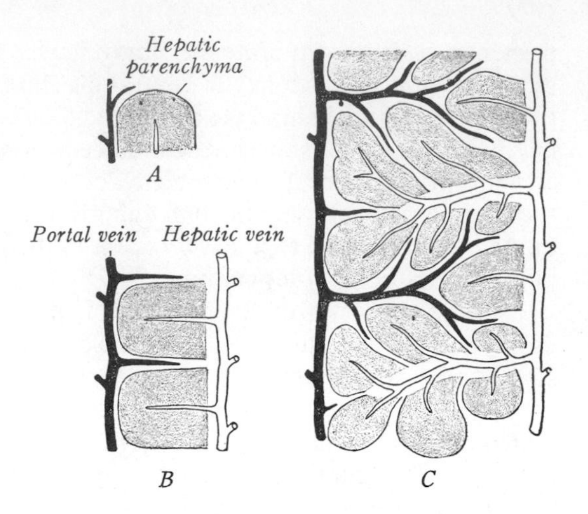

Figure 281. Diagrams to show the development of liver lobules. Masses of liver tissue (parenchyma) become clustered about branches of the excurrent hepatic vein; the portal vein branches (and those of the hepatic artery and bile duct, not shown) ramify external to the developing lobules. (From Arey, after Mall.)

Because of the dominance of metabolic over secretory functions, the structural arrangement of the liver is based on its relations to blood vessels rather than secretory ducts. A common pattern is that seen in Figure 280, and in a diagrammatic picture of embryonic development in Figure 281. The tissues are arranged in masses termed **lobules;** each has a central vein which carries blood to the hepatic veins and heart. External to the lobules (in addition to branches of the bile duct and arterioles) are vein branches from the portal system. Running inward, concentrically, toward each central vein are numerous sinusoids. The liver tissue between the sinusoids has the appearance of strands or cords of cells. Actually, however, these apparent cords are cut sections of plates of cells (Fig. 282) lying in a tangled network around the sinusoids. At the thickest, these plates are generally but two cells thick, so that every cell faces a sinusoid on one side or the other; in mammals they are but one cell thick.

THE PANCREAS (Figs. 283, 284). As we have seen, the intestine still produces in its walls (or in small and closely associated glands) certain of the enzymes utilized in digestion; however, much of the seat of enzyme production was transferred early in vertebrate history (with resulting greater efficiency) to a major external site, the pancreas. Stages in this transfer can be seen in amphioxus and cyclostomes. In the former there is no indication of a formed pancreas, but certain cells near the anterior end of the gut tube have the characteristics of pancreatic cells. In the lamprey (and also lungfish) these cells form clusters of small glands in this same region. In most other vertebrates the pancreas is a discrete (if somewhat amorphous) structure external to the intestine, lying in the dorsal mesentery. But even when well developed, the pancreas appears to form in many cases as several more or less separate outgrowths from the intestine (as, for example, in man, Fig. 284), and as a result of this multiple origin there tends to be considerable variation in the number and position of pancreatic ducts.

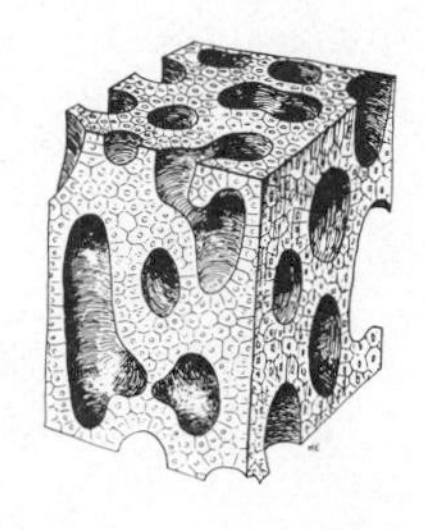
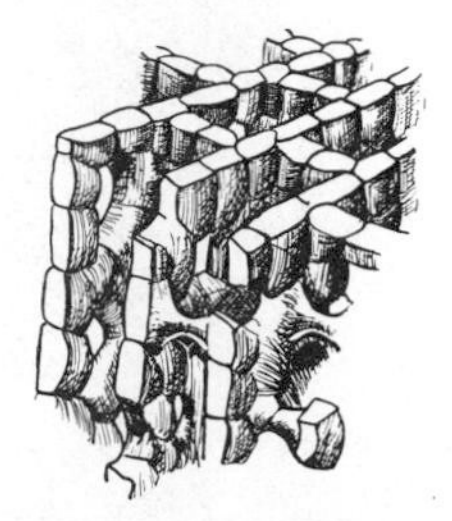

Figure 282. Diagrammatic enlargements of a small portion of a liver lobule (such as is shown in Fig. 280) to show the structure of the plates of liver cells. The plates are separated and perforated by channels which in life are occupied by sinusoids. *Left,* typical structure in a lower vertebrate; the plates are in general two cells thick. *Right,* mammalian structure; the plates are in general one cell in thickness. (After Elias.)

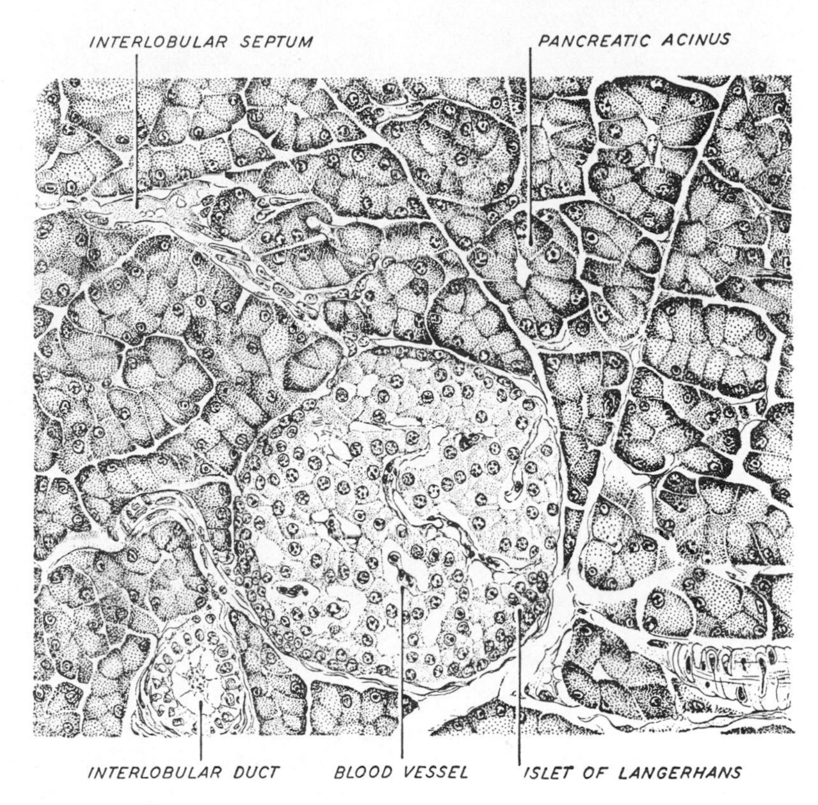

Figure 283. A section through the pancreas of a rat, to show both exocrine tissue (pancreatic acini) and a pancreatic island. A duct of the exocrine part of the gland and a connective tissue septum between lobes are also shown. (From Turner.)

The prominent function of the pancreas is that of an exocrine gland, which produces and pours into the intestine a series of enzymes (or rather proenzymes) which act upon all three major types of food materials and are responsible for a great part of the digestive activities of the gut. In addition, however, sections of the gland are endocrine in nature, as discussed in Chapter 17.

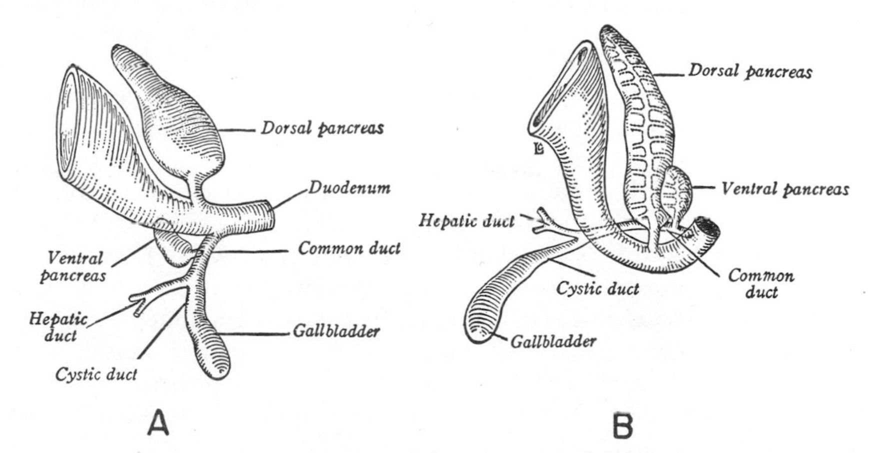

Figure 284. Diagrams showing two stages in the development of the human pancreas. *A*, An early stage, with both dorsal and (smaller) ventral pancreas developing. *B*, Later stage in which dorsal and ventral portions are beginning to fuse. (After Arey.)

Excretory and Reproductive Systems

13

From a functional point of view, a joint consideration of urinary and genital systems seems absurd, for excretion and reproduction have nothing in common. Morphologically, however, the two systems are closely associated and it is impossible to describe one without numerous cross references to the other. This association appears to be due to embryonic propinquity; the major organs of the two systems arise in areas of the mesoderm which lie close to one another in the walls of the trunk near the upper rim of the celomic cavity (Fig. 298).

URINARY ORGANS

KIDNEY TUBULE STRUCTURE AND FUNCTION. Paired kidneys, developed in varied form in all vertebrates, are the major organs of the urinary system. The basic kidney structure is the minute **kidney tubule, renal tubule,** or **nephron;** the numerous tubules connect with a duct system which eventually leads, posteriorly, to the body surface.

Among vertebrates as a whole the most generalized type of tubule is that shown diagrammatically in Figure 285 *A;* such tubules are present in forms as diverse as sharks, fresh-water and many marine teleosts, and amphibians. The proximal part is the spherical **renal corpuscle.** Its interior is a **glomerulus,** a compact cluster of blood vessels comparable to capillaries (Fig. 286). The outer part, the **capsule,** is a double-layered hemisphere which forms the proximal end of the tubule proper; its inner surface is in intimate contact with the walls of the blood vessels of the enclosed glomerulus. Its cavity is continued by that of the **convoluted tubule,*** along whose walls lies a network of capillary vessels; distinct proximal and distal sections of a tubule can usually be distinguished. Distally each tubule connects, in a fashion which varies in different groups, to ducts leading to the exterior of the body.

*Note that the term "tubule" is generally used in two ways: (1) as a synonym of nephron, i.e., a name for the entire nephric unit of renal corpuscle and convoluted tubule; and (2), more properly but more narrowly, for the latter structure only.

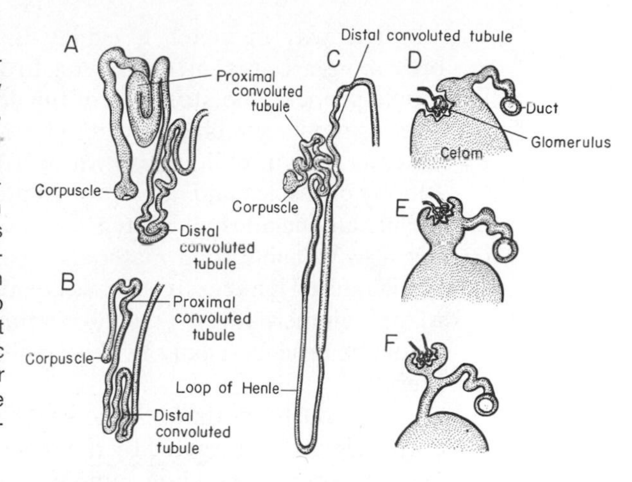

Figure 285. Tubule types. *A* to *C,* The three major types common in adult vertebrates. *A,* Presumably the most primitive, with corpuscle of good size, found in elasmobranchs, freshwater bony fishes, amphibians. *B,* Corpuscle reduced or absent, characteristic of marine teleosts, reptiles. *C,* Corpuscle large; a loop of Henle inserted; found in mammals, birds. *D* to *F,* Primitive tubule types found in lower vertebrates, principally in the embryo, and perhaps illustrating the early evolution of kidney tubules. *D,* Tubule runs from celom to kidney duct; glomerulus, if present in celom, not associated with tubule. *E,* Special small celomic chamber formed for glomerulus. *F,* This chamber has become the capsule of a renal corpuscle; tubule still connects with celom; closing the celomic opening leads to more progressive tubule type.

The result of nephric activity is the production of urine, destined for excretion and derived, obviously, from the associated circulatory vessels. Urine is composed mainly of water but contains in solution other substances, such as various salts and, particularly, nitrogenous wastes, generally urea or uric acid. The functions served are twofold: (1) elimination of waste and (2) regulation of the internal environment.

The destructive metabolism of carbohydrates and fats produces mainly carbon dioxide and water; disposal of these presents no serious problem. Proteins, however, contain nitrogen, and their residues include simple nitrogenous compounds, particularly ammonia, which are toxic to the animal. They are generally transformed rapidly in the liver into relatively harmless urea or uric acid, but even so, their removal is necessary; for this the kidneys bear the major responsibility, as they do for removing other harmful products which may be present in the body.

The cells of the body must live in a proper environment, one including, in solution in the fluids bathing them, appropriate amounts of specific simple salts. The maintenance of proper salt content demands a balance between intake (mainly via the intestine) and output. Excess salt intake requires excretory devices, particularly the kidney tubule; low salt content of the blood requires removal of excess water, likewise attainable through the work of the kidney.

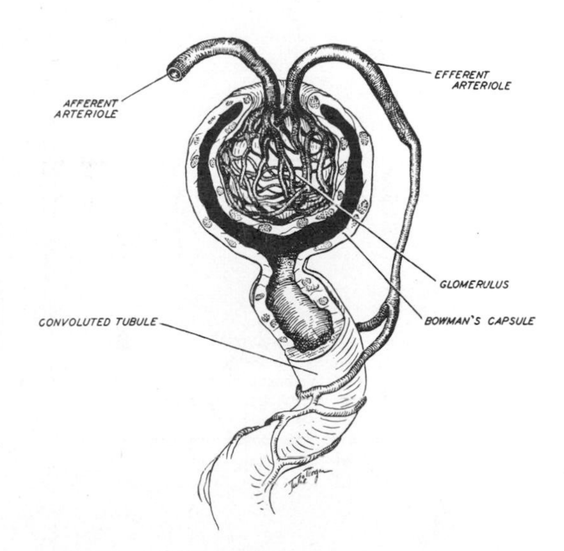

Figure 286. A mammalian renal corpuscle. Here the vessels around the tubule come from the efferent arteriole; in animals with a renal portal circulation, capillaries from this system surround the tube; they are frequently associated with ones from the efferent arteriole, but this is not always the case. (After Turner.)

The way in which a kidney functions is today fairly clear. Two distinct operations are present, one having to do with the renal corpuscle, the other with the tubule proper. The structure of the corpuscle suggests that we are dealing with a filtering device which draws off a filtrate from the blood plasma. This is indeed the case, as shown by liquid drawn from the capsule by micropipette in amphibians. Blood corpuscles and large protein molecules do not pass the filter, but otherwise the filtrate includes all contents of the blood, including not only water and wastes but also valuable food materials—particularly glucose. Further, the quantity of liquid filtered is excessive. It has been calculated that if all the liquid passed through a frog's glomeruli were actually eliminated from the body, nearly half a liter of urine would be produced daily by this small animal, and a man would produce about 200 liters!

Obviously, nothing of the sort happens. This excessive activity of the renal corpuscle is counteracted by the work of the convoluted tubule. There is a certain amount, sometimes a large amount, of excretion of waste substances from cells of the tubule into the urine as it passes by. The main tubule function, however, is to resorb, selectively, much of the filtrate; this includes much of the water (otherwise the animal would be rapidly dehydrated), but includes especially recapture of valuable materials, particularly glucose and salts, leaving the urine contents mainly water and nitrogenous wastes.

In a nephron the crude work is, so to speak, done by filtration from the glomerulus; the tubule proper adds the necessary refinement to the process.

TUBULE TYPES AND VERTEBRATE HISTORY. Whether fresh or salt water formed the original vertebrate home has been a problem of interest to students of kidney function as well as students of classification and phylogeny. The fossil record is ambiguous and open to controversy, but the earliest known vertebrates are found in marine deposits and probably, though less certainly, lived there. A study of kidney tubule structure and function can, however, lead to a very different conclusion.

Three types of nephric units may be found in the adult kidney of one group or another (Figs. 285–287). *(a)* In one type (Fig. 285 *A*), found characteristically developed in such varied forms as amphibians, fresh-water bony fishes, and elasmobranchs, there is a renal corpuscle of good size, and hence a high water output. *(b)* A second type (Fig. 285 *B*) is that found in many marine teleosts and

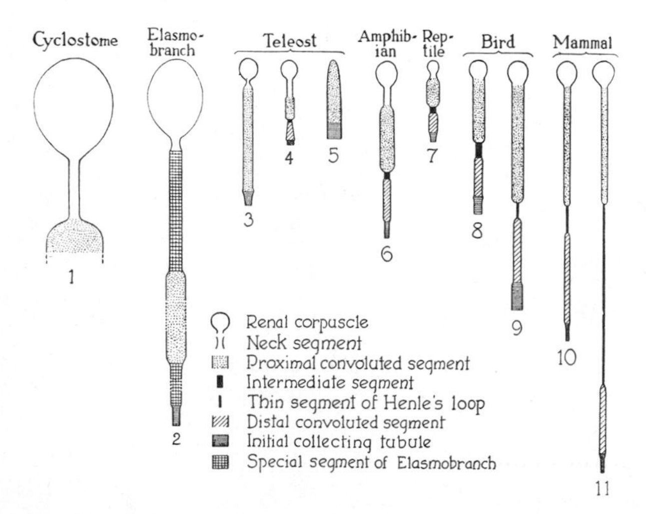

Figure 287. Diagram of kidney tubules of various vertebrates, all reduced on the same scale, to show the relative size of the components in the different groups. The glomeruli are at the upper end in each case, and the tubules are represented as if straightened out. The glomeruli are well developed in most groups and of enormous size in cyclostomes and elasmobranchs, but of reduced dimensions in reptiles, and are done away with in some marine teleosts (5). All have a proximal convoluted segment of the tubule; an intermediate segment, followed by a distal convoluted tubule, appears in some fishes and is present in all land forms. The intermediate segment becomes the loop of Henle in birds (in part) and mammals; this loop may be much elongated in the latter group. 1, Hagfish; 2, skate; 3, sculpin; 4, catfish; 5, toadfish; 6, frog; 7, painted turtle; 8 and 9, chicken; 10 and 11, rabbit. (After Marshall, Kempton, from Prosser.)

in reptiles; the corpuscle is small or absent, and hence water output is low. *(c)* A third type (Fig. 285 *C*) is seen in mammals, and, in a less extreme form, in birds. The glomerulus is large, but there is interjected into the middle of the convoluted tubule a long slim extra segment, the **loop of Henle.** This slim segment appears to be a powerful resorbant of water, and hence despite a plentiful output at the glomerulus, relatively little water reaches the bladder.

From the distribution of these types of tubule among vertebrates a consistent story can be constructed regarding the environmental history of vertebrates. It is assumed that type *(a)* is the primitive tubule, possessed by early fresh-water vertebrates and retained by forms which still inhabit such waters. Such an animal lives in a medium more dilute than its own body fluids, and hence is in danger of overdilution of these fluids (and consequent death) by osmosis through the surface of the body and of the gut. To prevent this, large amounts of water must be eliminated, and this is afforded by the presence of a large corpuscle. If, on the other hand, a fish enters the sea, it is liable to dehydration because of the greater salinity of the surrounding medium. Water must be conserved and much salt eliminated. In marine teleosts, the glomeruli are frequently reduced or absent and water output consequently reduced; further, salts and wastes are excreted by the cells of the gill membranes as well as by the kidneys.

Terrestrial vertebrates, living in a dry environment, have much the same problems as a marine fish. Water must be conserved. In modern reptiles this is accomplished by reduction in size of the renal corpuscles with a consequent decrease in water output. Birds and mammals have developed a different method of conservation. There is a normal glomerulus of large size and consequent high water output. The complex tubule, however, is an ''Indian giver'' and the presence of the loop of Henle results in absorption of much of the water; the product is a relatively concentrated urine.

The reasonable argument above leads to the conclusion that the presence of a large glomerulus is a primitive character, due to the need for an efficient water pump in a vertebrate living in fresh water. However, possibly vertebrates arose in the seas; the ancestral form could have been a marine fish, which had only a small glomerulus, and the development of a large glomerulus could have come later, with a move into fresh water.

Cartilaginous fish live in a medium with a higher salt concentration than their body fluids, yet pump water out through large glomeruli with as little concern as a fresh-water fish. They can do this without the danger of water loss through osmosis, for (in addition to salt elimination through a rectal gland) sharks have attained an osmotic pressure of their internal fluids equal to, or a bit higher than that of sea water without increase of their salt concentration. This is accomplished in peculiar fashion by retention in the blood stream, without apparent harm, of a large amount of urea, which raises the total concentration of materials in solution. We have thus in sharks and in salt-water teleosts two radically different kidney adaptations to salt water; this indicates that the two groups entered the seas independently, adapting themselves to the new environment in very different ways, and indicating, to many workers, that vertebrates must have arisen in fresh water.

Despite these arguments, most workers now agree that vertebrates originated in the oceans. All lower chordates are marine and almost all major groups of animals appear to have arisen in the seas. Certainly the structure and physiology of the kidneys suggest that both the Osteichthyes and the Chondrichthyes were originally inhabitants of fresh water (that this is true of the bony fish is also indicated by their possession of lungs, apparently as a primitive character). However, between the time of the earliest known vertebrates and the appearance of these groups in the fossil record, there is a gap of well over 50 million years—surely much could happen in that time. We know virtually nothing of the history of cyclostomes and can only speculate on their significance in this story. For that matter, there is no real reason to assume that the earliest known fossil vertebrates were actually the *first* vertebrates, so whether the former were marine or not probably is irrelevant. A marine origin seems inherently more probable, though our evidence is incomplete and contradictory.

PRIMITIVE TUBULE STRUCTURES. The types of nephric units described above are those most characteristic of adult vertebrate kidneys. There are, however, others which are more often found in embryos than in adults, particularly in the first units developed, and more often in lower than in higher vertebrates; there is hence ample reason to consider them primitive in nature. In these types (Fig. 285 *D–F*) there is always a ciliated funnel opening from the celomic cavity into the tubule; a typical glomerulus is sometimes present, but in some cases the glomerulus projects into the celomic cavity, and in still other cases it is absent altogether. Originally kidney tubules may have simply drained excess fluid and accumulated waste from the celomic cavities; the development of a glomerulus and its incorporation in the tubule was perhaps a later development.

Most invertebrate groups, and even the chordate amphioxus, have excretory structures of somewhat varied nature termed **nephridia.** These, however, are not homologous to the vertebrate nephron, which appears to have evolved independently, possibly as a water "pump" for life in fresh waters. Not improbably nephridia were present in ancestors of the vertebrates, as they are in amphioxus. But with the "invention" of the vertebrate kidney tubule, which could take care of wastes in addition to its major function of water elimination, nephridia, if present, would be redundant and could be abandoned.

ORGANIZATION OF THE KIDNEY SYSTEM. So far we have discussed merely the nature of the microscopic kidney units. Now to be considered is the organization of these secretory units and the duct systems leading from them to form the gross structures of the urinary system.

As seen in most vertebrates, the pattern of the kidney system appears fairly uniform, with paired, compact kidneys projecting into the abdominal cavity from its dorsal walls and paired ducts leading from them which may empty into a median bladder. Study, however, shows that there are basic differences from group to group; kidneys, ducts, and bladder are varied in nature and structure.

These variations have two major causes. (1) Unlike many organs, the kidney must begin to function at an early stage to take care of embryonic wastes; hence there must be rapidly developed a functioning embryonic kidney which, however, may be subject to modification or replacement in later embryonic and adult stages. (2) The genital organs lie adjacent to the kidneys, and—the testis particularly—tend to "invade" the urinary system, taking over part of its tubes and ducts for their products and in consequence causing marked modifications of the urinary organs.

We may begin our discussion of kidney organization by the description of the structure and development of an "idealized" primitive kidney which may be termed a **holonephros.** In our embryologic story we have noted the presence in the mesoderm on either flank of a band of kidney-forming tissue lying between the somites and the lateral plate, and frequently showing segmental division into a series of small **mesomeres** (or nephrotomes) (Figs. 75 *D*, 80 *C*, 288). Probably in the ancestral vertebrate each mesomere gave rise to a single renal tubule. As with the somites, the differentiation of mesomeres takes place in the embryo from front to

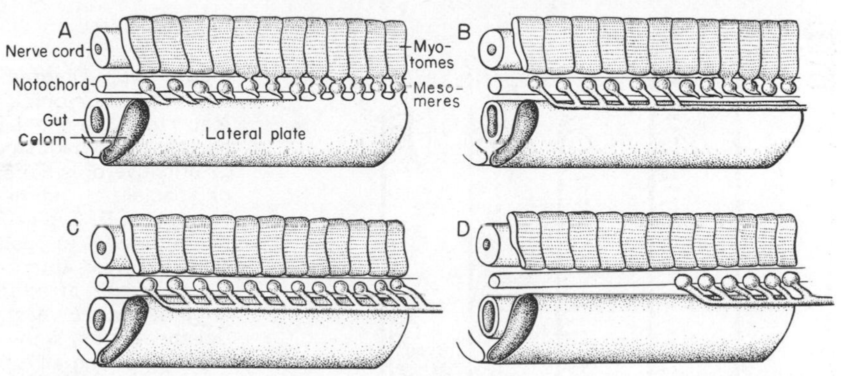

Figure 288. Diagrams of the anterior part of the trunk of an embryo (skin removed) to show the development of the archinephric duct. *A*, Most anterior—pronephric—mesomeres are budding out tubules which tend to fuse posteriorly. *B*, The pronephric tubules have formed the duct; some of the mesomeres farther posteriorly are forming tubules which are to enter the duct. *C*, The more posterior tubules have joined the duct. *D*, The pronephros lost, but the archinephric duct formed by it persists to drain the more posterior part of the kidney.

back, the oldest members of the tubule series being those at the front end, the last formed at the posterior end of the trunk.

In the embryo a longitudinal duct soon develops on either side, gathering the urine from the series of segmentally arranged units; the two ducts often unite before emptying to the exterior in the region of the cloaca. This primitive kidney duct is here termed the **archinephric duct.*** The duct, like the tubules, is of mesodermal origin. Generally it originates by a fusion of the tips of the most anterior and first-formed nephric units (Figs. 288 *A, B;* 289 *A*). It grows backward along the lateral surface of the mesomeres (or the band of mesomeric tissue), and as tubules develop farther back they grow outward to connect with it (Figs. 288 *C*, 289 *B*). The end result, ideally, is a **holonephros:** a kidney with a single nephric tubule in each trunk segment on either side of the body, the series draining though a pair of archinephric ducts.

Such an ideal holonephros is approached only in the larvae of hagfishes and gymnophionan amphibians. In adults of even the lowest of living vertebrates the most anterior and first-formed part of the kidney tubule system is specialized and degenerate. These anterior tubules are called the **pronephros** (Figs. 288 *B*, 289 *A*, 290 *A*). The remaining part of the kidney system, from which is formed, in one fashion or another, the kidney of adult living vertebrates, may be termed as a whole the **opisthonephros,** the "back kidney" (Fig. 289 *C, D*). This opisthonephros generally differs from the theoretic holonephros in three main particulars: (1) The anterior tubules (pronephros) are eliminated. (2) Above the hagfish level the simple segmental arrangement is lost; a variable (and often high) number of tubules may develop for each segment. (3) In most vertebrate groups the archinephric duct is used for sperm transport and a new urinary duct may develop.

AMNIOTE KIDNEY DEVELOPMENT. To show an extreme contrast with the ideal holonephros we will depart from a logical sequence and describe the development of the kidney in an amniote—specifically, a mammal. In mammals, as the mesoderm is undergoing differentiation, a short series of tubules at the back end of

---

*Wolffian duct, pronephric duct, mesonephric duct, and opisthonephric duct are all terms applied to this same duct at different developmental stages.

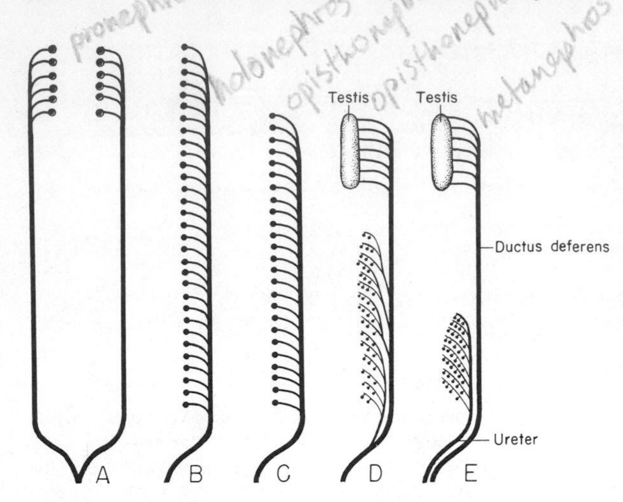

Figure 289. Diagrams of kidney types. *A*, Pronephros (embryonic); *B*, theoretical holonephros (each trunk segment with a single tubule), much as in a young hagfish or gymnophione amphibian; *C*, primitive opisthonephros: pronephros reduced or specialized, tubules segmentally arranged, as in hagfish. *D*, Typical opisthonephros: multiplication of tubules in posterior segments, testis usually taking over anterior part of system, trend for development of additional kidney ducts (most anamniotes). *E*, Metanephros of amniotes: an opisthonephros with a single additional duct, the ureter, draining all tubules. In *A*, both sides of the body are included; in *B* to *E*, one side only (cf. Fig. 305).

the head and future neck form as the pronephros, and in connection with them there develops an archinephric duct which grows rapidly back to the cloacal region (Fig. 290 *A*). These tubules function only briefly and then degenerate. Meanwhile, however, differentiation of tubules has continued backward without interruption to form a second embryonic nephric structure, the **mesonephros** (Fig. 290 *B*), which functions for much of the period of embryonic life in mammals and may persist until after birth in reptiles. The pronephric tubules are often rudimentary and without glomeruli; typical mesonephric elements are, in contrast, well formed. At first they are segmental in arrangement, but multiplication of tubules later takes place and the segmental condition is obscured.

As development continues, the mesonephric kidney degenerates in turn and its place is taken functionally by a **metanephros,** the functional kidney of the late

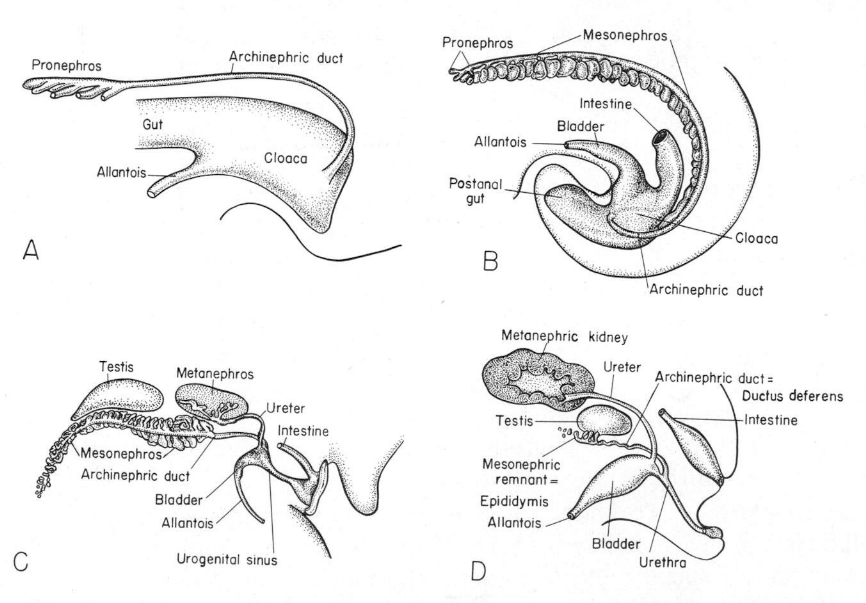

Figure 290. Diagrams to show formation of the metanephros of an amniote (male) embryo as seen from the left side. *A*, Pronephros and duct formed; *B*, mesonephros partly formed; *C*, pronephros reduced, posterior part of mesonephros functional, ureter formed, and metanephros beginning to differentiate; *D*, definitive stage; mesonephros reduced, and tubules and duct utilized only for sperm transport; metanephros, the functional kidney.

embryo and mature mammal. This arises (Fig. 290 *C, D*) from the most posterior part of the mesomeric tissue, which forms a compact mass in the roof of the lumbar region of the body cavity. In it are developed large numbers of kidney tubules. Unlike their predecessors, these do not open into the archinephric duct. Instead, they are drained by a new duct, the **ureter,** which buds off from the archinephric duct near its posterior end and grows upward and forward to enter the metanephric kidney and connect with its tubules.

We see in this story the development in the amniote embryo of three successive nephric structures—pronephros, mesonephros, metanephros. It is often stated or implied that these are distinct kidneys which have succeeded one another phylogenetically as they do embryologically. Upon consideration, however, it will be seen that there is little reason to believe this. The differences are readily explainable on functional grounds; the three appear to be regionally specialized parts of the original holonephros, which serve different functions.

The pronephric tubules are variable and often rudimentary in nature, but there is no sharp structural break between them and the mesonephric tubules which immediately succeed them. Their one distinctive feature is that they form the archinephric duct. But there is nothing really significant or mysterious about this; it is a practical matter. Once the first tubules are formed they begin to function. The formation of a urinary duct cannot wait until the entire kidney is formed; the anterior tubules just cannot wait that long!

In amniotes mesonephros and metanephros are readily distinguished. The latter is vastly greater in size and is served by a distinct duct. But it is, after all, derived from a part, even if a greatly expanded part, of the same band of tissue that forms, more anteriorly, the pronephros and mesonephros; the development of a distinct duct is presumably due to the impossibility of all of the multitude of tubules formed in this tissue draining directly into the archinephric duct; anyway the old duct is now used for sperm transport. Both mesonephros and metanephros are portions of the opisthonephros; the former is developed rapidly to function in the embryo during the period of the necessarily slower formation of the complex metanephros.

HEAD KIDNEY. In sharklike fishes and amniotes, the pronephros is an exceedingly short-lived structure. In contrast are conditions in other fishes and in amphibians (with small-yolked eggs) in which the embryo of necessity becomes an active food-seeking larva at an early stage. In such larvae the pronephros persists to satisfy excretory needs; it is, however, highly specialized and is frequently termed a **head kidney,** in reference to its anterior position. The number of tubules forming it is generally reduced to but one to three large convoluted tubules which frequently function in jointly draining liquids from a single large glomerulus situated in a special pocket of the celomic cavity. This larval head kidney may disappear in the adult but is commonly represented by a mass of lymphoid tissue; it persists throughout life in hagfishes and a few teleosts, in which it may function in blood formation.*

THE OPISTHONEPHROS OF ANAMNIOTES (Figs. 291, 292, 293). In lower vertebrates the distinction between mesonephric and metanephric portions of the kidney is never as distinct as in amniotes, and the entire structure is best termed an opisthonephros. In hagfishes the structure is a very simple one; the long slender kidney has a small number of tubules arranged in essentially segmental fashion the length of the trunk, and all are drained directly by the archinephric duct (Figs. 288

*More posterior parts of the kidney may also have this function—and unfortunately be given the same name.

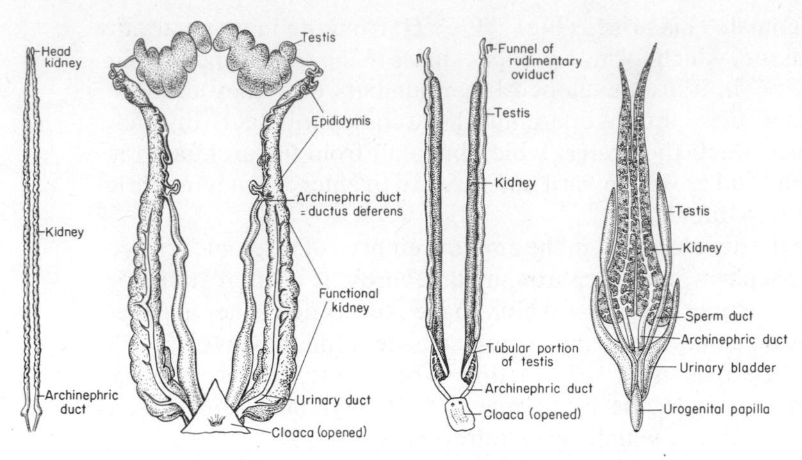

Figure 291. Urogenital systems in ventral view of males of *A*, the slime hag, *Bdellostoma; B,* the elasmobranch *Torpedo; C,* the lungfish *Protopterus; D,* a teleost, the seahorse *Hippocampus.* In *A* the testis, not shown, is pendent from a mesentery lying between the two kidneys and has no connection with them. In *B* the testis has appropriated the anterior part of the kidney as an epididymis, much as in most land vertebrates, and utilizes the entire length of the archinephric duct as a sperm duct. In *C* the sperm ducts drain, on the contrary, only into the posterior part of the kidney and thence to the archinephric duct. In *D* the sperm duct is entirely independent of the kidney system. (*A* after Conel; *B* after Borcea; *C* after Kerr, Parker; *D* after Edwards.)

C, 291 *A,* 297 *A*). Lampreys have grossly similar kidneys, but their histologic structure is unique with fusion of the capsules. In bony fishes the number of tubules may increase, but drainage is still directly by the original duct and the kidney is generally a long and slender structure (Figs. 291 *C, D;* 292 *B, C;* 297 *F, G*).

In other anamniotes, however, the structure is more complex (Figs. 289 *D;* 291 *B;* 292 *A;* 293; 297 *C, D*). The front part of the elongate opisthonephros tends to be reduced, while posteriorly it generally becomes much expanded, with a great

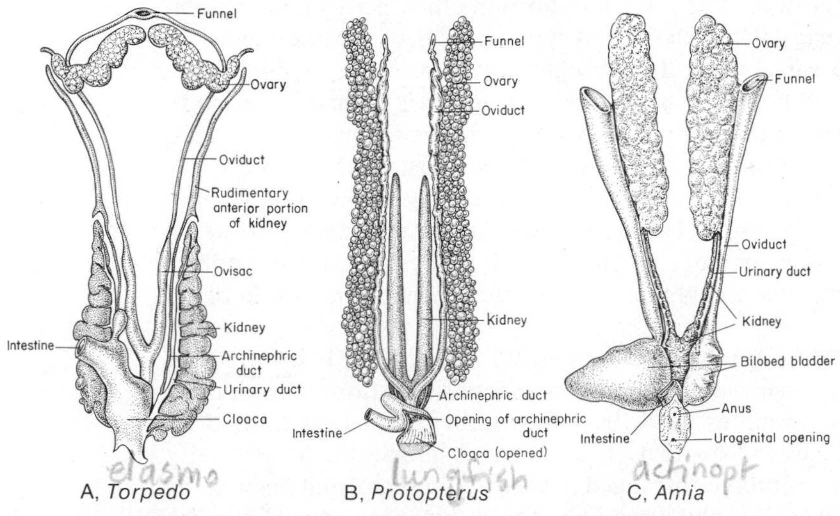

Figure 292. Urogenital systems in ventral view of females of *A,* the elasmobranch *Torpedo; B,* the lungfish *Protopterus;* and *C,* the primitive actinopterygian *Amia.* In *Torpedo* the shell gland is not developed. (*A* after Borcea; *B* after Parker, Kerr; *C* after Hyrtl, Goodrich.)

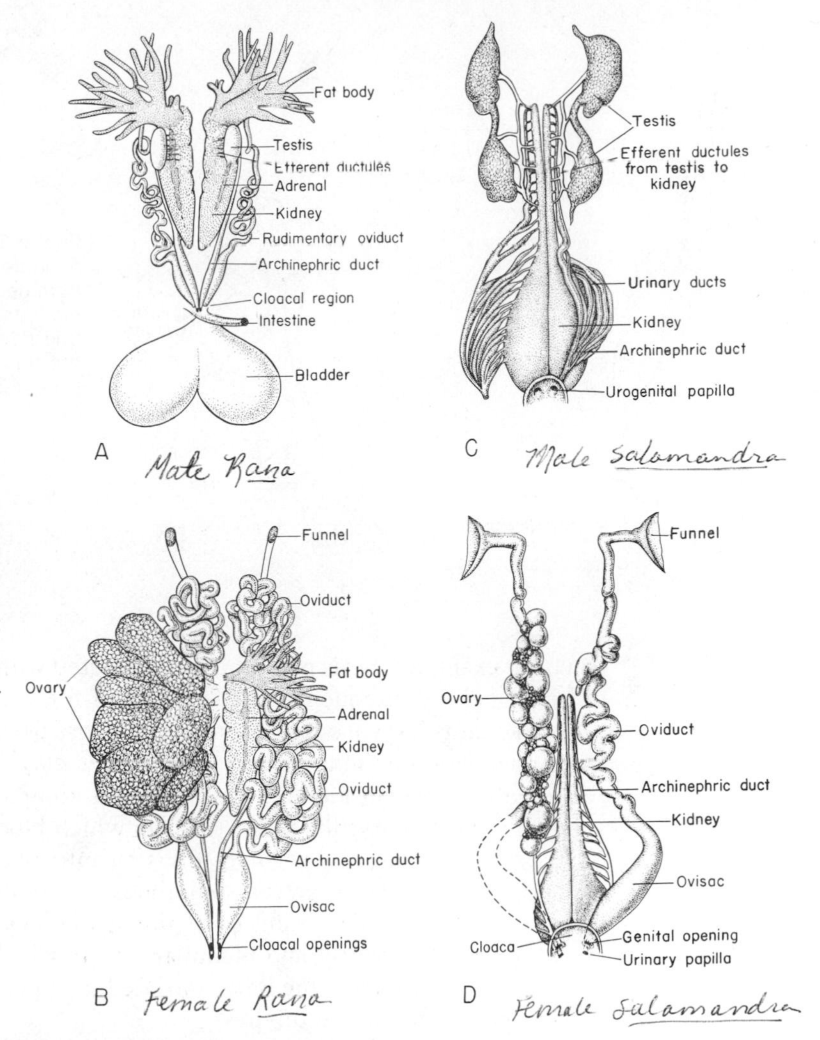

Figure 293. Urogenital system of amphibians. *A, B,* Male and female organs of a frog (*Rana*); in *B* the ovary (shown only on the right side of the body) is in a condition close to breeding maturity. The bladder and intestine are not shown in *B*. *C, D,* Male and female organs of the urodele *Salamandra*. In *C* the urinary ducts of the right side are detached and spread out to show their connections with the kidney. In *D* the ovary is shown only on the right side; the oviduct of the same side is partly removed to show the more posterior urinary ducts. Ventral views. (*A* and *B* after McEwen.)

multiplication of tubules, thus foreshadowing the amniote condition.* There is, further, an approach to the amniote condition in the duct system. Part of the urinary drainage may still be by way of the archinephric duct, but there is a strong trend in sharklike fishes and amphibians for the development of separate **accessory urinary ducts** which foreshadow the amniote development of a ureter and tend to leave the old duct free for sperm transport in the male.

THE AMNIOTE KIDNEY. As is indicated by this review of the phylogenetic development of the kidney in lower vertebrates and by our earlier description of mammalian development, the amniote kidney is a specialized end type in which the trend toward posterior concentration and toward development of a new duct system has attained a peak development with the formation of a definitive ureter. The anterior part of the old opisthonephros functions only in the embryo; the adult kidney is formed by a great expansion of the nephric tissue toward the back of the trunk (cf. Figs. 289 *E*, 297 *E*). In reptiles (Fig. 294) the metanephroi, often crenulated in appearance, contain in lizards a number of kidney tubules estimated in various forms to be from 3,000 to 30,000. In birds (Fig. 295) the kidney, likewise typically lobulated, contains a much larger number of tubules, some 200,000 in a

*Frogs are here (as often) exceptional for, in correlation with the greatly shortened body, the whole kidney is a short, compact structure.

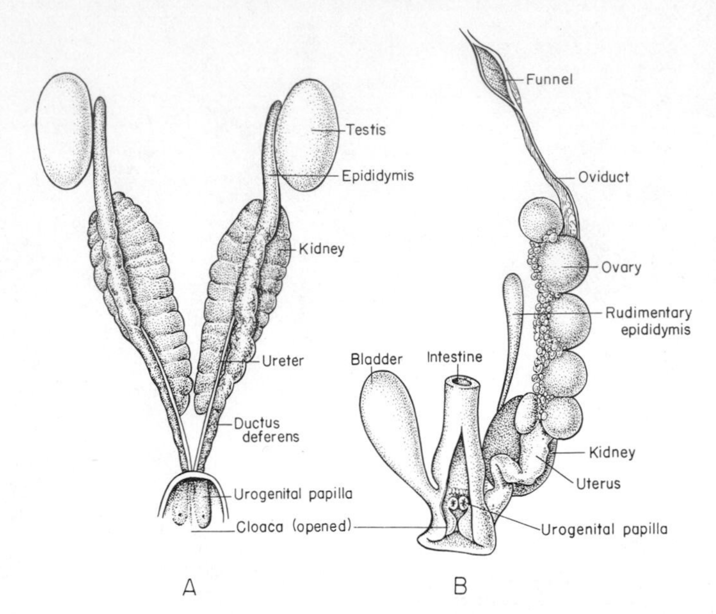

Figure 294. Urogenital organs in reptiles. *A*, Male organs of the lizard *Varanus*. *B*, Female organs of *Sphenodon*. In *A* the bladder is omitted; in *B* it is shown turned to one side. In *B* the organs of the left side only are figured. Ventral views. (*A* after Vandebroek; *B* after Osawa.)

fowl, for example; this is presumably correlated with the greater metabolic activity of birds and a consequent increase in need for waste disposal. In mammals the tubule count is likewise high; a mouse appears to have about 20,000, and in such large mammals as a man or cow the number may run into the millions.

The mammalian kidney (Fig. 296) is a compact structure, frequently bean-shaped, with a cavity, the **hilus,** through which blood vessels and ureter enter; in other cases the kidney is externally divisible into lobes. Within the kidney the ureter expands to form a **renal pelvis,** sometimes subdivided into a series of **calyces** into which collecting tubules drain. In section the kidney usually shows distinct "rind" and "marrow"—**cortical** and **medullary** portions, the former including glomeruli and convoluted tubules, the latter (striated in appearance) containing the loops of Henle and the collecting tubules.

We may parenthetically note here the nature of the blood supply to the kidney. The major function of filtration is performed by the glomeruli, which in all cases have a blood supply furnished by branches of the aorta. In cyclostomes, on the one hand, and mammals, on the other, the entire blood supply to the kidney is arterial. In all intermediate groups, however, from jawed fishes to reptiles and, to a very

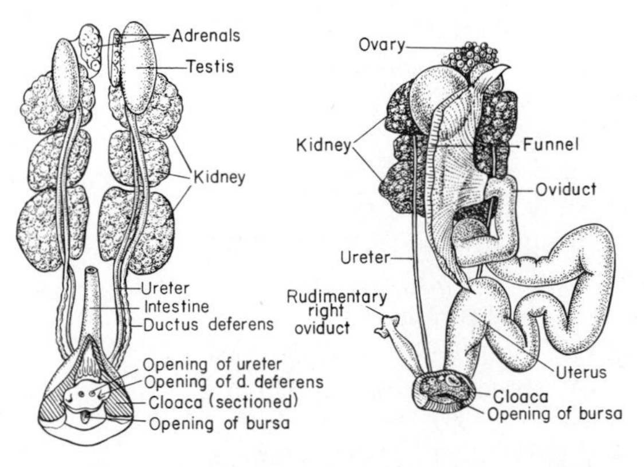

Figure 295. Urogenital organs of pigeon; *A*, male; *B*, female. The bursa (of Fabricius) is a pouch of lymphoid tissue opening dorsally into the cloaca of birds and discussed in the following chapter. (*A* after Röseler and Lamprecht; *B* after Parker.)

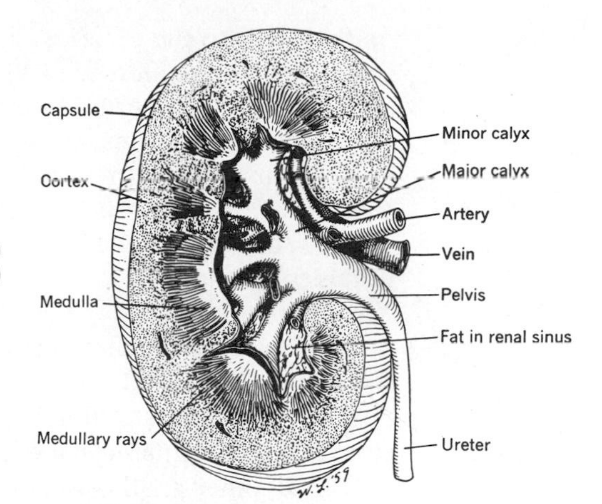

Figure 296. Section through a mammalian kidney. (From Windle, Textbook of Histology, The McGraw-Hill Company.)

slight degree, in birds, we find an accessory blood supply in the renal portal system (cf. p. 326). Venous blood in its course to the heart from the tail or hind legs, or both, may be forced to pass through a venous capillary system within the kidney which bathes the convoluted tubules, but is never concerned with the glomeruli.

EVOLUTION OF URINARY DUCTS (Figs. 297, 305). In the primitive vertebrate kidney, as seen in cyclostomes, a pair of simple archinephric ducts suffices for the drainage of urine; both sperm and eggs are shed freely into the peritoneal cavity

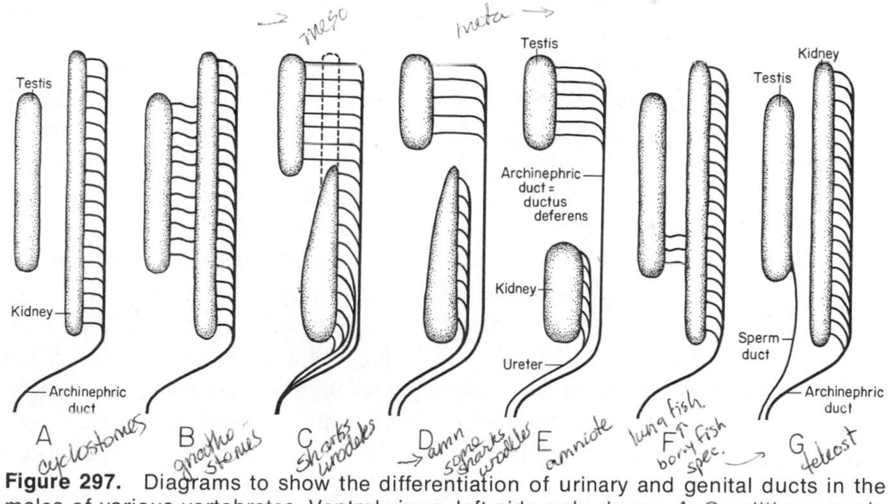

Figure 297. Diagrams to show the differentiation of urinary and genital ducts in the males of various vertebrates. Ventral views, left side only shown. *A,* Condition seen in cyclostomes; archinephric duct solely for urinary system; gonad not involved. *B,* Condition presumably primitive for gnathostomes, preserved in the sturgeon and gar pike. Testis connects at various points with kidney and thence with archinephric duct. *C,* Stage beyond *B* as seen in many sharks and urodeles. The testis has taken over the anterior part of the original kidney; the functional posterior part of the opisthonephros tends to drain by a series of ureter-like, accessory urinary ducts. *D,* Stage more advanced toward the amniote condition, found in some sharks and urodeles; the kidney drainage is by a single ureter-like duct. In females (not shown) of the types represented by males in *C* and *D,* the tendency toward development of new ducts for kidney drainage is generally not so marked as in the males; the condition shown in *D* is reached by few female forms below the amniote level. *E,* Amniote condition; a definite single ureter formed in both sexes. *F* and *G,* Lungfish and a teleost, showing a type of specialization peculiar to modern bony fishes. The testis tends to concentrate its connections toward the back end of the kidney (as in *F*), and in teleosts evolves a separate sperm duct, releasing the archinephric duct for its original urinary functions. Note that the two ducts for sperm and urine in *G* are *not* respectively homologous with ductus deferens and ureter, as might be thought at first sight.

on their way to the exterior. In gnathostomes, however, the seminiferous tubules come into connection with kidney tubules and utilize the original kidney duct as a passage for sperm. In a few modern vertebrates—the Australian lungfish, lower actinopterygians, the common frog and a few other amphibians—this duct serves equally for the conduction of sperm and urine. In general, however, this awkward dual function is unsatisfactory, and there appears to have been (so to speak) a struggle between urinary and genital systems for possession of the archinephric duct.

Among bony fishes the urinary system has been the winner. In African and South American lungfishes and *Polypterus* (Figs. 291 *C*, 297 *F*, 305 *c*) sperm enter the archinephric duct, but only near its posterior end; in teleosts there has developed a quite separate tube for sperm conduction and the archinephric duct is restored to its original urinary function (Figs. 291 *D*, 297 *G*, 305 *d*).

In all other gnathostomes the fight has gone the other way. The archinephric duct has been taken over in the male sex for sperm conduction and there is a development, partial or complete, of new ducts to care for the drainage of urine. In Figure 297 *C–D*, are shown diagrammatically stages in this process as seen in elasmobranchs, on the one hand, and amphibians, on the other. In both groups (as might be expected) development of new urinary ducts progresses more slowly, phylogenetically, in females, in which no competition exists, than in males. A first stage in differentiation is one in which there develop short urinary collecting tubes which empty into the posterior part of the archinephric duct. Beyond this, in many amphibians and female sharks, there is a stage in which the most posterior of these urinary tubes run separately to the cloaca but the more anterior ones still drain into the old duct. A final development, attained in male sharks and, in parallel fashion, in a few male amphibians, is one in which all ducts from the kidney unite to form a single duct leading independently to the cloaca, leaving the archinephric duct free for sperm transport. There is thus attained the development of a new duct, the accessory urinary duct, comparable to the ureter, which we have previously described as the functional duct in the amniotes.

URINARY BLADDERS. In a majority of vertebrates there is developed a bladder of some sort, a distensible sac in which urine may be stored. In female elasmobranchs and in some of the more primitive bony fishes (Figs. 292 *C;* 312 *D, E*) a small urinary bladder (or urinogenital sinus) may develop from the conjoined posterior ends of the archinephric ducts themselves, or at the posterior end of each duct. In male elasmobranchs the archinephric ducts transport sperm, and no bladder develops along their course, but small bladder-like expansions may occur in the accessory urinary ducts developed in this group. In cyclostomes and in teleosts (Figs. 291 *D*, 312 *A*) a bladder may develop from a pinched-off portion of the cloaca.

This last type of development is that which gives rise to the bladder of tetrapods. In them a bladder is useful both as a rudimentary sanitary measure and in some cases (such as frogs) as a source from which water can be resorbed to counteract desiccation under terrestrial conditions. In amphibians and some reptiles a bladder is prominently developed as an outgrowth from the floor of the cloaca (Figs. 293 *A*, 294 *B*, 313 *A*); this often lacks a direct connection with the ureters. Voiding of urine from the bladder takes place through the common cloacal outlet. In mammals the ureters come to enter directly into the bladder, so that no passage of urine through a cloacal cavity is necessary to reach it. Among reptiles the bladder is lost in some lizards, in snakes, in crocodilians; it is also lost in birds except in the ostrich. In its absence the urine is poured into the cloaca and may be mixed with the feces. We have noted in an earlier chapter (p. 112) that the tetrapod type of bladder plays an important part in amniote embryology, for the allantois, an important embryonic membrane, is elaborated as an outgrowth from the urinary bladder.

The tetrapod bladder is a highly distensible structure with stout walls endowed, particularly in mammals, with thick coats of smooth musculature. The lining is of a peculiar type, known (inappropriately) as **transitional epithelium.** When the bladder is empty, this appears to be of a thick, stratified nature; when distended, it is capable of thinning down to a layer or two of flat squamous cells.

GENITAL ORGANS

Sexual reproduction is almost universal in vertebrates (a few lizards are parthenogenetic) and in the vast majority of cases the two sexes are functionally separate. The basic reproductive structures are the **gonads**—ovary or testis. In these organs are produced the **gametes**—eggs (ova) or sperm—by the union of which the new generation is initiated. In all gnathostomes there are associated with the gonads tubes or ducts for the transport of gametes and in certain cases for the protection and nourishment of growing young within the female body. In various groups copulatory organs, aiding in internal fertilization of the eggs, may develop, and secondary sexual characters frequently affect general body size or proportions or such features as the plumage of birds, the mammary glands of mammals, and antlers and horns in ruminants.

SEX DEVELOPMENT. The sex of an individual depends basically upon the nature of its chromosomal inheritance, a balance between male and female potentialities received from the two parents. As discussed in Chapter 5, the early development of the embryo is due mainly to the organization already present in the unfertilized egg, and the influence of the sperm and the hereditary characters which it introduces are not appreciable until a relatively late stage. In consequence, the early embryo has, so to speak, no knowledge of which sex it is to become and must be prepared for either possibility. We thus find that for some time the sex organs of the embryo remain in an **indifferent stage,** during which gonads and their ducts proceed far in their development without showing indications of a trend specifically in either male or female direction (Figs. 298 *C,* 300). Eventually, there appears a definite sexual stage, presumably associated with the initiation of hormonal secretion. The gonads become definitely testes or ovaries, and only the ducts and other accessory structures appropriate to one sex or the other continue their development. Nonpertinent structures of the opposite sex cease to grow and may be resorbed, but are sometimes merely arrested in their growth, to persist as rudiments in the adult.

The mechanisms of sex determination, however, are so delicately balanced that in a variety of vertebrates the gonads hesitate (so to speak) between the two possible alternatives, and both eggs and sperm may tend to develop. Only in very exceptional cases (the Serranidae and three other families of marine teleosts) do both types of gametes come to maturity in the same individual at the same time of life, creating a functional hermaphrodite, but intersexual conditions of one sort or another are far from rare. A few hagfishes remain sterile intersexes throughout life; in some amphibians (as the common frog genus *Rana*) individuals may be functional females when young, but in old age shift to the male side of the balance and produce sperm.

GONADS. There is in vertebrates a great difference in the rapidity with which

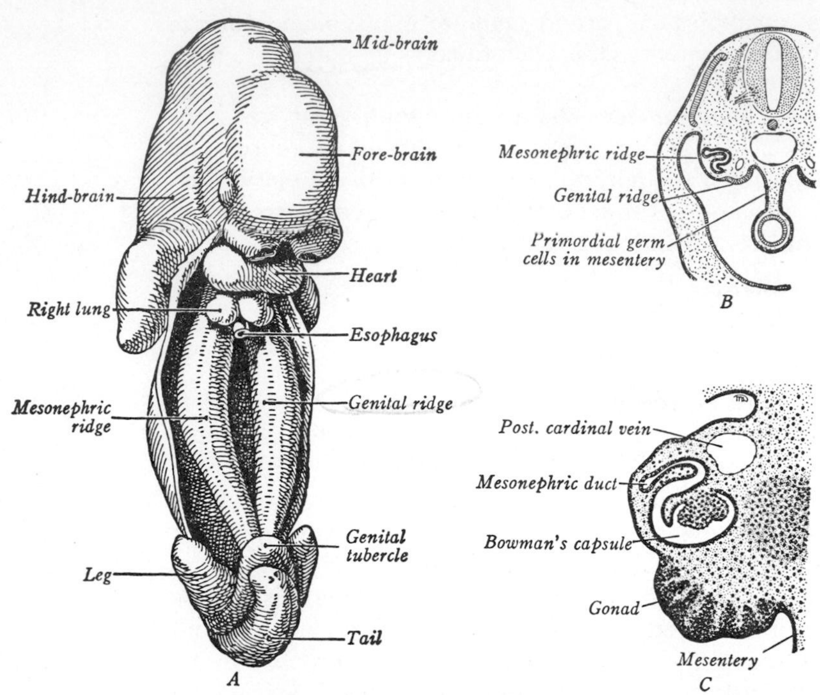

Figure 298. *A*, Ventral dissection of a human embryo of 9 mm., digestive tract removed, to show genital ridge and kidney (mesonephric ridge) projecting downward into the (opened) celomic cavity. *B*, Cross section of an embryo at an earlier stage (7 mm.); and *C*, a slightly later one (10 mm.). In the latter the primary sex cords are forming in the still "indifferent" gonad, and capsule and glomerulus are forming in the kidney tubules. (From Arey.)

different organ systems develop. The nervous system, for example, grows very rapidly in early stages; the genital organs are, on the other hand, slowest of any to develop. After all, whereas most bodily structures must be put to use at birth (or even earlier), the sex organs do not function until maturity of the individual has been attained.

The gonads make their appearance only at a stage when most of the other organ systems have been blocked out and the celomic cavities are well developed. Paired longitudinal **genital ridges** form along the roof of the celom, medial to the embryonic kidneys and on either side of the dorsal mesentery (Fig. 298). Elongate to begin with, the gonads developed from such ridges often become relatively short and compact in later stages, with a usual trend for anterior concentration of tissue. The **germinal epithelium** of the ridge, continuous with the mesodermal lining of the rest of the celom, forms the more important structural elements of the gonad (Figs. 298 *B, C;* 299 *A;* 300 *A*); mesenchyme lying beneath the epithelium forms connective tissue and in higher vertebrates, at least, gives rise to special **interstitial tissues** which are believed to be a source of gonad hormones.

Before the end of the indifferent stage the gonad generally develops into a swollen structure, extending downward into the celomic cavity from its dorsal wall. From the germinal epithelium covering its surface, fingerlike structures, the **primary sex cords** (Figs. 298 *C,* 299 *B,* 300 *B*), grow inward into the substance of the gonad. These cords contain, in addition to supporting elements, the germ cells from which eggs or sperm later develop.

It seems logical to assume that these germ cells arise locally, within the mesodermal epithelium of which they are a part. There is, however, a peculiar quirk

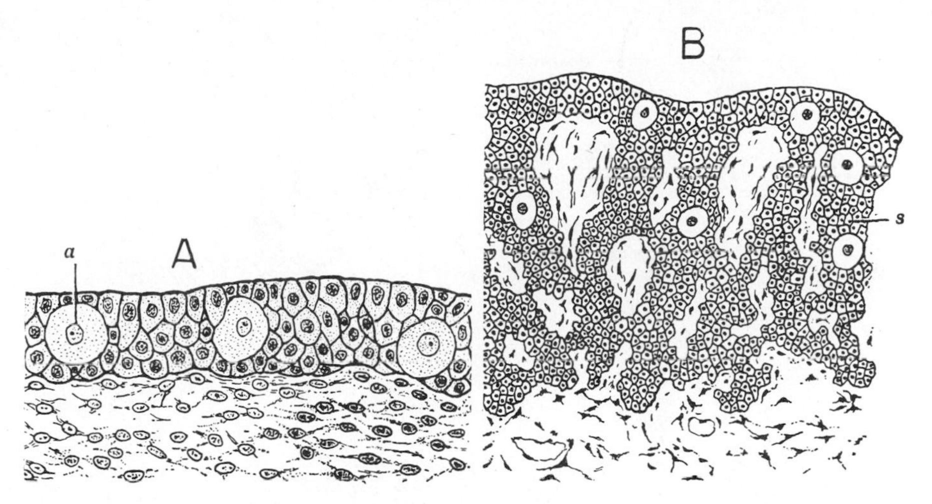

Figure 299. Early stages in mammalian gonad development. *A,* A section of the epithelium of the gonad at an early indifferent stage; primordial germ cells are present *(a),* but sex cords are not developed. *B,* A somewhat later stage, with primary sex cords *(s)* growing inward from the epithelial layer. (From Maximow and Bloom.)

in the story: the first germ cells to appear in ovary or testis seem actually to be derived from endoderm. In embryos of every vertebrate group from cyclostomes to mammals there have been observed, in the lining of the gut, cells which histologically are quite distinct from the ordinary cells of the digestive tract. These distinctive cells can be seen to leave the gut and migrate—either through the intervening tissues (Fig. 298 *B*) or by way of the blood stream—into the genital ridges to become the primary germ cells. It is generally believed that they produce at least a first generation of eggs or sperm. Whether or not, however, they are the ultimate source of all the eggs or sperm produced during the life of the individual is uncertain, and it is possible that their function may be merely the initiation and stimulation of the process of gamete formation.

Biologists have often emphasized that in animals in general the germ cells form an exceedingly independent tissue; the rest of the body is, from this point of view, merely a temporary structure shielding the potentially immortal germ plasm. In many invertebrates the future germ cells become distinct from the rest of the embryo at an early cleavage; the migration of vertebrate germ cells is, possibly, a demonstration of their equally distinctive nature.

OVARY: EGG PRODUCTION (Figs. 301, 302). In the development of an ovary beyond the indifferent stage the primary sex cords degenerate and there is generally a proliferation inward of **secondary sex cords** (Fig. 300 *D*). In these the ova arise from the germ cells after repeated divisions and by a complicated maturation process—the process of **oogenesis.** Each maturing egg cell may be surrounded by a cluster of other cells from the sex cords to form a **follicle,** and connective tissue cells may form a further external sheath. The follicular cells aid in the sustenance of the growing egg and produce sex hormones. In forms with large-yolked eggs the follicle becomes relatively enormous, forming a major bulge on the ovarian surface.

At seasons of reproductive activity, ripe follicles burst from the surface of the ovary into the surrounding celomic space—the process of **ovulation.** In most lower vertebrates the follicle is quickly resorbed, but in mammals and certain elasmobranchs it persists for some time, its cavity filled by a body of yellow material—the **corpus luteum,** which secretes the hormone **progesterone** (p. 416). The number of eggs in a mature state in the ovary at any one time is, in most groups, small—from

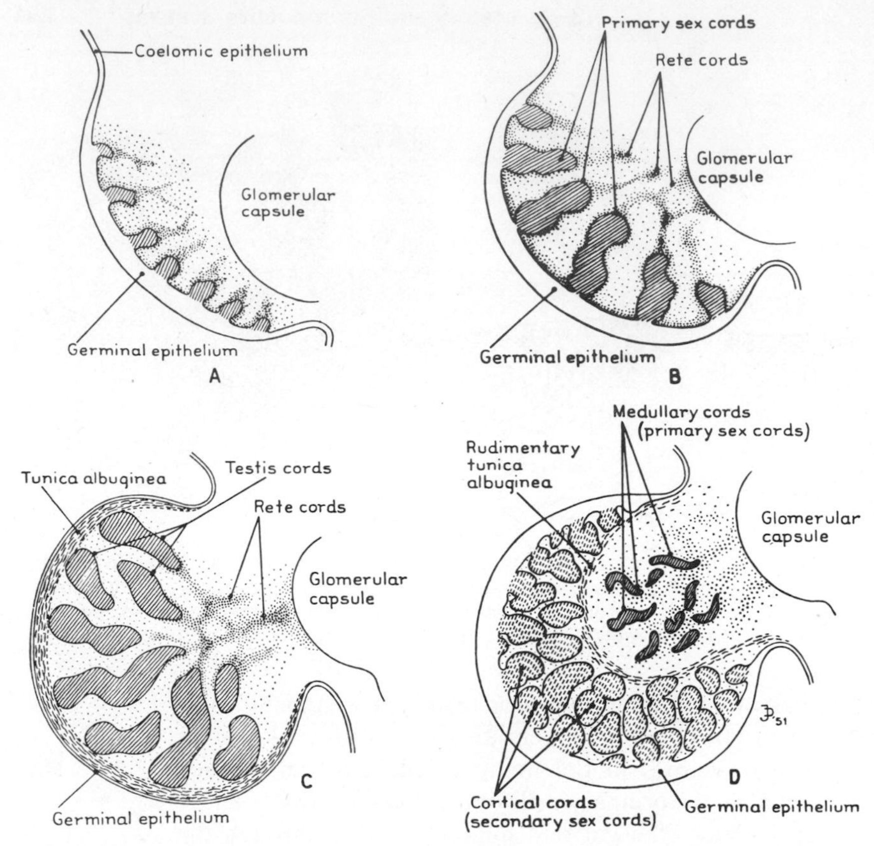

Figure 300. Development of testis and ovary of a mammal. *A,* Genital ridge, with incipient primary sex cords growing inward from the germinal epithelium; the stage is slightly earlier than Figure 299 *B. B,* Gonad still in the indifferent state; the primary sex cords are well developed, and cords are forming which will become the rete testis if the gonad becomes a testis. *C,* Development of the testis; the germinal epithelium degenerates, and is replaced by a sheath around the testis; testis tubules and rete continue development. *D,* Early development of an ovary, with reduction of primary sex cords and rete rudiments with, on the other hand, great development externally of secondary cords in which eggs develop. (From Burns, in Willier, Weiss and Hamburger, Analysis of Development.)

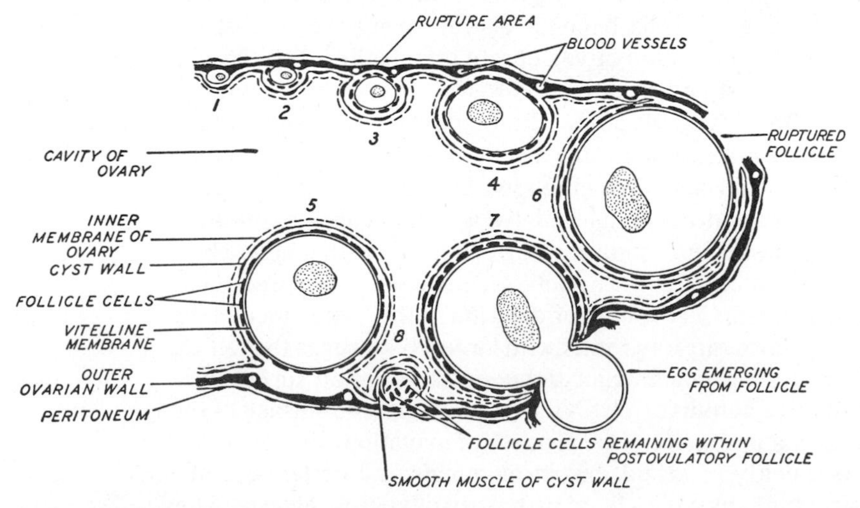

Figure 301. Diagrammatic section through a lobe of the frog's ovary. *1* to *5* illustrate stages in the growth of the follicle; *6* and *7,* rupture of a follicle and emergence of egg; *8,* postovulatory follicle. (From Turner.)

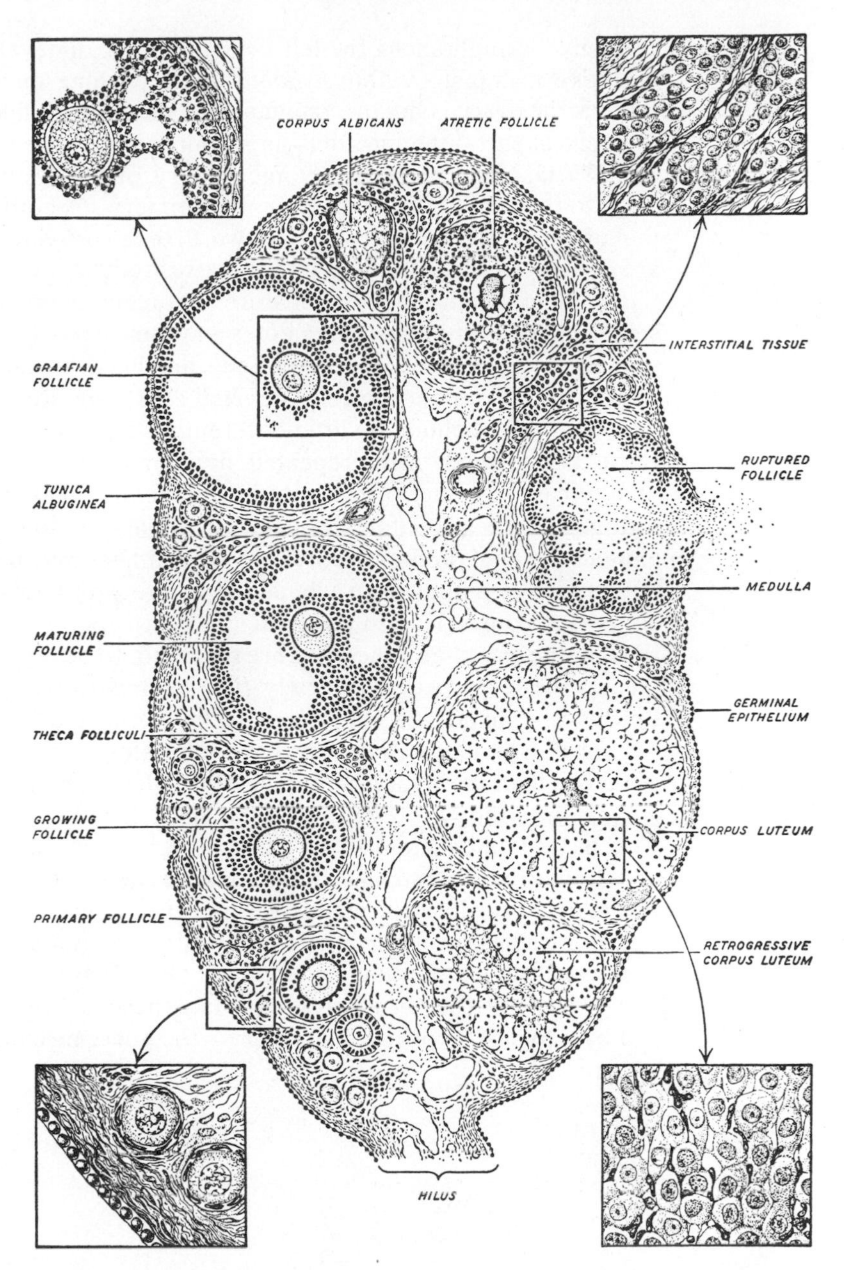

Figure 302. Diagram of a composite mammalian ovary. Progressive stages in the differentiation of a follicle are indicated on the left. The mature follicle may ovulate and form a corpus luteum *(right)* or degenerate without ovulation (atretic follicle). (From Turner.)

two to a dozen in many cases. In amphibians, however, hundreds or even thousands of ripe eggs may be present at breeding time, and in teleosts there may be hundreds of thousands or even millions of tiny eggs (the codfish is estimated to lay 4,000,000 in one season).

The ovary is usually a paired structure, often with a simple oval shape in a resting phase but frequently distended and irregular in outline at the breeding season. In cyclostomes and many teleosts the two ovaries are fused. In many teleosts the ovaries are hollow and connect directly with the oviducts (see p. 296).

In many elasmobranchs the left ovary remains undeveloped, in birds and in the primitive mammal *Ornithorhynchus* the left alone matures. In amphibians and reptiles the ovary is hollow, containing central lymph-filled cavities; in other cases the central part—the medulla—is a connective tissue structure.

TESTIS. The early embryonic history of the testis, through the indifferent stage to the time of the development of primary sex cords (Fig. 300 *C*) is similar to that of the ovary. From this point onward, the two diverge. No secondary sex cords are developed; instead the primary cords produce a series of hollow structures in the walls of which the sperm mature. In anamniotes these structures are usually small spherical **ampullae;** in amniotes and some teleosts there are, instead, elongate **seminiferous tubules** (Fig. 303). They are lined by an epithelium in which are relatively rare supporting cells—**Sertoli cells**—and the germinal elements. At the base of the epithelium are little-differentiated elements, the **spermatogonia.** From these are formed, after repeated divisions, the **spermatozoa;** when matured and released at the surface of the epithelium they include in their structure little except a head containing nuclear material and a long, mobile tail. Even in small animals the total production of these tiny gametes may be measured in hundreds of millions. The ampullae in which sperm are produced in lower vertebrates are somewhat comparable to egg follicles. When sperm are discharged at a breeding season, the ampullae concerned are resorbed, to be replaced by others which have meanwhile developed more slowly. The seminiferous tubules of higher vertebrates, on the other hand, are characteristically permanent structures.

The testis tends to be, on the whole, a more compact and regularly shaped structure than the ovary, and does not undergo as marked seasonal changes. In cyclostomes the two testes fuse, and there may be a partial fusion in sharks; in various birds and mammals the left testis tends to be slightly larger than the right.

In most vertebrates the adult gonads retain a position in the upper part of the celomic cavity. In most mammals, however, there occurs a **descent of the testis.** Paired pouches, projecting externally as the **scrotal sacs,** form in the floor of the abdominal cavity. During development (Fig. 304) the testes move backward and downward from their original position into these sacs, each accompanied by its duct and by a fold of its proper mesentery—the **gubernaculum.** In some cases the sacs

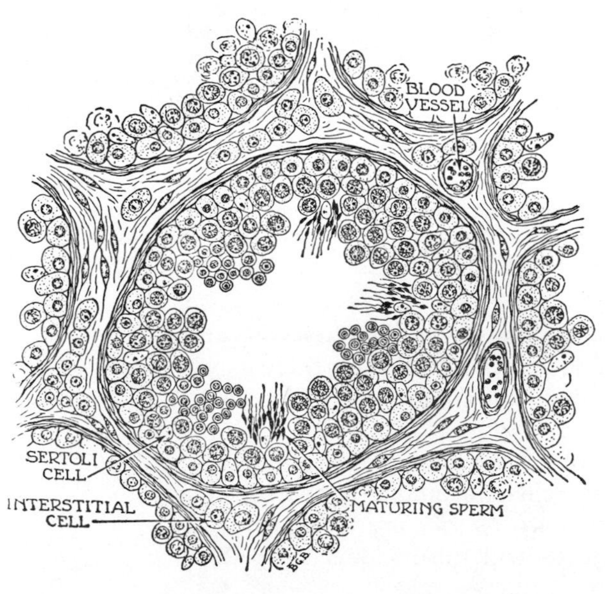

Figure 303. A small area of a mammalian testis, showing one tubule and parts of several others in cross section and the intertubular tissue. All stages in the development of sperm shown here would not normally appear at one time in any tubule. (From Hooker in Fulton-Howell.)

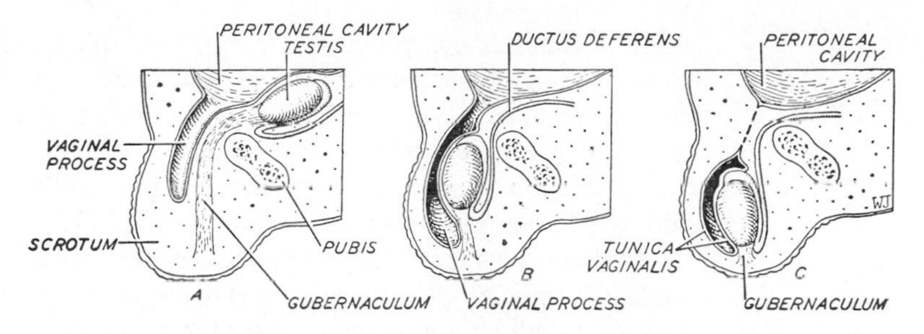

Figure 304. Descent of the testis in mammals; ventral surface of body at left. A vaginal process develops from the body cavity; its peritoneum forms the vaginal tunic of the scrotal sac. Broken line in *C*, position of inguinal canal in mammals in which sac is not completely closed. (From Turner.)

remain in open connection with the abdominal cavity, and the testes may be withdrawn into the body between breeding seasons. In other mammals the sacs may be permanently closed off, but there is, nevertheless, a weak spot here in the abdominal wall, rupture of which leads to the condition known in man as inguinal hernia. This unusual phenomenon of testicular descent appears to be associated with the fact that the internal temperatures of the mammal body are too high for the delicate process of sperm formation; the temperatures in the scrotal sacs are several degrees lower. However, in a few mammals, such as whales, the testes do not descend—but viable sperm are obviously produced.

THE OVIDUCT AND ITS DERIVATIVES IN LOWER VERTEBRATES (Figs. 292–294). In cyclostomes both eggs and sperm are shed into the celom and must find their own way to the posterior end of this cavity, whence a pair of **genital pores** afford them passage, in the breeding season, to the exterior. In all gnathostomes the sperm are conducted through closed tubes, but the eggs are still shed in most cases into the celom. They are not, however, really freed into this cavity for they are (except by accident) received through a funnel-like structure close beside the ovary, into a tube, the **oviduct** or **Müllerian duct.** Along the course of the oviduct there may be formed specialized enlargements for various purposes: storage of eggs before laying, deposition of a shell, or retention of the egg during embryonic development and a subsequent "live" birth—the **viviparous** condition.*

The oviduct parallels the archinephric duct in its embryonic course, and in elasmobranchs and urodeles it is formed by a longitudinal splitting in two of this duct. Probably the female duct (like the typical male sperm duct) was originally derived from the urinary system but in most tetrapods and in many fishes the oviduct forms independently, from mesodermal tissues, and thus shows no indication of such an origin. It is simple in structure in lungfishes and amphibians—forms which appear to be quite primitive in breeding habits. Beyond the proximal funnel, termed the **infundibulum,** the oviduct is a ciliated tube, small in diameter and relatively straight in resting phases; it may be expanded and highly convoluted at the breeding season. Its distal end may be especially enlarged as an **ovisac** for egg storage. Oviducts of this primitive type may open individually into the cloaca, or the two tubes may fuse at their posterior ends.

In many teleosts, there is an exceptional situation in that countless thousands or even millions of eggs may be released during a short breeding season. Under the

---

*In contrast to the primitive **oviparous** method, in which the egg is extruded and development takes place externally. Some distinguish from typical viviparity an **ovoviparous** method in which development is internal, but the young receive no nourishment from the mother.

normal system of egg reception through an open funnel, it is obvious that there would be the danger that the whole body cavity would be choked with eggs. In the teleosts this difficulty has been solved by sealing off, next to each ovary, a part of the celom into which the eggs are shed; from this the exit is into a funnel, probably not homologous to the true oviduct, leading to the exterior (Fig. 305 *g*). Other teleosts show a more normal condition (Fig. 305 *h*).

In sharklike fishes a large, shelled egg has evolved and a **shell gland** (nidamental gland) is formed as an expansion part-way down the oviduct. Here are present two types of glandular lining, one of which secretes an albuminous material—"egg

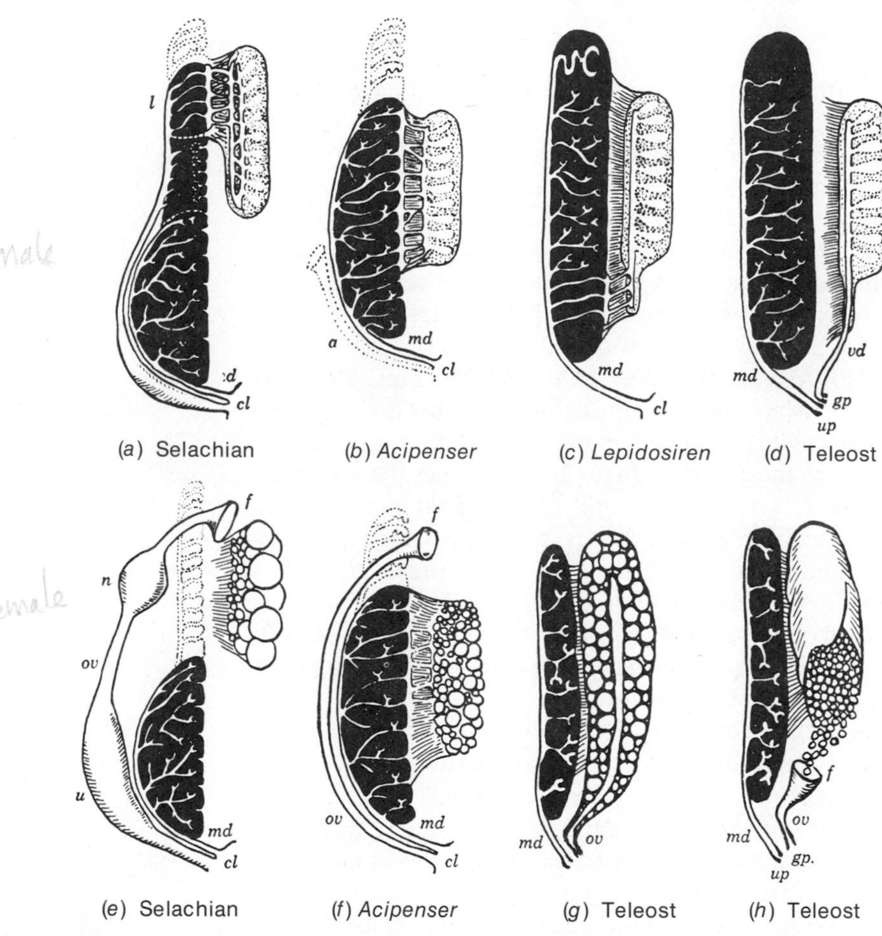

Figure 305. Representative types of urogenital systems in fishes. Upper series, males, lower series, females. *Acipenser* is a primitive actinopterygian, *Lepidosiren*, the South American lungfish. Black structures, opisthonephros; testis stippled; organs with circles (representing eggs) are ovaries; mesenteries hatched; vestigial pronephros, and vestigial male oviduct (in *b*) in stippled lines. *a*, Vestigial oviduct; *cl*, cloaca; *d*, special duct draining kidney in advanced selachians; *f*, open funnel of oviduct; *gp*, genital papilla and pore of teleosts; *l*, gland in anterior region of selachian kidney analogous to epididymis; *md*, archinephric duct as kidney duct; *n*, shell gland; *ov*, oviduct; *u*, selachian ovisac, or "uterus"; *up*, urinary pore of teleosts; *vd*, sperm duct of teleosts, analogous to ductus deferens of amniotes. In male selachians, as in amniotes, the testis tends to take over the original duct for sperm transport (cf. Fig. 297 *D*), but in other fishes this duct continues to drain the kidneys, and in teleosts a separate sperm duct *(vd)* develops. Specialized methods of egg transport are present in teleosts. (After Portman and Goodrich, from Hoar, in The Physiology of Fishes, M. E. Brown, editor, Academic Press.)

white"—about the egg; the other, in the lower part of the gland, forms a hard, horny egg case. Fertilization is internal in these cartilaginous fishes; the sperm travel "upstream" to fertilize the egg before it reaches the shell gland. Internal fertilization presumably developed because of the need for it in a form with a shelled egg. From this condition, however, it is easy to imagine the further step toward a viviparous condition, present in a considerable number of sharks and rays: the fertilized egg may be retained in the ovisac at the end of the duct until development is completed and the young born alive. In many elasmobranchs there is, still further, a development of methods of one sort or another whereby food materials may be furnished through the lining of the ovisac as a contribution to the nourishment of the young in a fashion paralleling functionally the mammalian placenta.

THE OVIDUCT IN AMNIOTES—UTERUS, VAGINA (Figs. 294 *B*, 295 *B*, 306, 309). In amniotes, as in sharks, a shelled egg has brought about specializations of the oviduct. In reptiles and birds the greater part of its length is the oviduct proper, or **uterine tube**—muscular, broad, and capable of great further distention at the breeding season. In amniotes, in contrast to sharks, the shell-forming gland is placed near the distal end of the tube, where the ovisac occurs in lower forms; it lies in the position of the **uterus** of mammals and is frequently called by that name. Birds are universally oviparous, but various lizards and snakes bear their young alive and in some cases have paralleled elasmobranchs and mammals in the development of structures through which nutriment may pass from mother to young. The two uteri of reptiles open separately into the cloaca; in birds the right oviduct, like the right ovary, is absent. In monotremes, which lay a shelled egg (although a relatively small one), the female organs are essentially similar to those of reptiles.

These organs, however, are markedly changed in typical mammals, in which the egg is tiny and development is viviparous. The uterine tube is a slender structure. The uterus, or womb, no longer produces a shell but is of the greatest importance as the place of development of the young. It is a thick-walled structure; its richly vascular epithelium forms, by union with the external membranes of the embryo, the placenta, through which maternal nourishment is afforded. In the most primitive mammals, the two uteri are still quite separate, the **duplex** condition (Fig. 307); in most, however, the distal ends of the two are fused to give a **bipartite** or **bicornuate** uterus; in higher primates there is complete union to a **simplex** type.

In reptiles and birds the most distal part of the oviducts, between uterus and cloaca, are short and undeveloped. In mammals, however, the terminal portions of the two ducts fuse to form a **vagina** for intromission of the male organ; the vagina opens into the urogenital sinus, derived from the cloaca. Marsupials show a curious (and surely aberrant) structure in which the vaginae are incompletely fused and may have a partly double—or even triple—construction (Fig. 308).

SPERM TRANSPORT: EPIDIDYMIS, DUCTUS DEFERENS. In cyclostomes the spermatozoa are discharged directly into the celom and, like the eggs, must make their own way to the outer world through genital pores at the back end of the celom. In all higher vertebrates this inefficient mode of transport has been abandoned and the male has developed a system of ejaculatory ducts which (in contrast with the female duct system) is closed throughout its course. In most groups these ducts are clearly "borrowed" from the urinary system. As the testis develops it lies close beside the embryonic kidney (Fig. 298). A short distance away are kidney tubules, from which the archinephric duct leads to the outer world. This short gap was bridged by the ancestral gnathostomes, and the sperm thus acquired a route to the exterior which could be followed in safety, avoiding the vicissitudes of travel through the celomic wilderness.

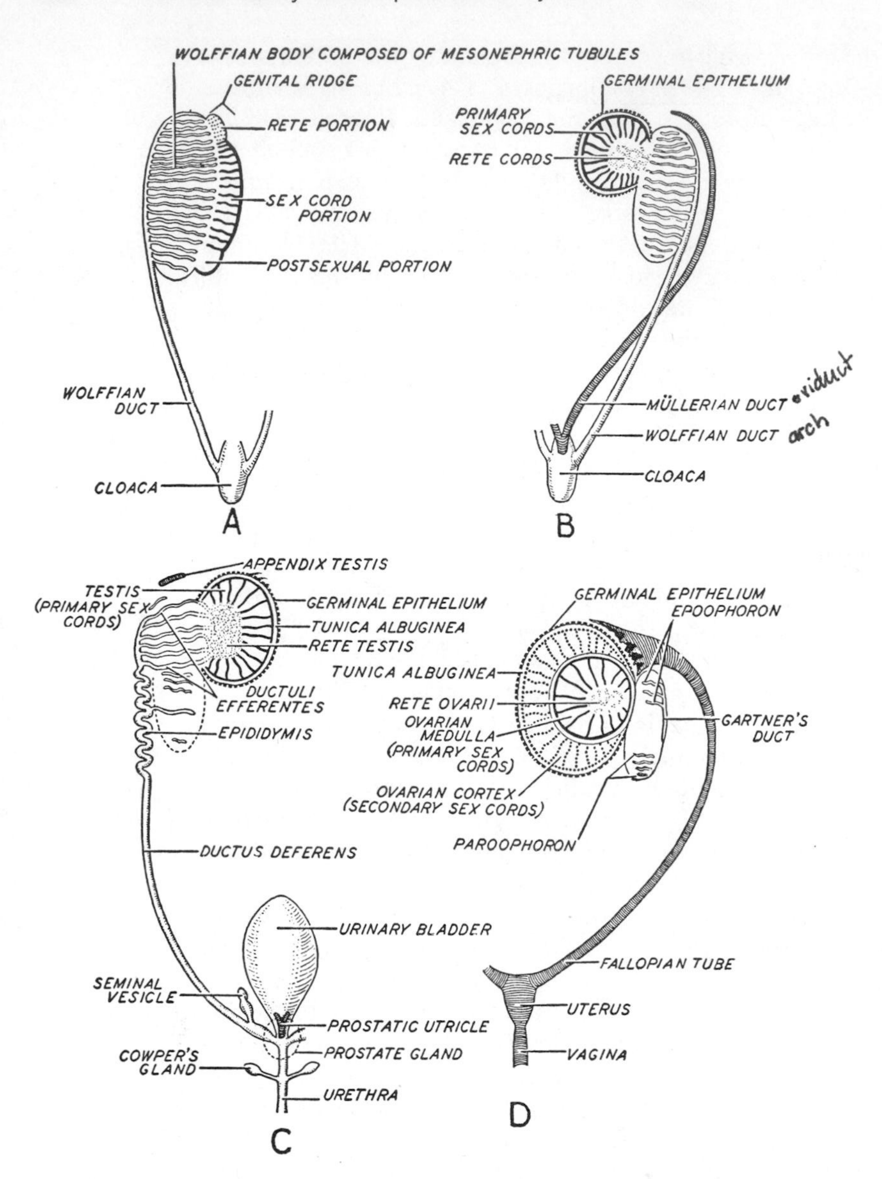

Figure 306. Embryonic development of the genital system of amniotes. *A,* Early indifferent stage, showing sex cords, developing rete testis, mesonephric kidney, and archinephric duct (wolffian body, wolffian duct). *B,* Somewhat later stage, in which the embryonic oviduct (müllerian duct) has appeared. *C,* The adult male (cf. Fig. 310). *D,* Adult female (cf. Fig. 309). (From Turner.)

Although there are variations of one sort or another, the connections between the seminiferous structures of the testis and the archinephric duct follow a basically similar pattern in most vertebrates (Fig. 311). Ripe ampullae or seminiferous tubules may be connected with one another by a **central canal** in the testis or a network of small canals, the **rete testis.** From this a number of parallel tubules extend across to the edge of the kidney. Here there may be a second longitudinal connection, or (as in mammals) the connecting tubules may pass directly to a series of erstwhile kidney tubules, which are termed **ductuli efferentes.**

These tubules emerge into the archinephric duct, originally designed for urine transport. As we have noted in the earlier part of the chapter, a dual function,

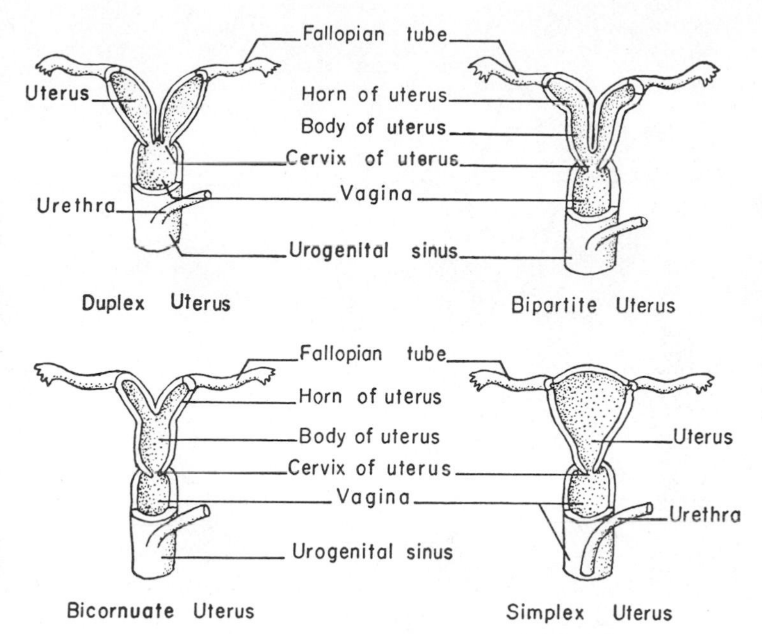

Figure 307. Diagrams to show the progressive fusion of the posterior ends of the oviducts (Fallopian tubes) in placental mammals. The uterus and part of the vagina have been cut open. (From Walker, after Wiedersheim.)

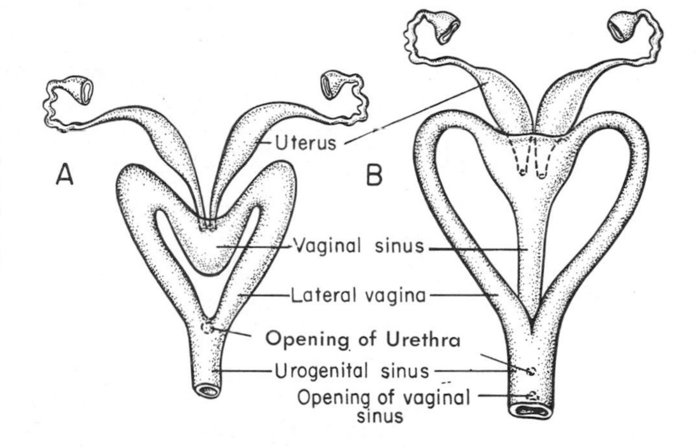

Figure 308. Female genital organs of marsupials. *A,* Opossum; *B,* kangaroo. From a median vaginal sinus there develop, in the opossum, a pair of lateral vaginae which unite distally in the urinogenital sinus. In the kangaroo figured the vaginal sinus has developed into a median vaginal tube. (After Vandebroek.)

conduction of both urine and semen, is functionally none too efficient and in the course of vertebrate history there has been a "struggle" between urinary and genital systems for use of the archinephric duct. As we have seen, the urinary system has tended to win out among bony fishes, and there has evolved a new sperm duct for sperm transport (Fig. 305 *d*). In most vertebrates, however, victory has gone to the genital system; the old archinephric duct has become a **ductus deferens,** serving for sperm transport alone.

The testis usually makes its connections with the anterior end of the kidney; as we have seen earlier in this chapter, urinary functions tend to be progressively concentrated posteriorly, and the front part of the old kidney system may become—most notably in mammals—a specialized region known as the **epididymis.** Even in chondrichthyeans (Fig. 291 *B*) the anterior end of the archinephric duct may become highly convoluted as the **ductus epididymis,** and back of the actual region of testis connection with the duct there is a part of the original kidney modified into glands secreting a fluid which is believed to stimulate the sperm. In mammals (Figs. 306, 310 *C,* 311 *B*) and in amniotes generally the epididymis,

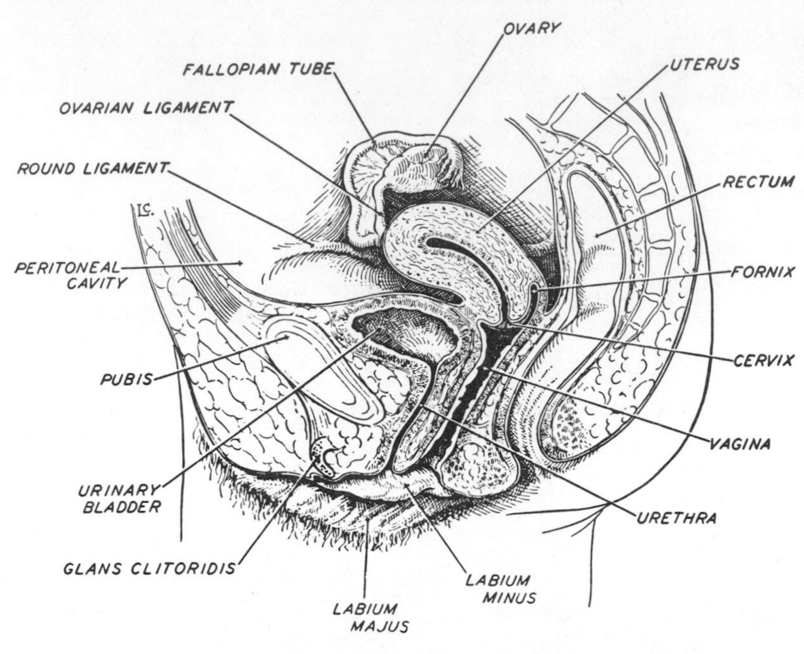

Figure 309. Female reproductive organs in *Homo*. (From Turner, General Endocrinology.)

containing the efferent ductules and convoluted duct, becomes a compact body resting close beside or upon the testis, as its name implies.

Distally the ductus deferens may expand into an ampulla for sperm storage in many groups and in mammals (Figs. 306 *C*, 310) there are further present other glands, including the **prostate gland** and **seminal vesicles,** which secrete liquid materials forming much of the seminal fluid.

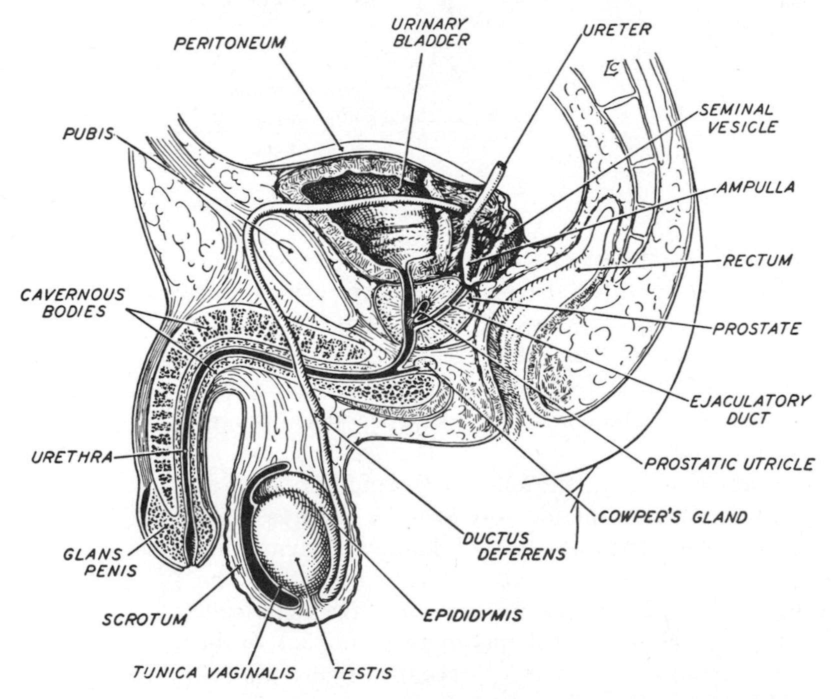

Figure 310. Male reproductive organs in *Homo*. (From Turner, General Endocrinology.)

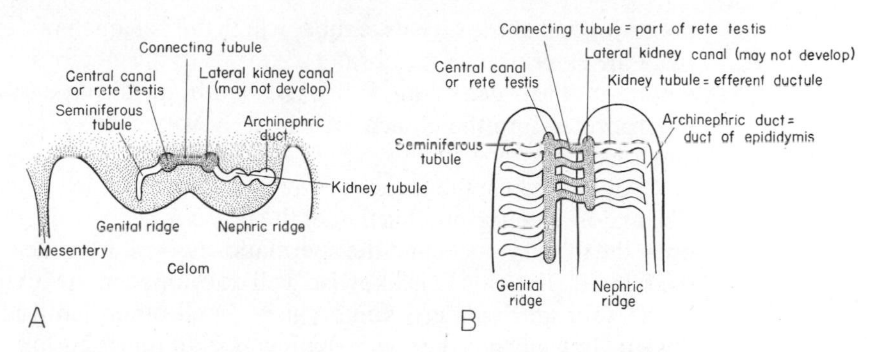

Figure 311. *A,* Cross section of an amniote embryo, to illustrate the fact that testis and kidney are adjacent (cf. Fig. 298) and may thus readily come into connection by the development of tubules bridging the gap between seminiferous and kidney tubules. *B,* Diagrammatic ventral view of a section of nephric and genital ridges to show mode of connection, usually with a central testis canal or rete testis between the sperm tubules, with connecting tubules, and frequently with a lateral kidney canal.

THE CLOACA AND ITS DERIVATIVES

In a great variety of vertebrates there is present at the back end of the trunk region a ventral pocket, opening to the exterior, in which are found the orifices of the digestive, genital, and urinary systems. This structure, appropriately termed the **cloaca** (the Roman name for a sewer), appears to have been a primitive vertebrate feature. We shall in this section follow, through the vertebrate groups, the history of the cloacal region and the varied disposition of the outlets of the systems concerned.

THE CLOACA IN FISHES AND LOWER TETRAPODS. Embryologically the cloaca has a twofold origin (cf. Fig. 314 *A*). Its major part consists of an expansion of the

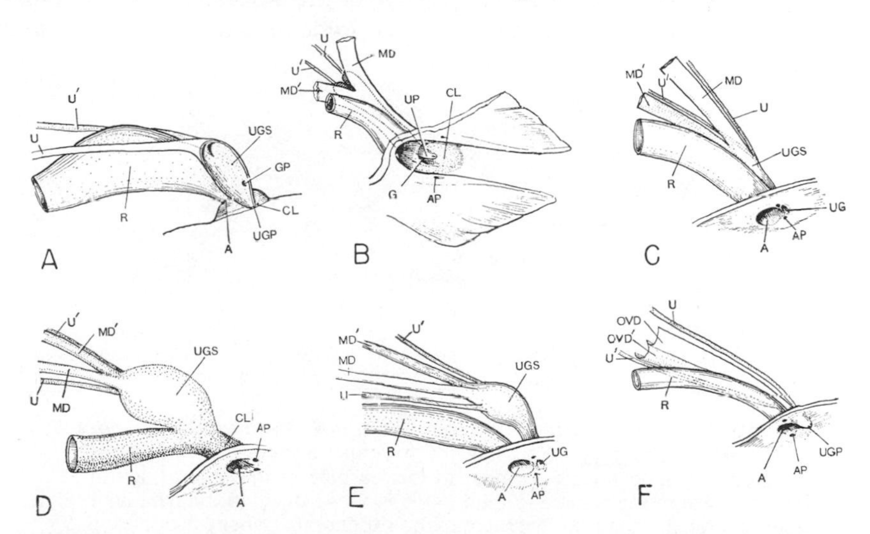

Figure 312. Cloacal and anal region in fishes. *A,* The lamprey, *Petromyzon; B,* a female shark; *C,* a young female chimaera; *D,* the Australian lungfish, *Epiceratodus; E,* a female sturgeon; *F,* a female salmon. Abbreviations: *A,* anus; *AP,* abdominal pore; *CL,* cloaca; *G,* genital opening; *GP,* genital pore; *MD, MD',* left and right oviducts (müllerian ducts); *OVD, OVD',* left and right oviducts of teleost; *R,* rectal region of intestine; *U, U',* left and right urinary ducts; *UG,* urinogenital opening; *UGP,* urinogenital papilla; *UGS,* urinogenital sinus; *UP,* urinary papilla. (From Dean.)

posterior end of the digestive tube, which during much of development is closed off from the exterior by a membrane. External to the membrane lies a depression of the ectoderm, the **proctodeum.** When the membrane disappears, this ectodermal area is incorporated into the cloaca, of which, however, it seems generally to form but a small part.

Among fishes, the cloaca is typically developed in elasmobranchs (Fig. 312 *B*). The major opening into it is that of the posterior end of the intestine; into it, further, open the urinary ducts and the spermatic ducts of the male or the paired oviducts of the female. The cloaca is likewise well developed in the lungfishes (Fig. 312 *D*) and in the sole surviving crossopterygian. In all other fish it is, however, reduced or absent. In hagfishes there is a shallow pocket representing a reduced cloaca, but in lampreys the anal opening is separate (Fig. 312 *A*). In lower ray-finned fishes and some teleosts urinary and reproductive tubes empty by a common sinus representing part of the cloaca, but the anus is distinct (Fig. 312 *E, F*). In most teleosts all three systems have separate openings, and the same is true of chimaeras (Fig. 312 *C*).

The primitive cloaca, however, was obviously present in the ancestral tetrapods, for it is found in all amphibians, reptiles, and birds, with products of all three systems entering it (Figs. 293–295, 313 *A*). Ventrally from the cloaca, we have noted, there develops in amphibians and many reptiles a large and distensible urinary bladder; this, however, generally has no direct connection with the ureters.

FATE OF THE CLOACA IN MAMMALS. In mammals the lowly monotremes (as the name indicates) still have a cloaca, but higher types have done away with this structure and have an anal opening separate from urinary and reproductive outlets. The monotreme cloaca (Fig. 313 *B*) shows a beginning of this subdivision. It is a unit for most of its extent, but the proximal part is subdivided into (1) a rectal region or **coprodeum** leading out from the intestine and (2) a more ventral **urodeum** which cares for both urinary and genital products. In marsupials there is a shallow pocket representing a last vestige of the cloaca. In placental mammals this ancient structure has vanished. The coprodeum simply becomes the rectal portion of the gut,

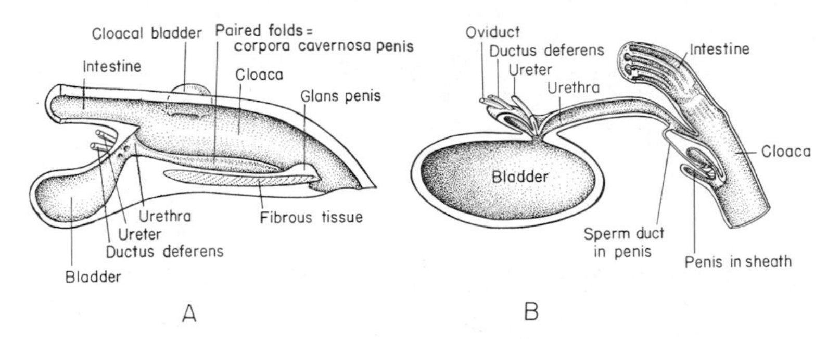

Figure 313. Section of the cloacal region of *A,* male tortoise, *B,* a monotreme mammal *(Echidna).* In *A* a penis-like structure is contained in the floor of the cloaca; paired folds may meet to form a tube at the time of sperm emission. In the monotreme a formed penis is present within the cloaca; it contains a tube, divided into several branches, for sperm transport, but urine passes out via the cloaca. In most reptiles the ureter opens dorsally into the cloaca at a point far from the bladder; in its chelonian position there is a closer approximation to the mammalian condition in the ventral shift of this opening. The tube labelled "urethra" in the two figures is equivalent to that structure in the female of higher mammals, but is here essentially a ventral proximal subdivision of the cloaca. A turtle specialization is the presence of a pair of small "supernumerary" bladder-like structures in the side walls of the cloaca. (*A* partly after Moens; *B* after Keibel.)

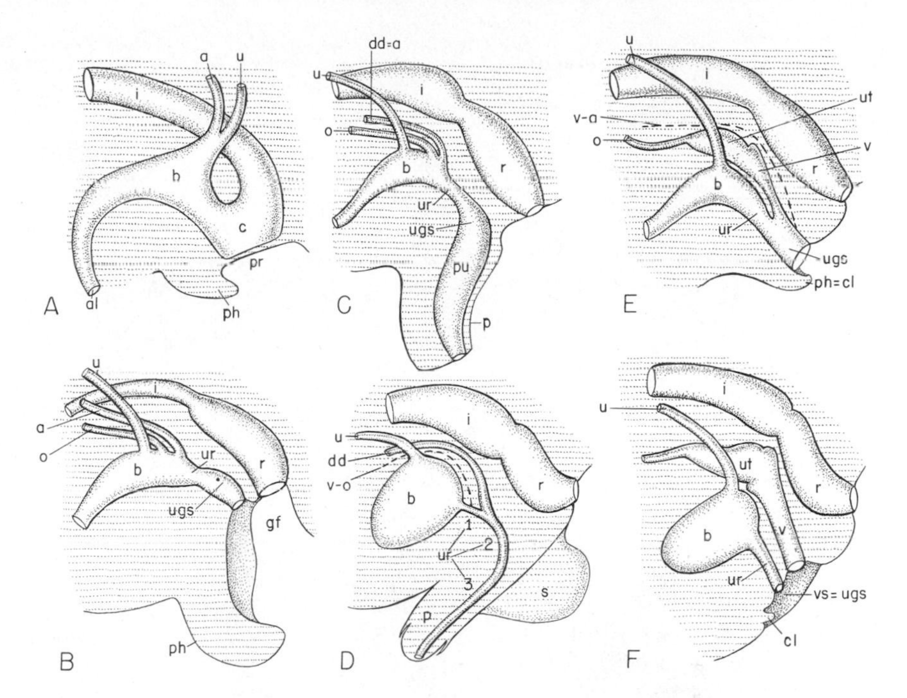

Figure 314. Embryology of the cloacal region of mammals; diagrammatic lateral views. *A,* Sexually indifferent stage, with intestine and allantois opening into undivided cloaca, archinephric duct and ureter opening together into base of allantois. *B,* Later indifferent stage; embryonic oviducts developed, ureter and archinephric duct separated, cloaca divided into rectum and urogenital sinus; phallic organ has begun development. *C,* Early stage of development of male, and *D,* adult male structure. In contrast to female, the oviduct disappears (broken line in *D*), the archinephric duct becomes the ductus deferens, and an extensive urethra includes, in addition, (1) the base of the allantois, (2) the urogenital sinus, and (3) a duct traversing the penis. *E,* Early stage of development of female, and *F,* adult female structure; disappearance of archinephric duct (broken line in *E*), development of bladder from base of allantois with short urethra distal to it, differentiation of uterus and vagina from embryonic oviduct, development of phallic organ as clitoris. *a,* Archinephric duct; *al,* stalk of allantois; *b,* bladder; *c,* cloaca; *cl,* clitoris; *dd,* ductus deferens; *gf,* genital fold; *i,* intestine; *o,* oviduct; *p,* penis; *ph,* phallic organ; *pr,* proctodeum; *pu,* penile urethra; *r,* rectum; *s,* scrotum; *u,* ureter; *ugs,* urogenital sinus; *ur,* urethra; *ut,* uterus; *v,* vagina; *v-a,* vestige of archinephric duct; *v-o,* vestige of oviduct; *vs,* vestibule.

opening at the anus. The urodeum, however, has a more complex history and differs much in the two sexes.

Conditions here are best understood by considering the developmental story, which recapitulates rather well the phylogenetic history (Fig. 314). In a placental mammal there is at an early stage a cloaca formed by a distal expansion of the gut and separated by a membrane from the proctodeal depression. The archinephric ducts and the oviducts empty into a ventral part of the cloaca, which extends outward to the allantois and in which the bladder is to develop; the ureters presently develop and open here as well.

While the embryo is still in a sexually indifferent stage, a horizontal septum extends outward to the closing membrane. This divides the cloaca into two chambers: a coprodeum, continuous with the gut above, and a urodeum, or **urogenital sinus,** below. Meanwhile, the bladder begins to expand, the ureters remaining in contact with it. Between bladder and urogenital sinus there is a relatively short narrow tube which is to form part or all of the **urethra** of the adult; the spermatic

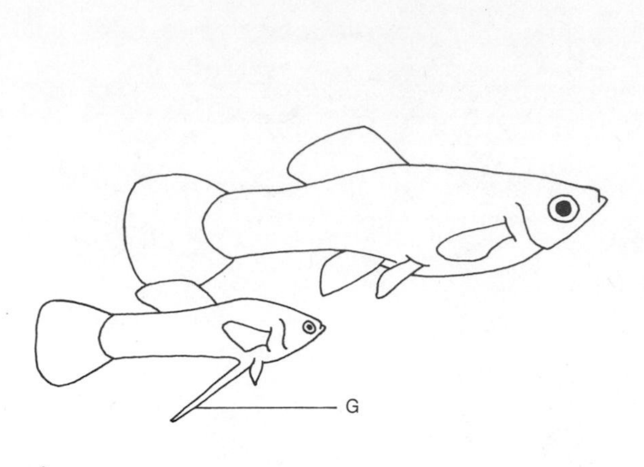

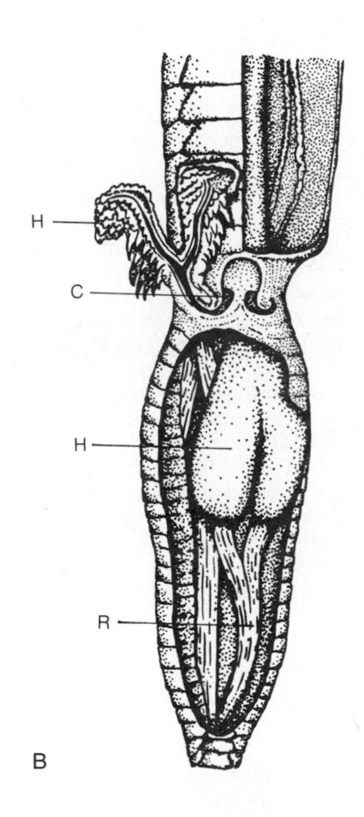

Figure 315. Copulatory organs of lower vertebrates. *A,* Poeciliid teleosts; the male has a gonopodium, formed from certain rays of the anal fin, which can be inserted into the female's reproductive tract. *B,* Rattlesnake; one of the hemipenes is everted from the cloaca and the other is in the resting position within the base of the tail. *C,* Cloaca; *G,* gonopodium; *H,* hemipenis; *R,* retractor muscles of hemipenis. (After Rosen and Gordon, and Hoffmann.)

ducts and oviducts come to terminate at the point where this tube opens into the sinus (Fig. 314 *B*).

Beyond this stage, conditions in the two sexes diverge. In the female (Fig. 314 *E, F*) the urogenital sinus becomes the **vestibule** of the urinary and genital systems; this may retain some depth (as in carnivores) or be a relatively shallow depression (as in primates). Into the vestibule open the conjoined distal ends of the oviducts as the vagina (the spermatic ducts, of course, degenerate); into the vestibule also opens a relatively short urethral tube from the bladder.

In the male the sinus has a different history. It becomes an elongate tube which continues on into the penis. Into its proximal end open the spermatic ducts or ducti

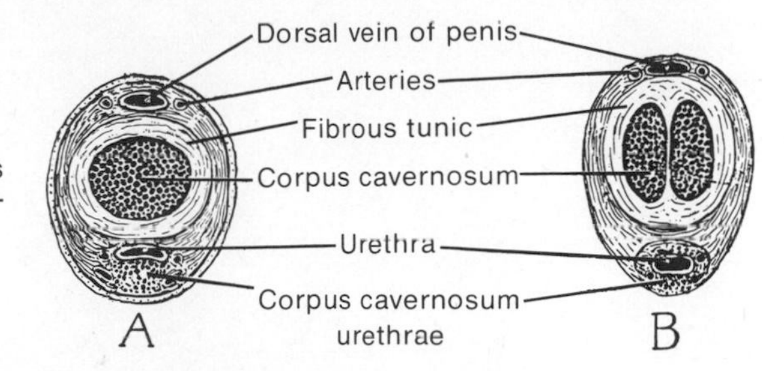

Figure 316. Sections through the penis of a rhesus monkey; *A*, distally, and *B*, proximally. (From Wislocki.)

deferentes (the oviducts degenerate) and the tube from the bladder. In the female this short proximal tube constitutes the entire urethra, but in the male the term urethra is applied to the whole extent of the tube from the bladder to the end of the penis. Female and male urethrae are, thus, not altogether comparable, for that of the male includes the homologues of both urethra and vestibule in the female.

EXTERNAL GENITALIA. In primitive water-dwelling vertebrates with shell-less eggs external fertilization is the general rule. But in forms with shelled eggs or with viviparous habits—including at least one placoderm, the Chondrichthyes, a few teleosts, and all amniotes—internal fertilization is a requirement, and special male structures are usually developed to facilitate the entrance of the sperm into the female genital tubes. In sharks, rays, and chimaeras these take the form of **claspers** extending from the pelvic fins (Fig. 144 *C*). These are inserted into the cloaca of the female; rolls of skin folded into tubes along the clasper form a channel for the sperm. In a number of viviparous teleosts a somewhat comparable median clasper or **gonopodium** is formed from the anal fin (Fig. 315 *A*).

In ancestral amniotes direct contact of male and female cloacae would seem to have been sufficient for the transfer of sperm, for *Sphenodon* lacks copulatory organs, as do most birds. In many reptiles, however, the male has some type of accessory organ, a **penis,** to aid in sperm transfer. The **hemipenes** of snakes and lizards form a pair of cloacal pockets, often containing thornlike spines; at the time of copulation these may be turned inside out, extruded, and inserted in the female cloaca (Fig. 315 *B*).

In turtles and crocodilians there are structures which may be morphologic forerunners of the penis of mammals (Fig. 313 *A*). Lying in the ventral wall of the cloaca, with a groove between them, is a pair of longitudinal ridges, the **corpora cavernosa penis,** composed of spongy tissue (as the name implies); a further spongy structure, the **glans penis,** lies at the outer end of the groove. On excitation these structures are distended with blood and the glans inserted into the female cloaca; the groove between the cavernous bodies closes into a tube which carries the sperm. In the female a comparable but smaller structure is the **clitoris.**

In monotremes there are somewhat comparable structures, and the clitoris of the female mammal is in general relatively undeveloped. In higher mammals, however, the penis becomes a discrete external organ. Glans and corpora cavernosa are retained; the groove between the latter structures is closed to become the distal portion of the urethra and is surrounded by an additional cavernous body, the **corpus cavernosum urethrae** (Fig. 316).

Circulatory System 14

In many small invertebrates there is no need for a circulatory system. Distances are short, and internal transportation of materials may be effected by diffusion and such flow of fluids as may result from body movement. With greater size and complexity, a circulatory system is required. A comparison with human communities is a fair analogy. In a village a transportation system is unnecessary; stores, school, church are all close to the homes they serve. With growth of the community this is no longer the case, and organized transportation systems are a necessity.

The simplest type of circulation seen among invertebrates is that of an open system, in which a heart forms a pump forcing blood out through a series of vessels—arteries—to various parts of the body. At the points where these vessels terminate, however, the blood is released into the tissue spaces to ooze back "on its own" to the heart. In amphioxus an advance is seen in that return vessels to the heart—the veins—are present; in mid-course, however, the blood is still in contact with the body cells which it serves. Vertebrates have achieved a completely closed system in which, between arteries and veins, the blood is enclosed in a network of tiny vessels—the capillaries—and is never in direct contact with tissues. Higher vertebrates have, in addition, evolved a set of vessels, the lymphatics, which return fluid from tissues to the heart.

FUNCTIONS. Foremost of functions served by the blood is the transport of materials to and from the cells (via the interstitial fluid). Oxygen must be constantly carried from gills, skin, or lungs, and a small but steady stream of food materials—mainly glucose, fats, and amino acids—must be supplied from the intestine or from storage and manufacturing centers, notably the liver. Conversely, wastes must be removed: carbon dioxide, destined for gills or lungs; nitrogenous wastes and excess metabolic water bound for the kidneys.

The maintenance of a stable and narrowly defined internal body environment is necessary for the welfare of the cells and tissues. The constant circulation of liquid throughout the body in the blood makes for uniformity of composition in the interstitial fluids of every region and aids in maintaining relatively uniform temperatures. Among further functions of the circulatory system are aid in the struggle against disease and in the repair of injuries and, through the circulation of hormones, the utilization of the blood stream as an accessory nervous system.

BLOOD

The blood, filling the vessels of the circulatory system, may be regarded as a tissue. Blood, like the connective and skeletal tissues, is derived from mesenchyme. All three consist of cellular elements lying in a "matrix." In bone and cartilage the matrix is a solid substance; in connective tissue it is gelatinous in consistency; in blood the matrix is a liquid, in which the cellular components float in free fashion.

BLOOD PLASMA. This liquid "matrix" of the blood, the **plasma,*** is a watery fluid of complex composition. The blood is essentially a part of the complex interstitial fluid enclosed within the walls of the blood vessels and both have the same salts. In addition, however, the blood contains materials peculiar to itself in the form of special **blood proteins**—albumin, globulins, and fibrinogen. These molecules are so large that they are normally unable to pass through capillary walls and thus leave the blood stream. The presence of the blood proteins raises the osmotic pressure of the blood above that of the interstitial fluids—a point of importance in capillary function. Further, the varied globulins play a variety of active roles, notably as antibodies effective against invading organisms or chemicals. Most of these proteins are formed by the liver, but many of the globulins are produced in lymphoid organs. Fibrinogen is the material responsible for clot formation when a vessel is cut.

As well as these stable and permanent plasma constituents, the blood contains materials in transit—food materials, notably glucose, en route to the cells, nitrogenous wastes (mainly as urea or uric acid), carbon dioxide, and minute amounts of hormones.

BLOOD CELLS (Fig. 317). Cellular blood components are absent in amphioxus but are present in all vertebrates. They normally include (1) red blood corpuscles or erythrocytes, (2) white blood corpuscles or leukocytes, and (3) thrombocytes.

Oxygen transportation is one of the most important functions of the circulatory system. Metallic compounds, particularly of iron or copper, are major aids in this regard and are found in many animals of various phyla, either free in the blood stream or contained in blood corpuscles. In the vertebrates the oxygen carrier is the iron compound **hemoglobin;** it is concentrated in the red blood corpuscles, the **erythrocytes.** In most vertebrates these are flattened oval structures which are

*The term "serum" refers to the liquid remaining after the protein clot material (fibrin) has been removed from clotted plasma.

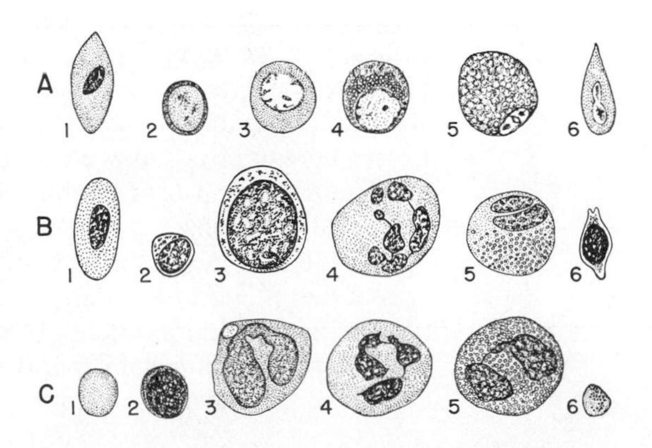

Figure 317. Blood cells of *A,* a teleost *(Labrax); B,* a frog *(Rana); C,* a mammal (man). *A1, B1, C1,* Erythrocytes; *A2, B2, C2,* lymphocytes; *A3, B4, C4,* neutrophilic granulocytes; *A4,* fine-grained acidophilic granulocyte; *A5,* coarse-grained acidophilic granulocyte; *A6, B6,* thrombocytes; *B3, C3,* monocytes; *C6,* blood platelet. *A,* × about 1800; *B, C,* × 1200. (*A* after Duthrie; *B* after Jordan; *C* after Maximow and Bloom.)

proper nucleated cells. In mammals, however, the mature erythrocyte sheds its nucleus; further, in nearly all mammals (camels and llamas are exceptions) the corpuscles are circular rather than oval in outline. There is considerable variation in their size; erythrocytes of mammals or birds may be but a few micrometers in diameter, but some amphibian corpuscles have a volume 100 times or more that of a typical mammalian corpuscle.

The white corpuscles, the **leukocytes,** are much fewer in number than the red corpuscles; they constitute only about 1 per cent of the blood cells in typical amniotes but may rise above 10 per cent in some fishes. Two main groups of white cells may be distinguished—lymphoid types, with a simple nucleus and clear cytoplasm, and granulocytes in which the nuclear materials are irregularly arranged and often subdivided and the cytoplasm is granular. Of the **lymphoid leukocytes** the common forms are the **lymphocytes,** small cells with a large nucleus and little cytoplasm. They derive their name from the fact that they abound in the lymph. Lymphocytes, of two distinct sorts, play an important role in the immunological reactions of the organism. To larger cells with a cytoplasm equally clear but more abundant, the term **monocytes** is applied.

The **granulocytes** or **polymorphonuclear leukocytes** are large cells with a nucleus that is irregular or lobate and a cytoplasm that is abundant and highly granular. A type in which the granules stain readily with acid dyes is reasonably termed **acidophilic** (eosinophilic); others which stain with basic dyes or respond in part to both types of dye are, respectively, **basophilic** and **heterophilic** (or neutrophilic). Heterophils of varied appearance are the most abundant granulocytes in all vertebrates (except reptiles); acidophils are uncommon, but widespread among vertebrate groups; basophils are still fewer in numbers and are seldom reported in fishes. Granulocytes accumulate rapidly in injured or infected tissues where some are phagocytes, "eating-cells"; further, other granulocytes, like lymphocytes, may play an important role in immunological protection of the organism and are involved in allergic reactions.

Thrombocytes are blood elements associated with the process of blood clotting. In most vertebrate classes they take the form of small, oval, pointed **spindle cells.** In mammals there are present instead tiny blood **platelets,** which lack a nucleus. The disintegration of thrombocytes releases materials involved in blood clotting.

BLOOD-FORMING TISSUES

In body tissues generally, most cellular differentiation commonly takes place in embryonic stages once and for all, and, furthermore, occurs in the place in which the mature cells are found. Not so with the blood. The life of most blood cells is measured in weeks or days, and these elements are constantly renewed. Further, the blood corpuscles are not fixed in position; once matured, they are free agents, which may circulate to any and every part of the body and carry with them no clue as to where or how they were formed. As a consequence, the study of the development and relationships of the blood cells is a difficult one, and there are many unsettled problems.

Little or none of the circulatory system is derived from any of the epithelial sheets laid down in the early embryo. Instead it is derived, like the connective tissues, from the mesenchyme cells. Correlated, it would seem, with this embryonic relationship is the fact that in many instances cells of one of these systems appear to be transformable into members of the other.

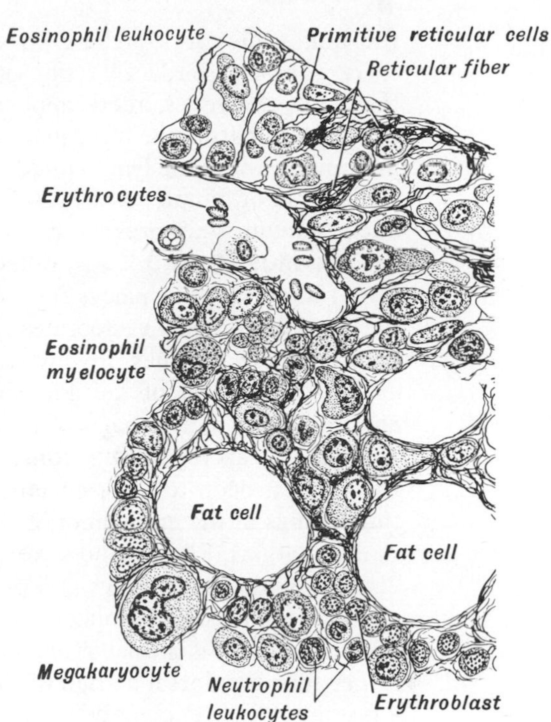

Figure 318. A blood-forming tissue–bone marrow from a mammalian femur. Much of the reticular framework and two reticular cells are seen, as well as two types of granular leukocytes in process of differentiation, and erythroblasts from which erythrocytes are formed. The megakaryocyte is a giant cell type from which it is believed blood platelets may be formed. (After Maximow and Bloom.)

Blood-forming or blood-storing tissues may be present in a number of areas in the body; these vary considerably from form to form and from embryonic sites to those present in the adult. In general, however, such sites have common structural features (Fig. 318). All are spaces which are enlargements of circulatory vessels or lie adjacent to such vessels. A reticular network of fibers forms the "skeleton" of the tissue. Enmeshed within this framework are found masses of blood cells in process of multiplication and differentiation. The seemingly simplest and least differential type of cells found in such tissues are **hemocytoblasts,** a basic type of primitive blood cell from which either red or white corpuscles may develop; lymphocytes may have a separate origin.

BLOOD-FORMING SITES. The first blood vessels formed in the embryo are those engaged in carrying food materials, and the first blood cells are erythrocytes formed in connection with them. These early cells are, in consequence, formed in mesenchyme in the yolk-filled belly floor in forms with a mesolecithal egg type, and in clusters of similar cells, termed **blood islands** (Fig. 268 *A*), on the surface of the yolk sac in large-yolked eggs. At a somewhat later stage blood cells may arise in a variety of regions from the mesenchyme or from the walls of blood vessels. Favored areas in the embryo include the kidney, liver, spleen, and pharyngeal tissues.

Even in the adult a great variety of organs may contain blood-forming centers in one group or another. In lampreys, elasmobranchs, many teleosts, and amphibians, the kidney continues to be important throughout life for blood-cell production, and in these forms and in turtles the liver contains blood-forming tissue. In sharks white cells are formed in the gonads. Lymphoid tissues continue to be present in the throat in various forms, from fish to mammals, as tonsil-like cell masses; lymphoid tissue is also common in the intestinal wall (Fig. 270). In higher vertebrates the **bone marrow** is a great center of blood formation; in this way the hollow interiors of long bones are put to positive use. In some frogs and in reptiles and birds all types of blood cells are produced in the marrow. In mammals, however, there appears to be little release of lymphocytes into the blood stream from the bone marrow. This cell type, instead, is found to be stored and multiplied in the **lymph nodes** (Fig. 321), small spherical organs situated along the course of lymph vessels. A few lymph nodes may be found in birds, but no such structures are present in lower vertebrate classes.

THYMUS GLAND. As noted above, tissues surrounding the pharynx are of importance as places for blood cell formation in the embryonic stage of all vertebrates. In most vertebrates much tissue of this sort is organized to form the **thymus gland.** In fishes, thymic material is generally present, deep to the surface, above most or all of the gill slits. In most tetrapods thymic tissue in varied shapes—often as two pairs of tissue masses—is present in the neck (Fig. 431 *D, E*); in mammals it generally consists of a single pair of glands at the anterior end of the thorax, deep to the sternum. The thymus is, in part at least, derived from thickenings of the epithelium of the embryonic gill pouches, generally from their dorsal margins. There is, however, much variation as to the pouches involved (Fig. 266); in fishes every typical gill pouch may produce thymic material, but in tetrapods only one or two pouches are involved. In mammals the second postspiracular gill pocket is the usual seat of thymic origin, and here it buds from the ventral rather than the dorsal margin of the pouch. The thymus tissue (Fig. 319) includes a framework of reticular cells, between which is a lymphoid tissue of white blood cells; in addition, the thymus contains minor amounts of other tissues (such as spherical corpuscles, one of which is shown in Figure 319). The reticular tissue of the gland seems definitely to originate from the epithelium, but the lymphoid cells may develop (as do other blood cells) from mesenchyme.

The gland grows rapidly during embryonic life, but before the adult stage is reached, the thymus ceases to grow. In tetrapods it may undergo degeneration, and in many adult mammals the thymus disappears entirely.

BURSA OF FABRICIUS. Another lymphoid organ quite similar to the thymus in many ways, the **bursa of Fabricius,** develops as a dorsal pouch off the cloaca of birds (Fig. 295). It resembles the thymus in its early development from near the end of the gut, rapid growth, production of lymphocytes in the embryo, and reduction or loss with maturity. Histologically, the lymphoid tissue is like that of the thymus, but the bursa also retains fairly normal mucosal epithelium believed to be the source of the lymphocytes, but (as usual) their origin is open to dispute. In birds the lymphocytes produced here appear to differ in their function in the immune system from the thymic lymphocytes. Mammals, which lack the bursa, produce, from an unknown source, lymphocytes of the bursal type.

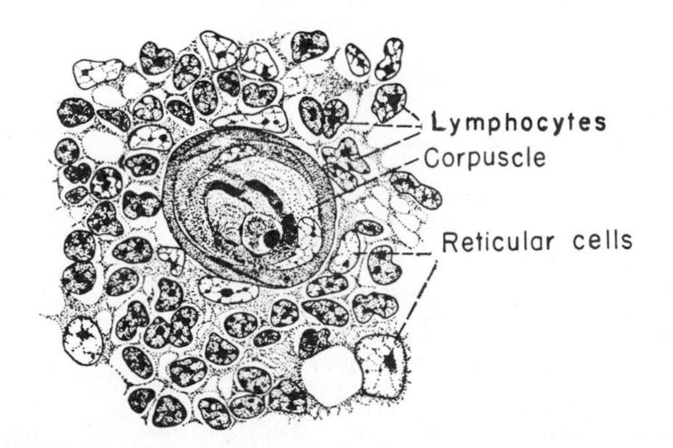

Figure 319. A small part of a mammalian thymus, showing the numerous lymphocytes, the cells forming the reticulum, and one of the peculiar corpuscles characteristic of the gland. (After Dahlgren and Kepner.)

SPLEEN. Only in the case of the spleen do we find tissues associated with blood-cell formation or storage forming a discrete major organ. In cyclostomes and lungfish even the spleen is present merely as a mass of reticular tissue surrounding part of the gut. But in every other vertebrate group it is, although adjacent to the gut, a distinct reddish structure lying in the dorsal mesentery. As may be seen from Figure 320, it may attain a complex structure. Within its reticular framework are packed masses of blood cells which in localized areas form either a **white pulp** consisting of leukocytes or a **red pulp** in which red cells predominate. The spleen is fed by an artery and drained by a vein, both of which may branch in complicated fashion; few lymphatic vessels are present.

The spleen is in every group an important center of blood cell formation. In the embryo, erythrocytes as well as granulocytes are formed there, and this function persists in the adult except in mammals, in which bone marrow has become the important seat of erythrocyte formation and white cells alone mature in the spleen. Red cells are, however, stored in great quantities in the spleen, in mammals as in lower groups, and destruction of such cells occurs here as well, to at least some degree.

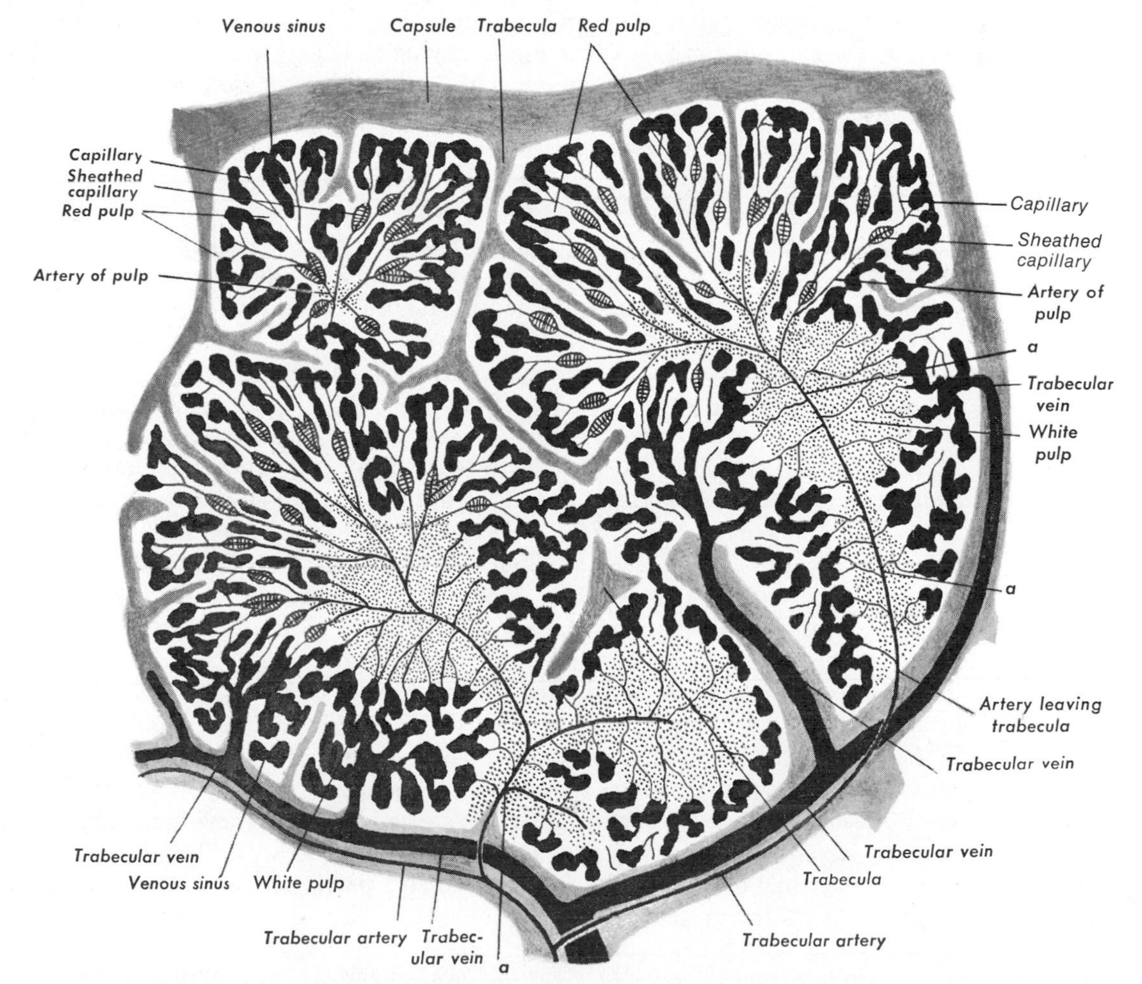

Figure 320. Diagram of a part of the mammalian spleen. Venous spaces in black; white pulp, heavy stipple; connective tissue capsule and trabeculae, light stipple; red pulp, unstippled. "Sheathed arteries" are surrounded by white pulp, here shown as hatched areas. (From Bloom and Fawcett.)

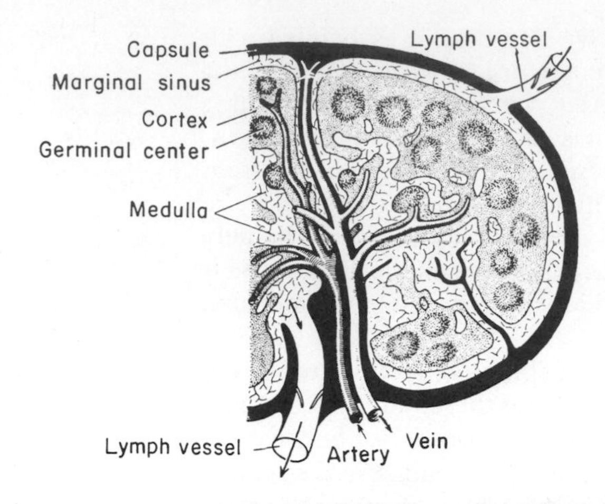

Figure 321. Diagram of the structure of a lymph node. In addition to the lymph vessels, a node is supplied by a small artery and vein. (After Portmann, Einführung in die vergleichende Morphologie der Wirbeltiere, Benno Schwabe & Co.)

CIRCULATORY VESSELS

The vessels of the circulatory system, like the blood cells, are derived from the embryonic mesenchyme. As food-containing liquid begins to flow through the body of the early embryo, adjacent mesenchyme cells gather about such channels and surround them with a continuous wall. All early formation of vessels takes place in this fashion; later in development and in adult life, new vessels may form (as tissues expand or injuries are repaired) by outgrowths from the lining of channels already established. The inner lining of blood vessels, termed an **endothelium,** consists of thin, leaf-shaped cells, continuous with one another at their margins. There are no openings except at the ultrastructural level and even these have a fine membrane over them; thus the circulating fluid is not normally in direct contact with the interstitial fluid or body cells, although most of the plasma contents can pass freely through the thin endothelial membrane.

The vessels of the circulatory system include (1) the heart; (2) arteries, by definition vessels carrying blood from heart to body tissues; (3) capillaries and comparable structures, typically small vessels connecting arteries and veins; (4) veins, returning blood to or toward the heart; and (5) lymphatics, auxiliary vessels prominent in higher vertebrates, which aid in the return of fluid from the tissues.

CAPILLARIES (Figs. 322, 325). These smallest of vessels, whose walls consist solely of a thin endothelial layer, generally have only the ''bore'' necessary to allow an erythrocyte to pass; they typically deploy from the ends of the arterial branches, wind among the tissues in such fashion that no cell is far from a capillary, and at their distal ends are re-collected into veins. Capillary networks, however, may be interpolated along the course of arterial or venous systems. In gill-bearing vertebrates the course of arterial flow from the heart is interrupted by a capillary system in the gills.* The return of venous blood from tissues to heart may be interrupted also by a forced passage, on the way, through a network of capillaries or similar vessels, as happens in the liver in all vertebrates and in the kidneys in many groups. An interjected system of veins which lead the blood to a capillary bed in some organ rather than directly to the heart is termed a **portal system** and named for the organ to which the system leads.

*We are accustomed to think, in visual images, of arterial blood as oxygenated and hence ''red''; but arterial blood between heart and gills in a fish is, of course, of the ''blue'' venous type; the arteries and veins leading to and from the tetrapod lung likewise have a reversal of the ''blue'' and ''red'' blood types.

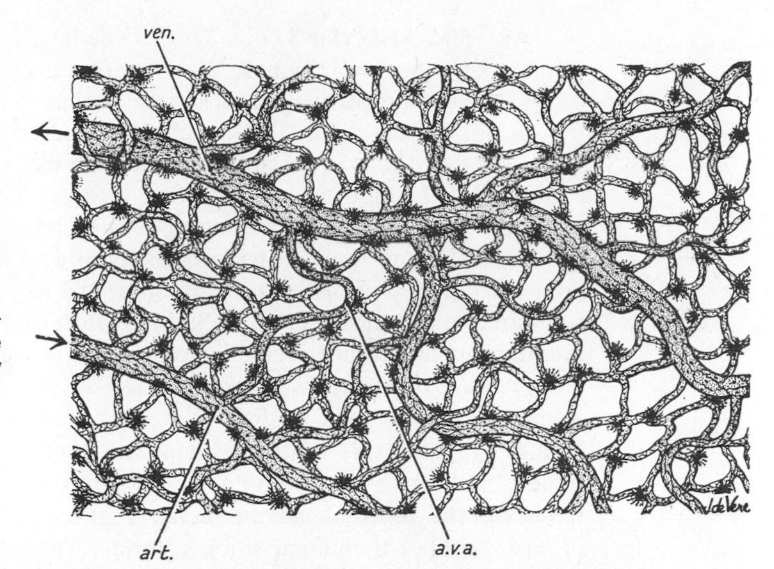

Figure 322. Portion of a capillary bed from the web of a frog's foot, showing small arteriole *(art.)* and venule *(ven.)*, a capillary network, and a direct arteriovenous anastomosis *(a.v.a.)*. (From Young, The Life of Mammals, Oxford University Press.)

Although capillaries are the major intermediaries between arteries and veins, there are other types of connections. There are sometimes found direct "short circuits" of larger caliber—**anastomoses**—between arteries and veins, or there may be substituted for capillaries small, thin-walled "ponds" of blood termed **sinusoids.** In some instances, a blood vessel breaks up into a complicated, coiled mass of tiny blood vessels reasonably termed "a marvellous network"—**rete mirabile;** a kidney glomerulus is such a structure. A special type of network may be present in the distal part of the limbs, particularly in wading birds, and the "flippers" of aquatic mammals. A maze of tiny channels from the arteries leading to the foot or "paddle" may surround the returning venous vessels. This appears to be a heat-conserving device, warmth being short-circuited from the arteries to the veins and back into the body, at the expense of the chilly limb. These, and many other retia, may consist of many parallel vessels and work to concentrate substances or conserve heat by a "countercurrent multiplier" system (Fig. 323).

Capillaries are too small to be dissected by ordinary means and hence are neglected from the point of view of gross anatomy. But it must never be forgotten that functionally the capillaries are the most important part of the circulatory system. Elsewhere blood is merely in transit; here it is at work, exchanging with the interstitial fluid, and through this with the cells, oxygen and food materials for carbon dioxide and wastes. At the proximal end of a capillary system the balance between hydrostatic pressure, tending to force materials out of the capillaries, and osmotic pressure is such that oxygen and other materials in solution pass out of the blood and to the tissues; at the far ends of the capillaries, hydrostatic pressure is, of course, lessened, and osmotic differences between blood and external fluid favor an inflow of carbon dioxide and wastes.

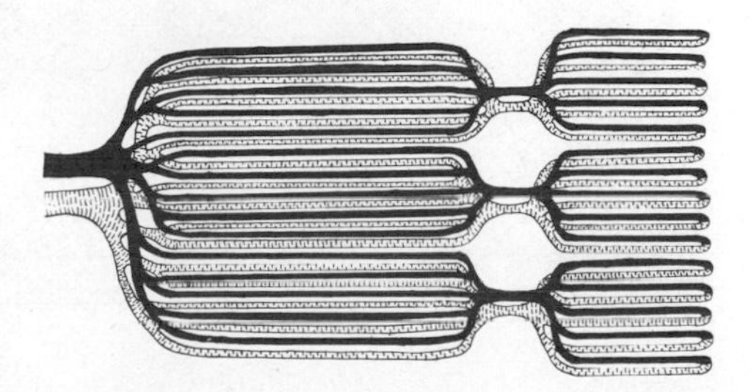

Figure 323. Diagram of a rete mirabile, from the red body of the swim bladder in an eel. The parallel arterial (lines) and venous (solid black) vessels provide a countercurrent multiplier, in this case one enabling the fish to secrete gas into the swim bladder at pressures well above those in the blood. (After Rietschel.)

ARTERIES AND VEINS (Fig. 324). The larger vessels of the body, **arteries** and **veins,** their smaller branches, the **arterioles** and **venules,** and likewise the major lymphatic vessels have external sheathing materials in their walls in addition to the ubiquitous endothelium. These include connective tissue fibers, elastic fibers, and smooth muscle cells in variable amounts; the walls of large arteries or veins contain small nutrient blood vessels for these tissues. The walls of the large vessels are customarily described as consisting of three layers or "tunics," the **tunica intima, tunica media,** and **tunica externa** or **adventitia.** As seen in small vessels the intima may consist only of the endothelium, but in a large artery there may be here also a thin sheet of connective tissue and a sheath of elastic tissue, the **internal elastic membrane.** The tunica media is dominantly a sheath of smooth muscle, which in large vessels is generally arranged in two layers, of circular and longitudinal fibers. Sometimes the tunica media is bounded externally by an **external elastic membrane.** Beyond this is the adventitia—connective tissue, often rather loose, which binds the vessel to adjacent structures. Arteries and veins have a similar build, but the veins (as may be seen in Figure 324) are typically thinner-walled and of larger caliber than comparable arteries. These differences are, of course, obviously related to the fact that arterial blood is flowing under higher pressure and at higher speed than that of the veins.

These functional differences account for further contrasts between arteries and veins. The "bore" of an artery tends to remain constant, diminishing only as branches are given off; a vein may expand along its course to form a large sac or **sinus.** The arterial system presents relatively few individual anomalies, but veins are highly variable. A fast-flowing mountain stream tends to take a direct and undeviating course, whereas a sluggish stream meanders, branches and re-forms, produces islands; so in the embryo a vein often appears as a network of variable channels (cf. Fig. 351); which channels are to form the definitive vein is not fixed, and hence the frequent anomalies. In veins there are often developed **valves,** commonly in pairs. These are folds of the intima, behind which lie pocket-like depressions. A backward flow of liquid in a sluggish vein is prevented by distention of the pockets and a consequent closing of the vessels. Almost never are such structures present in an artery, since arterial blood cannot ordinarily work back against the heart valves.

Lymphatic vessels are extremely thin-walled. In general they resemble veins and capillaries, but are even more irregular.

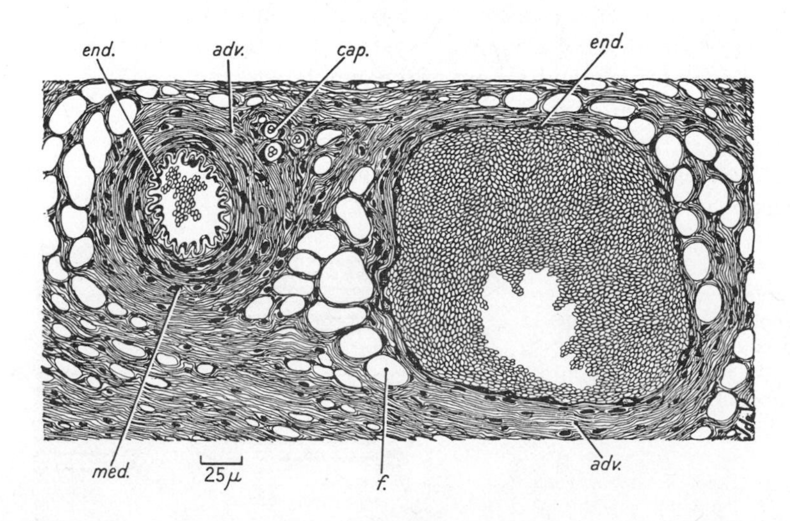

Figure 324. Section through a small artery and its accompanying vein, showing contrast in size and thickness of walls. *adv.,* Adventitia; *cap.,* capillary; *end.,* endothelium; *f,* fat cell; *med.,* media (muscle layer) of artery. (From Young, after Maximow and Bloom.)

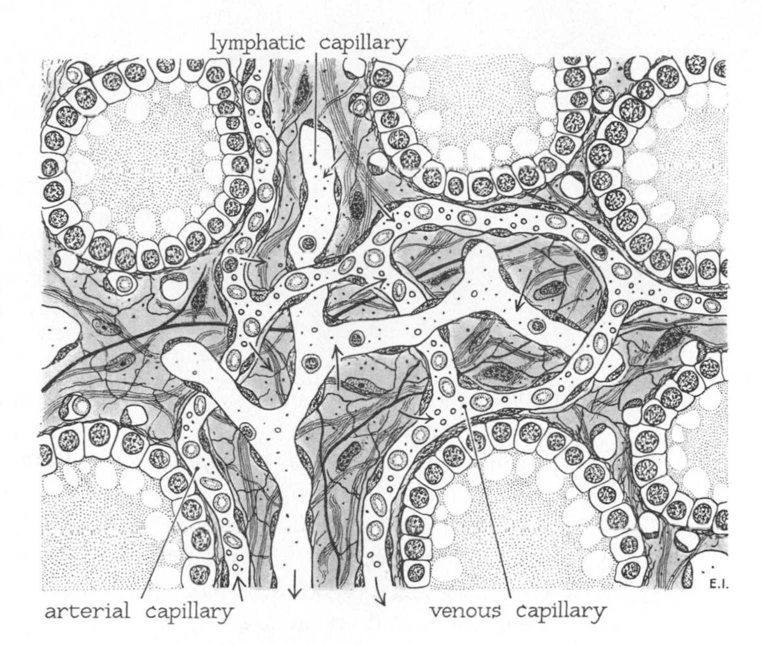

Figure 325. Diagram of blood and lymph capillaries, showing their relation in the mammalian thyroid. Arrows indicate the direction of flow of materials into and out of the capillaries. (From Kampmeier.)

ARTERIAL SYSTEM

AORTIC ARCH SYSTEM IN FISHES (Figs. 326, 327, 330). In primitive, gill-breathing vertebrates (and in amphioxus as well) all the blood from the heart courses forward in a **ventral aorta** lying in the floor of the throat. From this a series of **aortic arches*** curve upward on either side between successive gill slits. Each breaks up into capillaries in the gill membranes for aeration of the blood and is re-formed dorsally into further arterial vessels which pass to the tissues of the body and head. In tetrapods the gills are lost, but aortic arches are found in every embryo. Their history is one of the most interesting chapters in the structural evolution of the vertebrates.

In a diagrammatic primitive vertebrate (Fig. 330 *A*) we may picture the aortic circulation as including paired arches passing upward laterally in front of or between each gill slit or pouch and, beyond the capillary network, reaching the **dorsal aorta.** Posteriorly this major vessel is a single median trunk carrying blood backward to most of the body; anteriorly it consists of a pair of vessels, one on either side of the head. The number of aortic arches was presumably high and variable in the ancestral vertebrates (it is very high in amphioxus); apart from cyclostomes and a very few sharks, however, there are normally in vertebrates but five gill slits plus a spiracle and hence potentially six aortic arches, usually designated by Roman numerals. In vertebrate embryos these arches develop, in general, in regular order from front to back (Fig. 334). Until the gills begin to function the arches in fishes are continuous, uninterrupted vessels.

Vertebrates invariably refuse to adapt themselves to a man-made structural diagram, however, and in the aortic arches, as elsewhere, the "idealized" condition we have pictured is never preserved in the adult. There is considerable variation in the afferent and efferent portions of the arch when the gill capillaries develop in fishes. Diagrammatically, we have represented the arches as distributed so that one

*The triple use of the word "arch" in the gill region was commented on earlier; in this chapter *aortic* arch is implied throughout.

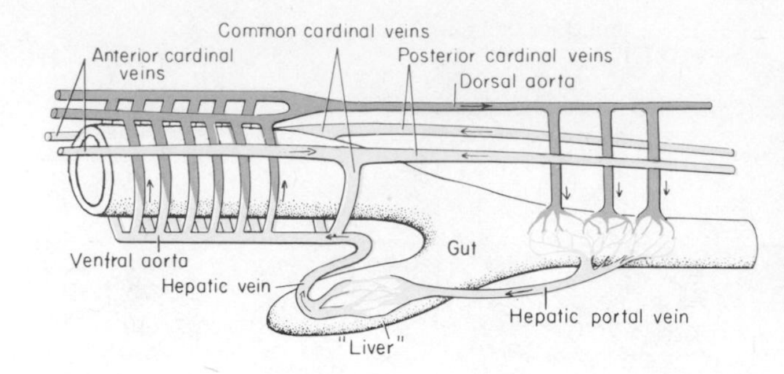

Figure 326. Diagram of the major circulatory vessels of amphioxus. Except for the absence of a heart at the posterior end of the ventral aorta, the system is closely comparable to that of vertebrates. Only a few of the numerous aortic arches are shown. The vessels are colored as if oxygen intake were entirely through gill walls; in fact much of it may be through the skin though most does occur at the gills.

supplies each gill (i.e., each gill bar); but in lampreys the arrangement is such that each arch supplies both front and back portions of a gill pouch, and in sharks the efferents are similarly placed opposite each gill slit. The manner in which an arch "breaks up" into capillary systems results in differences in the arrangement of afferent and efferent vessels; variations in different fish groups in this regard are shown in Figure 331. Other variations of a confusing sort lie in the presence of small efferent vessels (considered on p. 320) which may supply the lower jaw; still others lead to the throat and heart.

Most striking is the fact that in jawed fishes, in which six arches always develop in the embryo, certain of them are invariably lost or modified. This is the case with the mandibular arch. In living gnathostomes it frequently appears prominently in the embryo but never persists, except for its dorsal portion, which may aid in the blood supply to the head (Fig. 327, spiracular artery; cf. Fig. 330 *B*). The second aortic arch—the hyoidean—is well developed in the Chondrichthyes, but this too is lost in most ray-finned fishes and the lungfish *Epiceratodus*. Still another variant is that in the African lungfish *Protopterus*, which depends largely on lungs for breathing purposes, where arches III and IV run without break past the gill region. Similar reduction in gills occurs in a few teleosts.

AORTIC ARCHES IN AMPHIBIANS (Figs. 328, 330 *E*, 332 *A*). In the amphibian stage further changes occur in the arches, but in great measure the adult structure is not too far from that seen in the more advanced fishes—barring the fact that the vessels, with loss of functional gills, are continuous tubes. In the frog larva, gill capillaries are developed only to disappear later, but in urodeles, in which internal gills never develop, the arches remain as uninterrupted structures throughout all

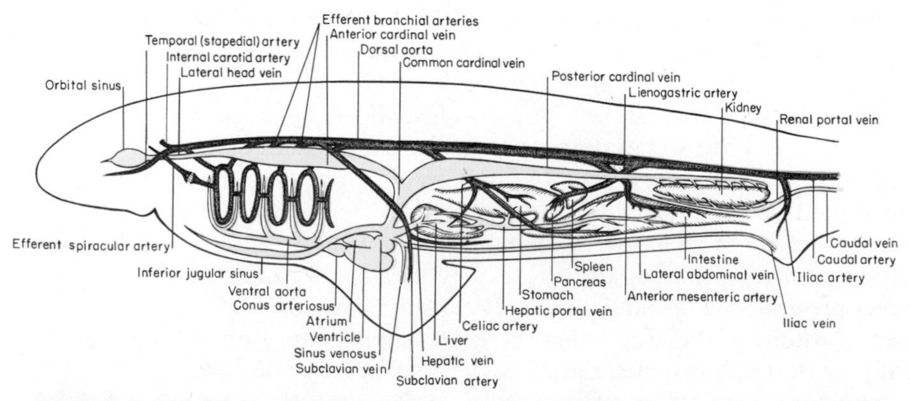

Figure 327. Diagram of the main blood vessels of a shark as seen in lateral view. The renal portal vein here (and in Fig. 328) should appear dorsal to the kidney; it is drawn ventrally to show the pattern more clearly.

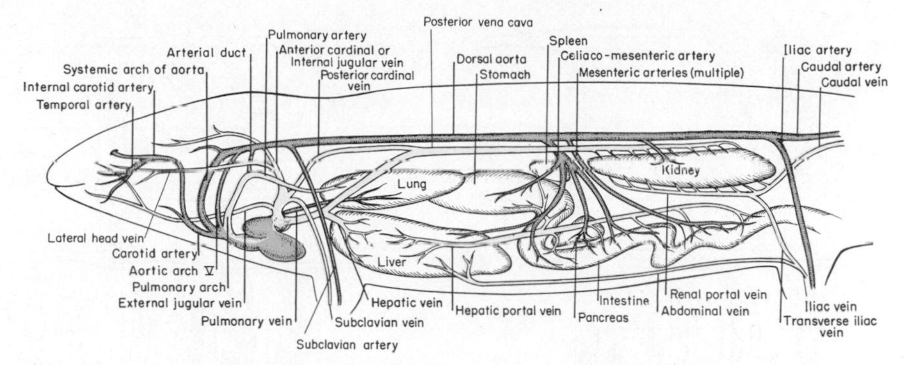

Figure 328. Diagram of the main blood vessels of a urodele amphibian as seen in lateral view.

stages. As in the more progressive fishes the first and second arches disappear during embryonic life. In many urodeles all the remaining four arches may persist, but arch V is absent in some urodeles and all anurans, leaving III, IV, and VI as the persistent members of the series.

Changes, however, occur in the dorsal connections of the arches. Even in fishes, blood flowing upward in arch III tends to pass forward toward the head rather than back toward the body; in tetrapods this arch and the continuation of the dorsal aorta forward from it become the **internal carotid artery.** A vessel running forward ventrally toward the tongue region from the back of the third arch develops as a **lingual artery,** and the end of the ventral aorta leading to both vessels is the **common carotid.** Back of the carotid, the segment of the dorsal aorta connecting it with the succeeding arches disappears in some amphibians and is absent in the great majority of amniotes; where it persists it is termed the **carotid duct.**

Arch IV is always a large paired vessel in lower tetrapods; it is termed the **systemic arch,** since it is the main channel for blood flowing from heart to body. Why this arch was selected as the channel, rather than the shorter route via arch V, is an unsolved puzzle.

The lung in tetrapods (and lungfishes) receives its blood from a **pulmonary artery** leading back from arch VI.* During the larval life of an amphibian, when the lung does not function, most of the blood in this arch travels straight up into the dorsal aorta as part of the main blood stream to the body. When air-breathing

*In modern amphibians, the skin has become a major breathing organ, thus reducing somewhat the importance of the lung supply. In frogs there is present a special branch from the pulmonary arch to the skin.

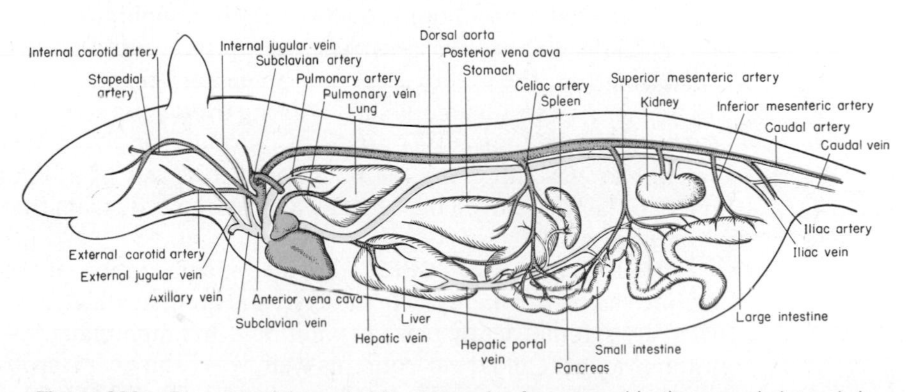

Figure 329. Diagram of the main blood vessels of a mammal (rat) as seen in lateral view.

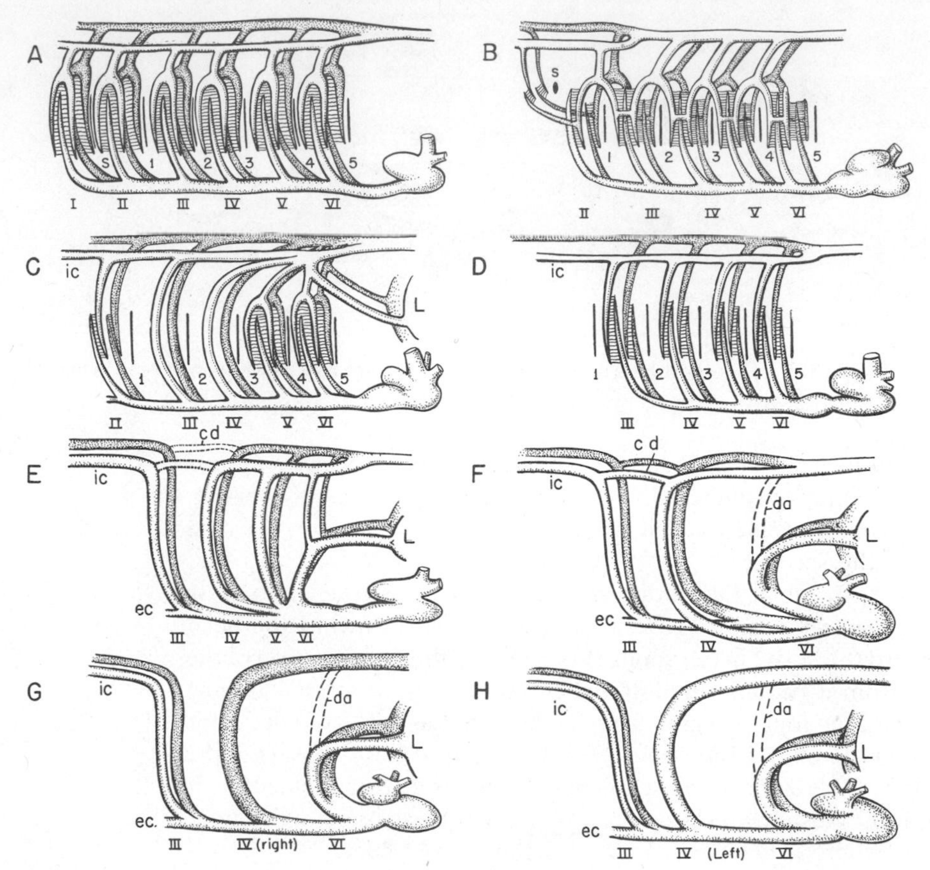

Figure 330. Diagrams of the aortic arches and derived vessels in various vertebrate types. *A,* Theoretical ancestor of the jawed vertebrates with six unspecialized aortic arches; *B,* typical fish condition as seen in a shark; *C,* the lungfish *Protopterus; D,* a teleost; *E,* a terrestrial salamander; *F,* a lizard; *G,* a bird; *H,* a mammal. Various accessory vessels are omitted. The vessels of the right side are heavily shaded. In terrestrial forms the position of the vessels is made (for purposes of the diagram) to correspond more or less to that of the arches from which they are derived. Aortic arches in Roman numerals; *s,* spiracular slit; following gill slits in Arabic numerals. *cd,* Carotid duct; *da,* embryonic ductus arteriosus; *ec,* external carotid artery; *ic,* internal carotid artery; *L,* lung. The carotid duct shown in the lizard is absent in other reptiles; in turtles the carotids arise by a separate stem directly from the heart. In *F-H* the embryonic arterial duct by-passing the lungs is shown in broken lines.

begins, much of the blood in this arch is diverted to the lung. The dorsal part of the arch, now unimportant, disappears in frogs and most amniotes; it persists, however, in reduced form among urodeles, the Gymnophiona, and a few reptiles, where it is called the **ductus arteriosus.** The base of the ventral aorta tends to divide, with separation from the main vessel of a **pulmonary trunk.**

In sum, we see in amphibians a strong trend toward separation of the old arch system into three parts: (1) a pair of carotids supplying the head, (2) a pair of systemic arches supplying the body, and (3) a pair of pulmonary arteries to the lungs; the last arise from the heart by a trunk separate from that leading to the other two groups.

AMNIOTE AORTIC ARCHES. Further specialization of the arch system in the amniotes has to do mainly with the systemic arch, in which asymmetry replaces the originally symmetrically paired conditions. In amphibians, as we have seen, both systemic arches, and the carotids as well, leave the heart through a common trunk. In living reptiles there are, instead, two trunks (Fig. 332 *C, D*). One—the smaller of

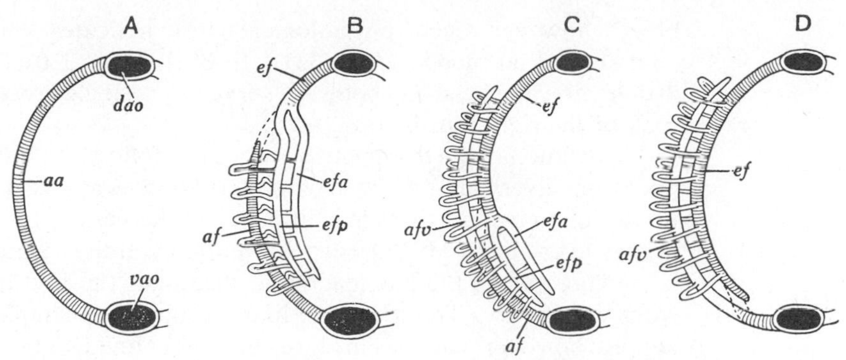

Figure 331. Diagram of circulation in a fish gill, left side, from behind. *A,* Embryonic condition, with continuous aortic arch, *aa,* from ventral aorta, *vao,* to dorsal aorta, *dao. B,* Shark condition; the afferent gill vessel, *af,* is formed from the aortic arch; the paired efferent vessels, *efa, efp,* are new formations (later embryonic changes in sharks make the efferent arteries appear to alternate with the afferents—as in Figure 330 *B*—but this is a secondary modification). *C,* Condition transitional to *D,* seen in sturgeons. *D,* Teleost condition. The embryonic arch gives rise to the efferent vessel, *ef;* the afferent vessel, *afv,* is a new formation. (After Sewertzoff, Goodrich.)

the two—leads only to the left systemic arch. The other and much larger vessel supplies both carotids and a right systemic arch which is larger than its mate. The base of the two openings from the heart are so situated that the larger opening appears to receive "fresh" blood from the lungs, and the left arch mainly venous

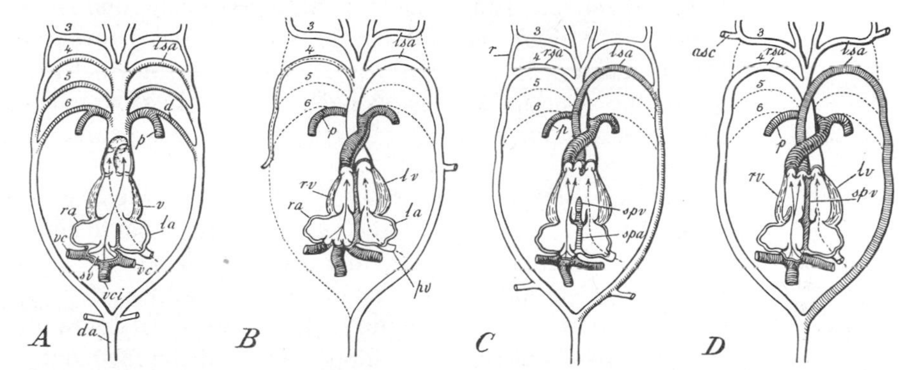

Figure 332. Diagram of the heart and aortic arches in tetrapods. *A,* Amphibian; *B,* a mammal; *C,* typical modern reptiles; *D,* crocodilian. Ventral views; the heart (sectioned) is represented as if the chambers were arranged in the same plane; the dorsal ends of the arches are arbitrarily placed at either side. Solid arrows represent the main stream of venous blood; arrows with broken line the blood coming from the lung. Vessels apparently carrying aerated blood are unshaded; those which appear to contain venous blood, hatched (physiological work, as noted in the text, shows the situation is not this simple). The two vessels at the top of each figure are the internal carotid (laterally) and external carotid (medially). In amphibians without a ventricular septum the two blood streams are somewhat mixed; subdivision of the arterial cone tends to bring about partial separation, but some venous blood is returned to the dorsal aorta. In mammals ventricular separation is complete, the arterial cone subdivided into two vessels, and the arches are reduced to the left systemic and pulmonary. The mammalian condition has apparently arisen directly from the primitive type preserved in the Amphibia, for in modern reptiles the conus arteriosus shows a division into three vessels, rather than two; one, returning venous blood back to the body, leads only to the left fourth arch. In crocodilians the ventricular septum is nearly complete, and the elimination of the left fourth arch would give the avian condition. *asc,* Anterior subclavian; *d,* ductus Botalli; *da,* dorsal aorta; *la,* left atrium; *lsa,* left systemic arch; *lv,* left ventricle; *p,* pulmonary artery; *pv,* pulmonary vein; *r,* portion of lateral aorta remaining open in some reptiles; *ra,* right atrium; *rsa,* right systemic arch; *rv,* right ventricle; *spa,* interatrial septum; *spv,* interventricular septum; *sv,* sinus venosus; *v,* ventricle; *vc,* anterior vena cava; *vci,* posterior vena cava. (From Goodrich.)

blood; however recent physiological work indicates that, in fact, both usually contain arterial blood (see p. 331). In birds (Fig. 330 *G*) the left systemic arch disappears, and head and body are served by a single great aorta which follows the path of the right fourth aortic arch.

Evolution of the mammalian arches has followed a different path. Mammalian ancestors diverged from those of modern reptiles at a very early stage, and there is no reason to believe that the split between the vessels leading to the two systemic arches ever occurred. But even so, a double arch is unnecessary and inefficient; somewhere along the line leading to mammals the fourth right arch disappeared from the picture. The mammal, like the bird, has simplified the systemic blood supply; however, whereas the birds have held to the right arch, in mammals it is the left member of the pair that has become the great arch of the aorta (Figs. 330 *H*, 332 *B*).

In mammals both carotids and both arteries to the pectoral limbs (the subclavians) are supplied with blood from this same great trunk. There are, however, great variations in their mode of branching, some of which are shown in Figure 333.

The embryonic development of the aortic arches of a mammal recapitulates to a considerable degree the phylogenetic story outlined above (Fig. 334). The first circulatory channel from heart to body is that of arch I. Gill pouches develop in back of this arch, and successive aortic loops—II, III, and IV—are formed between them; arch V appears as a transitory structure in some instances, and finally arch VI develops. As the more posterior arches form, the more anterior ones become reduced in importance; I and II disappear, and III becomes distinct as the carotid. The blood in arch VI in the embryo passes upward to the dorsal aorta and little of it enters the pulmonary artery until birth, at which time the upper part of this arch—the arterial duct—is occluded. Meanwhile the pulmonary trunk has separated from the aorta, the right fourth arch has disappeared, and the carotid duct has closed.

BLOOD SUPPLY TO THE HEAD (Fig. 335). In fishes the cranial region is supplied with arterial blood by the paired vessels which form the anterior end of the dorsal aorta, running forward along either side of the head; as we have seen, these become the internal carotid artery. The main trunk of this artery passes upward into the braincase in front of the pituitary. Before doing this, however, it gives off a major branch, an **orbital artery,** to supply much of the face and jaw region. This same situation holds true in most groups of tetrapods, including some mammals, where the orbital artery is usually called the **stapedial artery** because it often passes through an opening in this auditory ossicle.

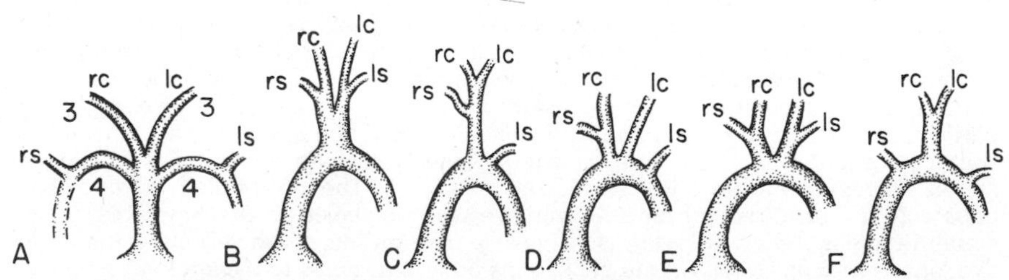

Figure 333. Diagrams, in ventral view, to show variations in the branching of the main blood vessels from the mammalian aortic arch. *A*, Embryonic condition, with ventral trunk of aorta and third (carotid) and fourth pairs of arches, of which the right fourth arch is later lost beyond the point of departure of the subclavian. By differential growth of the vessels the various arrangements shown in *B* to *F* are brought about (*D* is the human type). *lc*, Left carotid; *ls*, left subclavian; *rc*, right carotid; *rs*, right subclavian arteries. (After Hafferl, in part.)

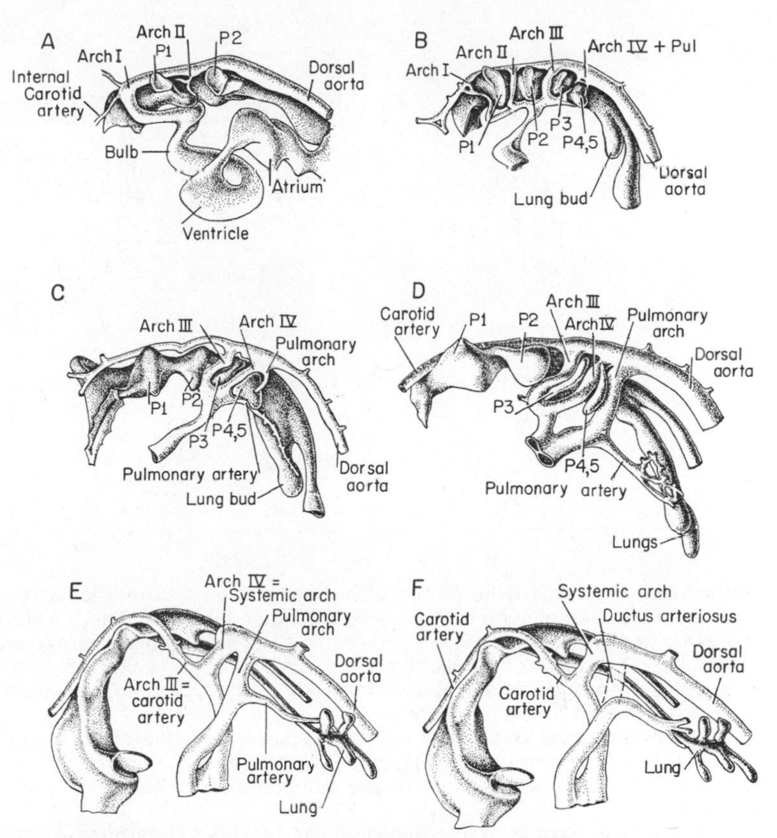

Figure 334. Development of the aortic arches of a mammal *(Homo).* The outline of the gut cavity, gill pouches, and lung buds are shown in addition to blood vessels; in *A,* the cavities within the heart are included. Arch I is developed. *B,* Arch I already reduced, II and III formed. *C,* Arch II reduced, IV (systemic) formed, VI (pulmonary) arch and pulmonary artery forming (Arch V does not develop in man). *D,* Pulmonary arch well developed. *E,* Carotid arch (III) separated dorsally from aorta, pulmonary arch becoming distinct at root from ventral aorta. *F,* Diagram to show reduction at birth of upper end of arch VI (ductus arteriosus). P^1 to P^5 = pharyngeal pouches. (After Streeter.)

We have, however, noted that in early tetrapods a small artery was present to supply blood to the tongue. In mammals generally (and to a lesser extent in many other forms) this becomes enlarged and elongated as the **external carotid artery.** It takes over the supply to the lower jaws, frequently in mammals that to the upper jaws, and in many forms (as in man) even supplies blood to the face, so that the stapedial artery is lost. The process is analogous to that of "stream piracy," whereby one river system taps the headwaters of another.

BLOOD SUPPLY TO THE BODY AND LIMBS (Figs. 327, 329, 336). In every vertebrate the major blood supply to the trunk, tail, and limbs is furnished by the dorsal aorta. Although paired anteriorly, the aorta in the trunk is a single median vessel, lying beneath the backbone and above the root of the mesentery. Three types of branches develop: (1) median ventral ones running down the mesenteries to the gut and its derivatives (liver and pancreas); (2) paired ventrolateral branches to the urogenital system; and (3) paired lateral branches to the muscles, skin, and other tissues of the body walls. The ventral—"splanchnic"— vessels are numerous in the embryo, but in adults are usually concentrated into a few main trunks, including a **celiac artery** to the stomach and liver and generally two **mesenteric arteries** to the

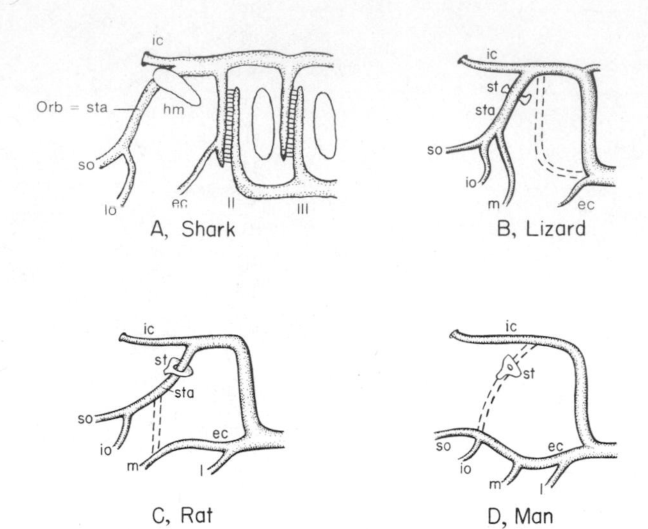

Figure 335. Diagrams of the left side of the head to show the evolution of the carotid system. To make homologies clear, all are shown as if no neck ever developed and the fish pattern were retained. In a relatively primitive fish stage *(A)* the direct forward continuation of the aorta is the internal carotid artery, which enters the braincase near the pituitary. This gives off a major branch, the orbital artery, which, passing close to the hyomandibular, supplies most of the more superficial features of the skull and upper jaw. An accessory blood supply to the head from the spiracle is omitted from the diagram.

In many tetrapods *(B)* a similar situation persists, the orbital artery being commonly called the stapedial, since it passes close to or through the stapes (= hyomandibular). However, the small external carotid present near the root of the carotid extends forward, and in mammals may take over part *(C)* or all *(D)* of the functions of the stapedial.

ec, External carotid; *hm,* hyomandibular; *ic,* internal carotid; *io,* infraorbital artery; *l,* lingual artery; *m,* mandibular artery; *orb,* orbital artery; *so,* supraorbital artery; *st,* stapes; *sta,* stapedial artery; II, III, second and third aortic arches. In *B* to *D* modified aortic root = common carotid.

intestine. Short lateral branches—the "visceral" branches—from the aorta reach the kidneys and gonads. Longer branches to the "outer tube" of the body were primitively segmental—actually intersegmental—and remain essentially in that

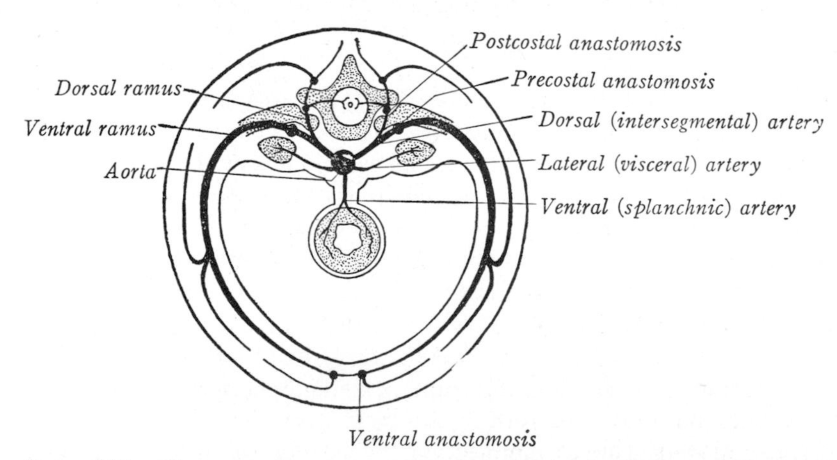

Figure 336. Diagrammatic cross section of the body of a higher vertebrate to show various types of branches which may be given off by the aorta. Most prominent are median ventral branches descending in the mesentery to the gut and associated structures and paired intersegmental arteries, the main ventral ramus of which descends the flanks between the myomeres or adjacent to successive ribs. Longitudinal anastomoses may occur between successive segments in various regions. (From Arey.)

condition in many lower vertebrates. In more advanced groups, however, we find that longitudinal connecting channels may form anastomoses at several points, both dorsally and ventrally. As a result there is a trend for reduction in the number of arteries coming from the aorta, leaving a relatively few large arteries, each of which may serve a considerable area of the back and flank.

In embryonic development a paired fin or leg is supplied by a network of small arteries (Fig. 351). During ontogeny one or another of these tends to become dominant and form a main channel from aorta to limb. In the pectoral appendage the main trunk is usually given its mammalian name of **subclavian artery** as it enters the limb; various other terms (as axillary and brachial) are applied to this same main vessel as it progresses distally. In the pelvic appendage the primitive trunk was the **ischiadic artery,** which emerged back of the pelvic girdle; in mammals, however, the major stem is the **iliac artery,** running out to the limb in front of the ilium (it is termed femoral, popliteal, and peroneal farther distally).

VENOUS SYSTEM

The veins—vessels bringing blood from capillary systems to the heart—have a complicated and variable arrangement. If, however, their embryonic history is studied, it is seen that they can be logically sorted out into a small number of systems. From this point of view we can distinguish (Fig. 337):

1. A **subintestinal system** flows forward beneath the gut in the embryo, and in the adult divides into the **hepatic portal system** running to the liver and the **hepatic veins** from the liver toward the heart.
2. Veins dorsal to the celom or gut carry blood toward the heart from the dorsal part of the body and the head (and generally from the paired limbs as well); they include the **cardinals** or the **venae cavae** which replace them, and their affluents.
3. A relatively minor group, the **abdominal vein** or veins, drain the ventral part of the body wall in most classes.
4. In lung-bearing forms, the **pulmonary veins** extend from lung to heart.

The first and fourth of these four components form the drainage of the gut and its outgrowths; they are essentially visceral venous systems. The second and third components are, on the contrary, mainly somatic venous elements, draining the outer wall of the body although the third becomes partly visceral in tetrapods.

HEPATIC PORTAL SYSTEM AND HEPATIC VEINS (Figs. 327, 329). The hepatic portal system, common to all vertebrates (and even amphioxus—see Fig. 326), is composed of veins which collect blood from the intestine and transport it to the sinusoids of the liver. It is functionally of the highest importance, since its presence guarantees that the liver had "first chance," for storage or transformation, at food materials absorbed by the intestinal capillaries.

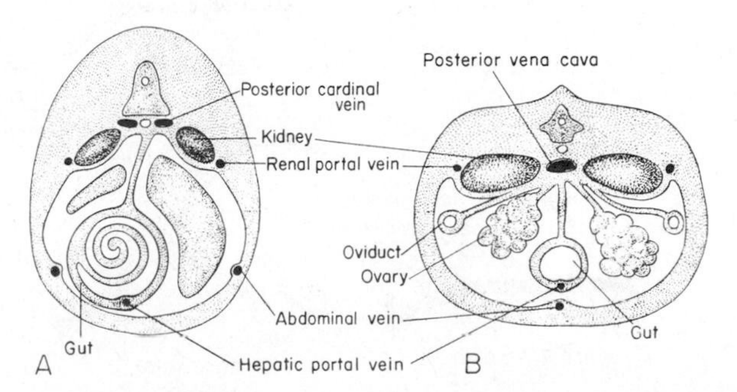

Figure 337. *A,* Cross section of the abdominal region of a shark, to show the position of the main veins. The renal portal vein is usually dorsal rather than lateral to the kidney. *B,* The same in a urodele amphibian.

Beyond the liver sinusoids the blood from the intestine is re-collected into a **hepatic vein** or veins. In most fishes this is a large vessel which empties directly into the heart. In sarcopterygian fishes and tetrapods, however, we find that, as discussed later (p. 326), part of the hepatic vein has been incorporated in the posterior vena cava. In consequence, the term "hepatic vein" is restricted in these forms to the vessel or its subdivisions which empty from the liver into the terminal part of the vena cava.

Generally the first blood vessels to appear in the embryo in forms with a mesolecithal type of egg are a pair of veins which form in the floor of the gut and coalesce into a single channel running forward ventrally as a **subintestinal vein** (Fig. 352). From the far anterior end of this trunk develop the heart and the ventral aorta—structures with which we are not concerned at this point. The remainder of this vessel gives rise to the hepatic and hepatic portal veins. For a time this runs without break from intestine to heart. Presently, however, the liver grows out ventrally from the gut. With its growth, liver and venous materials become intermingled; the vein breaks up into small vessels and finally into a hepatic sinusoid system, with the resulting formation of a separate portal trunk posteriorly and a hepatic vein anteriorly (Fig. 354). In large-yolked types a basically similar process of vein formation occurs, except that, as described in a later section, vitelline veins replace the subintestinal vessel in gathering food material from the yolk sac. In typical fishes no further important development occurs; in the embryo of lungfishes and of higher vertebrates a branch of the hepatic vein reaches dorsally along the mesenteries to tap the posterior cardinal system and form the anterior part of the posterior vena cava (p. 326). The hepatic portal vein remains a large and important vessel collecting blood not only from the intestine, but from the stomach, pancreas, and spleen as well, for conduction forward to the liver.

DORSAL VEINS—CARDINALS AND VENAE CAVAE. The principal blood drainage from the "outer tube" of the body is cared for by important longitudinal vessels situated dorsally above the gut and mesenteries. In lower vertebrates these veins are the cardinals; in higher forms major modifications of these vessels produce the venae cavae.

In the embryo of every vertebrate (and of amphioxus as well) paired veins appear at an early stage in the tissues above the celomic cavity, one on either side of the midline (Fig. 353). These are the primitive **cardinal veins.** The **posterior cardi-**

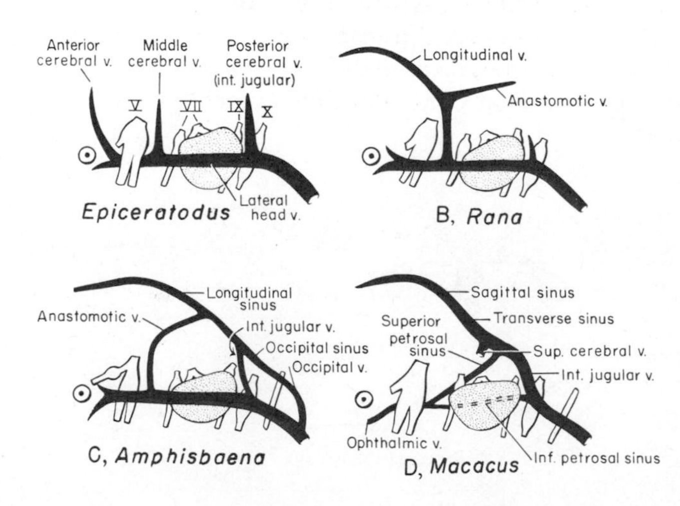

Figure 338. Diagrams of the left side of the head region to show stages in the evolution of the venous drainage. Roots of some cranial nerves are indicated (Roman numerals); the position of the eye is shown, and the otic capsule is stippled. In lower vertebrates the main drainage is by a lateral head vein, which forms in the orbital region and runs backward to become the anterior cardinal. This primitively receives several successive veins from the interior of the skull. A series of sinuses develops within the braincase; the lateral head vein is abandoned in mammals, and blood from the orbital region enters the sinus system, all finally draining from the skull as the internal jugular vein. *A*, Lungfish; *B*, frog; *C*, lizard; *D*, mammal (macaque monkey). (After van Gelderen.)

nals run forward along the trunk on either side of the dorsal aorta to the heart. Paired **anterior cardinals** begin as head veins on either side of the developing brain and run back dorsally above the gills or along the neck to meet their posterior mates. From their point of junction on either side a major vessel descends to enter the sinus venosus of the heart; this is the **common cardinal** (or duct of Cuvier). This characteristic cardinal system is retained in the adult of most fishes (Fig. 327). We may discuss separately the history of its anterior and posterior parts.

In all vertebrates except mammals the main stem of each anterior cardinal begins as a **lateral head vein** (Fig. 338 *A–C*) arising behind the orbit, receiving tributaries from face and brain; traveling back past the region of the ear to the level of the occiput, they are then termed anterior cardinals, and continue in typical fishes back to the common cardinals, receiving the veins from the pectoral appendages on the way. In lungfishes and tetrapods the posterior cardinals (as we shall see) are so modified and reduced that the common cardinals are merely continuations of the anterior trunks. With this modification the anterior cardinals come to resemble the vessels termed anterior venae cavae in mammals and are frequently called by this name.

In mammals (and, to a certain extent, crocodilians and birds) there is an important change in venous circulation in the head (Fig. 338 *D*). A sinus system is established within the expanded cranial cavity. Blood from much of the front part of the head enters the braincase to emerge posteriorly as the **internal jugular veins** and then, joining more superficial vessels, the **external jugulars,** to form **common jugulars;** the old lateral head vein is lost. After a junction with the vein from the front limb the vessel is termed in mammals an **anterior vena cava**. Despite the change in name, however, it is easily recognized that common jugular and anterior vena cava are the anterior cardinals of lower vertebrates. In many mammals (including man) there is a further change (Fig. 339) by which the blood from the left jugular (or vena cava) is shunted across to the right side, so that only a single vessel enters the heart. A comparable development is seen in birds.

The story of the **posterior cardinals** is more complex (Figs. 340, 341). It begins with a pair of simple dorsal vessels draining forward into the heart via the common cardinals; it ends in mammals with the draining of the same region by a single but complex vessel, the posterior vena cava. In between lies a considerable history.

In cyclostomes the posterior cardinals are simple paired vessels, receiving blood from the tail, kidneys, gonads, and dorsal parts of the body musculature and running forward uninterruptedly to the common cardinal veins. In the jawed fishes,

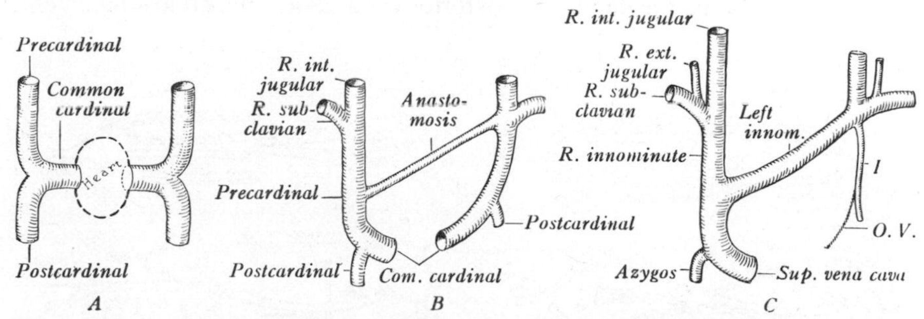

Figure 339. Ventral views of the veins anterior to the heart in successive developmental stages to show the formation, in man and certain other mammals, of a single anterior (or superior) vena cava from the two anterior cardinals (precardinals). An intercostal vein *(I)* and a small vein from the wall of the left atrium (*O.V.*, oblique vein) are persisting vestiges of the original left anterior cardinal. (From Arey.)

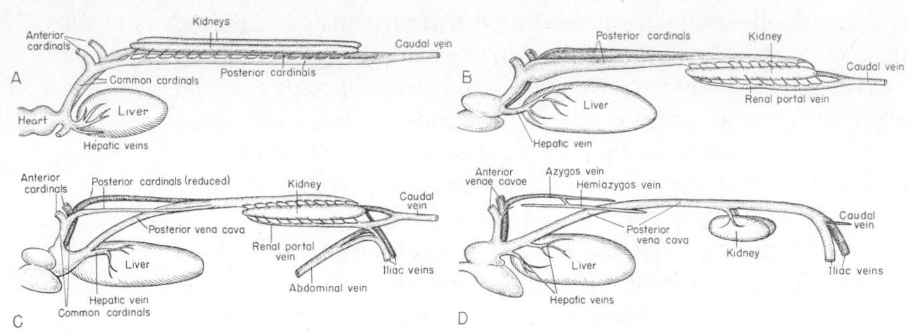

Figure 340. Diagrams in lateral view to show the evolution of the posterior cardinals and the development of a posterior vena cava. *A*, Lamprey (larva); *B*, typical fish condition; interjection of renal portal system. *C*, Lungfish or primitive tetrapod; a shortened route to the heart is established by utilizing part of the hepatic vein system in the initiation of a posterior vena cava. *D*, Mammal; the renal portal eliminated and the posterior cardinals separated from the posterior vena cava. Vessels of the right side are shown in deeper shading.

however, there develops a **renal portal system,** destined to persist upward into the avian stage. Blood from the posterior part of the trunk and the tail does not now go directly forward through the original cardinal channels, but is instead shunted through a network of capillaries around the kidney tubules (*not* through the glomeruli), whence it resumes its forward course, through the cardinals.

In the Sarcopterygii, as exemplified by the lungfishes, a second progressive change initiates the development of a **posterior vena cava.** A branch of the hepatic veins from the liver here extends up past that organ in a mesenteric fold to the dorsal wall of the body cavity and taps the right posterior cardinal. Once this connection is made, blood from this cardinal may follow this new circuit to the heart; further, since there are cross-connections between the two cardinals, the blood originally following the left cardinal anteriorly may likewise take this course. The old channels persist in lungfishes and urodeles, but in frogs and all higher forms are abandoned (leaving only variable stumps termed the **azygos veins**). The new major trunk from kidney region to heart may be properly termed the posterior vena cava.

In tetrapods the renal portal system shows some evidence of degeneration even in amphibians, more in reptiles; in birds it is in great measure abandoned and in mammals completely so. With this abandonment there occurs a third major stage in the development of the vena cava. The blood now passes directly forward from the back of the body past the kidneys, the whole channel along the trunk to the heart being the definitive posterior vena cava. This great single vein in adult mammals has

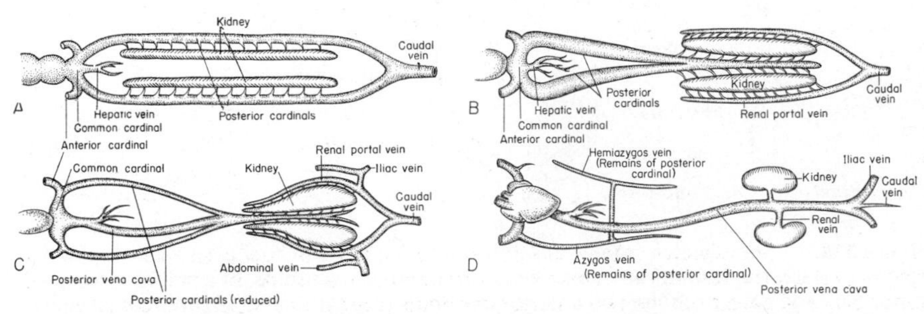

Figure 341. Diagrams in ventral view to show the evolution of the posterior cardinals and the development of a posterior vena cava. Stages as in Figure 340.

a seemingly simple structure. This is, however, deceptive. The phylogenetic story shows, as we have seen, that it is a composite patchwork of the old posterior cardinals and vessels replacing them and, anteriorly, of an enlarged part of the hepatic vein system. The embryonic development of the mammalian circulation recapitulates this evolutionary story in an elaborate (if somewhat variable) fashion, as shown in Figure 342. The complexities in embryology not reflecting phylogenetic events probably are required by the necessity of the system remaining functional throughout development.

ABDOMINAL VEINS. In the sharklike fishes paired **abdominal veins** run forward along the belly on either side, draining the flank musculature and paired appendages (Figs. 327 *A*, 337). They are absent in actinopterygians. In lungfishes we find instead of paired vessels a median abdominal element. This persists in amphibians and reptiles, but instead of entering the heart directly it joins the hepatic portal blood in the liver sinuses (Figs. 328, 337 *B*). The area drained by abdominal veins is sometimes extended to include the pelvic appendages and tail. The vein is absent in adult birds and mammals, but, as noted later, it is of interest in the embryo, where it is represented in the important umbilical veins (p. 337).

LIMB VEINS. In tetrapods pectoral and pelvic limbs are drained by large vessels termed the **subclavian** and **iliac veins.** The former enters the anterior cardinals or their replacements, the jugulars or anterior vena cava; the latter rather variably connects in lower tetrapods with the abdominal vein or the renal portal system or both; in mammals the two iliac veins are the main elements which join to form the posterior vena cava. Much smaller veins, with variable connections, drain the paired fins of fishes.

PULMONARY VEINS. These veins are, of course, absent in most living fishes, in which lungs are nonexistent. In actinopterygians veins from the lungs (*Polypterus*) or swim bladder enter the hepatic vein. In the lungfishes, however, the pulmonary trunk by-passes the sinus venosus and enters the atrial cavity of the heart directly. This separate course of the aerated blood from the lungs to the heart persists in all tetrapods.

LYMPHATICS

In most, if not all, vertebrates, we find, supplementing the venous system, a second series of vessels returning fluids from the tissues to the heart—the **lymphatic system.** Although paralleling the veins in many functions (and often paralleling them topographically) the lymphatics differ from them in major respects. A fundamental difference is that the lymphatics are not connected in any way with the arteries; they arise from capillaries, but these are blind at their tips (Fig. 325). There is thus no arterial pressure behind the fluid in the lymphatic vessels, and the flow of materials in them is generally sluggish. This contained liquid, the **lymph,** diffuses into the lymphatic vessels from the general tissue fluid and hence is generally similar in composition to this fluid and (apart from the absence of blood proteins) to the blood plasma. Except for white corpuscles which may enter by ameboid motion, there is, of course, no inflow of blood corpuscles. Abundant lymphocytes are found in the lymph nodes, and hence vessels, of mammals (p. 310), but these nodes, we have noted, are almost completely absent in other vertebrate groups. Related to the low pressure under which lymph travels, lymphatics are very thin walled, and even the largest of them are difficult to find and dissect unless specially injected. Cyclostomes and sharklike fishes possess a series of thin-walled sinusoids possibly repre-

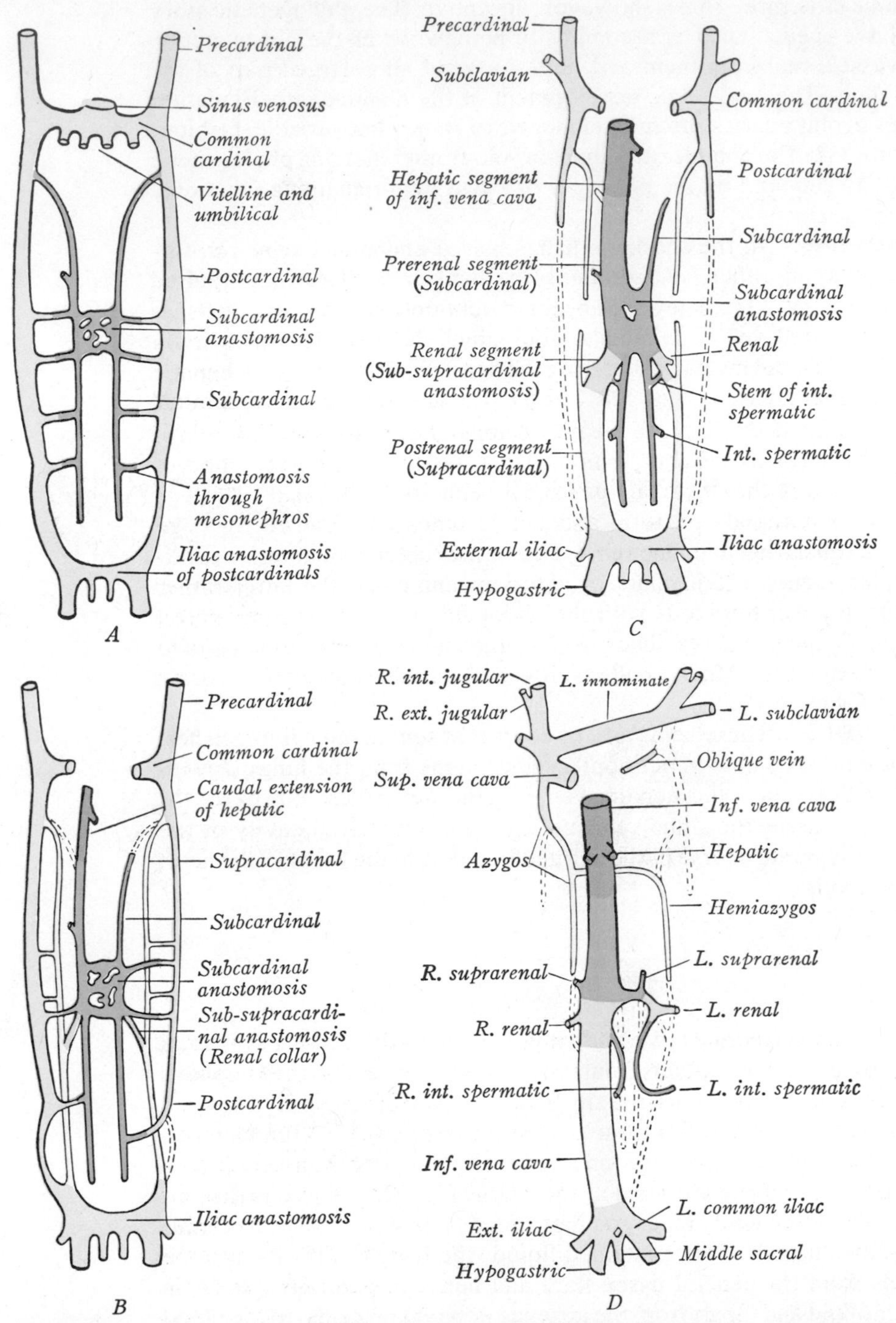

Figure 342. Development of the posterior vena cava of man, showing the embryological processes roughly paralleling the evolutionary history. The original cardinal system is shown in blue. Presently there develop subcardinal vessels (red), corresponding to the veins draining the kidney when a portal system is established in sharks. In *B*, this venous system is tapped (as in lungfishes) by a branch of the hepatic veins (purple). A third system of embryonic veins, the supracardinals (yellow) is not exactly paralleled in phylogeny. In green are vessels which develop in amniotes to bypass the kidney and eliminate the renal portal system. As seen in *D*, the definitive posterior vena cava includes fractions of all these structures. (From Arey, after McClure and Butler.)

senting a primitive lymphatic system. Typical lymphatics are present in bony fishes and highly developed in tetrapods—most notably in amphibians (Fig. 343 *A*). In this class lymph circulation is aided by the development of pulsating **lymph hearts**—small, two-chambered structures usually lying at points where lymph vessels enter the venous trunks. The major development of lymphatics in tetrapods may be due to the fact that blood pressure in the body capillaries is higher than in fishes; the lymphatics offer a relatively low pressure system of drainage of the tissues.

Lymphatics are prominently developed in the intestine, whence they carry (via the mesenteries) much of the absorbed fats in a milky liquid, the **chyle,** and in amphibians are particularly abundant in the subcutaneous tissues. Lymph vessels are absent from the central nervous system, liver, most skeletal tissues, and bone marrow. The arrangement of major lymph vessels varies greatly from group to group (Figs. 343, 344). They usually terminate at entrances into the cardinals or

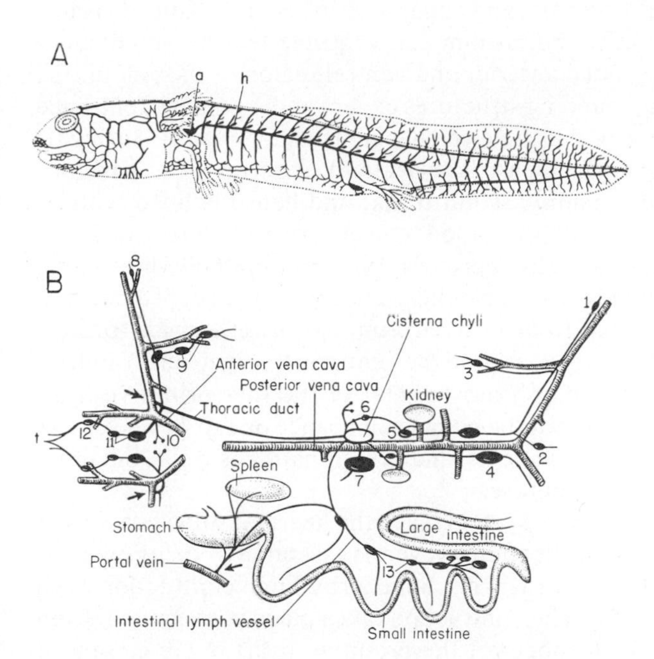

Figure 343. *A,* Side view of a salamander, showing superficial lymph vessels. Dorsal, lateral, and ventral longitudinal vessels are present; a series of lymph hearts *(h)* is present along the lateral vessel. Lymph from this vessel enters the venous circulation through an axillary sac *(a);* lymph from the ventral vessel enters through an inguinal sac. *B,* Diagram of the deep vessels of the lymphatic system of the rat, anterior end at left. Lymphatics in solid black; the neighboring veins are also shown. Nodes are numbered according to the region in which they lie; *1,* knee; *2,* tail; *3,* inguinal; *4,* lumbar; *5,* kidney; *6,* nodes about the cisterna chyli; *7,* intestinal node; *8,* elbow; *9,* axilla; *10,* thoracic; *11,* cervical; *12,* submaxillary; *13,* mesenteric nodes; *t,* plexus of lymphatics around tongue and lips. Arrows indicate the point of entrance of lymph into the veins near the junction of jugular and subclavian and into the portal vein. (*A* after Hoyer and Udziela; *B* after Job.)

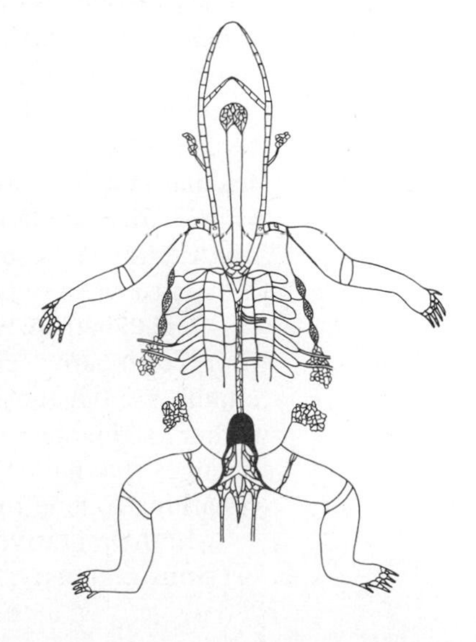

Figure 344. Diagram of the major lymphatic trunks in a crocodilian. This represents, despite its complexity, an extreme simplification. In this (and other reptiles) networks of lymphatics surround many blood vessels (the net shown in the midline is around the dorsal aorta), surround or lie in various organs (that in the throat is around the larynx), and underlie the skin (shown in the posterolateral part of the trunk). The lacteals and many other important parts of the system are omitted. (After Ottaviani and Tazzi.)

venae cavae; in higher tetrapods most or all of the lymph drains into the anterior venae cavae close to the heart (a point of lowest pressure in the circulatory system). The intestinal lymphatics usually run upward into a longitudinal **thoracic duct,** primitively paired but in mammals frequently reduced to a single vessel.

THE HEART

Some type of muscular pump is necessary for efficient circulation of the blood. Amphioxus has a whole series of tiny "heartlets," but in true vertebrates (aside from hagfish) the heart is a single structure, situated ventrally and well anteriorly in the trunk. It sucks in venous blood posteriorly from all regions of the body and pumps it anteriorly, in lower vertebrates, to the aortic arch system and the gill circulation. Primitively it consists of four successive chambers, termed, from back to front, sinus venosus, atrium, ventricle, and conus arteriosus; in advanced groups the first and last lose their identity, but atrium and ventricle tend to subdivide.

The heart is situated in a special anterior and ventral region of the celom, the pericardial cavity, free from surrounding structures except at the points of entrance and exit of vessels (Fig. 347); it is thus able to change its shape readily during its powerful pumping movements. The heart is essentially a series of expansions developed along the course of a main vascular trunk, and hence is lined with an endothelium continuous with that of the blood vessels, surrounding which are muscle and connective tissue; externally there is a thin outer epithelium as in the case of any other organ lying in the celomic cavities. The musculature, of a peculiar striated type, which we have noted to be derived from tissues otherwise forming smooth muscle fibers, is especially thick around the ventricle (or ventricles) and, on the other hand, is thin about the sinus venosus, first of the four primitive heart chambers. Between the chambers and at the points of entrance or exit of vessels are heart **valves,** basically similar to those in veins (and in lymphatics as well), but more powerful and usually of complex structure.

Fibers from the autonomic nervous system reach the heart (at sinus venosus or atrium) and may affect its rhythm; the heart, however, is essentially "on its own" as is shown by the fact that heart muscles will continue a rhythmic contraction even when cultured apart from the body. The contraction takes place in sequence, from back to front, through the four chambers of the primitive heart or, in advanced types, in the atria followed by the ventricles. The contraction of the musculature of one chamber (primitively the sinus venosus) induces stimulation, successively, of the muscle fibers of the remaining chambers of the heart. In amniotes—notably in birds and mammals—there develops a unique conducting mechanism, the **sinoventricular system** of specialized muscle fibers which simulates a local nervous system (Fig. 345). Stimulation of a **sinus node** in the right member of the pair of atria sets up contraction in these chambers; a second node is thereby stimulated and the impulse is carried via a bundle of fibers to the muscles of the ventricles.

THE PRIMITIVE HEART. The heart of typical fishes is a single tube consisting of four consecutive chambers. The heart of a bird or mammal likewise has four chambers, but these do not correspond to those of a primitive vertebrate; it is, rather, a double pump, with two chambers in each of its two parts. The great changes that have occurred in heart history are correlated with the shift from gill breathing to lung breathing.

In the primitive vertebrate heart (Figs. 346, 347 *A*) there are present, in order: (1) **sinus venosus,** a thin-walled sac into which blood enters from the cardinals and

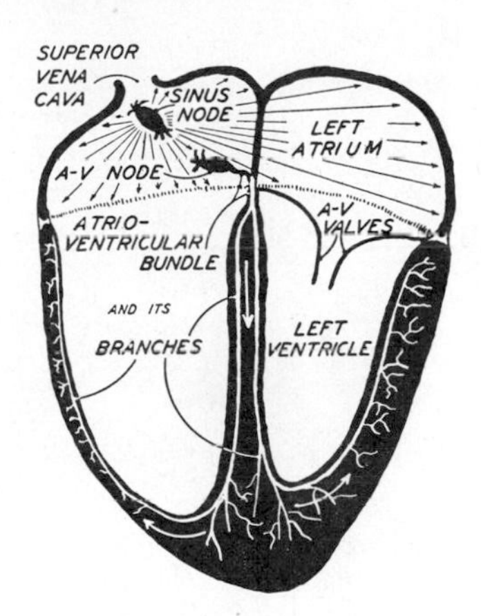

Figure 345. Diagram of a mammalian heart opened from the ventral surface to show the conducting system. (From Carlson and Johnson, The Machinery of the Body, University of Chicago Press.)

from the hepatic vein or veins; (2) **atrium** (or **auricle**),* still relatively thin-walled and distensible; (3) **ventricle,** the thick-walled major contractile portion of the heart; and (4) **conus arteriosus,** a narrow but stout tube leading to the ventral aorta, frequently furnished with several sets of valves. In the embryo these four chambers are arranged in an essentially straight line from back to front. But during development the front part of the heart tube tends to fold back ventrally in an S-shaped curve, thus combining length with compact structure (Fig. 347 *B–D*). As a result, even in a fish heart, the more "posterior" chambers tend to be situated dorsal, or even anterior, to the "anterior" ones. This makes visualization of heart construction difficult, and in diagrams (as Fig. 332) the heart is often represented as if "pulled out" into its embryonic longitudinal arrangement.

EVOLUTION OF THE DOUBLE HEART CIRCUIT (cf. Fig. 332). The primitive type of heart described above is present in most fishes. In lungfishes, however, and more fully in amphibians, a major difficulty arises with the substitution of lungs for gills as breathing organs. The heart now receives blood of two different types: "spent" blood from the body and "fresh," oxygenated blood from the lungs. The two streams should be kept separate, as far as possible, and sent to two different destinations—venous blood to the lungs, "fresh" blood to the body—by separate aortic trunks. But how to keep the two separate in a single-barrelled pump?

A perfect solution of this difficulty was not attained until the avian and mammalian stages were reached, but lungfishes and amphibians have made some progress toward separation of the blood streams. Even in lungfishes, the pulmonary vein does not enter the sinus venosus like the other venous trunks. Here and in amphi-

*Auricle, though often used as a synonym for atrium, refers properly to the supposedly earlike flaps on the mammalian atria.

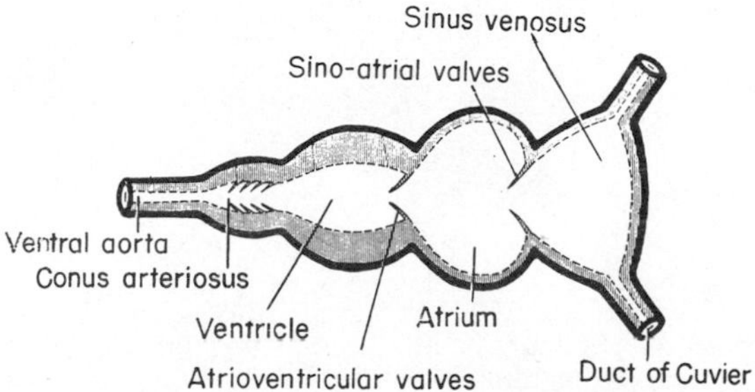

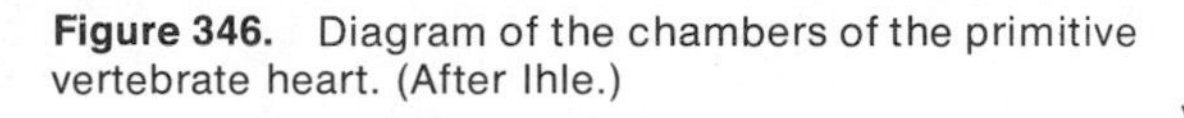
Figure 346. Diagram of the chambers of the primitive vertebrate heart. (After Ihle.)

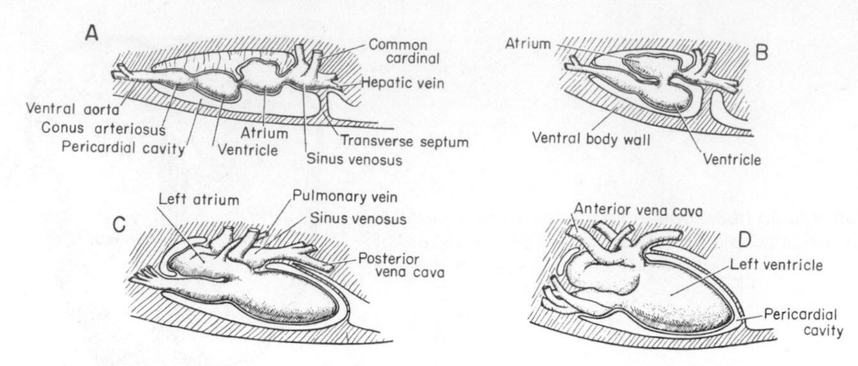

Figure 347. Diagrammatic views of left side of heart in various vertebrates, to show its position in the pericardial cavity and phylogenetic modification of the heart chambers. *A,* Hypothetic ancestral condition, found essentially repeated in embryos (cf. Fig. 346). The four primitive chambers are in line anteroposteriorly, and a dorsal mesentery is still present. *B,* Selachian stage; the mesentery is gone; the atrium has pushed forward above the ventricle, but the sinus venosus is still posteriorly placed. *C,* Amphibian stage; the sinus and accompanying blood vessels have moved anteriorly. *D,* Amniote stage; sinus and conus arteriosus have lost their identity; the heart attaches to the walls of the pericardium only anteriorly. (After Goodrich, 1930.)

bians (Fig. 348 *A*) the atrium is divided, more completely in amphibians than in lungfishes, into two halves, the pulmonary vein entering the left portion. The "spent" venous blood enters the right half of the atrium; the sinus venosus, which leads to this chamber, is reduced in size in some amphibians (anurans) and is merged with it in amniotes.*

But atrial separation is in vain if the two streams meet and mingle in the ventricle. A variety of adaptations in lungfishes and amphibians prevents complete mixture of the two, but much blending can nevertheless occur. It is only in amniotes that the two streams come to be effectively separated by subdivision of ventricle as well as atrium into two parts.

In most reptiles there is a ventricular septum, but it is incomplete and some

*Quite possibly the heart in modern amphibians has regressed somewhat in structure; the specialized skin-breathing of these forms reduces the importance of separating the two blood streams, since the oxygenated blood from the skin reaches the heart in common with the "spent" blood from the body.

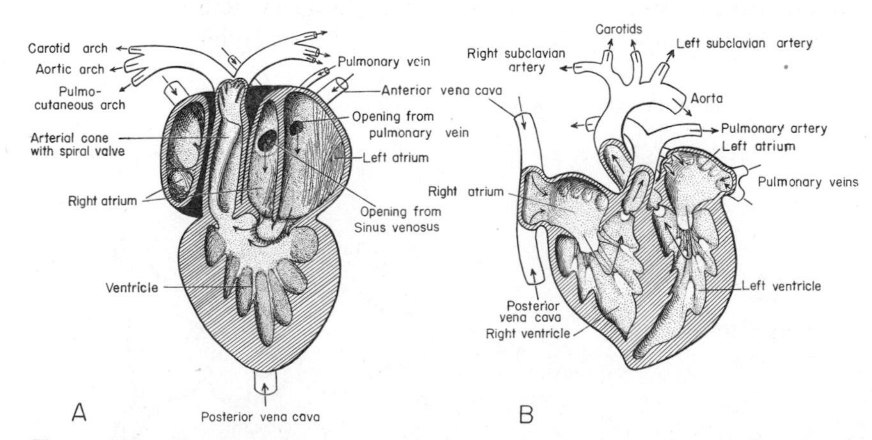

Figure 348. Diagrammatic section through the heart of *A,* a frog; *B,* a typical mammal. (Partly after Jammes.)

admixture may still take place. The structure of the reptilian ventricle is extremely complex—and difficult to show in a simple diagram such as Figure 332 *C*. It is divided into dorsal and ventral parts, with the former subdivided into right and left halves; the latter correspond, in some ways, to the separate ventricles of higher forms. Anatomic inspection suggests that the right systemic arch should receive oxygenated blood from the left side of the heart, while the left systemic arch, like the pulmonary, should receive nonoxygenated blood from the right. However, the situation is not that simple. Although the left systemic arch may, in some cases, receive "mixed" or even venous blood, this is rare—usually, it receives *only* oxygenated blood. As an aside, we may note that the idea that mixing is necessary because the pulmonary circulation lacks the capacity to handle half of the circulating blood is not substantiated by experimental work. In crocodilians the ventricular septum itself is complete but there is still a gap at the base of the arterial conus.

In lower tetrapods and even in lungfishes there is seen, however, some tendency for subdivision of the conus. In birds and mammals this division is completed; the conus is done away with as a separate structure and at long last in the evolutionary series the two blood streams of the heart are completely separated (Fig. 348 *B*).

To sum up, the introduction of the lung into the blood circuit in advanced fishes threw out of order the simple heart plan of primitive vertebrates and furnished a "problem" which advanced vertebrates found difficult to solve. Lungfishes, amphibians, and reptiles even today have not solved it completely, although their partial solutions are satisfactory enough to allow them to survive. There is no "perfect" solution for lungfishes or amphibians, which use lungs and either gills or the skin for respiration. In birds and mammals alone is the complete solution seen and complete separation of circuits attained. The result is surprising in its efficiency. The single pump of the original heart has become a double one; each half of the heart performs effectively its own distinct task.

HEART DEVELOPMENT (Fig. 349). We have noted that the first blood vessels to develop in the embryo form a subintestinal vein running forward from gut to gill arch region. The heart forms along the course of this vessel. Particularly in forms with a large yolk, in which the animal has at first no formed ventral surface, this vessel may long persist in the embryo in the form of paired vitelline veins, and the pulsating region which is to form the heart may at first be a paired structure (Fig. 349 *A*). About the early heart tube develops a portion of the celomic cavity, and in all vertebrates there presently occurs the S-shaped curvature of the heart and its subdivision into a series of chambers. The higher vertebrates show further de-

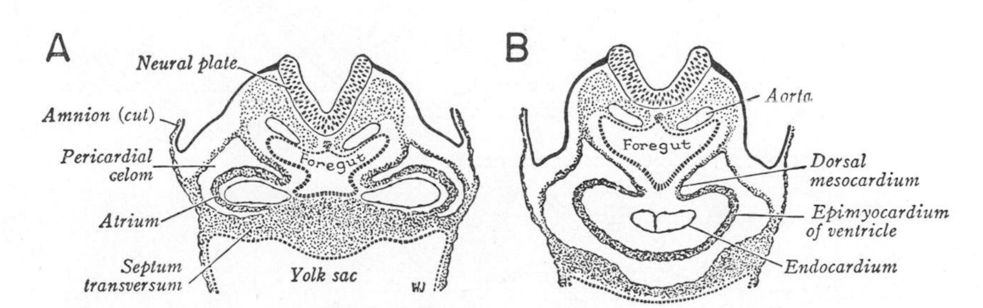

Figure 349. Cross sections of a mammalian embryo to show an early stage in heart formation, before the fusion of the two subintestinal vessels from which the heart forms. In the atrial region, *A*, the two are still widely separated; farther forward, in the ventricular region, *B*, the two tubes are apposed inside a single pericardial sac, but are not yet fused. (From Arey.)

velopment, with a gradual subdivision of atrium and ventricle and absorption into them of sinus venosus and conus arteriosus; these stages, as seen in a mammal, essentially recapitulate the phylogenetic history of the mammalian heart.

BLOOD CIRCUITS

In earlier sections of this chapter the components of the blood circuit have been described piecemeal. We shall here briefly review the general evolutionary history of the circulation as a whole with especial reference to blood pressures and capillary nets (Fig. 350).

As in any passage of liquid through tubes, the friction of the liquid on the walls tends to lessen the pressure given by the "pump"; the capillaries are, of course, the parts of the system in which the fall in pressure is the greatest. In fishes, in general, every drop of blood leaving the heart must pass through at least two capillary systems before returning to the heart—first in the gills, then in the general body tissues. Much of it, however, must pass through a third capillary net as well, for that which has gone to the gut must pass through the hepatic portal, that to the tail must on return pass through the renal portal system. The nature of the fish circulatory circuits thus makes heavy work of the maintenance of blood pressure. With the introduction of the pulmonary circuit and the abolition of gill capillaries in adult tetrapods, circulatory efficiency is greatly promoted. All body tissues are reached directly and with little loss of pressure. Two capillary systems only, not three, are encountered by blood traversing the hepatic portal or the renal portal of lower tetrapods, and the general body circulation passes but a single capillary net instead of two.* The substitution of lungs for gills has brought about, in the long run, not only an improved heart but also a more generally efficient circulatory system. With

*Minor portal systems, one in the pituitary, for example, do exist, but the amount of blood concerned is minimal.

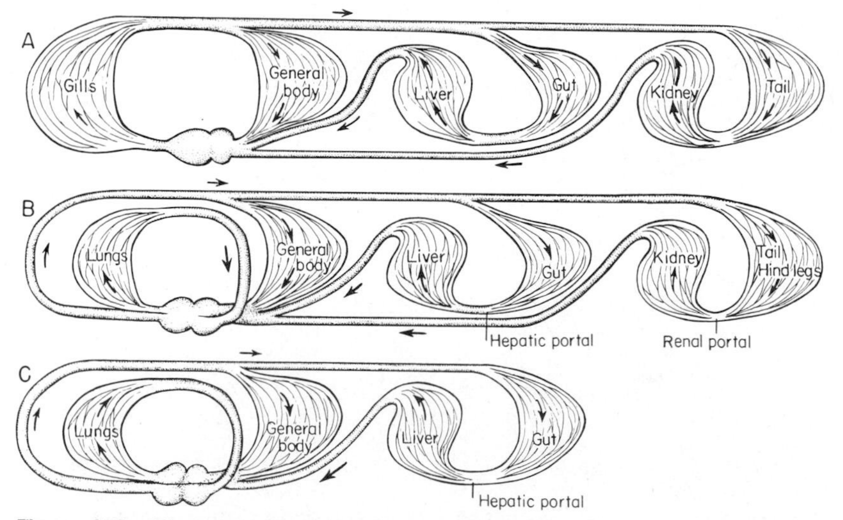

Figure 350. Diagrams showing the general nature of the blood circuits and capillary networks encountered in *A*, a typical fish; *B*, a terrestrial amphibian or a reptile, with elimination of gill circulation and introduction of pulmonary circuit; *C*, a mammal, with elimination of renal portal system and complete separation of the systemic and pulmonary circuits.

reduction in amniotes of the renal portal system there is further increased efficiency; in mammals it is only the intestinal circulation that encounters two capillary networks.

EMBRYONIC CIRCULATION

In earlier sections of this chapter we mentioned the formation of this or that blood vessel in the embryo. Although necessarily involving some repetition we may here attempt to gain a general picture of the development of the circulatory system, with attention to some of the vessels not present in the adult but necessary in the embryonic structures of forms with large-yolked eggs. It must be emphasized that the circulatory system cannot develop in the embryo merely with the "aim" of producing adult structures; it must be functionally effective at every moment of every embryonic or larval period. In general, too, we may note that in many areas (as in the limbs, Fig. 351) the circulation first develops in the form of a diffuse network, from which major vessels are "sorted out" only at a later stage.

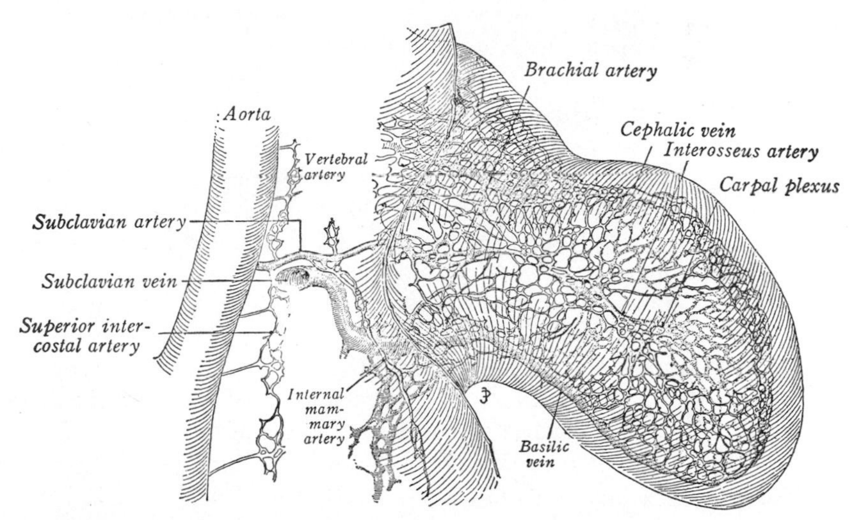

Figure 351. An early stage in the development of the forelimb of a pig embryo, to show the manner of formation of patterns in limb circulation. There is a network of interweaving small vessels from which the main adult vessels develop. Choice of one definitive channel or another allows for the occurrence of variants as anomalies. (From Woollard.)

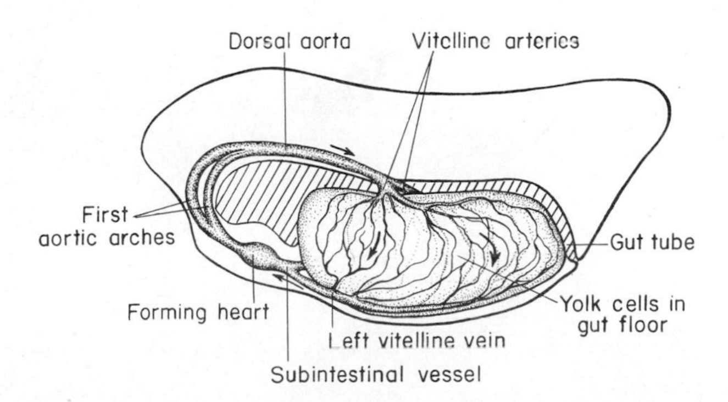

Figure 352. Diagram of the general circulation in a young frog tadpole. The food supply still lies in the yolky gut cells, and the vitelline circulation is of great importance.

The most generalized embryonic pattern is that developed in such forms as lungfishes and amphibians (Fig. 352), in which the picture is not complicated by excessive yolk or accessory membranes. Most of the nutriment on which the embryo must depend before feeding occurs lies in the yolky cells of the gut floor, and the first vessels arise there, forming a pair of **vitelline veins** which unite anteriorly into a subintestinal vessel. Along the course of this trunk the heart develops; the anterior end becomes the ventral aorta. Back of the heart the region of the subintestinal trunk is later invaded by liver tissue; the vessel is divided into the hepatic vein or veins anteriorly and the hepatic portal posteriorly, with a system of hepatic sinusoids between the two. At the front end of the body the trunk meanwhile has divided to curve upward on either side of the pharynx to form the first of the aortic arches; we have previously described the successive development of further arches more posteriorly and the usual degeneration of those first formed. Dorsally the major blood flow of the early embryo continues backward to form the dorsal aorta, paired anteriorly but further back forming a single median vessel below the notochord and above the gut and celom. Much of the blood from the aorta courses downward in **vitelline arteries,** corresponding in general to the celiac and mesenteric arteries of the adult, to reach the gut floor again and complete a primitive circuit.

But part of the aortic blood does not follow this path; instead, it leaves the aorta to supply the growing musculature of the body walls, the nervous system, and other structures of the outer tube of the body. A return system must be created for this blood. This is effected by the development of paired cardinals (Fig. 353) which develop the length of body and head, one on either side of the dorsal aorta. At a point above the heart each vessel sends a trunk, the common cardinal, down along the body wall to reach the heart. In all vertebrates above the cyclostome level the renal portal system is presently developed by interruption of the posterior cardinals. There is now established a general pattern of circulation which needs only few additions of any importance to attain adult conditions except for introduction of a circuit to and from the lungs in forms possessing these structures, and for development of circulation within the limbs.

The presence of a large amount of yolk and the development of a yolk sac in sharklike fishes and amniotes give a different appearance to the embryonic blood circuits (Fig. 353). The basic pattern is, however, the same as before. The nutrient

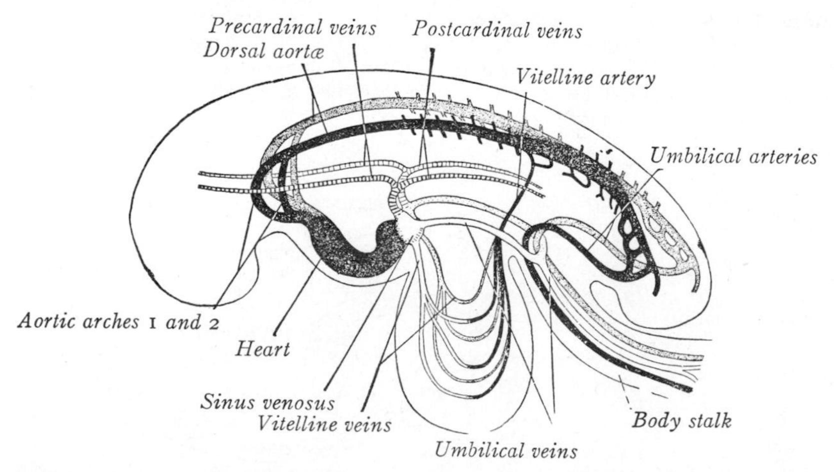

Figure 353. The circulatory vessels in a mammalian embryo. (From Arey, after Felix.)

yolk now lies not in the floor of a closed digestive tube but in a large pendant sac. In consequence the highly developed vitelline veins arise over the surface of the expanded yolk sac; the vitelline arteries are equally highly developed and extend outward onto this sac, paralleling the veins.

In amniotes the addition of the allantois makes a further complication in the picture. In reptiles and birds this forms (with the chorion) a respiratory organ, and vessels develop in the allantoic stalk for the necessary blood supply. These are the **allantoic arteries** and **allantoic veins;** the term **umbilical,** however, is generally used for them in mammalian studies, since they are the important vessels in the umbilical cord of the fetus. The allantoic arteries descend through the body walls to the allantoic stalk from the back end of the dorsal aorta. The corresponding veins (which may fuse for most of their course) do not, however, join the subintestinal vessels, as one might expect. Instead, they run forward in the lateral body wall on either side, and thus correspond to the abdominal veins of a shark. In early stages they enter the common cardinal or sinus venosus directly (Fig. 354 *A*). Later, however, they turn upward to pass through the liver; apparently the liver sinusoids are unable to care for the entire flow of blood and much of it in the embryo may pass through the liver tissue by way of a large duct—the **ductus venosus** (Fig. 354 *C, D*). In mammals these same vessels carry not only oxygen but also food from the placenta and hence are of the highest importance.

In all forms with a yolk sac, the vitelline vessels leading from it are resorbed

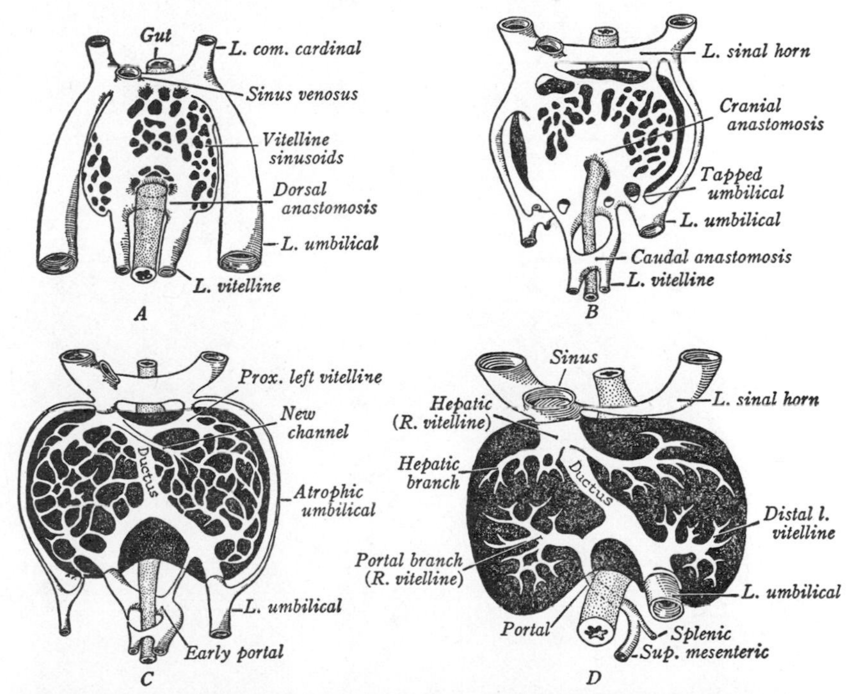

Figure 354. Diagrams of the liver region of human embryos at successive stages (4.5, 5, 6, and 9 mm in length), seen from the ventral surface, to show developmental changes in the vitelline and umbilical veins. The gut tube is stippled. In *A* the vitelline vessels from the yolk sac are well developed and pass through the liver tissue; in *B-D* their transformation into the portal system is seen. The umbilical veins (from the placenta) are already well developed in *A*, but run directly to the sinus venosus. In later stages this blood flow is diverted to the liver circulation, much of it flowing through this organ via a ductus venosus. The right umbilical vein is reduced; the left persists until birth, when this vessel and the ductus venosus undergo reduction. The posterior cardinal system would be dorsal to the liver and is not shown. (From Arey.)

when the sac contents are exhausted. In placental mammals, as we have noted, the yolk sac contains no food and the vitelline vessels disappear at an early stage. At hatching or at birth the allantoic (or umbilical) vessels likewise disappear.

Among tetrapods there is a marked circulatory change due to a change in the source of oxygen when lungs become functional—at metamorphosis, hatching, or birth. Up to that time there is little circulation of blood through the pulmonary arteries or veins. We have noted that on the arterial side of the circuit the retention of the ductus arteriosus allows blood in the pulmonary arch to bypass the lung. If, in amniotes, the two atria were separated in the embryo (as in the adult) by a partition, the left side of the heart would remain empty, since there is little blood entering from the pulmonary veins. In consequence, even in birds and mammals, gaps are present in the interatrial wall of the embryo's heart. When the lungs begin to function, there is a rapid and efficient shift in these features. The ductus arteriosus closes, the full pulmonary flow passes through the lungs and back to the heart, and, in birds and mammals, the gaps in the septum between the atria rapidly close.

15 Sense Organs

All cells, one may believe, are capable of receiving and responding to stimuli that measure, so to speak, some condition in the environment or a change in such condition. If, however, in a vertebrate the proper response to a sensory stimulus is one that should be performed by a distant region of the organism, or by the organism as a whole, reception of that stimulus is in vain unless there is some channel of communication between the sensory receptor and the organs—muscles or glands—which should make the appropriate response. Such communication may be made by hormonal action, but in general the mechanism used is that of the nervous system. The tips of nerve fibers are themselves capable of direct excitation, but more often the reception of sensation in vertebrates is the function of specialized **sensory cells,** mainly of ectodermal origin and generally grouped in organs of lesser or greater degree of complexity. These are attuned to physical or chemical stimuli of specific types and are associated with nerves which relay these stimuli to specific centers in the brain or to the nerve cord.

Anatomists divide such sensory nerves into two groups: the **somatic sensory nerves,** carrying impulses of a sort which in ourselves usually reach the level of consciousness, from the "outer tube" of the body—the skin and body surfaces and the muscles; and **visceral sensory nerves,** whose impulses, seldom reaching our consciousness, arise from the viscera. The physiologist customarily classifies sensory receptors in a fashion which fits readily into the neurologic scheme. **Exteroceptors** are those sensory structures of the skin and special senses which receive sensations from the outside world; **proprioceptors** include those situated in the striated voluntary muscles and tendons; **interoceptors** are those located in the internal organs. The first two of these correlate fairly well with the somatic system of sensory nerves, the third with the visceral sensory system of the anatomists.

SIMPLE SENSE ORGANS

One ubiquitous sensation—that of **pain**—seems not to need any special organ for its reception but may be produced by direct stimulation of the end fibers of sensory nerves. Particularly in lower vertebrates, other simple sensations may to some degree be received directly by excitation of nerve endings.

SENSE CORPUSCLES. A variety of sensations are received by small sensory structures, generally of microscopic size, which may be present in any part of the

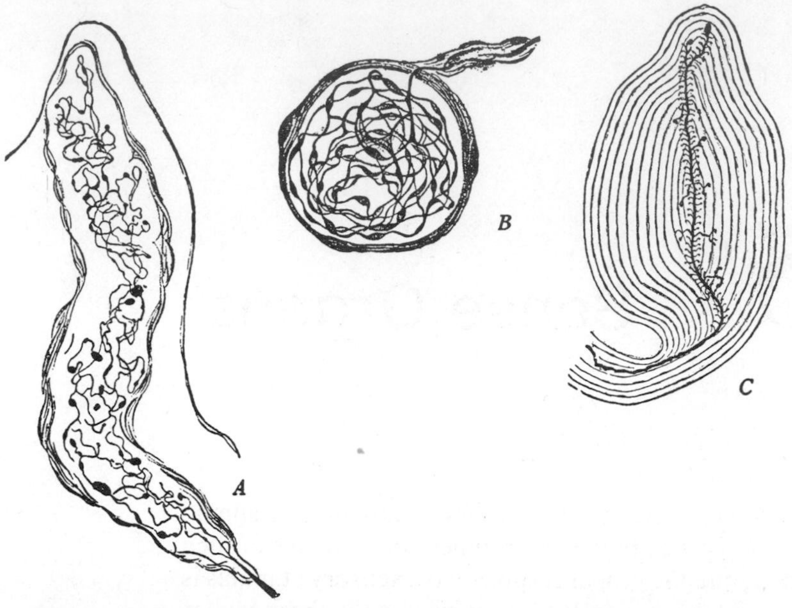

Figure 355. Some types of sensory organs from mammalian tissues. *A,* Tactile corpuscle (Meissner's corpuscle) from the connective tissue of the skin; *B,* an end bulb of Krause, sensitive to cold; *C,* a pacinian corpuscle, which registers pressure and tension. (From Ranson, after Dogiel, Sala, Böhm-Davidoff, Huber.)

body—in skin, muscles, or viscera. These are unknown in fishes, relatively rare in other lower vertebrates, but abundant in birds and mammals. Their appearance and structure vary greatly (Fig. 355). It is of course difficult to determine the specific functions of many of these structures, but from our own experience it seems that at least four types of simple sensations can be registered by such bodies—warmth, cold, touch, and pressure. Proprioceptive sensations, of the physiologists' terminology, include those of the **muscle spindles** (Fig. 356) and **tendon spindles.** In lower vertebrate classes terminal nerve fibrils may twist about individual muscle fibers or spread among tendon fibers. In mammals there are specialized receptors, the spindles, which consist of a group of small muscle fibers surrounded by a maze of sensory nerve endings and enclosed in a sheath. These receptors are the seat of "muscle sense." They not only register the state of contraction of the muscle concerned but (amazingly) give information as to the position in space of various parts of the body—information, as we are ourselves well aware, which can be furnished without the aid of other sensory structures, but strictly confined (in the absence of contact with other objects) to parts of the body containing striated muscles or tendons.

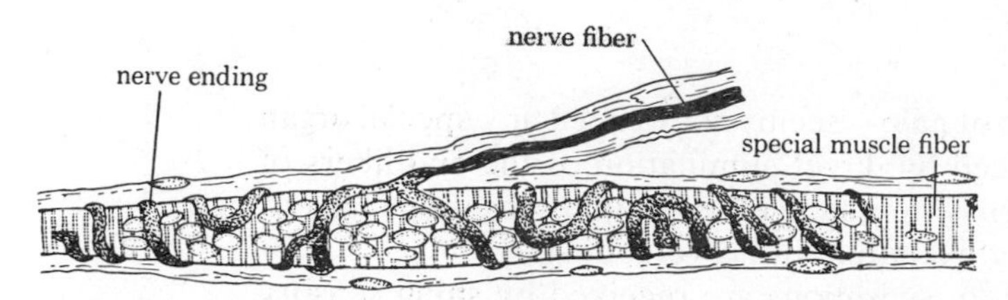

Figure 356. A muscle spindle. (After Windle, Textbook of Histology, McGraw-Hill.)

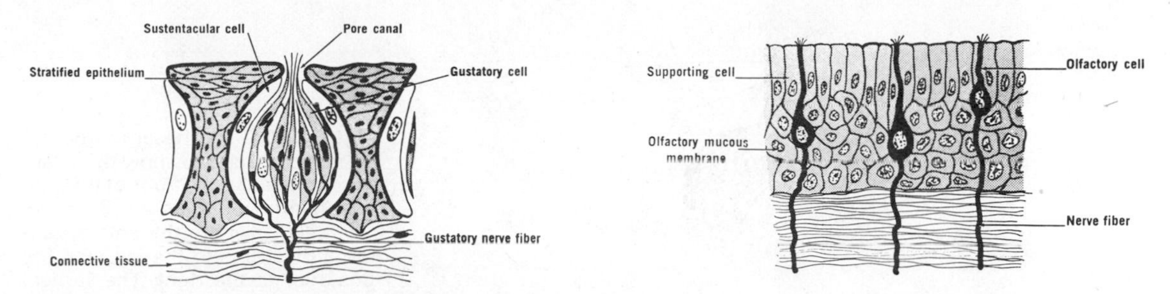

Figure 357. Microscopic sections through chemical sense organs. *A*, A taste bud; *B*, nasal mucosa. (From De Courcey, Medical Physiology.)

TASTE. In contrast to the simple senses named above, taste and smell are responses to chemical stimuli, the former received by **taste buds**—small barrel-like collections of elongate cells sunk within an epithelium (Fig. 357*A*). The "staves" of the barrel are supporting cells; in the center are the taste cells proper, elongate and tipped externally with a sensitive, hairlike process. Although for the most part confined to the mouth, and in mammals concentrated on the tongue, they may be more widely distributed, and in fishes and amphibians may be found in the skin. In some catfishes, for example, they are widespread over the entire surface of the body, giving a phenomenon possibly of pleasant (or unpleasant) gustatory sensations. It must be noted that much of what we casually think of as taste is actually a smelling of mouth contents (foods do not "taste" as well when a head cold clogs the nose). All taste buds look alike, but there appear to be four types, as regards reception, giving sensations of salty, sour, bitter, and sweet.

In sections which follow are described the more prominent of the complicated sensory structures of vertebrates—nose, eye, and ear, and the lateral line organs of fishes. Very probably, however, there is a variety of still further sensory structures, particularly in lower vertebrates, which give responses of types unfamiliar to us and hence difficult for us to understand. One such sense, about which we do have data, lies in the **pit organ** of the so-called pit vipers, such as the rattlesnakes. Placed between eye and nose, and filled with vascular tissue and nerve endings, the pit is highly sensitive to the movement nearby of a warm body of any sort—a sensory power extremely useful to such an animal which makes its living by capturing warmblooded rodents and is not too well endowed with "normal" sense organs.

THE NOSE

In tetrapods the nose has become associated with breathing, but its primary function is that of olfaction—the detection of minute amounts of chemical particles received from objects at a distance and the "sorting out" of these sensations into a variety of categories which are still poorly understood. In certain vertebrate groups smell is relatively unimportant; it is not generally highly developed in teleosts and is rather feeble in most birds, in marine mammals, and in higher primates, including man. In vertebrates generally, however, smell is in many ways the most important of all the senses; testimony to its importance is the fact that, as we shall later see, the most highly developed brain centers arise in an area primarily connected with smell.

In most fishes the nasal structures consist of a pair of pockets, placed well anteriorly in the head and without an internal opening to the mouth; each pocket has two openings, partially or completely separated, allowing a flow of water through

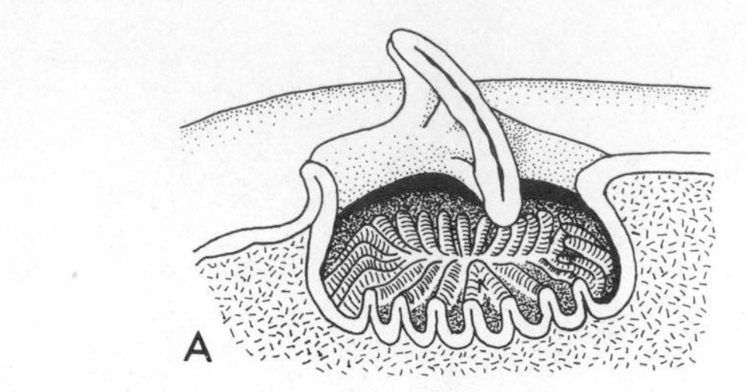

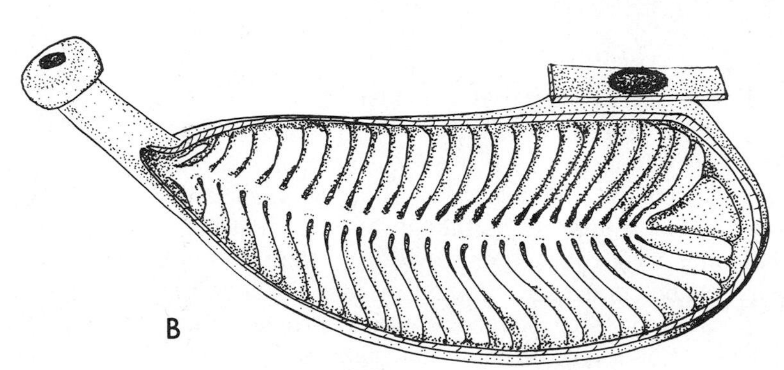

Figure 358. The nasal cavities of teleosts, opened to show the olfactory lamellae in the floor of the sac. *A,* A minnow *(Phoxinus); B,* an eel *(Anguilla);* in both the anterior external nostril is to the left, and the posterior to the right. The vertical flap in the minnow helps deflect a current of water through the nasal cavity. (After Rietschel and Liermann.)

the cavity beneath. In most bony fishes the nostrils are lateral in position; in sharks they lie beneath the snout. The floor of the nasal sac commonly bears a series of ridges, the **olfactory lamellae,** arranged in some type of rosette pattern (Fig. 358). These lamellae are typically covered by a simple columnar epithelium, containing **olfactory cells** interspersed with supporting elements (Fig. 357 *B*). On its exposed surface each olfactory cell bears a radiating brush of short, hairlike processes. In one remarkable feature—very probably primitive—these cells differ from typical vertebrate receptors. Others depend upon nerve fibers to relay inward the sensations received. The olfactory cells, on the contrary, do their own work; a long fiber extends from the cell itself inward to the brain.

The jawless vertebrates present a puzzling situation. In contrast to all other living vertebrates, the olfactory organ in cyclostomes is a single median pouch, opening at the tip of the snout (hagfishes) or high atop the head (lampreys); as a further peculiarity, it is combined with the hypophyseal sac (Figs. 17, 253). Is this condition primitive or specialized? The answer is none too clear. In some (but not all) of the most ancient fossil vertebrates, the nostril is comparable to that of a lamprey (Fig. 19 *A*); on the other hand, the lamprey nostril is in a more normal ventral position in the embryo and is bilobed, although not distinctly paired, in the lamprey larva.

In the typical crossopterygians there evolved a type of nostril with an opening into the roof of the mouth as well as to the exterior. In the tetrapods this passage is utilized as an adjunct to breathing. Here the olfactory pouch is filled with air rather than water, but nasal glands and liquid brought from the eye by the lacrimal (tear) duct keep the sensory epithelium moist and capable of functioning. In this new type of nasal apparatus the structure is at first simple. In amphibians (Fig. 359 *A, B*) an **external naris** leads into a somewhat elongated sac; posteroventrally, a large opening, the **choana,** or internal naris, opens directly from this into the front part of the

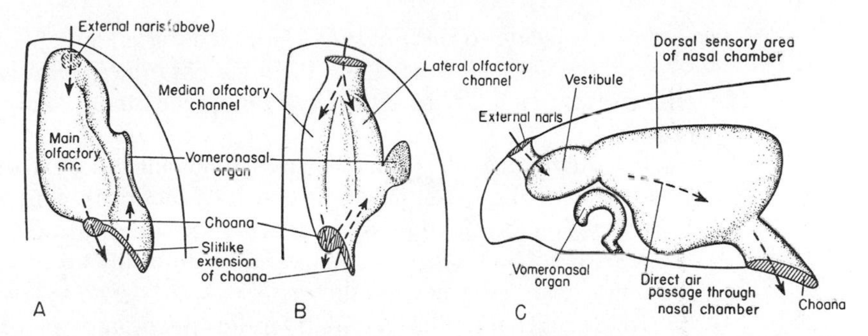

Figure 359. *A,* Ventral view of anterior part of the left side of the palatal region of the salamander, *Triturus,* with the nasal chambers shown as solid objects, the remainder as transparent structures (cf. Fig. 228). *B,* Similar view of the toad *Pipa. C,* Longitudinal section of the nasal region of a lizard, cut somewhat to the right of the midline, to show the cavities of the nasal apparatus. In the embryo lizard the vomeronasal organ was a medial pocket off the main nasal channel; in the adult (as in many mammals) this organ has separated to open independently into the roof of the mouth by a nasopalatine duct. Arrows show the main air flow inward in all figures and the outward flow toward the vomeronasal organ in the amphibians. (*A* after Matthes; *B* after Bancroft; *C* after Leydig.)

roof of the mouth. The inner surface of the sac is nearly smooth, and only part of its lining is sensory.

In typical reptiles the nasal region begins to assume a more complex structure (Fig. 359 *C*). The air passage is longer, there is usually a small but distinct **vestibule** anteriorly, and the sensory epithelium is confined to the upper part of the main chamber. Here there is typically developed one or several curved ingrowths from the lateral wall as **conchae** or **turbinals,** increasing the epithelial area. Beyond, a passage, the **nasopharyngeal duct,** short in lizards and snakes, leads back and down to the choana. In certain turtles, as we have noted previously, there is some development of a secondary palate; in crocodilians this is greatly elongated, with a related elongation of the duct to the choana. In birds, generally, smell appears to be of very little importance and the nasal structures, built on the general reptilian plan, are of modest size.

Among mammals, olfaction is reduced in whales and bats, but nasal structures reach the peak of their development (Fig. 229, 360) in typical members of the class. The nasal chamber may reach back to the orbits to occupy more than half the skull length. The conchae are usually highly developed, increasing the olfactory area in the upper part of the chamber and acting as air filters and "conditioners" in the direct air passage below. A long nasopharyngeal duct carries the air back above the

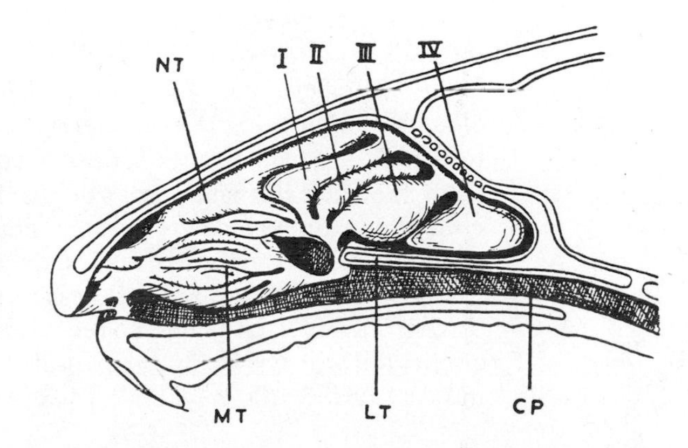

Figure 360. A section through the nasal region of a guinea pig, slightly to the right of the midline, to show the development of the turbinals, including the maxilloturbinal, *MT,* nasal turbinal, *NT,* and ethmoturbinals (I-IV). *CP,* choanal passage; *LT,* lamina transversalis, a remnant of the original primary palate, separating the sensory area of the nasal fossa from the choanal passage. (After Cave.)

secondary palate to the pharynx. In placental mammals there are present extensions of the nasal air spaces into the adjacent bones as the **sinus pneumatici** (plural) which effect a lightening of the skull (and have other effects all too familiar to some readers).

In numerous tetrapods there is a specialized part of the olfactory system termed the **vomeronasal organ** or organ of Jacobson which appears to have as its main function the picking up of olfactory sensations from the food in the mouth. In urodeles (Fig. 359 *A*) this organ is represented by an area of sensory epithelium lying in a channel somewhat distinct from that carrying inward the main current of air. In the other amphibians and in reptiles the organ is placed in a blind pouch lying to one side (Fig. 359 *B*). This appears to be the primitive reptile condition, as seen in *Sphenodon*. In lizards and snakes, however (Fig. 359 *C*), the vomeronasal organs occupy a pair of pouches which open quite separately into the roof of the mouth. A peculiar development is seen in some lizards and in snakes. The tongue, cleft into two prongs and darting in and out of the mouth, serves as an accessory olfactory organ. When the tongue is withdrawn, the tips are placed near the entrance to the vomeronasal pockets; chemical particles which adhere to them in the air are transferred to the moisture on the sensory epithelium of these pouches.

The organ is not well separated from the rest of the nose in turtles, and is vestigial to absent in crocodilians and birds, and also in some mammals (such as higher primates). Other mammals, however, have retained it; in many rodents it opens into the main nasal cavity, but in other forms which retain it, there are separate openings, as in lizards and snakes, into the roof of the mouth cavity.

THE EYE

The vertebrate body is subjected constantly to radiations which may vary from the extremely short but rapid waves of cosmic rays and those from atomic disintegration to the long, slow undulations utilized for radio transmission. Many of these affect protoplasm, but specific sensitivity appears to be limited to a narrow band part way between the two extremes; for knowledge of other wavelengths we must resort to mechanisms which transform their effects into terms receivable by our own limited senses. It is not unreasonable to find that the animal band of sensitivity corresponds in great measure to the range of radiations reaching the earth from the sun, since that body is the source of the vast bulk of the radiations that normally reach us. Of this band, the slower waves are received as heat, the faster, shorter waves perceived as light—the process of **photoreception.**

Specific sensitivity to light is widespread throughout the animal kingdom. In many simple invertebrates there develop "eye spots," clusters of sensitive cells, frequently with associated pigment, and often there is an evolutionary progress to an organized eye, with a lens for concentrating light upon sensitive cells in a closed chamber. Many simple eyes merely receive light "in bulk"; with better organization and the arrangement of sensory cells in a definite pattern so as to receive light from specific external areas, true vision results. Well developed eyes, with many common features, but surely independently evolved, are found in forms as far apart as molluscs, arthropods of various sorts, and vertebrates.

In the vertebrate eye, the essential structure (Figs. 361, 362) is the roughly spherical **eyeball,** situated in a recess, the **orbit,** on either side of the braincase, and connected with the brain by an **optic nerve,** contained in a stalk emerging from the internal surface of the eyeball. The eyeball has an essentially radial symmetry, with

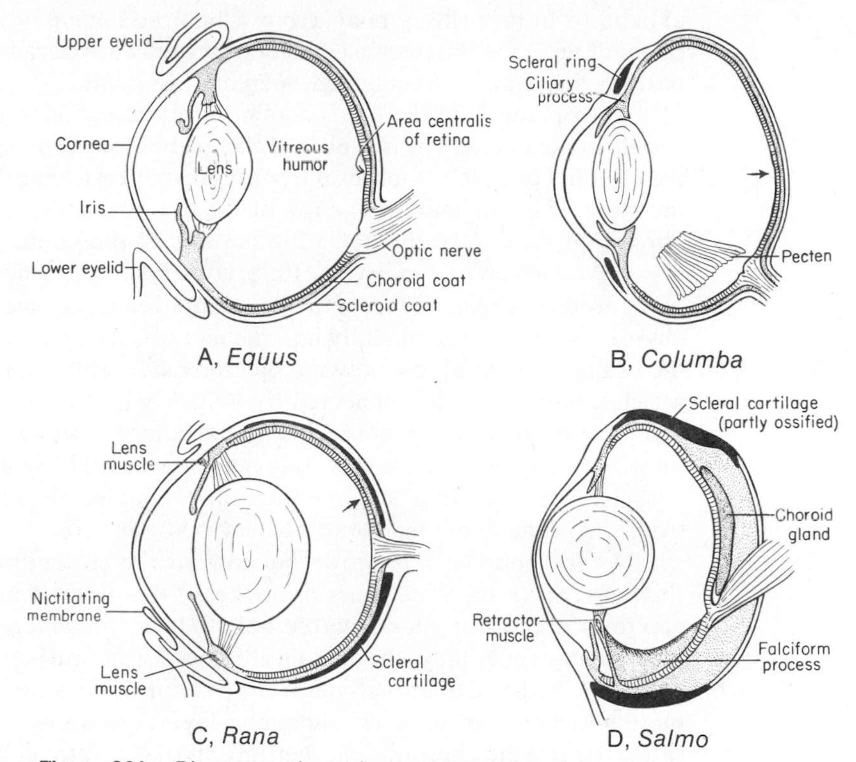

Figure 361. Diagrammatic vertical sections through the eye of *A*, a horse; *B*, a dove; *C*, a common frog; *D*, a teleost (salmon). Connective tissue of sclera and cornea unshaded; scleral ring or cartilage black; choroid, ciliary body, and iris stippled; retina hatched. Arrows point to the fovea. In *B* is shown the pecten, lying to one side of the midline. In *D* the section is slightly to one side of the choroid fissure through which the falciform process enters the eyeball. (After Rochon-Duvigneaud, Walls.)

a main axis running from inner to outer aspects. Within it is a set of chambers filled with watery or gelatinous liquids. In the interior, well toward the front, lies a **lens.** The walls of the hollow eyeball are formed basically of three layers, in order, from the outside inward, the **scleroid** and **choroid** coats and the **retina.** The scleroid is a complete sphere; choroid and retina are incomplete externally. The first two are of mesenchymal origin and are essentially supporting and nutritive in function; the retina (actually a double layer) includes the truly sensory part of the eye. At the outer end of the eyeball the scleroid coat is modified to form, with the overlying skin, the transparent **cornea.** Toward the exterior, choroid and retinal layers are fused and modified. Opposite the margins of the lens these conjoined layers usually

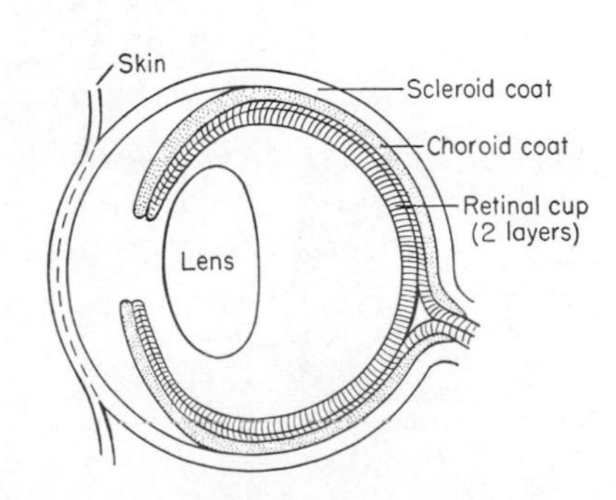

Figure 362. Diagrammatic section of an eye to show the arrangement of the successive embryonic layers.

expand to form a **ciliary body,** from which the lens may be suspended. Forward beyond this point the two fused layers curve inward parallel to the lens to form the **iris,** leaving a centrally situated opening, the **pupil.**

The operation of the eye is commonly (and reasonably) compared to that of a simple box camera. The chamber of the eyeball corresponds to the dark interior of the box; in both a lens focuses the light properly on a sheet of sensitive materials at the back of the chamber. The iris of the eye resembles the similarly named diaphragm of the camera in regulating the size of the pupil.

DEVELOPMENT (Fig. 363). Embryologically, the most important functional parts of the eyeball arise from the ectoderm (including neurectoderm), but mesenchyme also enters prominently into the picture. At about the time of completion of the brain tube there grow outward on either side from the forebrain spherical **optic vesicles,** which remain connected by a stalk with the brain. As each vesicle develops, its outer hemisphere folds into the inner to form a double-layered **optic cup** (in which, however, there may long persist a ventral fissure for entrance of blood vessels). The optic cup becomes the retina, which is thus a two-layered structure; from the retina come part of the ciliary body and iris.

As the optic vesicle grows out toward the surface, the overlying ectoderm thickens, and a spherical mass or pocket of this tissue sinks into the orifice of the cup to form the lens. In many (but not all) forms in which eye formation has been studied experimentally, the stimulus for lens formation is provided by the approach of the optic vesicle to the ectoderm. Further optic structures are contributed by the mesenchyme: first an inner sheathing layer, primarily vascular, surrounds the retina to form the choroid coat; then an external sheath of connective tissue forms a complete external sphere as the scleroid coat (sclera plus cornea).

SCLERA AND CORNEA. The scleroid coat is a stiff external structure which preserves the shape of the eyeball and resists pressures—internal or external—which might modify its shape. In cyclostomes, on the one hand, and mammals, on the other, it consists entirely of dense connective tissue, but in most other groups it is reinforced by cartilage or bone. Often there is a cartilaginous cup (ossified in some birds) enclosing much of the back of the eyeball, and further protection may be afforded by a **scleral ring** of bony plates lying in the sclera in front of the "equator" (Fig. 364). Fossil evidence shows that such a ring was present in many

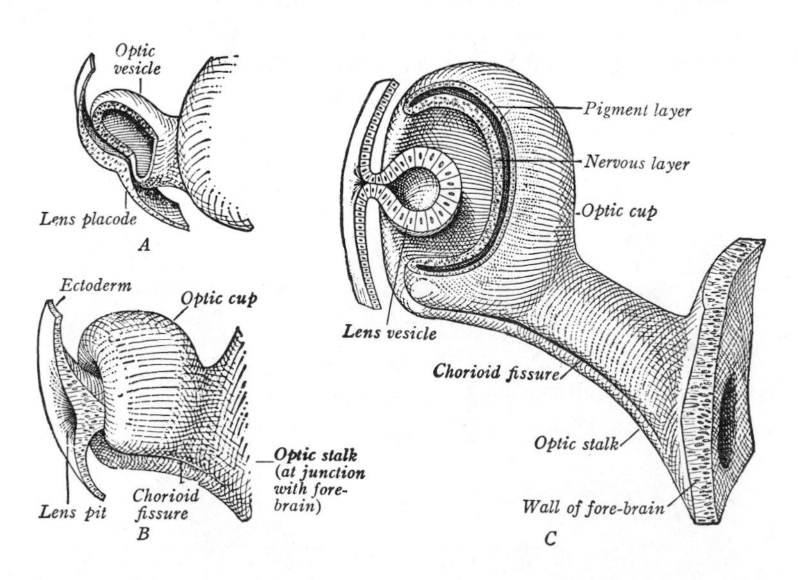

Figure 363. A series of diagrams to show the embryonic development of the optic cup and lens. (From Arey.)

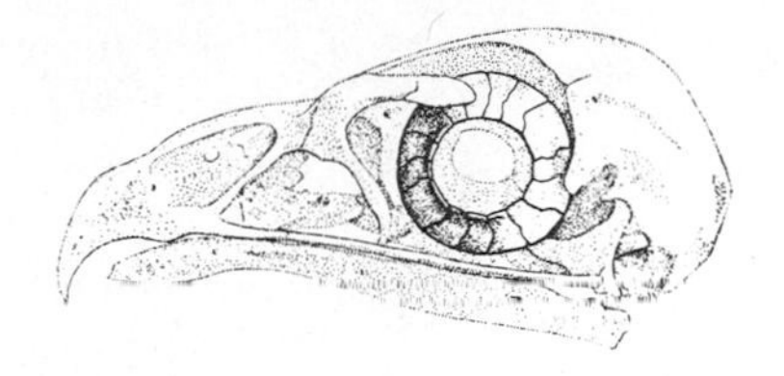

Figure 364. A skull of an eagle *(Aquila)*, showing the scleral ring in place. (From Edinger.)

primitive vertebrates, but it has persisted to the present only in actinopterygian fishes, reptiles, and birds. Primitively it appears to have consisted of four plates, but modern ray-finned fishes generally have but two, while reptiles and birds, on the other hand, have a much greater number of plates.

The superficial part of the eyeball is the translucent **cornea,** through which light enters the eyeball. In all vertebrates above the cyclostomes the scleral coat fuses inseparably here with the overlying skin; the cutaneous component of the cornea and the sensitive skin beneath the eyelids constitute the **conjunctiva.** The refractive index of the cornea—that is, its ability to deflect the course of light waves—is practically the same as that of water. In air, however, the cornea itself acts as a lens and relieves the true lens of much of the task of focusing; its importance is shown by the fact that in ourselves major defects calling for optical correction, such as astigmatism, are due to imperfection in corneal shape.

CHOROID. The inner mesodermal sheath of the eyeball contrasts strongly with the sclera, for it is a soft tissue, rich in blood vessels supplying the retina. Pigments in the choroid absorb most of the light reaching it after penetrating the retina. In addition, there often develops here a light-reflecting device, most familiar to us as seen in the ghostly eyes of a night-prowling cat illuminated by automobile headlights. This is the **tapetum lucidum,** which may be formed either of a layer of glistening connective tissue fibers or a sheet of tissue filled with crystals of guanine (a material also present in some chromatophores). In nocturnal forms or in fishes living deep in the water, this structure is of use in "conserving" the sparse light rays, turning those otherwise wasted back to the retina.

IRIS. This structure, found in all vertebrates, is formed by a combination of modified segments of both choroid and retinal layers, which, in attenuated form, join to furnish a pigmented diaphragm in front of the lens. This regulates the size of the pupil and, thus, the amount of light admitted through the lens to the retina. In some fishes the iris is fixed in dimensions, except when affected by movements of the underlying lens. But in sharks, some teleosts, and tetrapods, muscle cells are generally present; arranged in circular and radial patterns, they may expand or contract the pupillary opening, giving the effect of a camera diaphragm in opening widely in dim light, and "stopping down" the opening in bright light for better definition. Nocturnal forms frequently have a slit-shaped pupil which closes more readily to exclude bright light. The muscle fibers of the iris are formed from its retinal component, which is a derivative of the neural ectoderm of the embryo. Although muscle is normally formed by the mesoderm, these cells in the iris have all the attributes of muscle fibers, structurally and functionally.

THE LENS AND ACCOMMODATION. In tetrapods light rays are bent by the cornea, which consequently does much of the work of refraction, the lens merely acting as a "fine adjustment" for focusing. In fishes the cornea does not function in this way, and the entire task of focusing must be performed by the lens. To do this the lens of fishes is spherical and situated far forward in the eyeball, thus giving the lens maximum power and the longest possible distance for convergence of rays on to the retina; in tetrapods the lens is much less rounded and situated farther back in the eyeball. The lens is formed of elongated, rather simplified cells arranged in a complicated pattern of layers; it is completely transparent, firm in shape and, in lower vertebrates, resistant to distortion. In cyclostomes the lens has no peripheral attachments and is kept in place merely by pressure from the "vitreous humor" behind it and the cornea in front; in all other vertebrates it is attached peripherally by a **zonule,** which may be either a membrane or a radially arranged series of fibers.

As every user of a camera is aware, it is impossible to obtain exact "definition" of objects at varied distances without adjustment of the lens focus. Such adjustment in the eye is termed **accommodation.** The eyes of most vertebrates are capable of accommodation, but, curiously, it is attained in a different fashion in almost every major group. This suggests the probability that accommodation was not a property of the primitive vertebrate eye, and that different types of vertebrates have evolved this power independently of one another. The methods used may be broadly classified as follows:

A. Lens moved to achieve accommodation:
   1. "Resting" position for near vision; actively moved backward for distant objects (lampreys, teleosts)
   2. "Resting" position for far vision; actively moved forward for near objects (elasmobranchs, amphibians)

B. Shape of lens modified; "resting" form for distant objects, rounded shape for near objects (amniotes)

It would be impossible to give in any limited space an account of the various special muscular developments used in these methods of accommodation. Movements of the lens are accomplished by the pull of variously developed muscles. Among the amniotes, the lens is made more rounded in reptiles and birds by the push against its margin of pads developed by the ciliary processes. In mammals the lens is "normally" kept in a relatively flattened condition by a pull on its margins of the zonule fibers. These fibers are so connected to the ciliary muscles that, when the muscles relax, the elastic lens, released from tension, takes a more rounded shape (Fig. 365). The amniote methods are effective only if the lens retains its elasticity. In man, as older persons know, it stiffens with age, accommodation diminishes, and without artificial aids a book can be read only at arm's length, if at all.

CAVITIES OF THE EYEBALL. Much of the eyeball is, functionally, merely a blank space which need only be filled by liquid—"humors"—which will not block or distort light rays. The principal cavity of the eyeball, between lens and retina, is

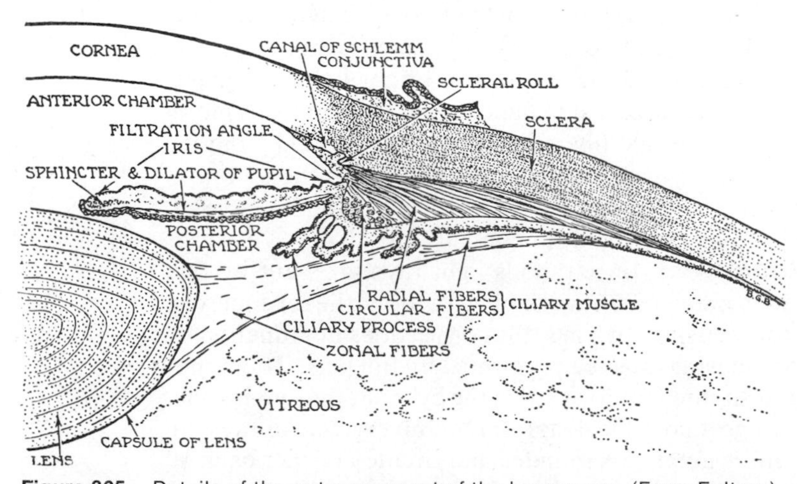

Figure 365. Details of the outer segment of the human eye. (From Fulton.)

filled by the thick, jelly-like **vitreous humor.** In front of the lens is the **aqueous humor.** The cavity filled by it between cornea and iris is termed the **anterior chamber** of the eye; the **posterior chamber** is not, as one might think, that filled by the vitreous humor, but the narrow area occupied by the aqueous humor between iris and lens (Fig. 365). A curious projection into the cavity of the vitreous humor in birds is the comblike **pecten** (Fig. 361 *B*). This structure is thought to have a nutritive function, but it has been suggested further that the "shadows" of the pecten ridges, falling on the retina, act as a grille ruling which enables a bird more readily to detect small or distant moving objects as their images pass from one component to another on this grille.

RETINA. All other parts of the eye are subordinate in importance to the retina; their duty is to see that light rays are brought in proper arrangement and focus to this structure for reception and transfer inward to the brain of visual stimuli. In the adult the two layers of the retinal cup are fused. The outer one is thin and contributes little but a set of pigment cells; the complex sensory and nervous mechanisms of the retina all develop from its inner portion. Details of retinal structure vary greatly from form to form and from one part to another of a single retina. Frequently, however, a sectioned retina has the general appearance seen at the left in Figure 366. Inside the pigment layer (which adjoins the choroid) is a

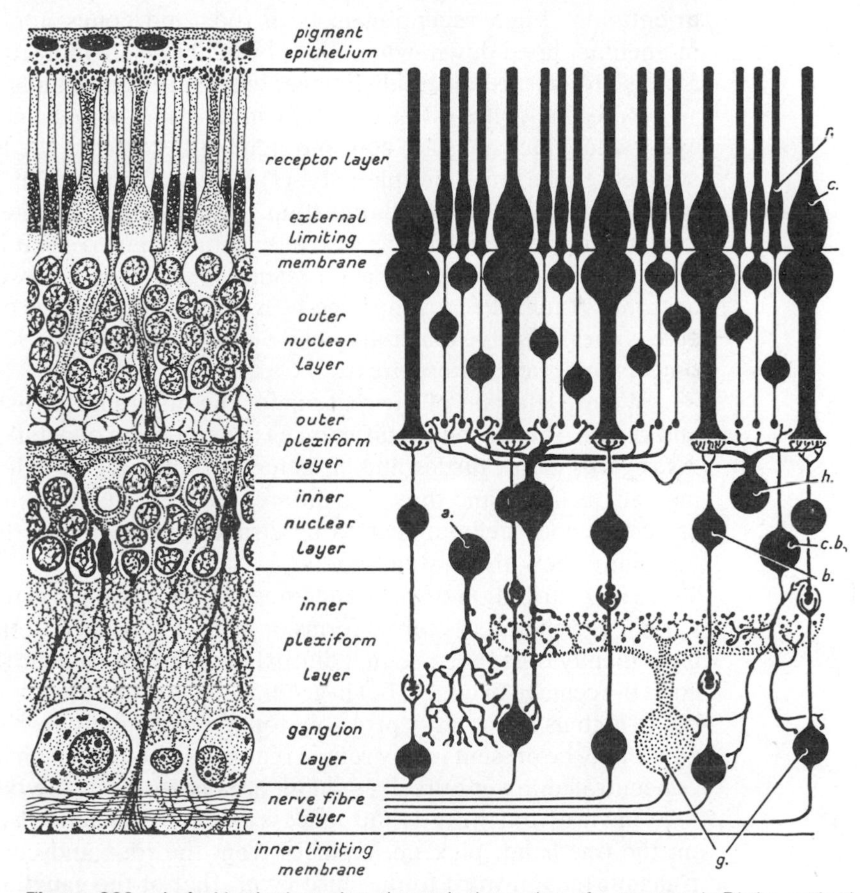

Figure 366. *Left,* Vertical section through the retina of a mammal; *Right,* retinal connections are revealed by silver impregnation. *a* and *h,* Nervous elements of more or less uncertain function. Other abbreviations: *b,* bipolar cells associated with single cones; *c,* cones; *cb,* bipolar cells, connecting with a series of rods; *g,* ganglion cells; *r,* rods. (From Walls, partly after Polyak.)

zone showing perpendicular striations; inside this again, are three distinct zones containing circular objects recognizable as cell nuclei. Appropriate staining reveals the true nature of these zones. The striated zone contains the elongated tips of the light-receiving cells, the rods and cones; the outer nuclear zone contains the cell bodies and nuclei of these structures. The middle nuclear zone is that of bipolar cells which transmit impulses inward from the rods and cones, and of accessory types of retinal nerve cells. The innermost nuclei are those of ganglion cells which pick up the stimuli from the bipolar elements and send fibers along the optic nerve to the brain.

The **rods** and **cones,** whose names derive from their usual shapes, are the actual photoreceptors. Each cell includes a sensory tip directed toward the choroid, a thickened section, and a basal piece containing the nucleus. One is immediately impressed by the fact that rod and cone cells in the vertebrate retina are *pointing the wrong way!* In a "logically" constructed retina, their tips should point, not away from, but toward the source of light. Some invertebrate (e.g., squid) eyes are so constructed; not those of vertebrates. This anomalous type of construction may be due to the retinal cells having first developed, phylogenetically, in the floor of a ventricle of the brain with their tips facing inward (and upward) to catch light rays entering the translucent body of a primitive chordate from above; when they were transferred to outfolded optic vesicles, these cells simply retained their original orientation. Light reception in both rods and cones appears to be due to the momentary breakdown, when "hit" by light rays, of molecules (rhodopsin, etc.) of a bluish to purplish or reddish tinge, which are compounds related to vitamin A.

Rods and cones differ markedly in function, as one can determine in his own eyes, where the cones are concentrated in the center of the field of vision, the rods situated, in the main, peripherally. (1) Rods are effective in faint light; cones come into play only with good illumination. At night one can often catch a glimpse of a faint star in the margin of the field of vision, but fail to see it if looking directly at it. (2) Cones as a group give good visual detail, rods a more blurred picture. For accurate vision we focus the cone-bearing center of the eye on the object. (3) Cones give color; rods give black-and-white effects only. In one's visual field, peripheral objects are gray and colorless.

We are ignorant of the reasons for the different "thresholds" to light of rods and cones. The reason for differences in perception of detail are pointed out below. As regards color, turtles and birds have colored oil globules in their cones which may act as filters and thus give differential reception to colors, but in other vertebrates all cone cells appear to be structurally identical; here again, we cannot "explain" how the system works.

The distribution of rods and cones in the various vertebrate groups is highly varied. In general, nocturnal forms or fishes living deep to the surface of the water have mainly rods in the retina; diurnal or forms living near the surface may have a good percentage of cones, but in general cones are relatively rare and color sensitivity as high as our own is probably found in relatively few vertebrates. Rods and cones may be present in any retinal region, but cones, when present, are commonly concentrated (as in ourselves) in an **area centralis** at the back of the retina.

Inward from the rods and cones is a layer of **bipolar cells,** with processes which, on the one hand, pick up impulses from the rods and cones and, on the other, transmit them inward to the third layer, that of the **ganglion cells** which form the optic nerve. It appears that often a single cone cell connects with a given bipolar cell, and only one such bipolar cell connects with an associated ganglion cell; thus each cone may have an individual pathway to optic nerve and brain. In contrast, a

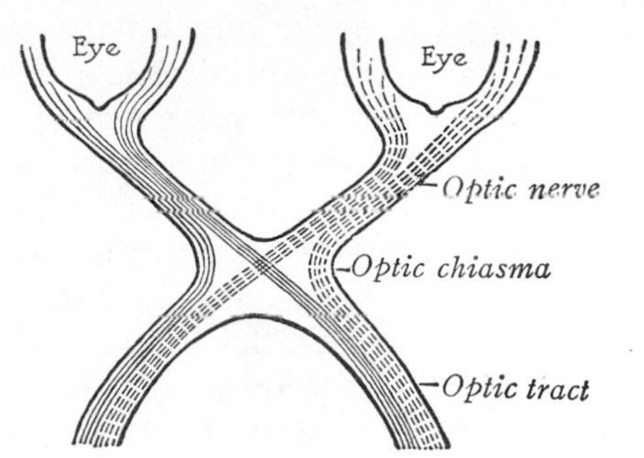

Figure 367. Diagram of the optic chiasma in a mammal with good stereoscopic vision. All fibers from the corresponding half of each eye pass to the same side of the brain. (From Arey.)

considerable number of rods converge into a single bipolar cell; in consequence, the brain obtains no information as to which of the group of rods has been stimulated, a condition which accounts for the lack of precision in vision using rods as compared with that of cones.

OPTIC NERVE. Although the nervous system is the subject of the next chapter, we may here discuss the central connections of the eye in relation to vision. The ganglion cells, situated on the inner surface of the retina, produce long fibers which converge to a point near the center of the retina, plunge through it (with the production of a **blind spot** because of absence here of rods and cones) and travel along the eye stalk to the brain as the optic nerve. While this is customarily called a nerve, it must be pointed out that, since the retina is, from an embryologic point of view, properly a part of the brain itself, the optic nerve is to be thought of not as a true, external nerve, but rather as a fiber tract connecting two regions of the brain.

As they reach the floor of the forebrain, the two optic nerves form the X-shaped **optic chiasma** (Fig. 367); in this crossroad, in most vertebrates, nearly all the fibers of the right optic nerve cross to the left side of the brain and vice versa; such a crossing of fibers is termed a **decussation.** In most vertebrate classes the two sets of fibers continue upward and backward to centers in the roof of the midbrain (tectum); in mammals almost all the fibers are, instead, relayed to a pair of special areas in the gray matter of the cerebral hemispheres.

In a majority of vertebrates the eyes are directed laterally, with nearly distinct fields of vision on either side; the brain builds up two separate pictures of two separate views. In a number of higher types, however—such as birds of prey and many mammals—the eyes are turned forward, the two visual fields overlap, and the two sets of impressions transmitted to the brain are more or less alike. In such cases the formation of two duplicate mental pictures seems an unnecessary procedure. Nevertheless this is done, as far as can be discovered, in nonmammalian forms, no matter how much overlap exists.

In many mammals, however, there appears **stereoscopic vision;** the two fields are mentally superimposed, with the result that such a form as man is aided, by the slight differences in point of view of the two eyes, in gaining effects of depth and three dimensional shape of objects, which are otherwise impossible of attainment. Associated with this new development is **incomplete decussation** at the chiasma. In mammals—and in mammals alone—we find that for those parts of the field which overlap, fibers from the areas in both retinas which view the same objects go to the same side of the brain. In consequence, certain groups of fibers do not cross (i.e., decussate), but turn a right angle at the chiasma to accompany their mates from the opposite eye. In man, for example, where the overlap of visual fields is nearly complete, practically all fibers from the left halves of both retinae enter the left side of the brain and those from the right halves enter the right side (Fig. 367). As a result, the visual area of each cerebral hemisphere builds up a half-picture of the total visual field as a "double exposure"; by further complicated interconnections between the hemispheres, the two halves of the picture are welded together to emerge into consciousness as a single stereoscopic view.

ACCESSORY STRUCTURES. We have earlier described the series of striated muscles which move the eyeball. Other external structures have to do chiefly with protection and care of the eye's external surface. Except in some sharks there is little development of eyelids in fishes. Lids are, however, developed in some fashion or other in most tetrapods; a dry cornea would become opaque, and closing the lids at intervals moistens and cleans the corneal surface. Upper and lower lids are always present and generally are opaque. In most groups the lower lid is the more prominent, but in mammals (and crocodilians) the reverse is true. In reptiles and birds generally and in many mammals there is a third eyelid, the **nictitating membrane,** a transparent fold of skin lying deep to the eyelids and drawn over the cornea from the anterior (or medial) to posterior margins.

No muscles were originally present in the eyelids, and various mechanisms have been evolved to move them. In some instances in lower tetrapods their movement is a passive one, opening and closing as the eyeball is pushed out or withdrawn; in other cases slips from the eyeball muscles extend forward to operate the eyelids. In mammals the facial muscles, grown forward over the head, form a ring of fibers which operate as a sphincter muscle in closing the eye (Fig. 213).

In terrestrial vertebrates are developed **lacrimal glands,** furnishing a "salt solution" to clean and moisten the cornea. In such primitive forms as the urodeles there is a row of small glands along the inside of the lower lid; in anurans, reptiles, and birds there tends to be a concentration of glandular development at the front (or medial) margin of the orbit; in mammals the lacrimal gland is usually developed at the outer or back corner of the eye. A useful adjunct, present in all tetrapods (except in turtles, where it is secondarily lost) is the **tear duct** (lacrimal duct), carrying surplus fluid from the corner of the eye to the nasal cavity (in lizards and most snakes surplus fluid is carried into the vomeronasal organ).

MEDIAN EYES. Ancestral vertebrates had a third eye, medially situated on the forehead and directed upward. In the oldest ostracoderms (as in Fig. 19) a socket which obviously contained such an eye is universally present and often conspicuous, although always much smaller than that of the paired eyes. This eye was generally present in the placoderms and present in Devonian bony fishes of all major groups. Further, it was universal in all the older land vertebrates—ancient amphibians and Paleozoic reptiles of all sorts (Figs. 176 *A, B;* 185 *A;* 186 *A*). By Triassic times, however, this accessory visual organ appears to have gone out of fashion. Today, median eyes are well developed only in lampreys, on the one hand, and *Sphenodon* and some lizards on the other (Figs. 368, 416); buried under the skin, they can do little more than detect the presence or absence of light, although a miniature cornea, lens and retina may be developed.

Like lateral eyes, these median eyes are brain outgrowths. There is, however, a curious complication in the story: the median eye, it appears, is not the same throughout, but may develop from either of two dorsal outpocketings of the brain roof, the **parietal organ** (parapineal organ) and **pineal organ.** Both develop eye structures in the lamprey; the former only in lizards and *Sphenodon*. The situation suggests that possibly the remote ancestor of the vertebrates may have had paired dorsal eyes as well as paired lateral ones. Despite loss of function, the pineal organ persists in higher vertebrates as a glandular structure having, as noted later, endocrine functions.

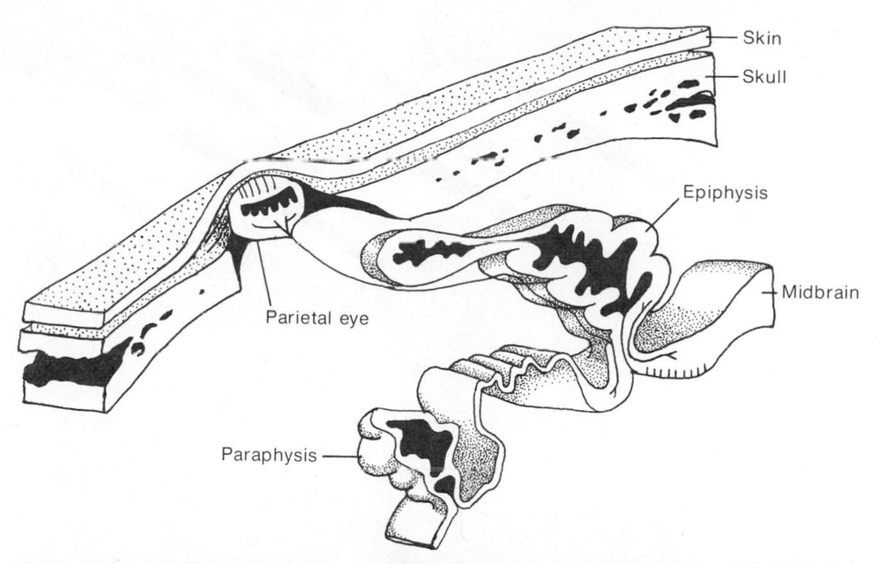

Figure 368. Median eye and associated structures in a lizard. The epiphysis (or pineal) and parietal organ (parapineal) are both dorsal outgrowths of the diencephalon which may, in different animals, form "eyes"; the paraphysis is another outgrowth, but never eyelike. In some lizards (and *Sphenodon*) the parietal eye lies within the parietal foramen in the skull and is a well-formed structure with a lens-like upper epithelium and retina-like lower part. It is attached to the diencephalic roof by a thin nerve. (After Wurtman, Axelrod, and Kelly.)

LATERAL LINE ORGANS

A highly developed sensory system of a type quite unknown in land dwellers is that of the lateral line organs (Figs. 369–371) found in fishes and larval (or aquatic) amphibians. The receptors are sensory cell clusters, the neuromasts; these are generally situated along a series of canals or grooves on the head and body. A main element of the system is the **lateral line** in the narrower sense of the term—a long canal or groove running the length of the trunk and tail. It continues forward onto the head, where similar canals form a complex pattern, typically including branches running forward both above and below the orbits and downward to run forward along the lower jaw. Isolated neuromasts, termed **pit organs,** may also be found on the head of fishes, and in modern amphibians the arrangement of the neuromasts on the head is discontinuous, with little evidence of linear arrangement. The canals in which the sensory organs are generally enclosed open to the surface at intervals by small pores; in a few fishes open grooves take the place of canals.

The sense organs of the lateral line system are the **neuromasts** (Fig. 369 *B*), consisting of bundles of cells which have much the appearance of taste buds. Each elongate sensory cell has a hairlike projection; invariably present above the "hairs" and enclosing their tips is a flexible mass of gelatinous material secreted by the neuromast cells, which may be termed a **cupula.** The neuromast organs are innervated by cranial nerves—most of those on the head by nerve VII, the entire body line by nerve X.

As we lack structures at all comparable to these ourselves (except, as we shall see, in the internal ear), the nature of the sensations registered by them has been difficult for us to determine. It appears that they respond, by movement of the cupula and consequent bending of the "hairs," to water vibrations or currents and thus aid the fish in locomotion through the water, where visible "landmarks" are

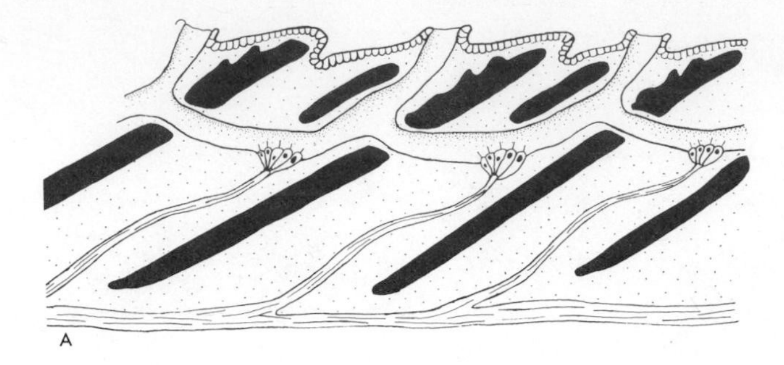

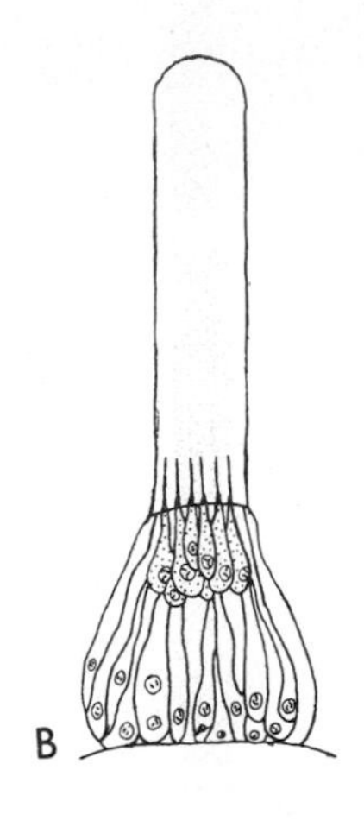

Figure 369. Lateral lines of fish. *A,* Part of the main lateral line canal of the perch *(Perca)* showing three pores to the surface and three neuromast organs. *B,* Diagram of a single neuromast organ showing the sensory processes and the gelatinous cupula (cf. Fig. 373). (After Goodrich and Yapp.)

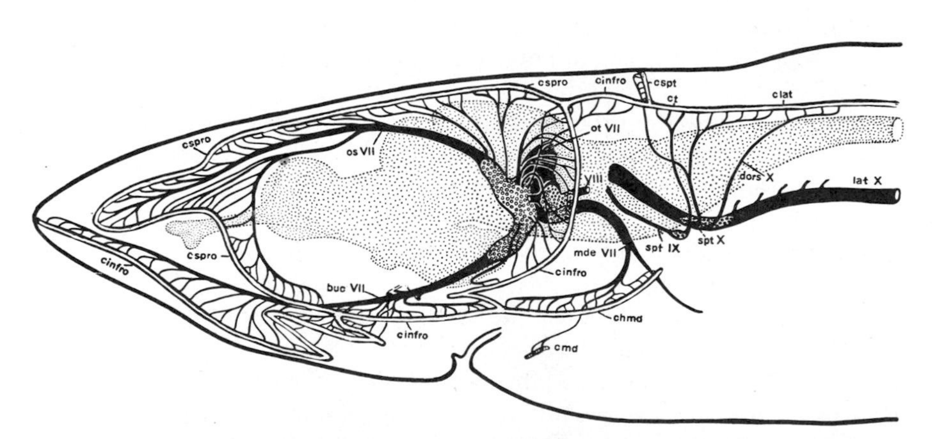

Figure 370. Left side of the head of a shark to show the lateral line canals (parallel lines) and the nerves (black) supplying them. Pit organs are not shown; *buc VII,* buccal ramus of nerve VII; *chmd,* hyomandibular canal; *cinfro,* infraorbital canal; *clat,* lateral line canal proper; *cmd,* mandibular canal; *cspro,* supraorbital canal; *cspt,* supratemporal (or occipital) canals; *ct,* temporal canal; *dors X,* dorsal ramus of nerve X; *lat X,* lateral line ramus of nerve X; *mde VII,* external mandibular ramus of nerve VII; *os VII,* superficial ophthalmic ramus of nerve VII; *ot VII,* otic ramus of nerve VII; *spt IX,* supratemporal ramus of nerve IX; *spt X,* supratemporal ramus of nreve X. The brain is stippled in and the ganglia of the cranial nerves shown by a pattern of small circles. (From Norris and Hughes.)

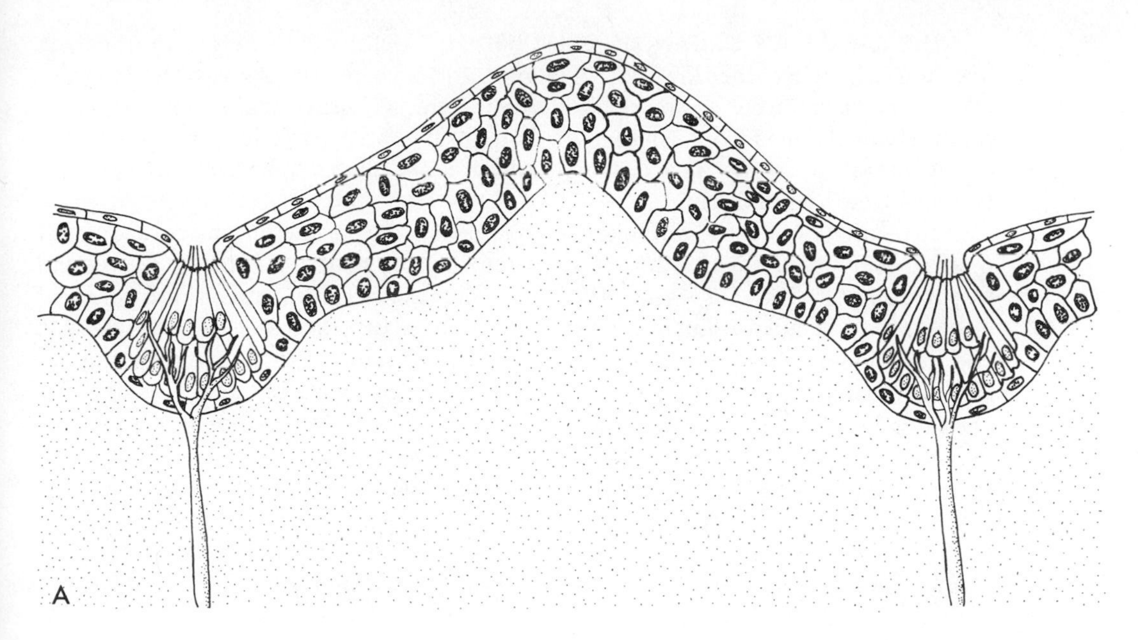

Figure 371. Lateral line organs of amphibians. *A,* Two neuromast organs in the skin of an aquatic frog *(Xenopus); B,* general pattern of neuromast organs in the skin of a urodele *(Pleurodeles).* The neuromasts are not enclosed within canals, but are separate organs on the skin; however, they do tend to form lines along the body and head. (After Murray and Escher.)

seldom present. Many reptiles and mammals have returned to an aquatic existence, but this useful sensory aid, once lost, never reappears.

Related to the lateral line system in development and probably in phylogeny, although differing in structure and probable function, are the **ampullae of Lorenzini,** clusters of jelly-filled little tubules scattered over the head region—particularly the snout—of elasmobranchs and containing at their bases sensory cells. Some may, like the proper lateral line organs, be sensitive to hydrostatic pressure; others, however, are highly sensitive receptors for temperature changes or for electrical charges.

THE EAR

First thoughts as to the primary anatomic or functional aspects of the vertebrate ear are liable to be misleading when based on familiar human features. One tends to think, when the word is mentioned, of the ornamental pinna of the mammalian ear, or, perhaps, of the middle ear cavity behind the drum, with its contained ear ossicles. These items, however, are entirely lacking in fishes; the basic ear structures of all vertebrates are those of the internal ear, the sensory structures buried deep within the otic capsule. We naturally think of hearing as the proper ear function. But in the ancestral vertebrates audition was apparently unimportant and perhaps absent; equilibrium was the primary sensory attribute of the "auditory" organ.

THE EAR AS AN ORGAN OF EQUILIBRIUM. Before considering the hearing function, which becomes increasingly important as we ascend the vertebrate scale, we may discuss the ear as an organ of equilibrium, a basic function which remains relatively unchanged from fish to man. Equilibrium is a type of sensation produced by the internal ear alone; all accessory structures are related to hearing and need not concern us at the moment.

In a variety of fishes, amphibians, and reptiles the paired internal ears are built upon a relatively uniform pattern in which most of the structures present are related to equilibration (Fig. 372 *A–D*). The **membranous labyrinth** consists of a series of closed sacs and canals lying within the ear capsules on either side of the braincase and containing a liquid, the **endolymph,** not dissimilar to that of the interstitial spaces. Two distinct major sac-like structures are generally present, the **utriculus** above and, more ventrally, the **sacculus.** A slender tube, the **endolymphatic duct,** usually extends upward and inward from them to terminate within the braincase in an **endolymphatic sac.** In both major ear vesicles there is found a large oval "spot" consisting of a sensory epithelium associated with branches of the auditory nerve; these are the **utricular macula** and the **saccular macula.** A pocket-like depression is formed in the floor of the sacculus. This is the **lagena,** which contains a small **lagenar**

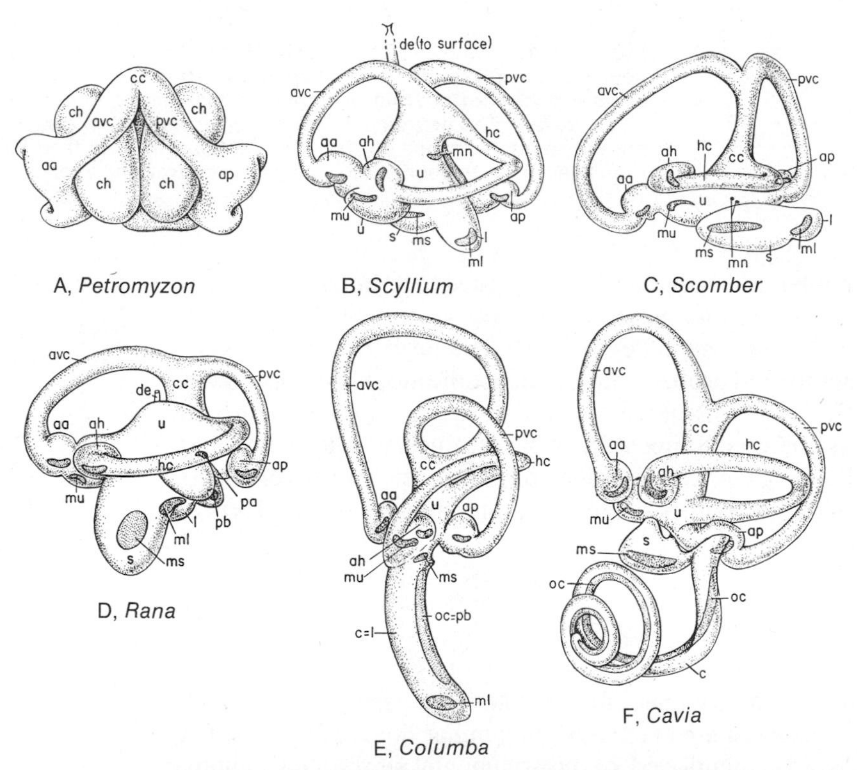

Figure 372. Membranous labyrinth of *A,* lamprey; *B,* shark; *C,* teleost; *D,* frog; *E,* bird; *F,* mammal; all external views of the left ear. Sensory areas are shown (except in *A*) as if the membrane were transparent. *aa,* Ampulla of anterior canal; *ah,* ampulla of horizontal canal; *ap,* ampulla of posterior canal; *avc,* anterior vertical canal; *c,* cochlear duct; *cc,* crus commune with which both vertical canals connect; *ch,* chambers in the lamprey ear lined with a ciliated epithelium; *de,* endolymphatic duct; *hc,* horizontal canal; *l,* lagena; *m,* macula of lagena; *mn,* macula neglecta; *ms,* macula of sacculus; *mu,* macula of utriculus; *oc,* organ of Corti; *pa,* papilla amphibiorum; *pb,* papilla basilaris; *pvc,* posterior vertical canal; *s,* sacculus; *u,* utriculus. (After Retzius.)

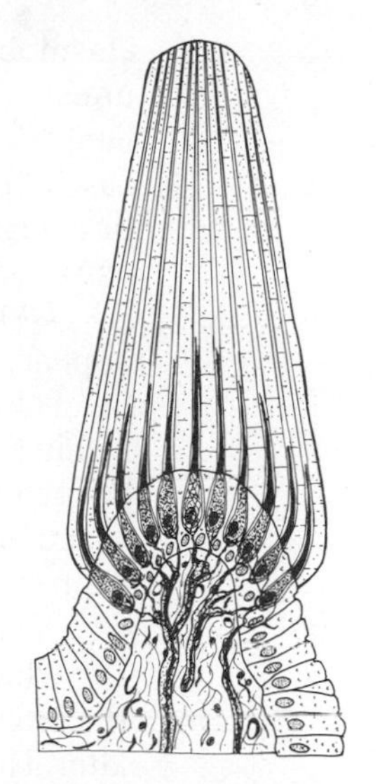

Figure 373. Crista (neuromast organ) from a human semicircular canal. Each sensory cell, here heavily stippled, has a hairlike process extending into the flexible, gelatinous cupula; bending of the cupula stimulates the cells. (After Bargmann.)

macula. The sensory cells of these maculae (and, indeed, those of the entire ear) are comparable to the neuromast cells of the lateral line system, having hairlike projections embedded in an overlying gelatinous material (Fig. 373). In the utricular and saccular maculae, and often that of the lagena as well, this material becomes a thickened structure in which are deposited crystals of calcium carbonate, forming an "earstone" or **otolith.** In ray-finned fishes the otoliths are massive structures which may fill almost the entire cavity of the vesicles.

The utricular macula, and to a much lesser degree, those of the sacculus and lagena, register, by the tilt of the otolith and bending of its sensory "hairs," the position of the head and linear acceleration; somewhat similar organs are found in a number of invertebrate types. They do not, however, furnish information as to turning movements. This is the function of another series of organs, the **semicircular canals.**

These tiny tubes spring out from the utriculus and connect with it at either end. In every jawed vertebrate three such canals are present, each running at right angles to the other, and thus representing the three planes of space. Two lie in vertical planes, the **anterior vertical canal** angling forward and outward from the upper surface of the utriculus, and the **posterior vertical canal,** running backward and outward; a **horizontal canal** extends laterally. Each canal has at one end a spherical expansion, an **ampulla.** The vertical canals bear these at their outer and lower ends, anteriorly and posteriorly; the horizontal canal (for no known reason in particular) has its ampulla anteriorly placed. Within each ampulla is a sensory area, usually elevated, termed a **crista** (Fig. 373). Here we find again the familiar neuromast "hair" cells, their tips embedded in a cupula. It seems clear that their function is to register turning movements in the several planes of space; displacement of liquid in one or more canals displaces the cupulae, with consequent bending of their sensory hairs.

Despite the fact that these organs of equilibrium are essentially uniform in basic structure in every jawed vertebrate, some variations occur here and there. For example, in sharks, sacculus and utriculus form but two parts of a common sac; in rays the canals are connected only by narrow ducts with the remainder of the system. The endolymphatic duct usually terminates in a sac of modest size within the cranial cavity but in frogs may expand the length of the spinal canal, and in elasmobranchs the duct extends upward to open on the top of the head. The most unusual conditions are found (as in other organ systems) in the cyclostomes. The lamprey has but two semicircular canals (Fig. 372 *A*); the hagfish but one. As in the case of the nose, we cannot be sure whether the cyclostome condition is a primitive or a degenerate one, although the presence of only two canals in some of the early ostracoderms suggests the former.

ORIGIN OF THE VERTEBRATE EAR. Embryologically, the internal ears first appear, like the lateral line organs, as ectodermal thickenings on either side of the head (Fig. 374). These sink inward to form a pair of sacs, which for some time may retain a connection with the exterior (the endolymphatic ducts in an adult shark are a retention of this condition). Typically, each sac then divides into utricular and saccular portions; from the former there separate off the semicircular canals, and from the latter arise further structures, discussed later.

This embryologic story—together with the nature of the sensory endings of the ear which closely resemble the externally placed neuromast organs—suggest that the internal ear originated phylogenetically as a specialized, deeply sunk part of the lateral line system. As will be noted in the next chapter, the nerves from the two sets of organs are closely associated. The two are closely associated functionally as well, furnishing a fish the major part of the data by which its locomotion is regulated.

HEARING IN FISHES. Although not the primary function of the internal ear, hearing is certainly present in many fishes, particularly among the teleosts, but it is uncertain as to which one or more of several sensory maculae are concerned. The saccular macula may be the main receptor in most cases, but possibly the lagenar and main utricular maculae may be involved and another possible suspect is the

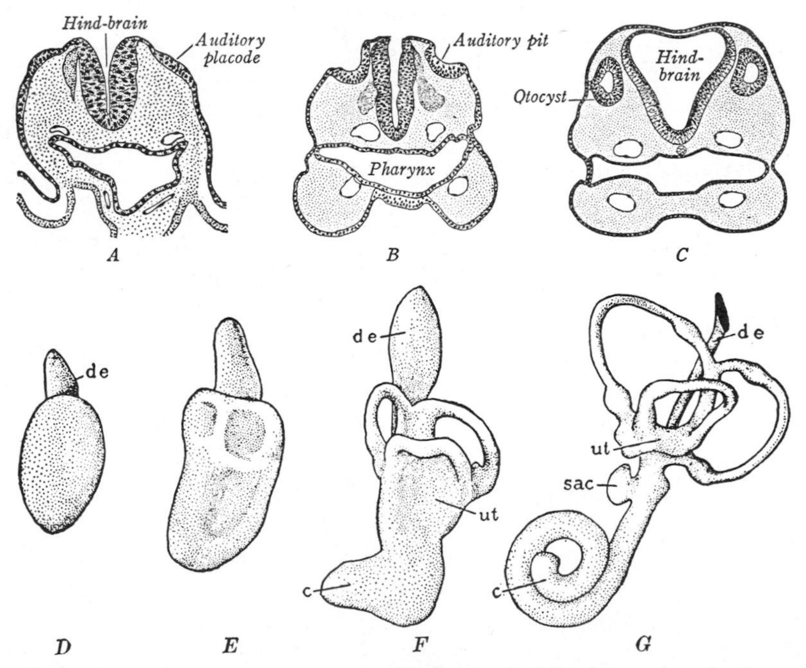

Figure 374. Diagrams to show the development of the internal ear in mammals. *A* to *C*, Cross sections of the head of early embryos; an ectodermal placode sinks inward on either side to form an otic vesicle. *D* to *G*, Successive stages in the development of the various parts of the membranous labyrinth from the otic vesicle: *c*, cochlear duct; *de*, endolymphatic duct; *sac*, sacculus; *ut*, utriculus. (*A* to *C* from Arey; *D* to *F* after His and Bremer.)

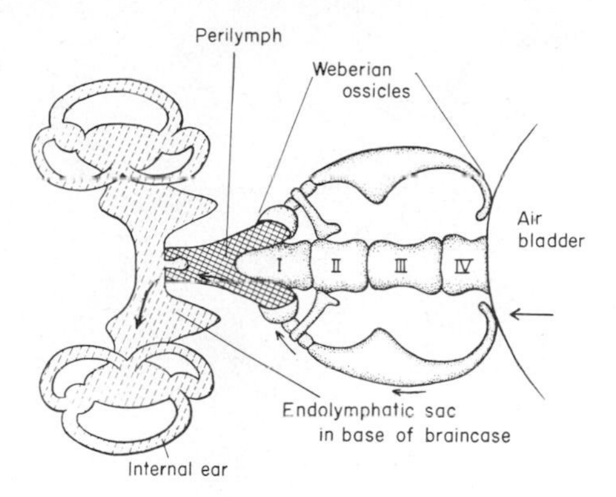

Figure 375. Diagrammatic horizontal section of the posterior part of the head and anterior part of the body of a teleost with weberian ossicles. Vibrations in an anterior subdivision of the air bladder set up corresponding vibrations in a series of small ossicles which in turn set up waves in a perilymphatic sac. This, again, sets up vibrations in an endolymphatic sac at the base of the braincase. Arrows show the course of transmission of the vibrations. Roman numerals indicate the vertebrae from which the weberian ossicles are derived. (After Chvanilov.)

macula neglecta (Fig. 372 *B, C*), a small sensory spot in the utriculus of many fishes and lower tetrapods.

Although fishes lack the middle ear apparatus utilized by terrestrial vertebrates in bringing external vibrations to the internal ear, several groups of teleosts have independently evolved comparable structures. Thus, a group including the catfishes, minnows, and carp (Ostariophysi) use the air bladder as a resonating chamber and carry vibrations forward to the ear by a series of small bones, termed the weberian ossicles, derived from the anterior vertebrae and ribs (Fig. 375).

THE MIDDLE AND EXTERNAL EAR IN AMNIOTES. Hearing is an important sense in tetrapods, but the sounds to be heard are relatively faint air waves which can ordinarily have little direct effect in setting up endolymphatic vibrations in the internal ear. Devices for amplification of these waves and their transmission to the internal ear are a necessity. Such devices, established, it seems, in the earliest tetrapods, are retained with little change in such reptiles as lizards (Fig. 376, 377 *A–C*). The spiracular gill cleft and the hyomandibular bone are the elements utilized. The spiracular pouch of the embryo never breaks through to the surface. The corresponding surface depression, where developed, is an **external auditory meatus;** the thin membrane between this depression and the pouch becomes the eardrum or **tympanic membrane,** which picks up air vibrations. The

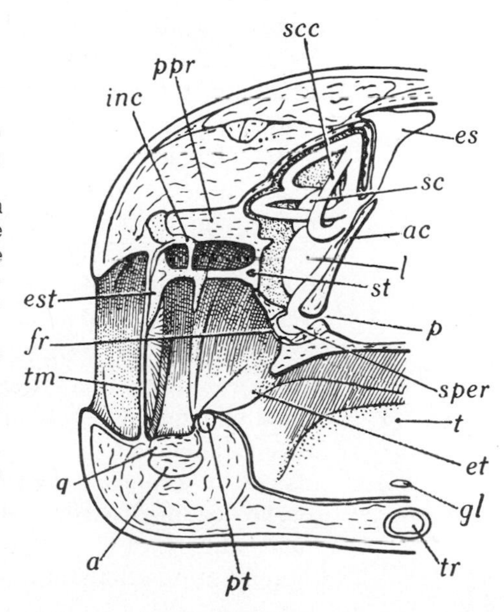

Figure 376. Posterior view of the left side of a lizard head to show the auditory apparatus. A shallow external depression leads to the ear drum *(tm)*. Internal to the drum, the stapes is seen, divided into two parts, the "extracolumella" *(est)* and columella or stapes proper *(st)*; processes from the former articulate above with the skull *(inc)* and below with the quadrate anterior to the middle ear cavity. This cavity opens by a broad eustachian tube *(et)* to the throat. The internal ear is shown in diagrammatic fashion. Other abbreviations: *a*, articular bone of lower jaw (= malleus); *ac*, inner wall of auditory capsule; *es*, endolymphatic sac; *fr*, fenestra rotunda; *gl*, glottis; *l*, lagena; *p*, perilymphatic duct connecting inner ear with cranial cavity; *ppr*, paroccipital process of otic region; *pt*, pterygoid; *q*, quadrate (= incus); *sc*, sacculus; *scc*, semicircular canals; *sper*, position of perilymphatic sac; *t*, tongue; *tr*, trachea. (From Goodrich, after Versluys.)

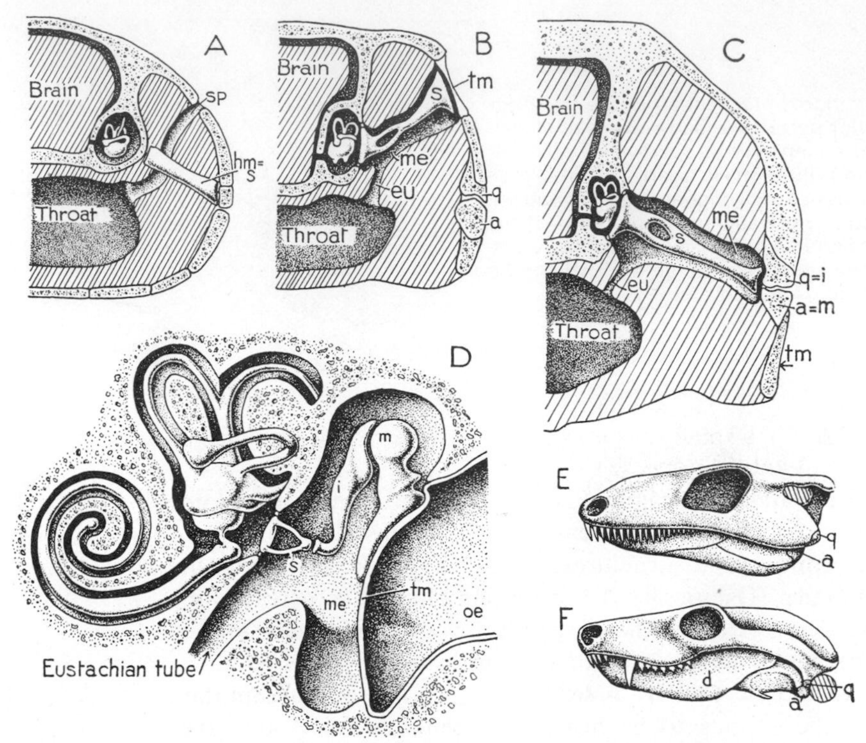

Figure 377. Diagrams to show the evolution of the middle ear and auditory ossicles. Diagrammatic sections through the otic region of the head of *A*, a fish; *B*, a primitive amphibian; *C*, a primitive reptile; *D*, a mammal (showing the ear region only); *E*, side view of the skull of a primitive land vertebrate; *F*, of a mammal-like reptile to show the shift of the eardrum from the otic notch of the skull to the region of the jaw articulation. *a*, Articular; *d*, dentary; *eu*, eustachian tube; *hm*, hyomandibular; *i*, incus; *m*, malleus; *me*, middle ear cavity; *oe*, outer ear cavity; *q*, quadrate; *s*, stapes; *sp*, spiracle; *tm*, tympanic membrane. (From Romer, Man and the Vertebrates, University of Chicago Press.)

pouch, enlarged to contain the hyomandibula, becomes a **middle ear cavity;** its connection with the throat is the **eustachian tube.** The fish hyomandibular bone changes its function to become a rodlike **stapes** (or columella); this crosses the middle ear cavity from the drum to an opening through the otic capsule into the internal ear, the **fenestra ovalis.** It thus transmits vibrations picked up by the drum to the liquids of the internal ear and, eventually, to its contained sensory structures. The availability of the hyomandibular for its new function was due, as discussed in an earlier chapter, to its being no longer required for support of the jaws; it may, however, retain a process connecting with the quadrate near the jaw articulation. As in the lizard shown in Figure 376, the stapes may form in two or more pieces and develop still further processes connecting with the skull or hyoid arch. The avian middle ear is similar to that of reptiles.

In mammals, external ear structures become prominent for the first time. There is a deep, tubular external meatus, and there is almost always a projecting **pinna** which may be of value as a collector of sound waves. A more fundamental change, however, is that seen within the middle ear cavity (Fig. 377 *D*). Here, instead of a single bone, there is present an articulated series of three **auditory ossicles** leading from eardrum to oval window—**malleus, incus,** and **stapes** (hammer, anvil, and stirrup). The origin of this series of ossicles was long debated. It was thought at one time that they might be due to a subdivision of the single reptilian element. Embryology, comparative anatomy, and paleontology combined have, however, revealed the true story. The inner element, the stapes, although much shortened, is equivalent to the whole stapedial apparatus of reptiles. The other two elements are homologous with the articular and quadrate bones which, in lower vertebrates, form the jaw joint. Mammals have evolved a new joint for the jaw; the structures forming the older joint have been put to new use. The reptilian eardrum lay close to the jaw joint; the articular has remained attached to it and become the malleus. The quadrate, connecting with the articular, on the one hand, and the stapes (the old fish hyomandibular) on the other, retains these connections as the incus. These bones, originally gill bar elements, afford a good example of the changes of function which homologous structures can undergo. Breathing aids have become feeding aids and, finally, hearing aids.

THE INTERNAL EAR IN REPTILES (Figs. 378, 379, 380 *A*). In tetrapods the parts of the internal ear devoted to equilibrium show little change; the auditory apparatus, however, gradually develops into structures which attain such size and importance that the older regions of the sacs and canals are often termed (rather slightingly) the **vestibule** of the internal ear.

It is the lagenar region in which this expansion takes place. The macula of the lagena persists in tetrapods, except in mammals above the monotreme level, but is unimportant. A second sensory area developed here—the **basilar papilla**—is the auditory organ of tetrapods, to which vibrations, brought in from without by the stapes, are received.

Between the walls of the otic capsule and the sacs and canals of the endolymph system are spaces crossed by connective tissue strands and filled by a second otic fluid, the **perilymph** (Fig. 379). In tetrapods there is organized in the perilymphatic spaces a conducting system which forms the last link in the transmission of vibrations to the basilar papilla. The stapes brings vibrations to the oval window. Inside this there is developed a large **perilymphatic cistern,** against which the stapes plays (Figs. 378, 379, 380 *A*). Vibrations received here are carried in a perilymph-filled duct around the lagena to its posterior border. At this point the duct lies beneath the basilar papilla and is separated from the under surface of its sensory

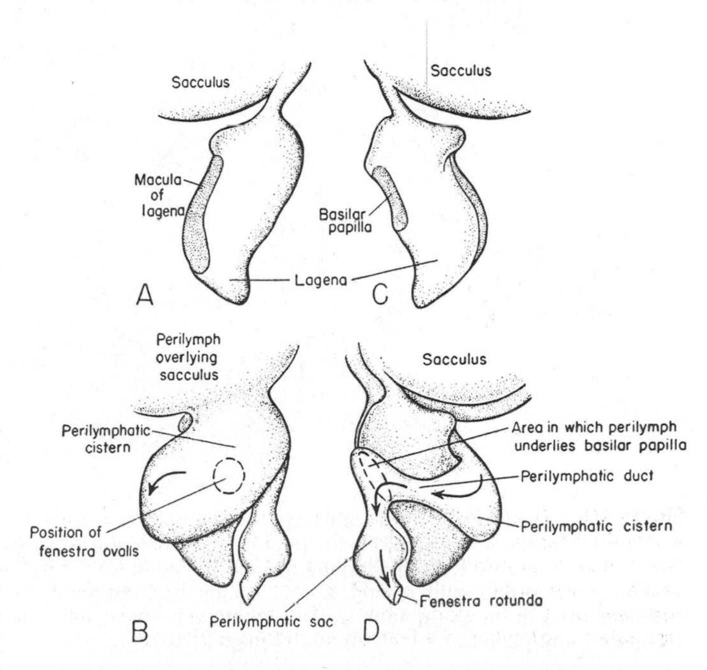

Figure 378. The ear of a late embryo of a lizard *(Lacerta)*. *A,* Left ear, lateral view of the membranous labyrinth in the floor of the sacculus and the lagena. *B,* The same with the perilymphatic system shown in addition. *C, D,* Medial views comparable to *A* and *B*, respectively. Arrows indicate course of conduction of vibrations from stapes to basilar papilla and on to the "round window" at the distal end, beyond the perilymphatic duct.

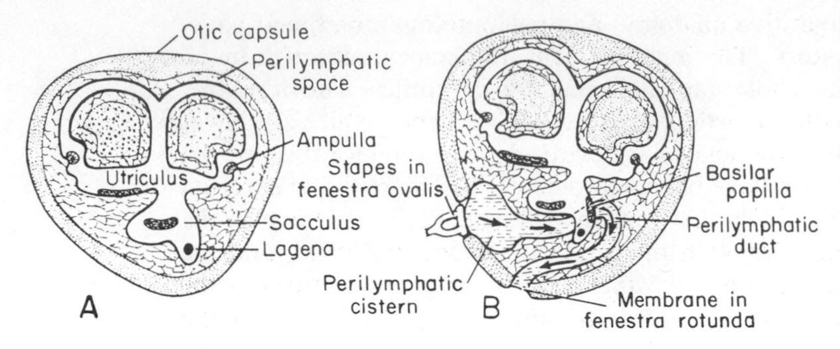

Figure 379. *A,* Schematic section through the ear capsule of a fish, to show the perilymphatic space surrounding the membranous labyrinth containing endolymph. *B,* Similar scheme of a tetrapod, in which a part of the perilymph area (arrows) is specialized to conduct sound from the fenestra ovalis to and past the auditory sensory area. The maculae are darkly shaded. (After de Burlet.)

cells only by an elastic **basilar membrane.** Vibrations of this membrane agitate the hair cells in the papilla, and at long last, in this roundabout fashion, the sensory organ is reached. This situation—an auditory sensory structure agitated by vibrations of the membrane at its base—is a fundamental feature in the construction of the apparatus in all amniotes. We shall find it repeated, and capable of description in much the same words, in birds and mammals.

As a final point here, it must be noted that a release mechanism for the vibrations carried must, of course, be set up at the far end of the perilymph duct. This mechanism was primitively a **perilymphatic sac,** projecting into the braincase. In most tetrapods, however, there is a further development. A "round window"—the **fenestra rotunda***—develops in the walls of the otic capsule; a membrane here vibrates in phase with the impulses received through the oval window at the other end of the perilymph system.

DEVELOPMENT OF THE COCHLEA (Fig. 380). Both birds and mammals have greatly refined their hearing ability by the development of a **cochlea.** The crocodilians demonstrate the manner of its development. Three structures are involved—the lagena, the perilymphatic duct, and the basilar papilla. The lagena expands into

*There is, unfortunately, also a foramen rotundum in the skull—they are *not* synonymous. The same unhappy situation occurs with the oval "hole" and "window" (foramen ovale and fenestra ovale).

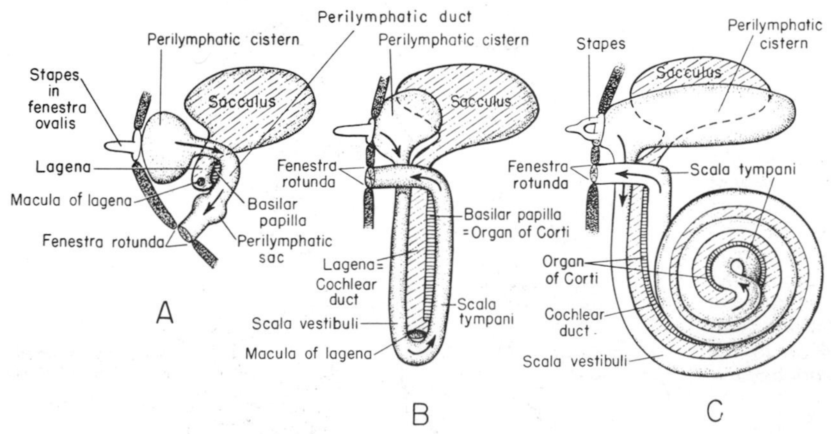

Figure 380. Diagrams of the saccular region to show the evolution of the cochlea. *A,* Primitive reptile with a small basilar papilla adjacent to the perilymphatic duct. *B,* The crocodile or bird type; the lagena has elongated to form a cochlear duct, the basilar sense organ with it, and a loop of the perilymphatic duct follows the cochlear duct in its elongation. *C,* The mammalian type; the cochlea is further elongated and coiled in a fashion economical of space.

a long tube, filled of course with endolymph; this is the **cochlear duct** (or scala media). The basilar papilla is likewise expanded into an elongate structure, running the length of the cochlear duct, as the **organ of Corti.** The perilymphatic duct, elongating beneath it in corresponding fashion, becomes a double loop. The part of the loop leading in from the oval window (in the vestibular part of the ear) is termed the **scala vestibuli;** the distal limb, leading to the round window (which is covered by a "tympanum") is the **scala tympani.** The three tubes—the endolymph-filled cochlear duct and the two perilymph-filled scalae—are closely pressed together and form a primitive cochlea.

The avian cochlea differs from that of a crocodilian mainly in greater elongation of the structure. In mammals there is still further lengthening of this triple tube system; as its name implies, it is in mammals coiled into a tidy spiral to keep it within the bounds of the otic capsule.

We cannot here discuss in detail the complicated microscopic structure of the organ of Corti (Figs. 380 *C*, 381) which includes a complex system of sensory and supporting cells and a covering membranous flap. An important element is the basilar membrane which underlies the organ. As in lower tetrapods, vibrations of this membrane by waves brought in by the perilymphatic system are responsible for stimulation of the organ of Corti. The functional "reason" for elongation of the basilar papilla appears to be discrimination of sounds of different pitch; the mem-

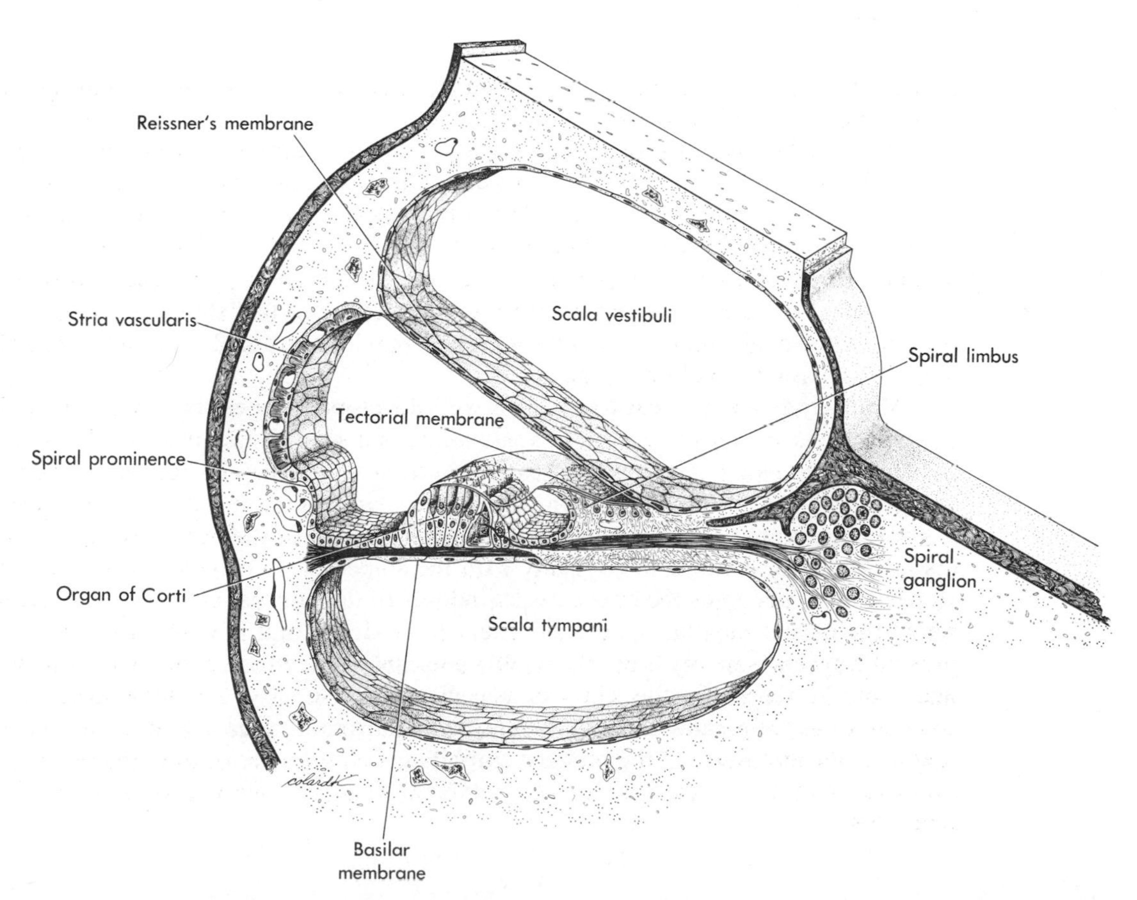

Figure 381. A much enlarged and schematic section through one turn of a mammalian cochlea to show the organ of Corti (cf. Fig. 380 *C*). (From Bloom and Fawcett.)

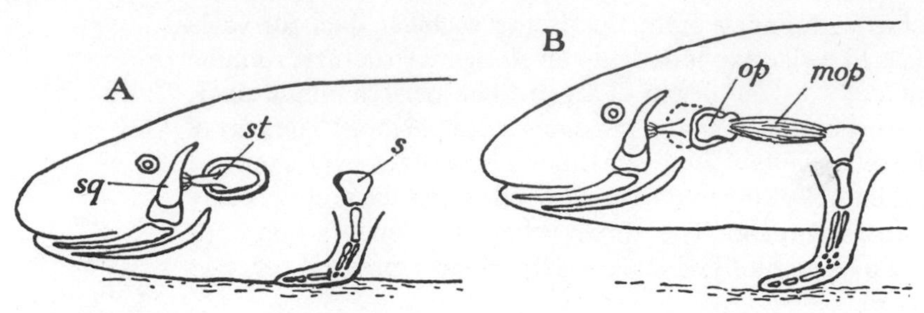

Figure 382. Diagram to show mechanism of communication between exterior and inner ear in urodeles. *A,* An aquatic form in which the stapes, or columella, picks up vibrations by a ligamentous attachment to the squamosal. *B,* A type in which the stapes is reduced, and the operculum picks up ground vibrations through a muscular connection with the shoulder girdle. *mop,* Opercular muscle; *op,* operculum; *s,* scapula; *sq,* squamosal; *st,* stapes. (After Kingsbury and Reed.)

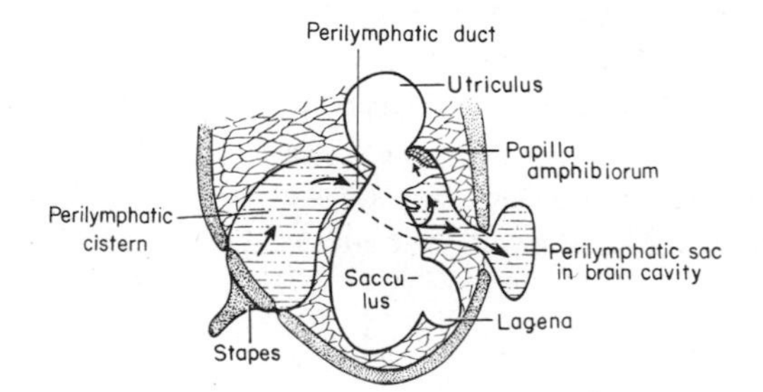

Figure 383. Schematic section of the internal ear in a salamander. The basilar papilla is absent here, but is found in addition to the papilla amphibiorum in anurans. (After de Burlet.)

brane grades in width along its length and is, in correlation, sensitive to different wave lengths at different parts of its extent.

THE EAR IN AMPHIBIANS. We have above omitted almost all reference to the ears of living amphibians for the reason that conditions in these forms are not, in general, primitive but specialized and seemingly degenerate in most cases and are, further, extremely varied. Four main points may be noted: (1) The drum and the middle ear cavity are often missing; (2) the stapes is frequently reduced or absent; (3) a second ossicle, the operculum (not the operculum of fishes), is frequently present in the oval window; and (4) the basilar papilla is often absent and a special amphibian papilla functions in its stead.

Many frogs have an eardrum and a well-developed stapes; but the drum and the middle ear cavity are absent in some anurans and in all members of the other two orders, and such hearing as is possible in these forms may be accomplished by using other types of "pickup," such as the two shown in Figure 382. The **operculum** is a flat plate, apparently a detached segment of the wall of the otic capsule, which fits into the oval window in company with the stapes or, in urodeles in which the stapes is lost, occupies the entire oval window. In the internal ear (Fig. 383) frogs retain the basilar papilla, but this is absent in urodeles, and in both cases there is present a special sensory area, the **papilla amphibiorum,** which appears to play the main role in hearing in this class of vertebrates. Evidently, auditory structures were in an experimental stage in the ancestral tetrapods. Here, as in many other features, the ancestors of the modern amphibians chose a series of paths less progressive than those chosen by their ancient relatives who gave rise to the amniotes.

16 The Nervous System

In a protozoan the single cell in itself receives sensations and responds to them. In higher, metazoan organisms there tends to be, to an increasing degree, a differentiation between cells specialized for the reception of sensations—**receptors**—and those which make the appropriate response—the **effectors.** In lowly forms the relations between these two types of cells may remain relatively simple; receptor cells may, by their physical and chemical activities, arouse their neighbors to respond. Even in vertebrates there is a retention of such a primitive method of stimulation in the circulation of hormones. But in most metazoans a means for more rapid and specific transmission of stimuli is present as the nervous system.

In primitive metazoans, such as coelenterates, this system may be merely a diffuse network of cells and fibers spread through the tissues. But in most animals of any degree of complexity the nervous system is more highly organized, with nerve trunks and centers where transfer of impulses between fibers takes place. In most groups a dominant center, a brain of some sort, makes its appearance. In the vertebrates the brain is situated anteriorly, close to the major sense organs, with a single dorsal hollow nerve cord—the spinal cord—running backward from this along the body. Brain and cord form the **central nervous system.** Running outward are numerous paired **nerves,** along which **ganglia**—clusters of nerve cells—may be found; these nerves and ganglia constitute the **peripheral nervous system.** We have noted that embryologically the nervous tissues are of ectodermal origin—mainly from the neurectoderm of the neural tube and neural crests, with additions from nearby ectodermal placodes.

STRUCTURAL ELEMENTS

THE NEURON. The nervous system contains numerous cell bodies, but more prominent are bundles of slender but elongate fibers which make up much of the bulk of the system. These fibers are, universally, processes of cells rather than independent structures. The basic units of the nervous system are **neurons;** each consists of a cell body and its processes, long or short.

Most of the cell bodies of the neurons are situated within the central nervous system. The shape is frequently stellate, because of the presence of multiple processes (Fig. 384); with appropriate stains, microscopic preparations show in the protoplasm various characteristic structures. Most notable are **Nissl bodies,** containing large amounts of RNA; this indicates that the cell body is the "manufacturing center" for the whole neuron, and the materials formed here flow out into the axon and other processes. In the adult there is very little evidence of cell division in the neurons, indicating, as a peculiarity of the nervous system, that a full complement of nerve cells has been attained by about the time of birth or hatching. In consequence (although cell processes may regenerate), destruction of a nerve cell through injury or disease is a permanent loss.

NERVE FIBERS. Extending outward from the cell body of the neuron are slender processes, whose distribution and lengths are quite variable. Most often thought of as typical are such neurons as those that innervate striated body muscles (Fig. 384 *B*). In these motor neurons short processes, slender, numerous, and branching, carry impulses inward toward the cell body; they are termed **dendrites,** from their treelike appearance. There is a single **axon,** a relatively stout and elongate process, which may be in large animals as much as several meters in length, carrying impulses away from the region of the cell body. A second common type is that of the afferent neurons that carry sensory stimuli inward to the central nervous system (Fig. 384 *A*); here the long process leading in from the point of reception to the cell body adjacent to the cord, as well as a second long process, which enters the cord, are comparable in structure to the axon of a motor neuron. Motor neurons may be termed **monopolar** and typical sensory neurons **bipolar** (**multipolar** neurons, as well, may be present in the central nervous system).

Functionally, the most important part of a major nerve fiber is its central structure, the **axis cylinder,** a strand of protoplasm continuous with that of the cell body. Its appearance is homogeneous in unstained materials; however, appropriate

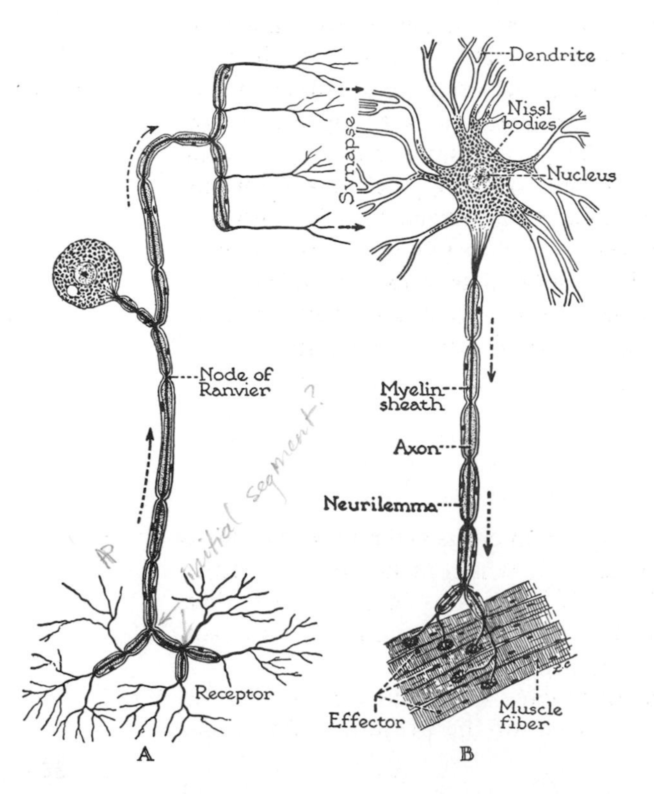

Figure 384. Two types of nerve cells. *A,* Afferent (sensory) neuron of spinal and cranial nerves; long axon-like processes run from sensory receptor to cell body in spinal ganglion and thence into cord, where ramification occurs. *B,* Efferent (motor) neuron, with cell body in cord and a long axon extending out to the effector (muscle fiber). In both parts the peripheral processes (axon of the motor cell) are greatly abbreviated; at this scale they should be several meters long. (From Millard, King and Showers, Human Anatomy and Physiology.)

stains or study under the electron microscope show the presence within it of numerous threadlike, longitudinal **neurofibrils.** The nerve fiber is covered by a very thin sheath proper to it. There are usually further coverings. Cyclostomes apart, most fibers except the very smallest, central or peripheral, are covered by a **myelin sheath** of fatty material. When the sheath is well formed, it gives a shiny, glistening appearance to the fibers. In some cases (as the postganglionic fibers of the autonomic system, described later) the sheath is thin, or a single sheath may be spread out to cover a series of small fibers. In such instances the sheath ordinarily is not observed and hence (incorrectly) assumed to be absent. In the case of fibers within the spinal cord or brain, a continuous sheath is laid down by adjacent, supporting, cellular elements. In peripheral fibers, the sheath is formed by special **sheath cells (Schwann cells),** which wrap themselves around segments of the axis cylinder in a fashion crudely comparable to a jelly roll (except that the turns are very numerous and very thin). In the intervals between the territory of two successive Schwann cells, the myelin sheath of peripheral fibers is interrupted and there is found a **node of Ranvier.** In fibers outside the central nervous system, there is a further and continuous covering, a tough, inelastic membrane, the **neurilemma,** also produced by the sheath cells.

If a fiber is cut, the portion distal to the cut degenerates and the proximal region and cell body may show evidence of damage. Peripheral fibers frequently regenerate, growing out from the stump connected with the cell body; they are aided, it seems, in seeking out their former paths by persistence of the sheath cells which surround the former axis cylinder. Experimental cutting of fiber bundles in the brain or cord, followed by differential staining of degenerating fibers, is a most important technique in the difficult task of determining the complicated "wiring" structure of the nervous system.

By analogy one tends to compare nerve transmission with an electric impulse, and it can be demonstrated that, as a nerve impulse travels along a fiber, there is a momentary change of the electric potential at the fiber surface, a "ripple of leakiness" traveling along the membrane. Rapid metabolic changes occur simultaneously in the axis cylinder—changes analogous to those in muscle contraction, making for rapid release of energy. However, a nerve impulse, though rapid, cannot be compared for speed with electricity. Even in the speediest mammalian fibers the rate is only about 130 meters per second; in lower groups the rates are on the average considerably slower. It is obvious that in the case of large animals (such as an elephant) the time lag between reception of a stimulus and reaction to it, even by the simplest reflex, may make coordination difficult.

The major qualities of nerve impulses may be briefly noted. An impulse is anonymous and nonspecific. The nature of a sensation "felt" in the brain depends upon the centers which receive it, not in any difference in the type of impulse received; could the "wiring" be changed, nerve impulses from the nose, for example, would give a sound effect if received in an auditory center. A nerve fiber is quite capable of transmitting an impulse in either direction. The unidirectional transmission normally found is due to the pattern of fiber connections; neurons are anatomically "polarized." As in the case of contraction of a muscle fiber, a nerve impulse is an "all or none" phenomenon. However, the strength of impulses along nerves as a whole may vary. There may be differences in the number of individual fibers stimulated; further, impulses do not ordinarily come singly, and a rapid sequence of impulses may have a cumulative effect on an effector.

THE SYNAPSE. Never does a single neuron span the entire distance between the sensory receptor initiating an impulse and the muscle or gland stimulated;

action takes place through a chain of neurons, always two and usually more. The point of transfer between successive neurons is termed a **synapse.** Typically the end of an axon breaks up into fine fibrils which lie close to the dendrites or cell body of the second neuron, but do not come into actual contact. The timing of nerve transmission shows that a distinct interval of a tiny fraction of a second is taken in bridging the gap at a synapse. The fibrils probably release minute amounts of a substance—usually acetylcholine—which stimulates the next fiber. In some peripheral situations it is definitely known that such chemical materials—**neurohumors**—are actually given off; the situation is less clear in the central nervous system.

THE REFLEX ARC (Figs. 385, 386 *A*). Before considering more complex structures, we may note the general nature of the simple type of nerve action known as a reflex, seen in such situations as the "automatic" withdrawal of a bare foot that has trod on a tack or of a finger that has touched a hot stove. A sensory stimulus picked up from receptor cells or fibril tips is carried toward the central nervous system by a long **afferent** or **sensory** nerve fiber. The cell body of the **sensory neuron** to which such a fiber belongs lies in a ganglion close to cord or brain, but the fiber continues directly past it into the central nervous system. Here it normally branches to synapse with and potentially stimulate a whole series of neurons. Conversely, each of these neurons may receive impulses from numerous afferent fibers, so that a considerable amount of interplay between receptors and effectors may take place.

In the simplest of reflexes the neurons here stimulated may be **efferent**—usually **motor neurons,** whose cell bodies lie in the cord or brain and whose long axons run out to effector organs (usually muscle fibers). But even a simple reflex is generally one stage more complicated and is a three-neuron chain. Afferent fibers usually do not synapse directly with motor cells but with **association neurons** contained entirely within the central nervous system; these, like the afferents, send out branching processes which connect with numerous motor cells. This further multiplies the number of possible responses to a sensory impulse and, conversely, the number of sensory impulses which may produce a given motor effect. Presumably no single afferent impulse is sufficient to activate an efferent neuron; the action

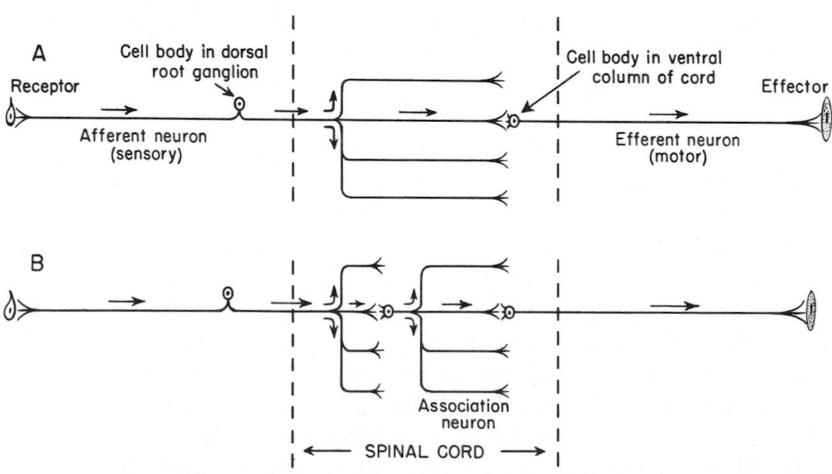

Figure 385. Diagrams to show simple reflexes. Area between broken lines is part of arc lying within the cord (cf. Fig. 386). *A*, Two-neuron reflex; *B*, association neuron interpolated, increasing the number of possible paths. This diagram is greatly oversimplified in that most efferent (and association) neurons would actually be receiving input from many different neurons, thereby producing convergence and summation.

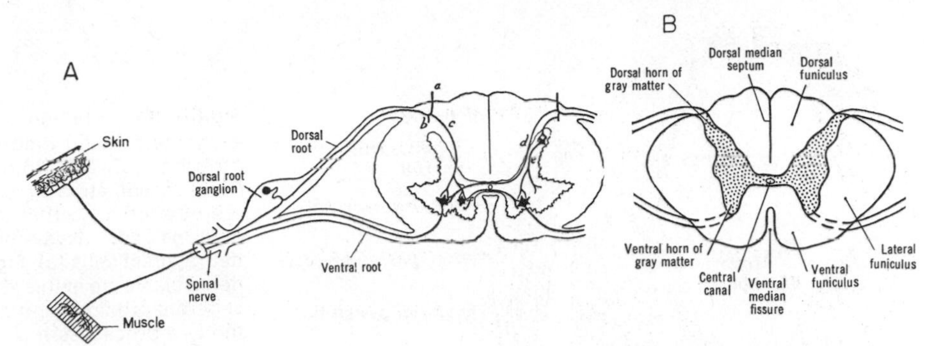

Figure 386. *A*, Diagram of mammalian spinal cord and nerve to show path of reflex arc. A sensory fiber entering by the dorsal root may send branches, *a, b,* up and down the cord. At various levels the sensory fiber may connect with motor neurons of the same side, *c,* or opposite side of the cord, *d;* or with association neurons, *e. B,* Diagrammatic section of a mammalian spinal cord, to show distribution of white and gray matter and funiculi. (After Gardner.)

is caused by a summation of stimuli received, giving rise to phenomena of "choice." We have here clues as to the way in which brain mechanisms more complicated than these simple reflexes may have been built up—namely through the development of higher association centers, with the interpolation of further series of association neurons, into which a wide variety of sensory impressions may drain and from which may come a wide variety of responses.

SPINAL NERVES

In general the peripheral nervous system is simply constructed. It consists essentially of the nerves which penetrate to almost every region of the body—groups of nerve fibers carrying afferent impulses from sensory endings to cord and brain and carrying outward efferent stimuli to muscles and glands. It includes, as well, the ganglia, found along the course of nerves, which contain cell bodies of sensory neurons.

Typical paired spinal nerves (Figs. 386 *A,* 387) are generally present in every body segment. There are two roots. The **ventral root** runs straight outward from the ventral margin of the side wall of the cord; the **dorsal root,** bearing a prominent ganglion, enters the cord higher up the side wall. In most vertebrates the two roots join to form a main trunk, from which various branches—**rami**—diverge. Neglecting for the time a ramus running toward the viscera, there is usually a **dorsal ramus** to the muscles and skin of the back and a **ventral ramus** to the more lateral and ventral parts of the body wall. The nerve trunk and its major branches carry both afferent and efferent fibers; the two roots, however, show a sharp division of functions. The ventral roots carry efferent, motor fibers whose cell bodies lie within the cord (Fig. 389). In typical spinal nerves of mammals the dorsal root contains only the afferent fibers, whose cell bodies lie in the dorsal root ganglion.

In many segments of the trunk each spinal nerve is a discrete structure, innervating the axial muscles formed from the myotome pertaining to its segment and, as well, a corresponding strip of skin. However, in certain regions, notably opposite the paired limbs, there is an interweaving of branches of spinal nerves to form a **plexus** (Fig. 388)—the **brachial plexus** and **lumbosacral plexus** for front and

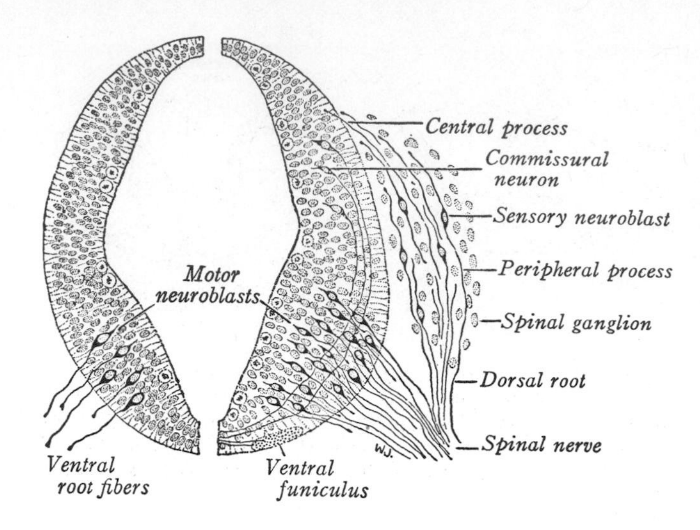

Figure 387. Sections of the spinal cord of early mammalian embryos. *Left,* axons are growing outward from motor nerve cells. *Right,* association or commissural neurons are developing within the cord and sensory neurons are developing externally from neural crest cells (cf. Fig. 77). These sensory neuroblasts are at this stage bipolar, i. e., with separate proximal and distal processes; later the two processes fuse proximally in higher vertebrates to give a unipolar condition to the mature ganglion cell (but one with a T-shaped process). (From Arey.)

hind legs, respectively. As a result we find that any muscle of a limb may be supplied by fibers from several spinal nerves.

CONSTANCY OF INNERVATION; NERVE GROWTH. Even in complicated limb plexuses we find that there is a high degree of constancy in the pattern of innervation of a given muscle in different animals, and there arose the doctrine, noted before (p. 205), that innervation is an absolutely constant feature—that a given muscle is always, over long evolutionary lines, innervated by the same nervous elements and by similar pathways. Actually, however, there are notable (although relatively rare) cases in which such a doctrine cannot be maintained. Any discussion of this problem leads to a consideration of nerve growth.

Of peripheral nerve fibers, the efferent elements grow outward toward the muscles which they innervate from cell bodies situated within the cord. Very likely in vertebrate ancestors the afferent cells were similarly contained in the cord. In most vertebrates, however, most such cell bodies are derived from the neural crest of the embryo (Fig. 77); they migrate downward to form the spinal ganglia and send fibers both inward to the cord and outward to the periphery. In the head part of the

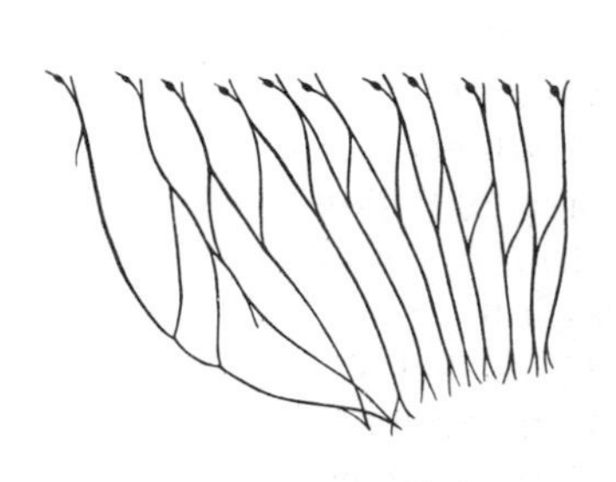

Figure 388. *Left,* nerve plexus supplying the left pelvic fin of a chimaera, showing interchange of fiber bundles between members of the series of spinal nerves concerned. The two roots and the dorsal ganglia of the nerves concerned are shown at the top of the figure; the base of the fin is at the bottom. *Right,* on a larger scale, the brachial plexus supplying the pectoral limb of a mammal. The nerve roots are not shown, and the branches of the plexus are cut short of their terminations; the largest trunk is that of the radial nerve supplying most of the forearm and front foot. The number of nerves involved is less than in the case of the fish fin, but the pattern of the plexus is complex. (Right figure, partly after Walker.)

ganglia are derived from thickenings—**placodes**—in the ectoderm lateral to the forming neural tube and crest.

How do these peripheral fibers grow outward from cord, brain or ganglia to reach their end organs, sensory or motor? Some investigators have assumed that there is a sort of specific, mystic affinity between a special nerve fiber and a special organ to which it attaches, so that the nerve fiber "finds its mate." That there is some degree of specific association is certain since, for example, efferent axons do not attach to sensory structures nor afferent fibers to muscle cells. Experimental work, such as that on transplantation of developing salamander limbs, shows that the presence of muscle does attract nerve fibers, but not necessarily those which would innervate them in a normal condition. Apparently fibers tend to push out from the nerve cord along paths of least resistance in the microscopic structural pattern of surrounding materials. Since in successive generations the topography of a region will tend to be the same, given nerve fibers will tend to follow similar courses and thus bring about an essential constancy of innervation without the necessity of postulating precise specificity.

NERVE COMPONENTS AND SPINAL NERVE COMPOSITION (Fig. 389). Highly useful in the study of both peripheral and central nervous systems is the doctrine of nerve components. This points out that both afferent (sensory) and efferent (motor) fiber types can be divided into **somatic** and **visceral** components. **Somatic afferent** fibers carry in sensations from the skin and muscles—the exteroceptive and proprioceptive groups of the physiologists. **Visceral afferents,** on the other hand, bring impulses from the interoceptive sensory structures of the gut and other internal structures. On the motor side, **somatic efferents** run to striated muscles of the somatic group in the "outer tube" of the body and limbs; **visceral efferents** supply the visceral musculature of the gut region and smooth muscles and glands in various positions. There are thus four components in any normal spinal nerve, typically the two sensory components traversing the dorsal root, the two motor elements in the ventral root; all four unite in the base of the nerve trunk (Fig. 389 *A*).

However, as we descend the vertebrate scale we find, increasingly, a departure from this two-and-two system. There is a tendency, representing (it would seem) a primitive condition, for visceral motor fibers to emerge via the dorsal root, thus giving a three-to-one situation (Fig. 389 *B*, *C*). Even in mammals there is evidence that some visceral motor fibers utilize the dorsal rather than the ventral root for emergence, and in amphibians and jawed fishes visceral motor fibers appear to be common in the dorsal root. Correlated with this, it would seem, are contrasts in the structure of spinal nerves. In most vertebrate groups dorsal and ventral roots are completely united distally and emerge at the same level of the cord. But in lower fishes the roots tend to alternate in position; in sharks and hagfishes the two roots are incompletely united, and in lampreys and amphioxus, finally, dorsal and ventral roots do not connect at all and are quite separate nerves (Fig. 390). This is probably the primitive condition and is a logical arrangement. The ventral roots here carry mainly or exclusively somatic motor fibers and lie opposite the myomeres which they supply; the dorsal roots carry all sensory fibers and most or all visceral motor elements and (reasonably) make their way outward between myomeres. Although this condition is, we have noted, much modified in most vertebrates by union of dorsal and ventral elements and transfer of the visceral component to the ventral root, we find today retention of this primitive system of dorsal and ventral nerves in the arrangement of the cranial nerves in all vertebrate groups.

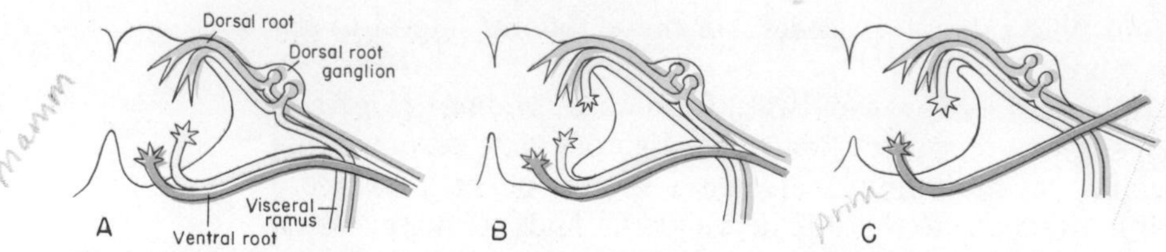

Figure 389. Diagrams show the distribution of nerve components in dorsal and ventral spinal roots. Somatic sensory, blue; visceral sensory, green; visceral motor, yellow; somatic motor, red. *A,* Mammalian condition; the dorsal root is almost purely sensory, and almost all motor fibers are in the ventral root. *B,* More primitive type, common in lower vertebrates; some visceral motor fibers emerge through the dorsal root. *C,* Probable primitive condition; dorsal and ventral roots are separate nerves, visceral motor fibers are part of the dorsal nerve, and the ventral nerve is purely somatic motor (cf. Fig. 390).

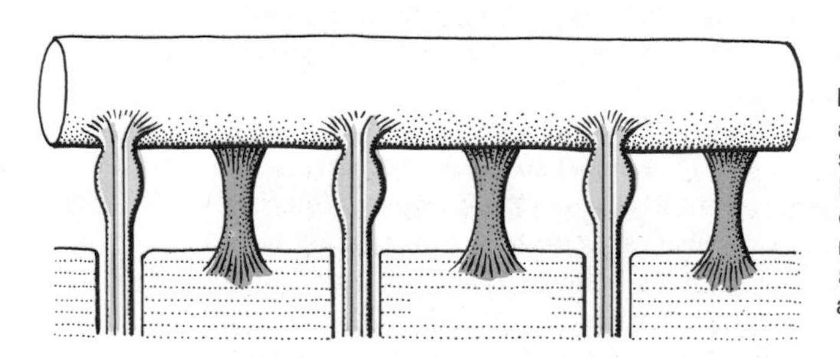

Figure 390. Diagram of the spinal cord and nerves of the left side of a lamprey seen in dorsal view (anterior end to the left), to show the alternating arrangement of separate dorsal and ventral spinal nerves, related to intermyotomic spaces and myotomes. Nerve components colored as in Figure 389.

VISCERAL NERVOUS SYSTEM

In the remote ancestors of the vertebrates the nervous system presumably consisted of two more or less discrete portions: one, a lowly organized set of superficial structures which responded to external stimuli; the other, a network of cells and fibers around the gut and other internal organs which enabled them to adjust directly to internal conditions. In vertebrates the more superficial portion of the nervous system has become highly organized and dominant. The old system around the gut persists, and there is evidence that to some degree the gut may still make local responses to internal stimuli. But with the rise of the external central nervous system of cord and brain, these structures have tended, so to speak, to conquer the nerve system of the viscera and in great measure to do away with its independence; the somatic animal takes over the visceral (Fig. 14). In higher vertebrates, particularly, effective connections have been established and much visceral activity is mediated by cord and hypothalamic centers in the brain. But, as we are well aware from personal experience, visceral sensations and motor responses to them are not generally associated with the higher centers of the brain; we "know" little of what our viscera feel and have little conscious control over them.

The afferent pathways of the visceral system call for little remark. Fibers from sensory endings in the gut ascend to spinal cord and brain through special visceral nerve trunks discussed below or through the vagus nerve of the cranial system, which extends much the length of the gut.

THE AUTONOMIC SYSTEM (Figs. 391–394). Certain specialized efferent visceral pathways from the brain to the striated muscles of the branchial arch system

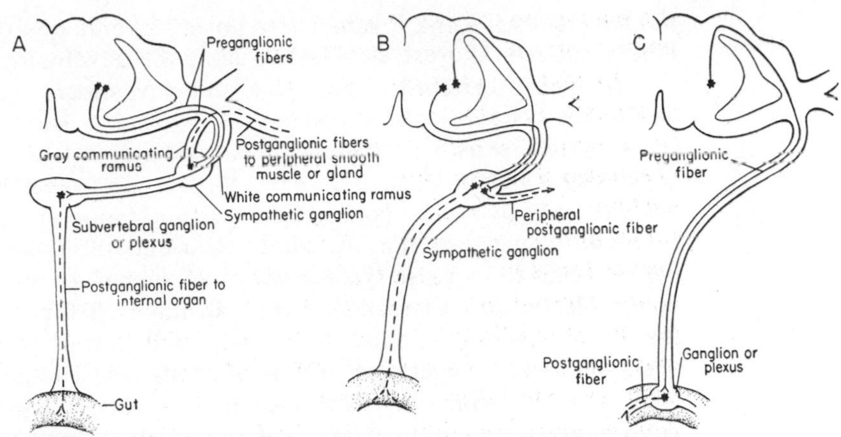

Figure 391. Diagrammatic cross sections to show the path of autonomic fibers. *A,* Sympathetic (thoracolumbar) distribution in a mammal, with autonomic ganglia both in a lateral chain and in a subvertebral position. Preganglionic fibers may be relayed in either position and run either to superficial structures via the major nerve trunks or to the viscera, in both cases with a long postganglionic neuron. *B,* Sympathetic distribution as found in many lower vertebrates; there is little development of a sympathetic chain, and no distinction of ganglia into two groups; fibers to peripheral structures course independently or with blood vessels, rather than with the major (somatic) nerve trunks. *C,* Course of parasympathetic fibers; the preganglionic fiber makes the entire run from cord or brain to a point in or near the organ concerned, where there is a relay to a short postganglionic neuron.

will be discussed later. The remainder of the visceral efferent fibers innervate the smooth muscles and glands of the body. These fibers constitute the **autonomic system,** in reference to the generally self-governing nature of their reflexes. A classification is shown below.*

- Visceral nervous system
  - Afferent
  - Efferent
    - Special branchial
    - Autonomic
      - Sympathetic
      - Parasympathetic

The course and nature of the efferent pathways from the cord to the end organs differ notably from that described earlier for somatic motor fibers. The visceral impulse utilizes two neurons in succession. The first, the **preganglionic neuron,** is comparable to a somatic element; its cell body lies in the cord; its axon is well myelinated. In a typical body segment this axon, after leaving the cord and entering the nerve trunk, leaves it shortly to descend ventrally in a **visceral ramus** (or communicating ramus; Figs. 389, 391). It extends, however, only part way to the effector organ, muscular or glandular, for at some point along its course it enters a ganglion of the autonomic system. Here the motor impulses are relayed to a second, **postganglionic neuron,** whose axon (with little or no myelin in its sheath) completes

*There is wide variation in the use of the terms concerned. The term sympathetic has been defined by various authors *(a)* narrowly, as here; *(b)* as equivalent to autonomic; *(c)* as equivalent to the entire visceral system, both afferent and efferent. As still another confusing variant in terminology, autonomic is sometimes used for the entire visceral system, afferent as well as efferent.

the passage to the end organ. These postganglionic neurons are derived from the embryonic neural crest, descending along the developing nerves.

In higher vertebrates anatomic and physiologic features indicate that the autonomic system can be sorted out into two subdivisions, termed (1) **sympathetic** (in a narrow sense) or **thoracolumbar system** and (2) the **parasympathetic** or **craniosacral system** (Figs. 392, 393). Most organs receive innervation from both systems. The two differ both functionally and topographically (although the functional differences are not absolutely clear-cut). Stimulation of true sympathetic nerves tends to increase the activity of an animal, speed up circulation and slow down digestive processes, and, in general, make it fit for fight or frolic. The action of the parasympathetic, on the other hand, tends to slow down activity and promote digestion and a "vegetative" phase of existence. The postganglionic neurons of both systems stimulate their muscular or other end organs by giving off neurohumors from the fiber tips. In the sympathetics proper the materials produced are **noradrenalin** (norepinephrine) and **adrenalin** (epinephrine), hormones also produced (as noted in Chapter 17) by the adrenal gland; in the case of the parasympathetics **acetylcholine** is produced. Still further points of contrast are anatomic. In a mammal the sympathetic outflow is from the thoracic and lumbar regions of the cord; parasympathetic fibers are associated with cranial nerves—notably the vagus—plus a second outflow in the sacral region. Another anatomic difference lies in the fact that in the sympathetics the relay to the second neuron takes place in ganglia close to the backbone or, at the farthest, in the dorsal mesentery not far below it, whereas in the parasympathetic system the first neuron runs all the way from brain or cord to a ganglion in, or close to, the organ concerned (Fig. 391).

Most of the development of the complex autonomic system seen in mammals appears to have taken place gradually in the upward course of vertebrate evolution. In teleosts and all tetrapods the visceral rami of the trunk nerves enter the ganglia of

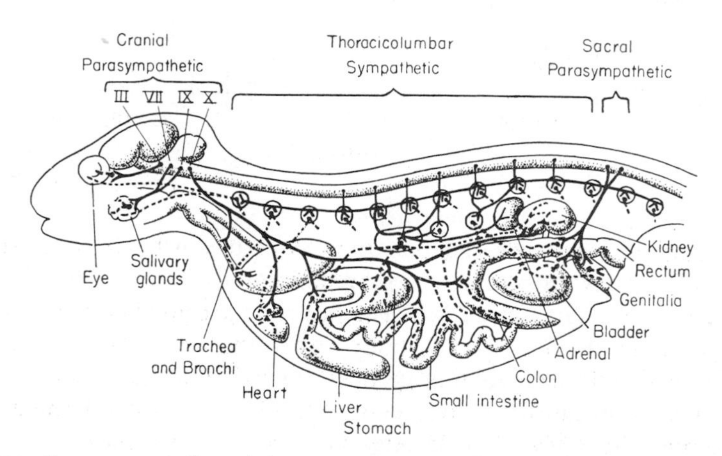

Figure 392. Diagrammatic representation of the autonomic system of a mammal. Only a fraction of the true number of trunk segments is represented. A sympathetic chain is developed, allowing exchange of fibers between segments. Sympathetic ganglia are represented by circles; short nerves projecting from them represent gray rami rejoining the main segmental nerve trunk and running to peripheral structures. There is here a regional sorting out of autonomic nerves into parasympathetic elements associated with cranial and sacral nerves, and sympathetic nerves arising from trunk segments. The two systems nearly completely overlap, both reaching nearly every organ. In the sympathetic system, the relay to postganglionic fibers takes place in the sympathetic ganglia for peripheral structures and those in the head and chest; the relay to abdominal viscera occurs in a series of ganglia–celiac, superior mesenteric, inferior mesenteric–which are more ventrally situated. Preganglionic nerves or fibers in full line; postganglionic in broken line.

the **sympathetic chain** which runs along either side of the backbone and gives off secondary neurons to peripheral structures; no such structures exist in lower fishes. In lower vertebrates, such as sharks (Fig. 394), there is no regional sorting-out of sympathetic and parasympathetic types and little of the system of double innervation.

CRANIAL NERVES

In the head there is present a special series of varied nerves which are, particularly at first sight, difficult to compare with those of the body (Fig. 395). They were first studied in man and given names and numbers based on their mammalian functions and positions. Although, as we shall see, the human arrangement does not hold throughout, we shall introduce the study of cranial nerves by listing them:*

I. *Olfactory:* sensory, from the olfactory epithelium.
II. *Optic:* sensory, from the eye.
III. *Oculomotor:* innervates four of the six eye muscles.
IV. *Trochlear:* to the superior oblique muscle of the eye.
V. *Trigeminal:* a large nerve with three main branches, mainly bringing in somatic sensations from the head, with motor fibers to the jaw muscles.
VI. *Abducens:* to the posterior rectus muscle (which abducts the eye).
VII. *Facial:* partly sensory, but mainly important in mammals as supplying the muscles of the face.
VIII. *Acoustic* (=**Auditory** or **Stato-acoustic**): sensory, from the internal ear.
IX. *Glossopharyngeal:* a small nerve, mainly sensory, and innervating (as the name implies) much of the tongue and pharynx.
X. *Vagus:* a large nerve, both sensory and motor, which (as the name suggests) does not restrict itself to the head but runs backward to innervate much of the viscera—heart, stomach, and so forth.
XI. *Accessory* (=**Spinal accessory**): a motor nerve accessory to the vagus.
XII. *Hypoglossal:* a motor nerve to the muscles of the tongue.

One can brutally memorize such a list of cranial nerves and their functions, but no one interested in the nervous system can stop at this point. We have here a series of nerves which are amazing in variety and seemingly haphazard in distribution, and one cannot but attempt to "make sense" out of them. Is there any logic in their distribution? Can they be grouped in any sort of natural categories?

A clue to classification lies in a consideration of nerve components (cf. Table 3). We have noted that in the postcranial region, nerve components of four types are present. These are represented in the cranial region as well, but in addition, nerves to nose, eye and ear form a **special somatic sensory** group, and on the visceral side there are special types among both sensory and motor components. Taste fibers are considered as a **special visceral sensory** component, and the nerve supply to the striated visceral muscles of the jaws and gills is quite unlike the autonomic system, forming a **special visceral motor** category. We thus have seven types of components, as listed in the table. We can group cranial nerves into three types which (as indicated by the double rulings) show essentially clear-cut distinctions as to the

*The initial letters of the names of the cranial nerves are the initial letters of the words of the following choice bit of poetry: "On old Olympus' towering top a Finn and German viewed a hop."

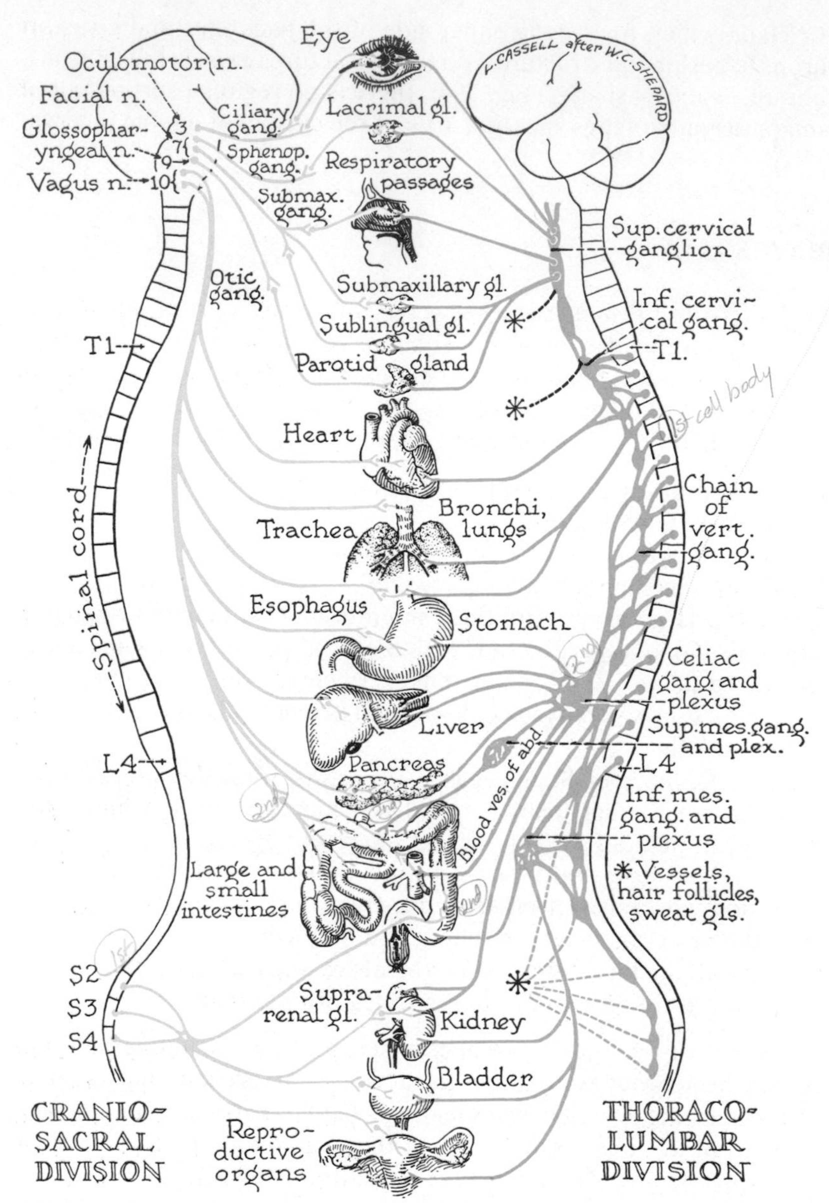

Figure 393. The human autonomic system. At the left, the parasympathetic system; right, the sympathetic system. (From Millard, King and Showers, Human Anatomy and Physiology.)

components present. These are *(a)* **special sensory nerves** of somatic type—I, II, VIII, and lateral line nerves; *(b)* **dorsal root** or **branchial nerves,** containing sensory components and special visceral motor components associated with the branchial region; and *(c)* **ventral root nerves** containing almost exclusively somatic motor fibers. The first category is peculiar to the cranial region; the other two are comparable to the dorsal and ventral roots of spinal nerves of lower vertebrates, especially to the separate dorsal and ventral nerves seen in lampreys and amphioxus.

SPECIAL SENSORY NERVES. In all vertebrates the three main sense organs—nose, eye, and ear—are innervated by special nerves; in primitive aquatic vertebrates we find also special nerve trunks for the lateral line organs.

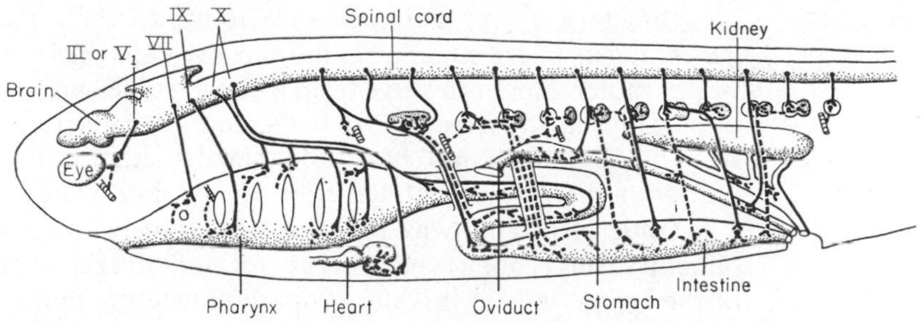

Figure 394. Diagrammatic representation of the autonomic system of a shark. As in Figure 392, only a fraction of the true number of trunk segments is represented, and "samples" only of types of innervation of abdominal viscera and blood vessels are given. Trunk sympathetic ganglia are developed–indicated by white circles–and are usually associated with bodies of chromaffin tissue (stippled). There is no development of a sympathetic chain, no development of gray rami to peripheral structures, and no regional sorting out of sympathetic and parasympathetic systems (cf. Fig. 392). Preganglionic autonomic nerves in full line, postganglionic nerves or fibers in broken line. (After J. Z. Young, modified).

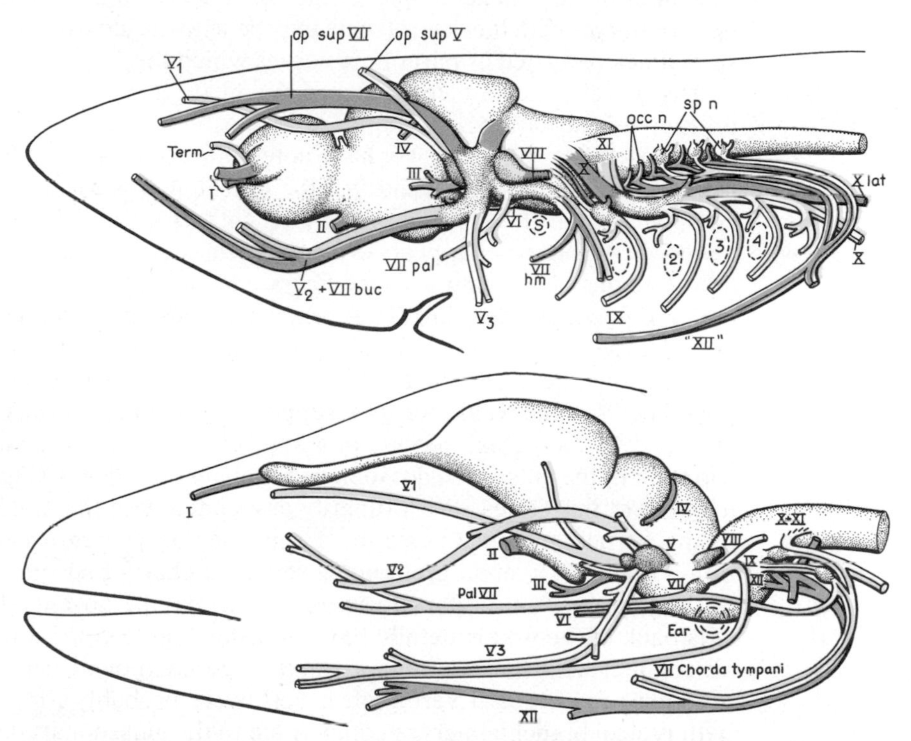

Figure 395. *Above,* diagram of the distribution and components of the cranial nerves of a shark *(Squalus). Below,* diagram of the distribution and components of the cranial nerves of a lizard *(Anolis).* Special somatic sensory nerves (I, II, lateral line, VIII), purple; somatic sensory components, blue; visceral sensory, green; visceral motor, yellow; somatic motor, red. Roman numerals refer to the cranial nerves. *buc,* Buccal; *hm,* hyomandibular; *lat,* lateral line trunk; *occ n,* occipital nerves; *op sup,* superficial ophthalmic; *pal,* palatine; *S,* position of spiracle; *sp n,* anterior spinal nerves; *Term,* terminal nerve; *1* to *4,* position of gill slits; V_1, ophthalmic (profundus) ramus of trigeminal; V_2, maxillary division; V_3, mandibular division of trigeminal; *Pal* VII, palatine ramus of facial. "XII," Trunk of conjoined occipital and anterior spinal nerves corresponding to amniote hypoglossal XII. In *Squalus,* special somatic sensory components of VII and X = lateral line nerves. (*Squalus* after Norris and Hughes; *Anolis* data from Willard, Watkinson.)

Olfactory (I). As mentioned previously (p. 342), the olfactory is not a typical nerve, for its fibers come from the sensory cells of the nose and run inward to the brain, rather than both ways from a cell in some ganglion. In most vertebrates the olfactory is not a formed nerve but a number of discrete fiber bundles passing back from nasal epithelium to brain. In animals with a well-formed vomeronasal organ (p. 344) a discrete branch develops for its innervation.

Optic (II). This was described previously (p. 351). As in the case of the olfactory, this is not a typical nerve, for its fibers run inward from the ganglion cells of the retina; indeed it is not properly a true nerve but a specialized brain tract.

Acoustic (VIII). A more normal nerve, supplying the internal ear, its fibers arise from true ganglion cells, although these may be in part situated well peripherally, close to the sensory structures.

Lateral Line Nerves (cf. Fig. 370). We have noted that the lateral line organs are intimately related to the auditory sense, and hence it is reasonable to find that the nerves to these structures are closely associated with the acoustic nerve. In fishes there are two major lateral line nerves, arising from the medulla anterior and posterior to the acoustic. The anterior nerve supplies most of the lateral line organs of the head; the posterior supplies the neuromasts of the occipital region and trunk. The anterior element accompanies the facial nerve, and most or all of the posterior nerve emerges with the vagus (some may be with the glossopharyngeal); the two are sometimes reckoned as part of the nerves which they accompany, but the association is merely one of "convenience," and the lateral line nerves are basically independent structures.

BRANCHIAL NERVES. We have noted that dorsal and ventral nerve roots were primitively distinct nerves and that the dorsal nerves carried not only all sensory components but the visceral motor elements as well. Table 3 shows that a large series of cranial nerves appears to belong to this dorsal category—they lack somatic motor elements, contain sensory fibers and for the most part include a visceral motor component as well. This series includes the terminal nerve, profundus nerve, trigeminus proper, facial, glossopharyngeal, and vagus (Fig. 396). The dorsal root series of cranial nerves, however, differs in one major respect from dorsal roots of the trunk, for they supply the region of the gills and are primitively arranged, as branchial nerves, in a segmental pattern corresponding to the distribution of the gills. The glossopharyngeal nerve of fishes (Fig. 397) is a "model" member of this series. It is primarily associated with the first typical gill slit, and its main trunk descends posterior to this opening as a **post-trematic ramus.** In addition there are a small **pretrematic ramus,** a **pharyngeal ramus** reaching the roof of the pharynx, and a **dorsal ramus** (which may be absent) to the skin. The gill slits back of the first generally have, in fishes, quite comparable nerves, but they all connect with the brain via a single large compound nerve, the vagus. More anteriorly in ancestral vertebrates there were probably three "normal" gill slits with typical branchial nerves comparable to the glossopharyngeal, but in all living forms the peculiar modifications associated with the formation of jaws have brought about marked specializations of the nerves in this region.

Terminal Nerve. In many members of every vertebrate class except agnathans (at least cyclostomes) and birds there is found a tiny nerve which runs from brain to nasal cavity but is not olfactory, although apparently sensory in nature. Possibly it is a remnant of a most anterior member of this series which primitively innervated the mouth region.

Profundus Nerve (V_1). The **ophthalmicus profundus** nerve is a stout trunk which receives somatic sensations from the snout. In most vertebrates it is intimately associated with the trigeminus and counted as the first of its three trunks, but in lower vertebrates it is often quite independent. It appears to have been in the ancient ostracoderms a typical "complete" branchial nerve, associated with a gill slit which was lost when the mouth expanded.

Trigeminal Nerve (V_2, V_3). The trigeminal nerve proper is believed to have been associated with a second gill slit, present in ostracoderms, but, like the first, lost when the mouth expanded its gape as jaws developed. In contrast to the profundus, however, it has remained highly developed and innervates the jaw muscles as well as having a somatic sensory component. There are two principal branches, the **maxillary** and **mandibular rami,** somewhat comparable to pre- and post-trematic rami of more posterior gill nerves.

Facial Nerve (VII). This is the nerve proper to the spiracular gill slit, and in fishes it usually has a fairly normal branchial nerve construction. We have noted earlier that in mammals the muscles proper to this arch have spread over head and face as the muscles of expression; the facial nerve owes its name to their innervation.

Glossopharyngeal Nerve (IX). As noted, this small nerve is associated in fishes with the first gill slit; it is persistently small and unimportant in tetrapods.

Vagus Nerve (X, XI). This is the largest and most versatile of the cranial nerves; the separately named **accessory nerve** of amniotes is essentially a posterior motor root of the vagus. There is usually a small cutaneous sensory branch, but the vagus is essentially a visceral nerve. In fishes the vagus supplies all the branchial arches behind that of the first typical gill; in addition a powerful **visceral** (or **abdominal) ramus** extends backward along the gut as a major element of the autonomic system.

SOMATIC MOTOR NERVES. The nerves in this category—III, IV, VI, and XII of the human series—are highly comparable in most regards to primitive ventral roots of spinal nerves. They are almost entirely composed of somatic motor fibers and innervate striated muscles derived from somites.

Eye Muscle Nerves (Oculomotor, Trochlear, Abducens) (III, IV, VI). These small nerves innervate the eye muscles derived from the three pre-otic somites (cf. p. 211). The little trochlear is unusual in that it curves upward within the substance of the brain stem and emerges dorsally to supply the muscle of the opposite side of the head. Autonomic fibers frequently accompany the oculomotor nerve, but are not really part of it.

Hypoglossal (XII). In fishes (particularly among sharks) the posterior end of the skull—hence of the cranial nerve series—is not a fixed point, for the occiput is formed by a variable number of vertebrae. There is, hence, a variable number of **spino-occipital nerves,** which are essentially anterior members of the trunk series but tend to lose their dorsal roots, and mainly supply muscles formed from the somites of the occipital region. In amniotes the condition is stabilized; back of the vagus (and accessory) there is a final cranial nerve, the hypoglossal, usually formed by three ventral roots which presumably represent three body segments fused into the occiput. There is no hypoglossal nerve in modern amphibians, but fossil evidence shows it to have been present in early amphibians as a direct inheritance from the occipital nerves of the fish; in this regard, as in many others, modern amphibians are degenerate rather than primitive. We have noted that the myotomes of the general occipital region migrate in the embryo backward and downward around the gills to form the hypobranchial musculature of fishes and the tongue musculature of tetrapods (Fig. 199). Trunks of occipital nerves in fishes follow the muscles in this movement (and are hence often termed the hypo-

TABLE 3. **Table of Nerve Components of Cranial Nerves**

Nerve Types	***Special Sensory***	***Branchial (Dorsal)***					***Ventral***
Components	*Special Somatic Sensory*	*General Somatic Sensory*	*General Visceral Sensory*	*Special Visceral Sensory*	*Special Visceral Motor*	*Visceral Motor (Autonomic)*	*Somatic Motor*
O. Terminalis		X					
I. Olfactory	X						
II. Optic	X						
III. Oculomotor						(X)	X
IV. Trochlear							X
V_1. Profundus		X				(X)	
$V_{2,3}$. Trigeminal proper		X			X		
VI. Abducens							X
VII. Facial	L	(X)	X	X	X	X	
VIII. Acoustic	A						
IX. Glossopharyngeal	L	(X)	X	X	X	X	
X and XI. Vagus (and accessory)	L	X	X	X	X	X	
XII. Hypoglossal							X

Proprioceptive fibers (muscle sense) are not included, but are present in all somatic motor nerves. *L,* Lateralis sensory components of lower vertebrates (in X, the vagus, alone in amphibians); *A,* acoustic components of lateralis-acoustic system. Components in parentheses: variable or negligible. The three areas between vertical double-ruled lines indicate the components proper to each of the three nerve types. Except for the usual presence of autonomic fibers accompanying the oculomotor nerve, the distinctions are clear-cut.

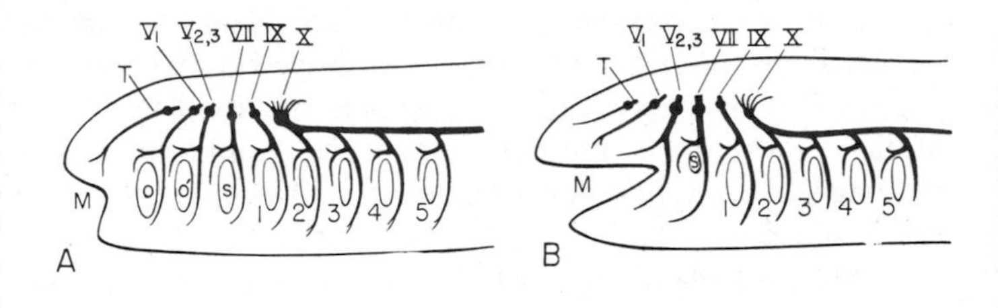

Figure 396. Diagrams showing the distribution of branchial (dorsal root) cranial nerves. *A,* Hypothetic primitive condition, with typical nerves to each of two anterior gill slits lost in jawed vertebrates, and a terminal nerve to anterior end of head. *B,* Condition in jaw-bearing fishes. *M,* Mouth; *O, O',* anterior gill slits lost in gnathostomes; *S,* spiracular slit; *T,* terminal nerve; *1* to *5,* typical gill slits of gnathostomes.

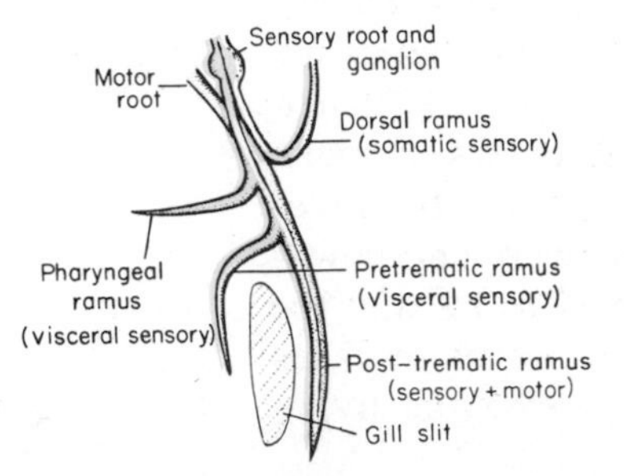

Figure 397. Diagram showing the composition of a typical branchial nerve, such as the fish glossopharyngeal. The three nerve components colored as in Figures 389 and 395.

branchial nerve), and in amniotes the hypoglossal nerve follows a similar path, back, down, and then forward around the pharyngeal region to innervate the tongue musculature.

SEGMENTATION OF THE HEAD

It is convenient, at this point, to break off our discussion of the nervous system to consider a separate problem, but one in which much of the evidence derives from the study of the cranial nerves. Obviously the trunk of vertebrates is segmentally arranged, but such clear division into repeated units is not seen anteriorly in the head. Nevertheless many workers have tried to demonstrate that the segmentation, however masked, is actually there. Certain cranial nerves seem comparable to dorsal or ventral roots of the spinal nerves in primitive chordates—and spinal nerves are clearly segmentally arranged. No completely convincing demonstration of such a pattern can be made for the head, but we can present a more or less plausible story (Fig. 398 and Table 4).

Ventral root nerves contain only somatic motor fibers and innervate somatic muscles which develop from the myotomes of the somites; somites are the primary segmental structures in vertebrates. Therefore ventral roots should be segmental. In the head, as elsewhere, these nerves do in fact line up with the myotomes. Three small pre-otic somites give rise to the extrinsic muscles of the eye; each has its own nerve, numbers III, IV, and VI. The more posterior somites in the head give rise to the hypobranchial muscles and are innervated by the hypobranchial or hypoglossal nerve; this nerve goes to muscles formed from several somites and arises from several separate roots. The segmental pattern is maintained. Unfortunately the ear, as it develops, occupies the space that would normally contain somites, and the sequence of myotomes is interrupted. At least two pairs of somites may start to form and then be eliminated here. We will assume that only two were lost—which, as we shall soon see, makes things balance fairly well.

Dorsal root, or branchial, nerves also form a series. However this series lines up not with myotomes or other somatic structures, but with the skeletal and muscular elements of the gills. Thus the trigeminal is the nerve of the mandibular arch, the facial is the nerve of the hyoid, and so forth. The pattern is not completely simple: the vagus has separate branchial branches to successive branchial arches; it also arises as a series of separate roots from the brain. Anteriorly we may (as in Table 4) assume only a single premandibular visceral arch and consider the terminalis as the nerve of the region anterior to the gills. It is possible that there were two premandibular arches and that the terminalis was the nerve of the more anterior one. In any event the relation between these nerves and the visceral arches is clear.

But is there any reason to believe that dorsal and ventral roots must line up neatly here as they do in the trunk? Not really. It seems plausible—after all, if the nerves are similar, the whole pattern may be similar. Anyway it seems tidier and simpler if both repetitive series, that of myotomes and that of visceral arches, possess the same interval between successive units. Another (rather weak) argument is that it appears to work—we can draw pictures like Figure 398 and make charts like Table 4. However, remember we only *assume* two somites are lost as the ear develops. Also different vertebrates include a variable number of somatic segments within the head (that is, the number of occipital arches varies)—again we can juggle things so that they come out even. Nevertheless, we do suspect that the

TABLE 4. **Segmentation of the Head**

Segment	***0 = Snout***	***1***	***2***	***3***	***4***	***5***	***6***	***7***	***8***	***9***
Myotome	(none)	First pre-otic	Second pre-otic	Third pre-otic	(crowded out by developing ear)		First post-otic	Second postotic	Third postotic	Fourth post-otic
Ventral Root Nerve	(none)	Oculomotor	Trochlear	Abducens	←(lost)→		←Hypobranchial = Hypoglossal→			Ventral root of second spinal nerve
Dorsal Root Nerve	Terminalis	Profundus	Trigeminal	Facial	Glosso-pharyngeal	←Vagus (including Accessory)→				Dorsal root of second spinal nerve
Visceral Arch	(none)	Premandibular	Mandibular	Hyoid	First branchial	Second branchial	Third branchial	Fourth branchial	Fifth branchial	(none)
Branchial Opening	(none)	(lost with development of large, jawed mouth)		Spiracle	First branchial	Second branchial	Third branchial	Fourth branchial	Fifth branchial	(none)

Compare with Figure 398 and description in the text. This table is based largely on a dogfish, in which the first spinal nerve is largely vestigial (although fibers that "should" be in it contribute to the hypobranchial and vagus nerves).

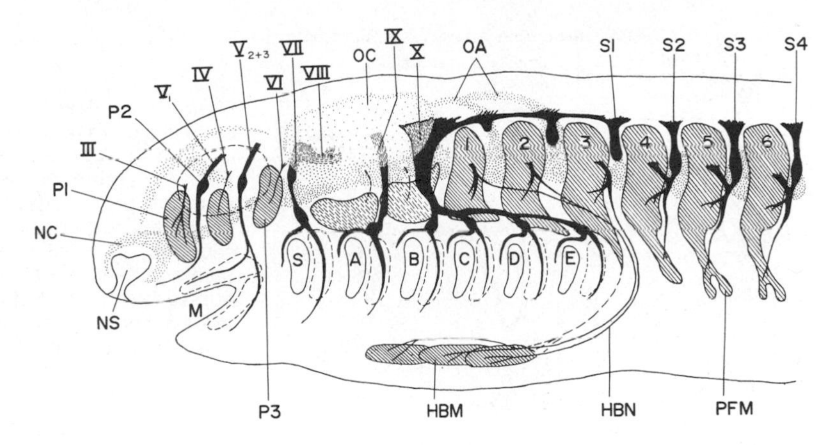

Figure 398. Diagram of the embryo of a shark *(Scyllium)* to indicate the segmental structure of a vertebrate head. Cartilage of axial skeleton is stippled; nerves are shown in solid black (the parts behind the otic capsule are indicated by fine dashed lines); somatic muscles are hatched (parts that degenerate in the embryo are in dashed hatching); the orbit and the visceral cartilages are outlined by dashed lines. *A–E,* The five normal branchial openings; *HBM,* hypobranchial muscles; *HBN,* hypobranchial nerve (= hypoglossal of amniotes); *M,* mouth; *NC,* nasal capsule; *NS,* nasal sac; *OA,* occipital arches; *OC,* otic capsule; *PFM,* muscle buds for pectoral fin muscles; *P1–P3,* the three pre-otic somites; *S,* spiracle; *S1–S4,* the first four spinal nerves; *1–6,* the first six post-otic somites; III–X, the cranial nerves (numbered in the usual fashion). Note that this diagram shows two somites and their ventral root nerves "crowded out" by the developing otic capsule and generally is in agreement with the scheme presented in Table 4. (After Goodrich.)

head was originally segmented in much the same fashion as the trunk, and that this pattern was obscured by the various specializations, formation of major sense organs, and the like, that occurred there.

CENTRAL NERVOUS SYSTEM—ACCESSORY ELEMENTS

Ordinary connective tissue is not present in the central nervous system, but as differentiation of the neural tube occurs in the embryo, some of the cells there become specialized for supporting rather than nervous functions, substituting for the connective tissue found in most organs. A fraction of them remain around the borders of the cavities of the brain and cord and retain an epithelial formation, but most are small star-shaped cells termed **neuroglia** scattered among the neurons.

Brain and cord are not merely protected by the braincase and the neural arches of the vertebrae, but are further ensheathed in one or more wrappings of true connective tissue, the **meninges** (Fig. 399). In most fishes there is but a single meninx of complex structure, but in all tetrapods at least two are present. The outer, the **dura mater,** is a stout sheath, connected by slender filaments with a softer inner membrane, which is closely applied to brain or cord. In mammals this inner membrane is divided into two delicate structures, an outer **arachnoid** and an inner **pia mater,** the two separated by a fluid-filled **subarachnoid space** crossed by a cobweb of delicate tissue threads.

The tube present in the neural canal of the embryo persists in the adult as the ventricles of the brain and central canal of the spinal cord. These cavities (and the

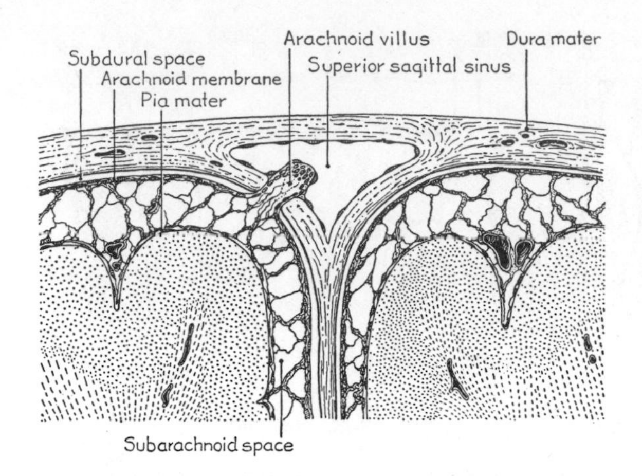

Figure 399. Transverse section of a portion of the brain of a mammal to show the meninges. The section is taken through the partition (falx cerebri) between the two cerebral hemispheres. The superior sagittal sinus is a prominent venous channel following a longitudinal course backward between the hemispheres; sections of smaller blood vessels are seen in meninges and brain. The subarachnoid space is occupied by cerebrospinal fluid; the arachnoidal villi offer a minor means of transfusion of material between this fluid and the blood. (After Weed.)

subarachnoid space in the meninges) are filled by a clear liquid, the **cerebrospinal fluid,** secreted by the choroid plexuses described below and similar in composition to the interstitial fluid or the perilymphatic liquid of the ear. Materials may reach it from the blood through the medium of special vascular structures, notably the choroid plexuses of the brain (p. 388).

SPINAL CORD

The **spinal cord** (Figs. 386, 389, 400), which runs the length of the body, is a little-modified adult representative of the nerve tube formed in the early embryo. A small liquid-filled central canal persists. Suboval or circular in lower vertebrates, the cord tends to expand in bilateral fashion in higher forms. Two layers can be readily distinguished, a central area of **gray matter,** mainly composed of cell bodies, and a peripheral **white matter,** formed of countless myelinated fibers coursing up and down the cord. The gray substance was, it seems, primitively arranged in a fairly even fashion about the central cavity, but in most vertebrates it has a symmetric arrangement which in section has an H-shape or that of a butterfly's wings. There thus appear to be a pair of ''horns'' on either side; actually, of course, each ''horn'' is merely a section of a longitudinal structure, and we should speak rather of a dorsal column and a ventral column.

The **ventral column** is the seat of the cell bodies of efferent neurons of the spinal nerves. Their numbers will vary, of course, in any given part of the column with the volume of musculature at that level, and in tetrapods this column is much expanded in the regions supplying the limbs. The visceral efferent neurons are situated above and lateral to those of the somatic type and sometimes are distinguished as a **lateral column.**

The **dorsal column** is associated with the dorsal, sensory nerve roots and is the seat of the cell bodies of association neurons through which impulses brought in from sense organs may be relayed and distributed. The arrangement of various clusters of association neurons is complex and variable, but in some cases (particularly in certain embryos) we can distinguish a larger series associated with somatic sensory reception, situated dorsally and medially, and a smaller visceral sensory group placed more ventrally and laterally. There thus appear to be in the gray matter four areas on either side related to the four major nerve components, being in sequence from dorsal to ventral: somatic sensory, visceral sensory,

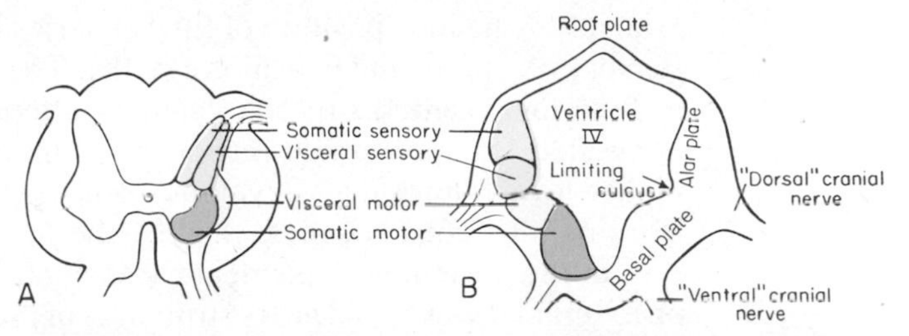

Figure 400. Diagrams showing the distribution of sensory and motor columns. Somatic sensory association column, blue; visceral sensory, green; visceral motor column, yellow; somatic motor, red. *A,* The spinal cord of the adult of certain lower vertebrates. *B,* The embryonic medulla oblongata; the embryonic spinal cord shows a similar arrangement of the columns. The plate of tissue lying below the limiting sulcus is termed the basal plate; from this, motor centers arise. The sensory region above is the "wing" or alar plate. (Partly after Herrick.)

visceral motor, and somatic motor. The same arrangement is found in the gray matter of the brainstem (Fig. 400 *B*).

The **white matter** is composed of ascending and descending fibers of sensory nerve cells, of similar fibers from the association cells, of fibers which carry sensory stimuli forward to the brain, and of fibers returning from the brain to act on motor neurons. The last two categories are especially abundant in higher vertebrates, in which the trunk is more completely under the influence of the brain than it is in lower forms. Topographically the "horns" divide the white matter into dorsal, lateral and ventral **funiculi;** more important are restricted areas within the funiculi occupied by specific **fiber tracts** with a given type of function and connections; these vary too much from group to group to be described here in detail, although in general the dorsal funiculi carry mainly ascending tracts, the ventral funiculi descending motor tracts.

THE BRAIN

In all vertebrates, as well as in the more highly organized invertebrates, we find a concentration of nervous tissue at the anterior end of the body in the form of a brain. Such a concentration is to be expected, for in an active bilaterally symmetric animal it is this region which first makes contact with environmental situations to which response must be made, and in which, hence, sense organs and associated nervous structures are most advantageously situated.

BRAIN ARCHITECTURE. Primitively, we may believe, the brain was merely a modestly developed anterior region of the neural tube where, in addition to local reflexes, special sensory stimuli were assembled and "referred for action" to the semiautonomous body via the spinal cord; such is the situation in amphioxus. Within the vertebrates, however, there has occurred a strong trend for the concentration in the brain of command over bodily functions, with the development of many complex centers. We have noted, in discussing the elementary composition of the nervous system, how the intercalation of association neurons into the simple reflex arcs greatly broadens the field of possible responses to a sensory stimulus and, conversely, greatly increases the variety of stimuli which may excite a specific motor response. The brain pattern is essentially an elaboration of this principle—the interposition of further series of neurons between primary areas of sensory reception and final motor paths. These intermediate neurons are clustered in functional **centers.** In such centers afferent impulses may be correlated and integrated for appropriate responses or motor mechanisms coordinated; on still higher levels there may develop association centers of whose activity memory, learning, and consciousness may be the products.

In the present very elementary account our attention will be mainly centered on external features and gross structures of the sort seen in Figures 404 through 412. But while such superficial aspects of brain anatomy are significant, an adequate understanding of the working of the brain can no more be gained from them than a knowledge of a telephone system can be had from an acquaintance with the external appearance and room plan of the telephone exchange. What is important in a telephone system is the wiring arrangements and switchboards; in a brain the centers act as switchboards, in a sense, and tracts of fibers form the wiring between them.

It may well be that the brain "wiring" was primitively much like that of the spinal cord—a general crisscross of fibers interconnecting all areas. To some degree there does persist in the brain a certain amount of seemingly random distribution of fibers, and there is further a persistence of a primitive condition in the **reticular formation,** a system of interlacing cells and fibers carrying motor impulses along the motor columns of the brain stem (seen diagrammatically in Figures 424 through 426). In general, however, there is a strong tendency for the clustering together of nerve cells in centers and the assembling of fibers with like connections into definite bundles. Although certain special centers have special names, most are termed **ganglia** or **nuclei** (making an unfortunate duplicate biologic use of the latter word). Fiber bundles connecting nuclei with one another or with the cord are in general termed **tracts;** the fibers of a tract are, of course, axons of neurons whose cell bodies lie in the nucleus of origin. The brain is constructed on an essentially bilaterally symmetric pattern; in consequence, cross-connections of fiber bundles, termed **commissures,** must be present between the two sides in order that the animal may not, literally, have a dual personality.

BRAIN DEVELOPMENT (Fig. 401). The general topography of the brain and its parts is best understood through a consideration of its development. The brain develops rapidly in the embryo—much more rapidly than almost any other organ—and there is early established a generalized structural pattern upon which the numerous variations seen in the adult brains of different groups are superposed. In early stages the future brain is merely an expanded front part of the neural tube. Presently its anterior end tends to fold downward, producing a **cephalic flexure,** and a bit later there is a constriction more posteriorly at a point known as the **isthmus.** There is thus established a division into three major regions of the brain termed, in front-to-back order, the **prosencephalon, mesencephalon** and **rhombencephalon**—in plain English, forebrain, midbrain, and hindbrain. Although various specialized outgrowths are later added, the original "tubular" portions of these three "segments" are still recognizable in the adult, where they are collectively known as the **brain stem;** here are persistently located centers for many simple but basically important neural functions. The lengthwise division of the brain stem into three portions is correlated with the fact that in most vertebrates each of the three is associated with one of the three major sense organs: nose, eye, and ear and lateral line (Fig. 402). In each region there develops from the stem a dorsal outgrowth of layered "gray matter" which is primitively associated with one of these sensory structures. These are, in anteroposterior order: the cerebral hemispheres of the

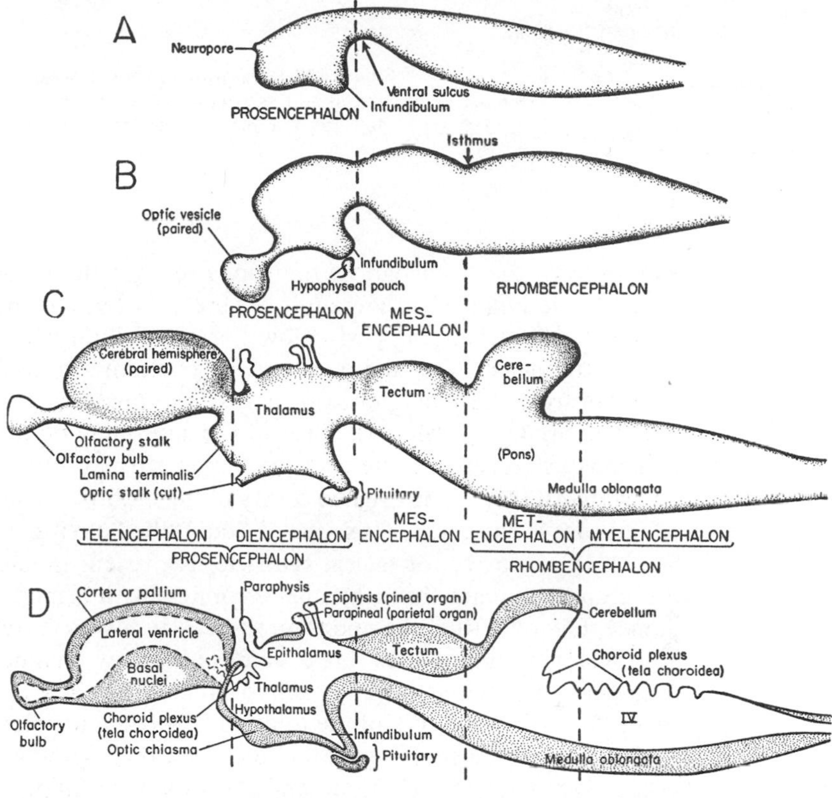

Figure 401. Diagrams to show the development of the principal brain divisions and structures. *A,* Only prosencephalon (primitive forebrain) distinct from remainder of neural tube. *B,* Three main divisions established. *C,* More mature stage in lateral view. *D,* The same in median section. (Partly after Bütschli.)

prosencephalon, primarily associated with smell; the roof of the midbrain—tectum—associated with vision; and the cerebellum, an outgrowth of the hindbrain associated with the ear and lateral line.

By the time the three primary subdivisions are established special structures have begun to make their appearance in the forebrain. The optic vesicles, discussed in the last chapter, push out from its floor. More posteriorly, there is a down-growing median projection, the **infundibulum;** concomitantly, a pocket of epithelium, the **hypophyseal pouch** (Rathke's pouch), grows upward from the roof of the mouth. In later stages modified infundibular tissues and those derived from the pouch come to form the **pituitary gland,** discussed in the next chapter. Dorsally there grows from the roof of the forebrain a series of median processes and a median "eye stalk" (sometimes two). Particularly striking and important is the development more anteriorly of paired dorsal outgrowths of the forebrain. These are hollow pockets of tissue which extend forward toward the nasal region; from them develop the **cerebral hemispheres** and, still farther anteriorly, the **olfactory bulbs.** These structures constitute the **telencephalon,** the anterior terminal segment of the brain; the unpaired part of the forebrain is the **diencephalon.**

Back of the forebrain region there are relatively few major modifications as development proceeds. The midbrain shows only a pair of dorsal swellings which form the **tectum,** prominent in lower vertebrates. In the hindbrain a dorsal out-

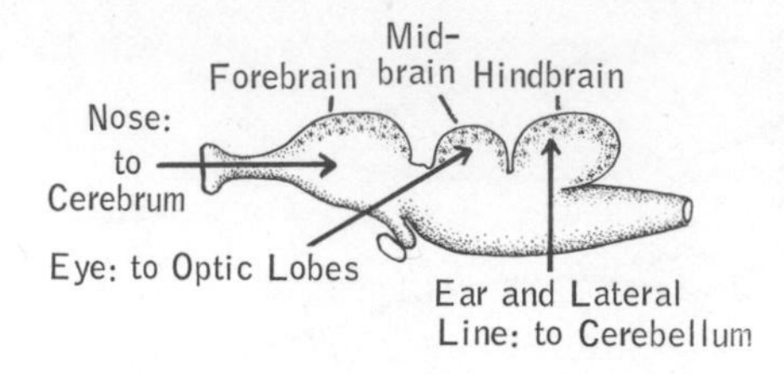

Figure 402. Diagram to show the relation in lower vertebrates of the three major sense organs to the three dorsal areas of gray matter in the three major subdivisions of the brain.

growth from the front part of the roof becomes the **cerebellum.** The brain stem here is little changed in the adult, where it is termed the **medulla oblongata.** In mammals, however, the part below the cerebellum is expanded into a structure termed the **pons.** Pons and overlying cerebellum are distinguished as the **metencephalon** from the most posterior part of the medulla, the **myelencephalon.**

The principal brain structures of the adult may be tabulated according to the divisions established in the embryo shown at the bottom of this page.

VENTRICLES. The original cavity of the embryonic neural tube persists in the adult brain in the form of a series of liquid-filled cavities and passages (Figs. 403, 407, 421). A cavity, or **lateral ventricle,** is present in each cerebral hemisphere; these communicate through small foramina with a median **third ventricle** in the diencephalon. Within the midbrain there is in lower vertebrates a well-developed ventricle; but in amniotes this becomes a narrow channel, the **cerebral aqueduct** (aqueduct of Sylvius), leading back to a **fourth ventricle** in the medulla. Commonly there develop in the roof of the third and fourth ventricles areas of thin and highly folded vascular tissue, the **choroid plexuses,** through which exchange of materials takes place between the blood and cerebrospinal fluid.

MEDULLA OBLONGATA. Approach to the study of brain architecture is best made by first considering those parts which are simplest in construction and most closely resemble the spinal cord. The brain stem is simpler than its specialized dorsal outgrowths, and in that part of it which lies in the hindbrain, the medulla oblongata, we find a structure basically similar to the cord.

It is from the medulla (and the adjacent part of the midbrain) that there arise all the cranial nerves except the atypical ones from the nose and eye and the terminalis. The medulla itself is basically similar to a section of the spinal cord, except that the

Prosencephalon	Telencephalon	Cerebral hemispheres, including olfactory lobes, basal nuclei (corpus striatum) and cerebral cortex (pallium); olfactory bulbs
	Diencephalon	Epithalamus; thalamus; hypothalamus; various appendages
Mesencephalon		Tectum, including optic lobes (corpora quadrigemina in mammals); tegmentum; crura cerebri (cerebral peduncles) in mammals
Rhombencephalon	Metencephalon	Part of medulla oblongata; cerebellum; pons of mammals
	Myelencephalon	Most of medulla oblongata

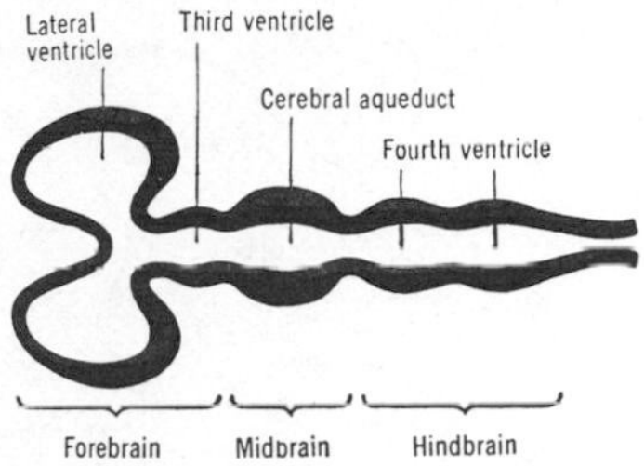

Figure 403. Diagram showing position of brain ventricles. (From Gardner.)

central canal is greatly enlarged to form the fourth ventricle, and its membranous roof is infolded to form the posterior choroid plexus. As a consequence the columns of gray matter are pushed apart to lie on either side of the ventricle. These columns (Figs. 400 *A*, 415) are basically the four we have already seen present in the cord and arranged in the same order, with a horizontal sulcus separating the sensory columns above from the motor columns below. In the embryo the columns are simple; in the adult, however, they tend (particularly in higher vertebrates) to break up into a series of nuclei of specialized nature, as indicated in Figure 415 *B*. These nuclei, mostly associated with specific cranial nerves, give us all the elements required for reflex circuits between sensory reception and the responding effector organs of the head and branchial region. In addition, however, we find at the upper margin of the medulla a special area or set of nuclei which serve for primary reception of sensations from the ear and the primitively associated lateral line structures. In mammals, we have noted, there is a further specialization of the medulla in the development, anteriorly, of a swollen region, the **pons,** containing a great mass of neurons that, as noted later, relay impulses from hemispheres to cerebellum.

In lower vertebrates trunk and tail are to a considerable degree independent of

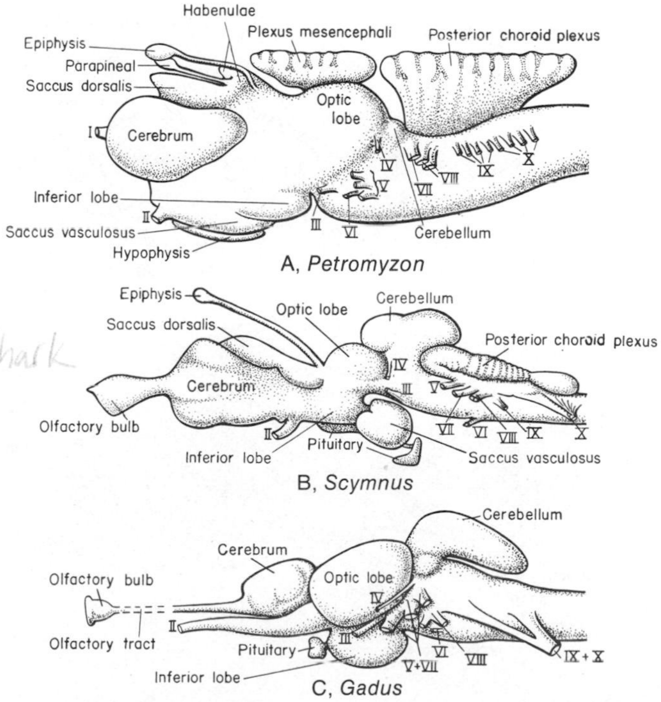

Figure 404. Lateral views of brain of *A,* a lamprey; *B,* a shark; *C,* a codfish. In the lamprey an exceptional condition is the development of a vascular choroid area, the plexus mesencephali, on the roof of the midbrain. Roman numerals indicate cranial nerves. (After Bütschli, Ahlborn.)

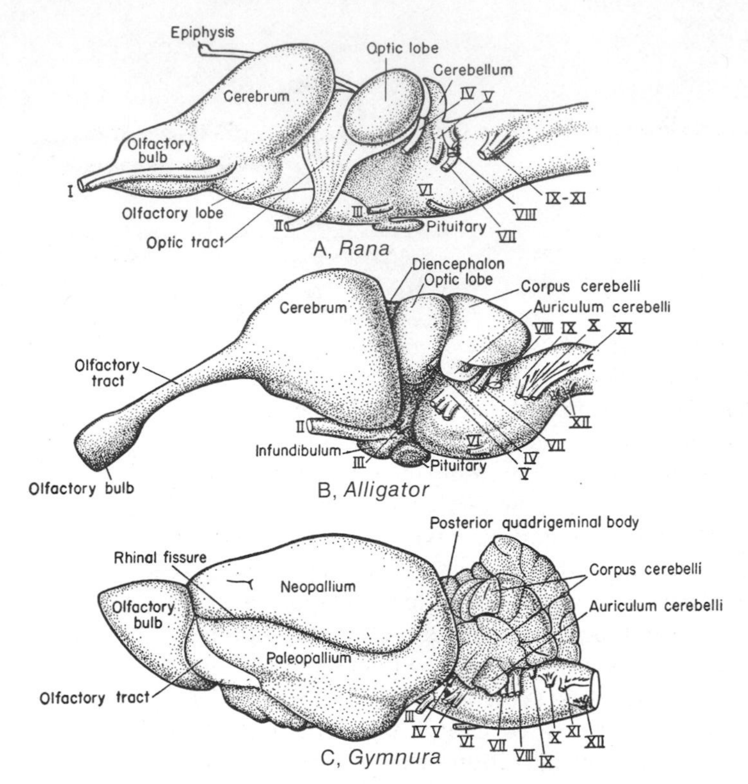

Figure 405. Lateral views of brain of *A,* a frog; *B,* an alligator; *C,* an insectivore representing a primitive mammalian type. In normal head posture the front end of the alligator brain is tilted upward. Roman numerals indicate cranial nerves. (After Bütschli, Clark, Crosby, Gaupp, Wettstein.)

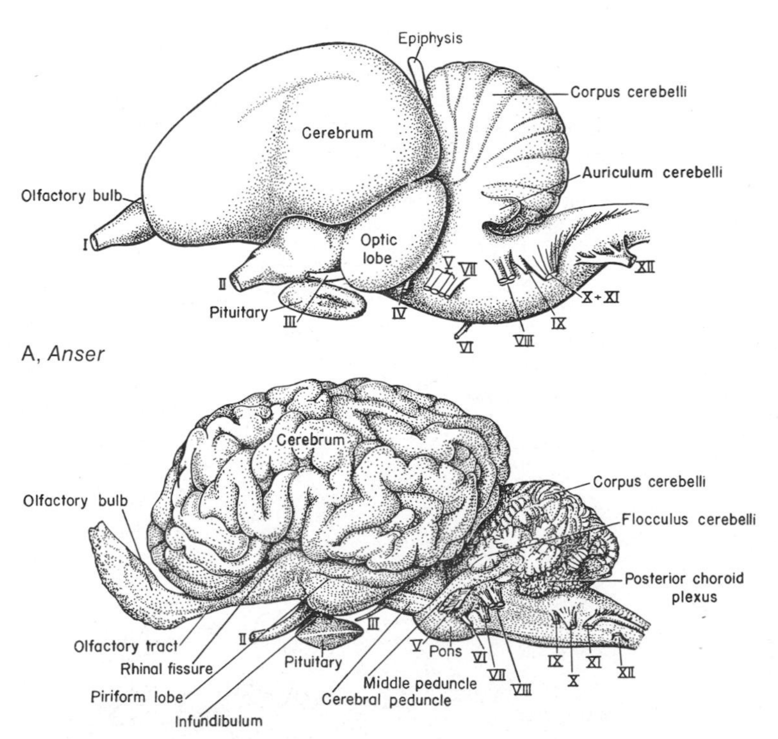

Figure 406. Lateral views of brain of *A,* a goose; *B,* a horse. The goose brain, like that of the alligator, is tilted upward anteriorly in life. Roman numerals indicate cranial nerves. (After Bütschli, Kuenzi, Sisson.)

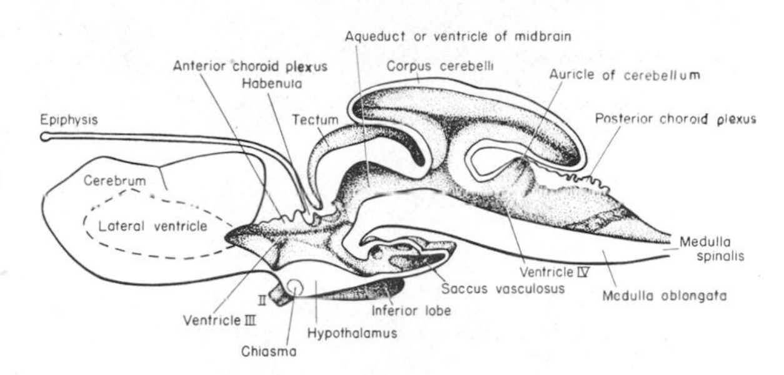

Figure 407. Right half of the brain of a shark *(Scyllium)* in median aspect. Unshaded areas are those sectioned. (After Haller, Bruckhardt.)

the brain in their activities. In fishes and tailed amphibians, however, the medulla contains the cell bodies of a pair of spectacular **giant cells of Mauthner,** whose large axons extend the length of the cord and exercise control over the rhythmic movements of the trunk and tail, important in fish locomotion.

CEREBELLUM (Figs. 413, 414). Rising above the brain stem at the anterior end of the medulla is the cerebellum, a brain center, often of large size, which is of extreme importance in the coordination and regulation of motor activities and the maintenance of posture. It acts in a passive, essentially reflex fashion in equilib-

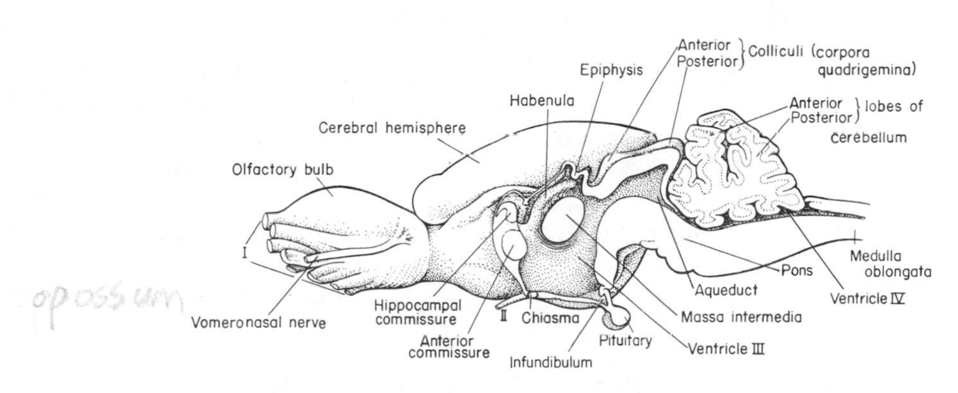

A, *Didelphys*

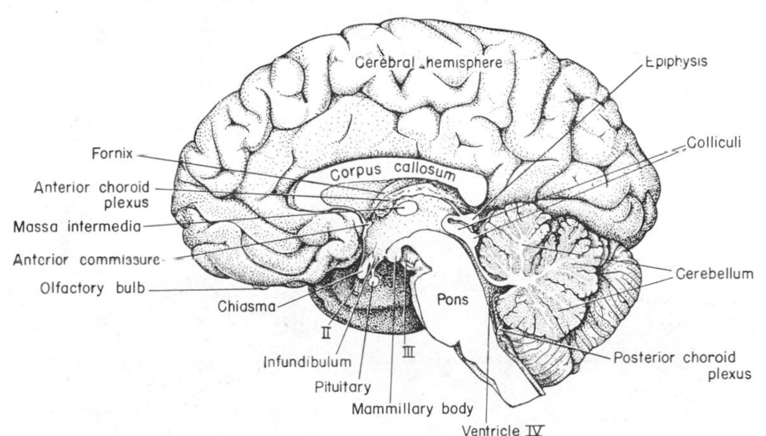

B, *Homo*

Figure 408. Right half of the brain, in median aspect, of *A*, an opossum; *B*, man. Unshaded areas are those sectioned. The internally bulging side walls of the diencephalon may meet and fuse in the midline, forming a "massa intermedia," which, however, has no functional importance. (*A* after Loo.)

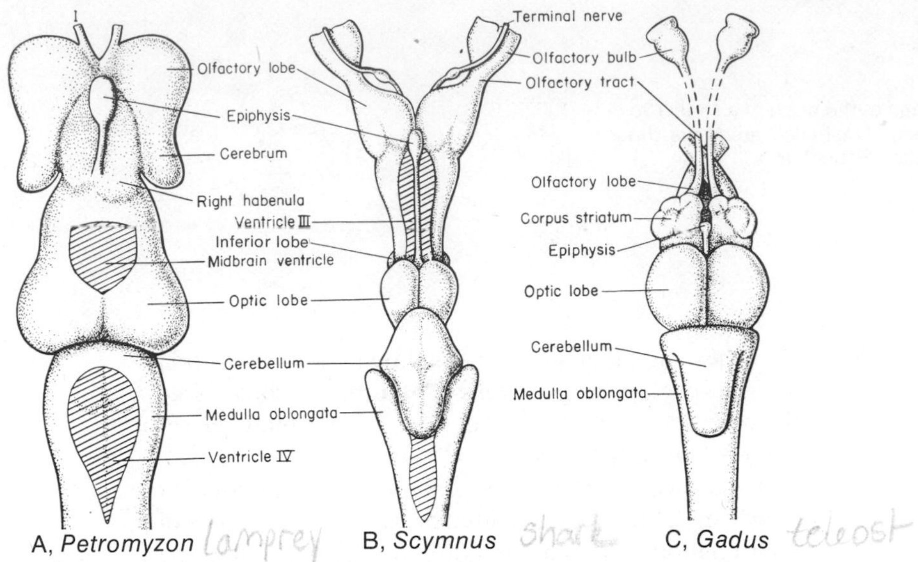

Figure 409. Dorsal views of the brain of *A*, a lamprey; *B*, a shark; *C*, a teleost (codfish). Hatched areas are those in which a choroid plexus has been removed, exposing the underlying ventricle. (After Bütschli, Ahlborn.)

rium. Its function in regulating muscular activity may be compared to that of "staff work" in the movement of an army. To carry out the general orders of an army commander it is necessary that there be on hand information as to the position, current movements, condition, and equipment of the bodies of troops concerned. Similarly, a "directive" from higher brain centers for muscular action—say the movement of a limb—cannot be carried out efficiently unless there are available data as to the current position and movement of the limb, the state of relaxation or contraction of the muscles involved, and the general position of the body. Such data

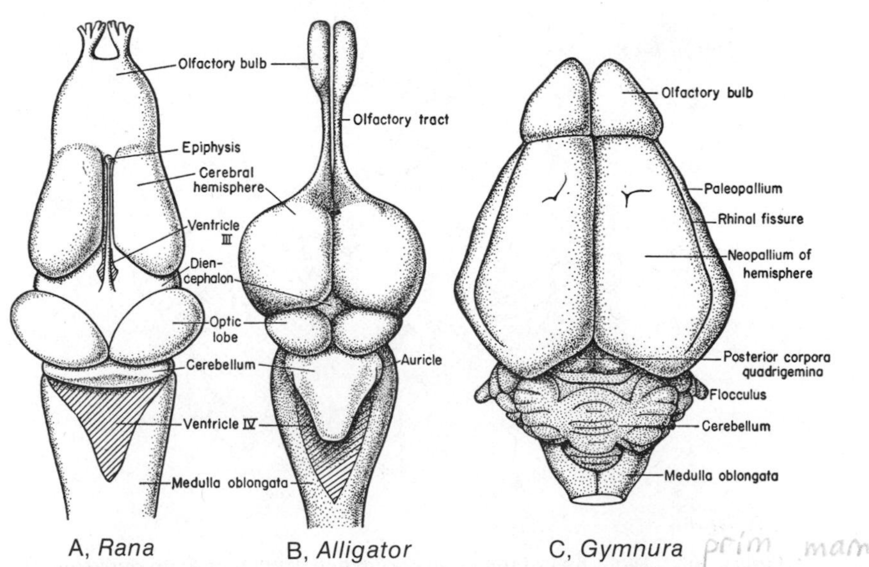

Figure 410. Dorsal views of the brain of *A*, a frog; *B*, an alligator; *C*, an insectivore. Hatched areas are those in which a choroid plexus has been removed, exposing the underlying ventricle. (After Gaupp, Crosby, Wettstein, Clark.)

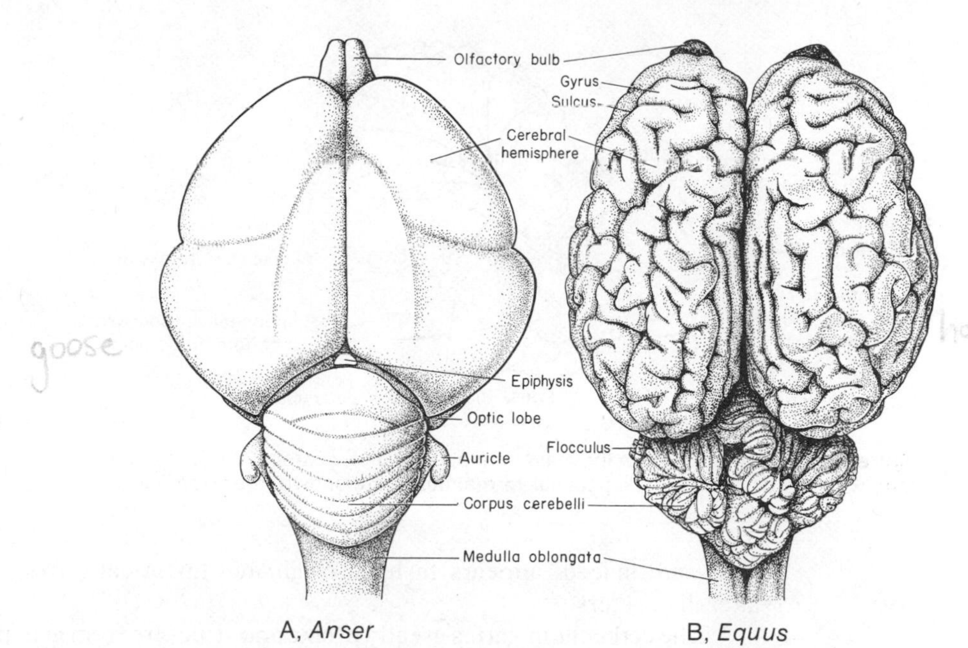

Figure 411. Dorsal views of the brain of *A*, a goose; *B*, a horse. (After Bütschli, Kuenzi, Sisson.)

are assembled in the cerebellum and synthesized there, and resulting "orders" issued by efferent pathways render the movement effective. Although there are connections with various sensory centers, the data utilized by the cerebellum in primitive forms are derived mainly from two sources, the proprioceptive structures in the muscles and tendons, and the equilibratory apparatus of the ear plus lateral line organs. We have noted that these last sense organs have a primary reception center in the upper margin of the medulla; the cerebellum rises upward from this

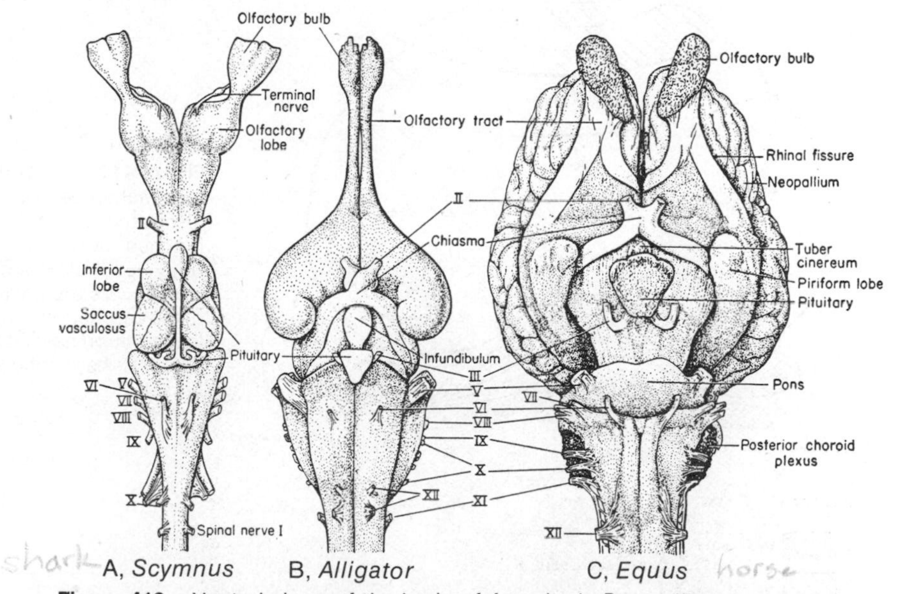

Figure 412. Ventral views of the brain of *A*, a shark; *B*, an alligator; *C*, a horse. Roman numerals indicate cranial nerves. (After Bütschli, Wettstein, Sisson.)

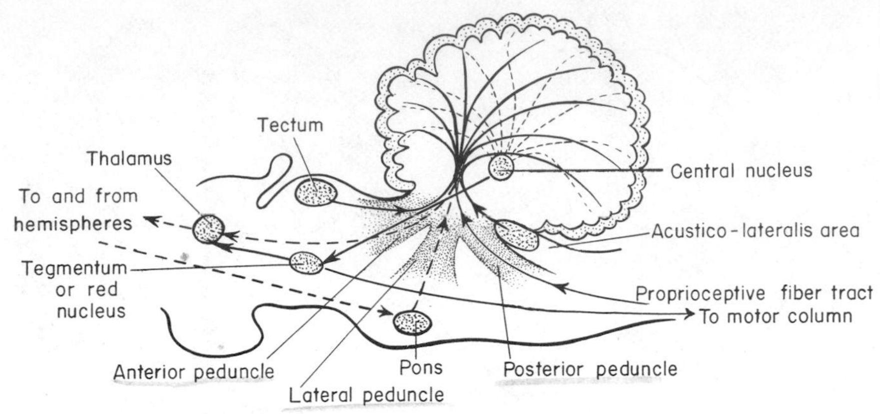

Figure 413. Diagram to show the main connections of the cerebellum. The connections with the cerebral cortex, peculiar to mammals, are shown in broken lines.

area and, indeed, appears to have originated historically from these acoustico-lateralis centers.

The cerebellum varies greatly in size and structure from group to group; its size is generally correlated with the locomotor agility of the animal. It is seen at the height of its development in birds and mammals, as shown in Figure 414. Its most ancient portions, phylogenetically, are the **auricles,** or **flocculi,** especially concerned with equilibrium and closely connected with the inner ear. In contrast to every other area of the brain except the cerebral hemispheres and the midbrain roof, the cerebellum is a region in which there develops a **cortex,** a layered superficial sheet of gray cellular material, complex in structure and often highly convoluted. The cerebellar cortex is connected with other brain areas by stout bundles of fibers

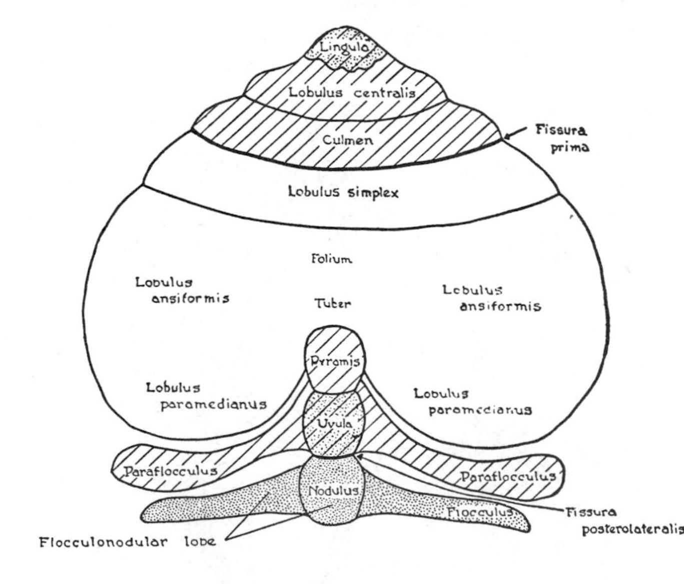

Figure 414. Diagram of a surface view of a mammalian cerebellum (showing details not discussed in the text). The stippled and hatched portions, associated with equilibrium (vestibular) and with muscle sensations (spinal), are the phylogenetically oldest parts of the cerebellum; the white area is a mammalian addition associated with the cortex of the cerebral hemispheres. (From Fulton, after Larsell.)

which form the **cerebellar peduncles,** shown in Figure 413. In mammals, but not in other classes, there are strong connections in both directions between the cerebellum and the cerebral hemispheres; the swollen pons is a relay point in the path from hemispheres to cerebellum.

MIDBRAIN AND DIENCEPHALON. In contrast with the posterior part of the brain stem, the mesencephalon and diencephalon show specialized features in vertebrates of all classes.

The lateral walls of the midbrain, termed the **tegmentum,** function mainly as the seat of centers and tracts carrying motor impulses down the brain stem from "higher" centers. The midbrain roof has had a checkered career. In all vertebrates except mammals the fibers of the optic nerve, which enter the brain in the diencephalic region, do not terminate there but run upward and backward to the mesencephalic roof, the **tectum.** This is an area of gray matter which in many vertebrates is highly developed. Primarily it is a visual center, but to it are attached fiber paths from other sensory centers—from those of the ear and lateral line, from the somatic sensory areas and from the nose via the cerebral hemispheres. As a result, sensory stimuli from all somatic sources are here associated and synthesized, and motor responses originated. The tectum thus appears to be in fishes and amphibians (where cerebral hemispheres are little developed) the true "heart" of the nervous system—the center which wields the greatest influence on body activity. In reptiles and birds the tectum is still an area of great importance but is rivalled, and in birds overshadowed, by developments within the hemispheres.

In mammals the tectum has undergone a startling reduction in importance; most of its functions have been transferred to the gray matter of the cerebral hemispheres, and most of the sensory stimuli which are integrated in the midbrain in lower vertebrates are in mammals projected, instead, to the cerebral cortex. Not even visual sensations are received here; the once important tectum is represented only by two pairs of small swellings, the **corpora quadrigemina,** which function only for visual reflexes and as a relay station for auditory stimuli on their way to the cerebral hemispheres.

The **diencephalon,** the region surrounding the third ventricle, has dorsal and ventral outgrowths of interest (Fig. 416). We have noted in the last chapter the frequent presence of median eye structures or their glandular representatives, such as the **pineal organ.** Here, too, are located the **anterior choroid plexus** and in some cases a thin-walled sac, the **paraphysis,** of almost unknown function. In the floor of the diencephalon is the optic chiasma, and in most fishes there is frequently a large vascular sac of uncertain function. Most important of diencephalic appendages is the **pituitary gland** or **hypophysis cerebri,** the major endocrine structure of the body, described in the next chapter.

The walls of the diencephalon are termed the **thalamus,** and this is in turn divided into the epithalamus, thalamus proper, and hypothalamus. The **epithalamus** is of relatively little importance. The **hypothalamus** contains a number of nuclei which are the highest centers of the visceral nervous system. The range of their functions is incompletely known; it is, however, of interest that (for example) temperature regulation in birds and mammals and sleep in the latter class are controlled by the hypothalamus.

The **thalamus** proper is in lower vertebrates an area of modest importance but is in every case a relay area for impulses passing to and from the cerebral hemispheres. The ventral part of the thalamus is a forward outpost of the motor columns of the brain stem and cord, and functions as a relay center for part, at least, of the motor impulses travelling downward from the hemispheres. The dor-

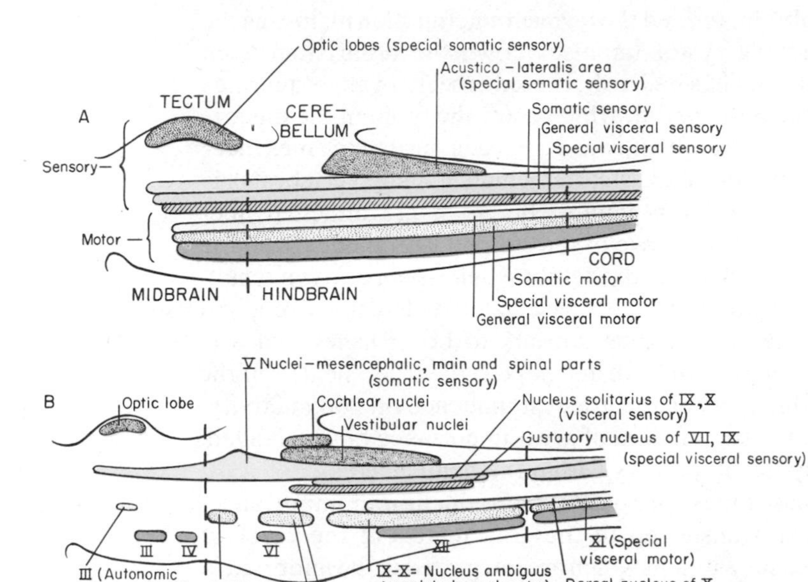

Figure 415. Diagrams of midbrain and hindbrain regions in lateral view to show the arrangement of sensory and motor nuclei. Somatic sensory, blue; special somatic sensory, stippled; visceral sensory, green; special visceral sensory, hatched; visceral motor, yellow; special visceral motor, stippled; somatic motor, red. *A,* Hypothetic primitive stage, in which brain stem centers were continuous with one another and with the columns of the cord. Even at such a stage, however, it would be assumed that special somatic centers would have developed for eye and ear. The brain includes a special visceral motor column for the branchial muscles. *B,* Comparable diagram of the mammalian situation. The somatic sensory column is still essentially continuous (almost entirely associated with nerve *V*), but the other columns are broken into discrete nuclei. The visceral sensory column includes both a general visceral nucleus (mainly for afferent fibers from the viscera via the vagus) and a special nucleus for the important sense of taste. Of efferent visceral nuclei, there are small anterior ones for autonomic eye reflexes and the salivary glands, and a large nucleus for parasympathetic fibers to the viscera via the vagus. There are important branchial motor nuclei for *V, VI,* and *IX, X* (ambiguus). The somatic motor column includes small nuclei for eye muscles anteriorly and a hypoglossal nucleus posteriorly.

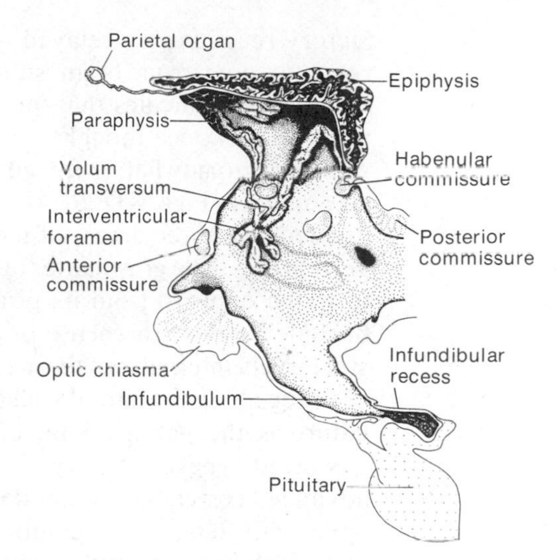

Figure 416. Sagittal section through the diencephalon of *Sphenodon*, showing the various structures projecting into and out of the third ventricle. Anterior is to the left. (After Gisi.)

sal part of the thalamus is a sensory relay center for impulses travelling upward to the cerebral hemispheres. In lower vertebrates, where the hemispheres are relatively undeveloped, this region is of no great importance. But in higher vertebrate groups in which, as we shall see, the hemispheres are the dominant association centers, the functions of the dorsal thalamus are pronounced. Cutaneous and auditory stimuli are relayed upward through dorsal thalamic nuclei; and, still further, fibers of the optic nerve, which in most vertebrates plunge through the diencephalon to the mesencephalic roof, in mammals are mostly relayed here to the gray matter of the hemispheres.

CEREBRAL HEMISPHERES. The evolution of the cerebral hemispheres is one of the most spectacular stories in comparative anatomy. These paired outgrowths of the forebrain began, it would seem, simply as loci of olfactory reception. Early in tetrapod history they became large and important centers of sensory correlation; by the time the mammalian stage is reached, the greatly expanded surfaces of the hemispheres have become the dominant association centers, seat of the highest mental faculties. The development of such centers in this area emphasizes the importance of the sense of smell in vertebrates; as we have seen, the acousticolateralis system and vision are senses upon which important correlation mechanisms were erected early in vertebrate history, but in the long run smell has proved dominant. Smell is of little account in higher primates, such as ourselves, but in most vertebrate groups it has been and is a major channel through which information about the outside world is received. It is thus but natural that its brain centers should form a base for the erection of higher correlative and associative mechanisms.

The most anterior outposts of the brain are paired **olfactory bulbs,** in which fibers from the olfactory cells of the nose are received and relayed backward through an **olfactory tract** to the cerebral hemispheres. These structures are universally present but in fishes usually relatively small and incompletely differentiated, and only the anterior parts of the hemispheres are paired. Primitively, as seen in cyclostomes, the hemispheres function mainly in a fashion preserved in higher vertebrates only in that portion of the structures termed the olfactory lobes (Figs. 417, 418 *A*). In such areas olfactory sensations are assembled and learned olfactory reactions are relayed to the more posterior centers; few fibers travel the reverse route from brain stem to hemispheres for correlation there, although recent work indicates that the hemispheres are not as completely olfactory structures as was once thought.

In a somewhat more advanced type of hemisphere such as that seen in amphibians (Figs. 417 *B,* 418 *B*) most of the tissues of the hemispheres can be divided into three areas, of interest because of their history in more progressive vertebrates. The gray matter of much of the hemispheres tends in these higher types to move outward from its primitive internal position to the surface and thus to become the **cerebral cortex,** or **pallium** (''cloak''). In amphibians the gray matter is still largely internal, but these terms may be used in the light of later history. A band of tissue along the lateral wall of the hemisphere of a persistently simple olfactory nature is the **paleopallium,** destined to form the olfactory (pyriform) lobes of advanced types. Dorsally and medially is the **archipallium,** a somewhat more advanced correlation center destined to become the hippocampus in mammals and apparently related to ''emotional'' behavior. Ventrally lies a large area of gray matter which persistently remains internal in the higher development of hemisphere structures; this forms the **basal nuclei,** the corpus striatum of mammals.

An aberrant, ''everted,'' type of forebrain is seen in most bony fishes (Fig. 419). In these fishes there is none of the trend toward outward movement of the gray matter seen in higher tetrapods; on the contrary, the outer walls of the hemispheres are thin membranes and the cellular material is crowded into a mass bulging into the ventricles from below. This development is typical of actinopterygians. As one might expect, since tetrapods do not show this structure, lungfishes have ''normal,'' not ''everted,'' hemispheres; however, the only living coelacanth, the rather aberrant *Latimeria,* displays an odd, rather intermediate condition.

In reptiles (Figs. 417 *C,* 418 *C*) the hemispheres are advanced over those of amphibians both in size and complexity. Some of the gray matter is trending toward a superficial position. The basal nuclei have moved inward to occupy a considerable area of the floor and are far from being purely olfactory in nature. Strong fiber bundles project upward to them from the thalamus and back from them to the brain stem; the basal nuclei are obvious correlation centers of importance. In birds (Fig. 420 *A*) this trend for development of the basal nuclei has progressed further. The hemispheres are large, but their development is due to enormous expansion of the basal nuclei or **corpus striatum,** and the outer walls of the hemispheres are little developed. It is obvious that the basal nuclei form a dominant association center in which, one may believe, are concentrated the mechanisms which evoke the complex ''instinctive'' action patterns seen in birds. In addition, they form a **hyperstriatum,** which may include higher centers.

In mammals cerebral evolution has taken quite another course from that found in birds (Figs. 417 *E–F,* 418 *E–F;* an ancestral reptilian stage is shown in Figs. 417 *D,* 418 *D*). The basal nuclei are moderately well developed in the interior of the hemisphere as the **corpus striatum.** The old-fashioned olfactory cortex area, the paleopallium, persists as the relatively small **pyriform lobes;** the archipallium persists likewise as the **hippocampus,** a small area tucked away on the medial aspect of the hemisphere. In mammals emphasis is placed on a new type of surface gray matter, the **neopallium** or **neocortex.** This may develop to some slight degree in reptiles; in mammals the neopallium takes over the greater part of the expanded and generally highly convoluted surface of the hemispheres. This new covering is from the first a highly developed type of association center, with four to six layers of cells present throughout its extent. It receives, like the corpus striatum, fibers which relay to it sensory stimuli from the brain stem. As it has developed in mammalian evolution it has come to take over all the higher mental functions present in either the tectum or the basal nuclei in other groups and has become not only the major directive center of the animal's activities, but also the seat of memory and of such qualities as intelligence and consciousness. Other, older centers, such as the tectum and basal nuclei, may retain some control over muscular activity through various relays; the mammalian neopallium has developed a powerful **pyramidal tract** of fibers which extends directly from the cortex to "voluntary" motor regions of stem and cord.

With expansion, the neopallium of the hemispheres tends to cover and envelop other brain structures in more progressive mammals, as may be seen by comparing more primitive mammalian brains such as those of Figures 405 *C*, 408 *A*, 410 *C* with Figures 406 *B*, 408 *B*, 411 *B;* in this process there is much shifting and distortion of older hemispheric areas, and the ventricles assume a complex and distorted structure (Fig. 421). Since the neopallium is essentially a sheet rather than a solid mass of material, simple increase in size of the hemispheres would be an unsatisfactory means of growth, and in advanced mammalian types the cortex is highly convoluted and thrown into folds or **gyri,** with intervening furrows, the **sulci.** The fossil record indicates that this folded pattern arose independently in many lines of mammals (Fig. 422).

The mammalian cortex is often described as composed of frontal, parietal, occipital, and temporal lobes. These terms, however, are merely topographic and have no precise meaning as regards the architecture or functioning of cortical areas. A complex "wiring" system connects all parts of the cortex with one another, suggesting that the gray matter is essentially a single unit. Experiments on laboratory animals and study of the results of disease or injury on human brains show that this is true to a considerable extent. On the other hand, it is clear that certain cortical areas are normally associated with specific functions (Fig. 423). The front part of the neopallium includes a motor area; the posterior part is associated with sensory perception; special regions are associated with eye and ear; in a general "somatic sensory" region are definite areas for reception of cutaneous and proprioceptive sensations from various specific parts of the body. In man particularly, however, we find that these specific functional areas of the neopallium occupy only a relatively small part of the surface. Between them are found large "blank areas" of gray matter, most conspicuously one occupying much of the frontal lobe. Obviously these regions are far from blank; they are association areas of the highest and most generalized type, the seat of such mental properties as learning ability, initiative, foresight, and judgment.

BRAIN PATTERNS—SUMMARY. We may summarize here some of the main features of brain structure described in brief fashion above.

Much of the brain stem is a persistently primitive region, with motor and sensory columns and centers rather closely comparable to those of the spinal cord. The history of brain evolution has been mainly one of the development, above and in front of the medulla, of higher centers; most prominent of such centers are dorsal outgrowths of laminated gray matter which have grown up in areas concerned with the three special senses. In such centers sensory data are assembled and synthesized, and resultant motor stimuli sent out to the motor areas of brain stem and cord.

1. The primary center of reception for stimuli from the ear and lateral line lies in the medulla; above this developed the cerebellum, which initiates no bodily movement, apart from posture reflexes, but insures that motor directives initiated in other centers be carried out in proper fashion. The principal information upon

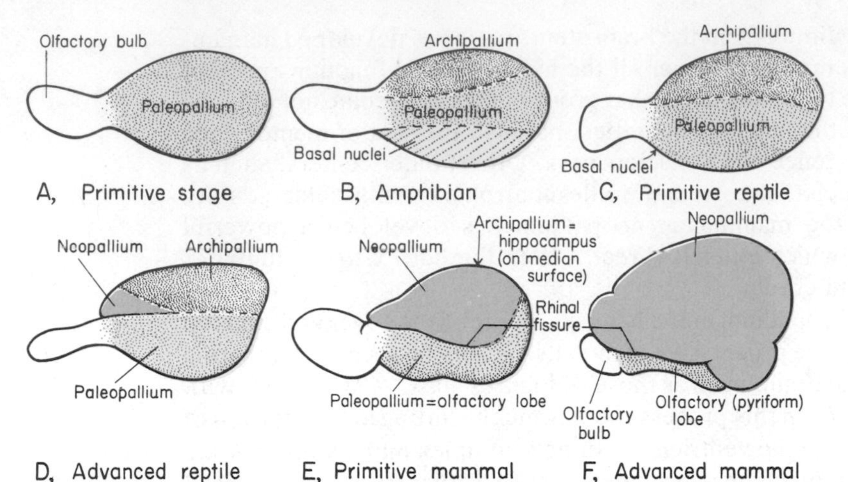

Figure 417. Diagrams to show progressive differentiation of the cerebral hemisphere (cf. Fig. 418). Lateral views of left hemisphere and olfactory bulb. In *A* the hemisphere is merely an olfactory lobe. *B,* Dorsal and ventral areas, archipallium (= hippocampus) and basal nuclei (corpus striatum) are differentiated. *C,* The basal nuclei have moved to the inner part of the hemisphere. *D,* The neopallium appears as a small area (in many reptiles). *E,* The archipallium is forced to the median surface, but the neopallium is still of modest dimensions, and the olfactory areas are still prominent below the rhinal fissure (as in primitive mammals). *F,* The primitive olfactory area is restricted to the ventral aspect, and the neopallial areas are greatly enlarged (as in advanced mammals). The various cellular components of the hemispheres are distinguished by color.

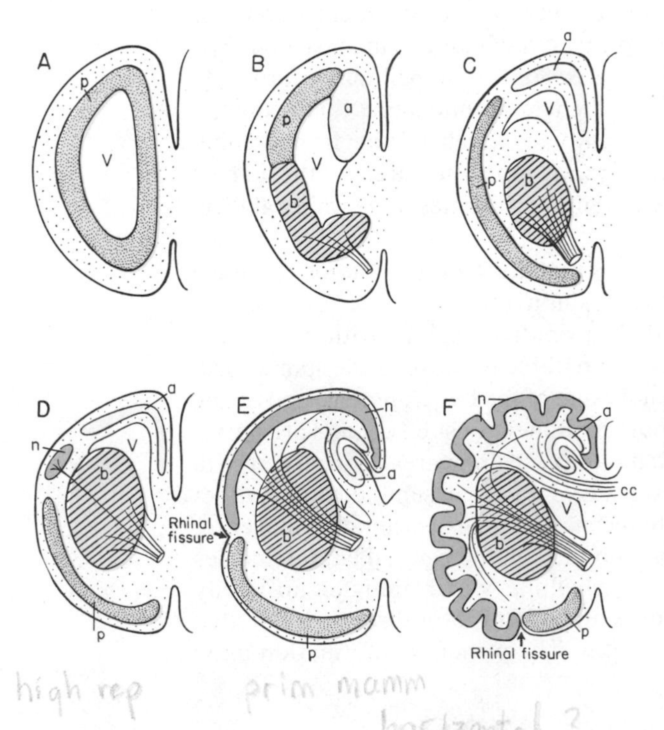

Figure 418. Diagrammatic cross sections of left cerebral hemisphere to show stages in the evolution of the corpus striatum and cerebral cortex. *A,* Primitive stage, essentially an olfactory lobe; gray matter internal and little differentiated. *B,* Stage seen in modern amphibians. Gray matter still deep to surface, but differentiated into paleopallium (= olfactory lobe), archipallium (= hippocampus), and basal nuclei (= corpus striatum), the last becoming an association center, with connections from and to the thalamus (indicated by lines representing cut fiber bundles). *C,* More progressive stage, in which basal nuclei have moved to interior, and pallial areas are moving toward surface. *D,* Advanced reptilian stage; beginnings of neopallium. *E,* Primitive mammalian stage; neopallium expanded, with strong connections with brain stem; archipallium rolled medially as hippocampus; paleopallial area still prominent. *F,* Progressive mammal; neopallium greatly expanded and convoluted; paleopallium confined to restricted ventral area as pyriform lobe. The corpus callosum developed as a great commissure connecting the two neopallial areas. *a,* Archipallium; *b,* basal nuclei; *cc,* corpus callosum; *n,* neopallium; *p,* paleopallium; *V,* ventricle. The different types of "gray matter" colored as in Figure 417.

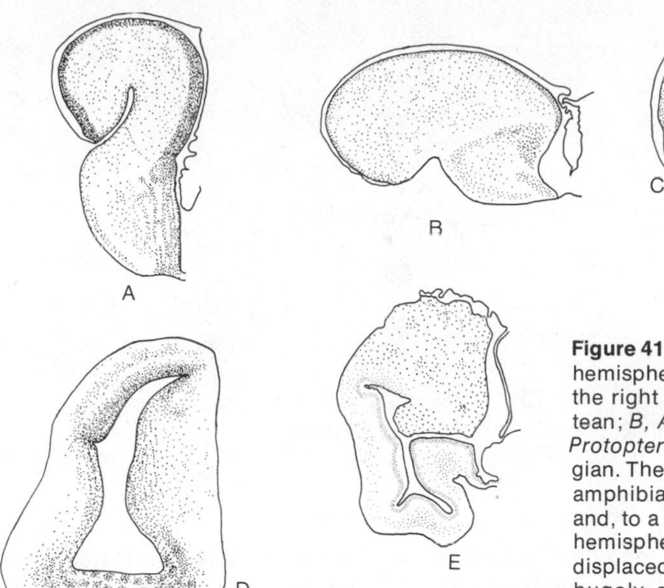

Figure 419. Transverse sections through one cerebral hemisphere of various Osteichthyes; the midline is on the right of each drawing. *A, Polypterus,* a chondrostean; *B, Amia,* a holostean; *C, Carassius,* a teleost; *D, Protopterus,* a lungfish; *E, Latimeria,* a crossopterygian. The hemisphere of the lungfish resembles that of amphibians (cf. Fig. 418B); in all the actinopterygians and, to a lesser extent, *Latimeria,* the dorsal wall of the hemisphere is membranous, the ventricle is dorsally displaced, and the ventral position represented by hugely expanded basal nuclei—a pattern termed "everted." (After Nieuwenhuys and Kuhlenbeck.)

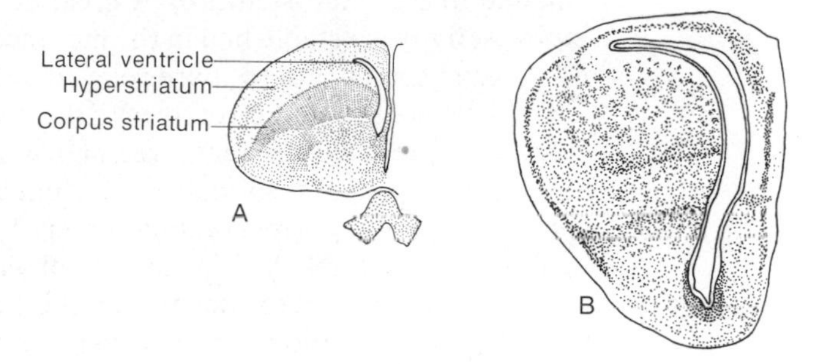

Figure 420. Transverse sections through the cerebral hemisphere of a sparrow, *A,* and an alligator, *B.* In birds, the basal nuclei form a greatly expanded corpus striatum and, above it, hyperstriatum; there is essentially no cortex. This is not an archosaurian characteristic, as the alligator has a relatively "normal" forebrain. (After Kappers, Huber, and Crosby.)

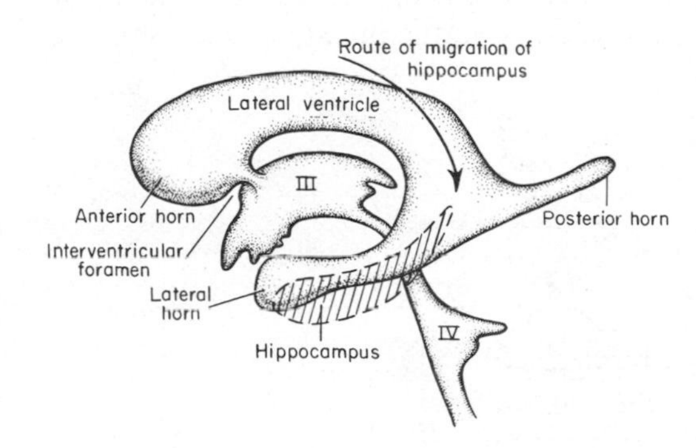

Figure 421. The brain ventricles of an advanced mammalian type *(Homo)* in lateral view from the left. The ventricles are represented as solid objects, the brain tissue being removed. With expansion of the cerebral hemisphere, the lateral ventricle has expanded backward to a posterior horn in the occipital lobe, and downward and forward laterally to a lateral horn in the temporal lobe. With this backward and downward expansion, various shifts in position of brain parts occurred. The hippocampus, which developed dorsally on the median surface of the hemisphere (cf. Fig. 418 *F*) has been rotated, in advanced mammals, backward and downward into a ventral position near the midline.

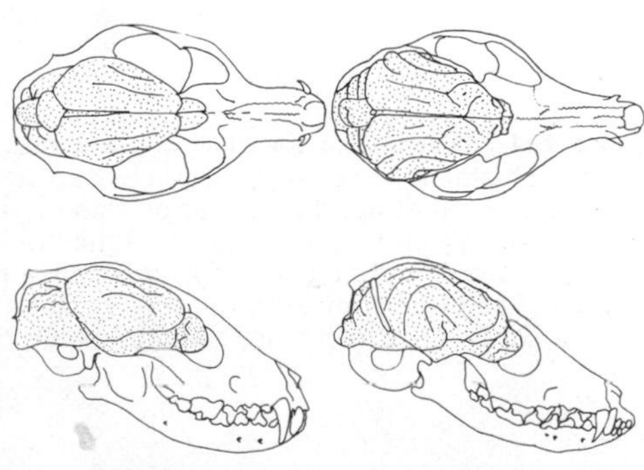

Figure 422. Brains (stippled) in skulls of a fossil and a Recent dog to show the increase in size and complexity of the brain, especially the cerebral hemisphere. *Hesperocyon,* on the left, is an Oligocene form (roughly 30 million years old); *Fennecus,* on the right, is a modern foxlike form of about the same size. The actual brain of the fossil is, of course, not preserved, but the form of the cranial cavity reflects its structure in considerable detail. (After Radinsky.)

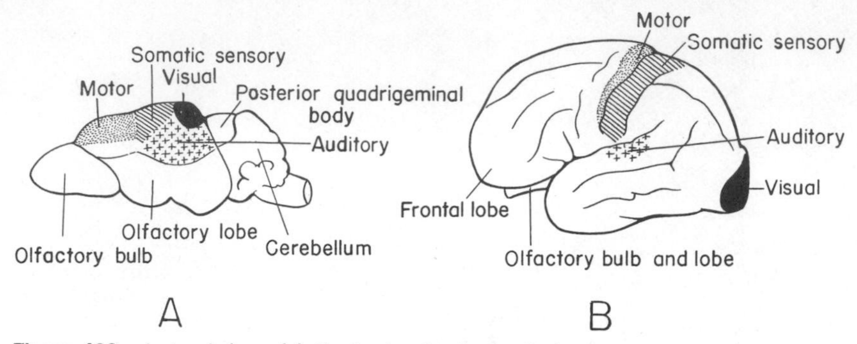

Figure 423. Lateral view of *A*, the brain of a shrew; *B*, the cerebrum of man; to show cortical areas.

which it acts comes from the adjacent centers for the ear and lateral line and from the proprioceptive system of the muscles and tendons. In mammals intimate connections are established between cerebellum and cerebral hemispheres.

2. In lower vertebrates the main centers dominating nervous activity are situated in the brain stem. *(a)* A great center of coordination and of initiation of motor activity is established in the mesencephalic tectum (Fig. 424); this is primarily a visual center, but to it are relayed stimuli of all other somatic sensory types, and from it are sent out directives to the motor centers and columns. As we ascend the vertebrate scale the tectal region becomes rivalled and then exceeded by the hemispheres, and in mammals the tectum is of small importance. *(b)* The tectum is somatic in nature; corresponding centers for visceral sensations and visceral motor responses are established in the hypothalamus.

3. In higher vertebrates the cerebral hemispheres, originally a center largely for olfactory sensation, have become more and more important as association centers. *(a)* First of cerebral areas to gain importance is that of the basal nuclei or corpus striatum, to which, in all amniotes, tracts from the thalamus relay somatic

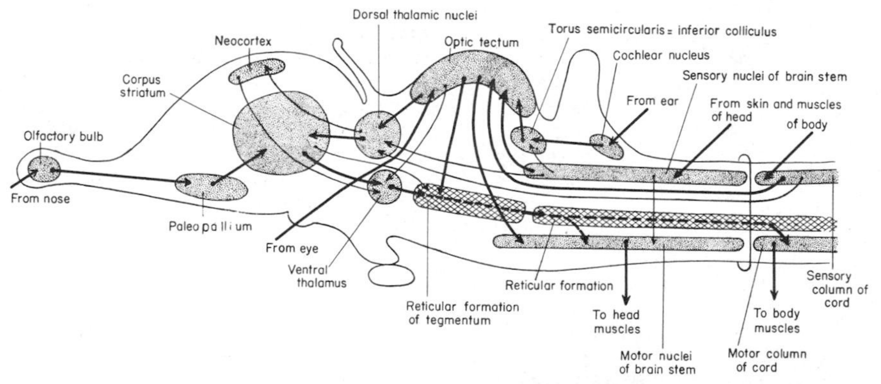

Figure 424. Diagram of the main centers and "wiring" arrangement of a reptile, in which the tectal region of the midbrain plays a dominant role; the corpus striatum (basal ganglia) is of some importance as a correlation center, but the neocortex (neopallium) is unimportant. The reticular formation of the brain stem (cross hatched) is important in carrying motor impulses to nuclei of the stem and cord. In this oversimplified diagram only a limited number of paths between somatic receptors and effectors are included; visceral centers and paths are omitted, as are cerebellar connections (shown in Fig. 413).

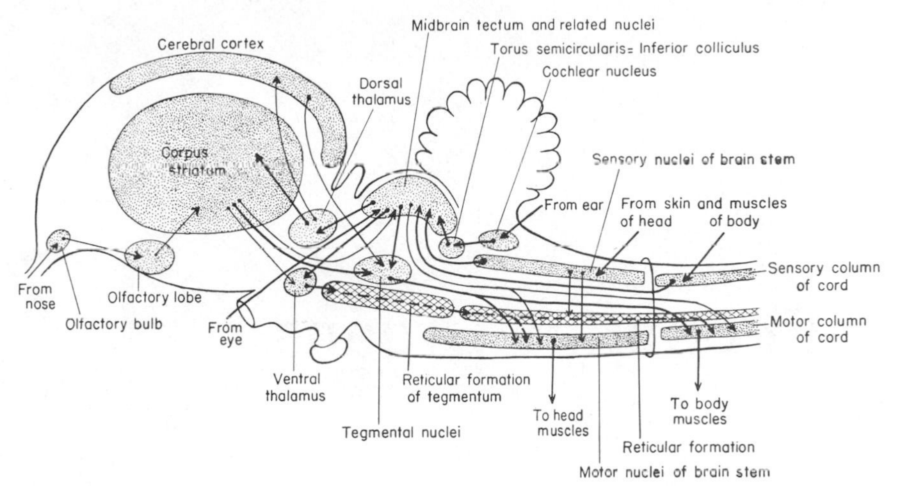

Figure 425. A "wiring diagram" of a bird brain, comparable to that of Figure 424. The midbrain tectum is still of importance, but the corpus striatum is the dominant center in many regards.

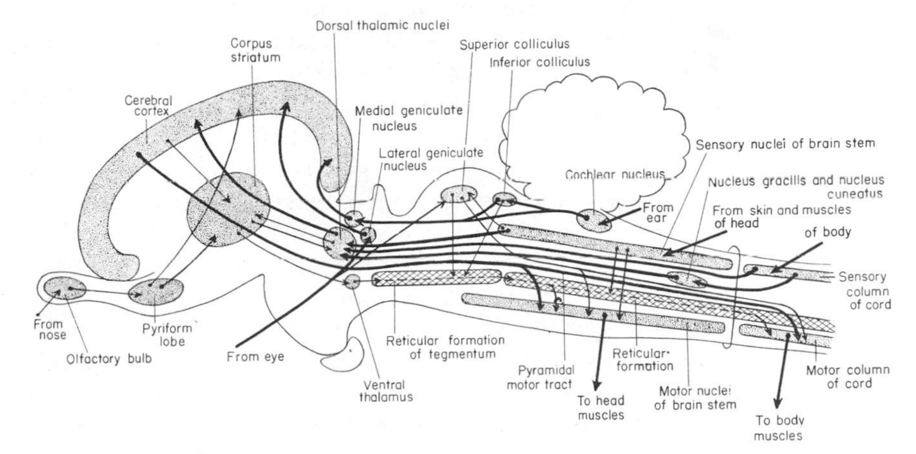

Figure 426. A "wiring diagram" of a mammalian brain comparable to Figures 424 and 425. The midbrain tectum is reduced to a minor reflex center, and the corpus striatum is relatively unimportant; most sensory impulses are projected "upward" to the cerebral cortex, whence a direct motor path (pyramidal tract) extends to the motor centers of brain stem and cord.

sensations and from which descending fibers carry motor stimuli to midbrain and motor columns. In reptiles the corpus striatum rivals the older tectum in importance and in birds it is a dominant center (Fig. 425). *(b)* In mammals, however, a different development has occurred (Fig. 426). In the gray matter of the cortex there develops a new, greatly expanded correlation and association center, the neopallium. This assumes the greater part of the higher functions once concentrated in the tectum or corpus striatum, gains a complete array of somatic sensory data through projection fibers from the thalamus, and develops direct motor paths to the motor columns of the brain stem and spinal cord.

Endocrine Organs 17

In the last chapter we described, in the nervous system, an exceedingly complicated but highly efficient method of coordinating bodily activities by ''messages'' received from and sent to specific areas of the body with speed and precision. We shall here consider a second integrative system, under which information and directives are carried through the blood by chemical ''messengers,'' the **hormones** produced by **endocrine glands**—glands, that is, which have no ducts but instead pass their products into circulatory vessels. This method of transmission is, of course, slower than transmission by nerve impulse, and hormonal effects are often broadly distributed over the body to a variety of organs and tissues, in contrast to the ''pinpointing'' possible in the nervous system. But despite these apparent drawbacks, many of the hormones are not only important but are absolutely essential for the maintenance of the life of the organism.

For convenience, we have gathered in this chapter data on all the known hormone-producing structures of the body, although they do not actually form an organ system, but are scattered here and there throughout the body—literally from stem to stern—and may derive from a variety of sources. The situation is somewhat similar to that of the blood-forming organs: just as it makes no difference in which part of the body blood corpuscles are produced, so too the area in which hormones are produced is inconsequential, as long as they can be passed into some element of the circulatory system and thence be distributed throughout the body.

Neural and hormonal systems of communication, although distinct, are far from independent of one another. Directly or indirectly, the nervous system may be powerfully affected by hormones. On the other hand, the ''master gland'' of the endocrine system, the hypophysis, is strongly influenced by the adjacent hypothalamus, and some of its hormones are actually produced by ganglia there. Again, the adrenal medulla, although an endocrine organ, is composed of modified nerve cells.

Which is the older regulatory system, nervous or endocrine? There is no clear answer to this; probably both evolved in parallel fashion. Elementary nervous systems are present in some of the most primitive metazoan animals. Hormonal systems are known in numerous invertebrates, and undoubtedly many more await discovery.

THE HYPOPHYSIS

Below the diencephalic region of the brain lies a small but essential structure, the major endocrine organ of the body, the **pituitary gland** or **hypophysis cerebri** (Figs. 229, 404, 408, 427, 430). In most vertebrates the pituitary tissues form a single compact mass, contained in a pocket (the **sella turcica**) in the floor of the braincase. Actually, however, the gland is a dual structure, its two portions having very different embryological origins and functioning in different fashions (Fig. 428).

Downward from the embryonic diencephalon extends a hollow, fingerlike process, the **infundibulum.** Upward from the embryonic mouth there grows an ectodermal pocket, the **hypophyseal pouch** (Rathke's pouch). From both these embryonic structures there proliferate masses of tissues, which unite to form the adult pituitary, or hypophysis.

Although various terminologies have been (and are) applied to subdivisions of the gland, it is best considered as consisting of two parts or lobes in accordance with its embryonic origins: the **adenohypophysis,** derived from the hypophyseal pouch, and the **neurohypophysis,** formed from brain tissue. The major portion of the adenohypophysis—in fact, the major portion of the whole gland—is the **pars distalis.** In addition there may be distinguished, particularly in mammals, a **pars tuberalis,** growing up around the stalk of the infundibulum, and a **pars intermedia,** which may fuse with the neural part of the gland. The neurohypophysis consists mainly of the **lobus nervosus,** or neural lobe; but the **infundibulum,** from the bottom of which the lobe develops, may also be considered as part of the neurohypophysis. And as we shall see, a fraction of the hypothalamus is functionally also a part of the neurohypophyseal apparatus.

There is considerable variation in pituitary structure among some of the lower vertebrates. In the lamprey, for example, there is no formed neural lobe, its homologue being simply a plate of tissue in the floor of the diencephalon, and the adenohypophysis is formed of tissue derived from a tube leading backward beneath the brain from the nasal opening (Fig. 253 *B*). In cartilaginous fishes, there is found a distinct infundibulum and a well-formed mass of neural lobe tissue. Here, and in almost all higher types, the hypophyseal pouch is closed and there is a well-developed adenohypophysis, but in shark-like fishes and ray-finned types there are no well-marked subdivisions of that part of the gland. In lungfishes and most

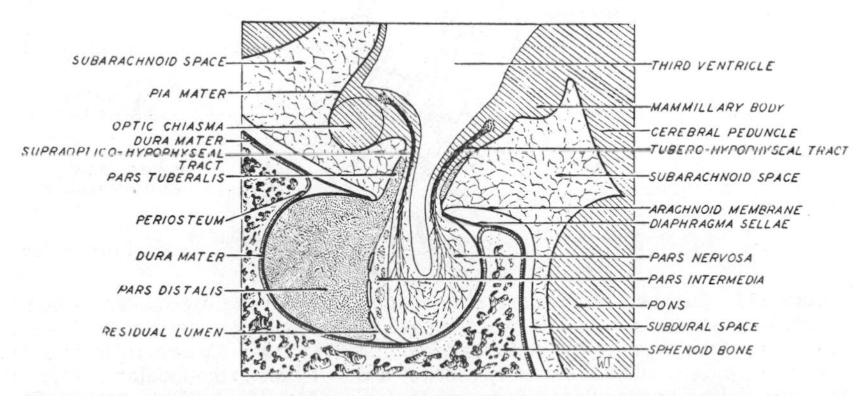

Figure 427. Section through a human pituitary and the adjacent structures at the base of the brain. (From Turner, General Endocrinology, W. B. Saunders Company.)

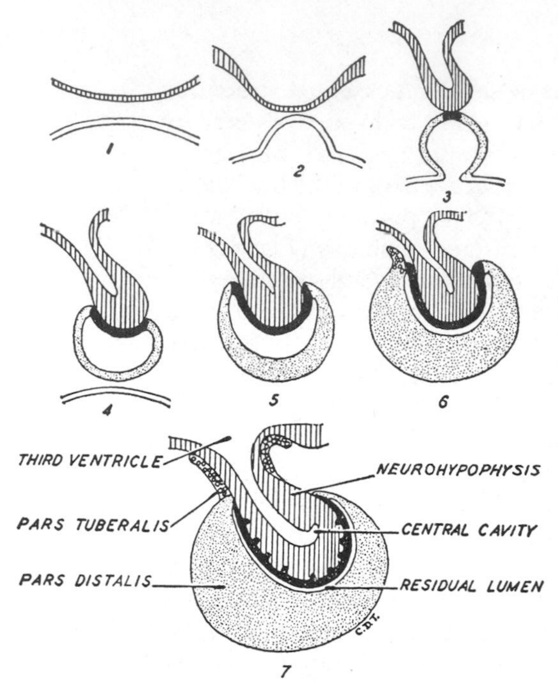

Figure 428. Diagrams showing stages in the embryonic development of the mammalian pituitary from brain tissue (hatched) and from the hypophyseal (Rathke's) pouch (the pars intermedia in black). As will be seen, only the neurohypophysis is formed from neural material; the three other parts, constituting the adenohypophysis, are derived from the pouch epithelium. (After Turner.)

tetrapods there is a distinct pars intermedia in addition to the pars distalis, but a pars tuberalis is a relatively uncommon development.

It was originally assumed that the hormones given off by both parts of the hypophysis were produced by the cells located in the organ itself. In recent years, however, it has been discovered that the neural lobe is merely a storage area for the

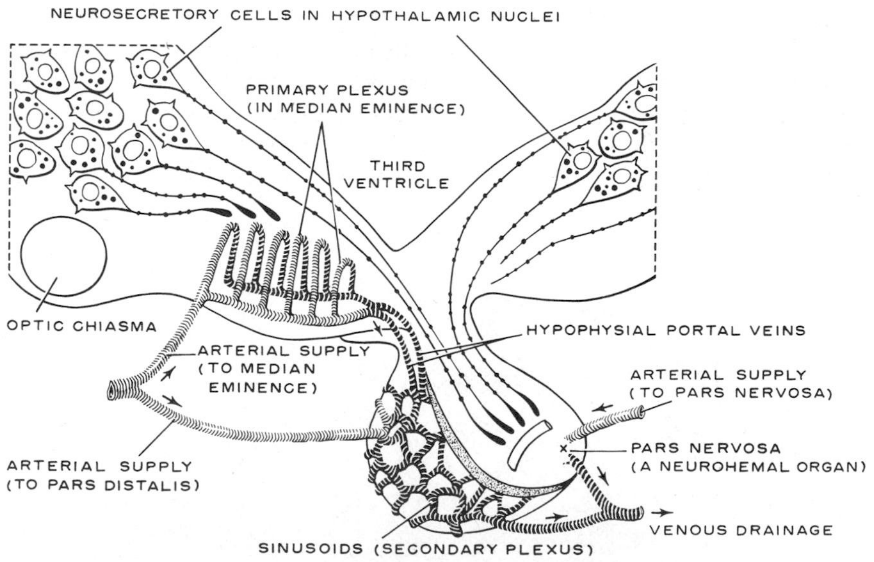

Figure 429. Diagram of anatomic connections between the hypothalamus and the pituitary gland. Cells in certain hypothalamic nuclei secrete materials which are the hormones of the pars nervosa and are picked up by blood vessels passing through that part of the pituitary. Other secretory cells release materials in the median eminence at the base of the diencephalon; they appear to be picked up here by the primary plexus of the hypophyseal portal system and, passing to the sinusoids in the pars distalis of the adenohypophysis, stimulate hormone production there. (From Turner, General Endocrinology, W. B. Saunders Company.)

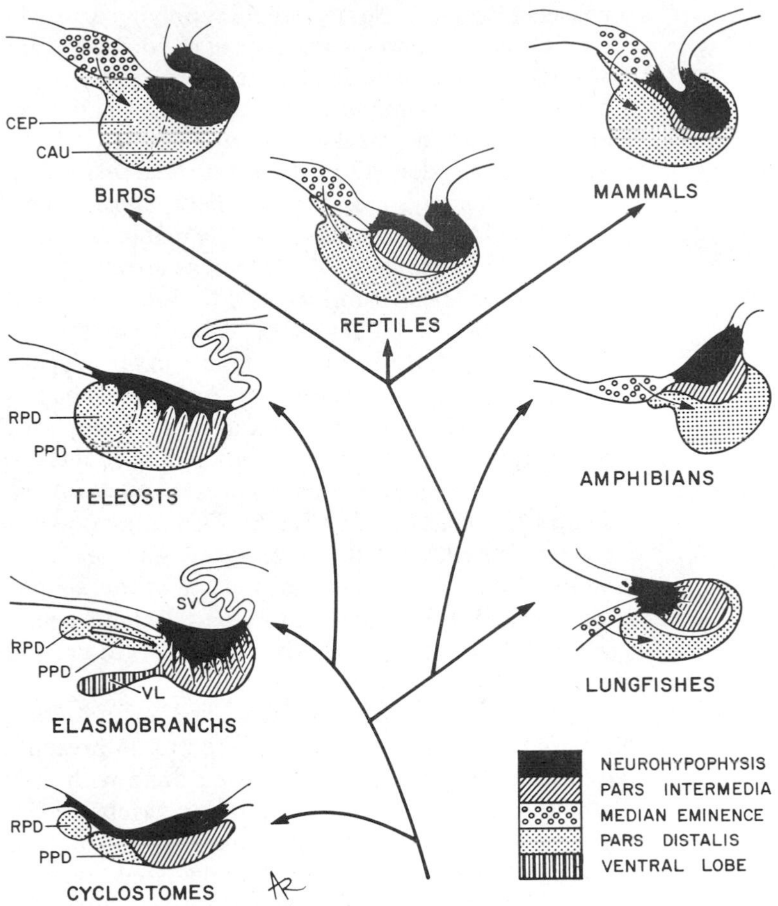

Figure 430. Schema to show probable evolutionary changes in the vertebrate pituitary gland. Arrows extending from the median eminence to the pars distalis represent the hypophyseal portal system. *CAU,* caudal division of the avian anterior lobe; *CEP,* cephalic division of the same; *PPD,* proximal pars distalis; *RPD,* rostral pars distalis; *VL,* ventral lobe of the elasmobranch pituitary; *SV,* saccus vasculosus. (From Turner, General Endocrinology, W. B. Saunders Company.)

hormones that it passes into the blood. The hormones are actually formed in the cell bodies of neurons of the brain, located in the **supraoptic** and **paraventricular nuclei** of the hypothalamus. The materials secreted by these cells pass down their axons into the neural lobe for storage and eventual entry into the blood.

In contrast, the more numerous hormones of the adenohypophysis, both pars distalis and pars intermedia, are produced by the cells actually present in these structures. The adenohypophysis is strongly influenced by the brain. The means by which this influence is exerted is a matter of interest. In many fishes nerve fibers pass into the adenohypophysis, thus affording a reasonable path by which the brain can affect glandular activity there. But in lungfishes and tetrapods this is not the case. Nerve fibers may enter the pars intermedia, but (except for some which appear to be autonomic fibers for blood vessels) none penetrate to the pars distalis, where almost all the hormones are formed. In this condition, how does the brain influence the gland?

Apparently indirectly, through a curious local portal system of tiny blood vessels (Fig. 429). Some vessels supplying arterial blood to the adenohypophysis pass close to or through the floor of the diencephalon anterior to the infundibulum. A tract carrying secretions to the neural lobe passes through this same area, which, particularly in mammals, tends to form a small swelling on the lower surface of the brain, termed the **median eminence.** In most fishes and all tetrapods the blood vessels to the adenohypophysis enter the region of the median eminence, break up into a capillary system, then re-collect as a set of small portal veins to pass on to the pars distalis. In passage through the median eminence the blood presumably picks up neurohumoral materials that act as agents transmitting "information" from the brain to the gland. This seems a curiously roundabout way of accomplishing an important function, but no other effective method is known or has been suggested.

Attempts have been made to identify predecessors of the hypophysis, or of its components, in lower chordates. Amphioxus has a group of cells in the roof of the mouth and the floor of the neural tube above, which have been compared to the hypophyseal pouch and infundibulum respectively, but the homology is very uncertain. In adult tunicates there is present a **neural gland,** which is in certain regards suggestive of a pituitary (Fig. 6). This gland opens into the morphologically dorsal side of the entrance to the pharynx, and lies close to the ganglion which is the nearest approach to a brain present in the simple nervous system of the adult tunicate. Morphologically, the neural gland could reasonably be considered as antecedent to the hypophysis, but there is at present no positive evidence that it is an endocrine-producing structure.

Some nine or ten hormones are known to be produced by the pituitary. Still others have been, or are, suspected to be present there. The greater part of the work on these hormones has been done with mammals, but a large part of the whole series is found throughout the vertebrate classes. Most pituitary hormones are produced by the adenohypophysis, principally by the pars distalis. All are proteins or polypeptides. They may be listed and briefly described:

GROWTH HORMONE OR SOMATOTROPHIN (STH or GH). This has a very broad influence on growth and metabolism in general, with marked influence on growth of skeleton and muscle, metabolism of fats and carbohydrates, and protein synthesis; it further acts to enhance the effect of other hormones on the activity of thyroid, adrenal cortex, and reproductive organs.

CORTICOTROPHIN (ACTH). Vital for adrenal cortical activity in secretion of hormones and exerts some metabolic influence in other regards.

THYROTROPHIN (TSH). Essential for the stimulation of the thyroid to form and release its hormones.

PROLACTIN (LTH). This is mainly associated with sexual structures and activities and hence with the next two hormones is termed **gonadotrophic.** These hormones are best known (and named) from their observed effects in mammals. Prolactin stimulates milk secretion and, in some mammals, prolongs the functional life of the corpus luteum (and consequent continued secretion of progesterone; cf. p. 416). The functions of prolactin in lower vertebrates are far from completely known. In pigeons it stimulates the gland in the crop that produces "pigeon milk"; in some urodeles it causes the animal to return to the water for reproduction; in fish it may be involved in osmoregulation, serve as a "growth hormone," or be involved in the regulation of pigmentation. Prolactin-producing cells appear to predominate in the anterior pars distalis of many teleosts.

LUTEINIZING OR INTERSTITIAL CELL-STIMULATING HORMONE (LH or ICSH). Influences maturation of the gonads and production of sex hormones, acts in formation of the corpora lutea and secretion of progesterone in the ovary, and stimulates the interstitial cells of the testis, promoting the production of male sex hormones and the maturation of sperm.

FOLLICLE-STIMULATING HORMONE (FSH). Stimulates the growth of ovarian follicles, promotes spermatogenesis and, in conjunction with the luteinizing hormone, promotes estrogen secretion and ovulation.

INTERMEDIN OR MELANOCYTE-STIMULATING HORMONE (MSH). A polypeptide which acts to disperse pigment granules in melanophores, to aggregate reflecting organelles in iridophores, and to darken the skin. In contrast to the hormones listed so far, this (as its name indicates) is produced in several varieties by the intermediate portion of the gland when this part is distinctly developed. It has no clearly known function in birds and mammals; however, it has been implicated in several effects on the brain.

In contrast to the wealth of hormones produced by the adenohypophysis, the secretions given off by the neurohypophysis consist of a limited number of related polypeptides, prominently **vasopressin,** or antidiuretic hormone (ADH), and **oxytocin.** Vasopressin is principally associated with increasing blood pressure through contraction of arterioles and controlling water output or intake in various fashions in the different vertebrates. Oxytocin is best known from its effects on the female mammal in promoting contraction of uterine muscle, and development of the mammary gland and ejection of milk after the birth of young; it appears, however, that this hormone has sexual effects on at least certain other vertebrates—spawning in minnows, for example. It seems clear that, in contrast to the importance of the numerous hormones of the adenohypophysis, those of the neural lobe play a relatively modest part in the body economy.

PARATHYROID GLANDS

Among the glands derived in tetrapods from the gill pouch region of the embryo are small structures, usually two pairs, termed the **parathyroid glands** (Figs. 266, 267). In the adult these are situated at somewhat variable positions in the neck. In man, they are embedded in the thyroid tissues. Attention was long ago called to these glands through the discovery that extirpating parts of the human thyroid containing them resulted in the death of the patient, because the parathyroid hormone, **parathormone,** is deeply involved in the metabolism of calcium and, to a lesser degree, of phosphorus. Parathyroid glands as such are absent in fishes, but there is evidence to suggest that the **ultimobranchial bodies,** budded from the last gill pouch of the embryo, are partial equivalents in fish. These glands (and equivalent cells in mammals) produce the hormone **calcitonin,** which lowers the level of calcium. Small bodies, termed the **corpuscles of Stannius,** found in teleosts, have been compared to parathyroids, but are certainly not homologous—they form from urogenital ducts.

THYROID GLAND

Also derived from the throat, but (in contrast to the parathyroid) budded from the floor of the embryo pharynx rather than its walls, is the **thyroid gland** (Figs. 431, 432). In adult fishes it lies below the gill chamber, and in tetrapods generally ventral to the windpipe at some point along the length of the neck. In many vertebrates it is a single structure, although often bilobed; in birds, lizards, amphibians,

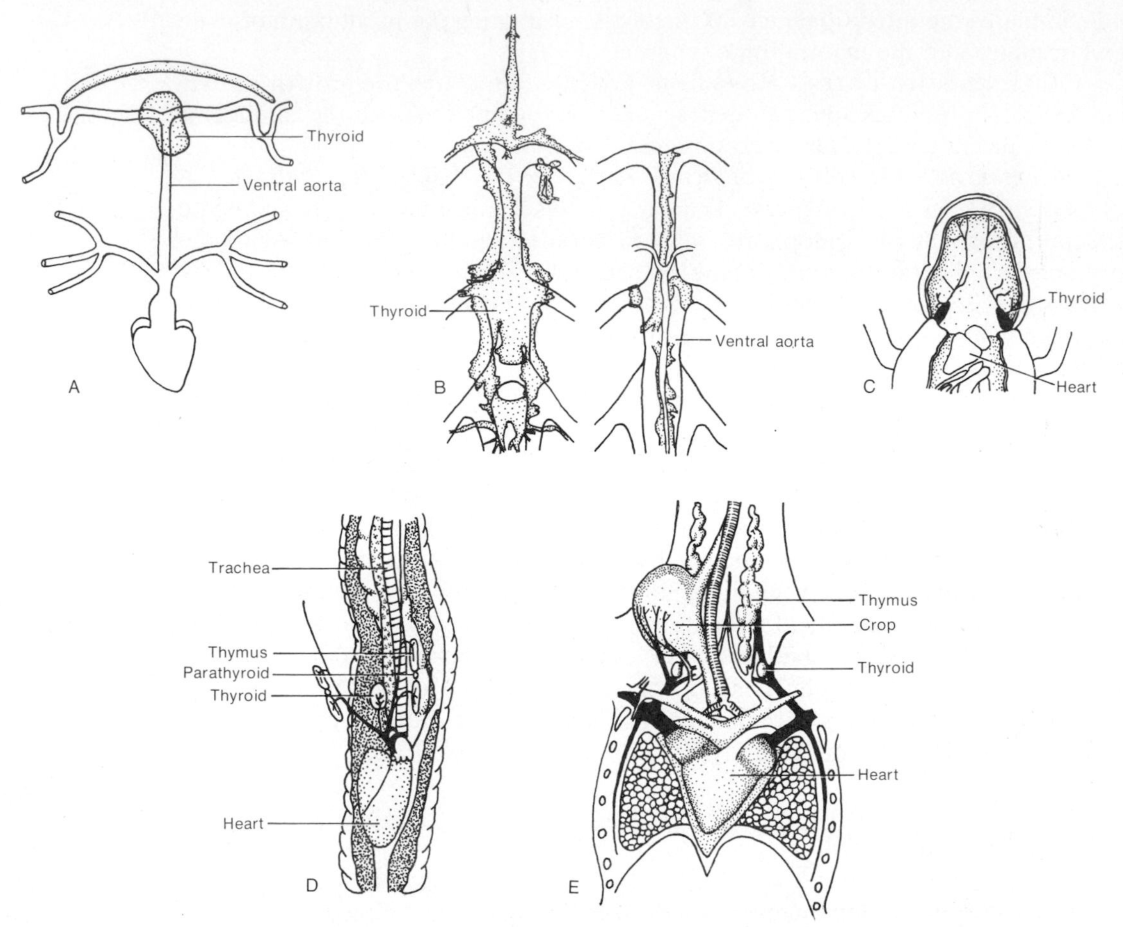

Figure 431. Thyroid and adjacent organs in various vertebrates. All are in ventral view except *B* in which the ventral view is left and a dorsal view on the right. *A, Raja,* a ray; *B, Salmo,* a salmon; *C, Ambystoma,* a urodele; *D, Natrix,* a snake; *E, Gallus,* a chicken. (After Ferguson, Hoar, Weichert, Clark, and Ede.)

and teleosts it is usually paired in the adult. In teleosts, and to a lesser extent in other vertebrate types, small detached masses of thyroid tissue may be found in places far removed from the main gland. The thyroid tissues consist of numerous small spherical follicles, bounded by a secretory epithelium that discharges into a central cavity filled with a gelatinous **colloid substance.** In this colloid are found, in storage, quantities of an iodine-bearing protein from which the iodine-bearing hormones, mainly **thyroxine,** are formed and discharged into the blood. The thyroid products are highly important in maintaining tissue metabolism, and are concerned with reproductive functions and growth phenomena. Most spectacular of the thyroid functions is control of metamorphosis in amphibians.

The thyroid has a pedigree that stretches far back in chordate history. Both amphioxus and tunicates have ciliated and partly glandular channels in the floor of the pharynx which are concerned in the feeding mechanism. Such a channel is termed the **endostyle** (Figs. 4–6). In both amphioxus and tunicate endostyles there are produced, as well as mucus, iodine compounds that are carried on into the digestive tract together with food materials. Ammocoete larvae of lampreys have a similar feeding habit and ciliated grooves much like those of amphioxus and tunicates. Here the ventral groove terminates posteriorly in a deep pouch in the

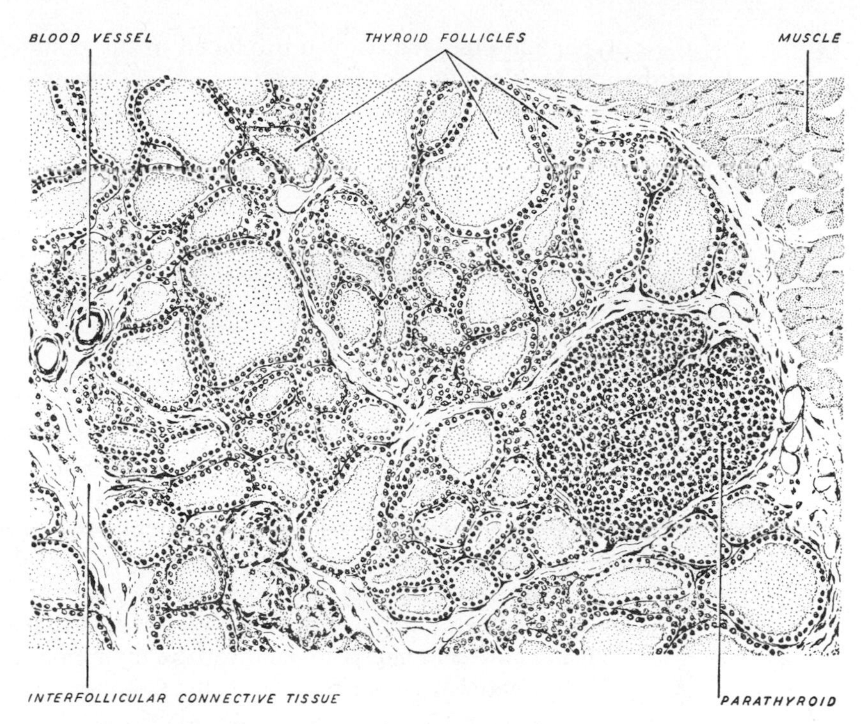

Figure 432. Thyroid and parathyroid tissues of the rat. (From Turner.)

pharyngeal floor, where there is produced an iodine-bearing material that is carried on down the digestive tract (Fig. 253 *A*).

The endostyle, including the ammocoete gland, is a median ventral pharyngeal structure, comparable in position to the thyroid gland. Are they truly homologous? The lamprey gives us a positive, conclusive answer. At metamorphosis the larval endostylar pouch closes off from the gut and breaks up into a series of follicles that are indisputably thyroid in structure. The thyroid, it would seem, was originally an exocrine gland, whose products were taken into the digestive tract. It has become an endocrine organ; but it is of interest that, alone of hormones, those of the thyroid are still effective when taken by mouth.

PANCREATIC ISLANDS

Although the greater part of the glandular tissue of the pancreas is devoted to the production of enzymes that pass through ducts to the intestine, areas of tissue of another type can be seen distributed through the gland as isolated islands (Fig. 283). These **islets of Langerhans** consist of at least two types of cells, glandular in nature but not furnished with ducts; they thus obviously form an endocrine organ, sending secretions into the blood. The insular material is usually diffused among the ordinary tissues of the formed pancreas. In teleosts, however, small clusters of islet cells are spread here and there in the general region of the gut, and in a few fishes these tissues make up a small special organ of their own.

The islands produce a specific protein hormone, **insulin.** This material has an important regulatory action on metabolism, particularly of carbohydrates; interruption of its supply brings on the disease diabetes mellitus. A second substance, the polypeptide **glucagon,** is also produced in the pancreatic islands, notably in birds and reptiles; this tends to increase blood sugar by the breaking down of glycogen in liver storage.

INTERRENAL TISSUES AND THE ADRENAL CORTEX

In most tetrapods there is found, adjacent to the kidneys and often capping them, a pair of endocrine structures termed the **adrenal glands** (or epinephric glands, suprarenal glands; Figs. 293, 295). Microscopic examination shows the presence in them of two different types of tissues, intermingled or juxtaposed in lower tetrapods, but formed into distinct cortical and medullary layers of a single compact organ in mammals (Fig. 433). Both parts are endocrine glands, but of very different sorts. The medullary tissue is a modified part of the nervous system, but the cortical substance is of quite another nature. Fishes almost never have formed adrenals (Fig. 434), but in sharks the two components are quite distinct. In other fishes diffuse cell masses representing both components can be found between and around the kidneys and along the course of the major blood vessels dorsal to the celom. The cortical materials are termed **interrenal tissues.***

That the cortical material is vital for the maintenance of life was recognized more than a century ago, for human deaths from an ailment known as Addison's disease were invariably associated with deterioration of the adrenal cortex. The cortical cells have since been discovered to secrete a series of steroid hormones that have widespread influence over bodily functions. The broad functions of the

*Bodies within the kidneys of some teleosts, the corpuscles of Stannius, are sometimes said to be interrenal; however they differ in origin and histology.

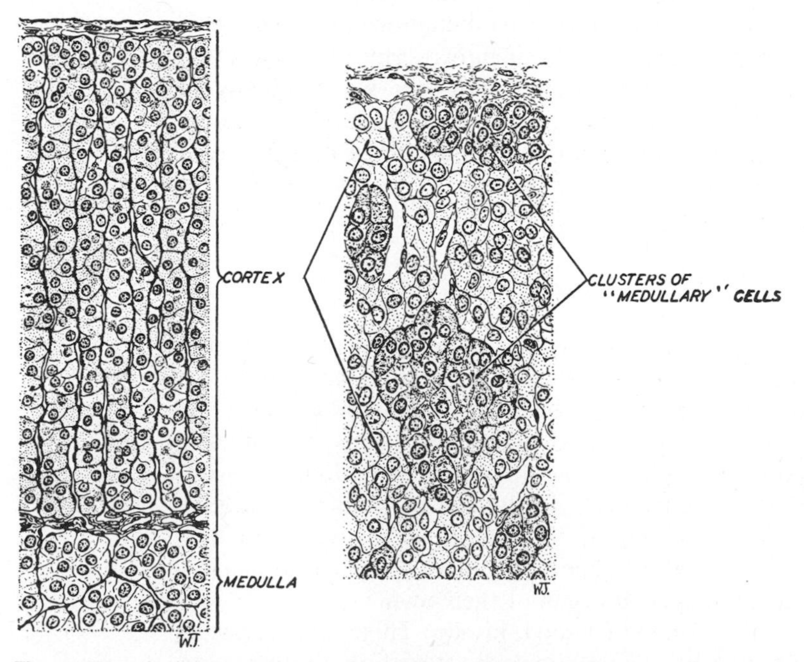

Figure 433. A section through part of the adrenal gland (outer surface above) in a mammal (rat), with division into cortical and medullary layers, and a reptile *(Heloderma),* in which the two tissues are intermingled. (From Turner.)

cortical hormones lie in aiding the body to meet long-continued environmental stresses, in contrast to the function of the adrenal medulla in coping with brief emergencies (as discussed below). More specific effects of the cortical hormones are, among others, on regulation of the salt and water balance in cells and body fluids, and on metabolism.

The importance of cortical hormones in water and salt regulation suggests a relationship of some sort between the cortical materials and the kidneys. That this physiologic association and the close topographic relationship of kidney and adrenal are not accidental, but have historic significance, is indicated by the embryologic origin of the cortical tissues. They appear as strands of cells, which bud off from the epithelium of the roof of the celom medial to the developing kidney tubules and lateral to the gonads. Kidney and adrenal cortex are thus derived from adjacent regions of the embryonic mesoderm as are also the gonads, which produce similar steroid hormones.

CHROMAFFIN TISSUES AND THE ADRENAL MEDULLA

Very different in origin and function from the cortical portion of the adrenal gland is its medullary portion and the structures antecedent to it in lower verte-

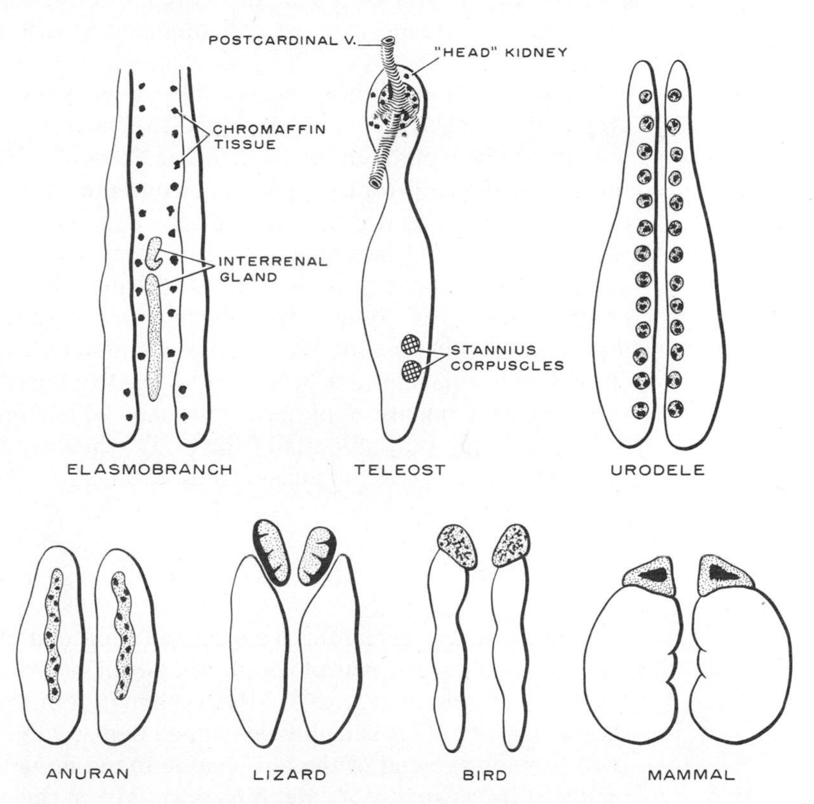

Figure 434. Comparative morphology of the adrenal tissues of vertebrates. The kidneys are shown in outline; solid black indicates chromaffin (medullary) tissues; stippling indicates interrenal (cortical) tissues. The two tissue types are typically separated spatially in elasmobranchs. In bony fishes the two tissues tend to be concentrated in the head kidneys, around branches of the postcardinal veins; they are frequently intermingled. (Based on Gorbman, A. and Bern, H. A.: Textbook of Comparative Endocrinology, New York, John Wiley & Sons, 1962.)

brates. Here we are dealing with a portion of the nervous system which has been modified to perform an endocrine function. We have seen that the visceral motor nerve supply to the internal organs of the body is of a peculiar type in which the impulses do not directly reach the smooth muscles or glands concerned, but are relayed through a series of postganglionic neurons, which give off neurohumors at their tips. The cells of the adrenal medulla and homologous structures in lower vertebrates are much modified postganglionic neurons.

In various instances there are described clusters of cells found throughout the vertebrate body, particularly along the region near the dorsal aorta or adjacent to sympathetic ganglia, which are termed **chromaffin cells** because of their reaction to certain stains. Embryologically, they arise from cells migrating downward along the path of the visceral nerve rami from the neural crest, and are thus identical in origin with the postganglionic neurons of the sympathetic system. In fishes (Fig. 434), small masses of such cells, often associated with interrenal tissues, are present between the kidneys and along the dorsal wall of the body cavity. They are appropriately termed **paraganglia,** since they are embryologically identical with the sympathetic ganglia, which they may adjoin, particularly in sharks (Fig. 394). In tetrapods occasional small cell clusters of this type persist, but, except in urodeles, the chromaffin material is concentrated into a compact mass of tissue, which forms part of the adrenal gland capping the kidney. In mammals the chromaffin cells are concentrated in the center of the adrenal body, forming its medulla; in lower tetrapods they are more diffuse and interspersed with the cortical component (Fig. 433).

These cell masses are innervated by preganglionic autonomic nerve fibers; on stimulation they secrete into the blood two related chemicals, which are identical with those given off by the postganglionic fibers of the sympathetic system: **adrenalin** and **noradrenalin** (or **epinephrine** and **norepinephrine**). Here, however, adrenalin is the more abundant product. The cells of the adrenal medulla do not look like nerve cells, for they lack fibers. But since they are homologous with postganglionic sympathetic neurons, it is not surprising that they produce comparable neurohumors. The contrast is that the true postganglionic sympathetic cell produces only a tiny amount of adrenalin-like material, which affects only structures immediately adjacent to it, whereas the mass of adrenal cells is capable of rapidly releasing large quantities of these materials, which may have a strong, immediate, "shotgun" effect on all parts of the body when carried about by the circulatory system, "bracing" the organism to meet emergencies.

UROPHYSIS

Just as in all vertebrates a special anterior area of the central nervous system may develop as the neurohypophysis, so in most, if not all, fishes a posterior secretory system may arise. To this system (not unreasonably) the term **urophysis** has been given. This is highly developed in many teleosts (Fig. 435). In the nerve cord, toward the end of the tail, can be found large cells, obviously secretory in nature. Fibers—axons—extend backward from these cells and may terminate in bulbous tips, which are filled with secreted material. These bulbs are most commonly clustered together on the underside of the cord, causing a slight swelling or a wartlike structure, often readily visible to the naked eye upon dissection. The nature of the secretion, mainly a protein, is as yet imperfectly known, but it appears to influence regulation of the salt content of the blood.

SEX HORMONES

Far more than any other activity, reproduction in vertebrates is powerfully influenced by hormones, with regard to both anatomic structures and behavior patterns. The subject merits a volume in itself; here we will but briefly note the aspects of this complex picture that are of morphological significance. The sex hormones, as noted earlier in this chapter, are strongly influenced by the gonadotrophic hormones of the pituitary.

With one exception,* the hormones produced by the gonads are steroids, termed **androgens** when produced primarily by the testis, and **estrogens** when produced by the ovary. These steroids are very similar in chemical composition to the hormones of the adrenal cortex. The relationships are, in fact, so close that the gonads and cortex are, as it were, unable completely to sort out their respective roles as hormone producers. A limited amount of sex hormones may be detected among the cortical secretions, and a small fraction of the gonadal production is of substances proper to the adrenal cortex. Still further, male and female gonads are not completely differentiated in their products, for the male gonads are found to secrete a certain amount of female hormone, and vice versa.

This similarity of hormone products between adrenal cortex and gonads correlates with the similarity in origin of the cells that produce them. Apart from the actual germ cells (which are not involved in hormone production), all the materials making up the gonads are derived, we have seen, from the mesoderm lining the dorsal rim of the celomic cavity on either side of the midline, and the cortical adrenal cells are derived from the adjacent region of the mesoderm, between gonad and kidney (p. 290).

The major androgens of the male gonad, where analysis has been made, are the steroids **testosterone** and **androstenedione;** in the female the major steroids are the estrogens, **estradiol** and the much less potent **estrone.** What cellular elements in the gonads produce these sex hormones? The major source in the testis appears to be distinctive **interstitial cells,** which are quite separate from the formed seminiferous tubules or ampullae and lie (together with connective tissue) in the interstices between them (Fig. 303). In certain cases, however, a second source of androgens

*The protein **relaxin** is a female hormone that relaxes the pelvic symphysis and facilitates birth of offspring in mammals and also has functions aiding reproduction in lower vertebrates.

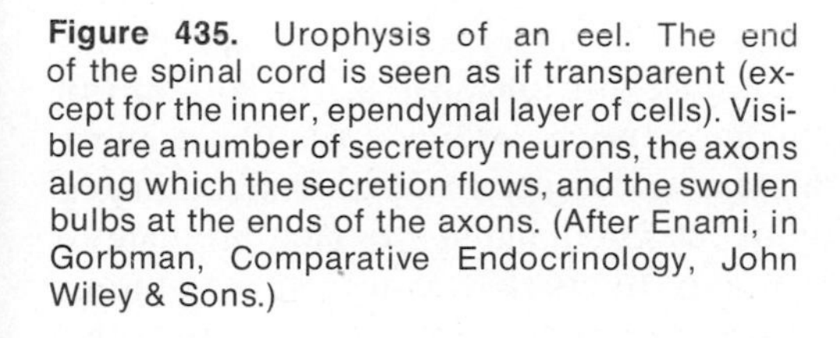

Figure 435. Urophysis of an eel. The end of the spinal cord is seen as if transparent (except for the inner, ependymal layer of cells). Visible are a number of secretory neurons, the axons along which the secretion flows, and the swollen bulbs at the ends of the axons. (After Enami, in Gorbman, Comparative Endocrinology, John Wiley & Sons.)

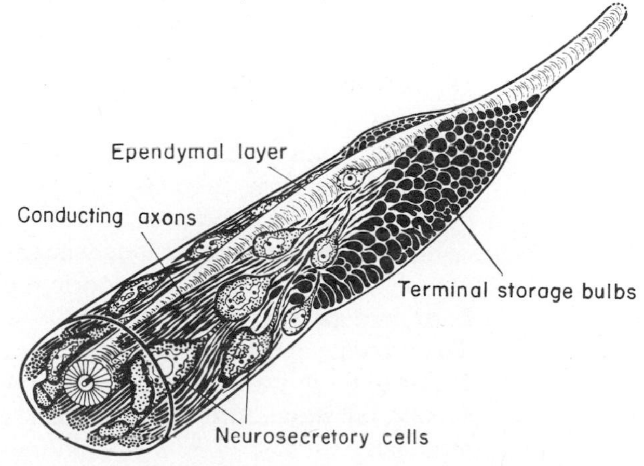

appears to be present. In the lining of the tubules or ampullae there are found, beside the sperm-producing elements, supporting cells (**Sertoli cells**) that have the same embryonic origin as the interstitial cells and may have similar hormone-producing potentialities (Fig. 303). Ovarian cellular materials are of common embryonic origin with interstitial cells and there may be a certain amount of interstitial tissue in the ovary. Most ovarian cells, however, are engaged as **follicle cells,** furnishing sustenance to the developing eggs. In addition to this function, the follicle cells are the major producers of estrogens, although minor amounts of interstitial tissues present may also be concerned in this process (cf. p. 291, Fig. 302).

Apart from the production of the primary sex hormones, we find that in the ovary a second type of steroid hormone formation may occur. When an egg bursts from the ovary, one would expect the follicle in which it has been enclosed to degenerate. This appears to be the case in many groups; but in mammals, most notably, there is no immediate degeneration. Instead, the follicle cells long persist, forming a yellow-colored tissue, the **corpus luteum,** and filling the empty follicle. The corpus luteum in mammals secretes an important steroid hormone, **progesterone;** this prepares the uterine epithelium for implantation of the ovum and, if fertilization and implantation are accomplished, stimulates the development of a placenta. But although major interest in the corpus luteum and its hormonal function has been centered on mammals, with their placental type of development, a corpus luteum, likewise apparently productive of progesterone, develops after the bursting of the ovarian follicle in elasmobranchs. Still further, although no typical corpus luteum is formed in other vertebrate groups, progesterone has nevertheless been discovered in the ovaries of certain other vertebrate types, notably birds. Some elasmobranchs and a few lower tetrapods bear their young alive, but in general no major specific function is known for progesterone in lower vertebrates. This suggests that we have in progesterone an example of the process of hormonal evolution: a chemical product given off by a tissue that originally may have had little positive function, but later came to serve a useful role in the economy of the body and in mammals has at last become an important hormone.

The nature and functions of the mammalian placenta are part of the story of vertebrate embryology rather than vertebrate anatomy. We may, however, note here that during the course of pregnancy, the mammalian placenta itself becomes an endocrine-producing organ, secreting a number of steroids, including not only estrogen and progesterone but also a special hormone of its own, **gonadotrophin,** which is important in maintaining pregnancy.

OTHER HORMONES

Hormones so far discussed have been produced by specific endocrine glands and in general form an interlocking system, acting with and against one another. In contrast to this is a series of **gastrointestinal hormones,** produced from the ordinary epithelium of the gut and independent of other hormones in their production and action. The mucosa of the pyloric region of the mammalian stomach produces a hormone, **gastrin,** which affects the secretion of hydrochloric acid in the fundus. The proximal portion of the small intestine produces **secretin** when food enters this region of the gut; this stimulates the flow of pancreatic juice. There is evidence of a second intestinal hormone, **pancreozymin-cholecystokinin,** that stimulates pancreatic secretion and effects evacuation of the gallbladder when fats enter the intestine. Still further hormonal agents have been postulated on the basis of their physiological activity, but, unlike those already mentioned, have not been defined chemically.

In other areas of bodily activity there are evidences of influences caused by chemical materials carried through the blood. In most such cases, however, there is little evidence of production of such chemicals by specific "endocrine" structures and it is difficult to know where to draw a line.

There are a number of body structures whose tissues have a glandular appearance and hence have been suspected, despite the absence of convincing evidence, of having endocrine properties. Most often cited in this category are the pineal and parapineal structures, which tend to persist even when their original visual function has been lost as in the mammalian pineal organ (cf. p. 352). In such forms as sharks and frogs the pineal, although not open to the surface, contains sensory cells, and still acts as a photoreceptor. In mammals this is no longer the case. The pineal, however, appears to receive data on the presence or absence of light by way of nerves from the autonomic system, and contains secretory cells which produce a hormone **melatonin.** This receives its name from the fact that when injected into tadpoles it causes lightening of the skin owing to concentration of pigment within the melanophores; more than that, however, the pineal in mammals appears, through its secretion, to show anti-gonadal activity; the active agent is not certainly known. We probably have much more to learn in the future about the functions of this supposedly "vestigial" structure.

We have noted earlier that the functions of the thymus are not too well known; endocrine activity is strongly implied, and the thymus may be the source of a factor inducing the differentiation of lymphoid cells. Again, ultimobranchial bodies, budded off from the last gill pouch, have been thought to have some possible hormonal nature. In mammals they become embedded in the thyroid and parathyroid and are thought to give rise to C-cells, which are the source of calcitonin. C-cells are present in the ultimobranchial bodies of lower forms, some of which also have parathyroids. The secretory cells of ultimobranchial bodies, unlike the rest of the organ, appear to come from neural crest cells.

appendix 1

a synoptic classification of chordates

The classification given here is presented primarily for the purpose of allowing the student to place in their proper position the forms discussed in the text. In consequence, no attempt is made to list the genera or even families of vertebrates, and in many cases suborders and orders are neglected when such subdivisions of groups lack interest for present purposes. Most of the major fossil forms are mentioned, although our anatomic knowledge of them is practically confined to the skeletal system.

In addition to the major terms in the classification given below, (1) the first three subphyla are often termed **Protochordata**, lower chordates, in contrast to the **Vertebrata**; (2) among the vertebrates, the term **Gnathostomata**, jawed vertebrates, may be used to contrast all other higher vertebrate groups with the class **Agnatha**; (3) **Tetrapoda** is frequently used for amphibian and higher four-footed types in contrast with **Pisces**, fishes in a broad sense; (4) **Amniota**, referring to features of embryonic development found in reptiles, birds, and mammals, may be used for these three classes, all fishes and the amphibians as well being grouped as **Anamniota**.

PHYLUM CHORDATA

SUBPHYLUM HEMICHORDATA

Little or no development of notochord or dorsal nerve cord; often considered to be a separate phylum, distinct from but closely related to the Chordata.

CLASS PTEROBRANCHIA

Simple, sessile, plantlike animals, which gather food with ciliated lophophores (Figs. 8 *A, B,* p. 24).

CLASS ENTEROPNEUSTA

Acorn worms; worm-shaped burrowers with well developed pharyngeal gills (Fig. 8 *F,* p. 24).

SUBPHYLUM UROCHORDATA

Tunicates and related forms, with notochord and nerve cord well developed in the larvae in many cases, but adults sessile or floating organisms, consisting mainly of an elaborate branchial apparatus (Figs. 6, 8 *C, D, E,* pp. 22, 24).

SUBPHYLUM CEPHALOCHORDATA

Amphioxus *(Branchiostoma)* with notochord, nerve cord, and pharyngeal gills all well developed in the adult stage (Fig. 4, p. 19).

SUBPHYLUM VERTEBRATA

Generally with developed backbone, skull, brain, and kidneys.

CLASS AGNATHA

Jawless vertebrates.

†ORDER OSTEOSTRACI.
†ORDER ANASPIDA.
†ORDER HETEROSTRACI.
†ORDER COELOLEPIDA.

"Ostracoderms"; usually armored forms from the Ordovician, Silurian, and Devonian Periods; gills well developed in the well-known osteostracans, like *Cephalaspis,* which were probably bottom-dwelling filter-feeders (Figs. 18, 19, 21, pp. 39, 40, 42).

ORDER PETROMYZONTIA. Lampreys; freshwater or marine forms "parasitic" or predaceous on other fish; these are often lumped with the following order as the Cyclostomata (Fig. 16 *C,* p. 38).

ORDER MYXINOIDEA. Hagfishes; marine scavengers (Figs. 16 *A, B,* p. 38).

CLASS ELASMOBRANCHIOMORPHI

Cartilaginous fish and certain more primitive and bony relatives.

†**Subclass Placodermi.** Jawed fishes, usually with heavy bony armor and often with peculiarly constructed fins; almost all Devonian in age.

ORDER ARTHRODIRA. Predaceous forms, often with a well developed joint between the skull and thoracic armor (Figs. 22 *A, B,* p. 44).

ORDER PTYCTODONTIDA. Mollusc-crushing forms resembling the cartilaginous Holocephali (Fig. 22 *F,* p. 44).

ORDER PHYLLOLEPIDA.
ORDER PETALICHTHYIDA.
ORDER RHENANIDA.

Odd, usually flattened forms that tend to have a reduced amount of bone and superficially resemble the modern Batoidea (Figs. 22 *C, D, E,* p. 44).

† Indicates extinct.

ORDER ANTIARCHI. Small bottom dwellers with bony armor covering the pectoral fins (Fig. 22 *G*, p. 44).

Subclass Chondrichthyes. Cartilaginous fishes.

INFRACLASS ELASMOBRANCHII. Sharks and related forms; upper jaw not fused to braincase and gills with separate external openings.

†ORDER CLADOSELACHII. Primitive Paleozoic sharks (Fig. 24 *A*, p. 46).

†ORDER PLEURACANTHODII. Odd, freshwater, Paleozoic sharks (Fig. 24 *B*, p. 46).

ORDER SELACHII. Typical sharks, Paleozoic to Recent, with claspers and narrow-based fins (Fig. 24 *C*, p. 46).

ORDER BATOIDEA. Skates and rays; flattened forms basically similar to sharks (Fig. 24 *D*, p. 46).

INFRACLASS HOLOCEPHALI. Chimaeras and related forms; upper jaw fused to braincase, and gills covered by an operculum.

Various extinct orders are very poorly known (Fig. 25 *A*, p. 47).

ORDER CHIMAERIFORMES. Chimaeras or ratfish (Fig. 25 *B*, p. 47).

CLASS OSTEICHTHYES

Generally called bony fish; however, many other fish also contain bone (Fig. 28, p. 49).

†**Subclass Acanthodii.** "Spiny sharks"; Paleozoic forms of doubtful relationships, once placed with Placodermi but probably related to the ancestral Osteichthyes (Fig. 27, p. 48).

Subclass Actinopterygii. Ray-finned fishes.

Superorder Chondrostei. Primitive ray-finned fishes, with heterocercal tails; represented by the varied and mainly Paleozoic †palaeoniscoids and the living *Polypterus*, sturgeons, and paddlefish (Figs. 32, 33, p. 53).

Superorder Holostei. Dominant ray-finned forms of the Mesozoic, with abbreviated heterocercal tails; the only living forms are *Amia* and *Lepisosteus* (Fig. 34, p. 54).

Superorder Teleostei. Dominant fishes of the Cenozoic and Recent, with homocercal tails; they include many thousands of forms classified in about 6 to nearly 50 orders (Figs. 36, 38, pp. 56, 57).

Subclass Sarcopterygii. Fleshy-finned forms (sometimes called Choanichthyes).

ORDER CROSSOPTERYGII. Predaceous forms with the upper jaw not fused to the braincase.

†Suborder Rhipidistia. Paleozoic forms ancestral to tetrapods (Fig. 29 *A*, p. 50).

Suborder Coelacanthiformes. Aberrant marine forms, including the living *Latimeria* (Fig. 29 *B*, p. 50).

ORDER DIPNOI. Lungfish; Paleozoic to Recent, mollusc-crushing forms with the upper jaw fused to the braincase and, in most cases, large toothplates (Fig. 30, p. 51).

CLASS AMPHIBIA

Tetrapods, but without the development of an amniote egg (Fig. 39, p. 58).

†**Subclass Labyrinthodontia.** Primitive forms in which the vertebral centra are formed by pleurocentra and intercentra.

ORDER ICHTHYOSTEGALIA. The earliest amphibians, from the late Devonian and Mississippian, which retained a fishlike tail (Fig. 41 *A*, p. 60).

ORDER TEMNOSPONDYLI. The common larger amphibians of the Carboniferous, Permian, and Triassic; the intercentrum is larger than the pleurocentrum (Figs. 41 *B, C,* p. 60).

ORDER ANTHRACOSAURIA. Relatively rare Paleozoic amphibians, in which the pleurocentrum is larger than the intercentrum; they include the ancestors of reptiles (Fig. 41 *D,* p. 60).

†**Subclass Lepospondyli.** Paleozoic forms in which the vertebral centra formed as single, often spool-shaped structures.

ORDER AISTOPODA. Limbless, snakelike forms (Fig. 40 *A,* p. 59).

ORDER NECTRIDEA. Salamander-like forms, some with "horns" (Fig. 40 *C,* p. 59).

ORDER MICROSAURIA. More salamander-like forms (Fig. 40 *B,* p. 59).

ORDER LYSOROPHIA. Forms with reduced limbs and highly modified skulls.

Subclass Lissamphibia. The modern, smooth-skinned amphibians.

ORDER ANURA. Frogs and toads; specialized for hopping.

ORDER URODELA. Salamanders and newts; a primitive body form, but many degenerate characters.

ORDER GYMNOPHIONA. Wormlike burrowing forms from the tropics; also called **Apoda**.

CLASS REPTILIA

Amniotes, but without advanced avian or mammalian characters (i.e., no feathers or hair) (Fig. 42, p. 63).

Subclass Anapsida. Forms with no temporal fenestrae.

†ORDER COTYLOSAURIA. Primitive "stem reptiles" of the late Paleozoic and Triassic (Fig. 43 *A,* p. 63).

ORDER TESTUDINES. Turtles and tortoises, with a bony shell; also called **Chelonia** or **Testudinata** (Figs. 43 *B, C,* p. 63).

Subclass Lepidosauria. Primitively diapsid reptiles without archosaur specializations.

†ORDER EOSUCHIA. Permian and Triassic ancestral diapsids.

ORDER RHYNCHOCEPHALIA. The living *Sphenodon* of New Zealand and fossil relatives (Fig. 46 *A,* p. 66).

ORDER SQUAMATA. Forms with reduced temporal arches.

Suborder Lacertilia. Lizards; primitive and very diverse forms, including numerous limbless ones (Fig. 46 *B,* p. 66).

Suborder Amphisbaenia. Specialized, usually limbless, burrowing forms (Fig. 46 *C,* p. 66).

Suborder Serpentes. Snakes; limbless forms (Fig. 46 *D,* p. 66).

Subclass Archosauria. "Ruling reptiles"; diapsid forms, usually with extra cranial fenestrae and adaptations for bipedal locomotion (Fig. 47, p. 68).

†ORDER THECODONTIA. Triassic ancestors of dinosaurs, birds, and others (Fig. 48 *A,* p. 69).

ORDER CROCODILIA. Crocodiles and alligators; amphibious survivors of the primitive archosaurs (Fig. 48 *B,* p. 69).

†ORDER PTEROSAURIA. Flying reptiles of the Mesozoic with membranous wings; often called pterodactyls (Fig. 48 *E*, p. 69).

†ORDER SAURISCHIA. "Reptile-like" dinosaurs with a triradiate pelvis (Fig. 48 *C*, p. 69).

Suborder Theropoda. The bipedal, carnivorous dinosaurs.

Suborder Sauropodomorpha. The largest of the quadrupedal herbivores.

†ORDER ORNITHISCHIA. "Birdlike" dinosaurs with a tetraradiate pelvis; all were herbivores (Fig. 48 *D*, p. 69).

Suborder Ornithopoda. Bipedal, unarmored forms.

Suborder Stegosauria. Quadrupeds with odd plates down the back.

Suborder Ankylosauria. Another group of armored quadrupeds.

Suborder Ceratopsia. The horned dinosaurs; rhinoceros-like forms.

†**Subclass Euryapsida.** Diverse, possibly unrelated reptiles with a single dorsal temporal fenestra on each side; their classification is much debated at present; also called **Parapsida** and **Synaptosauria**.

ORDER ARAEOSCELIDIA. Various obscure terrestrial forms, mostly rather lizard-like; also called **Protorosauria** (Fig. 45 *A*, p. 65).

ORDER SAUROPTERYGIA. Plesiosaurs and relatives; marine forms of the Mesozoic with the limbs transformed into powerful paddles (Fig. 45 *C*, p. 65).

ORDER PLACODONTIA. Armored marine mollusc-eaters of the Triassic (Fig. 45 *B*, p. 65).

ORDER ICHTHYOSAURIA. Ichthyosaurs, forms paralleling the mammalian porpoises in their marine adaptations (Fig. 45 *D*, p. 65).

†**Subclass Synapsida.** Mammal-like reptiles with a single lateral temporal fenestra on each side.

ORDER PELYCOSAURIA. Primitive Carboniferous and Permian forms, often very similar to cotylosaurs, but with a temporal fenestra (Fig. 52 *A*, p. 74).

ORDER THERAPSIDA. Advanced, often very mammal-like forms from the later Permian and Triassic (Fig. 52 *B*, p. 74).

CLASS AVES

Birds; winged descendants of the archosaurs, with feathers and temperature control.

†**Subclass Archaeornithes.** *Archaeopteryx*, the Jurassic bird with a basically reptilian skeleton, plus feathers (Fig. 49 *A*, p. 71).

Subclass Neornithes. All other birds; skeleton is "modernized" (Fig. 51, p. 73).

†**Superorder Odontognathae.** Toothed birds of the Cretaceous (Fig. 49 *B*, p. 71).

Superorder Palaeognathae. The ostrich-like birds or ratites, usually with reduced wings and relatively primitive skeletons; probably an artificial group (Fig. 50, p. 72).

Superorder Neognathae. All other birds; placed in a large number of separate orders, but all essentially similar in most anatomical features.

CLASS MAMMALIA

Animals with hair; the habit of nursing their young; a single bone on each side of the lower jaw (Fig. 53, p. 76).

Subclass Prototheria. Primitive mammals defined by certain technical characters, such as a small alisphenoid bone and no tritubercular teeth.

INFRACLASS ALLOTHERIA. Forms with widened braincases and no jugals.

ORDER MONOTREMATA. The duckbill and spiny anteaters of Australia and New Guinea.

†ORDER MULTITUBERCULATA. Jurassic to Eocene forms, perhaps comparable in habits to the later rodents.

†INFRACLASS EOTHERIA.

ORDER TRICONODONTA. } Various poorly known, small, primitive
ORDER DOCODONTA. } mammals.

Subclass Theria. Normal mammals with well developed alisphenoids.

†INFRACLASS PATRIOTHERIA. Small, primitive, ancestral forms.

ORDER SYMMETRODONTA. Forms with primitive teeth.

ORDER PANTOTHERIA. Forms with tritubercular teeth; ancestral to higher mammals.

INFRACLASS METATHERIA.

ORDER MARSUPIALIA. The pouched mammals; young are born alive, but at an immature stage (Fig. 54, p. 78).

Suborder Polyprotodonta. Primitive, mainly carnivorous forms, including the American opossum and many Australian forms.

Suborder Peramelida. Bandicoots; Australian omnivorous forms.

Suborder Caenolestoidea. Carnivorous South American forms, mainly extinct.

Suborder Diprotodonta. Herbivorous Australian forms such as wombats and kangaroos.

INFRACLASS EUTHERIA. The higher mammals, with an efficient placenta (Fig. 55, p. 79).

ORDER INSECTIVORA. Primitive eutherians; usually small and insectivorous.

Suborder Proteutheria. Very primitive extinct forms and a few living groups, including the tree shrews (Fig. 56 *A*, p. 79).

Suborder Macroscelidea. The African elephant shrews (Fig. 56 *B*, p. 79).

Suborder Dermoptera. The "flying lemur."

Suborder Lipotyphla. Moles, shrews, hedgehogs, and relatives (Fig. 56 *C, D, F,* p. 79).

Suborder Zalambdodonta. A small group of insectivores with odd teeth (Fig. 56 *E*, p. 79).

†ORDER TILLODONTIA. Large, early, and aberrant herbivores.

†ORDER TAENIODONTA. Very large, archaic, somewhat rodent-like forms.

ORDER CHIROPTERA. Bats; the only true flying mammals.

Suborder Megachiroptera. Large, fruit-eating forms.

Suborder Microchiroptera. Ordinary small bats.

Order Primates (Fig. 57, p. 81).

†Suborder Plesiadapoidea. Early Tertiary forms with rodent-like adaptations; poorly understood.

Suborder Lemuroidea. Lemurs and relatives.

Suborder Tarsioidea. *Tarsius* and extinct relatives; transitional between lemurs and monkeys.

Suborder Platyrrhini. South American monkeys and marmosets; the nostrils open to the sides.

Suborder Catarrhini. Old World monkeys, apes, and men; the nostrils open downward.

†Order Creodonta. Archaic carnivores.

Suborder Deltatherida. Small, insectivore-like forms.

Suborder Hyaenodontia. Large carnivorous forms.

Order Carnivora. The true carnivores (Fig. 58, p. 82).

Suborder Fissipedia. Terrestrial carnivores.

†Infraorder Miacoidea. Extinct ancestors of the modern forms.

Infraorder Aeluroidea. Cats and their relatives, including civets, mongooses, hyenas, and the like; also called **Feloidea**.

Infraorder Arctoidea. Dogs and their relatives, including weasels, skunks, otters, raccoons, bears, and the like; also called **Canoidea**.

Suborder Pinnipeda. Marine carnivores; seals, sea lions, and walruses.

†Order Condylarthra. Primitive forms, probably ancestral to the ungulates and many related forms (Fig. 59 *A*, p. 84).

†Order Pantodonta.
†Order Dinocerata.
†Order Embrithopoda.
†Order Notoungulata.
Suborder Notioprogonia.
Suborder Toxodontia.
Suborder Typotheria.
†Order Xenungulata.
†Order Pyrotheria.
†Order Astrapotheria.
†Order Litopterna.

All of these are generally archaic ungulates in a very broad sense; the first two orders are basically characteristic of the Northern Hemisphere, the third is African, and the remainder are South American (Fig. 59 *B–E*, p. 84).

Order Hyracoidea. The conies of Africa and the Near East; rabbit-like in general appearance, but actually ungulates; this and the next two orders appear related and of African origin.

Order Proboscidea. The elephants and their relatives.

†Suborder Moeritherioidea. An early African relative of elephants.

Suborder Euelephantoidea. Elephants, mammoths, and mastodons.

†Suborder Deinotherioidea. Elephant-like forms with recurved lower tusks.

Order Sirenia. The sea cows, manatees, and dugongs; an aquatic offshoot of some ungulate stock.

†ORDER DESMOSTYLIA. Forms vaguely like sirenians.

ORDER PERISSODACTYLA. Odd-toed ungulates (Fig. 60, p. 85).

Suborder Hippomorpha. Horses plus various fossils, including titanotheres—large, ungainly, horned ungulates.

†Suborder Ancylopoda. Chalicotheres; odd ungulates, generally horselike but with claws.

Suborder Ceratomorpha. Tapirs, rhinoceroses, and the like.

ORDER ARTIODACTYLA. Even-toed ungulates (Fig. 61, p. 86).

†Suborder Palaeodonta. Ancestral forms.

Suborder Suina. Relatively primitive forms with simple stomachs and bunodont teeth; pigs, and relatives such as hippos.

Suborder Ruminantia. Advanced forms with complex stomachs and selenodont teeth.

Infraorder Tylopoda. Camels and various extinct groups; they are intermediate between the Suina and Pecora in some ways.

Infraorder Pecora. Advanced ruminants, often with horns or antlers; deer, giraffes, pronghorn, bison, sheep, goats, antelopes, and the like.

ORDER EDENTATA. So-called "toothless" mammals; a South American group also called the Xenarthra.

Suborder Pilosa. The hairy edentates.

†Infraorder Gravigrada. Ground sloths.

Infraorder Tardigrada. Tree sloths.

Infraorder Vermilingua. South American anteaters.

Suborder Loricata. The armored edentates; armadillos and the extinct glyptodonts.

ORDER PHOLIDOTA. The Old World pangolin, another "anteater."

ORDER TUBULIDENTATA. The aardvark of Africa, still another type of anteater.

ORDER CETACEA. The whales and porpoises; they may represent two independent groups.

†Suborder Archaeoceti. Probably ancestral forms.

Suborder Odontoceti. The toothed whales, porpoises, and dolphins.

Suborder Mysticeti. Whalebone whales; the big filter-feeders.

ORDER RODENTIA. Gnawing animals (except the rabbits); their classification is extremely poorly understood and that given here is conservative and oversimplified.

Suborder Protrogomorpha. Primitive forms including the living mountain beaver (which is not a beaver).

Suborder Sciuromorpha. Squirrels and their allies.

Suborder Myomorpha. Rats, mice, and many more.

Suborder Caviomorpha. Guinea pig and many other South American forms such as porcupines; certain African and other Old World forms (more porcupines) may well be related.

ORDER LAGOMORPHA. Rabbits and a few relatives; gnawing forms, but not closely related to the rodents.

appendix 2

scientific terminology

In anatomic terminology common Latin (or Greek) words are used as such for any part of the body for which the ancients had a name. For numerous other structures, scientific names have been invented (1) by using, in a new sense, some classical word which seemed to be descriptive of the part concerned, or (2) commonly, by combining Greek or Latin roots to form a new compound term. The student frequently attempts to memorize such terms without understanding their meaning and with consequent mental indigestion. We give here the roots from which many of these descriptive terms and compounds are derived, as an aid to comprehension. As will be seen, some names formed by the anatomists are rather fanciful or farfetched; some are none too appropriate. This list is not intended, of course, as a glossary or dictionary of scientific words. We have not, for instance, included common names of bones and muscles. Most of the terms used in this book are defined or discussed in the text. For a wider vocabulary, use of a standard biologic or medical dictionary* is recommended, but the larger editions of Webster, Oxford, and others are satisfactory in most regards. Abbreviations: *F.*, French; *G.*, Greek; *L.*, Latin; *NL.*, "New" Latin; *Sp.*, Spanish.

A-, ab. L. prefix implying separation.
Abdomen. L., from *abdere* (?), to hide.
Abducens. L., *ab*, away, + *ducens*, leading.
Abductor. L., *ab*, away, + *ducere*, to lead.
Accessory. L., *accessorius*, supplementary.
Acelous. G., *a*, not, + *koilos*, hollow.
Acetabulum. L., *acetabulum*, vinegar cup.
Acoustic. G., *akoustikos*, pertaining to hearing.
Acrania. G., *a*, without, + *krania*, heads.
Acrodont. G., *akron*, height or extremity, + *odous*, tooth.
Acromion. G., *akron*, height or extremity, + *omos*, shoulder.
Actino-. G., a ray.
Ad. L. prefix, to, toward, at, or near.
Adductor. L., *ad*, to, + *ducere*, to lead.
Adrenal. L., *ad*, near, + *renes*, kidneys.
Alisphenoid. L., *ala*, wing, + G., *sphen*, wedge, + G., *eidos*, form.
Allantois. G., *allas*, sausage, + *eidos*, form, appearance.
Alveolus. L., *alveolus*, little cavity.
Ambiguus. L., *ambiguus*, uncertain, changeable.
Amnion. G., *amnion*, fetal membrane.
Amphi-. G. prefix, on both sides; hence, around or double.
Amphibia. G., *amphi*, double, + *bios*, life.
Amphicelous. G., *amphi*, both, + *koilos*, hollow.
Amphioxus. G., *amphi*, both, + *oxys*, sharp.
Amphiplatyan. G., *amphi*, both, + *platys*, flat.
Amphistylic. G., *amphi*, both, + *stylos*, pillar.
Ampulla. L., *ampulla*, a flask or vessel swelling in the middle.
A-, an-. G. prefix (alpha privative), without or not.
Ana-. G. prefix, on, upward, throughout, frequently, or reinforcing a meaning.
Analogy. G., *ana*, according to, + *logos*, due ratio; hence, proportionate.
Anamniota. G., *an*, without, + *amnion*, fetal membrane.
Anapsid. G., *an*, not, + *apsis*, arch.

*Such as Dorland's American Illustrated Medical Dictionary. 25th ed. Philadelphia, W. B. Saunders Company, 1974; or A Dictionary of Scientific Terms by Henderson and Henderson. 7th ed. Princeton, Van Nostrand, 1960.

Anastomosis. G., *anastomoein,* to bring to a mouth, cause to communicate.
Andro-. G., *andros,* a male.
Ankylosis. G., *ankylosis,* stiffening of the joint.
Annulus. L., *anulus (annulus),* a ring.
Anura. G., *an,* without, + *oura,* tail.
Anus. L., *anus,* fundament.
Apo-. G. prefix, from.
Apoda. G., *a,* without, + *poda,* feet.
Aponeurosis. G., *apo,* from, + *neuron,* tendon.
Apophysis. G., *apo,* from, + *physis,* growth.
Appendicular. L., *appendere,* to hang upon.
Apsid. G. suffix, from *apsis,* an arch.
Arachnoid. G., *arachnes,* spider, + *eidos,* shape or likeness.
Arch-, archi-. G. prefix, first or chief; hence, primitive or ancestral.
Archenteron. G., *arch,* first, + *enteron,* intestine, gut.
Archipallium. G., *archi,* first, + *pallium,* cloak.
Archipterygium. G., *archesthai,* to begin, + *pterygion,* a little wing.
Arcualia (pl.). L., *arcualis,* bowshaped.
Arrector. L., *arrigere,* to raise.
Arytenoid. G., *arytana,* jug, + *eidos,* shape or likeness.
Astragalus. G., *astragalos,* ankle bone, used as a die (commonly pl.).
Atrium. L., *atrium,* court, entrance hall.
Auditory. L., *audire,* to hear.
Auricle. L. dim., *auricula,* external ear.
Auto-. G. prefix, *autos,* self.
Autonomic. G., *autos,* self, + *nomos,* law.
Autostylic. G., *autos,* self, + *stylos,* pillar.
Axial. L., *axis,* axle of a wheel, the line about which any body turns.
Azygos. G., *a,* not, + *zygon,* yoke; hence, unpaired.

Basal. L., *basis,* footing or base.
Basi-. L. prefix, pertaining to the base.
Basibranchial. L., *basis,* base, + *branchiae,* gills.
Basihyoid. L., *basis,* base, + G., *hyoeides,* Y-shaped.
Bi-. L. prefix, two, twice, or double.
Biceps. L., *bis,* twice, + *caput,* head.
Bilateral. L., *bi,* two, + *latus,* side.
Blasto-. G. prefix, bud, germ, or sprout.
Blastocele. G., *blastos,* germ, + *koilos,* hollow.
Blastoderm. G., *blastos,* germ, + *derma,* skin.
Blastodisc. G., *blastos,* germ, + *diskos,* a round plate, quoit.
Blastomere. G., *blastos,* germ, + *meros,* part.
Blastopore. G., *blastos,* germ, + *poros,* passage, opening.
Blastula. L. dim. of G. *blastos,* germ.
Brachial. L., *brachialis,* belonging to the arm.
Brachium. (pl. **-ia**). L., *brachium,* arm, especially the forearm.
Brachy-. G. prefix from *brachus,* short.
Branchial. L., *branchiae,* or G., *branchia,* gills.
Branchiostegal. G., *branchia,* gills, + *stegein,* to cover.
Bronchus. G., *bronchia,* end of windpipe.
Buccal. L., *bucca,* cheek.
Bulbus. L., *bulbus,* bulb, swollen root.
Bulla. L., a bubble, hence spherical in shape.
Bunodont. G., *bounos,* mound, + *odous,* tooth.
Bursa. L., a purse, hence purse-shaped object.

Calcaneum. L., *calcaneum,* heel.
Callosum. L., *callosus, -a, -um,* thick-skinned.
Calyx (pl. **calices**). L., *calyx,* husk, cup-shaped protective covering.
Caninus. L., *caninus,* pertaining to a dog.
Capillary. L., *capillaris,* pertaining to the hair.
Capitulum. L. dim. *(caput),* small head.
Caput (pl. **capita**). L., *caput,* head.
Carapace. NL., *carapax,* bony or chitinous covering.
Cardiac. G., *kardiakos (kardia),* pertaining to the heart.
Cardinal. L., *cardinalis,* pertaining to a door hinge.
Carina. L., keel.
Carnassial. F., *carnassier,* carnivorous (L., *caro, carnis,* flesh).
Carnivorous. L., *carno,* flesh, + *vorare,* to devour.
Carnosus. L., *carnosus,* fleshy.
Carpus. G., *karpos,* wrist.
Cartilago (pl. **-agines**). L., *cartilago,* gristle, cartilage.
Caudal. L., *cauda,* tail.
Cava. L., *cavus, -a, -um,* hollow.
Cavernosus. L., *caverna,* a hollow or cave.
Cecum. L., *caecus, -a, -um,* blind.
Celiac. G., *koilia,* belly.
Celom (e). G., *koiloma,* a hollow.
Cephalic. G., *kephale,* head.
Cephalo-. G. combining form, *kephale,* head.
Ceratobranchial. G., *keras,* horn, + *branchia,* gills.
Ceratotrichia. G., *keras,* horn, + *thrix,* hair.
Cercal. G., *kerkos,* a tail.
Cerebellum. L. dim., *cerebrum,* brain.

Cerebrum. L., *cerebrum,* brain.
Cervical. L., *cervix,* neck.
Chiasma. G., *chiasma,* figure of X.
Chiro-. G., *cheir,* a hand.
Choana (pl. **-ae**). G., *choane,* funnel.
Choledochus. G., *chole,* bile, + *dochos (dechomai),* container.
Chondrichthyes. G., *chondros,* cartilage, + *ichthys,* fish.
Chondro-. G., combining form, cartilaginous.
Chondroblast. G., *chondros,* cartilage, + *blastos,* shoot or germ.
Chondroclast. G., *chondros,* cartilage, + *klaein,* to break.
Chorda. G., *chorde,* string of gut, cord.
Chorion. G., *chorion,* skin.
Choroid. G., *chorion,* skin, + *eidos,* likeness.
Chromaffin. G., *chroma,* color, + L., *affinis,* showing affinity for.
Chromatophore. G., *chroma,* color, + *pherein,* to bear.
Chromosome. G., *chroma,* color, + *soma,* body.
Chyme. G., *chymos,* juice.
Ciliary. L., *cilium* (pl. *cilia),* eyelash.
Circum-. L. prefix, around.
Clavicle. L., *clavicula,* a little key.
Cleithrum. G., *kleithron,* a door latch or bar.
Cloaca. L., *cloaca,* sewer, drain.
Cnemial. G., *kneme,* lower leg.
Coccyx. L., a cuckoo, hence a structure shaped like a cuckoo's bill.
Cochlea. L., *cochlea* (G., *kochlias),* spiral, snail shell.
Collagen. G., *kolla,* glue, + L., *gen,* begetter of.
Colon. L., *colon,* great gut.
Columella. L. dim. *(columma),* little pillar.
Commissure. L., *commissura (cum + mittere),* connection.
Concha. L., *concha,* bivalve, oyster shell.
Condyle. G., *kondylos,* knuckle.
Conjunctiva. L., *conjunctivus,* connecting.
Constrictor. L., *constringere,* to draw together.
Coprodeum. G., *kopros,* dung, + *hodaios,* pertaining to a way.
Coraco-, corono-. G., *korax* or *corone,* crow; hence crowlike.
Corium. G., *chorion,* skin, leather.
Cornea. L., *corneus,* horny.
Coronary. L., *coronarius,* pertaining to a wreath or crown; hence, encircling.
Corpus (pl. **corpora**). L., body.
Cortex. L., *cortex,* bark, rind.
Cortical. L., *cortex, -icis,* bark, rind.
Cosmin. G., *kosmos,* orderly arrangement, + *eidos,* form.
Costa. L., *costa,* rib.
Costal. L., *costa,* rib.
Cranial. G., *kranion,* skull.
Cribriform. L., *cribrum,* sieve, + *forma,* form.
Cricoid. G., *krikos,* ring, + *eidos,* form.
Crista. L., *crista,* crest.
Crus (pl. **crura**). L., *crus,* leg.
Ctenoid. G., *kteis* (gen. *ktenos),* a comb, + *eidos,* form.
Cuneiform. L., *cuneus,* wedge, + *forma,* shape.
Cupula. L., a small cask or cup.
Cutis. L., *cutis,* skin.
Cycloid. G., *kyklos,* circle, + *eidos,* form.
Cyno-. G., *kuon, kunos,* dog.
Cystic. G., *kystis,* the bladder, a bag or pouch.
Cytoplasm. G., *kytos,* cell, + *plasma,* plasma, anything molded.

Dactyl. G., *daktylos,* finger.
De-. L. prefix signifying down, away from, deprived of.
Deciduous. L., *deciduus (de + cado),* falling off.
Decussatio. L., *decussatio,* crosswise intersection.
Decussation. L., *decussatio,* intersection of two lines, as in Roman X.
Deferens. L., *de,* away, + *ferens,* carrying.
Dendr-. G., *dendros,* tree.
Dens, Dentis. L., tooth.
Depressor. L., *de,* down, + *premere,* to press.
Dermal. G., *derma,* skin.
Dermatome. G., *derma,* skin, + *temnein,* to cut.
Di-. G., *dis,* twice; hence, twofold or double.
Di-, dia-. G. prefix, through, between, apart, across.
Diaphragm. G., *diaphragma,* partition, midriff.
Diaphysis. G., *dia,* between, + *physis,* growth.
Diapophysis. G., *dia,* apart, + *apophysis,* outgrowth.
Diapsid. G., *di,* double, + *apsis,* arch.
Diarthrosis. G., *dia,* through, + *arthroun,* to fasten by a joint.
Diastema (pl. **-ata**). G., *diastema,* interval.
Digit. L., *digitus,* finger.
Diphycercal. G., *diphyes,* twofold, + *kerkos,* tail.
Diplospondylous. G., *diploos,* double, + *spondyle,* vertebra.
Dipnoi. G., *di,* double, + *pnein,* to breathe.
Distal. L., *distare,* to stand apart.
Dorsal. L., *dorsum,* back.
Duct. L., *ducere,* to lead or draw.

Duodenum. L., *duodeni,* twelve (meaning twelve fingerbreadths).
Dura. L., hard.

E-, ex-. L. prefix, out, out of, from.
Ectepicondyle. G., *ek,* out, + *epi,* upon, + *kondylos,* knuckle, knob.
Ectoderm. G., *ektos,* outside, + *derma,* skin.
Effector. L., *efficere,* to bring to pass.
Efferent. L., *ex,* out, + *ferre,* to bear.
Ejaculatory. L., *e,* out, + *jacere,* to throw.
Ek-, ekto-. G. prefix, out of, from, outside.
Elasmo-. G., *elasma,* a flat plate or strap.
Embolomerous. G., *en,* in, + *ballein,* to throw, + *meros,* portion or part.
En-, endo-. G. prefix, in, within.
Endocardium. G., *endon,* within, + *kardia,* heart.
Endochondral. G., *endon,* within, + *chondros,* cartilage.
Endocrine. G., *endon,* within, + *krinein,* to separate.
Endoderm. G., *endon,* within, + *derma,* skin.
Endolymph. G., *endon,* within, + L., *lympha,* water.
Endometrium. G., *endon,* within, + *metra,* womb.
Endoneurium. G., *endon,* within, + *neuron,* nerve.
Endoskeleton. G., *endon,* within, + *skeletos,* dried up.
Endostyle. G., *endon,* within, + *stylos,* pillar.
Endothelium. G., *endon,* within, + *thele,* nipple.
Entepicondyle. G., *entos,* in, + *epi,* upon, + *kondylos,* knuckle.
Enzyme. G., *en,* in, + *zyme,* leaven.
Epaxial. G., *epi,* on, + *axis,* center line.
Ependyma. G., *ependyma,* upper garment.
Epibranchial. G., *epi,* on, + *branchia,* gills.
Epicardium. G., *epi,* on, + *kardia,* heart.
Epicondyle. G., *epi,* on, + *kondylos,* knuckle.
Epidermis. G., *epi,* on, + *derma,* skin.
Epididymis. G., *epi,* on, + *didymoi,* testicles.
Epiglottis. G., *epi,* on, + *glotta,* tongue.
Epimere. G., *epi,* on, + *meros,* part.
Epineurium. G., *epi,* on, + *neuron,* nerve.
Epiphysis. G., *epi,* on, + *physis,* growth.
Epiploic. G., *epiploon,* caul, omentum.
Epithalamus. G., *epi,* on, + *thalamos,* chamber.
Epithelium. G., *epi,* on, + *thele,* nipple.
Erythrocyte. G., *erythros,* red, + *kytos,* cell.
Esophagus. G., *oisein (phero),* to carry, + *phagein,* to eat.
Ethmoid. G., *ethmos,* sieve, + *eidos,* form.
Eu-. G., adv., well; the better of alternates.
Eury-. G., broad.
Excretion. L., *(excretus) ex,* out, + *cernere,* to sift.
Exocrine. G., *ex,* out, + *krinein,* to separate.
Extensor. L., *ex,* out, + *tendere,* to stretch.
Extrinsic. L., *extrinsecus,* on the outside.

Facialis. L., *facies,* face.
Falciform. L., *falx,* sickle, + *forma,* shape.
Falx. L., *falx, falcis,* sickle.
Fascia (pl. **-iae**). L., *fascia,* band.
Fenestra. L., window.
Fer. L. suffix, carrier of.
Fiber. L., *fibra,* string, thread.
Fibril. NL., *fibrilla,* a little thread.
Fibula. L., brooch, pin.
Filoplume. L., *filum,* thread, + *pluma,* soft feather, down.
Filum. L., *filum,* thread.
Fimbria. L., *fimbriae,* threads, fringe.
Firmisternal. L., *firmus,* steadfast, strong, + G., *sternon,* chest.
Fissure. L., *fissura (findo),* a cleft.
Flagellum (pl. **-a**). L., *flagellum,* a little whip.
Flexor. L., *flexus,* bent.
Flocculus. NL. dim. *(floccus),* a tuft of wool.
Follicle. L., *folliculus,* a small bag.
Fontanelle. F., a small fountain.
Foramen. L., hole.
Form. L., *forma,* shape.
Fornix. L., *fornix,* arch or vault.
Fossa. L., ditch, channel, something dug.
Fovea. L., *fovea,* small pit.
Frontal. L., *frons, frontis,* forehead, brow.
Fundus. L., *fundus,* bottom.
Funiculus. L., *funiculus,* a slender rope.

Gametes. G., *gametes,* spouse.
Gan-. G., *ganos,* brightness.
Ganglion. G., *ganglion,* a swelling under the skin.
Gastralia (pl.). G., *gaster,* belly.
Gastrula. NL. dim. (from G., *gaster,* stomach).
Gen-. L., race, begotten.
Geniculate. L., *geniculatus,* with bent knee.
Genital. L., *genitalis (gigno),* pertaining to birth.
Geo-. G., *ge,* the earth.

Germinal. L., *germen,* bud, germ.
Germinative. L., *germen,* bud, germ.
Glans. L., *glans,* acorn.
Glenoid. G., *glene,* socket, + *eidos,* form.
Gli-, glia. G., *gloia,* glue.
Glomerulus. L. dim. of *glomus,* a ball.
Glomus (pl. **glomera**). L., *glomus,* ball.
Glossopharyngeus. G., *glossa,* tongue, + *pharynx,* throat.
Glottis. G., *glottis,* mouth of the windpipe.
Gluteus. G., *gloutos,* rump.
Gnathos. G., *gnathos,* jaw.
Gnathostomata (pl.). G., *gnathos,* jaw, + *stoma,* mouth.
Gonad. G., *gone,* seed.
Grad. L., *gradi,* walk.
Granulocytes. L., *granulum,* small grain, + G., *kytos,* cell.
Granulosus. L., *granulosus,* full of grains.
Granulum. L., *granulum,* small grain.
Guanin (e). Sp., *guano,* dung of sea fowl.
Guanophore. Sp. from Peruvian, *huanu,* dung, + G., *pherein,* to bear.
Gubernaculum. L., *gubernaculum,* helm.
Gular. L., *gula,* throat.
Gymno-. G., naked.
Gyrus (pl. **gyri**). G., *gyros,* a turn.

Habenula. L., *habena,* strap.
Haemal, hemal. G., *haima,* blood.
Hamatum. L., *hamatus,* hook-shaped.
Helico-. G., *helix,* twisted.
Hemi-. G. prefix, signifying half.
Hemibranch. G., *hemi,* half, + *branchia,* gills.
Hemichordata. G., *hemi,* half, + *chorde,* string.
Hemipenis. G., *hemi,* half, + L., *penis,* penis.
Hemisphere. G., *hemi,* half, + *sphaira,* ball.
Hemocytoblast. G., *haima,* blood, + *kytos,* cell, + *blastos,* germ.
Hemoglobin. G., *haima,* blood, + L., *globus,* sphere.
Hemopoietic. G., *haima,* blood, + *poietikos,* creative.
Hepatic. L., *hepar,* liver.
Hetero-. G. combining form signifying other, different.
Heterocelous. G., *heteros,* other, + *koilos,* hollow.
Heterocercal. G., *heteros,* other, + *kerkos,* tail.
Heterodont. G., *heteros,* other, + *odous,* tooth.
Heterotopic. G., *heteros,* other, + *topos,* place.
Hippocampus. G., *hippos,* horse, + *kampos,* sea monster.
Histology. G., *histos,* web, + *logos,* discourse, account.
Holo-. G. combining form signifying whole.
Holoblastic. G., *holos,* whole, + *blastos,* germ.
Holobranch. G., *holos,* whole, + *branchia,* gills.
Holocephali. G., *holos,* whole, + *kephale,* head.
Holonephros. G., *holos,* whole, + *nephros,* kidney.
Holostei. G., *holos,* whole, + *osteon,* bone.
Homo-. G. combining form signifying one and the same.
Homocercal. G., *homos,* same, + *kerkos,* tail.
Homoiothermous. G., *homos,* same, + *thermos,* hot.
Homolecithal. G., *homos,* same, + *lekithos,* yolk.
Homology. G., *homos,* same, + *logos,* ratio.
Hormone. G., *hormaein,* to excite.
Humor. L., *humor,* moisture, fluid.
Hyaline. G., *hyalos,* glass.
Hyoid. G., *hyoeides,* Y-shaped.
Hyomandibular. G., *upsilon* (Y-shaped letter), + L., *mandibula,* jaw.
Hyostylic. G., *upsilon* (Y-shaped letter), + *stylos,* pillar.
Hypaxial. G., *hypo,* under, + L., *axis,* center line, axis.
Hyper-. G., above, over.
Hypo-. G. prefix signifying under, below.
Hypobranchial. G., *hypo,* below, + *branchia,* gills.
Hypoglossal. G., *hypo,* under, + *glossa,* tongue.
Hypomere. G., *hypo,* under, + *meros,* part.
Hypophysis. G., *hypo,* under, + *physis,* growth.
Hypothalamus. G., *hypo,* under, + *thalamos,* chamber, couch.
Hypsodont. G., *hypsos,* height, + *odous,* tooth.
Hypural. G., *hypo,* under, + *oura,* tail.

Ichthy-. G., *ichthys,* fish.
Ileum. G., *eilein,* to wind or turn.
In-. L. prefix signifying not; also signifying in, into, within, toward, on.
Incisor. L., *incisus (incidere),* cut.
Incus. L., *incus,* anvil.
Inductor. L., *inducere,* to lead on, excite.
Infra-. L. prefix signifying below, lower than.
Inframeningeal. L., *infra,* below, + G., *meninx,* membrane.
Infraparietal. L., *infra,* below, + *paries,* wall.

Infraspinous. L., *infra,* below, + *spina,* spine.
Infundibulum. L., *infundibulum,* a funnel.
Inguinal. L., *inguina,* groin.
Integument. L., *in,* over, + *tegere,* to cover.
Inter-. L. prefix signifying between, among.
Intercalated. L., *inter,* between, + *calare,* to call.
Intercostal. L., *inter,* between, + *costa,* rib.
Intermaxillary. L., *inter,* between, + *maxilla,* jaw.
Interrenal. L., *inter,* between, + *renes,* kidneys.
Interstitial. L., *inter,* between, + *sistere,* to set.
Intervertebral. L., *inter,* between, + *vertebra,* joint.
Intestine. L., *intus,* within.
Intrinsic. L., *intrinsecus,* inward.
Invagination. L., *in,* in, + *vagina,* sheath.
Invertebrate. L., *in,* not, + *vertebratus,* jointed.
Iridocyte. G., *iris, -idos,* rainbow, + *kytos,* cell.
Iris. G., *iris,* rainbow.
Ischiofemoral. G., *ischion,* hip, + L., *femur,* thigh.
Ischium (pl. **-ia**). G., *ischion,* hip.
Iso-. G. prefix signifying equal.
Isolecithal. G., *isos,* equal, + *lekithos,* yolk.
Isomer. G., *isos,* equal, + *meros,* part.
Iter. L., way, passage.

Jejunum. L., *jejunus,* empty (of food).
Jugal. L., *jugum,* yoke.
Jugular. L., *jugularis (jugulum),* pertaining to the neck.

Keratin. G., *keras,* horn.
Kinetic. G., *kinesis,* movement.

Labial. L., *labialis (labia),* pertaining to the lips.
Lacerate. L., *lacerare,* to tear.
Lacrimal. L., *lacrima,* tear.
Lagena. L., *lagena,* flask.
Lamina (pl. **-ae**). L., *lamina,* thin plate.
Larva (pl. **-ae**). L., *larva,* ghost, mask.
Larynx. G., *larynx,* upper part of windpipe.
Lateral. L., *lateralis (latus),* pertaining to a side.
Lemma. G., skin.
Lepidotrichia. G., *lepis,* scale, + *thrix,* hair.
Leukocyte. G., *leukos,* white, + *kytos,* cell.
Levator. L., *levare,* to raise.
Lieno-. L., *lien,* spleen.
Ligamentum. L., *ligamentum,* a bandage.
Ling-. L., *lingua,* tongue.
Lipid. G., *lipos,* fat, + *eidos,* resemblance.
Lipo-. G. combining form signifying fat.
Lipophore. G., *lipos,* fat, + *phoros (phero),* bearing.
Lith-. G., *lithos,* stone.
Lobus. G., *lobos,* lobe.
Lophodont. G., *lophos,* ridge, + *odous,* tooth.
Lucidum. L., *lucidus (lux),* full of light, clear.
Lumbar. L., *lumbare,* apron for the loins.
Lumen. L., light, opening.
Luteum. L., *luteus,* yellow.
Lymphocyte. L., *lympha,* water, + G., *kytos,* cell.

Macrophage. G., *makros,* large, + *phagein,* to eat.
Macula. L., *macula,* spot, stain.
Magnus, -a, -um. L., large.
Malar. L., *mala,* upper jaw.
Malleus. L., *malleus,* hammer.
Mammillary. L., *mammillaris (mamma, -ae),* of or in the breast.
Marginal. L., *marginalis (margo),* bordering.
Marsupium. L., *marsupium,* pouch.
Mastoid. G., *mastos,* breast, + *eidos,* resemblance.
Matrix. L., *matrix (mater),* womb, groundwork, or mold.
Maxilla. L., originally jaw, now bone of upper jaw.
Meatus (pl. **-us**). L., *meatus,* passage.
Medial. L., *medialis (medius),* pertaining to the middle.
Mediastinum. L., *mediastinus,* servant, drudge.
Medulla. L., *medulla,* marrow, pith.
Mega-. G., *megas,* great.
Melanin. G., *melas, melanos,* black.
Membrane. L., *membrana,* skin.
Meninx (pl. **meninges**). G., *meninx,* membrane.
Ment-. L., *mentum,* chin.
Mere. G., *meros,* part.
Mes-, meso-. G. prefix signifying middle.
Mesencephalon. G., *mesos,* middle, + *en,* in, + *kephale,* head.
Mesenchyme. G., *mesos,* middle, + *en,* in, + *chymos,* juice.
Mesentery. G., *mesos,* midway between, + *enteron,* gut.
Mesocardium. G., *mesos,* middle, + *kardia,* heart.
Mesoderm. G., *mesos,* middle, + *derma,* skin.
Mesolecithal. G., *mesos,* middle, + *lekithos,* yolk.

Mesonephros. G., *mesos,* middle, + *nephros,* kidney.
Mesopterygium. G., *mesos,* middle, + *pterygion,* little wing.
Mesorchium. G., *mesos,* middle, + *orchis,* testis.
Mesovarium. G., *mesos,* middle, + L., *ovarium,* ovary.
Meta-. G. prefix meaning after, next; denoting change of time or situation.
Metabolic. G., *metabole,* change.
Metacarpus. G., *meta,* after, + L., *carpus,* wrist.
Metamere. G., *meta,* after, + *meros,* part.
Metamorphosis. G., *meta,* signifying change, + *morphe,* form.
Metanephros. G., *meta,* after, + *nephros,* kidney.
Metapleura. G., *meta,* after, + *pleura,* side.
Metapodial. G., *meta,* after, + *pous, podos,* foot.
Metatarsus. G., *meta,* after, + L., *tarsus,* ankle.
Metencephalon. G., *meta,* after, + *enkephalos,* brain.
Micro-. G., *mikros,* small.
Molar. L., *molaris (mola),* pertaining to a millstone.
Monocyte. G., *monos,* single, + *kytos,* cell.
Morph-. G., *morphe,* shape.
Mucus. L., *mucus,* snivel, slippery secretion.
Multangulum. L., *multus,* many, + *angulus,* angle.
Muscle. L., *musculus,* little mouse.
Myelencephalon. G., *myelos,* marrow, + *enkephalos,* brain.
Myelin. G., *myelos,* marrow.
Myo-. G. combining form signifying muscle.
Myocardium. G., *mys, myos,* muscle, + *kardia,* heart.
Myocomma. G., *mys, myos,* muscle, + *komma,* implying separation.
Myodome. G., *mys, myos,* muscle, + L., *domus,* house.
Myomere. G., *mys, myos,* muscle, + *meros,* part.
Myotome. G., *mys, myos,* muscle, + *tome,* a cutting.

Naris (pl. **-es**). L., *naris,* nostril.
Neopallium. G., *neos,* youthful, new, + L., *pallium,* cloak.
Nephridia. G., *nephridios,* belonging to the kidneys.
Nephrotome. G., *nephros,* kidney, + *tome,* a cutting.
Neural. G., *neuron,* nerve.
Neurenteric. G., *neuron,* nerve, + *enteron,* gut.
Neurilemma. G., *neuron,* nerve, + *lemma,* husk, sheath.
Neuro-. G. combining form signifying nerve.
Neuroglia. G., *neuron,* nerve, + *gloia,* glue.
Neurohumor. G., *neuron,* nerve, + L., *humor,* fluid.
Neuromast. G., *neuron,* nerve, + *mastos,* round hill.
Neuron. G., *neuron,* sinew, tendon; equivalent of L. *nervus;* whence, nerve.
Neuropil. G., *neuron,* nerve, + *pilos,* felt.
Nictitating. L., *nictare,* to wink.
Nidamental. L., *nidamentum,* materials for a nest.
Node. L., *nodus,* knot.
Notochord. G., *noton,* back, + *chorde,* cord.
Nuchal. L., *nucha,* nape of the neck.

Obliquus. L., *obliquus,* slanting.
Oculomotor. L., *oculus,* eye, + *motor,* mover.
Odontoblast. G., *odous,* tooth, + *blastos,* shoot, germ.
Oid. G., *eidos,* like (used as suffix).
Olecranon. G., *olekranon,* point of the elbow.
Olfactory. L., *olere,* to smell, + *facere,* to make.
Omasum. L., *omasum,* paunch.
Omentum. L., *omentum,* adipose membrane enclosing the bowels.
Omni-. L., *omnis,* all.
Omo-. G., *omos,* shoulder.
Omphalo-. G. combining form *(omphalos)* signifying the navel.
Ontogeny. G., *onta,* things that exist, + *gennan,* to beget.
Oo-. G., *oon,* egg.
Operculum. L., *operculum,* lid.
Ophthalmic. G., *ophthalmos,* eye.
Opistho-. G. prefix signifying backward, behind.
Opisthocelous. G., *opisthe,* behind, + *koilos,* hollow.
Opisthonephros. G., *opisthe,* behind, + *nephros,* kidney.
Optic. G., *opsis,* sight.
Or-. L., *os, oris,* mouth.
Ortho-. G., *orthos,* straight, direct.
Osseous. L., *os* (pl. *ossa*), bone.
Ossicle. L., *ossiculum,* small bone.
Osteoblast. G., *osteon,* bone, + *blastos,* germ.
Osteocyte. G., *osteon,* bone, + *kytos,* cell.
Ostraco-. G., *ostracon,* pot, shell.
Otic. G., *otikos,* belonging to the ear.

Otolith. G., *ous, otos,* ear, + *lithos,* stone.
Ovum. (pl. **ova**). L., ovum, egg.
Oxyphil. G., *oxys,* acid, + *philos,* friend.

Paleontology. G., *palaios,* ancient, + *onta,* existing things, + *logos,* science.
Paleopallium. G., *palaios,* ancient, + L., *pallium,* covering.
Pallium. L., *pallium,* cloak.
Palma. L., *palma,* the (open) hand.
Palpebra. L., *palpebra,* eyelid.
Pancreas. G., *pan,* all, + *kreas,* flesh.
Papilla. L., *papilla,* pimple.
Para-. G. prefix signifying alongside of, near.
Parabronchii. G., *para,* beside, + *bronchos,* windpipe.
Paracentrum. G., *para,* near, + *kentron (L. centrum),* center.
Parachordal. G., *para,* beside, + *chorde,* cord.
Paraganglion. G., *para,* along, beside, + *ganglion,* knot.
Paraphysis. G., *para,* beside, + *physis,* growth.
Parapsid. G., *para,* beside, + *apsis,* arch.
Parathyroid. G., *para,* near, + *thyreos,* oblong shield, + *eidos,* form.
Parietal. L., *paries,* wall.
Parotid. G., *para,* near, + *ous, otos,* ear.
Pecten. L., *pecten,* comb.
Pectoral. L., *pectoralis (pectus),* belonging to the breast.
Pedunculus. L., *pediculus,* a little foot.
Pelvic. L., *pelvis,* basin.
Peri-. G. prefix meaning around.
Pericardial. G., *peri,* around, + *kardia,* heart.
Perichondrium. G., *peri,* around, + *chondros,* cartilage.
Perichordal. G., *peri,* around, + *chorde,* cord.
Perilymph. G., *peri,* around, + L., *lympha,* fluid.
Perimysium. G., *peri,* around, + *mys,* muscle.
Periosteum. G., *peri,* around, + *osteon,* bone.
Peristalsis. G., *peristaltikos,* clasping and compressing.
Peritoneum. G., *peritonaion,* membrane containing the lower viscera.
Phallic. G., *phallikos,* pertaining to the penis.
Pharynx. G., *pharynx,* throat.
Phore. G., *phoreus,* bearer.
Photophore. G., *phos, photos,* light, + *pherein,* to bear.
Phrenic. G., *phren,* diaphragm.
Phylogeny. G., *phylon,* race, + *gennan,* to beget.
Physis. G., *phyein,* to generate, hence an outgrowth.
Pia-. L., soft.
Pineal. L., *pinea,* pine cone.
Pinna (pl. **-ae**). L., *penna, pinna,* feather; hence, wing.
Pisiform. L., *pisum,* pea, + *forma,* shape.
Pituitary. L., *pituita,* slime, phlegm.
Placenta. L., *placenta,* a flat cake.
Placode. G., *plax,* plate, + *eidos,* likeness.
Planta. L., *planta,* sole of the foot.
Plastron. F., *plastron,* breastplate.
Platybasic. G., *platys,* broad, flat, + L., *basis,* base.
Plectrum. G., *plektron,* hammer.
Pleuro-. G. combining form signifying the side.
Pleurocentrum. G., *pleura,* side, + *kentron* (L., *centrum*), center.
Pleurodont. G., *pleura,* ribs, side, + *odous,* tooth.
Plexus (pl. **-us**). L., *plexus,* plaiting, braid.
Pneumatic. G., *pneumatikos,* pertaining to breath.
Pod. G., *pous, podos,* foot.
Poikilothermous. G., *poikilos,* changeful, + *thermos,* heat.
Pons. L., *pons, pontis,* bridge.
Portal. L., *porta* (pl. *-ae*), gate.
Porus. G., *poros,* passage.
Post-. L. prefix signifying behind, after.
Prae-, pre-. L. prefix signifying before, in front.
Premolar. L., *pre,* in front, + *molaris,* molar.
Prepuce. L., *praeputium,* foreskin.
Primordial. L., *primordium,* beginning.
Pro-. G. or L. prefix signifying before, in front of, or prior.
Procelous. G., *pro,* in front, + *koilos,* hollow.
Proctodeum. G., *proktos,* anus, + *hodaios,* pertaining to a way.
Profundus. L., *profundus,* deep.
Pronator. L., *pronare,* to bend forward.
Pronephros. G., *pro,* before, + *nephros,* kidney.
Proprioceptor. L., *proprius,* special, + *capere,* to take.
Prosencephalon. G., *pros,* before, + *enkephalos,* brain.
Prostate. L., *pro,* in front, + *stare,* to stand.
Protonephros. G., *protos,* first, + *nephros,* kidney.
Protoplasm. G., *protos,* first, + *plasma,* form.
Proximal. L., *proximus,* next.
Psalterium. L., stringed instrument; book of psalms.
Pseudobranch. G., *pseudes,* false, + *branchia,* gills.
Pterygoid. G., *pteryx,* wing, + *eidos,* likeness.

Pterylae. G., *pteron,* feather, + *hyle,* a wood.
Pubis (pl. **-es**). L., *pubis,* mature.
Pulmonary. L., *pulmo,* lung.
Pygal. G., *pyge,* rump.
Pygostyle. G., *pyge,* rump, + *stylos,* pillar.
Pylorus. G., *pylouros,* gate-keeper.
Pyriform. L., *pirum,* pear, + *forma,* shape.

Quadriceps. L., *quattuor,* four, + *caput,* head.
Quadrigeminus. L., *quadrigeminus,* fourfold, four.

Radial. L., *radius,* rod, spoke.
Ramus. L., *ramus,* branch.
Re-. L., again, backward.
Receptor. L., *recipere,* to take back, receive.
Rectum, rectus. L., straight.
Remiges (pl.). L., *remex,* rower.
Renal. L., *renes,* kidneys.
Rete. L., *rete,* network.
Reticulum. L., *reticulum,* a little net.
Retina. L., *rete,* net.
Retractor. L., *retrahere,* to draw back.
Retrices. L., *retro,* back, + *cedere,* to go.
Rhabdo-. G., *rhabdos,* rod.
Rhachitomous. G., *rhachis,* spine, + *temnein,* to cut.
Rhinal. G., *rhis,* nose.
Rhombencephalon. G., *rhombos,* kind of parallelogram, + *enkephalos,* brain.
Rhyncho-. G., *rhynchos,* snout, beak.
Rostrum (pl. **-a**). L., *rostrum,* beak.
Rotator. L., *rotare,* to whirl about.
Ruminate. L., *ruminare,* to chew the cud.

Sacculus. L., *sacculus,* a little bag.
Sacrum. L., *sacer,* sacred.
Sagittal. L., *sagitta,* arrow.
Salpinx. G., *salpinx,* trumpet.
Sarcolemma. G., *sarx,* flesh, + *lemma,* husk, skin.
Saur. G., *sauros,* lizard, hence reptile.
Scala. L., *scala,* staircase.
Sclera. G., *skleros,* hard.
Sclerotic. G., *skleros,* hard.
Sclerotome. G., *skleros,* hard, + *temnein,* to cut.
Scrotum. L., *scrotum,* skin.
Sebaceous. L., *sebum,* tallow, grease.
Selenodont. G., *selene,* moon (hence, crescent), + *odous,* tooth.
Seminiferous. L., *semen,* seed, + *ferre,* to bear.
Septum. L., *saeptum,* fence.
Ser-. L., *serum,* whey, watery liquid.
Sinus (pl. **-us**). L., *sinus,* curve, cavity, bosom.
Somatic. G., *soma* (pl. *somata*), body.
Somatopleure. G., *soma,* body, + *pleura,* side.
Somite. G., *soma,* body, + suffix *-ite,* indicating origin.
Spermatozoon (pl. **-a**). G., *sperma,* seed, + *zoon,* animal.
Sphenoid. G., *sphen,* wedge, + *eidos,* likeness.
Sphincter. G., *sphingein,* to bind tight.
Spina. L., *spina,* thorn.
Spiracle. L., *spiraculum,* air hole.
Splanchnic. G., *splanchna,* viscera.
Splanchnopleure. G., *splanchna,* viscera, + *pleura,* side.
Splenial. L., *splenium,* patch.
Squamo-. L., *squama,* scale.
Stapes. L., *stapes,* stirrup.
Stato-. G., *status,* standing.
Stego-. G., *stege,* roof.
Stereospondylous. G., *stereos,* solid, + *spondylos,* vertebra.
Sternum. G., *sternon,* chest.
Stomodeum. G., *stoma,* mouth, + *hodaios,* pertaining to a way.
Stratum (pl. **strata**). L., *stratum,* layer.
Striatum. L., *striatus,* grooved, streaked.
Styloid. G., *stylos,* pillar, + *eidos,* likeness.
Sub-. L. prefix signifying under, beneath, near.
Subcostal. L., *sub,* under, + *costa,* rib.
Sublingual. L., *sub,* under, + *lingua,* tongue.
Subunguis. L., *sub,* under, + *unguis,* nail.
Subvertebral. L., *sub,* under, + *vertebra,* joint.
Sulcus. L., *sulcus,* furrow.
Supinator. L., *supinare,* to bend backward.
Supracostal. L., *supra,* above, + *costa,* rib.
Supraspinatus. L., *supra,* above, + *spina,* thorn.
Sym-, syn-. G. prefix signifying with or together.
Sympathetic. G., *syn,* together, + *pathein,* to suffer.
Symphysis. G., *syn,* together, + *physis,* growth.
Synapse. G., *syn,* together, + *haptein,* to fasten.
Synarthrosis. G., *syn,* together, + *arthron,* joint.
Synsacrum. G., *syn,* together, + L., *sacer (os sacrum),* sacrum.
Syrinx. G., *syrinx,* pipe.

Tabular. L., *tabula,* board, table.
Talonid. L., *talus,* heel, + G., *eidos,* form.
Tapetum. L., *tapete,* carpet.
Tarsus. G., *tarsos,* sole of the foot.

Tectum. L., *tectum (tego),* roof.
Tegmentum. L., *tegumentum,* a covering.
Tela. L., *tela,* web.
Telencephalon. G., *telos,* end, + *enkephalos,* brain.
Telolecithal. G., *telos,* end, + *lekithos,* yolk.
Temporal. L., *temporalis,* belonging to time.
Temporal. L., *tempora,* the temples.
Tendon. L., *tendere,* to stretch.
Tentorium. L., *tentorium,* tent.
Terminalis. L., *terminare,* to limit.
Testis. L., *testis,* testicle.
Tetrapod. G., *tetra,* four, + *pous,* foot.
Thalamus. G., *thalamos,* chamber or couch.
Thecodont. G., *theke,* case, sheath, + *odous,* tooth.
Ther-. G., beast, mammal.
Thorax (pl. **thoraces**). G., *thorax,* breastplate, breast.
Thrombocytes. G., *thrombos,* clot, + *kytos,* cells.
Thymus. G., *thymos,* sweetbread.
Thyroid. G., *thyreos,* shield, + *eidos,* form.
Tome. G., cut.
Trabecula (pl. **-ae**). L., *trabecula,* a little beam.
Trachea. G., *tracheia,* windpipe.
Trans-. L., across.
Triceps. L., *tres,* three, + *caput,* head.
Trichia. G., *thrix, trichos,* hair.
Trigeminus. L., *trigeminus,* born three together.
Triplo-. G., *triploos,* triple.
Triquetrum. L., *triquetrus,* three-cornered, triangular.
Trochanter. G., *trochos,* wheel, pulley.
Trochlea. G., *trochilia,* pulley.
Trophoblast. G., *trophe,* nourishment, + *blastos,* shoot, germ.
Tropibasic. G., *trope,* a turning, + L., *basis,* base.
Tuberculum. L., *tuberculum,* a small hump.
Tunica. L., *tunica,* undergarment.
Turbinal. L., *turbo,* a top, anything that spins or shows turning.
Tympanic. L., *tympanum,* drum.
Ula, -ulus. L., diminutive.
Ultimo-. L., furthest, last.
Umbilical. L., *umbilicus,* navel.
Unciform. L., *uncus,* hook, + *forma,* shape.
Uncinate. L., *uncinatus,* furnished with a hook.
Unguli-. L., *unguis,* fingernail, toenail.
Urea. G., *ouron,* urine.
Urodela. G., *oura,* tail, + *delos,* evident.
Urodeum. G., *ouron,* urine, + *hodaios,* pertaining to a way.
Urogenital. G., *ouron,* urine, + L., *genitalis,* genital.
Uropygial. G., *orros,* end of os sacrum, + *pyge,* rump.
Urostyle. G., *oura,* tail, + *stylos,* pillar.
Uterus. L., *uterus,* womb.
Utriculus. L., *utriculus,* small skin or leather bottle.
Vagus. L., *vagus,* wandering.
Valvula. L., *valvula,* a little fold or valve.
Vas. L., *vas,* vessel.
Vascular. L., *vasculum,* small vessel.
Velum. L., *velum,* veil.
Ventral. L., *venter,* belly.
Ventricle. L., *ventriculus,* little cavity, loculus.
Vermiform. L., *vermis,* worm, + *forma,* shape.
Vertebra. L., *vertebra,* joint.
Vesicle. L., *vesicula,* a small bladder.
Vestibulum. L., *vestibulum,* entrance court.
Vibrissa (pl. **-ae**). L., *vibrissa,* hair in the nostril.
Villus (pl. **villi**). L., *villus,* shaggy hair.
Visceral. L., *viscera,* entrails, bowels.
Vitelline. L., *vitellus,* yolk of egg.
Vitreus. L., *vitreus,* of glass; hence, transparent.
Viviparous. L., *vivus,* living, + *parere,* to beget.
Vomer. L., *vomer,* ploughshare.
Vore, -vorous. L., *vorare,* to eat.
Xiphiplastron. G., *xiphos,* sword, + F., *plastron,* breastplate.
Zygapophysis. G., *zygon,* yoke, + *apophysis,* process of a bone.
Zygomatic. G., *zygoma,* yoke.

LATIN WORD ENDINGS

Although scientific terms are often used in English form, some knowledge of the use of these words in Latin form is desirable. Latin is a highly inflected language, with a variable series of terminations for nouns and adjectives expressing not only singular and plural numbers, but also genders (of an artificial nature) and a variety of cases; still further, there are several different systems of forming such terminations ("declensions"). Fortunately, however, nearly all use of scientific terms involves only two cases—nominative and genitive. Fewer than a score of endings affixed to the root word will cover most instances.

Adjectives (which must agree in gender, number, and case with their nouns) are "declined" according to one of the two following schemes, for each of which a common adjective is used as an example (the ending, attached to the root, is in boldface).

Most nouns follow one of these same schemes. Thus, *fibula* is a feminine noun of the first declension and is declined *fibula, fibulae, fibulae, fibularum; humerus* is a masculine noun of the second declension, declined *humerus, humeri, humeri, humerorum; sternum, sterna, sterni, sternorum,* a neuter noun of the second declension; *cutis, cutes, cutis, cutium* (skin), a feminine noun of the third declension.

There are, however, two complications: (1) in the third declension most nouns have a short form for the nominative singular, a longer root for the other case endings. Thus *femur* (third declension neuter) becomes *femora,* and so on, in other cases; other typical examples are *meninx, meninges; foramen, foramina; caput, capita*. (2) A few nouns used anatomically belong to a further declension—a fourth declension. Of masculine words of this gender—*plexus* and *meatus* are examples—the plural spelling is the same as the singular; hence the English form is preferable for common use. A common neuter noun of this declension is *cornu* (horn), declined *cornu, cornua, cornus, cornuum*.

First and Second Declension (Combined)

	Masculine	*Neuter*	*Feminine*
Nominative singular	magn**us**	magn**um**	magn**a**
Nominative plural	magn**i**	magn**a**	magn**ae**
Genitive singular	magn**i** (Masculine and Neuter)		magn**ae**
Genitive plural	magn**orum** (Masculine and Neuter)		magn**arum**

Third Declension

	Masculine and Feminine	*Neuter*
Nominative singular	grand**is**	grand**e**
Nominative plural	grand**es**	grand**ia**
Genitive singular	grand**is** (all genders)	
Genitive plural	grand**ium** (all genders)	

appendix 3

references

A few of the more useful general works, or works on special topics or animal types, review articles, and a limited number of original research papers and monographs are listed here. To look further into the literature of any special topic, these two publications are most useful:

Zoological Record, 1864–date. London. Each annual volume lists all papers published during the year concerning each class of vertebrates, and follows this with classified lists of those papers that deal with various topics in anatomy, embryology, and so forth.

Biological Abstracts, 1926–date. Philadelphia. A voluminous journal which attempts to abstract and index all papers published in any field of biology.

GENERAL

Alexander, R. McN. 1975. The Chordates. London, Cambridge University Press.

Bolk, L., et al., eds. 1931–1939. Handbuch der vergleichenden Anatomie der Wirbeltiere. 6 vols. Berlin and Vienna, Urban und Schwartzenberg. The classic reference for comparative anatomy, with summaries of all that was known to its date.

Bronn, H. G., et al. 1874–date. Klassen und Ordnungen des Thier-Reichs. Leipzig and Heidelberg, Winter. A voluminous work by various authors, published in parts, some old, some new, some as yet incomplete, which gives great attention to the anatomy of the various vertebrate groups as well as to classification and distribution.

Cuvier, G. 1805. Leçons d'Anatomie Comparée. 5 vols. Paris, Baudouin. The first great comparative anatomy.

Giersberg, H., and Rietschel, P. 1967–1968. Vergleichende Anatomie der Wirbeltiere. 2 vols. Jena, Gustav Fischer. The first two volumes of a proposed three; an excellent and comprehensive work.

Goodrich, E. S. 1930. Studies on the Structure and Development of Vertebrates. London, The Macmillan Company. A stimulating discussion of many anatomic problems by a first rate authority. Reprinted by Dover Publications, New York, 1958.

Grassé, P.-P., ed. 1948–date. Traité de Zoologie, Anatomie, Systématique, Biologie. Paris, Masson et Cie. Not yet complete. Vol. XI treats of lower chordates; vols. XII–XVII of vertebrates.

Gregory, W. K. 1951. Evolution Emerging. 2 vols. New York, The Macmillan Company. Extremely valuable illustrations although now somewhat out of date. Reprinted by Arno Press, New York, 1974.

Hildebrand, M. 1968. Anatomical Preparations. Berkeley and Los Angeles, University of California Press. An excellent book on how to make all sorts of preparations.

Kükenthal, W., and Krumbach, T., eds. 1923–date. Handbuch der Zoologie. Berlin and Leipzig, W. de Gruyter and Company. A huge compendium like the Bronn, also incomplete.

Marinelli, W., and Strenger, A. 1954–1973. Vergleichende Anatomie und Morphologie der Wirbeltiere. Wien, Franz Deu-

ticke. Parts so far issued treat of cyclostomes, *Squalus,* and *Acipenser.*
Orr, R. T. 1976. Vertebrate Biology. 4th ed. Philadelphia, W. B. Saunders Company. More natural history than anatomy.
Owen, R. 1866–1868. On the Anatomy of Vertebrates. 3 vols. London, Longmans, Green. A classic, full of original observations.
Young, J. Z. 1963. The Life of Vertebrates. 2nd ed. London and New York, Oxford University Press. An excellent, group by group account, not only of structure but of life habits and functions of the vertebrates.
Young, J. Z. 1957. The Life of Mammals. London, Oxford University Press. A companion to the preceding, stressing anatomy and physiology, and including material on lower vertebrates.

FUNCTIONAL MORPHOLOGY

Denison, R. H., et al. 1961. Evolution and dynamics of vertebrate feeding mechanisms. Am. Zool. *1*:177–234.
Gans, C. 1974. Biomechanics. An Approach to Vertebrate Biology. Philadelphia, J. B. Lippincott Company. A brief description of well selected examples by a leader in the field.
Gans, C., and Bock, W. J. 1965. The functional significance of muscle architecture —a theoretical analysis. Ergeb. Anat. Entwickl. *38*:116–142.
Gray, J. 1953. How Animals Move. London, Cambridge University Press.
Gray, J. 1968. Animal Locomotion. London, Weidenfeld and Nicolson. An extensive and detailed summary of the work done in this area by the most active group.
Hildebrand, M. 1974. Analysis of Vertebrate Structure. New York, John Wiley and Sons. A basic comparative anatomy text with extensive discussions of functional morphology, especially of locomotor mechanisms.
Howell, A. B. 1944. Speed in Animals. Their Specializations for Running and Leaping. Chicago, University of Chicago Press. Reprinted by Hafner, New York, 1965.
Liem, K. F. 1970. Comparative functional anatomy of the Nandidae (Pisces: Teleostei). Fieldiana: Zool. *56*:1–166.
Nursall, J. R., et al. 1962. Vertebrate locomotion. Am. Zool. *2*:127–208.
Smith, J. M., and Savage, R. J. C. 1956. Some locomotory adaptations in mammals. J. Linnean Soc. (London) *42*:603–622. An older paper, but a classic in the field.
Thompson, D'A. W. 1942. On Growth and Form. 2nd ed. Cambridge, England, Cambridge University Press. (An abbreviated edition was published in 1961.) A classic by a master mathematician, zoologist, and Greek scholar.
Tricker, R. A. R., and Tricker, B. J. K. 1967. The Science of Movement. New York, American Elsevier.

PHYSIOLOGY

Brobeck, J. K., ed. 1973. Best and Taylor's Physiological Basis of Medical Practice. 9th ed. Baltimore, Williams & Wilkins.
Buddenbrock, W. von. 1950–1961. Vergleichende Physiologie. 5 vols. Basel, Birkhäusen.
Field, J., et al., eds. 1959–date. Handbook of Physiology. Washington, American Physiological Society. A huge, still incomplete survey in many volumes; largely but far from entirely human.
Florey, E. 1977. General and Comparative Animal Physiology. 2nd ed. Philadelphia, W. B. Saunders Company.
Gordon, M. S., et al. 1972. Animal Physiology: Principles and Adaptations. 2nd ed. New York, Macmillan, Inc.
Guyton, A. C. 1976. Textbook of Medical Physiology. 5th ed. Philadelphia, W. B. Saunders Company.
Hoar, W. S. 1975. General and Comparative Physiology. 2nd ed. Englewood Cliffs, N.J., Prentice-Hall, Inc. This and the following include much data on vertebrate physiology and anatomy.
Prosser, C. L. 1973. Comparative Animal Physiology. 3rd ed. Philadelphia, W. B. Saunders Company.
Ruch, T. C., and Patton, H. D. 1973. Medical Physiology and Biophysics. (20th edition of Howell's Textbook of Physiology.) Philadelphia, W. B. Saunders Company. Essentially human physiology alone.
Vander, A. J., et al. 1975. Human Physiology. The Mechanisms of Body Function. 2nd ed. New York, McGraw-Hill.

EVOLUTIONARY THEORY

The references here are only a few of the major sources in this field; there are also many textbooks, including several in paperback editions.

Darwin, C. 1859. On the Origin of Species. London, John Murray. (Reprinted in many versions by many companies.) The original classic—and still well worth reading.

Dobzhansky, T. 1970. Genetics of the Evolutionary Process. New York, Columbia University Press.

Mayr, E. 1963. Animal Species and Evolution. Cambridge, The Belknap Press of Harvard University Press.

Rensch, B. 1960. Evolution above the Species Level. New York, Columbia University Press.

Simpson, G. G. 1953. The Major Features of Evolution. New York, Columbia University Press.

Simpson, G. G. 1961. Principles of Animal Taxonomy. New York, Columbia University Press.

PALEONTOLOGY

Colbert, E. H. 1969. Evolution of the Vertebrates. 2nd ed. New York, John Wiley and Sons.

Piveteau, J. 1952–1968. Traité de Paléontologie. 7 vols. Paris, Masson et Cie. A comprehensive work, four volumes of which treat of vertebrates.

Romer, A. S. 1966. Vertebrate Paleontology. 3rd ed. Chicago, University of Chicago Press. The standard text on this topic.

Romer, A. S. 1971. The Vertebrate Story. (Rev. ed.) Chicago, University of Chicago Press. An elementary account of vertebrate evolution.

Stahl, B. J. 1974. Vertebrate History: Problems in Evolution. New York, McGraw-Hill. An interesting book that stresses disagreements and unsettled questions rather than simply presenting the "party line."

LOWER CHORDATES

Barrington, E. J. W. 1965. The Biology of the Hemichordata and Protochordata. Edinburgh, Oliver and Boyd; San Francisco, W. H. Freeman and Company.

Barrington, E. J. W., and Jefferies, R. P. S., eds. 1975. Protochordates. Symposia of the Zoological Society of London, No. 36.

Berrill, N. J. 1950. The Tunicata. London, The Ray Society.

Berrill, N. J. 1955. The Origin of the Vertebrates. London, Oxford University Press.

Garstang, W. 1928. The morphology of Tunicata. Quart. J. Microsc. Sci. *72*:51–187.

Grassé, P. -P., ed. 1948. Traité de Zoologie, Tome XI. Échinodermes-Stomochordés-Prochordés. Paris, Masson et Cie. Contains a comprehensive account of lower chordates by Dawydoff, Brien, Drach, and others.

FISHES

Alexander, R. McN. 1967. Functional Design in Fishes. London, Hutchinson and Company.

Allis, E. P., Jr. 1897. The cranial muscles and cranial and first spinal nerves in *Amia calva*. J. Morphol. *12*:487–808. This and further works by Allis cited below are well illustrated accounts of cranial anatomy.

Allis, E. P., Jr. 1903. The skull and cranial and first spinal muscles and nerves in *Scomber scomber*. J. Morphol. *18*:45–328.

Allis, E. P., Jr. 1909. The cranial anatomy of the mail-cheeked fishes. Zoologica (Stuttgart) *22*:1–219.

Allis, E. P., Jr. 1922. Cranial anatomy of *Polypterus*. J. Anat. *56*:189–294.

Allis, E. P., Jr. 1923. The cranial anatomy of *Chlamydoselachus anguineus*. Acta Zoologica *4*:123–221.

Berg, L. S. 1947. Classification of Fishes, Both Recent and Fossil. Ann Arbor, Edwards Brothers. A translation of a Russian original. A new edition in Russian published in 1949, translated into German in 1958.

Breder, C. M. 1926. The locomotion of fishes. Zoologica (New York) *4*:159–297.

Brodal, A., and Fänge, R., eds. 1963. The Biology of *Myxine*. Oslo, Universitetsforlaget.

Brown, M. E., ed. 1957. The Physiology of Fishes. 2 vols. New York, Academic Press. Despite the limitation of the title, gives in the main a comprehensive account of fish biology and anatomy, although now largely replaced by Hoar and Randall (see below).

Budker, P. 1971. The Life of Sharks. London, Weidenfeld and Nicolson.

Daniel, J. F. 1934. The Elasmobranch Fishes. 3rd ed. Berkeley, University of California Press. Shark anatomy.

Dean, B. 1895. Fishes, Living and Fossil. New York, Macmillan. Old, but still valuable for Recent forms.

Dean, B. 1906. Chimaeroid fishes. Carnegie Institution of Washington, Publication 32.

Dean, B. 1916–1923. A Bibliography of Fishes. 3 vols. New York, American Museum of Natural History.

Gans, C., and Parsons, T. S. 1964. A Photographic Atlas of Shark Anatomy. New York, Academic Press.

Gilbert, P. W., Mathewson, R. F., and Rall, D. P. 1967. Sharks, Skates, and Rays. Baltimore, Johns Hopkins Press.

Goodrich, E. S. 1909. A Treatise on Zoology, edited by E. Ray Lankester. Part IX. Vertebrata Craniata, Fascicule I. "Cyclostomes and Fishes." London, The Macmillan Company. A mine of data on fish anatomy; badly indexed, however.

Greenwood, P. H., et al., eds. 1973. Interrelationships of Fishes. (Supplement No. 1 to the Zoological Journal of the Linnean Society, Vol. 53, 1973.) London, Academic Press. A fairly recent survey with papers by many leading workers.

Harder, W. 1964. Anatomie der Fische. Handbuch der Binnenfischerei Mitteleuropas, Band IIA. Stuttgart, E. Schweizerbart.

Hardisty, M. W., and Potter, I. C., eds. 1971–1972. The Biology of Lampreys. 2 vols. London, Academic Press.

Hoar, W. S., and Randall, D. J., eds. 1969–1971. Fish Physiology. 6 vols. New York, Academic Press. Mainly physiological, but a mine of information on many aspects of the biology of fishes.

Lagler, K. F., et al. 1962. Ichthyology. New York, John Wiley and Sons, Inc.

Marshall, N. B. 1965. The Life of Fishes. London, Weidenfeld and Nicolson.

Millot, J. 1954. Le troisième coelacanthe. Le Naturaliste Malagache. 1er Supplement. Superficial structures of *Latimeria*.

Millot, J. 1955. The coelacanth. Sci. Am. *193*(6):34–39.

Millot, J., and Anthony, J. 1958, 1965. Anatomie de *Latimeria chalumnae*. I. Squelette, muscles et formations de soutien. II. Système nerveux et organs des sens. Paris, Centre National de la Recherche Scientifique.

Moy-Thomas, J. A., and Miles, R. S. 1971. Palaeozoic Fishes. 2nd ed. Philadelphia, W. B. Saunders Company. An excellent and up-to-date summary.

Norman, J. R. 1963. A History of Fishes. 3rd ed. London, Ernest Benn, Ltd. Life history, habits, as well as structure.

Romer, A. S. 1946. The early evolution of fishes. Quart. Rev. Biol. *21*:33–69.

Thomson, K. S. 1969. The biology of lobe-finned fishes. Biol. Rev. *44*:91–154.

AMPHIBIA

Ecker, A., Wiedersheim, R., and Gaupp, E. 1888–1904. Anatomie des Frosches. 3 vols., 2nd ed. Braunschweig, Friedrich Viewig und Sohn. A thorough account of frog anatomy, which has passed through the hands of three successive authors.

Francis, E. T. B. 1934. The Anatomy of the Salamander. London and New York, Oxford University Press.

Goin, C. J., and Goin, O. B. 1971. Introduction to Herpetology. 2nd ed. San Francisco, W. H. Freeman and Company.

Holmes, S. J. 1927. The Biology of the Frog. 4th ed. New York, Macmillan Company.

Moore, J. A., ed. 1964. Physiology of the Amphibia. New York, Academic Press.

Noble, G. K. 1931. The Biology of the Amphibia. New York, McGraw-Hill. Reprinted by Dover Publications, New York, 1954. Still probably the best general reference on the modern amphibians.

Parsons, T. S., and Williams, E. E. 1963. The relationships of the modern Amphibia: a re-examination. Quart. Rev. Biol. *38*:26–53.

Porter, K. R. 1972. Herpetology. Philadelphia, W. B. Saunders Company.

Romer, A. S. 1947. Review of the Labyrinthodontia. Bulletin, Museum of Comparative Zoology, Harvard, *99*:1–368.

Schmalhausen, I. I. 1968. The Origin of Terrestrial Vertebrates. New York, Academic Press. Translated from the Russian by L. Kelso; edited by K. S. Thomson.

Špinar, Z. V. 1972. Tertiary Frogs from Central Europe. The Hague, Dr. W. Junk N.V. Some of these excellent fossils have preserved soft parts.

Vial, J. L., ed. 1973. Evolutionary Biology of the Anurans. Columbia, University of Missouri Press. An excellent summary of many aspects.

Wiedersheim, R. 1879. Die Anatomie der Gymnophionen. Jena, Gustav Fischer.

REPTILES

See also the references by Goin and Goin and by Porter, listed above under Amphibia.

Bellairs, A. d'A. 1969. The Life of Reptiles. 2 vols. London, Weidenfeld and Nicolson.

Bellairs, A. d'A., and Underwood, G. 1951. The origin of snakes. Biol. Rev. *26*:193–237.

Bojanus, L. H. 1819–1821. Anatome Testudinis Europaeae. Vilnae, Josephi Zawadzki. Reprinted by the Society for the Study of Amphibians and Reptiles. Essentially a series of plates, probably the best yet made of any submammalian form.

Carr, A. 1952. Handbook of Turtles. Ithaca, N.Y., Cornell University Press.

Colbert, E. H. 1961. Dinosaurs, Their Discovery and Their World. New York, E. P. Dutton and Company.

Gans, C., et al., eds. 1969–date. Biology of the Reptilia. London, Academic Press. A large series, still far from complete, covering almost all aspects of reptilian biology.

Pope, C. H. 1955. The Reptile World. New York, Alfred A. Knopf.

Romer, A. S. 1956. Osteology of the Reptiles. Chicago, University of Chicago Press. The basic reference for reptilian classification, as well as osteology.

Underwood, G. 1967. A Contribution to the Classification of Snakes. London, Trustees of the British Museum (Natural History).

Williston, S. W. 1914. Water Reptiles of the Past and Present. Chicago, University of Chicago Press.

BIRDS

Bowman, R. I. 1961. Morphological differentiation and adaptation in the Galápagos finches. University of California Publications in Zoölogy, Vol. 58.

Bradley, O. C. 1960. The Structure of the Fowl. 4th ed. Edinburgh and London, Oliver and Boyd, Ltd.

Chamberlain, I. W. 1943. Atlas of Avian Anatomy. East Lansing, Michigan State College, Agricultural Experiment Station.

De Beer, G. 1954. *Archaeopteryx lithographica*. London, British Museum (Natural History).

Dorst, J. 1974. The Life of Birds. 2 vols. New York, Columbia University Press. Mainly natural history.

Farner, D. S., et al., eds. 1971–1975. Avian Biology. 5 vols. New York, Academic Press. The major modern summary of anatomy and physiology.

Fürbringer, M. 1888. Untersuchungen zur Morphologie und Systematik der Vögel. Zugleich ein Beitrag zur Anatomie des Stütz- und Bewegungsorgane. 2 vols. Amsterdam and Jena, Gustav Fischer. Old but still basic work, including excellent comparative anatomical data.

Heilmann, G. 1926. The Origin of Birds. New York, D. Appleton-Century Company. Reprinted by Dover Publications, New York, 1971.

Holmgren, N. 1955. Studies on the phylogeny of birds. Acta Zoologica *36*:243–328.

Koch, T. 1973. Anatomy of the Chicken and Domestic Birds. Ames, Iowa State University Press.

Lucas, A. M., and Stettenheim, P. R. 1972. Avian Anatomy. Integument. Washington, D.C., U.S. Dept. of Agriculture Handbook, 362.

Marshall, A. J., ed. 1960–1961. Biology and Comparative Physiology of Birds. 2 vols. New York, Academic Press. Still very useful, though partly replaced by Farner, et al. (see above).

Newton, A., and Gadow, H. 1893–1896. A Dictionary of Birds. London, Adam and Charles Black. Old but still useful.

Pycraft, W. P. 1910. A History of Birds.

London, Methuen and Company. Includes anatomy.
Shufeldt, R. W. 1890. The Myology of the Raven. London, Macmillan and Company.
Strong, R. M. 1939–1959. A bibliography of birds. Publication Field Museum of Natural History, Zoology, *25*.
Sturkie, P. D. 1965. Avian Physiology, 2nd ed. Ithaca, N.Y., Comstock Publishing Associates.
Thomson, A. L., ed. 1964. A New Dictionary of Birds. London, Thomas Nelson & Sons.
Webb, M. 1957. The ontogeny of the cranial bones, cranial peripheral and cranial parasympathetic nerves, together with a study of the visceral muscles of *Struthio*. Acta Zoologica *38*:81–203.
Welty, J. C. 1975. The Life of Birds. 2nd ed. Philadelphia, W. B. Saunders Company. Probably the best single volume on all aspects of birds.

MAMMALS

Anderson, S., and Jones, J. K., Jr., eds. 1967. Recent Mammals of the World—A Synopsis of Families. New York, The Ronald Press.
Baum, H., and Zietzschmann, O. 1936. Handbuch der Anatomie des Hundes. Berlin, P. Parey.
Bensley, B. A., and Craigie, E. H. 1948. Practical Anatomy of the Rabbit. 8th ed. Philadelphia, Blakiston Company.
Bourlière, F. 1954. The Natural History of Mammals. New York, Alfred A. Knopf.
Bradley, O. C., and Grahame, T. 1943. Topographical Anatomy of the Dog. 5th ed. New York, The Macmillan Company.
Crouch, J. E. 1969. Text-Atlas of Cat Anatomy. Philadelphia, Lea & Febiger.
Davis, D. D. 1964. The giant panda. A study of evolutionary mechanisms. Fieldiana, Zoology, Mem. 3.
Davison, A., and Stromsten, F. A. 1937. Mammalian Anatomy, with Special Reference to the Cat. 7th ed. Philadelphia, Blakiston Company.
Ewer, R. F. 1973. The Carnivores. Ithaca, N.Y., Cornell University Press.
Field, H. E., and Taylor, M. E. 1969. An Atlas of Cat Anatomy. (Rev. ed.) Chicago, University of Chicago Press.
Flower, W. H., and Lydekker, R. 1891. An Introduction to the Study of Mammals, Living and Extinct. London, Adam and Charles Black. Old, but still useful.
Gerhardt, U. 1909. Das Kaninchen. Leipzig, H. E. Ziegler and R. Woltereck.
Getty, R. 1975. Sisson & Grossman's Anatomy of the Domestic Animal. 5th ed., 2 vols. Philadelphia, W. B. Saunders Company. Comprehensive accounts of horse, ox, sheep, goat, dog, cat, and domestic birds.
Goffart, M. 1971. Function and Form in the Sloth. Oxford, Pergamon Press.
Goss, C. M., ed. 1973. Gray's Anatomy. 29th ed. Philadelphia, Lea & Febiger. One of several standard human anatomies; others are those of Morris and Cunningham.
Greene, E. G. 1935. Anatomy of the rat. Tr. Am. Philosophical Soc. (n.s.) *27*:1–370. Reprint by Hafner, New York, 1971.
Griffiths, M. 1968. Echidnas. Oxford, Pergamon Press.
Gunderson, H. L. 1976. Mammalogy. New York, McGraw-Hill.
Harrison, R. J. 1972. Functional Anatomy of Marine Mammals. London, Academic Press.
Hartman, C. G., and Straus, W. L., Jr., eds. 1933. The Anatomy of the Rhesus Monkey. Baltimore. Williams & Wilkins.
Hill, W. C. O. 1953–1974. Primates: Comparative Anatomy and Taxonomy. 8 vols. Edinburgh, University Press.
Hofer, H., Schultz, A. H., and Starck, D., eds. 1956–1958. Primatologia. 4 vols. Basel, Karger. Not yet complete.
Howell, A. B. 1930. Aquatic Mammals. Springfield, Charles C Thomas. Reprinted by Dover, 1970.
Le Gros Clark, W. E. 1926. On the anatomy of the pen-tailed tree-shrew (*Ptilocercus lowii*). Proc. Zool. Soc. London *1926*: 1179–1309.
Le Gros Clark, W. E. 1934. Early Forerunners of Man. Baltimore, Williams & Wilkins. A discussion of the anatomy of lower primates.
Le Gros Clark, W. E. 1958. History of the Primates. 6th ed. London, British Museum (Natural History).
Le Gros Clark, W. E. 1960. The Antecedents of Man. Chicago, Quadrangle Books.
Matthews, L. H. 1969. The Life of Mammals. 2 vols. London, Weidenfeld and Nicolson.
Miller, M. E., et al. 1964. Anatomy of the

Dog. Philadelphia, W. B. Saunders Company. One of the most detailed descriptions of a mammal other than man.
Nickel, R., Schummer, A., and Seiferle, E. 1960, 1961. Lehrbuch der Anatomie der Haustiere. Bd. 1: Bewegungsapparat. Bd. 2: Eingeweide. 2nd ed. Berlin and Hamburg, Parey.
Norris, K. 1966. Whales, Dolphins, and Porpoises. Berkeley, University of California Press.
Reighard, J. E., and Jennings, H. S. 1935. Anatomy of the Cat. 3rd ed. New York, Henry Holt and Company, Inc.
Schultz, A. H. 1969. The Life of Primates. London, Weidenfeld and Nicolson.
Sharman, G. B. 1970. Reproductive physiology of marsupials. Science *167*:1221–1228.
Simpson, G. G. 1945. The principles of classification and a classification of mammals. Bulletin, American Museum of Natural History *88*:1–350. The standard reference, currently being revised.
Slijper, E. J. 1962. Whales. London, Hutchinson and Company.
Thenius, E. 1972. Grundzüge der Verbreitungsgeschichte der Säugetiere. Jena, Gustav Fischer.
Vaughan, T. A. 1972. Mammalogy. Philadelphia, W. B. Saunders Company.
Walker, E. P., et al. 1975. Mammals of the World. 3rd ed. Baltimore, The Johns Hopkins Press.
Weber, M., Burlet, H. M. de, and Abel, O. 1927–1928. Die Säugetiere. 2 vols., 2nd ed. Jena, Gustav Fischer. A standard work on mammalian anatomy and classification.
Wimsatt, W. A., ed. 1970. Biology of Bats. 2 vols. New York, Academic Press.
Woollard, H. H. 1936. The anatomy of *Tarsius spectrum*. Proc. Zool. Soc. London *70*:1071–1184.
Zietzschmann, H. C. O., et al., eds. 1943. Ellenberger-Baum: Handbuch der vergleichenden Anatomie der Haustiere. 18th ed. Berlin, Springer-Verlag. The standard anatomy of domestic animals on which almost all others are based.

CELLS AND TISSUES

Andrew, W., and Hickman, C. P. 1974. Histology of the Vertebrates. A Comparative Text. St. Louis, C. V. Mosby Company.
Bloom, W., and Fawcett, D. 1975. Textbook of Histology. 10th ed. Philadelphia, W. B. Saunders Company. One of the standard medical texts.
Brachet, J., and Mirsky, A. E., eds. 1964. The Cell: Biochemistry, Physiology, Morphology. 6 vols. New York, Academic Press.
Ham, A. W. 1974. Histology. 7th ed. Philadelphia, Lippincott Company.
Kurtz, S. M., ed. 1964. Electron Microscopic Anatomy. New York, Academic Press.
Leeson, C. R., and Leeson, T. S. 1976. Histology. 3rd ed. Philadelphia, W. B. Saunders Company.
LeGros Clark, W. E. 1971. The Tissues of the Body. 6th ed. London, Oxford University Press. A text which stresses aspects of interest to the gross anatomist.
Patt, D. I., and Patt, G. R. 1969. Comparative Vertebrate Histology. New York, Harper & Row.
Symposium on the cell, by various authors. 1961. Sci. Am. *205*(3):50–52, 100–238.

EMBRYOLOGY

Arey, L. B. 1974. Developmental Anatomy. 7th ed.(rev.). Philadelphia, W. B. Saunders Company. Entirely human.
Balinsky, B. I. 1975. An Introduction to Embryology. 4th ed. Philadelphia, W. B. Saunders Company. Probably the best standard text.
Bellairs, R. 1971. Developmental Processes in Higher Vertebrates. London, Logo Press Ltd.
Bodemer, C. W. 1968. Modern Embryology. New York, Holt, Rinehart and Winston.
Conklin, E. G. 1932. The embryology of amphioxus. J. Morphol. *54*:69–151.
Damas, H. 1944. Recherches sur le développement de *Lampetra fluviatilis* L. Contribution à l'étude de la céphalogenèse des vertébrés. Arch. Biol. *55*:1–284.
DeBeer, G. R. 1958. Embryos and Ancestors. 3rd ed. London and New York, Oxford University Press. A masterful and classic statement of the relationships among anatomy, embryology, and phylogeny.

De Haan, R. L., and Ursprung, H., eds. 1965. Organogenesis. New York, Holt, Rinehart and Winston.
Hamilton, W. J., Boyd, J. D., and Mossman, H. W. 1972. Human Embryology. 4th ed. Baltimore, Williams & Wilkins.
Hörstadius, S. 1950. The Neural Crest. London, Oxford University Press.
Lillie, F. R. 1952. Development of the Chick. Revised and edited by H. L. Hamilton. New York, Holt.
Nelsen, O. 1953. Comparative Embryology of Vertebrates. New York, Blakiston Company.
Patten, B. M. 1948. Early Embryology of the Pig. 3rd ed. New York, McGraw-Hill.
Patten, B. M. 1971. Early Embryology of the Chick. 5th ed. New York, McGraw-Hill.
Patten, B. M., and Carlson, B. M. 1974. Foundations of Embryology. 3rd ed. New York, McGraw-Hill.
Romanoff, A. L. 1960. The Avian Embryo. New York, The Macmillan Company.
Saunders, J. W. 1970. Patterns and Principles of Animal Development. New York, Macmillan, Inc.
Starck, D. 1955. Embryologie. Stuttgart, Georg Thieme Verlag. Some emphasis on man, but includes vertebrates generally.
Torrey, T. W. 1971. Morphogenesis of the Vertebrates. 3rd ed. New York, John Wiley and Sons.
Waddington, C. H. 1952. The Epigenetics of Birds. London, Cambridge University Press.
Willier, B. H., Weiss, P. A., and Hamburger, V., eds. 1955. Analysis of Development. Philadelphia, W. B. Saunders Company. Reprinted by Hafner, New York, 1971.

SKIN

Bagnara, J. T., and Hadley, M. E. 1973. Chromatophores and Color Change. Englewood Cliffs, N.J., Prentice-Hall, Inc.
Elias, H., and Bortner, S. 1957. On the phylogeny of hair. Am. Mus. Novit. No. 1820:1–15.
Fingerman, M. 1965. Chromatophores. Physiol. Rev. *45*:296–339.
Hardy, J. D. 1961. Physiology of temperature regulation. Physiol. Rev. *41*:521–606.
Harvey, E. N. 1952. Bioluminescence. New York, Academic Press.
Kon, S. K., and Cowie, A. T. 1961. Milk: The Mammary Gland and its Secretion. New York, Academic Press.
Lillie, F. R. 1942. On the development of feathers. Biol. Rev. *17*:247–266.
Maderson, P. F. A., ed. 1972. The vertebrate integument: Symposium. Am. Zool. *12*:12–171.
Montagna, W., and Parakkal, P. F. 1974. The Structure and Function of Skin. 3rd ed. New York, Academic Press. (See also Sci. Am. *212*[2]:56–69, 1965.)
Paris, P. 1914. Recherches sur la gland uropygienne des oiseaux. Arch. Zool. Exp. Gén. *53*:132–276.
Rook, A. J., and Walton, G. S., eds. 1965. Comparative Physiology and Pathology of the Skin. Philadelphia, F. A. Davis Company.
Schaffer, J. 1940. Die Hautdrüsenorgane der Säugetiere. Berlin and Wien, Urban und Schwarzenberg.

SKELETON

Brien, P. 1962. Etude de la formation, de la structure des écailles des Dipneustes actuels et de leur comparaison avec les autres types d'écailles des poissons. Annalen Koninklijk Mus. Midden-Afrika, Ser. 8°, Zool. Wetensch., No. 108, pp. 53–125.
Denison, R. H. 1951. The exoskeleton of early Osteostraci. Fieldiana, Geol. *11*:197–218.
Fisher, H. I. 1946. Adaptations and comparative anatomy of the locomotor apparatus of New World vultures. Am. Midland Nat. *35*:545–727.
Flower, W. H. 1885. An Introduction to the Osteology of the Mammalia. 3rd ed. London, The Macmillan Company. An old but useful little book. Reissued by Dover Press, New York, 1962.
Goodrich, E. S. 1904. On the dermal fin-rays of fishes—living and extinct. Quart. J. Microsc. Sci. *47*:465–522.
Goodrich, E. S. 1908. On the scales of fish, living and extinct, and their importance in classification. Proc. Zool. Soc. London, *1908*:751–774.
Gregory, W. K., Miner, R. W., and Noble,

G. K. 1923. The carpus of *Eryops* and the primitive cheiropterygium. Bull. Am. Mus. Nat. Hist. *48*:279–288.
Gregory, W. K., and Raven, H. C. 1944. Studies on the origin and early evolution of paired fins and limbs. Ann. N.Y. Acad. Sci. *42*:273–360.
Haines, R. W. 1942. The evolution of epiphyses and of endochondral bone. Biol. Rev. *17*:267–292.
Jayne, H. 1898. Mammalian Anatomy. Part I. The Skeleton of the Cat. Philadelphia, Lippincott Company.
Kerr, T. 1952. The scales of primitive living actinopterygians. Proc. Zool. Soc. London *122*:55–78.
Kummer, B. 1959. Bauprinzipien des Säugerskeletes. Stuttgart, Georg Thieme Verlag.
McLean, F. C., and Urist, M. R. 1968. Bone. 3rd ed. Chicago, University of Chicago Press.
Moss, M. L. 1968. Comparative anatomy of vertebrate dermal bone and teeth. Acta Anat. (Basel) *71*:178–208.
Murray, P. D. F. 1936. Bones. A Study of the Development and Structure of the Vertebrate Skeleton. London, Cambridge University Press.
Ørvig, T. 1951. Histological studies of placoderms and fossil elasmobranchs. I. The endoskeleton, with remarks on the hard tissues of lower vertebrates in general. Arkiv f. Zool. (2)*2*:321–454.
Parker, W. K. 1868. A Monograph on the Structure and Development of the Shoulder Girdle and Sternum. London, Ray Society.
Piiper, J. 1928. On the evolution of the vertebral column in birds. Philos. Trans. R. Soc. Lond. (B) *216*:285–351.
Reynolds, S. H. 1913. The Vertebrate Skeleton. 2nd ed. Cambridge, England, Cambridge University Press.
Romer, A. S. 1942. Cartilage: an embryonic adaptation. Am. Nat. *76*:394–404.
Schaeffer, B. 1941. The morphological and functional evolution of the tarsus in amphibians and reptiles. Bull. Am. Mus. Nat. Hist. *78*:395–472.
Schmalhausen, J. J. 1912–1913. Zur Morphologie der unpaaren Flossen. Ztschr. Wissensch. Zool. *100*:509–587; *104*:1–80.
Shufeldt, R. W. 1909. Osteology of birds. Bull. New York State Mus. *130*:5–381.
Wake, D. B. 1970. Aspects of vertebral evolution in the modern Amphibia. Forma et Functio *3*:33–60.
Watson, D. M. S. 1917. The evolution of the tetrapod shoulder girdle and fore-limb. J. Anat. *52*:1–63.
Westoll, T. S. 1958. The lateral fin-fold theory and the pectoral fins of ostracoderms and early fishes. In Westoll, T. S., ed.: Studies on Fossil Vertebrates. London, University of London, pp. 180–211.
Wilder, H. H. 1903. The skeletal system of *Necturus maculatus* Rafinesque. Mem. Boston Soc. Nat. Hist. *5*:357–439.
Williams, E. E. 1959. Gadow's arcualia and the development of tetrapod vertebrae. Quart. Rev. Biol. *34*:1–32.

SKULL

Bellairs, A. d'A. 1949. The anterior braincase and interorbital septum of Sauropsida with a consideration of the origin of snakes. J. Linnean Soc. London, Zool. *41*:482–512.
Brock, G. T. 1929. On the development of the skull of *Leptodeira hotamboia*. Quart. J. Microsc. Sci. *73*:289–334.
Crompton, A. W. 1953. The development of the chondrocranium of *Spheniscus demersus* with special reference to the columella auris of birds. Acta Zoologica *34*:71–146.
Crompton, A. W. 1963. The evolution of the mammalian jaw. Evolution *17*:431–439.
DeBeer, G. R. 1937. The Development of the Vertebrate Skull. London and New York, Oxford University Press. Publication preceded by a series of detailed papers on various forms by DeBeer and colleagues. Good bibliography.
Frazzetta, T. H. 1962. A functional consideration of cranial kinesis in lizards. J. Morphol. *111*:287–319.
Frazzetta, T. H. 1968. Adaptive problems and possibilites in the temporal fenestration of tetrapod skulls. J. Morphol. *125*:145–158.
Gaupp, E. 1900. Das Chondrocranium von *Lacerta agilis*. Anat. Hefte (Arb.) *15*:433–595.
Gregory, W. K. 1933. Fish skulls: A study of the evolution of natural mechanisms. Tr. Am. Philosophical Soc. *23*:75–481.
Hofer, H. 1955. Neuere Untersuchungen zur Kopfmorphologie der Vögel. Basel, Acta 11th Congrès International d'Ornithologie 104–137.

Jollie, M. T. 1957. The head skeleton of the chicken and remarks on the anatomy of this region in other birds. J. Morphol. *100*:389–436.

Jollie, M. T. 1960. The head skeleton of the lizard. Acta Zoologica *41*:1–64.

Kampen, P. N. van. 1905. Die Tympanalgegend des Säugetierschädels. Morphol. Jahrbuch *34*:321–722.

Lakjer, T. 1927. Studien über die Gaumenregion bei Sauriern im Vergleich mit Anamniern und primitiven Sauropsiden. Zool. Jahrbücher (Anat.) *49*:57–356.

Lang, C. 1956. Das Cranium der Ratiten mit besonderer Berücksichtigung von *Struthio camelus*. Ztschr. Wissensch. Zool. *159*:165–224.

Parker, W. K.: Structure and development of the skull. A long series of papers on the following forms: Ostrich, Fowl, *Rana*, Batrachia, Salmon, Pig, Urodela, *Tropidonotus*, Lacertilia, *Acipenser*, *Lepidosteus*, Edentata and Insectivora, Birds, Sharks and Skates, Crocodilia, *Opisthocomus*, in the following journals: Philos. Trans. Roy. Soc. London *(B) 156, 159, 161, 163, 164, 167, 169, 170, 173, 176*, 1866–1885; Tr. Zool. Soc. London *9, 10, 11, 13*, 1875–1891; Tr. Linnean Soc. London, Zool. *1, 2*, 1875–1888. Old but well illustrated and valuable.

Peyer, B. 1912. Die Entwicklung des Schädelskelettes von *Vipera aspis*. Morphol. Jahrbuch *44*:563–621.

Starck, D. 1941. Zur Morphologie des Primordialcraniums von *Manis javanica* Desm. Morphol. Jahrbuch *86*:1–122. One of a series of mammal skull studies by Starck and his students.

Versluys, J. 1912. Das Streptostylie-Problem und die Bewegungen im Schädel bei Sauropsiden. Zool. Jahrbücher Suppl. *15*(2):545–714.

MUSCLES

Bourne, G. H., ed. 1972–date. The Structure and Function of Muscle. 4 vols. New York, Academic Press.

Braus, H. 1901. Die Muskeln und Nerven der Ceratodusflosse. Semon's Zoologische Forschungsreisen in Australien *1*:137–300.

Cheng, C. C. 1955. The development of the shoulder region of the opossum, *Didelphys virginiana*, with special reference to the musculature. J. Morphol. *97*:415–471.

Drüner, L. 1902–1904. Zungenbein-, Kiemenbogen- und Kehlkopf-Muskeln der Urodelen. Zool. Jahrbücher (Anat.) *15*:435–622; *19*:361–690.

Dunlap, D. G. 1960. The comparative myology of the pelvic appendage in the Salientia. J. Morphol. *106*:1–76.

Edgeworth, F. H. 1935. The Cranial Muscles of Vertebrates. London, The Macmillan Company.

Elliott, D. H. 1965. Structure and function of mammalian tendon. Biol. Rev. *40*:392–421.

Fürbringer, M. Zur vergleichenden Anatomie des Brustschulterapparates und der Schultermuskeln. Jena. Ztschr. Naturwiss. *7*:237–320, 1873; *8*:175–280, 1874; *34*:215–718, 1900; *36*:289–736, 1902; Morphol. Jahrbuch *1*:636–816, 1876.

George, J. C., and Berger, A. J. 1966. Avian Myology. New York, Academic Press.

Gilbert, P. W. 1957. The origin and development of the human extrinsic ocular muscles. Carnegie Inst. Washington, Contrib. Embryol. *36*:59–78.

Grundfest. H. 1960. Electric fishes. Sci. Am. *203*(4):115–124.

Haas, G. 1931. Die Kiefermuskulatur und die Schädelmechanik der Schlangen in vergleichender Darstellung. Zool. Jahrbücher (Anat.) *53*:127–198. See also *ibid. 52*:1–218, 1930.

Hofer, H. 1950 Zur Morphologie der Kiefermuskulatur der Vögel. Zool. Jahrbücher (Anat.) *70*:427–556.

Howell, A. B. 1937. Morphogenesis of the shoulder architecture: Aves. Auk *54*:363–375.

Huber, E. 1931. Evolution of Facial Musculature and Facial Expression. Baltimore, Johns Hopkins University Press.

Hudson, G. E. 1937. Studies on the muscles of the pelvic appendage in birds. Am. Midland Nat. *18*:1–108.

Huxley, H. E. 1965. The mechanism of muscular contraction. Sci. Am. *213* (6):18–27.

Kerr, N. S. 1955. The homologies and nomenclature of the thigh muscles of the opossum, cat, rabbit, and rhesus monkey. Anat. Rec. *121*:481–493.

Konigsberg, I. R. 1964. The embryological origin of muscle. Sci. Am. *211*(2):61–66.

Lakjer, T. 1926. Studien über die Trigeminus-versorgte Kaumuskulatur der Sauropsiden. Copenhagen, C. A. Reitzel.

Lissmann, H. W. 1958. On the function and

evolution of electric organs in fish. J. Exper. Biol. *35*:151–191.

Maurer, F. 1898. Die Entwicklung der ventralen Rumpfmuskulatur bei Reptilien. Morphol. Jahrbuch *26*:1–60.

Nursall, J. R. 1956. The lateral musculature and the swimming of fish. Proc. Zool. Soc. London *126*:127–143.

Romer, A. S. 1927. The development of the thigh musculature of the chick. J. Morphol. *43*:347–385.

Romer, A. S. 1944. The development of tetrapod limb musculature—the shoulder region of *Lacerta*. J. Morphol. *74*:1–41.

Schumacher, G. H. 1961. Funktionelle Morphologie der Kaumuskulatur. Jena, Gustav Fischer.

Sewertzoff, A. N. 1907. Studien über die Entwickelung der Muskeln, Nerven und des Skeletts der Extremitäten der niederen Tetrapoda. Bull. Soc. Impériale Naturalistes Moscou (n.s.) *21*:1–430.

Starck, D., and Barnikol, A. 1954. Beiträge zur Morphologie der Trigeminusmuskulatur der Vögel (besonders der Accipitres, Cathartidae, Striges und Anseres). Morphol. Jahrbuch *94*:1–64.

Straus, W. L., and Rawles, M. E. 1953. An experimental study of the origin of the trunk musculature and ribs in the chick. Am. J. Anat. *92*:471–510.

Sullivan, G. E. 1962. Anatomy and embryology of the wing musculature of the domestic fowl *(Gallus)*. Aust. J. Zool. *10*:458–518.

Sy, M. 1936. Funktionell-anatomische Untersuchungen am Vogelflügel. J. Ornithologie *84*:199–296.

Wilder, H. H. 1912. The appendicular muscles of *Necturus maculosus*. Zool. Jahrbücher Suppl. *15*(2):383–424.

CELOM

Butler, G. W. 1889, 1892. On the subdivision of the body cavity in lizards, crocodiles, and birds. Proc. Zool. Soc. London *1889*: 452–474; snakes, *1892*:477–498.

Keith, A. 1905. The nature of the mammalian diaphragm and pleural cavities. J. Anat. Physiol. *39*:243–284.

Mall, F. P. 1897. Development of the human coelom. J. Morphol. *12*:395–453.

Wells, L. J. 1954. Development of the human diaphragm and pleural sacs. Carnegie Inst. Washington, Contrib. Embryol. *35*:107–134.

MOUTH, PHARYNX, LUNGS

Adams, W. E. 1939. The cervical region of the Lacertilia. J. Anat. *74*:57–71.

Appelbaum, E. 1942. Enamel of sharks' teeth. J. Dent. Res. *21*:251–257.

Ballantyne, F. M. 1927. Air bladder and lungs; a contribution to the morphology of the air bladder of fish. Tr. Roy. Soc. Edinburgh *55*:371–394.

Bijtel, J. H. 1949. The structure and the mechanism of movement of the gill filaments in Teleostei. Arch. Néerl. Zool. *8*:267–288.

Comroe, J. H., Jr. 1966. The lung. Sci. Am. *214*(2):57–68.

Dahlberg, A. A., ed. 1971. Dental Morphology and Evolution. Chicago, University of Chicago Press.

Edmund, A. G. 1960. Tooth Replacement Phenomena in the Lower Vertebrates. Contribution 52, Life Sciences Division, Royal Ontario Museum, Toronto.

Gans, C. 1970. Strategy and sequence in the evolution of the external gas exchangers of ectothermal vertebrates. Forma et Functio *3*:61–104.

Gaunt, A. S., ed. 1973. Vertebrate sound production. Am. Zool. *13*:1139–1255.

Gibbs, S. P. 1956. The anatomy and development of the buccal glands of the lake lamprey (*Petromyzon marinus* L.) and the histochemistry of their secretion. J. Morphol. *98*:429–470.

Gregory, W. K. 1934. A half century of trituberculy. The Cope-Osborn theory of dental evolution, with a revised summary of molar evolution from fish to man. Proc. Am. Philosophical Soc. *73*:169–317.

Hughes, G. M. 1963. Comparative Physiology of Vertebrate Respiration. Cambridge, Mass., Harvard University Press.

Jones, F. R. H., and Marshall, N. B. 1953. The structure and functions of the teleostean swim-bladder. Biol. Rev. *28*:16–83.

King, A. S. 1966. Avian lungs and air sacs. Int. Rev. Gen. Exp. Zool. *2*:171–267.

Klapper, C. E. 1946. The development of

the pharynx of the guinea pig with special emphasis on the fate of the ultimobranchial body. Am. J. Anat. *79*:361–397.

Locy, W. A., and Larsell, O. 1916. The embryology of the birds' lung. Am. J. Anat. *19*:447–501.

Marshall, N. B. 1960. Swimbladder structure of deep-sea fishes in relation to their systematics and biology. Discovery Reports *31*:1–122.

Miles, A. E. W., ed. 1967. Structural and Chemical Organization of Teeth. 2 vols. New York, Academic Press.

Miller, W. S. 1947. The Lung. 2nd ed. Springfield, Ill., Charles C Thomas.

Moss, M. 1970. Enamel and bone in shark teeth; with a note on fibrous enamel in fishes. Acta Anat. *77*:161–187.

Müller, B. 1908. The air sacs of the pigeon. Smithsonian Miscellaneous Collections *1*:365–414.

Owen, R. 1840–1845. Odontography—A Treatise on the Comparative Anatomy of the Teeth. London, Hippolyte Bailliere. Despite its antiquity, a valuable comprehensive account; cf. the next.

Peyer, B. 1968. Comparative Odontology. Chicago, University of Chicago Press.

Poll, M. 1962. Étude sur la structure adulte et la formation des sacs pulmonaires des protoptères. Annalen Koninklijk Mus. Midden-Afrika, Ser. 8°, Zool. Wetensch., No. 108, pp. 131–171.

Schmidt-Nielsen, K. 1971. How birds breathe. Sci. Am. *225*(6):72–79.

Scott, J. H., and Symons, N. B. 1974. Introduction to Dental Anatomy. 7th ed. Edinburgh and London, E. and S. Livingstone.

Sonntag, C. F. 1920–1924. The comparative anatomy of the tongues of the Mammalia. Proc. Zool. Soc. London *1920*:115–129; *1921*:1–29, 277–322, 497–521, 741–767; *1922*:639–657; *1923*:129–153, 515–529; *1924*:725–755.

Woodland, W. N. F. 1911. On the structure and function of the gas glands and retia mirabilia associated with the gas bladder of some teleostean fishes. Proc. Zool. Soc. London *1911*:183–248.

Woskoboinikoff, M. 1932. Der Apparat der Kiemenatmung bei den Fischen. Zool. Jahrbücher (Anat.) *55*:315–488.

DIGESTIVE SYSTEM

Barrington, E. J. W. 1945. The supposed pancreatic organs of *Petromyzon fluviatilis* and *Myxine glutinosa*. Quart. J. Microsc. Sci. *85*:391–417.

Blake, I. H. 1930, 1936. Studies on the comparative histology of the digestive tube of certain teleost fishes. J. Morphol. *50*:39–70; *60*:77–102.

Burger, J. W. 1962. Further studies on the function of the rectal gland in the spiny dogfish. Physiol. Zool. *35*:205–217.

Calhoun, M. L. 1954. Microscopic Anatomy of the Digestive System of the Chicken. Ames, Iowa State College Press.

Cornselius, C. 1925. Morphologie, Histologie und Embryologie des Muskelmagens der Vögel. Morphol. Jahrbuch *54*:507–559.

Elias, H. 1955. Liver morphology. Biol. Rev. *30*:263–310.

Gorham, F. W., and Ivy, A. C. 1938. General function of the gall bladder from the evolutionary standpoint. Field Museum of Natural History, Zoology Series, *22*:159–213.

Greene, C. W. 1912. Anatomy and histology of the alimentary tract of the king salmon. Bull. U.S. Bureau Fisheries, *32*:73–100.

Hill, W. C. O. 1926. A comparative study of the pancreas. Proc. Zool. Soc. London *1926*:581–631.

Hirsch, G. C. 1950. Magenlose Fische. Zool. Anzeiger, Ergänz. *145*:302–326.

Hopkins, G. S. 1895. On the enteron of American ganoids. J. Morphol. *11*:411–442.

Jacobshagen, E. 1913. Untersuchungen über das Darmsystem der Fische und Dipnoer. II. Jena. Ztschr. Naturw. *49*:373–810.

Kaden, L. 1936. Über Epithel und Drüsen des Vogelschlunds. Zool. Jahrbücher (Anat.) *61*:421–466.

Mitchell, P. C. 1901. On the intestinal tract of birds; with remarks on the valuation and nomenclature of zoological characters. Tr. Linnean Soc. London, Zool. *8*:173–275.

Mitchell, P. C. 1906. On the intestinal tract of mammals. Tr. Zool. Soc. London *17*:437–536.

Neumayer, L. 1930. Die Entwicklung des Darms von *Acipenser*. Acta Zoologica *11*:39–150.

Oguri, M. 1964. Rectal glands of marine and fresh water sharks, comparative histology. Science *144*:1151–1152.

Pernkopf, E. 1930. Beiträge zur ver-

gleichende Anatomie des vertebraten Magens. Ztschr. Anat. *91*:329–390.
Peterson, H. 1908. Beiträge zur Kenntniss des Baues und der Entwickelung des Selachierdarmes. Jena. Ztschr. Naturw. *43*:619–652; *44*:123–148.
Rogers, T. A. 1958. The metabolism of ruminants. Sci. Am. *198*(2):34–38.
Rouiller, C., ed. 1963–1964. The Liver. 2 vols. New York, Academic Press.
Slijper, E. J. 1946. Die physiologische Anatomie der Verdauungsorgane bei den Vertebraten. Tabulae Biologicae *21*:1–81.

EXCRETORY AND REPRODUCTIVE SYSTEMS

Bentley, P. J., and Follett, B. K. 1963. Kidney function in a primitive vertebrate, the cyclostome *Lampetra fluviatilis*. J. Physiol. *169*:902–918.
Borcea, J. 1906. Recherches sur le système urogenital des elasmobranchs. Arch. Zoologie exp. et gén. *4*(4):199–484.
Boyden, E. A. 1922. The development of the cloaca in birds. Am. J. Anat. *30*:163–201.
Buchanan, G., and Fraser, E. A. 1918. The development of the urinogenital system in the Marsupialia with special reference to *Trichosurus vulpecula*. Part I. J. Anat. *53*:35–95.
Chase, S. W. 1923. The mesonephros and urogenital ducts of *Necturus maculosus* Rafinesque. J. Morphol. *37*:457–532.
Conel, J. L. 1917. The urogenital system of myxinoids. J. Morphol. *29*:75–164.
Edwards, J. G. 1928, 1929. Studies on aglomerular and glomerular kidneys. Am. J. Anat. *42*:75–108; Anat. Rec. *44*:15–28.
Everett, H. B. 1945. The present status of the germ-cell problem in vertebrates. Biol. Rev. *20*:45–55.
Fox, H. 1963. The amphibian pronephros. Quart. Rev. Biol. *38*:1–25.
Fraser, E. A. 1950. The development of the vertebrate excretory system. Biol. Rev. *25*:159–187.
Grady, H. G., and Smith, D. E., eds. 1963. The Ovary. Baltimore, Williams & Wilkins.
Gray, P. 1930–1936. The development of the amphibian kidney. Quart. J. Microsc. Sci. *73*:507–546; *75*:425–466; *78*:445–473.
Holmgren, N. 1950. On the pronephros and the blood in *Myxine glutinosa*. Acta Zool. *31*:233–348.
Huber, G. C. 1917. On the morphology of the renal tubules of vertebrates. Anat. Rec. *13*:305–339.
Kempton, R. T. 1943, 1953. Studies on the elasmobranch kidney. J. Morphol. *73*: 247–263; Biol. Bull. *104*:45–56.
Kindahl, M. 1938. Zur Entwicklung der Exkretionsorgane von Dipnoërn und Amphibien. Acta Zool. *19*:1–190.
Leigh-Sharpe, W. H. 1920–1926. The comparative morphology of the secondary sexual characters of elasmobranch fishes. J. Morphol. *34*:245–265; *35*:359–380; *36*:221–243; *42*:307–308.
Maschkowzeff, A. 1934–1935. Zur Phylogenie der Geschlechtsdrüsen und der Geschlechtsausfuhrgänge bei den Vertebrata auf Grund von Forschungen betreffend die Entwicklung des Mesonephros und der Geschlechtsorgane bei den Acipenseriden, Salmoniden und Amphibien. Zool. Jahrbücher (Anat.) *59*:1–68, 201–276.
Meyer, D. B. 1964. The migration of primordial germ cells in the chick embryo. Develop. Biol. *10*:154–190.
Moore, C. R. 1926. The biology of the mammalian testis and scrotum. Quart. Rev. Biol. *1*:4–50.
Schmidt-Nielsen, K. 1959. Salt glands. Sci. Am. *200*(1):101–116.
Semon, R. 1892. Studien über den Bauplan des Urogenitalsystems der Wirbeltiere. Dargelegt an der Entwickelung dieses Organsystems bei *Ichthyophis glutinosus*. Jena. Ztschr. Naturw. *26*:89–203.
Smith, H. W. 1932. Water regulation and its evolution in fishes. Quart. Rev. Biol. *7*:1–26.
Smith, H. W. 1951. The Kidney. London and New York, Oxford University Press.
Smith, H. W. 1953. From Fish to Philosopher. Boston, Little, Brown and Company. Vertebrate evolution with kidney evolution as the leitmotif.
Witschi, E. 1948. Migration of the germ cells of human embryos from the yolk sac to the primitive gonadal folds. Carnegie Inst. Washington, Contrib. Embryol. *32*:67–80.
Young, W. C., ed. 1961. Sex and Internal Secretion. Baltimore, Williams & Wilkins.

CIRCULATORY SYSTEM

Abramson, D. I., ed. 1962. Blood Vessels and Lymphatics. New York, Academic Press.

Adolph, E. F. 1967. The heart's pacemaker. Sci. Am. *216*(3):32–37.

Barclay, A. K., Franklin, K. J., and Pritchard, M. M. L. 1945. The Foetal Circulation. Springfield, Ill., Charles C Thomas.

Barnett, C. H., Harrison, R. J., and Tomlinson, J. D. W. 1958. Variations in the venous systems of mammals. Biol. Rev. *33*:442–487.

Benninghoff, A. 1921. Beiträge zur vergleichenden Anatomie und Entwicklungsgeschichte des Amphibienherzens. Morphol. Jahrbuch *51*:354–412.

Bugge, J. 1961. The heart of the African lungfish, *Protopterus*. Vidensk. Medd. fra Dansk. Naturhist. Foren. *123*:193–210.

Burnet, M. 1962. The thymus gland. Sci. Am. *207*(5):50–57.

Butler, E. G. 1927. The relative role played by the embryonic veins in the development of the mammalian vena cava posterior. Am. J. Anat. *39*:267–353.

Chardon, M. 1962. Contribution à l'étude du système circulatoire lié à la respiration des Protopteridae. Annalen Koninklijk Mus. Midden-Afrika, Ser. 8°, Zool. Wetensch., No. 108, pp. 53–99.

Chèvremont, M. 1948. Le système histiocytaire ou réticulo-endothélial. Biol. Rev. *23*:267–295.

Chiodi, V., and Bortolami, R. 1966. The Conducting System of the Vertebrate Heart. Bologna, Edizioni Calderini.

Congdon, E. D. 1922. Transformation of the aortic arch during the development of the human embryo. Carnegie Inst. Washington, Contrib. Embryol. *14*:47–110.

Cooper, E. L., ed. 1975. Developmental immunology. Am. Zool. *15*:1–213.

Davis, D. D., and Storey, H. E. 1943. The carotid circulation in the domestic cat. Publ. Field Museum of Natural History, Zool. *28*:5–47.

DeLong, K. T. 1962. Quantitative analysis of blood circulation through the frog heart. Science *138*:693–694.

Foxon, G. E. H. 1955. Problems of the double circulation in vertebrates. Biol. Rev. *30*:196–228.

Goodrich, E. S. 1919. Note on the reptilian heart. J. Anat. *53*:298–304.

Greil, A. 1903. Beiträge zur vergleichenden Anatomie und Entwicklungsgeschichte des Herzens und des Truncus arteriosus der Wirbeltiere. Morphol. Jahrbuch *31*:123–310.

Greil, A. 1908–1913. Entwickelungsgeschichte des Kopfes und des Blutgefässsystems von *Ceratodus forsteri*. Semon's Zoologische Forschungsreise in Australien *1*:661–1492.

Heuser, C. H. 1923. The branchial vessels and their derivatives in the pig. Carnegie Inst. Washington, Contrib. Embryol. *15*:121–139.

Hill, W. C. O. 1953. The blood-vascular system of *Tarsius*. Proc. Zool. Soc. London, *123*:655–692.

Hochstetter, T. 1906. Beiträge zur Anatomie und Entwickelungsgeschichte des Blutgefässsystemes der Krokodile. Voeltzkow, A., Reise in Ostafrika, *4*:1–139.

Huntington, G. S., and McClure, C. F. W. 1920. The development of the veins in the domestic cat. Anat. Rec. *20*:1–31.

Kampmeier, O. F. 1969. Evolution and Comparative Morphology of the Lymphatic System. Springfield, Ill., Charles C Thomas.

Kern, A. 1926. Das Vogelherz. Morphol. Jahrbuch *56*:264–315.

Krogh, A. 1929. The Anatomy and Physiology of Capillaries. New Haven, Yale University Press.

Mathur, P. N. 1944. The anatomy of the reptilian heart. Part I. *Varanus monitor* (Linné). Proc. Ind. Acad. Sci., B, *20*:1–29.

Mayerson, H. S. 1963. The lymphatic system. Sci. Am. *208*(6):80–90.

Miller, J. F. A. P. 1964. The thymus and the development of immunological responsiveness. Science *144*:1544–1551.

Mossman, H. W. 1948. Circulatory cycles in the vertebrates. Biol. Rev. *23*:237–255.

O'Donoghue, C. H. 1912. The circulatory system of the common grass snake (*Tropidonotus natrix*). Proc. Zool. Soc. London *1912*:612–647.

O'Donoghue, C. H. 1920. The blood-vascular system of the tuatara, *Sphenodon punctatus*. Philos. Tr. Roy. Soc. London *(B)210*:175–252.

O'Donoghue, C. H., and Abbott, E. 1928. The blood-vascular system of the spiny dogfish, *Squalus acanthias* Linné, and *Squalus sucklii* Gill. Tr. Roy. Soc. Edinburgh *55*:823–890.

Padget, D. H. 1957. The development of the cranial venous system in man, from the viewpoint of comparative anatomy. Carnegie Inst. Washington, Contrib. Embryol. *36*:79–140.

Parsons, T. S., organizer. 1968. Functional

morphology of the heart of vertebrates. Am. Zool. *8*:177–229.
Quiring, D. P. 1949. Collateral Circulation. Philadelphia, Lea and Febiger.
Regan, F. P. 1929. A century of study upon the development of the eutherian vena cava inferior. Quart. Rev. Biol. *4*:179–212.
Robertson, J. I. 1913. The development of the heart and vascular system of *Lepidosiren paradoxa*. Quart. J. Microsc. Sci. *59*:53–132.
Rusznyak, I., et al. 1960. Lymphatics and Lymph Circulation: Physiology and Pathology. Oxford, Pergamon Press Ltd.
Satchell, G. H. 1971. Circulation in Fishes. Cambridge, England, Cambridge University Press.
Shearer, E. M. 1930. Studies on the embryology of circulation in fishes. Am. J. Anat. *46*:393–459.
Shindo, T. 1914. Zur vergleichenden Anatomie der arteriellen Kopfgefässe der Reptilien. Anat. Hefte *51*:267–356.
Wiggers, C. J. 1957. The heart. Sci. Am. *196*(5):75–87.
Wood, J. E. 1968. The venous system. Sci. Am. *218*(1):86–96.
Yoffey, J. M., and Courtice, F. C. 1956. Lymphatics, Lymph, and Lymphoid Tissue. Cambridge, Mass., Harvard University Press.

SENSE ORGANS

Allis, E. P., Jr. 1934. Concerning the course of the laterosensory canals in recent fishes, pre-fishes, and *Necturus*. J. Anat. *68*:361–415.
Allison, A. C. 1953. The morphology of the olfactory system in vertebrates. Biol. Rev. *28*:195–244.
Atz, J. W. 1952. Narial breathing in fishes and the evolution of internal nares. Quart. Rev. Biol. *27*:366–377.
Baldauf, R. J., and Heimer, L., eds. 1967. Vertebrate olfaction. Am. Zool. *7*:385–432.
Baradi, A. F., and Bourne, G. H. 1953. Gustatory and olfactory epithelia. Internat. Rev. Cytol. *2*:289–330.
Bellairs, A. d'A., and Boyd, J. D. 1947, 1950. The lachrymal apparatus in lizards and snakes. Proc. Zool. Soc. London *117*:81–101; *120*:269–309.
Burne, R. H. 1909. The anatomy of the olfactory organ of teleostean fishes. Proc. Zool. Soc. London *1909*:610–662.
Cahn, P. H. 1967. Lateral Line Detectors. Bloomington, Ind., Indiana University Press.
Chranilov, N. S. 1927, 1929. Beiträge zur Kenntniss der Weber'schen Apparates der Ostariophysi. Zool. Jahrbücher (Anat.) *49*:501–597; *51*:323–462.
Dempster, W. T. 1930. The morphology of the amphibian endolymphatic organ. J. Morphol. *50*:71–120.
Detweiler, S. R. 1943. Vertebrate Photoreceptors. New York, The Macmillan Company.
Dijkgraaf, S. 1952. Bau und Funktionen der Seitenorgane und des Ohrlabyrinths der Fische. Experientia *8*:205–216.
Disler, N. N. 1960. Lateral Line Sense Organs and Their Importance in Fish Behaviour. Moskva, Izdatelstvo Akademii Nauk SSSR.
Eakin, R. M. 1973. The Third Eye. Berkeley, University of California Press.
Fänge, R., Schmidt-Nielsen, K., and Osaki, H. 1958. The salt gland of the herring gull. Biol. Bull. *115*:162–171.
Frisch, K. von. 1936. Über den Gehörs in der Fische. Biol. Rev. *11*:210–243.
Gaupp, E. 1913. Die Reichertsche Theorie (Hammer- Amboss- und Kieferfrage). Arch. Anat. Physiol. Supplement *V*:1–417. On evolution of middle ear ossicles.
Guggenheim, L. 1948. Phylogenesis of the Ear. Culver City, Calif., Murray and Gee.
Haagen-Smit, A. J. 1952. Smell and taste. Sci. Am. *186*(3):28–32.
Holmgren, N. 1942. General morphology of the lateral sensory line system of the head in fishes. Kungl. Svenska Vetenskapsakad. Handl. (3)*20*(1):1–46.
Kleerekoper, H. 1969. Olfaction in Fishes. Bloomington, Ind., Indiana University Press.
Lowenstein, O. 1936. The equilibrium function of the vertebrate labyrinth. Biol. Rev. *11*:113–145.
Noble, G. K., and Schmidt, A. 1937. The structure and function of the facial and labial pits of snakes. Proc. Am. Philosophical Soc. *77*:263–288.
Parker, G. H. 1912. Smell, Taste, and Allied Senses in the Vertebrates. Philadelphia, Lippincott Company.
Parsons, T. S. 1959. Studies on the comparative embryology of the reptilian nose. Bull. Mus. Comp. Zool., Harvard, *120*:104–277.
Parsons, T. S., ed. 1966. The vertebrate ear. Am. Zool. *6*:368–466.
Polyak, S. 1957. The Vertebrate Visual Sys-

tem. Chicago, University of Chicago Press.

Pumphrey, R. J. 1948. The sense organs of birds. Ibis, *90*:171–199; Annual Report of the Smithsonian Institution 305–330.

Reed, H. D. 1920. The morphology of the sound-transmitting apparatus in caudate Amphibia. J. Morphol. *33*:325–375.

Retzius, G. 1881–1884. Das Gehörorgan der Wirbelthiere. Morphologischhistologische Studien. 2 vols. Stockholm, Samson and Wallin.

Rochon-Duvigneaud, A. 1943. Les Yeux et la Vision des Vertébrés. Paris, Masson et Cie.

Stensiö, E. A. 1947. The sensory lines and dermal bones of the cheek in fishes and amphibians. Kungl. Svenska Vetenskaps-akad. Handl. (3)*24*(3):1–195.

Versluys, J. 1898. Die mittlere und äussere Ohrsphäre der Lacertilia und Rhynchocephalia. Zool. Jahrbücher (Anat.) *12*:161–406. See also *ibid. 18*:107–188, 1902.

Walls, G. L. 1942. The vertebrate eye and its adaptive radiation. Cranbrook Institute of Science, Bulletin No. 19.

Werner, S. C. 1960. Das Gehörorgan der Wirbeltiere und des Menschen. Leipzig, George Thieme.

NERVOUS SYSTEM

Addens, J. L. 1933. The motor nuclei and roots of the cranial and first spinal nerves of vertebrates. Part I. Introduction, Cyclostomes. Zeitschr. Ges. Anat. Abt. I, *101*:307–410.

Aronson, L. R. 1963. The central nervous system of sharks and bony fishes, with special reference to sensory and integrative mechanisms. *In* Sharks and Survival, P. W. Gilbert, ed., 165–241. Boston, D. C. Heath.

Bartelmez, G. W. 1915. Mauthner's cell and the nucleus motorius tegmenti. J. Comp. Neurol. *25*:87–128.

Bass, A. D. 1959. Evolution of nervous control from primitive organisms to man. Am. Assoc. Adv. Sci. Publ. 52.

Boeke, J. 1935. The autonomic (enteric) nervous system of *Amphioxus lanceolatus*. Quart. J. Microsc. Sci. *77*:623–658.

Bullock, T. H. 1945. The anatomical organization of the nervous system of Enteropneusta. Quart. J. Microsc. Sci. *86*:55–111.

Campenhout, E. van. 1930. Historical survey of the development of the sympathetic nervous system. Quart. Rev. Biol. *5*:23–50, 217–234.

Causey, G. 1960. The Cell of Schwann. Edinburgh and London, E. and S. Livingstone Ltd.

Cobb, S. 1963. Notes on the avian optic lobe. Brain *86*:363–372.

Cobb, S. 1966. The brain of the emu, *Dromaeus novaehollandiae*. II. Anatomy of the principal nerve cell ganglia and tracts. Breviora, Mus. Comp. Zool., No. 250:1–27.

Cole, F. J. 1897. On the cranial nerves of *Chimaera monstrosa*. Tr. Roy. Soc. Edinburgh *38*:631–680.

Conel, J. L. 1929, 1931. The development of the brain of *Bdellostoma stouti*. I. External growth changes. II. Internal growth changes. J. Comp. Neurol. *47*:343–403; *52*:365–499.

Eccles, J. 1965. The synapse. Sci. Am. *212*(1):56–69.

Franz, V. 1923. Nervensystem der Akranier. Jena. Ztschr. Naturw. *59*:401–526.

Gans, C., and Crosby, E. C., eds. 1964. Recent advances in neuroanatomy. Am. Zool. *4*:4–96.

Goodrich, E. S. 1937. On the spinal nerves of the Myxinoidea. Quart. J. Microsc. Sci. *80*:153–158.

Hassler, F., and Stephan, H., eds. 1967. Evolution of the Forebrain. New York, Plenum Publishers.

Herrick, C. J. 1899. The cranial nerves of the bony fishes. J. Comp. Neurol. *9*:153–455. Cf. also *10*:265–322, 1900; *11*:177–249, 1901.

Herrick, C. J. 1926. Brains of Rats and Men. Chicago, University of Chicago Press.

Herrick, C. J. 1931. An Introduction to Neurology. 5th ed. Philadelphia, W. B. Saunders Company.

Herrick, C. J. 1948. The Brain of the Tiger Salamander. Chicago, University of Chicago Press.

Hughes, A. 1960. The development of the peripheral nerve fiber. Biol. Rev. *35*:283–323.

Igarashi, S. and Kamiya, T. 1972. Atlas of the Vertebrate Brain. Tokyo, University of Tokyo Press.

Johnels, A. G. 1956. On the peripheral autonomic system of the trunk region of *Lampetra planeri*. Acta Zool. *37*:251–285.

Johnston, J. B. 1901. The brain of *Acipenser;* a contribution to the morphology of the vertebrate brain. Zool. Jahrbücher (Anat.) *15*:59–263.

Johnston, J. B. 1906. The Nervous System

of Vertebrates. Philadelphia, Blakiston Company.

Kappers, C. U. A., Huber, G. C., and Crosby, E. C. 1936. The Comparative Anatomy of the Nervous System of Vertebrates, Including Man. 2 vols. New York, The Macmillan Company. A mine of information on comparative neurology, but difficult to work with for one not a neurologist.

Kingsbury, B. F. 1895. On the brain of *Necturus maculatus*. J. Comp. Neurol. *5*:139–205.

Krieg, W. J. S. 1942. Functional Neuroanatomy. Philadelphia, Blakiston Company.

Kuhlenbeck, H. 1967–1975. The Central Nervous System of Vertebrates. 4 vols. New York, Academic Press. A still incomplete and rather idiosyncratic series with much interpretation.

Kuntz, A. 1953. The Autonomic Nervous System. 4th ed. Philadelphia, Lea and Febiger.

Larsell, O. 1967–1972. The Comparative Anatomy and Histology of the Cerebellum. 3 vols. Minneapolis, University of Minnesota Press.

Lindström, T. 1949. On the cranial nerves of the cyclostomes with special reference to the N. trigeminus. Acta Zool. *30*:315–458.

Llinas, R., ed. 1969. Neurobiology of Cerebellar Evolution and Development: Proceedings. Chicago, American Medical Association.

Mitchell, G. A. G. 1953. Anatomy of the Autonomic Nervous System. Edinburgh and London, E. and S. Livingstone, Ltd.

Nicol, J. A. C. 1952. Autonomic nervous systems in lower chordates. Biol. Rev. *27*:1–49.

Nieuwenhuys, R. 1962. Trends in the evolution of the actinopterygian forebrain. J. Morphol. *111*:69–88.

Norris, H. W. 1913. Cranial nerves of *Siren lacertina*. J. Morphol. *24*:245–338.

Norris, H. W., and Hughes, S. P. 1920. The cranial, occipital and anterior spinal nerves of the dogfish. J. Comp. Neurol. *31*:293–395.

Papez, J. W. 1929. Comparative Neurology. New York, Thomas Y. Crowell Company. Reprinted in 1961. Out of date, especially on fiber connections, but still useful for comparative approach.

Petras, J. M., and Noback, C. R., eds. 1969. Comparative and evolutionary aspects of the vertebrate central nervous system. Ann. N.Y. Acad. Sci. *167*(1):1–513. Papers by many of the leading workers in one of the few truly comparative surveys.

Portmann, A. 1946, 1947. Etudes sur la cérebralisation chez les oiseaux. Alauda *14*:2–20; *15*:1–15.

Ransom, S. W., and Clark, S. L. 1959. The Anatomy of the Nervous System. 10th ed. Philadelphia, W. B. Saunders Company.

Retzlaff, E. 1957. A mechanism for excitation and inhibition of the Mauthner's cells in teleosts. J. Comp. Neurol. *107*:209–225.

Silén, L. 1950. On the nervous system of *Glossobalanus marginatus* Meek. Acta Zool. *31*:149–175.

Snider, R. S. 1958. The cerebellum. Sci. Am. *199*(2):84–90.

Stefanelli, A. 1951. The mauthnerian apparatus in the Ichthyopsida. Quart. Rev. Biol. *21*:17–34.

Stensiö, E. A. 1963. The brain and the cranial nerves in fossil, lower craniate vertebrates. Skrifter Norske Videnskaps-Akademi I Oslo I. Mat.-Naturv. Klasse. Ny Serie. No. *13*:1–120.

Stettner, L. J., and Matyniak, K. A. 1968. The brain of birds. Sci. Am. *218*(6):64–77.

Strong, O. S. 1895. The cranial nerves of Amphibia. J. Morphol. *10*:101–230.

Tretjakoff, D. 1927. Das periphere Nervensystem des Flussneunauges. Ztschr. Wiss. Zool. *129*:359–952.

Watkinson, G. B. 1906. The cranial nerves of *Varanus bivittatus*. Morphol. Jahrbuch *35*:450–472.

Weed, L. W. 1917. The development of the cerebrospinal spaces in pig and in man. Carnegie Inst. Washington, Contrib. Embryol. *5*:3–116.

Weiss, P. A. 1934. In vitro experiments on the factors determining the course of the outgrowing nerve fiber. J. Exp. Zool. *68*:393–448.

Willard, W. A. 1915. The cranial nerves of *Anolis carolinensis*. Bull. Mus. Comp. Zool., Harvard *59*:17–116.

Worthington, J. 1906. The descriptive anatomy of the brain and cranial nerves of *Bdellostoma dombeyi*. Quart. J. Microsc. Sci. *57*:137–181.

Yntema, C. L., and Hammond, W. S. 1947. The development of the autonomic nervous system. Biol. Rev. *22*:344–359.

Young, J. Z. 1931. On the autonomic nervous system of the teleostean fish, *Uranoscopus scaber*. Quart. J. Microsc. Sci. *74*:492–525.

Young, J. Z. 1933. The autonomic system of selachians. Quart. J. Microsc. Sci. *75*:571–624.

Zeman, W., and Innes, J. R. M. 1963. Craigie's Neuroanatomy of the Rat. New York, Academic Press.

ENDOCRINE ORGANS

Bargmann, W. 1960. The neurosecretory system of the diencephalon. Endeavour *19*:125–133.

Barrington, E. J. W. 1975. An Introduction to General and Comparative Endocrinology. 2nd ed. Oxford, Oxford University Press.

Bern, H. A. 1967. Hormones and endocrine glands of fishes. Science *158*:455–462.

Boyd, J. D. 1950. The development of the thyroid and parathyroid glands and the thymus. Ann. R. Coll. Surg. Engl. *7*:455–471.

Chester Jones, I. 1957. The Adrenal Cortex. London, Cambridge University Press.

Dodd, J. M., and Kerr, T. 1963. Comparative morphology and histology of the hypothalamo-neurohypophysial system. Symposium 9, Zool. Soc. London, 9–27.

Fields, W. S., Guillemin, R., and Carton, C. A., eds. 1956. Hypothalamic-hypophysial Interrelationships. A Symposium. Houston, Baylor University College of Medicine.

Goldsmith, E. D. 1949. Phylogeny of the thyroid: descriptive and experimental. Ann. N.Y. Acad. Sci. *50*:282–316.

Gorbman, A., ed. 1967. Recent developments in endocrinology. Am. Zool. *7*:81–169.

Gorbman, A., and Bern, H. A. 1962. A Textbook of Comparative Endocrinology. New York, John Wiley and Sons.

Green, J. D. 1951. The comparative anatomy of the hypophysis, with special reference to its blood supply and innervation. Am. J. Anat. *88*:225–311.

Grossman, M. I. 1950. Gastrointestinal hormones. Physiol. Rev. *30*:33–90.

Holmes, R. L., and Ball, J. N. 1974. The Pituitary Gland: A Comparative Account. Cambridge, England, Cambridge University Press.

Lynn, G. W., and Wachowski, H. E. 1951. The thyroid gland and its functions in cold-blooded vertebrates. Quart. Rev. Biol. *26*:123–168.

Macchi, I. A., and Gapp, D. A., eds. 1973. Comparative aspects of the endocrine pancreas. Am. Zool. *13*:565–709.

Marshall, F. H. A. 1960. The Physiology of Reproduction. London, Longmans, Green.

Matty, A. J. 1966. Endocrine glands in lower vertebrates. Int. Rev. Gen. Exp. Zool. *2*:44–138.

Norris, H. W. 1941. The Plagiostome Hypophysis, General Morphology and Types of Structure. Lancaster, Science Press.

Pickford, G. E., et al. 1973. The current status of fish endocrine systems. Am. Zool. *13*:710–936.

Scharrer, E., and Scharrer, B. 1963. Neuroendocrinology. New York, Columbia University Press.

Turner, C. D., and Bagnara, J. T. 1976. General Endocrinology. 6th ed. Philadelphia, W. B. Saunders Company.

Villee, D. B. 1975. Human Endocrinology: A Developmental Approach. Philadelphia, W. B. Saunders Company.

Von Euler, U. S., and Heller, H. S., eds. 1963. Comparative Endocrinology. New York, Academic Press.

Watzka, M. 1933. Vergleichende Untersuchungen über den ultimobranchialen Körper. Z. Mikrosk. Anat. Forsch. *34*:485–533.

Wilkins, L. 1960. The thyroid gland. Sci. Am. *202*(3):119–129.

Wurtman, R. J., and Axelrod, J. 1965. The pineal gland. Sci. Am. *213*(1):50–60.

index

B

C

D

H

I

N

O

T

U

-8 81

2195
WHEATON
COLLEGE